The Theory of Open Quantum Systems

The Theory of Open Quantum Systems

Heinz-Peter Breuer
Albert-Ludwigs-Universität Freiburg, Physikalisches Institut

Francesco Petruccione
University of KwaZulu-Natal, School of Physics - Quantum Research Group

CLARENDON PRESS · OXFORD

This book has been printed digitally and produced in a standard specification
in order to ensure its continuing availability

OXFORD
UNIVERSITY PRESS

Great Clarendon Street, Oxford OX2 6DP

Oxford University Press is a department of the University of Oxford.
It furthers the University's objective of excellence in research, scholarship,
and education by publishing worldwide in
Oxford New York

Auckland   Cape Town Dar es Salaam   Hong Kong   Karachi
Kuala Lumpur Madrid Melbourne Mexico City   Nairobi
New Delhi Shanghai Taipei Toronto
With offices in
Argentina Austria Brazil Chile Czech Republic France Greece
Guatemala Hungary Italy Japan South Korea Poland Portugal
Singapore Switzerland Thailand Turkey Ukraine Vietnam

ISBN 978-0-19-921390-0

Printed and bound in Great Britain by CPI Antony Rowe,
Chippenham and Eastbourne

A Gerardo Marotta
Presidente dell' Istituto Italiano per gli Studi Filosofici
per avere promosso e sostenuto questa ricerca

Für Heike und Valeria
für unendliche Geduld, Verständnis und Hilfe

PREFACE

Quantum mechanics is at the core of our present understanding of the laws of physics. It is the most fundamental physical theory, and it is inherently probabilistic. This means that all predictions derived from quantum mechanics are of a probabilistic character and that there is, as far as we know, no underlying deterministic theory from which the quantum probabilities could be deduced. The statistical interpretation of quantum mechanics ultimately implies that predictions are being made about the behaviour of ensembles, i.e. about collections of a large number of independent, individual systems, and that the statements of quantum theory are tested by carrying out measurements on large samples of such systems.

Quantum mechanical systems must be regarded as open systems. On the one hand, this is due to the fact that, like in classical physics, any realistic system is subjected to a coupling to an uncontrollable environment which influences it in a non-negligible way. The theory of open quantum systems thus plays a major rôle in many applications of quantum physics since perfect isolation of quantum systems is not possible and since a complete microscopic description or control of the environmental degrees of freedom is not feasible or only partially so. Most interesting systems are much too complicated to be describable in practice by the underlying microscopic laws of physics. One can say even more: Not only is such a microscopic approach impossible in practice, it also does not provide what one really wants to know about the problem. Even if a solution of the microscopic evolution equations were possible, it would give an intractable amount of information, the overwhelming part of which is useless for a reasonable description.

Practical considerations therefore force one to seek for a simpler, effectively probabilistic description in terms of an open system's dynamics. The use of probability theory allows the treatment of complex systems which involve a huge or even an infinite number of degrees of freedom. This is achieved by restricting the mathematical formulation to an appropriate set of a small number of relevant variables. Experience shows that under quite general physical conditions the time evolution of the relevant variables is governed by simple dynamical laws which can be formulated in terms of a set of effective equations of motion. The latter take into account the coupling to the remaining, irrelevant degrees of freedom in an approximate way through the inclusion of dissipative and stochastic terms.

There is another reason for invoking the notion of an open system in quantum theory which is of more fundamental origin. Quantum theory introduces a deterministic law, the Schrödinger equation, which governs the dynamics of the probability distributions. This equation describes the evolution of chance, that is the dynamics of ensembles of isolated systems. However, as a probabilistic theory quantum mechanics must also encompass the random occurrence of definite events which are the realizations of the underlying probability distributions. In order to effect the occurrence of chance events a quantum system must be subjected to interactions with its surroundings. Any empirical test of the statistical predictions on a quantum system requires one to couple it to a measuring apparatus which generally leads to non-negligible influences on the quantum object being measured. Thus, quantum mechanics in itself involves an intimate relationship to the notion of an open system through the action of the measurement process.

This book treats the central physical concepts and mathematical techniques used to study the dynamics of open quantum systems. The general approach followed in the book is to derive the open system's dynamics either from an underlying microscopic theory by the elimination of the environmental degrees of freedom, or else through the formulation of specific measurement schemes in terms of a quantum operation. There is a close physical and mathematical connection between the evolution of an open system, the state changes induced by quantum measurements, and the classical notion of a stochastic process. The book provides a detailed account of these interrelations and discusses a series of physical examples to illustrate the mathematical structure of the theory.

To provide a self-contained presentation Part I contains a survey of the classical theory of probability and stochastic processes (Chapter 1), and an introduction to the foundations of quantum mechanics (Chapter 2). In addition to the standard concepts, such as probability space, random variables and the definition of stochastic processes, Chapter 1 treats two topics, which are important for the further development of the theory. These are piecewise deterministic processes and Lévy processes. In Chapter 2 the emphasis lies on the statistical interpretation of quantum mechanics and its relationship to classical probability theory. As a preparation for later chapters, we also discuss composite quantum systems, the notion of entangled states, and quantum entropies. A detailed account of the quantum theory of measurement within the framework of quantum operations and effects is also included.

The fundamentals of the description of the quantum dynamics of open systems in terms of quantum master equations are introduced in Part II, together with its most important applications. In Chapter 3, special emphasis is laid on the theory of quantum dynamical semigroups which leads to the concept of a quantum Markov process. The relaxation to equilibrium and the multi-time structure of quantum Markov processes are discussed, as well as their irreversible nature which is characterized with the help of an appropriate entropy functional. Microscopic derivations for various quantum master equations are presented, such as the quantum optical master equation and the master equation for quantum Brownian motion. The influence functional technique is investigated in the context of the Caldeira–Leggett model. As a further application, we derive the master equation which describes the continuous monitoring of a quantum ob-

ject and study its relation to the quantum Zeno effect. Chapter 3 also contains
a treatment of non-linear, mean field quantum master equations together with
some applications to laser theory and super-radiance.

In Chapter 4 we study the important field of environment-induced decoherence and the transition to the classical behaviour of open quantum systems. A
number of techniques for the determination of decoherence times is developed.
As specific examples we discuss experiments on the decoherence of Schrödinger
cat-type states of the electromagnetic field, the destruction of quantum coherence in the Caldeira–Leggett model, and the environment-induced selection of a
pointer basis in the quantum theory of measurement.

While Parts I and II mainly deal with the standard aspects of the theory,
Parts III–V provide an overview of more advanced techniques and of new developments in the field of open quantum systems. Part III introduces the notion
of an ensemble of ensembles and the concept of stochastic wave functions and
stochastic density matrices. The underlying mathematical structure of probability distributions on Hilbert or Liouville space and of the corresponding random
state vectors is introduced in Chapter 5. These concepts are used to describe the
dynamics of continuous measurements performed on an open system in Chapter 6. It is shown that the time evolution of the state vector, conditioned on
the measurement record, is given by a piecewise deterministic process involving continuous evolution periods broken by randomly occurring, sudden quantum jumps. This so-called unravelling of the quantum master equation in the
form of a stochastic process is based on a close relation between quantum dynamical semigroups and piecewise deterministic processes. The general theory
is illustrated by means of a number of examples, such as direct, homodyne and
heterodyne photodetection.

The general formulation in terms of a quantum operation is given in Chapter
8, where we also investigate further examples from atomic physics and quantum
optics, e.g. dark state resonances and laser cooling of atoms. In particular, the
example of the sub-recoil cooling dynamics of atoms nicely illustrates the interplay between incoherent processes and quantum interference effects which leads
to the emergence of long-tail Lévy distributions for the atomic waiting time.

The numerical simulation of stochastic processes on high-performance computers provides an efficient tool for the predictions of the dynamical behaviour
in physical processes. The formulation of the open system's dynamics in terms of
piecewise deterministic processes or stochastic differential equations in Hilbert
space leads to efficient numerical simulation techniques which are introduced and
examined in detail in Chapter 7.

Part IV is devoted to the basic features of the more involved non-Markovian
quantum behaviour of open systems. Chapter 9 gives a general survey of the
Nakajima–Zwanzig projection operator methods with the help of which one derives so-called generalized master equations for the reduced system dynamics. In
the non-Markovian regime, these master equations involve a retarded memory
kernel, that is a time-convolution integral taken over the history of the reduced

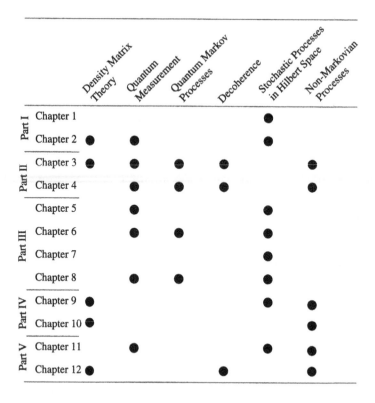

FIG. 0.1. Possible pathways through the book for a first reading. The heavy dots indicate those chapters in which the basic information on central concepts of the book may be found.

system. In general, one is thus confronted with the difficult task of treating an integro-differential equation for the open system's density matrix. We therefore develop in Chapter 9 a method of particular relevance based on an equation of motion which is local in time and which is known as the time-convolutionless projection operator method. This method serves as a starting point for a systematic perturbation expansion around the Markovian limit and for a numerical treatment. A number of applications to non-Markovian dynamics in physical systems is investigated in Chapter 10, such as the Jaynes–Cummings model, quantum Brownian motion and the spin-boson model.

The final Part V is concerned with the relativistic formulation of the dynamics of open quantum systems and of quantum measurement theory. Chapter 11 deals with the relativistic formulation of the state reduction postulate in quantum measurement theory. This postulate is used to study a number of applications to local and non-local measurements and to the restrictions on the measurability of non-local quantities imposed by the causality principle. The relativistic formulation allows us to discuss several important experiments from a unified perspective, such as EPR-type experiments, measurements of Bell state operators, exchange measurements, and quantum teleportation. The relativistic density matrix theory of quantum electrodynamics is developed in Chapter 12 using functional methods from field theory, path integral methods, and the influence functional formulation. As an important example a detailed theory of decoherence in quantum electrodynamics is presented.

The cross-disciplinary nature of the field of open quantum systems necessarily requires the treatment of various different aspects of quantum theory and of diverse applications in many fields of physics. We sketch in Fig. 0.1 various possible pathways through the book which may be followed in a first reading.

The book addresses undergraduate and graduate students in physics, theoretical physics and applied mathematics. As prior knowledges only a basic understanding of quantum mechanics and of the underlying mathematics, as well as an elementary knowledge of probability theory is assumed. The chapters of the book are largely written as self-contained texts and can be employed by lecturers as independent material for lectures and special courses. Each chapter ends with a short bibliography. Since most chapters deal with a rapidly evolving field it was sometimes impossible to give a complete account of the literature. Instead we have tried to give some important examples of original papers and introductory review articles that cover the subjects treated; personal preferences strongly influenced the lists of references and we apologize to those authors whose work is not cited properly. A number of excellent general textbooks and monographs from which we learned a great deal is listed in the following bibliography.

Bibliography

Alicki, R. and Lendi, K. (1987). *Quantum Dynamical Semigroups and Applications*, Volume 286 of *Lecture Notes in Physics*. Springer-Verlag, Berlin.

Braginsky, V. B. and Khalili, F. Ya. (1992). *Quantum Measurement*. Cambridge University Press, Cambridge.

Carmichael, H. (1993). *An Open Systems Approach to Quantum Optics*, Volume m18 of *Lecture Notes in Physics*. Springer-Verlag, Berlin.

Cohen-Tannoudji, C., Dupont-Roc, J. and Grynberg, G. (1998). *Atom–Photon Interactions*. John Wiley, New York.

Davies, E. B. (1976). *Quantum Theory of Open Systems*. Academic Press, London.

Gardiner, C. W. and Zoller, P. (2000). *Quantum Noise* (second edition). Springer-Verlag, Berlin.

Giulini, D., Joos, E., Kiefer, C., Kupsch, J., Stamatescu, I.-O. and Zeh, H. D. (1996). *Decoherence and the Appearence of a Classical World in Quantum Theory*. Springer-Verlag, Berlin.

Louisell, W. (1990). *Quantum Statistical Properties of Radiation*. John Wiley, New York.

Mandel, L. and Wolf, E. (1995). *Optical Coherence and Quantum Optics*. Cambridge University Press, Cambridge.

Nielsen, M. A. and Chuang, I. L. (2000). *Quantum Computation and Quantum Information*. Cambridge University Press, Cambridge.

Scully, M. O. and Zubairy, M. S. (1997). *Quantum Optics*. Cambridge University Press, Cambridge.

Walls, D. F. and Milburn, G. J. (1994). *Quantum Optics*. Springer-Verlag, Berlin.

Weiss, U. (1999). *Quantum Dissipative Systems*, Volume 2 of *Series in Modern Condensed Matter Physics*. World Scientific, Singapore.

ACKNOWLEDGEMENTS

It is a pleasure to thank the Istituto Italiano per gli Studi Filosofici for promoting the research programme which led to this book. The Istituto supported a series of fruitful workshops on the theory of open quantum systems. In particular we owe thanks to its President Avv. Gerardo Marotta, who gave continual help and encouragement throughout the research and the preparation of the manuscript.

We have received a great deal of help from friends and colleagues. Our thanks to Robert Alicki, Francois Bardou, Thomas Filk, Domenico Giulini, Gerard Milburn and Ludger Rüschendorf for critically reading parts of the manuscript and for making several suggestions for improvements. We have profited much from lively discussions with them.

We are also indebted to our students Peter Biechele, Kim Boström, Uwe Dorner, Jens Eisert, Daniel Faller, Wolfgang Huber, Bernd Kappler, Andrea Ma, Wolfgang Pfersich and Frithjof Weber. They developed with us parts of the material presented in the book and contributed with several ideas.

The excellent cooperation with the staff of Oxford University Press deserves special thanks. In particular, we express our gratitude to Sönke Adlung for the professional guidance through all stages of the preparation of the work.

CONTENTS

II DENSITY MATRIX THEORY

3 Quantum master equations

Part I

Probability in classical and quantum physics

1

CLASSICAL PROBABILITY THEORY AND STOCHASTIC PROCESSES

This chapter contains a brief survey of classical probability theory and stochastic processes. Our aim is to provide a self-contained and concise presentation of the theory. We concentrate on those subjects which will be important for the developments of the following chapters. More details and many interesting examples and applications of classical probability theory may be found, e.g. in the excellent textbooks by Feller (1968, 1971) and Doob (1953) for the more mathematically oriented readers, and by Gardiner (1985), van Kampen (1992) and Reichl (1998) for readers who are more interested in physical applications.

1.1 The probability space

The fundamental concept of probability theory is the probability space. It consists of three basic ingredients, namely a sample space of elementary events, a σ-algebra of events, and a probability measure on the σ-algebra. These notions will be introduced and explained below. We shall follow here the axiomatic approach to probability which is mainly due to Kolmogorov (1956).

1.1.1 *The σ-algebra of events*

The formal objects to which we want to attribute probabilities are called events. Mathematically, these events are subsets of some basic set Ω, the *sample space*, or space of events. The subsets of Ω containing just one element $\omega \in \Omega$ are referred to as *elementary events.*

Given some sample space Ω one is usually not interested in all possible subsets of Ω (this may happen, for example, if Ω is infinite and non-countable), that is, we need to specify which kind of subsets $A \subset \Omega$ we would like to include in our theory. An important requirement is that the events form a so-called *σ-algebra*, which is a system $\mathcal{A}$ of subsets of Ω with the following three properties.

1. The sample space itself and the empty set belong to the system of events, that is $\Omega \in \mathcal{A}$ and $\emptyset \in \mathcal{A}$.
2. If $A_1 \in \mathcal{A}$ and $A_2 \in \mathcal{A}$, then also the union $A_1 \cup A_2$, the intersection $A_1 \cap A_2$, and the difference $A_1 \setminus A_2$ belong to the system $\mathcal{A}$.
3. If we have a countable collection of events $A_1, A_2, \ldots, A_n, \ldots \in \mathcal{A}$, then also their union $\cup_{n=1}^{\infty} A_n$ belongs to $\mathcal{A}$.

We shall always write $A \in \mathcal{A}$ to express that the subset $A \subset \Omega$ is an event of our theory. The above requirements ensure that the total sample space Ω and the empty set $\emptyset$ are events, and that all events of $\mathcal{A}$ can be subjected to the logical operations 'AND', 'OR' and 'NOT' without leaving the system of events. This is why $\mathcal{A}$ is called an algebra. The third condition is what makes $\mathcal{A}$ a σ-algebra. It tells us that any countable union of events is again an event.

1.1.2 *Probability measures and Kolmogorov axioms*

The construction of the probability space is completed by introducing a probability measure on the σ-algebra. A probability measure is simply a map $\mu : \mathcal{A} \longrightarrow \mathbb{R}$ which assigns to each event A of the σ-algebra a real number $\mu(A)$,

$$A \mapsto \mu(A) \in \mathbb{R}. \tag{1.1}$$

The number $\mu(A)$ is interpreted as the probability of the event A. The probability measure μ is thus required to satisfy the following Kolmogorov axioms:

1. For all events $A \in \mathcal{A}$ we have

$$0 \leq \mu(A) \leq 1. \tag{1.2}$$

2. Probability is normalized as

$$\mu(\Omega) = 1. \tag{1.3}$$

3. If we have a countable collection of disjoint events

$$A_1, A_2, \ldots, A_n, \ldots \in \mathcal{A}, \quad \text{with } A_i \cap A_j = \emptyset \text{ for } i \neq j, \tag{1.4}$$

then the probability of their union is equal to the sum of their probabilities,

$$\mu\left(\cup_{n=1}^{\infty} A_n\right) = \sum_{n=1}^{\infty} \mu(A_n). \tag{1.5}$$

On the basis of these axioms one can build up a consistent probability theory. In particular, the Kolmogorov axioms enable one to determine the probabilities for all events which arise from logical operations on other events. For example, one finds

$$\mu(A_1 \cup A_2) = \mu(A_1) + \mu(A_2) - \mu(A_1 \cap A_2). \tag{1.6}$$

Summarizing, a probability space consists of a sample space Ω, a σ-algebra $\mathcal{A}$ of events, and a probability measure μ on $\mathcal{A}$. This concept of a probability space constitutes the axiomatic basis of classical probability theory. Of course, from a physical viewpoint one has to relate these abstract notions to experience and to specific theoretical models of reality, which may be a non-trivial task.

1.1.3 *Conditional probabilities and independence*

An important concept of probability theory is the notion of statistical independence. This concept is often formulated by introducing the *conditional probability* $\mu(A_1|A_2)$ of an event A_1 under the condition that an event A_2 occurred,

$$\mu(A_1|A_2) = \frac{\mu(A_1 \cap A_2)}{\mu(A_2)}. \tag{1.7}$$

Of course, both events A_1 and A_2 are taken from the σ-algebra and it is assumed that $\mu(A_2) > 0$. These events are said to be statistically independent if

$$\mu(A_1|A_2) = \mu(A_1), \tag{1.8}$$

or, equivalently, if

$$\mu(A_1 \cap A_2) = \mu(A_1) \cdot \mu(A_2). \tag{1.9}$$

This means that the probability of the mutual occurrence of the events A_1 and A_2 is just equal to the product of the probabilities of A_1 and A_2.

   If we have several events $A_1, A_2, \ldots, A_n$ the condition of statistical independence is the following: For any subset $(i_1, i_2, \ldots, i_k)$ of the set of indices $(1, 2, \ldots, n)$ we must have

$$\mu\left(A_{i_1} \cap A_{i_2} \cap \ldots \cap A_{i_k}\right) = \mu(A_{i_1})\mu(A_{i_2}) \ldots \mu(A_{i_k}), \tag{1.10}$$

which means that the joint occurrence of any subset of the events A_i factorizes. As simple examples show (Gardiner, 1985), it is not sufficient to check statistical independence by just considering all possible pairs A_i, A_j of events.

   An immediate consequence of definition (1.7) is the relation

$$\mu(A_1|A_2) = \mu(A_2|A_1)\frac{\mu(A_1)}{\mu(A_2)}, \tag{1.11}$$

which is known as Bayes's theorem.

1.2   Random variables

The elements ω of the sample space Ω can be rather abstract objects. In practice one often wishes to deal with simple numbers (integer, real or complex numbers) instead of these abstract objects. For example, one would like to add and multiply these numbers, and also to consider arbitrary functions of them. The aim is thus to associate numbers with the elements of the sample space. This idea leads to the concept of a *random variable*.

1.2.1   *Definition of random variables*

A random variable X is defined to be a map

$$X : \Omega \mapsto \mathbb{R}, \tag{1.12}$$

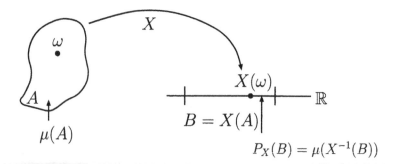

FIG. 1.1. Illustration of the definition of a random variable. A random variable
X is a map from the sample space to the set of real numbers. The probability
that the random number falls into some Borel set B is equal to the probability
measure $\mu(A)$ of the event $A = X^{-1}(B)$ given by the pre-image of B.

which assigns to each elementary event $\omega \in \Omega$ a real number $X(\omega)$. Given some
ω the value

$$x = X(\omega) \tag{1.13}$$

is called a *realization* of X. In the following we use the usual convention to denote
random numbers by capital letters, whereas their realizations are denoted by the
corresponding lower case.

Our definition of a random variable X is not yet complete. We have to impose
a certain condition on the function X. To formulate this condition we introduce
the σ-algebra of Borel sets[1] of $\mathbb{R}$ which will be denoted by $\mathcal{B}$. The condition on
the function X is then that it must be a measurable function, which means that
for any Borel set $B \in \mathcal{B}$ the pre-image $A = X^{-1}(B)$ belongs to the σ-algebra $\mathcal{A}$
of events. This condition ensures that the probability of $X^{-1}(B)$ is well defined
and that we can define the *probability distribution of* X by means of the formula

$$P_X(B) = \mu\left(X^{-1}(B)\right). \tag{1.14}$$

A random variable X thus gives rise to a probability distribution $P_X(B)$ on the
Borel sets B of the real axis (see Fig. 1.1).

Particular Borel sets are the sets $(-\infty, x]$ with $x \in \mathbb{R}$. Consider the pre-images
of these set, that is the sets

$$A_x \equiv \{\omega \in \Omega | X(\omega) \leq x\}. \tag{1.15}$$

By the condition on X these sets are measurable for any $x \in \mathbb{R}$ which enables
one to introduce the function

$$F_X(x) \equiv \mu(A_x) = \mu\left(\{\omega \in \Omega | X(\omega) \leq x\}\right). \tag{1.16}$$

---

[1]The σ-algebra of Borel sets of $\mathbb{R}$ is the smallest σ-algebra which contains all subsets of the
form $(-\infty, x)$, $x \in \mathbb{R}$. In particular, it contains all open and closed intervals of the real axis.

For a given x this function yields the probability that the random number X takes on a value in the interval $(-\infty, x]$. $F_X(x)$ is referred to as the *cumulative distribution function* of X. One often employs the following shorthand notation,

$$F_X(x) \equiv \mu(X \leq x). \tag{1.17}$$

As is easily demonstrated, the cumulative distribution function has the following properties:

1. $F_X(x)$ increases monotonically,

$$F_X(x_1) \leq F_X(x_2), \quad \text{for } x_1 < x_2. \tag{1.18}$$

2. $F_X(x)$ is continuous from the right,

$$\lim_{\varepsilon \to +0} F_X(x + \varepsilon) = F_X(x). \tag{1.19}$$

3. $F_X(x)$ has the following limits,

$$\lim_{x \to -\infty} F_X(x) = 0, \quad \lim_{x \to +\infty} F_X(x) = 1. \tag{1.20}$$

The random variable X is said to have a *probability density* $p_X(x)$ if the cumulative distribution function can be represented as

$$F_X(x) = \int_{-\infty}^{x} dx\, p_X(x). \tag{1.21}$$

If $F_X(x)$ is absolutely continuous we get the formula

$$p_X(x) = \frac{dF_X(x)}{dx}. \tag{1.22}$$

In the following we often represent distribution functions by their densities $p_X(x)$ in this way, as is common in the physics literature. This is permissible if we allow $p_X(x)$ to involve a sum of δ-functions and if we exclude certain singular distribution functions (Feller, 1971). In terms of the density of X we may also write eqn (1.14) as

$$P_X(B) = \int_{B} dx\, p_X(x), \tag{1.23}$$

where the integral is extended over the Borel set B.

We have just considered a single random variable. One can, of course, study also an arbitrary collection

$$X = (X_1, X_2, \ldots, X_d) \tag{1.24}$$

of random variables which are defined on the same probability space. The vector-valued function $X : \Omega \mapsto \mathbb{R}^d$ is called a multivariate random variable or a random vector, which simply means that each component X_i is a real-valued random variable. For a given $\omega \in \Omega$ the quantity $x = X(\omega) = (X_1(\omega), \ldots, X_d(\omega))$ is a realization of the multivariate random variable. The joint probability density of a multivariate random variable is denoted by $p_X(x)$. The probability for the variable to fall into a Borel set $B \subset \mathbb{R}^d$ is then given by

$$P_X(B) = \mu\left(X^{-1}(B)\right) = \int_B d^d x \, p_X(x). \tag{1.25}$$

In accordance with our former definition, two random variables X_1 and X_2 on the same probability space are said to be statistically independent if

$$\mu(X_1 \leq x_1, X_2 \leq x_2) = \mu(X_1 \leq x_1) \cdot \mu(X_2 \leq x_2) \tag{1.26}$$

for all x_1, x_2. Here, the left-hand side is the shorthand notation for the probability

$$\mu(X_1 \leq x_1, X_2 \leq x_2) \equiv \mu\left(\{\omega \in \Omega | X_1(\omega) \leq x_1 \text{ and } X_2(\omega) \leq x_2\}\right). \tag{1.27}$$

The joint statistical independence of several random variables is defined analogously to definition (1.10).

1.2.2 Transformation of random variables

Given a d-dimensional random variable X we can generate new random variables by using appropriate transformations. To introduce these we consider a Borel-measurable function

$$g : \mathbb{R}^d \longrightarrow \mathbb{R}^f. \tag{1.28}$$

Such a function is defined to have the property that the pre-image $g^{-1}(B)$ of any Borel set $B \subset \mathbb{R}^f$ is again a Borel set in $\mathbb{R}^d$. Thus, the equation

$$Y = g(X) \tag{1.29}$$

defines a new f-dimensional random variable Y. If P_X is the probability distribution of X, then the probability distribution of Y is obtained by means of the formula

$$P_Y(B) = P_X(g^{-1}(B)). \tag{1.30}$$

The corresponding probability densities are connected by the relation

$$p_Y(y) = \int d^d x \, \delta^{(f)}(y - g(x)) p_X(x), \tag{1.31}$$

where $\delta^{(f)}$ denotes the f-dimensional δ-function. This formula enables the determination of the density of $Y = g(X)$. For example, the sum $Y = X_1 + X_2$ of two random variables is found by taking $g(x_1, x_2) = x_1 + x_2$. If X_1 and X_2 are independent we get the formula

$$p_Y(y) = \int dx_1 \, p_{X_1}(x_1) p_{X_2}(y - x_1), \tag{1.32}$$

which shows that the density of Y is the convolution of the densities of X_1 and of X_2.

1.2.3 *Expectation values and characteristic function*

An important way of characterizing probability distributions is to investigate their expectation values. The *expectation value* or mean of a real-valued random variable X is defined as[2]

$$\mathrm{E}(X) \equiv \int_{-\infty}^{+\infty} x dF_X(x) = \int_{-\infty}^{+\infty} dx\, x p_X(x). \qquad (1.33)$$

Here, the quantity $dF_X(x)$ is defined as

$$dF_X(x) \equiv F_X(x + dx) - F_X(x) = \mu(x < X \le x + dx). \qquad (1.34)$$

Correspondingly, the integrals in (1.33) are regarded as Lebesgue–Stieltjes integrals (Feller, 1971). More generally, the expectation value of some measurable function $g(X)$ of X is defined to be

$$\mathrm{E}(g(X)) = \int_{-\infty}^{+\infty} g(x) dF_X(x) = \int_{-\infty}^{+\infty} dx\, g(x) p_X(x). \qquad (1.35)$$

Particularly important expectation values are the moments of order m:

$$\mathrm{E}(X^m) = \int_{-\infty}^{+\infty} x^m dF_X(x) = \int_{-\infty}^{+\infty} dx\, x^m p_X(x). \qquad (1.36)$$

The variance of a random variable X is defined by

$$\mathrm{Var}(X) \equiv \mathrm{E}\left([X - \mathrm{E}(X)]^2\right) = \mathrm{E}(X^2) - \mathrm{E}(X)^2. \qquad (1.37)$$

The significance of the variance stems from its property to be a measure for the fluctuations of the random variable X, that is, for the extent of deviations of the realizations of X from the mean value $\mathrm{E}(X)$. This fact is expressed, for example, by the Chebyshev inequality which states that the variance controls the probability for such deviations, namely for all $\varepsilon > 0$ we have

$$\mu(|X - \mathrm{E}(X)| \ge \varepsilon) \le \frac{1}{\varepsilon^2} \mathrm{Var}(X). \qquad (1.38)$$

In particular, if the variance vanishes then the random number X is, in fact, deterministic, i.e. it takes on the single value $x = \mathrm{E}(X)$ with probability 1. The variance plays an important rôle in the statistical analysis of experimental data

---

[2] We follow the usual convention of mathematical probability theory and denote the expectation value of a classical random variable by the symbol E, in order to distinguish it from the quantum mechanical expectation value which will be denoted by angular brackets.

(Honerkamp, 1998), where it is used, for example, to estimate the standard error of the mean for a sample of realizations obtained in an experiment.

For a multivariate random variable $X = (X_1, X_2, \ldots, X_d)$ one defines the matrix elements of the *covariance matrix* by

$$\text{Cov}(X_i, X_j) \equiv \text{E}\left([X_i - \text{E}(X_i)][X_j - \text{E}(X_j)]\right). \quad (1.39)$$

The $d \times d$ matrix with these coefficients is symmetric and positive semidefinite. As is well known, the statistical independence of two random variables X_1, X_2 implies that the off-diagonal element $\text{Cov}(X_1, X_2)$ vanishes, but the converse is not true. However, the off-diagonal elements are a measure of the linear dependence of X_1 and X_2. To see this we consider, for any two random variables with non-vanishing variances, the correlation coefficient

$$\text{Cor}(X_1, X_2) \equiv \frac{\text{Cov}(X_1, X_2)}{\sqrt{\text{Var}(X_1)\text{Var}(X_2)}} \quad (1.40)$$

which satisfies $|\text{Cor}(X_1, X_2)| \leq 1$. If the absolute value of the correlation coefficient is equal to one, $|\text{Cor}(X_1, X_2)| = 1$, then there are constants a and b such that $X_2 = aX_1 + b$ with unit probability, i.e. X_2 depends linearly on X_1.

Let us finally introduce a further important expectation value which may serve to characterize completely a random variable. This is the *characteristic function* which is defined as the Fourier transform of the probability density,

$$G(k) = \text{E}\left(\exp[ikX]\right) = \int dx \, p_X(x) \exp(ikx). \quad (1.41)$$

It can be shown that the characteristic function $G(k)$ uniquely determines the corresponding probability distribution of X. Under the condition that the moments of X exist the derivatives of $G(k)$ evaluated at $k = 0$ yield the moments of X,

$$\text{E}(X^m) = \frac{1}{i^m} \left. \frac{d^m}{dk^m} \right|_{k=0} G(k). \quad (1.42)$$

For this reason $G(k)$ is also called the *generating function*. For a multivariate random variable the above expression for the characteristic function is readily generalized as follows,

$$G(k_1, k_2, \ldots, k_d) = \text{E}\left(\exp\left[i\sum_{j=1}^{d} k_j X_j\right]\right). \quad (1.43)$$

An important property of the characteristic function is the following. As we have already remarked, if X and Y are two independent random variables the probability density of their sum $Z = X + Y$ is the convolution of the densities of X and Y. Consequently, the characteristic function of Z is the product of the characteristic functions of X and Y.

1.3 Stochastic processes

Up to now we have dealt with random variables on a probability space without explicit time dependence of their statistical properties. In order to describe the dynamics of a physical process one needs the concept of a *stochastic process* which is, essentially, a random variable whose statistical properties change in time. The notion of a stochastic process generalizes the idea of deterministic time evolution. The latter can be given, for example, in terms of a differential equation which describes the deterministic change in time of some variable. In a stochastic process, however, such a deterministic evolution is replaced by a probabilistic law for the time development of the variable.

This section gives a brief introduction to the theory of stochastic processes. After a formal definition of a stochastic process we introduce the family of joint probability distributions which characterizes a stochastic process in a way that is fully sufficient for practical purposes. In a certain sense, the differential equation of a deterministic theory is replaced by such a family of joint probability distributions when one deals with stochastic processes. This is made clear by a theorem of Kolmogorov which is also explained in this section.

1.3.1 *Formal definition of a stochastic process*

In mathematical terms, a stochastic process is a family of random variables $X(t)$ on a common probability space depending on a parameter $t \in T$. In most physical applications the parameter t plays the rôle of the time variable. The parameter space T is therefore usually an interval of the real time axis.

Corresponding to this definition, for each fixed t the quantity $X(t)$ is a map from the sample space Ω into $\mathbb{R}$. A stochastic process can therefore be regarded as a map

$$X : \Omega \times T \longrightarrow \mathbb{R}, \tag{1.44}$$

which associates with each $\omega \in \Omega$ and with each $t \in T$ a real number $X(\omega, t)$. Keeping ω fixed, we call the mapping

$$t \mapsto X(\omega, t), \quad t \in T, \tag{1.45}$$

a *realization*, *trajectory*, or *sample path* of the stochastic process.

In the above definition the map (1.44) can be quite general. Because of this the notion of a stochastic process is a very general concept. We need, however, one condition to be satisfied in order for $X(t)$ to represent a random variable for each fixed t. Namely, for each fixed t the function $X(t)$ which maps Ω into $\mathbb{R}$ must be measurable in the sense that the pre-images of any Borel set in $\mathbb{R}$ must belong to the algebra of events of our probability space.

A multivariate stochastic process $X(t)$ is defined similarly: It is a vector-valued stochastic process $X(t) = (X_1(t), X_2(t), \ldots, X_d(t))$, each component $X_i(t)$, $i = 1, 2, \ldots, d$, being a real-valued stochastic process. Thus, formally a multivariate stochastic process can be regarded as a map

$$X : \Omega \times T \longrightarrow \mathbb{R}^d. \tag{1.46}$$

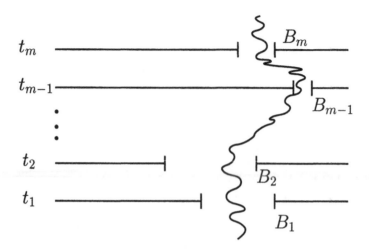

FIG. 1.2. A sample path $X(t, \omega)$ of a stochastic process which passes at times t_1, t_2, ..., t_{m-1}, t_m the sets B_1, B_2, ..., B_{m-1}, B_m, respectively. The probability for such a path to occur is given by the joint probability $P(B_1, t_1; \ldots; B_m, t_m)$.

1.3.2   *The hierarchy of joint probability distributions*

What characterizes a stochastic process is the way in which the random variables $X(t)$ for different times t are related to each other, that is, the degree of statistical dependencies between the random variables of the family.

According to the definition given above, a stochastic process is, formally speaking, nothing but a time-dependent random variable. It can therefore be characterized uniquely if one constructs a probability space and a map of the form (1.46) on it. This is however not the way one characterizes a stochastic process in practice. In most applications one tries to construct an appropriate process by the observation of the statistical correlations between the random variables $X(t_\nu)$ at a finite set of discrete times t_ν. This is done on the basis of experimental data, employing some phenomenological model, or with the help of some underlying microscopic physical theory.

Thus, the physical theory usually provides a so-called *family of finite joint probability distributions* which is defined as follows (see Fig. 1.2). We take a set $t_1, t_2, \ldots, t_m$ of discrete times and Borel sets $B_1, B_2, \ldots, B_m$ in $\mathbb{R}^d$, and consider for a multivariate stochastic process $X(t)$ the quantity

$$P(B_1, t_1; B_2, t_2; \ldots; B_m, t_m) \equiv \mu\left(X(t_1) \in B_1, X(t_2) \in B_2, \ldots, X(t_m) \in B_m\right). \tag{1.47}$$

This quantity is a joint probability distribution of order m. It gives the probability that the process $X(t)$ takes on some value in B_1 at time t_1, some value in B_2 at time t_2, ..., and some value in B_m at time t_m. The set of all joint probability distributions for all $m = 1, 2, \ldots$, all discrete times t_ν, and all Borel sets B_ν is called the family of finite joint probability distributions of the stochastic process.

Each stochastic process gives rise to such a family of joint probabilities. It follows immediately from the above definition that the joint probabilities satisfy the Kolmogorov consistency conditions:

$$P(\mathbb{R}^d, t) = 1, \tag{1.48}$$

$$P(B_1, t_1; \ldots; B_m, t_m) \geq 0, \tag{1.49}$$

$$P(B_1, t_1; \ldots; B_{m-1}, t_{m-1}; \mathbb{R}^d, t_m) = P(B_1, t_1; \ldots; B_{m-1}, t_{m-1}), \tag{1.50}$$

$$P(B_{\pi(1)}, t_{\pi(1)}; \ldots; B_{\pi(m)}, t_{\pi(m)}) = P(B_1, t_1; \ldots; B_m, t_m). \tag{1.51}$$

The first two conditions state that the distributions must be non-negative and that the probability for the sure event $X(t) \in \mathbb{R}^d$ is normalized to 1. The third condition asserts that for $\nu > 1$ the sure event $X(t_\nu) \in \mathbb{R}^d$ can always be omitted from the set of arguments, whereas the fourth condition means that the joint probability distribution is invariant under all permutations π of its arguments.

Of course, for a given stochastic process $X(t)$ the consistency conditions (1.48)–(1.51) are a trivial consequence of the definition (1.47). The important point to note is that the following non-trivial theorem is also true: Suppose that a family of functions is given which satisfy conditions (1.48)–(1.51). Then there exists a probability space and a stochastic process $X(t)$ on this space such that the family of joint probabilities pertaining to $X(t)$ coincides with the given family.

This is the theorem of Kolmogorov. It ensures that for any consistent family of joint probabilities there exists a stochastic process $X(t)$ on some probability space. It should be noted, however, that the process $X(t)$ is not unique; for a given family of joint probability distributions there may exist different stochastic processes, where the term *different* means that these processes may be different on events with non-zero measure.

In practice, the non-uniqueness of $X(t)$ usually does not cause any difficulty since the family of finite-dimensional joint probabilities uniquely determines the probabilities of all events which can be characterized by any finite number of random variables. Thus, if we assess, for example, some stochastic model by comparison with a set of experimental data, which is always finite, no problem due to the non-uniqueness of the process will ever be encountered.

1.4 Markov processes

Markov processes play an important rôle in physics and the natural sciences. One reason for this fact is that many important processes, as for example the processes arising in equilibrium statistical mechanics, are Markovian provided one chooses an appropriate set of variables. Another reason is that many types of stochastic processes become Markovian by an appropriate extension of the state space. Finally, Markov processes are relatively easy to describe mathematically. In this section we define and classify the most important Markov processes and briefly review their properties.

1.4.1 *The Chapman–Kolmogorov equation*

Essentially, a Markov process is a stochastic process $X(t)$ with a short memory, that is a process which rapidly forgets its past history. This property is what makes a Markov process so easy to deal with, since it ensures that the whole hierarchy of joint probabilities introduced in the preceding section can be reconstructed from just two distribution functions.

The condition for the rapid decrease of memory effects can be formulated in terms of the conditional joint probabilities as follows,

$$\mu(X(t) \in B|X(t_m) = x_m, \dots, X(t_1) = x_1) = \mu(X(t) \in B|X(t_m) = x_m). \quad (1.52)$$

This is the Markov condition. It is assumed to hold for all $m = 1, 2, 3, \dots$, for all ordered sets of times,

$$t_1 < t_2 < \dots < t_m < t, \quad (1.53)$$

for all Borel sets B and all $x_1, x_2, \dots, x_m \in \mathbb{R}^d$. The Markov condition states that the probability for the event $X(t) \in B$, conditioned on m previous events $X(t_1) = x_1, \dots, X(t_m) = x_m$, only depends on the latest event $X(t_m) = x_m$.

In the following we shall discuss the consequences of the Markov condition by introducing the joint probability densities

$$P(B_m, t_m; \dots; B_1, t_1) = \int_{B_m} dx_m \dots \int_{B_1} dx_1 \, p_m(x_m, t_m; \dots; x_1, t_1) \quad (1.54)$$

and the corresponding conditional probability densities

$$p_{l|k}(x_{k+l}, t_{k+l}; \dots; x_{k+1}, t_{k+1}|x_k, t_k; \dots; x_1, t_1) \equiv \frac{p_{k+l}(x_{k+l}, t_{k+l}; \dots; x_1, t_1)}{p_k(x_k, t_k; \dots; x_1, t_1)} \quad (1.55)$$

in terms of which the Markov condition takes the form

$$p_{1|m}(x, t|x_m, t_m; \dots; x_1, t_1) = p_{1|1}(x, t|x_m, t_m). \quad (1.56)$$

This equation demonstrates that the quantity $p_{1|1}(x, t|x', t')$ plays a crucial rôle in the theory of Markov processes.

For any stochastic process (not necessarily Markovian) $p_{1|1}(x, t|x', t')$ is equal to the probability density that the process takes on the value x at time t under the condition that the process took the value x' at some prior time t'. This conditional probability is therefore referred to as the *conditional transition probability* or simply as the *propagator*. We introduce the notation

$$T(x, t|x', t') \equiv p_{1|1}(x, t|x', t') \quad (1.57)$$

for the propagator. As follows directly from the definition, the propagator fulfils the relations,

$$\int dx\, T(x,t|x',t') = 1, \tag{1.58}$$

$$\lim_{t \to t'} T(x,t|x',t') = \delta(x - x'). \tag{1.59}$$

The first equation expresses the fact that the probability for the process to take any value at any fixed time is equal to 1, and the second equation states that with probability 1 the process does not change for vanishing time increment.

The probability density $p_1(x,t)$, which is simply the density for the unconditioned probability that the process takes on the value x at time t, will be denoted by

$$p(x,t) \equiv p_1(x,t). \tag{1.60}$$

The density $p(x,t)$ is connected to the initial density at some time t_0 by the obvious relation

$$p(x,t) = \int dx'\, T(x,t|x',t_0)p(x',t_0). \tag{1.61}$$

A stochastic process is called *stationary* if all joint probability densities are invariant under time translations, that is if for all τ

$$p_m(x_m, t_m + \tau; \ldots ; x_1, t_1 + \tau) = p_m(x_m, t_m; \ldots ; x_1, t_1). \tag{1.62}$$

In particular, stationarity implies that the probability density p is independent of time, $p(x,t) = p(x)$, and that the propagator $T(x,t|x',t')$ depends only on the difference $t - t'$ of its time arguments. With the help of stationary processes one describes, for example, equilibrium fluctuations in statistical mechanics.

A process is called *homogeneous* in time if the propagator depends only on the difference of its time arguments. Thus, a stationary process is homogeneous in time, but there are homogeneous processes which are not stationary. An example is provided by the Wiener process (see below).

The great simplification achieved by invoking the Markov condition derives from the fact that the total hierarchy of the joint probabilities can be reconstructed from an initial density $p(x,t_0)$ and an appropriate propagator. According to eqn (1.61) the density $p(x,t)$ for later times $t > t_0$ is obtained from the initial density and from the propagator. Thus, also the joint probability distribution $p_2(x,t;x',t')$ is known, of course. By virtue of the Markov condition all higher-order distribution functions can then be constructed, provided the propagator fulfils a certain integral equation which will now be derived.

To this end, we consider three times $t_1 < t_2 < t_3$ and the third-order distribution p_3 and invoke the definition of the conditional probability and the Markov condition to obtain

$$\begin{aligned} p_3(x_3,t_3;x_2,t_2;x_1,t_1) &= p_{1|2}(x_3,t_3|x_2,t_2;x_1,t_1)p_2(x_2,t_2;x_1,t_1) \quad (1.63) \\ &= p_{1|1}(x_3,t_3|x_2,t_2)p_{1|1}(x_2,t_2|x_1,t_1)p_1(x_1,t_1). \end{aligned}$$

We integrate this equation over x_2,

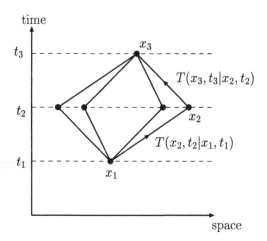

FIG. 1.3. Illustration of the Chapman–Kolmogorov equation (1.66).

$$p_2(x_3, t_3; x_1, t_1) = p_1(x_1, t_1) \int dx_2 p_{1|1}(x_3, t_3|x_2, t_2) p_{1|1}(x_2, t_2|x_1, t_1), \quad (1.64)$$

and divide by $p_1(x_1, t_1)$,

$$p_{1|1}(x_3, t_3|x_1, t_1) = \int dx_2 p_{1|1}(x_3, t_3|x_2, t_2) p_{1|1}(x_2, t_2|x_1, t_1). \quad (1.65)$$

On using our notation (1.57) for the propagator we thus have

$$T(x_3, t_3|x_1, t_1) = \int dx_2\, T(x_3, t_3|x_2, t_2) T(x_2, t_2|x_1, t_1), \quad (1.66)$$

which is the *Chapman–Kolmogorov equation.*

The Chapman–Kolmogorov equation admits a simple intuitive interpretation, which is illustrated in Fig. 1.3: Starting at the point x_1 at time t_1, the process reaches the point x_3 at a later time t_3. At some fixed intermediate time t_2 the process takes on some value x_2. The probability for the transition from (x_1, t_1) to (x_3, t_3) is obtained by multiplying the probabilities for the transitions $(x_1, t_1) \longrightarrow (x_2, t_2)$ and $(x_2, t_2) \longrightarrow (x_3, t_3)$ and by summing over all possible intermediate positions x_2.

Once we have a propagator $T(x, t|x', t')$ and some initial density $p(x, t_0)$ we can construct the whole hierarchy of joint probability distributions. As we have already seen, the propagator and the initial density yield the time-dependent density $p(x, t)$. It is easy to verify that with these quantities all m-th order joint probability densities are determined through the relation

$$p_m(x_m, t_m; x_{m-1}, t_{m-1}; \ldots; x_1, t_1) = \prod_{\nu=1}^{m-1} T(x_{\nu+1}, t_{\nu+1}|x_\nu, t_\nu) p(x_1, t_1), \quad (1.67)$$

where $t_0 \leq t_1 \leq t_2 \leq \ldots \leq t_m$.

In summary, to define a stochastic Markov process we need to specify a propagator $T(x,t|x',t')$ which obeys the Chapman–Kolmogorov equation (1.66), and an initial density $p(x,t_0)$. The classification of Markov processes therefore amounts, essentially, to classifying the solutions of the Chapman–Kolmogorov equation.

1.4.2 *Differential Chapman–Kolmogorov equation*

The Chapman–Kolmogorov equation (1.66) is an integral equation for the conditional transition probability. In order to find its solutions it is often useful to consider the corresponding differential form of this equation, the *differential Chapman–Kolmogorov equation*.

We suppose that the propagator $T(x,t|x',t')$ is differentiable with respect to time. Differentiating eqn (1.66) we get the differential Chapman–Kolmogorov equation,

$$\frac{\partial}{\partial t}T(x,t|x',t') = \mathcal{A}(t)T(x,t|x',t'). \qquad (1.68)$$

Here, $\mathcal{A}$ is a linear operator which generates infinitesimal time translations. It is defined by the action on some density $\rho(x)$,

$$\mathcal{A}(t)\rho(x) \equiv \lim_{\Delta t \to 0} \frac{1}{\Delta t} \int dx' \left[T(x,t+\Delta t|x',t) - \delta(x-x') \right] \rho(x')$$

$$= \lim_{\Delta t \to 0} \frac{1}{\Delta t} \left[\int dx' \, T(x,t+\Delta t|x',t)\rho(x') - \rho(x) \right]. \qquad (1.69)$$

In the general case the operator $\mathcal{A}$ may depend on t. However, for a homogeneous Markov process the propagator $T(x,t+\Delta t|x',t)$ for the time interval from t to $t+\Delta t$ does not depend on t and, thus, the generator is time independent in this case.

For a homogeneous Markov process we can write the propagator as $T_\tau(x|x')$ where $\tau = t-t' \geq 0$ denotes the difference of its time arguments. The Chapman–Kolmogorov equation can then be rewritten as

$$T_{\tau+\tau'}(x|x') = \int dx'' \, T_\tau(x|x'')T_{\tau'}(x''|x'), \quad \tau, \tau' \geq 0. \qquad (1.70)$$

Once the generator $\mathcal{A}$ is known, the solution of the Chapman–Kolmogorov equation for a homogeneous Markov process can be written formally as

$$T_\tau(x|x') = \exp(\tau\mathcal{A}) \delta(x-x'), \quad \tau \geq 0. \qquad (1.71)$$

These equations express the fact that the one-parameter family $\{T_\tau|\tau \geq 0\}$ of conditional transition probabilities represents a dynamical semigroup. The term *semi*-group serves to indicate that the family $\{T_\tau|\tau \geq 0\}$ is, in general, not a full group since the parameter τ is restricted to non-negative values.

From a physical viewpoint the semigroup property derives from the irreversible nature of stochastic processes: Suppose that at time t_0 an initial density

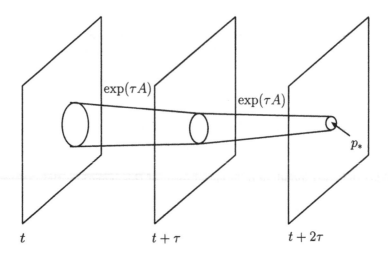

FIG. 1.4. Schematic picture of the irreversible nature of a dynamical semigroup.
The figure indicates the contraction of the range of the propagators T_τ for
increasing τ.

$p(x, t_0)$ is given. The above family of conditional transition probabilities then
allows us to propagate uniquely this initial density to times $t = t_0 + \tau > t_0$.
With the help of eqns (1.61) and (1.71) we get

$$p(x, t) = \exp(\tau A)\, p(x, t_0). \tag{1.72}$$

However, the resulting process is, in general, not invariant with respect to time
inversions, which means that it is not possible to find for *each* $p(x, t_0)$ a probability density $p(x, t)$ at some earlier time $t < t_0$ which evolves into $p(x, t_0)$.
Mathematically, this means that the range of the operator $\exp(\tau A)$ is contracting for increasing τ, that is, this operator is not invertible in the total space of
all probability distributions (Fig. 1.4). This is why irreversible processes enable
one to distinguish the future from the past.

The above situation occurs, for example, if the process relaxes to a unique
stationary state $p_*(x)$ in the limit of long times,

$$\lim_{t \to +\infty} p(x, t) = p_*(x). \tag{1.73}$$

Such processes arise in statistical mechanics, for example, when one studies the
relaxation of closed physical systems to an equilibrium state. Then, clearly, the
stationary density $p_*(x)$ must be a zero mode of the generator,

$$A p_*(x) = 0. \tag{1.74}$$

Consequently, also backward propagation leaves $p_*(x)$ invariant, and there is no
way of obtaining any specific distribution other than $p_*(x)$ at any prior time.

In the following sections we shall introduce three basic types of Markov processes which can be distinguished by the form of their generator or, equivalently, by the short-time behaviour of their propagator.

1.4.3 Deterministic processes and Liouville equation

The simplest example of a Markov process is provided by a deterministic process. It is defined by some initial density $p(x, t_0)$ and by a propagator which describes a deterministic time evolution corresponding to a system of ordinary differential equations

$$\frac{d}{dt}x(t) = g(x(t)), \quad x(t) \in \mathbb{R}^d. \tag{1.75}$$

Here, $g(x)$ denotes a d-dimensional vector field. For simplicity we assume this system to be autonomous, i.e. that the vector field $g(x)$ does not depend explicitly on time, such that the resulting process becomes homogeneous (see below). The most prominent examples of this type of process are the processes that arise in equilibrium statistical mechanics, in which case eqn (1.75) represents the Hamiltonian equations of motion in phase space.

To such an ordinary differential equation there corresponds the phase flow which is denoted by $\Phi_t(x)$. This means that for fixed x the phase curve

$$t \mapsto \Phi_t(x) \tag{1.76}$$

represents the solution of eqn (1.75) corresponding to the initial value $\Phi_0(x) = x$.

The sample paths of the deterministic process are given by the phase curves (1.76). Thus, the propagator for such a process is simply

$$T(x, t|x', t') = \delta\left(x - \Phi_{t-t'}(x')\right). \tag{1.77}$$

This equation tells us that the probability density for the process to reach the point x at time t, under the condition that it was in x' at time t', is different from zero if and only if the phase flow carries x' to x in the time interval from t' to t, that is, if and only if $x = \Phi_{t-t'}(x')$.

As is easily verified the propagator (1.77) fulfils the relations (1.58) and (1.59). On using the group property of the phase flow, which may be expressed by

$$\Phi_t(\Phi_s(x)) = \Phi_{t+s}(x), \tag{1.78}$$

one can also show that (1.77) satisfies the Chapman–Kolmogorov equation. Thus we have constructed a solution of the Chapman–Kolmogorov equation and defined a simple Markov process.

In order to find the infinitesimal generator $\mathcal{A}$ for a deterministic process we insert expression (1.77) into definition (1.69):

$$
\begin{aligned}
\mathcal{A}\rho(x) &= \lim_{\Delta t \to 0} \frac{1}{\Delta t} \int dx' \left[\delta(x - \Phi_{\Delta t}(x')) - \delta(x - x') \right] \rho(x') \\
&= \frac{d}{dt}\bigg|_{t=0} \int dx' \delta(x - \Phi_t(x'))\rho(x') = \int dx' g_i(x') \left[\frac{\partial}{\partial x'_i} \delta(x - x') \right] \rho(x') \\
&= -\frac{\partial}{\partial x_i} \left[g_i(x)\rho(x) \right].
\end{aligned}
\tag{1.79}
$$

Here, the $g_i(x)$ denote the components of the vector field $g(x)$, and summation over the index i is understood. Thus, the generator of the deterministic process reads

$$
\mathcal{A} = -\frac{\partial}{\partial x_i} g_i(x),
\tag{1.80}
$$

and the differential Chapman–Kolmogorov equation takes the form

$$
\frac{\partial}{\partial t} T(x, t | x', t') = -\frac{\partial}{\partial x_i} \left[g_i(x) T(x, t | x', t') \right].
\tag{1.81}
$$

This is the Liouville equation for a deterministic Markov process corresponding to the differential equation (1.75). Of course, the density $p(x, t)$ fulfils an equation which is formally identical to eqn (1.81).

1.4.4   Jump processes and the master equation

The deterministic process of the preceding subsection is rather simple in that it represents a process whose sample paths are solutions of a deterministic equation of motion; only the initial conditions have been taken to be random. Now we consider processes with discontinuous sample paths which follow true probabilistic dynamics.

1.4.4.1   Differential Chapman–Kolmogorov equation

We require that the sample paths of $X(t)$, instead of being smooth solutions of a differential equation, perform instantaneous jumps. To formulate a differential Chapman–Kolmogorov equation for such a jump process we have to construct an appropriate short-time behaviour for its propagator.

To this end we introduce the transition rates $W(x|x', t)$ for the jumps which are defined as follows. The quantity $W(x|x', t)\Delta t$ is equal to the probability density for an instantaneous jump from the state x' into the state x within the infinitesimal time interval $[t, t+\Delta t]$, under the condition that the process is in x' at time t. Given that $X(t) = x'$, the total rate for a jump at time t is therefore

$$
\Gamma(x', t) = \int dx W(x|x', t),
\tag{1.82}
$$

which means that $\Gamma(x', t)\Delta t$ is the conditional probability that the process leaves the state x' at time t by a jump to some other state.

An appropriate short-time behaviour for the propagator can now be formulated as

$$T(x, t + \Delta t | x', t) = W(x|x', t)\Delta t + (1 - \Gamma(x', t)\Delta t)\,\delta(x - x') + \mathcal{O}(\Delta t^2). \quad (1.83)$$

The first term on the right-hand side gives the probability for a jump from x' to
x during the time interval from t to $t + \Delta t$. The prefactor of the delta function of
the second term is just the probability that no jump occurs and that, therefore,
the process is still in x' at time $t + \Delta t$. As it should do, for $\Delta t \to 0$ the propagator
approaches the delta function $\delta(x - x')$ (see eqn (1.59)). In view of (1.82) the
propagator also satisfies the normalization condition (1.58).

It is now an easy task to derive the differential Chapman–Kolmogorov (1.68)
equation for the jump process. Inserting (1.83) into (1.69) we find for the generator of a jump process

$$
\begin{aligned}
\mathcal{A}(t)\rho(x) &= \lim_{\Delta t \to 0} \frac{1}{\Delta t} \int dx'\, [T(x, t + \Delta t | x', t) - \delta(x - x')]\,\rho(x') \\
&= \lim_{\Delta t \to 0} \frac{1}{\Delta t} \int dx'\, [W(x|x', t)\Delta t - \Gamma(x', t)\Delta t \delta(x - x')]\,\rho(x') \\
&= \int dx'\, W(x|x', t)\rho(x') - \Gamma(x, t)\rho(x) \\
&= \int dx'\, [W(x|x', t)\rho(x') - W(x'|x, t)\rho(x)]. \quad (1.84)
\end{aligned}
$$

In the last step we have used definition (1.82) for the total transition rate. This
immediately leads to the equation of motion for the propagator,

$$\frac{\partial}{\partial t} T(x, t | x', t') = \mathcal{A}(t)T(x, t | x', t') \quad (1.85)$$

$$= \int dx''\, [W(x|x'', t)T(x'', t | x', t') - W(x''|x, t)T(x, t | x', t')].$$

This is the differential Chapman–Kolmogorov equation for the jump process. It
is also called the *master equation*. The same equation holds, of course, for the
density $p(x, t)$,

$$\frac{\partial}{\partial t} p(x, t) = \mathcal{A}(t)p(x, t)$$

$$= \int dx'\, [W(x|x', t)p(x', t) - W(x'|x, t)p(x, t)]. \quad (1.86)$$

This equation is also referred to as the master equation. It must be kept in mind,
however, that the master equation is really an equation for the conditional transition probability of the process. This is an important point since a time-evolution
equation for the first-order density $p(x, t)$ is not sufficient for the definition of a
stochastic Markov process.

The master equation (1.86) has an intuitive physical interpretation as a balance equation for the rate of change of the probability density at x. The first term on its right-hand side describes the rate of increase of the probability density at x which is due to jumps from other states x' into x. The second term is the rate for the loss of probability due to jumps out of the state x.

Note that in the above derivation we have not assumed that the process is homogeneous in time. In the homogeneous case the transition rates must be time independent, $W(x|x',t) = W(x|x')$. Then also the total transition rate $\Gamma(x')$ is time independent, of course.

For the case of an integer-valued process, which may be univariate or multivariate, we write $X(t) = N(t)$. The first-order probability distribution for such a discrete process is defined by

$$P(n,t) = \mu\left(N(t) = n\right),\tag{1.87}$$

whereas the propagator takes the form

$$T(n,t|n',t') = \mu(N(t) = n|N(t') = n').\tag{1.88}$$

The corresponding master equation for the probability distribution then takes the form

$$\frac{\partial}{\partial t}P(n,t) = \sum_{n'=-\infty}^{+\infty} \left[W(n|n',t)P(n',t) - W(n'|n,t)P(n,t)\right],\tag{1.89}$$

with a similar equation for the propagator.

1.4.4.2   *The homogeneous and the non-homogeneous Poisson process*   Let us discuss two simple examples for an integer-valued jump process, namely the homogeneous and the non-homogeneous Poisson process. These examples will be important later on.

As a physical example we consider a classical charged matter current described by some current density $\vec{j}(\vec{x},t)$ of the form

$$\vec{j}(\vec{x},t) = \vec{j}(\vec{x})e^{-i\omega_0 t} + \vec{j}(\vec{x})^* e^{i\omega_0 t}.\tag{1.90}$$

The current density is assumed to be transverse, that is $\vec{\nabla}\cdot\vec{j} = 0$. As is known from quantum electrodynamics (Bjorken and Drell, 1965) such a current creates a radiation field, emitting photons of frequency ω_0 at a certain rate γ. An explicit expression for this rate can be derived with the help of the interaction Hamiltonian

$$H_I(t) = -\frac{e}{c}\int d^3x\,\vec{j}(\vec{x},t)\cdot\vec{A}(\vec{x})\tag{1.91}$$

which governs the coupling between the classical matter current $\vec{j}(\vec{x},t)$ and the quantized radiation field $\vec{A}(\vec{x})$. Treating $H_I(t)$ as a time-dependent interaction and applying Fermi's golden rule one obtains an explicit expression for the rate of photon emissions

$$\gamma = \frac{e^2 \omega_0}{2\pi \hbar c^3} \int d\Omega \sum_{\lambda=1,2} |\vec{\epsilon}(\vec{k}, \lambda) \cdot \vec{j}(\vec{k})|^2, \qquad (1.92)$$

where $\vec{j}(\vec{k})$ denotes the Fourier transform of the current density,

$$\vec{j}(\vec{k}) = \int d^3 x \, \vec{j}(\vec{x}) e^{-i\vec{k}\cdot\vec{x}}. \qquad (1.93)$$

Obviously, the relation (1.92) involves an integral over the solid angle $d\Omega$ into the direction $\vec{k}$ of the emitted photons and a sum over the two transverse polarization vectors $\vec{\epsilon}(\vec{k}, \lambda)$.

The rate γ describes the probability per unit of time for a single photon emission. We are interested in determining the number $N(t)$ of photon emissions over a finite time interval from 0 to t. To this end we assume that the current, acting as a fixed classical source of the radiation field, is not changed during the emission processes. We then have a constant rate γ for all photon emissions and each emission process occurs under identical conditions and independently from all the previous ones. The number $N(t)$ then becomes a stochastic Markov process, known as a (homogeneous) Poisson process. It is governed by the following master equation for the propagator,

$$\frac{\partial}{\partial t} T(n, t|n', t') = \gamma T(n-1, t|n', t') - \gamma T(n, t|n', t'). \qquad (1.94)$$

In the following we study the deterministic initial condition $N(0) = 0$, that is

$$P(n, 0) = \delta_{n,0}. \qquad (1.95)$$

The Poisson process provides an example for a one-step process for which, given that $N(t) = n$, only jumps to neighbouring states $n \pm 1$ are possible. The Poisson process is even simpler for only jumps from n to $n + 1$ are possible and the jump rate is a constant γ which does not depend on n. It should thus be no surprise that the master equation for the Poisson process can easily be solved exactly. This may be done by investigating the characteristic function

$$G(k, t) = \sum_{n=n'}^{+\infty} e^{ik(n-n')} T(n, t|n', t'), \qquad (1.96)$$

with n' and t' held fixed. We have extended here the summation from $n = n'$ to infinity to take into account that

$$T(n, t|n', t') = 0 \quad \text{for } n < n'. \qquad (1.97)$$

This is a direct consequence of the fact that the jumps only increase the number $N(t)$. Inserting the characteristic function (1.96) into the master equation (1.94) we obtain

$$\frac{\partial}{\partial t}G(k,t) = \gamma \left[e^{ik} - 1\right] G(k,t),$$ (1.98)

which is immediately solved to yield

$$G(k,t) = \exp\left[\gamma(t - t')\left(e^{ik} - 1\right)\right],$$ (1.99)

where we have used the initial condition $G(k, t = t') = 1$ which, in turn, follows from $T(n, t'|n', t') = \delta_{nn'}$. On comparing the Taylor expansion of the characteristic function,

$$G(k,t) = \sum_{n=0}^{+\infty} e^{ikn} \frac{[\gamma(t - t')]^n}{n!} e^{-\gamma(t-t')}$$

$$= \sum_{n=n'}^{+\infty} e^{ik(n-n')} \frac{[\gamma(t - t')]^{n-n'}}{(n - n')!} e^{-\gamma(t-t')}$$ (1.100)

with definition (1.96) we obtain for the propagator of the Poisson process

$$T(n, t|n', t') = \frac{[\gamma(t - t')]^{n-n'}}{(n - n')!} e^{-\gamma(t-t')}, \quad n \geq n'.$$ (1.101)

As should have been expected the process is homogeneous in time. It is also homogeneous in space in the sense that $T(n, t|n', t')$ only depends on the difference $n - n'$. The corresponding distribution $P(n, t)$ becomes

$$P(n,t) = \frac{(\gamma t)^n}{n!} e^{-\gamma t}, \quad n \geq 0,$$ (1.102)

which represents a Poisson distribution with mean and variance given by

$$E(N(t)) = \text{Var}(N(t)) = \gamma t.$$ (1.103)

Figure 1.5 shows a realization of the Poisson process.

The homogeneous Poisson process describes the number $N(t)$ of independent events in the time interval from 0 to t, where each event occurs with a constant rate γ. Let us assume that for some reason this rate may change in time, that is $\gamma = \gamma(t)$. The resulting process is called a *non-homogeneous Poisson process*, since it is no longer homogeneous in time. As one can check by insertion into the master equation,

$$\frac{\partial}{\partial t}T(n, t|n', t') = \gamma(t)T(n - 1, t|n', t') - \gamma(t)T(n, t|n', t'),$$ (1.104)

the propagator of the non-homogeneous Poisson process is given by

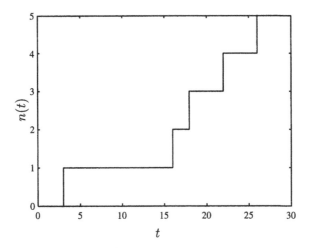

FIG. 1.5. Sample path of the Poisson process which was obtained from a numerical simulation of the process.

$$T(n,t|n',t') = \frac{[\mu(t,t')]^{n-n'}}{(n-n')!}e^{-\mu(t,t')}, \quad n \geq n', \tag{1.105}$$

and by $T(n,t|n',t') = 0$ for $n < n'$. Here, we have set

$$\mu(t,t') \equiv \int_{t'}^{t} ds\, \gamma(s). \tag{1.106}$$

The probability $P(n,t)$ is accordingly found to be

$$P(n,t) = \frac{[\mu(t,0)]^{n}}{n!}e^{-\mu(t,0)}, \quad n \geq 0. \tag{1.107}$$

By means of the above explicit expressions we immediately get the generalization of the relations (1.103) for the non-homogeneous Poisson process,

$$E(N(t)) = \text{Var}(N(t)) = \mu(t,0). \tag{1.108}$$

Let us also determine the two-time correlation function which is defined by

$$E(N(t)N(s)) = E(N(s)N(t))$$
$$= \sum_{n_1,n_2} n_2 n_1 T(n_2,t|n_1,s)T(n_1,s|0,0), \tag{1.109}$$

where in the second line we have assumed, without restriction, that $t \geq s$. Differentiating with respect to t and invoking the master equation we find

$$\frac{\partial}{\partial t}\mathrm{E}(N(t)N(s))$$

$$= \gamma(t) \sum_{n_1,n_2} \left[n_2 T(n_2 - 1, t|n_1, s) - n_2 T(n_2, t|n_1, s) \right] n_1 T(n_1, s|0, 0)$$

$$= \gamma(t) \sum_{n_1,n_2} n_1 T(n_2, t|n_1, s) T(n_1, s|0, 0)$$

$$= \gamma(t) \sum_{n_1} n_1 T(n_1, s|0, 0) = \gamma(t)\mathrm{E}(N(s))$$

$$= \gamma(t)\mu(s, 0). \tag{1.110}$$

In the second step we have performed the shift $n_2 \to n_2 + 1$ of the summation variable n_2 in the gain term of the master equation. Hence, on using the initial condition $\mathrm{E}(N(s)N(s)) = \mu(s, 0)^2 + \mu(s, 0)$ we find the following expression for the two-time correlation function of the non-homogeneous Poisson process,

$$\mathrm{E}(N(t)N(s)) = \mu(t, 0)\mu(s, 0) + \mu(s, 0). \tag{1.111}$$

1.4.4.3 *Increments of the Poisson process*   It will be of interest for later applications to derive the properties of the increment $dN(t)$ of the non-homogeneous Poisson process. For any time increment $dt \geq 0$ the corresponding increment of the Poisson process is defined by

$$dN(t) \equiv N(t + dt) - N(t), \quad dt \geq 0. \tag{1.112}$$

Note that we do not assume at this point that dt is infinitesimally small. The probability distribution of the increment $dN(t)$ may be found from the relation

$$\mu(dN(t) = dn) = \sum_{n=0}^{\infty} T(n + dn, t + dt|n, t)P(n, t). \tag{1.113}$$

Since the process is homogeneous in space the sum over n can immediately be carried out and using eqn (1.105) we find

$$\mu(dN(t) = dn) = T(dn, t + dt|0, t) = \frac{[\mu(t + dt, t)]^{dn}}{dn!} e^{-\mu(t+dt,t)}, \tag{1.114}$$

where $dn = 0, 1, 2, \ldots$. The first and second moments of the Poisson increment are therefore given by

$$\mathrm{E}(dN(t)) = \mu(t + dt, t), \tag{1.115}$$

$$\mathrm{E}(dN(t)^2) = \mu(t + dt, t) + \mu(t + dt, t)^2. \tag{1.116}$$

If we now take dt to be infinitesimally small we see that, apart from terms of order $\mathcal{O}(dt^2)$,

$$\mathrm{E}(dN(t)^2) = \mathrm{E}(dN(t)) = \gamma(t)dt. \tag{1.117}$$

In addition, we also observe that this last relation becomes a deterministic relation in the limit $dt \to 0$. Namely, according to eqn (1.114) the probability for the event $dN(t) \geq 2$ is a term of order dt^2, and, consequently

$$\mu \left[dN(t)^2 = dN(t) \right] = 1 + \mathcal{O}(dt^2). \tag{1.118}$$

Thus, apart from terms of order dt^2, the equation

$$dN(t)^2 = dN(t) \tag{1.119}$$

holds with probability 1 and can be regarded as a deterministic relation in that limit.

These results are easily generalized to the case of a multivariate Poisson process $N(t) = (N_1(t), N_2(t), \ldots, N_d(t))$. We take the components $N_i(t)$ to be statistically independent, non-homogeneous Poisson processes with corresponding rates $\gamma_i(t)$. It follows that the relations

$$\mathrm{E}(dN_i(t)) = \gamma_i(t)dt, \tag{1.120}$$
$$dN_i(t)dN_j(t) = \delta_{ij}dN_i(t), \tag{1.121}$$

hold in the limit $dt \longrightarrow 0$, where terms of order $\mathcal{O}(dt^2)$ have been neglected. These important results will be useful later on when we construct the stochastic differential equation governing a piecewise deterministic process.

1.4.5 Diffusion processes and Fokker–Planck equation

Up to now we have considered two types of stochastic Markov processes whose sample paths are either smooth solutions of a differential equation or else discontinuous paths which are broken by instantaneous jumps. It can be shown that the realizations of a Markov process are continuous with probability one if

$$\lim_{\Delta t \to 0} \frac{1}{\Delta t} \int\limits_{|x-x'|>\varepsilon} dx \, T(x, t + \Delta t | x', t) = 0 \tag{1.122}$$

for any $\varepsilon > 0$. This equation implies that the probability for a transition during Δt whose size is larger than ε decreases more rapidly than Δt as Δt goes to zero.

Of course, deterministic processes must fulfil this continuity condition. In fact, we have for a deterministic process (θ denotes the unit step function)

$$\lim_{\Delta t \to 0} \frac{1}{\Delta t} \int\limits_{|x-x'|>\varepsilon} dx T(x, t + \Delta t | x', t)$$

$$= \lim_{\Delta t \to 0} \frac{1}{\Delta t} \int\limits_{|x-x'|>\varepsilon} dx \delta(x - \Phi_{\Delta t}(x'))$$

$$= \lim_{\Delta t \to 0} \frac{1}{\Delta t} \theta(|x' - \Phi_{\Delta t}(x')| - \varepsilon)$$

$$= 0. \tag{1.123}$$

In the case of a jump process the continuity condition is clearly violated if the jump rate $W(x|x',t)$ allows jumps of size larger than some $\varepsilon > 0$,

$$\lim_{\Delta t \to 0} \frac{1}{\Delta t} \int_{|x-x'|>\varepsilon} dx T(x, t + \Delta t | x', t) = \int_{|x-x'|>\varepsilon} W(x|x',t) > 0. \qquad (1.124)$$

There exists, however, a further class of stochastic Markov processes, the diffusion processes, which are not deterministic processes but also satisfy the continuity condition. We derive the differential Chapman–Kolmgorov equation for a multivariate diffusion process $X(t) = (X_1(t), X_2(t), \ldots, X_d(t))$ by investigating a certain limit of a jump process which is described by the master equation (1.86). To this end, we write the transition rate as

$$W(x|x',t) = f(x', y, t), \qquad (1.125)$$

where $y = x - x'$, that is as a function f of the starting point x', of the jump increment y, and of time t. Inserting (1.125) into the master equation (1.89) we find

$$\frac{\partial}{\partial t} p(x,t) = \int dy f(x - y, y, t) p(x - y, t) - p(x, t) \int dy f(x, y, t). \qquad (1.126)$$

The fundamental assumption is now that $f(x', y, t)$ varies smoothly with x', but that it is a function of y which is sharply peaked around $y \approx 0$. In addition it is assumed that $p(x, t)$ varies only slowly with x on scales of the order of the width of f. This enables one to expand the gain term $f(x - y, y, t) p(x - y, t)$ to second order in y,

$$\frac{\partial}{\partial t} p(x,t) = \int dy f(x, y, t) p(x, t) - \int dy\, y_i \frac{\partial}{\partial x_i} [f(x, y, t) p(x, t)] \qquad (1.127)$$
$$+ \int dy\, \frac{1}{2} y_i y_j \frac{\partial^2}{\partial x_i \partial x_j} [f(x, y, t) p(x, t)] - p(x, t) \int dy f(x, y, t).$$

The indices i, j label the different components of the multivariate process and a summation over repeated indices is understood. Taking fully into account the strong dependence of $f(x', y, t)$ on y, we have *not* expanded with respect to the y-dependence of the second argument of $f(x - y, y, t)$. We see that the first term on the right-hand side cancels the loss term. Thus, introducing the first and the second moment of the jump distribution as

$$g_i(x, t) \equiv \int dy\, y_i f(x, y, t), \quad D_{ij}(x, t) \equiv \int dy\, y_i y_j f(x, y, t), \qquad (1.128)$$

we finally arrive at the differential Chapman–Kolmgorov equation for a diffusion process,

$$\frac{\partial}{\partial t} p(x,t) = -\frac{\partial}{\partial x_i} [g_i(x, t) p(x, t)] + \frac{1}{2} \frac{\partial^2}{\partial x_i \partial x_j} [D_{ij}(x, t) p(x, t)]. \qquad (1.129)$$

This is the famous *Fokker–Planck equation* for a diffusion process. A formally identical equation holds for the propagator of the process.

Again, this equation admits an obvious interpretation as a continuity equation for the probability density if we rewrite it as

$$\frac{\partial}{\partial t}p(x,t) + \frac{\partial}{\partial x_i}J_i(x,t) = 0, \qquad (1.130)$$

where we have introduced the probability current density

$$J_i(x,t) \equiv g_i(x,t)p(x,t) - \frac{1}{2}\frac{\partial}{\partial x_j}\left[D_{ij}(x,t)p(x,t)\right]. \qquad (1.131)$$

As we already know, the first term on the right-hand side of the Fokker–Planck equation describes a deterministic drift which corresponds to the differential equation with vector field g,

$$\frac{d}{dt}x(t) = g(x(t),t). \qquad (1.132)$$

The second term on the right-hand side of the Fokker–Planck equation describes the diffusion of the stochastic variable $X(t)$. According to its definition the matrix $D(x,t)$, known as the diffusion matrix, is symmetric and positive semidefinite.

The most prominent example of a diffusion process is obtained by considering a one-dimensional diffusion process, setting the drift equal to zero, $g(x,t) \equiv 0$, and by taking $D(x,t) \equiv 1$. This leads to the Gaussian propagator

$$T(x,t|x',t') = \frac{1}{\sqrt{2\pi(t-t')}}\exp\left(-\frac{(x-x')^2}{2(t-t')}\right), \qquad (1.133)$$

showing that the process is both homogeneous in time and in space. This process is often referred to as a *Brownian motion* process. If we further take the initial density

$$p(x,t=0) = \delta(x) \qquad (1.134)$$

the process $X(t)$ becomes the famous *Wiener process* $W(t)$.

Another important example of a diffusion process is obtained by adding to a Brownian motion process with constant diffusion coefficient D a deterministic drift corresponding to the differential equation $dx/dt = -kx$ with a constant coefficient $k > 0$. The corresponding propagator then takes the form

$$T(x,t|x',t') = \sqrt{\frac{k}{\pi D\left[1-e^{-2k(t-t')}\right]}}\exp\left(-\frac{k(x-e^{-k(t-t')}x')^2}{D\left[1-e^{-2k(t-t')}\right]}\right), \qquad (1.135)$$

and the stationary first-order distribution is given by

$$p(x) = \sqrt{\frac{k}{\pi D}}\exp\left(-\frac{kx^2}{D}\right). \qquad (1.136)$$

This defines the *Ornstein–Uhlenbeck process*. Its physical significance is due to the fact that, up to trivial linear transformations, it is essentially the only process

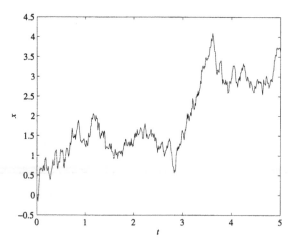

FIG. 1.6. Sample path of the Wiener process.

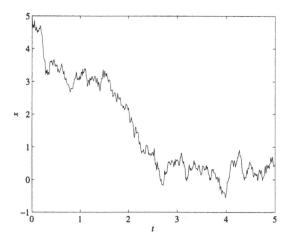

FIG. 1.7. Sample path of the Ornstein–Uhlenbeck process.

which is stationary, Gaussian and Markovian. This assertion is known as Doob's theorem. A Gaussian process is defined to be a process whose joint probability distributions are Gaussian functions. The only other process with the above three properties is the so-called completely random process, for which all m-th order joint probability distributions are products of first-order Gaussian distributions. We show in Figs. 1.6 and 1.7 realizations of the Wiener and of the Ornstein–Uhlenbeck process.

With the help of the explicit expression (1.133) for the conditional transition probability of the Wiener process we can calculate all relevant quantities of interest. For example we find for the mean and the variance of the Wiener process

$$E(W(t)) = \int dw\, wT(w, t|0, 0) = 0, \tag{1.137}$$

$$E(W(t)^2) = \int dw\, w^2 T(w, t|0, 0) = t. \tag{1.138}$$

The second relation shows the well-known linear increase of the variance with time t. The two-time correlation function of the Wiener process will also be of interest. It is defined by

$$E(W(t)W(s)) = E(W(s)W(t)) \tag{1.139}$$
$$= \int dw_2 \int dw_1\, w_2 w_1 T(w_2, t|w_1, s) T(w_1, s|0, 0),$$

where the second equation presupposed that $t \geq s$. The integrals can easily be evaluated with the help of (1.133). The result is

$$E(W(t)W(s)) = \min(t, s), \tag{1.140}$$

which holds for $t \geq s$ and for $t < s$.

On using these results we can also study the properties of the increment $dW(t)$ of the Wiener process which is defined for any $dt \geq 0$ by

$$dW(t) \equiv W(t + dt) - W(t). \tag{1.141}$$

From eqns (1.137) and (1.140) we get

$$E[dW(t)] = 0, \tag{1.142}$$
$$E\left[dW(t)^2\right] = dt. \tag{1.143}$$

For a multivariate Wiener process with statistically independent components $W_i(t)$ we have the important formulae:

$$E[dW_i(t)] = 0, \tag{1.144}$$
$$E[dW_i(t)dW_j(t)] = \delta_{ij} dt, \tag{1.145}$$
$$E\left[(dW_i(t))^{2k}\right] = \mathcal{O}(dt^k), \tag{1.146}$$
$$E\left[(dW_i(t))^{2k+1}\right] = 0. \tag{1.147}$$

The last two relations hold for all positive integers k, and may be shown directly using the Gaussian propagator for the Wiener process.

1.5   Piecewise deterministic processes

We are now in a position to introduce a certain type of stochastic process which plays an important rôle in this book, namely the so-called piecewise deterministic processes (PDPs). Essentially, such a process is obtained by combining a deterministic time-evolution with a jump process.

Piecewise deterministic processes have a large variety of applications in the natural sciences and technology. A number of examples and a mathematically rigorous treatment of PDPs may be found in the book by Davis (1993). In the theory of open quantum systems PDPs provide the appropriate mathematical concept to describe the evolution of the state of a continuously monitored system (see Chapter 6). We shall present in this section three different mathematical formulations of a piecewise deterministic process, namely the Liouville master equation for its propagator, the path integral representation, and the stochastic differential equation for the random variable.

1.5.1   The Liouville master equation

The sample paths of a PDP consist of smooth deterministic pieces which are given by the solution of some differential equation and which are interrupted by instantaneous jumps. On the basis of our previous results it is now easy to write down an appropriate expression for the short-time behaviour of the propagator of a PDP,

$$T(x, t + \Delta t | x', t) = (1 - \Gamma(x')\Delta t)\, \delta\, (x - x' - g(x')\Delta t)$$
$$+ W(x|x')\Delta t + \mathcal{O}(\Delta t^2). \tag{1.148}$$

The δ-function represents the short-time behaviour which is characteristic of a deterministic process and which is obtained by expanding the flow of the deterministic differential equation (1.75),

$$\Phi_{\Delta t}(x') = x' + g(x')\Delta t + \mathcal{O}(\Delta t^2). \tag{1.149}$$

The prefactor of the δ-function involves the total rate $\Gamma(x') = \int dx W(x|x')$ for jumps out of the state x'. The factor $(1 - \Gamma(x')\Delta t)$ is therefore the probability that no jump occurs in the interval Δt.

The second term on the right-hand side of eqn (1.148) is the probability for a jump from x' to x within time Δt. Note that the conditions (1.58) and (1.59) are satisfied. Note further that we consider here, for simplicity, only processes which are homogeneous in time since the generalization to the inhomogeneous case is obvious.

On account of the above short-time behaviour we can immediately write down a differential Chapman–Kolmogorov equation for the propagator,

$$\frac{\partial}{\partial t} T(x, t | x', t') = -\frac{\partial}{\partial x_i} [g_i(x) T(x, t | x', t')] \tag{1.150}$$

$$+ \int dx'' \left[W(x|x'') T(x'', t | x', t') - W(x''|x) T(x, t | x', t') \right].$$

The first term on the right describes the deterministic drift, whereas the second part leads to the jumps $x' \rightarrow x$ with rate $W(x|x')$. Accordingly, we call eqn (1.150) the *Liouville master equation*.

1.5.2 *Waiting time distribution and sample paths*

Let us determine a central quantity for the description of PDPs, i.e. the waiting time distribution $F(\tau|x',t')$. This quantity is defined to be the probability for the next jump to occur during the time interval $[t',t'+\tau]$ under the condition that we are in the state x' at time t'. To find this quantity we observe that

$$F(\tau+d\tau|x',t') - F(\tau|x',t') \equiv dF(\tau|x',t') \tag{1.151}$$

is just the probability for the next jump to occur in the time interval $[t'+\tau,t'+\tau+d\tau]$. If we divide this probability by the probability $1-F(\tau|x',t')$ that no jump occurred in the previous interval from t' to $t'+\tau$, we get the conditional probability for a jump over the interval $d\tau$. The latter quantity must be equal to $d\tau$ times the total rate $\Gamma(\Phi_\tau(x'))$ for a jump out of the state we are presently in, namely $\Phi_\tau(x')$ (since, if the jump occurs at time $t'+\tau$, the deterministic evolution has carried us from the state x' to $\Phi_\tau(x')$). Hence we have

$$\frac{dF(\tau|x',t')}{1-F(\tau|x',t')} = \Gamma(\Phi_\tau(x'))d\tau. \tag{1.152}$$

This equation shows that $F(\tau|x',t')$ does not, in fact, depend on the time t' and we shall omit this argument in the following (note however that for a PDP which is not homogeneous in time, e.g. for a non-autonomous differential equation, it does depend on t').

The above relation leads to the differential equation

$$\frac{d}{d\tau}\ln\left[1-F(\tau|x')\right] = -\Gamma(\Phi_\tau(x')) \tag{1.153}$$

which is easily solved to yield

$$F(\tau|x') = 1 - \exp\left[-\int_0^\tau ds\,\Gamma(\Phi_s(x'))\right], \tag{1.154}$$

where we have used the initial condition $F(\tau=0|x')=0$. Given that we are in the state x', expression (1.154) is the cumulative distribution function for the random waiting time τ of the PDP.

For a pure jump process $(g=0)$ we simply have

$$F(\tau|x') = 1 - \exp\left[-\Gamma(x')\tau\right], \tag{1.155}$$

demonstrating that for $\Gamma(x') > 0$ pure jump processes have an exponentially distributed waiting time. For $\Gamma(x')=0$ we have $F\equiv 0$ which means that the waiting time is infinite: The process never leaves the state x' which is called a trapping state of the process.

For a true PDP $(g\neq 0)$ the waiting time distribution is, in general, not an exponential function. As eqn (1.154) shows this is a direct consequence of the deterministic time evolution between the jumps as a result of which the total transition rate can change in a non-trivial way. Since $\Gamma(x')$ is non-negative the function

$$I(\tau|x') = \int_0^\tau ds\, \Gamma(\Phi_s(x')) \qquad (1.156)$$

increases monotonically with τ. Thus, we can basically distinguish two cases:

1. If the function $I(\tau|x')$ is not bounded it follows that

$$\lim_{\tau\to\infty} I(\tau|x') = +\infty, \qquad (1.157)$$

and, therefore, we have

$$\lim_{\tau\to\infty} F(\tau|x') = 1. \qquad (1.158)$$

This implies that the process jumps with certainty after some time. One can obtain a realization of the waiting time τ for example by making use of the inversion method. To this end, one first draws a random number η which is uniformly distributed over the interval $[0,1]$. The random waiting time may then be determined by solving the implicit equation

$$F(\tau|x') = \eta \qquad (1.159)$$

for τ (see Fig. 1.8).

2. If, on the other hand, the function $I(\tau|x')$ is bounded it follows that the limit

$$\lim_{\tau\to\infty} I(\tau|x') = I(\infty|x') < \infty \qquad (1.160)$$

exists. In this case we find

$$\lim_{\tau\to\infty} F(\tau|x') = 1 - \exp[-I(\infty|x')] \equiv 1 - q < 1. \qquad (1.161)$$

This means that with a finite probability q, where $0 < q \leq 1$, the process no longer jumps. The quantity q is called the *defect*. Formally, this situation can be dealt with by defining the state space of the random waiting time to be $\mathbb{R}_+ \cup \{\infty\}$. Then q is the probability for the event $\tau = \infty$. Again a realization of τ can be obtained with the help of the inversion method (see Fig. 1.8). Having drawn a uniformly distributed random number $\eta \in [0,1]$ one first decides whether $\eta \geq 1 - q$ or $\eta < 1 - q$. In the former case one sets $\tau = \infty$, in the latter case τ is to be determined by eqn (1.159).

We have just described how to determine a realization of the random time intervals τ between the jumps of a PDP. The above procedure allows us to design a simple algorithm for the generation of a sample path $x(t)$ of the PDP:

1. Assume that at some time t_0 the realization has reached the state x_0 through the preceding jump. In the case that t_0 is the initial time, x_0

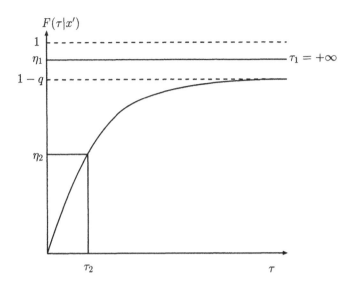

FIG. 1.8. The figure shows a typical waiting-time distribution $F(\tau|x')$ with defect q and illustrates how a realization of τ is obtained by the inversion method. η_1 and η_2 are realizations of a uniform random number $\eta \in [0,1]$. For η_1 the corresponding realization of the waiting time is infinite, $\tau_1 = +\infty$, while for η_2 the realization of the waiting time is equal to τ_2.

must be drawn from the initial probability density $p(x_0, t_0)$. In any case we have

$$x(t_0) = x_0. \tag{1.162}$$

2. Draw a random waiting time τ from the distribution function $F(\tau|x_0)$ by employing the procedure described above. In the time interval from t_0 to $t_0 + \tau$ the realization $x(t)$ then follows the solution of the deterministic equation, that is,

$$x(t_0 + s) = \Phi_s(x_0) \tag{1.163}$$

for $0 \leq s \leq \tau$.

3. At time $t_0 + \tau$ the sample path performs an instantaneous jump into some state z which is a random variable to be drawn from the probability density

$$Q(z|x(t_0 + \tau)) = \frac{W(z|x(t_0 + \tau))}{\Gamma(x(t_0 + \tau))}. \tag{1.164}$$

Note that $Q(z|x)$ is just the conditional probability density for a jump into the state z, given that we already know that a jump takes place out of the state x. Note further that $Q(z|x)$ is correctly normalized as

$$\int dz Q(z|x) = 1. \tag{1.165}$$

Finally, we set

$$x(t_0 + \tau) = z \tag{1.166}$$

and proceed with the first step above.

This algorithm can easily be translated into a numerical computer program. Using such a program we can generate a sample of realizations of the process and then estimate all quantities of interest as sample averages. This is the essence of the stochastic simulation technique, which will be illustrated with the help of a number of examples in Chapters 6, 7 and 8.

1.5.3   Path integral representation of PDPs

The above algorithm for the generation of the sample paths of a PDP yields the structure of the possible paths of the process and the probability of their occurrence. In mathematical terms this means that the propagator $T(x, t|x', t')$ of the PDP may be represented in terms of those sample paths that start in x' at time t' and end in x at time t. Each of these paths contributes with a certain probability, and summing over all paths with an appropriate weighting we get the propagator. This path integral representation of PDPs will be derived in the present section.

Up to now we have characterized a PDP by its differential Chapman–Kolmogorov equation (1.150), that is by its short-time asymptotics. There is, however, an equivalent integral representation for the propagtor of a PDP which is given by

$$T(x, t|x', t') = T^{(0)}(x, t|x', t') \tag{1.167}$$

$$+ \int_{t'}^{t} ds \int dy \int dz \, T^{(0)}(x, t|y, s)W(y|z)T(z, s|x', t').$$

Here, we have introduced the quantity

$$T^{(0)}(x, t|x', t') = [1 - F(t - t'|x')] \, \delta \left(x - \Phi_{t-t'}(x') \right) \tag{1.168}$$

which is the conditional probability density for the process to reach the state x at time t without any jump, given that it was in x' at time t'. As is easily checked, the solutions of (1.167) fulfil eqns (1.58) and (1.59). We may also verify by an explicit calculation that the solutions of (1.167) satisfy the differential Chapman–Kolmogorov equation (1.150).

We will not go into the details of these calculations since (1.167) allows a simple interpretation (see Fig. 1.9). The propagator is written as a sum of two terms, corresponding to two possibilities for the process to proceed from (x', t') to (x, t): The first possibility is that no jump occurs in the time interval from t' to t. The corresponding conditional probability density is given by the quantity $T^{(0)}(x, t|x', t')$ introduced above. The second possibility is that at least one jump occurs. The corresponding contribution is the second term on the right-hand

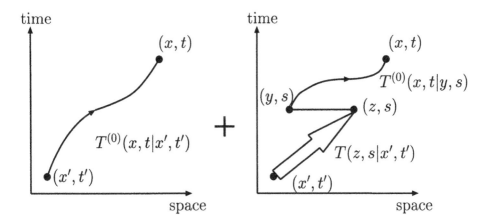

FIG. 1.9. Graphical representation of the Kolmogorov forward equation (1.167).

side of eqn (1.167). The variable s denotes the instant of the last jump.[3] Then $T(z, s|x', t')$ is the probability to be in some state z immediately before the jump, and $W(y|z)ds$ is the probability for a jump from z to y during ds. Multiplying this by $T^{(0)}(x, t|y, s)$ we get the probability that the process reaches the state z at time s, that it then jumps into the state y, and that it reaches the state x in the remaining time interval without any further jumps. The total contribution to $T(x, t|x', t')$ is found by integrating over all possible jump times s and over all intermediate states z and y.

Equation (1.167) is the *Kolmogorov forward equation*. It is an integral representation for the propagator of a PDP which directly leads to the path integral representation. Our above interpretation makes it natural to decompose the propagator of the process into separate terms $T^{(N)}(x, t|x', t')$, each of which represents the contribution from those paths with exactly N jumps:

$$T(x, t|x', t') = \sum_{N=0}^{\infty} \varepsilon^N T^{(N)}(x, t|x', t'). \tag{1.169}$$

The first term in this sum has already been defined above. We have introduced a formal expansion parameter ε which will be set equal to 1 afterwards. We introduce this parameter also into the integral equation by regarding the rates $W(y|z)$ formally as quantities of order ε. In this way the decomposition (1.169) becomes a perturbative expansion in powers of the transition rates.

If we insert eqn (1.169) into eqn (1.167) and collect terms of the same order in ε we find the following recursion relation which is valid for $N \geq 1$,

---

[3]We exclude the possibility that an infinite number of jumps accumulates at a point of the time interval.

$$T^{(N)}(x,t|x',t') = \int_{t'}^{t} ds \int dy \int dz \, T^{(0)}(x,t|y,s)W(y|z)T^{(N-1)}(z,s|x',t').$$

$$(1.170)$$

On iterating this equation one obtains $T^{(N)}(x,t|x',t')$ for all N. Substituting the result into (1.169) and setting $\varepsilon = 1$ we finally get the path integral representation of the propagator,

$$T(x,t|x',t') = T^{(0)}(x,t|x',t') \tag{1.171}$$

$$+ \sum_{N=1}^{\infty} \int_{t'}^{t} ds_N \int_{t'}^{s_N} ds_{N-1} \ldots \int_{t'}^{s_2} ds_1 \int dy_N dz_N \int dy_{N-1} dz_{N-1} \ldots \int dy_1 dz_1$$

$$T^{(0)}(x,t|y_N,s_N)W(y_N|z_N)T^{(0)}(z_N,s_N|y_{N-1},s_{N-1})$$

$$\times W(y_{N-1}|z_{N-1})T^{(0)}(z_{N-1},s_{N-1}|y_{N-2},s_{N-2})\ldots W(y_1|z_1)T^{(0)}(z_1,s_1|x',t').$$

We see that the sum over paths involves a sum over the number N of jumps, a multiple integral over all intermediate jump times $s_1 < s_2 < \ldots < s_N$, and a multiple integral over the intermediate states $y_1, z_1, y_2, z_2, \ldots, y_N, z_N$. Here, z_ν is the state just before the ν-th jump, whereas y_ν is the state immediately after the ν-th jump. Finally, the integrand in eqn (1.171) is the corresponding probability for a specific path given by the number N, the s_ν and the y_ν, z_ν.

1.5.4 *Stochastic calculus for PDPs*

In the preceding sections we have formulated the dynamics of a stochastic Markov process $X(t)$ with the help of appropriate time evolution equations for the conditional propagator of the process. Alternatively, one can also describe the process in terms of an evolution equation for the random variables itself. Of course, such an equation is, in general, not a deterministic differential equation, but must be an equation of motion involving random coefficients, that is, a stochastic differential equation. This way of describing a stochastic process is well known for the case of diffusion processes. We first review briefly this case as a motivation for our discussion of piecewise deterministic processes.

1.5.4.1 *Itô calculus for diffusion processes*   Consider a multivariate diffusion process $X(t) = (X_1(t), X_2(t), \ldots, X_d(t))$ which is governed by the Fokker–Planck equation (1.129). Taking an arbitrary differentiable scalar function f one finds for the expectation value of $f(X(t))$ by employing the Fokker–Planck equation

$$\frac{\partial}{\partial t}\mathrm{E}\left[f(X(t))\right] = \int dx \, f(x)\frac{\partial}{\partial t}p(x,t)$$

$$= \mathrm{E}\left[\frac{\partial f}{\partial x_i}g_i(X(t)) + \frac{1}{2}\frac{\partial^2 f}{\partial x_i \partial x_j}D_{ij}(X(t))\right]. \tag{1.172}$$

By use of the increment $dX(t) = X(t+dt) - X(t)$ of the process we may rewrite this as follows,

$$E\left[f(X(t) + dX(t)) - f((X(t))\right] \tag{1.173}$$
$$= E\left[\frac{\partial f}{\partial x_i} g_i(X(t)) + \frac{1}{2}\frac{\partial^2 f}{\partial x_i \partial x_j} D_{ij}(X(t))\right] dt + \mathcal{O}(dt^2).$$

Now, let us take f to be one of the coordinate functions, $f(x) = x_i$. This yields

$$E\left[dX_i(t)\right] = E\left[g_i(X(t))\right] dt + \mathcal{O}(dt^2). \tag{1.174}$$

To leading order the expectation value of the increment $dX_i(t)$ is proportional to dt, where the coefficient is given by the expectation value of $g_i(X(t))$ describing the deterministic drift. Next, we take $f(x)$ to be the function $f(x) = x_i x_j$ for a fixed pair of indices (i, j). The result is then

$$E\left[dX_i(t)dX_j(t)\right] = E\left[D_{ij}(X(t))\right] dt + \mathcal{O}(dt^2). \tag{1.175}$$

For small dt the covariance matrix of the increments $dX_i(t)$ is given by dt times the expectation of the diffusion matrix.

The important point to notice is that the behaviour of the process expressed by eqns (1.174) and (1.175) can be reproduced by means of an appropriate linear combination of Wiener increments $dW_i(t)$, $i = 1, 2, \ldots, d$. These increments are supposed to be mutually independent and independent of $X(t)$. If we set

$$d\tilde{X}_i(t) \equiv g_i(X(t))dt + B_{ij}(X(t))dW_j(t), \tag{1.176}$$

with an appropriate matrix $B_{ij}(x)$, we find with the help of eqns (1.144) and (1.145) that

$$E\left[d\tilde{X}_i(t)\right] = E\left[g_i(X(t))\right] dt, \tag{1.177}$$

and

$$E\left[d\tilde{X}_i(t)d\tilde{X}_j(t)\right] = E\left[D_{ij}(X(t))\right] dt + \mathcal{O}(dt^2). \tag{1.178}$$

The last equation holds provided the matrix $B_{ij}(x)$ is related to the diffusion matrix $D_{ij}(x)$ by (remember that the summation convention is in force)

$$B_{ik}(x)B_{jk}(x) = D_{ij}(x). \tag{1.179}$$

This shows that the mean and the covariance of the increments $dX_i(t)$ and $d\tilde{X}_i(t)$ coincide to order dt. Therefore we see that on small time scales the increments $dX_i(t)$ of the diffusion process behave as the increments $d\tilde{X}_i(t)$ which are obtained by adding an appropriate linear combination of Wiener increments to the deterministic drift. Thus, one expects that the increments $dX_i(t)$ obey a *stochastic differential equation* of the form

$$dX_i(t) = g_i(X(t))dt + B_{ij}(X(t))dW_j(t). \tag{1.180}$$

It can be shown that the diffusion process $X(t)$, which was defined originally in terms of the Fokker–Planck equation for its conditional propagator, is indeed equivalent to the stochastic differential equation (1.180). A rigorous mathematical treatment of stochastic differential equations requires the introduction of the concept of stochastic integration which enables the development of a stochastic calculus for diffusion processes, known as *Itô calculus*. Equation (1.180) is then to be interpreted as a stochastic differential equation in Itô form. We do not enter here into a detailed discussion of the stochastic calculus, which is the subject of several excellent textbooks (Gardiner, 1985; Doob, 1953; Kloeden and Platen, 1992; Arnold, 1974).

Let us, however, demonstrate that the increments $d\tilde{X}_i(t)$ defined by eqn (1.176) reproduce correctly, within the desired order, all expectation values of the form (1.173) as they are determined through the Fokker–Planck equation. To show this we have to determine the expectation value

$$
\mathrm{E}\left[f(X(t) + d\tilde{X}(t)) - f(X(t))\right]
$$
$$
= \mathrm{E}\left[f(X_i(t) + g_i(X(t))dt + B_{ij}(X(t))dW_j(t)) - f(X(t))\right]. \quad (1.181)
$$

On using eqns (1.144)–(1.147) we get by expanding f around $X(t)$ and disregarding terms of order dt^2,

$$
\mathrm{E}\left[f(X(t) + d\tilde{X}(t)) - f(X(t))\right]
$$
$$
= \mathrm{E}\left[\frac{\partial f}{\partial x_i}(g_i dt + B_{ij} dW_j) + \frac{1}{2}\frac{\partial^2 f}{\partial x_i \partial x_j} B_{ik} B_{jl} dW_k dW_l\right] + \mathcal{O}(dt^2)
$$
$$
= \mathrm{E}\left[\frac{\partial f}{\partial x_i} g_i(X(t)) + \frac{1}{2}\frac{\partial^2 f}{\partial x_i \partial x_j} D_{ij}(X(t))\right] dt + \mathcal{O}(dt^2). \quad (1.182)
$$

This is seen to coincide within order dt with the expectation value (1.173) determined by the Fokker–Planck equation, which proves our statement.

A similar calculation leads to the following formula for the increment of an arbitrary function $\phi(X(t))$ of the stochastic variable, known as the Itô formula,

$$
d\phi(X(t)) = \left[\frac{\partial \phi}{\partial x_i} g_i(X(t)) + \frac{1}{2}\frac{\partial^2 \phi}{\partial x_i \partial x_j} D_{ij}(X(t))\right] dt + \frac{\partial \phi}{\partial x_i} B_{ij}(X(t))dW_j(t).
$$
$$
(1.183)
$$

We observe that this expression for the increment of ϕ may be obtained directly by expanding the difference $d\phi(X(t)) = \phi(X(t) + dX(t)) - \phi(X(t))$ in powers of $dX(t)$ and by employing the following rules of the Itô calculus:

$$
dW_i(t)dW_j(t) = \delta_{ij}dt, \quad (1.184)
$$
$$
dW_i(t)dt = 0, \quad (1.185)
$$
$$
[dW_i(t)]^k = 0, \quad \text{for } k > 2. \quad (1.186)
$$

In summary, these rules can be stated as follows: (i) Pick up all terms of order $dt^{1/2}$, and dt, whereby $dW_i(t)$ is treated as a quantity of order $dt^{1/2}$, and (ii) discard all terms of order $dt^{3/2}$ and higher.

1.5.4.2 *Itô calculus for PDPs* Let us now develop an analogous calculus for piecewise deterministic processes. To this end, we first consider a real-valued PDP $X(t)$ and again define the increments by $dX(t) \equiv X(t+dt) - X(t)$, $dt \geq 0$. Our aim is to construct, in analogy to the procedure of the foregoing subsection, a stochastic differential equation for the PDP which is of the following form,

$$dX(t) = g(X(t))dt + dJ(X(t)). \tag{1.187}$$

Obviously, the first term describes the deterministic drift of the PDP. The second term $dJ(X(t))$ represents the jump part of the process. It is intuitively clear that this term is not small in the usual sense as dt goes to zero, since $X(t)$ undergoes instantaneous jumps of finite size. This fact makes the stochastic calculus for PDPs very different from ordinary calculus as we shall see below.

Our task is to find the statistical properties of the jump part $dJ(X(t))$. Let us assume that the state $X(t)$ is given. According to the Liouville master equation (1.150) the system may perform an instantaneous jump from this given state to some other state z with a rate $W(z|X(t))$. If no jump occurs in the time interval dt we must have $dJ(x) = 0$, of course, which means that only the deterministic drift is present.

Consider now the quantity $dN_z(t)dz$ which is defined to be the number of jumps from the given state $X(t)$ into the volume element dz around z, taking place during the interval dt. For small dt the average of the random number $dN_z(t)dz$ must be equal to dt times the rate $W(z|X(t))dz$ for jumping into the element dz, that is,

$$\mathrm{E}\,[dN_z(t)] = W(z|X(t))dt. \tag{1.188}$$

This equation shows that the expectation value on the left-hand side is a conditional expectation, namely it is conditioned on the given state $X(t)$.

If the random number $dN_z(t)dz$ takes on the value 1 we get the new state $X(t+dt) = z$, that is the increment $dX(t)$ is equal to $z - X(t)$. Thus we are led to the following ansatz for the stochastic jump term,

$$dJ(X(t)) = \int (z - X(t))dN_z(t)dz, \tag{1.189}$$

such that the stochastic differential equation for the PDP can be written as

$$dX(t) = g(X(t))dt + \int (z - X(t))dN_z(t)dz. \tag{1.190}$$

We have not yet fully specified the statistics of the random quantities $dN_z(t)$, only their expectation values have been defined in (1.188). Since the PDP involves a jump rate which increases linearly with dt one expects that for small dt at most one jump can occur, that is, we expect that the relation

$$dN_{z'}(t) \cdot dN_z(t) = \delta(z' - z)dN_z(t) \tag{1.191}$$

holds.

According to eqn (1.191) the random numbers $dN_z(t)$ provide a *field* of independent Poisson increments as may be seen as follows. We suppose that, starting from the given state $X(t)$, the process may perform a discrete set of transitions

$$X(t) \longrightarrow z_\alpha = z_\alpha(X(t)) \tag{1.192}$$

labelled by an index α. The rate for a particular transition α will be denoted by $\gamma_\alpha(X(t))$. This rate as well as the final state $z_\alpha(X(t))$ of the transition may depend on the initial state $X(t)$, of course. Hence we have

$$W(z|X(t)) = \sum_\alpha \gamma_\alpha(X(t))\delta\left[z - z_\alpha(X(t))\right]. \tag{1.193}$$

We further take a disjoint partition of the state space into small elements Δz_α in such a way that each z_α is contained in precisely on Δz_α. The random quantities

$$dN_\alpha(t) \equiv \int_{\Delta z_\alpha} dN_z(t)dz \tag{1.194}$$

therefore denote the number of jumps into the element Δz_α during the time interval dt. By virtue of eqns (1.188) and (1.193) their expectation values are given by

$$\mathrm{E}(dN_\alpha(t)) = \gamma_\alpha(X(t))dt. \tag{1.195}$$

Using eqn (1.191) we also find

$$dN_\alpha(t)dN_\beta(t) = \delta_{\alpha\beta}dN_\alpha(t). \tag{1.196}$$

This shows that the $dN_\alpha(t)$ behave for small dt as independent Poisson increments (compare with eqns (1.120) and (1.121)). Finally, with these definitions the stochastic differential equation (1.190) reduces to the particularly transparent form

$$dX(t) = g(X(t))dt + \sum_\alpha \{z_\alpha(X(t)) - X(t)\}\, dN_\alpha(t). \tag{1.197}$$

This equation has an obvious interpretation: In view of (1.196) either all $dN_\alpha(t)$ are zero, or $dN_\alpha(t) = 1$ for precisely one index α. In the former case $X(t)$ follows the deterministic drift, in the latter case $X(t)$ performs a particular jump given by the index α.

Equations (1.190) or (1.197) provide the desired formulation of a PDP in terms of a stochastic differential equation. Let us check that they lead to the correct dynamical behaviour as dictated by the Liouville master equation (1.150). From the latter we find for the expectation value of the increment of any function f

$$\mathrm{E}\left[f(X(t) + dX(t)) - f(X(t))\right] \tag{1.198}$$

$$= \mathrm{E}\left[\frac{\partial f}{\partial x}g(X(t)) + \int \{f(z) - f(X(t))\}\, W(z|X(t))dz\right]dt + \mathcal{O}(dt^2).$$

Our aim is to show that the same equation is obtained if one substitutes the right-hand side of eqn (1.190) for $dX(t)$ and uses the properties (1.188) and

(1.191). To this end we first determine the powers of the jump part $dJ(t)$ in the stochastic differential equation. Invoking eqn (1.191) it is easy to verify that for all $k = 1, 2, 3, \ldots$ we have

$$[dJ(X(t))]^k = \int \{z - X(t)\}^k \, dN_z(t)dz. \tag{1.199}$$

This means that, after averaging, all powers of $dJ(X(t))$ are of the same order dt. Consider now the expectation value

$$E \equiv \mathrm{E}\left[f(X(t) + g(X(t))dt + dJ(X(t))) - f(X(t))\right] \tag{1.200}$$

which must be demonstrated to be equal to the right-hand side of eqn (1.198) up to terms of order dt^2. This may be achieved by first expanding with respect to the drift term, keeping the jump term to all orders, which leads to

$$E = \mathrm{E}\left[\frac{\partial f}{\partial x}(X + dJ)g dt + f(X + dJ) - f(X)\right] + \mathcal{O}(dt^2). \tag{1.201}$$

In the first term the derivative of f is evaluated at the point $X + dJ$. However, since this term is already of order dt we can evaluate it at the point X, thereby making only an error of order dt^2, due to the above property of dJ. Hence we write

$$E = \mathrm{E}\left[\frac{\partial f}{\partial x}g dt + f(X + dJ) - f(X)\right] + \mathcal{O}(dt^2). \tag{1.202}$$

The second term must be treated, however, in an entirely different way, since, on averaging, all powers of dJ are of the same order dt, that is, we must evaluate the difference $f(X + dJ) - f(X)$ exactly. This is obviously due to the fact that the process performs instantaneous jumps of finite size, such that any approximation of this difference by a finite Taylor expansion fails. Hence we write, denoting by $f^{(k)}$ the k-th derivative of f,

$$\begin{aligned}
\mathrm{E}\left[f(X + dJ) - f(X)\right] &= \mathrm{E}\left[\sum_{k=1}^{\infty} \frac{1}{k!} f^{(k)}(X) \left[dJ\right]^k\right] \\
&= \mathrm{E}\left[\sum_{k=1}^{\infty} \frac{1}{k!} f^{(k)}(X) \int \{z - X\}^k \, dN_z(t)dz\right] \\
&= \mathrm{E}\left[\int \{f(z) - f(X)\} \, dN_z(t)dz\right] \\
&= \mathrm{E}\left[\int \{f(z) - f(X)\} \, W(z|X)dz\right] dt. \tag{1.203}
\end{aligned}$$

Thus we have

$$E = \mathrm{E}\left[\frac{\partial f}{\partial x}g + \int \{f(z) - f(X)\} \, W(z|X)dz\right] dt + \mathcal{O}(dt^2), \tag{1.204}$$

which coincides with the right-hand side of (1.198) and proves that the stochastic differential equation (1.190) correctly reproduces the expectation values of the increments of all functions f.

Finally we note that stochastic differential equations similar to (1.190) or (1.197) also hold in the case of a multivariate process. Explicitly, we may write

$$dX_i(t) = g_i(X(t))dt + \int \{z_i - X_i(t)\} \, dN_z(t)dz \qquad (1.205)$$

for the stochastic differential equation of a multivariate PDP with components $X_i(t)$.

1.6   Lévy processes

In the preceding sections our investigation of Markov processes was based mainly on the differential Chapman–Kolmogorov equation. An alternative approach is to search directly for solutions of the corresponding integral equation. Such a strategy is indeed possible if certain symmetry and invariance properties are imposed. An example is the requirement that the process is both homogeneous in space and time which leads to the important class of Lévy processes (Itô, 1993; Bertoin, 1996). As will be discussed in this section it is possible to give a complete characterization of this class of stochastic processes. Lévy processes have a large number of interesting applications (Klafter, Shlesinger and Zumofen, 1996). An example will be given in Chapter 8, where it is shown that certain stable Lévy distributions arise in the stochastic representation of the dynamics of open quantum systems.

1.6.1   *Translation invariant processes*

A Lévy process is defined to be a stochastic process $X(t)$, $t \geq 0$, which is both homogeneous in space and in time. For simplicity we assume that the process is real valued (the generalization to d dimensions is straightforward), and that the process starts at time $t_0 = 0$ at the point $X(0) = 0$. The propagator of such a process depends only on the differences of its time and space arguments,

$$T(x,t|x',t') = T_{t-t'}(x - x'), \qquad (1.206)$$

which means that the propagator is invariant with respect to space-time translations.

Alternatively, translational invariance may be formulated as follows. We take any ordered set of times,

$$t_0 \equiv 0 < t_1 < t_2 < \ldots < t_m. \qquad (1.207)$$

Spatial homogeneity then implies that the m random variables

$$X(t_\nu) - X(t_{\nu-1}), \quad \nu = 1, 2, \ldots, m, \qquad (1.208)$$

are mutually independent. Such a process is also called an *additive process*, or a process with independent increments. Homogeneity in time means that the random variable $X(t+s) - X(t'+s)$, with $t > t'$, follows the same distribution as the random variable $X(t) - X(t')$ for all $s > 0$. Such processes are called processes with stationary increments. A Lévy process is thus a stochastic process with independent and stationary increments.

Employing translational invariance we write the Chapman–Kolmogorov equation as follows,

$$T_{t+t'}(y) = \int dy' \, T_t(y-y')T_{t'}(y'), \qquad (1.209)$$

which means that the propagator over the time $t+t'$ is equal to the convolution of the propagators over the times t and t'. It is thus advantageous to consider the characteristic function

$$G(k,t) \equiv \mathrm{E}\left[e^{ikX(t)}\right] = \int dy \, e^{iky}T_t(y), \qquad (1.210)$$

such that the Chapman–Kolmogorov equation leads to

$$G(k, t_1 + t_2) = G(k, t_1)G(k, t_2). \qquad (1.211)$$

This equation shows that the logarithm $\ln G(k,t)$ of the characteristic function (the cumulant generating function) is linear in t and that we can write the characteristic function as

$$G(k,t) = \exp\left[t\Psi(k)\right]. \qquad (1.212)$$

The quantity $\Psi(k)$ is called the *characteristic exponent* of the process.

The characteristic exponent $\Psi(k)$ is directly related to the generator $\mathcal{A}$ of the Lévy process. We write the propagator as

$$
\begin{aligned}
T_t(y) &= \int \frac{dk}{2\pi} e^{-iky} G(k,t) \\
&= \int \frac{dk}{2\pi} e^{-iky} \left(1 + t\Psi(k)\right) + \mathcal{O}(t^2) \\
&= \delta(y) + t \int \frac{dk}{2\pi} e^{-iky} \Psi(k) + \mathcal{O}(t^2).
\end{aligned}
\qquad (1.213)
$$

Hence we get by the definition of the generator

$$
\begin{aligned}
\mathcal{A}\rho(x) &= \lim_{t\to 0} \frac{1}{t} \int dy \left[T_t(y) - \delta(y)\right] \rho(x-y) \\
&= \int dy \left[\int \frac{dk}{2\pi} e^{-iky} \Psi(k)\right] \rho(x-y) \\
&\equiv \int dy \mathcal{A}(y)\rho(x-y).
\end{aligned}
\qquad (1.214)
$$

Thus we observe that the integral kernel $\mathcal{A}(y)$ of the generator is equal to the Fourier transform of the characteristic exponent $\Psi(k)$.

1.6.2   *The Lévy–Khintchine formula*

According to eqn (1.211) the characteristic function $G(k,t)$ is the m-th power of $G(k,t/m)$. This implies that the propagator over time t is equal to the m-th convolution power of the propagator over time t/m, that is,

$$T_t = T_{t/m} * T_{t/m} * \ldots * T_{t/m}, \tag{1.215}$$

where $*$ denotes the convolution and the right-hand side contains m factors. A distribution with this property is called infinitely divisible. One can see this more directly as follows. We set $t_\nu = \nu t/m$ for $\nu = 0, 1, 2, \ldots, m$. By means of the identity

$$X(t) = \sum_{\nu=1}^{m} \left(X(t_\nu) - X(t_{\nu-1}) \right), \tag{1.216}$$

the random variable $X(t)$ is represented as a sum of m independent and identically distributed random numbers. This is precisely the definition for a random number $X(t)$ (and its corresponding distribution T_t) to be infinitely divisible.

It is possible to give a complete characterization of all possible Lévy processes by making use of the property that their propagators are infinitely divisible. This is achieved by a fundamental theorem which is due to Lévy and Khintchine. Applied to the present case this theorem asserts that the most general form of the characteristic exponent of a Lévy process is given by the following equation:

$$\Psi(k) = igk - \frac{D}{2}k^2 + \int \left[e^{iky} - 1 - iky\theta(h - |y|) \right] W(y)dy. \tag{1.217}$$

Here, g and D are real constants with $D \geq 0$. $\theta(x)$ is the unit step function such that

$$\theta(h - |y|) = \begin{cases} 1, & |y| < h, \\ 0, & |y| \geq h. \end{cases} \tag{1.218}$$

This function cuts off the last term in the integral over y at some length scale h, the significance of which will be discussed below. Finally, $W(y)$ is some non-negative measure, the Lévy measure, defined on $\mathbb{R} \setminus \{0\}$ with the properties

$$\int_{|y|<h} y^2 W(y)dy < \infty, \tag{1.219}$$

$$\int_{|y|\geq h} W(y)dy < \infty. \tag{1.220}$$

According to the Lévy–Khintchine theorem any translational invariant process $X(t)$ has a characteristic exponent $\Psi(k)$ of the form (1.217). For a given length scale h, the constants g and D, and the measure $W(y)$ in this representation are uniquely determined by the process.

To get a feeling for the Lévy–Khintchine representation let us consider some simple examples. First we take $W(y) \equiv 0$. Then we have

$$\Psi(k) = igk - \frac{D}{2}k^2, \tag{1.221}$$

and, consequently,

$$G(k,t) = \exp\left[\left(igk - \frac{D}{2}k^2\right)t\right]. \tag{1.222}$$

On Fourier transforming this characteristic function we immediately get

$$T_t(y) = \frac{1}{\sqrt{2\pi Dt}}\exp\left\{-\frac{(y-gt)^2}{2Dt}\right\}, \tag{1.223}$$

which is recognized as the propagator of a Gaussian diffusion process with constant drift coefficient g (corresponding to a linear drift) and constant diffusion coefficient D. Of course, g and D must be constant since the process is assumed to be homogeneous.

Now, consider the case $g = D = 0$ and take the Lévy measure

$$W(y) = \gamma\delta(y - y_0), \tag{1.224}$$

with some positive constant γ and $y_0 > h$. The characteristic exponent is then found to be

$$\Psi(k) = \gamma\left(e^{iky_0} - 1\right), \tag{1.225}$$

such that

$$G(k,t) = \exp\left[\gamma t\left(e^{iky_0} - 1\right)\right]. \tag{1.226}$$

Thus, $X(t)$ is a Poisson process with rate γ and step size y_0 (compare with eqn (1.99)) or, equivalently, a Poisson process $N(t)$ with unit step size times y_0, i.e. $X(t) = y_0 N(t)$.

This example is readily generalized to the case that the Lévy measure is a discrete sum of δ-functions,

$$W(y) = \sum_\alpha \gamma_\alpha\delta(y - y_\alpha), \tag{1.227}$$

where $\gamma_\alpha > 0$ and $y_\alpha > h$. The characteristic function of the corresponding process is in this case given by

$$G(k,t) = \exp\left[\sum_\alpha \gamma_\alpha t\left(e^{iky_\alpha} - 1\right)\right] = \prod_\alpha \exp\left[\gamma_\alpha t\left(e^{iky_\alpha} - 1\right)\right]. \tag{1.228}$$

The result is that the Lévy process $X(t)$ is now a discrete sum of independent Poisson processes with rates γ_α and step sizes y_α. Introducing the corresponding independent Poisson processes $N_\alpha(t)$ with unit step size we thus have

$$X(t) = \sum_\alpha y_\alpha N_\alpha(t). \tag{1.229}$$

Hence,

$$dX(t) = \sum_\alpha y_\alpha dN_\alpha(t), \tag{1.230}$$

and

$$\mathrm{E}\left[dN_\alpha(t)\right] = \gamma_\alpha dt, \quad dN_\alpha(t)dN_\beta(t) = \delta_{\alpha\beta}dN_\alpha(t). \tag{1.231}$$

This corresponds exactly to our former results (1.195) and (1.196) for the special case of a PDP without drift and with Poisson increments $dN_\alpha(t)$ the statistics of which does not depend on $X(t)$, which is due to the translational invariance of the process. Accordingly, the Lévy measure $W(y)$ is equal to $W(x + y|x) = W(x|x - y)$, which is just a discrete distribution of jump rates.

In the general case we expect the Lévy measure $W(y)$ to represent a continuous sum of independent Poisson processes indexed by their jump sizes y. This can be seen directly from the general form of the generator $\mathcal{A}$ as follows with the help of expression (1.214) and the Lévy–Khintchine formula:

$$\mathcal{A}\rho(x) = \int dy \int \frac{dk}{2\pi} \left[igk - \frac{D}{2}k^2\right] e^{-iky}\rho(x - y) \tag{1.232}$$

$$+ \int dy \int \frac{dk}{2\pi} \int dz \left[e^{ikz} - 1 - ikz\theta(h - |z|)\right] e^{-iky}W(z)\rho(x - y).$$

The first term is easily seen to be the generator of a linear diffusion process, whereas the second term is determined as

$$\int dy \int dz \left[\delta(z - y) - \delta(y) + z\theta(h - |z|)\delta(y)\frac{\partial}{\partial x}\right] W(z)\rho(x - y)$$

$$= \int dz\, W(z) \left[\rho(x - z) - \rho(x) + z\theta(h - |z|)\frac{\partial}{\partial x}\rho(x)\right]. \tag{1.233}$$

Thus we find for the most general form of the generator of a Lévy process:

$$\mathcal{A}\rho(x) = \left[-g\frac{\partial}{\partial x} + \frac{D}{2}\frac{\partial^2}{\partial x^2}\right]\rho(x) \tag{1.234}$$

$$+ \int dy\, W(y) \left[\rho(x - y) - \rho(x) + y\theta(h - |y|)\frac{\partial}{\partial x}\rho(x)\right].$$

Apart from the additional term under the integral which is proportional to $\theta(h - |y|)$ this is just the generator of a process obtained by adding a jump process to a Gaussian diffusion process. Of course, due to the requirement of translational invariance of the process, drift and diffusion coefficient must be constant, and the jump rate $W(y)$ in the master equation part in eqn (1.234) only depends on the difference y of the positions of the initial and final states.

What is the significance of the additional term $y\theta(h - |y|)\partial\rho/\partial x$ which obviously depends on the arbitrary length scale h? To answer this question we take a closer look at the conditions (1.219) and (1.220) imposed on the Lévy measure. These conditions control the behaviour of $W(y)$ at zero and infinity. Up to now in this section, and also in the preceding ones, we have always tacitly assumed that the total transition rate

$$\Gamma = \int dy W(y) \qquad (1.235)$$

is finite. In this case both conditions (1.219), (1.220) are satisfied for any $h > 0$. But also the quantity $\int_{|y|<h} dy \, yW(y)$ is then finite and we may absorb the last term in (1.234) into the drift coefficient, which yields

$$\mathcal{A}\rho(x) = \left[-g'\frac{\partial}{\partial x} + \frac{D}{2}\frac{\partial^2}{\partial x^2}\right]\rho(x) + \int dy W(y)\left[\rho(x-y) - \rho(x)\right], \qquad (1.236)$$

where we have introduced a new drift coefficient by

$$g' = g - \int_{|y|<h} dy \, yW(y). \qquad (1.237)$$

Thus we see that for a finite total transition rate Γ the scale h is completely arbitrary and the generator of the Lévy process takes on precisely the form encountered before.

The important point to note is, however, that the conditions (1.219) and (1.220) do not exclude the possibility that the total transition rate diverges as a result of a singularity of $W(y)$ at $y = 0$. The reason for this divergence is that a singularity of the Lévy measure at $y = 0$ can lead to an accumulation of an infinite number of jumps in a finite time interval. As an example take $W(y)$ to be proportional to $|y|^{-2}$. Such a measure decreases sufficiently rapidly for the integral (1.220) to converge, whereas the singularity at $y = 0$ is weak enough to satisfy also (1.219). But the total transition rate given by the integral (1.235) clearly diverges. In spite of this, the generator (1.234) and the characteristic exponent $\Psi(k)$ are well defined and finite.

Let us rewrite the general form (1.234) of the generator as follows,

$$\mathcal{A}\rho(x) = \left[-g\frac{\partial}{\partial x} + \frac{D}{2}\frac{\partial^2}{\partial x^2}\right]\rho(x) + \int_{|y|\geq h} dy W(y)\left[\rho(x-y) - \rho(x)\right]$$

$$+ \int_{|y|<h} dy W(y)\left[\rho(x-y) - \rho(x) + y\frac{\partial}{\partial x}\rho(x)\right]. \qquad (1.238)$$

The second term on the right-hand side yields the big jumps of size $|y| \geq h$, while the third term involves the small jumps. If we apply the generator to a smooth probability density $\rho(x)$ which varies significantly on a typical length scale $L \gg h$, we may expand the integrand in the third term around x to obtain the expression

$$\frac{1}{2} \int_{|y|<h} dy \, y^2 W(y) \frac{\partial^2}{\partial x^2} \rho(x). \tag{1.239}$$

By condition (1.219) this expression is, for any fixed $h > 0$, well defined and finite. We thus see that the contribution (1.239) from the small jumps leads to diffusion-type behaviour of the process on small length scales.

1.6.3   Stable Lévy processes

In the preceding subsection we have discussed the most general form of the generator of a Lévy process. A question that arises from the foregoing discussion is, are there interesting Lévy measures for which the total transition rate diverges? As we shall see below there are such measures the existence of which is closely related to the notions of scale invariance and stability of the corresponding processes.

1.6.3.1   Stable distributions and scaling relations

A Lévy process $X(t)$ is called *stable*[4] if it has the following scaling property. For any $u > 0$ there is a corresponding $\lambda(u) > 0$ such that the random number $X(ut)$ follows the same distribution as $\lambda(u)X(t)$. This means that a change of the time scale by a factor of u yields the same distribution as a change of the length scale by an appropriate factor $\lambda(u)$ (here and in the following we shall assume, to be definite, that the variable X has the dimension of length). The stability of the process thus expresses a certain scale invariance, which is often referred to as self-similarity. We write the stability condition symbolically as

$$X(ut) \sim \lambda(u)X(t), \tag{1.240}$$

whereby the symbol $\sim$ serves to indicate that the left- and right-hand sides of the expression follow the same distribution.

One can show that a Lévy process is stable if and only if the corresponding propagator represents a stable distribution. Let us first define stable distributions. Consider some random number Z with distribution $p(z)$, and mutually independent copies $Z_1, Z_2, \ldots, Z_\nu, \ldots$ of Z all with the same distribution $p(z)$. Then $p(z)$ is defined to be a stable distribution if for each positive integer m there exists a factor $\lambda(m)$ such that the sum $\sum_{\nu=1}^{m} Z_\nu$ has the same distribution as the random number $\lambda(m)Z$,

$$\sum_{\nu=1}^{m} Z_\nu \sim \lambda(m)Z. \tag{1.241}$$

Thus, a distribution p is stable if all m-th convolution powers of p are equal to p up to an appropriate change of the scale. A prominent example for a stable

---

[4]More precisely, our definition is that of stability in the strict sense.

distribution is the normal distribution which obviously has this property with
the scaling factor $\lambda(m) = \sqrt{m}$.

It is easy to show that the propagator $T_t(y)$ of a stable Lévy process $X(t)$
must be a stable distribution: For any m we can write

$$X(mt) = \sum_{\nu=1}^{m} \left(X(\nu t) - X([\nu - 1]t) \right). \tag{1.242}$$

The left-hand side in this equation has the same distribution as $\lambda(m)X(t)$,
whereas the right-hand side is the sum of m independent, identically distributed random numbers $X(\nu t) - X([\nu - 1]t)$ which follow the same distribution
as $X(t)$. This shows that the distribution of $X(t)$, namely $T_t(y)$ must be stable.
The converse is also true: If we take some stable distribution, then it yields the
propagator of a stable Lévy process.

Next we investigate what stability means for the characteristic exponent
$\Psi(k)$. The relation $X(ut) \sim \lambda(u)X(t)$ leads to the scaling relation

$$T_{ut}(y) = \frac{1}{\lambda(u)} T_t \left(\frac{y}{\lambda(u)} \right) \tag{1.243}$$

for the propagator. The corresponding relation for the characteristic functions
becomes

$$\Psi(k) = \frac{1}{u} \Psi \left(\lambda(u)k \right). \tag{1.244}$$

If we look at the Lévy–Khintchine formula (1.217) we get the following obvious
solutions of the scaling condition (1.244): For $D = W = 0$ we have $\lambda(u) =
u$ which is the scaling for a linear drift, corresponding to the singular stable
distribution $T_t(y) = \delta(y - gt)$. For $g = W = 0$ one finds $\lambda(u) = \sqrt{u}$ which is
the scaling for the Brownian motion case, corresponding to the stable Gaussian
propagator

$$T_t(y) = \frac{1}{\sqrt{2\pi Dt}} \exp \left[-\frac{y^2}{2Dt} \right]. \tag{1.245}$$

The question is now, are there non-trivial Lévy measures that lead to stable
Lévy processes? To answer this question we put $D = 0$, since we already know
that $D > 0$ for Brownian motion. We keep however a non-vanishing g for reasons
that will become clear later on. On using the Lévy–Khintchine formula the scaling
condition (1.244) then takes the form

$$igk + \int \left[e^{iky} - 1 - iky\theta(h - |y|) \right] W(y)dy \tag{1.246}$$

$$= i \left(\frac{\lambda}{u} \right) gk + \int \left[e^{iky} - 1 - iky\theta(\lambda h - |y|) \right] \frac{1}{u\lambda} W \left(\frac{y}{\lambda} \right) dy.$$

This equation suggests that we can satisfy the scaling relation by taking Lévy
measures with the property

$$W(y) = \frac{1}{u\lambda} W\left(\frac{y}{\lambda}\right), \tag{1.247}$$

provided the scale dependence through the term $\theta(\lambda h - y)$ can be removed in some way (it can!). Ignoring this term for a moment we are thus led to (1.247), which, of course, could have been obtained directly by dimensional analysis, since $W(y)$ has the dimension (length·time)$^{-1}$. Thus we find the stable Lévy measures

$$W_\alpha(y) = \begin{cases} C_+ y^{-1-\alpha}, & y > 0, \\ C_- |y|^{-1-\alpha}, & y < 0, \end{cases} \tag{1.248}$$

with some non-negative constants $C_\pm$. These measures lead to the scaling relation

$$\lambda(u) = u^{1/\alpha}, \tag{1.249}$$

that is we have $X(ut) \sim u^{1/\alpha} X(t)$.

The possible values for the exponent α, the *scaling exponent*, are restricted by the conditions (1.219) and (1.220) which must be satisfied by any Lévy measure. Condition (1.219) leads to $\alpha < 2$, while condition (1.220) yields $\alpha > 0$, and hence,

$$0 < \alpha < 2. \tag{1.250}$$

Recall that the case $\alpha = 2$, corresponding to the scaling relation $\lambda(u) = \sqrt{u}$, is obtained for a non-vanishing diffusion coefficient D and by setting $g = W = 0$.

The above scaling relation has interesting consequences for a sequence of independent random numbers $X, X_1, X_2, \ldots, X_\nu, \ldots$, following the corresponding stable distribution. Consider the case $0 < \alpha < 1$. Then we have

$$\sum_{\nu=1}^m X_\nu \sim m^{1/\alpha} X, \tag{1.251}$$

with $1/\alpha > 1$. This shows that the sum grows faster than the number m of terms in it. In other words, the maximal term of the sum is likely to become very large and to dominate the value of the sum. It is clear that the central limit theorem of probability theory is therefore violated which is related to the fact that the stable distributions do not have a finite variance for $\alpha < 2$. Indeed, it can be shown that a stable distribution with finite variance must necessarily be Gaussian (or concentrated at a single point). This is seen from the fact that the density $p(y)$ of a stable distribution with exponent $\alpha \neq 2$ behaves for large y as $|y|^{-1-\alpha}$.

1.6.3.2 *Characteristic exponents of stable Lévy processes* Our analysis is not yet complete since we have neglected above the scale-dependent term in eqn (1.246) and since we do not yet know the characteristic exponents $\Psi_\alpha(k)$ of the stable processes.

As we have seen, if one changes the length scale h to some other value h', the corresponding change of the characteristic exponent $\Psi_\alpha(k)$ can be compensated by an appropriate shift of the drift coefficient which is given by

$$g \longrightarrow g' = g + \int dy\, y W_\alpha(y)\left[\theta(h' - |y|) - \theta(h - |y|)\right]. \qquad (1.252)$$

Henceforth, this new coefficient g' will be called the renormalized drift coefficient. Thus we see that the choice of the length scale is to a large extent arbitrary and we can employ this freedom to find the characteristic exponents of the stable distributions. We distinguish three cases.

1. Consider first the case $0 < \alpha < 1$. In this case the integral

$$\int\limits_{|y|<h} dy\, y W_\alpha(y) \qquad (1.253)$$

converges. Thus, we can let h' go to zero in eqn (1.252). We further put the renormalized drift coefficient equal to zero, since a non-zero drift term would violate the scaling relation (see eqn (1.246)). The characteristic exponent then reduces to

$$\Psi_\alpha(k) = \int\limits_{-\infty}^{+\infty} \left(e^{iky} - 1\right) W_\alpha(y) dy$$

$$= C_+ \int\limits_{0}^{+\infty} \left(e^{iky} - 1\right) \frac{dy}{y^{1+\alpha}} + C_- \int\limits_{-\infty}^{0} \left(e^{iky} - 1\right) \frac{dy}{|y|^{1+\alpha}}. \qquad (1.254)$$

The integrals can be determined explicitly and yield the final result for the characteristic exponent of the α-stable distributions

$$\Psi_\alpha(k) = \left(-C_0 + i\frac{k}{|k|}C_1\right)|k|^\alpha. \qquad (1.255)$$

The new constants C_0 and C_1 are related to the constants $C_\pm$ of the Lévy measure through

$$C_0 = -(C_+ + C_-)\Gamma(-\alpha)\cos\left(\frac{\pi\alpha}{2}\right), \qquad (1.256)$$

$$C_1 = -(C_+ - C_-)\Gamma(-\alpha)\sin\left(\frac{\pi\alpha}{2}\right). \qquad (1.257)$$

Here Γ denotes the gamma function. As it should be, we have $C_0 \geq 0$ since $\Gamma(-\alpha) < 0$. C_0 is the *scale parameter* and C_1 the *symmetry parameter*; for $C_1 = 0$ the corresponding stable distribution is obviously symmetric.

2. Consider now the case $1 < \alpha < 2$. We cannot follow the above procedure in the present case for the integral (1.253) now diverges. However, for $1 < \alpha < 2$ the integral

$$\int_{|y|>h} dy \, y W_\alpha(y) \tag{1.258}$$

is finite. This allows us to take the limit $h' \longrightarrow \infty$ in eqn (1.252). Putting again the renormalized drift coefficient equal to zero we now have

$$\Psi_\alpha(k) = \int_{-\infty}^{+\infty} \left(e^{iky} - 1 - iky\right) W_\alpha(y) dy \tag{1.259}$$

$$= C_+ \int_0^{+\infty} \left(e^{iky} - 1 - iky\right) \frac{dy}{y^{1+\alpha}} + C_- \int_{-\infty}^0 \left(e^{iky} - 1 - iky\right) \frac{dy}{|y|^{1+\alpha}}.$$

Again the integrals can be evaluated explicitly and the result is precisely the same as above, that is, we again get eqn (1.255) with the coefficients given by (1.256) and (1.257). Note that again $C_0 \geq 0$ since now $\Gamma(-\alpha) > 0$ and $\cos(\pi\alpha/2) < 0$.

3. The case $\alpha = 1$ requires special treatment which involves a non-zero drift coefficient. Note that the drift term satisfies the scaling condition (see eqn (1.246)) since $\alpha = 1$ gives $\lambda(u) = u$. It can be shown that for the process to be strictly stable the Lévy measure $W_1(y)$ must be taken to be symmetric. Thus in the special case $\alpha = 1$ we write

$$W_1(y) = \frac{C_0}{\pi} \frac{1}{|y|^2}, \quad C_0 \geq 0. \tag{1.260}$$

The characteristic exponent is then given by

$$\Psi_1(k) = iC_1 k + \frac{C_0}{\pi} \int_{-\infty}^{+\infty} \left(e^{iky} - 1 - \frac{iky}{1+(y/h)^2}\right) \frac{dy}{y^2}. \tag{1.261}$$

Note that we have included here a finite drift term with coefficient C_1 and that we have changed the cutoff function from $\theta(h-|y|)$ to the function $1/(1+(y/h)^2)$. This is permissible since this change also amounts to a finite renormalization of the drift coefficient. The integral in eqn (1.261) can be evaluated explicitly using the method of residues which yields the h-independent value $-\pi|k|$. As a result one finds the same formula as for the other cases, namely

$$\Psi_1(k) = iC_1 k - C_0|k| = \left(-C_0 + i\frac{k}{|k|}C_1\right)|k|. \tag{1.262}$$

On Fourier transforming this result for $C_1 = 0$ we immediately get the propagator

$$T_t(y) = \int \frac{dk}{2\pi} e^{-C_0|k|t - iky} = \frac{1}{\pi} \frac{C_0 t}{(C_0 t)^2 + y^2}, \tag{1.263}$$

representing the symmetric Cauchy process (Fig. 1.10).

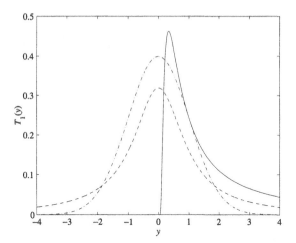

FIG. 1.10. Plots of the propagators $T_1(y)$ for three α-stable Lévy processes: the Gaussian propagator (1.245) with $\alpha = 2$ and $D = 1$ (dashed-dotted line), the Cauchy propagator (1.263) with $\alpha = 1$ and $C_0 = 1$ (dashed line), and the propagator (1.264) corresponding to $\alpha = 1/2$ (solid line).

We have obtained above the most general form for the characteristic exponents of stable Lévy processes with scaling exponent α. In addition we have explicit expressions for the corresponding Lévy measures. However, the stable distribution functions obtained by Fourier transforming the characteristic functions are known explicitly only in very special cases. For $\alpha = 1$ we have already seen that one gets the Cauchy distribution. Another case is provided by the propagator given by

$$T_t(y) = \frac{t}{\sqrt{2\pi y^3}} \exp\left\{ -\frac{t^2}{2y} \right\}, \quad y > 0, \tag{1.264}$$

and by $T_t(y) = 0$ for $y \leq 0$ (see Fig. 1.10). This case corresponds to a stable distribution with scaling exponent $\alpha = 1/2$, scale parameter $C_0 = 1$, and symmetry parameter $C_1 = 1$. Hence we have $\lambda(u) = u^2$. We shall discuss in Chapter 8 a physical example where this stable distribution emerges in the distribution of the waiting times of a PDP.

References

Arnold, L. (1974). *Stochastic Differential Equations*. John Wiley, New York.

Bertoin, J. (1996). *Lévy Processes*. Cambridge University Press, Cambridge.

Bjorken, J. D. and Drell, S. D. (1965). *Relativistic Quantum Fields*. McGraw-Hill, New York.

Davis, M. H. A. (1993). *Markov Models and Optimization*. Chapman & Hall, London.

Doob, J. L. (1953). *Stochastic Processes*. John Wiley, New York.

Feller, W. (1968). *An Introduction to Probability Theory and Its Applications*, Volume I. (third edition). John Wiley, New York.

Feller, W. (1971). *An Introduction to Probability Theory and Its Applications*, Volume II. (second edition). John Wiley, New York.

Gardiner, C. W. (1985). *Handbook of Stochastic Methods for Physics, Chemistry and the Natural Sciences* (second edition). Springer-Verlag, Berlin.

Honerkamp, J. (1998). *Statistical Physics*. Springer-Verlag, Berlin.

Itô, Kiyosi (ed.) (1993). *Encyclopedic Dictionary of Mathematics* (second edition). The MIT Press, Cambridge, Massachusetts.

Klafter, J., Shlesinger, M. F. and Zumofen, G. (1996, February). Beyond Brownian motion. *Phys. Today*, **49**, 33–39.

Kloeden, P. E. and Platen, E. (1992). *Numerical Solutions of Stochastic Differential Equations*. Springer-Verlag, Berlin.

Kolmogorov, A. N. (1956). *Foundations of the Theory of Probability*. Chelsea Publishing Company, New York.

Reichl, L. E. (1998). *A Modern Course in Statistical Physics* (second edition). John Wiley, New York.

van Kampen, N. G. (1992). *Stochastic Processes in Physics and Chemistry* (second edition). North-Holland, Amsterdam.

2

QUANTUM PROBABILITY

Having outlined the classical theory of probability and stochastic processes we now turn to the notion of probability in quantum mechanics. The emphasis in this chapter lies on the standard formulation of quantum mechanics as may be found in much more detail in many excellent textbooks (von Neumann, 1955; Landau and Lifshitz, 1958; Cohen-Tannoudji, Diu and Laloë, 1977; Bohm, 1993).

In the first section we concentrate on those aspects which clarify the relation between quantum mechanics and classical probability theory. Most importantly, we discuss that, although quantum mechanics is an intrinsically probabilistic theory, the application of probabilistic concepts to quantum mechanics is quite different from that of the classical theory. It turns out that the notion of a probability space together with the corresponding space of random variables is *not* applicable to quantum mechanics. On the contrary, this concept is incompatible with the basic structures of quantum mechanics, namely a Hilbert space of state vectors and the corresponding algebra of observables.

We further introduce in this chapter the concept of a composite quantum system which will be important for the study of open systems. A section on quantum entropies is also included. The description of measurements performed on a quantum system provides an important part of the statistical interpretation of quantum mechanics. We therefore present in this chapter the foundations of the generalized theory of quantum measurement which is based on the notions of operations and effects. These concepts will play a significant rôle in the development of a stochastic theory of open quantum systems.

2.1 The statistical interpretation of quantum mechanics

In this section we outline the basic mathematical notions used to formulate quantum mechanics and its statistical interpretation. We start with some formal concepts from functional analysis and then present in the following subsections the fundamentals of their physical and statistical significance. Further mathematical details on the probabilistic interpretation of quantum mechanics may be found in Holevo (1982).

2.1.1 *Self-adjoint operators and the spectral theorem*

In quantum mechanics the state of a closed physical system is described by a state vector[5], or wave function ψ which is an element of some Hilbert space $\mathcal{H}$.

---

[5]We shall use both the notation ψ and the 'ket' $|\psi\rangle$ for an element of the Hilbert space $\mathcal{H}$. The symbol $\langle\psi|$ denotes an element of the dual space $\mathcal{H}^* \cong \mathcal{H}$, i.e. a 'bra' vector.

The scalar product in the Hilbert space will be denoted by angular brackets, that is we write $\langle \phi | \psi \rangle$ for the scalar product of two state vectors ψ, $\phi \in \mathcal{H}$. Accordingly, the norm of ψ is given by

$$||\psi|| \equiv \sqrt{\langle \psi | \psi \rangle}. \tag{2.1}$$

The Hilbert space is assumed to be separable, which means that there exists a finite or countable dense orthonormal basis $\{\varphi_\alpha\}$ of state vectors satisfying

$$\langle \varphi_\alpha | \varphi_\beta \rangle = \delta_{\alpha\beta}, \tag{2.2}$$

such that each state vector $|\psi\rangle$ has a unique decomposition of the form

$$|\psi\rangle = \sum_\alpha |\varphi_\alpha\rangle \langle \varphi_\alpha | \psi \rangle. \tag{2.3}$$

The measurable quantities, or *observables*, of the closed physical system are represented by linear, self-adjoint operators in the Hilbert space. A linear operator $\hat{R}$,

$$\hat{R} : \mathcal{D}(\hat{R}) \longrightarrow \mathcal{H}, \tag{2.4}$$

is self-adjoint if its domain $\mathcal{D}(\hat{R})$ coincides with the domain $\mathcal{D}(\hat{R}^\dagger)$ of the adjoint operator $\hat{R}^\dagger$ and if $\hat{R}\psi = \hat{R}^\dagger\psi$ on the common domain.

A theorem of fundamental importance in connection with self-adjoint operators is the spectral theorem (Akhiezer and Glazman, 1981). It states that for any self-adjoint operator $\hat{R}$ there exists a unique spectral family E_r such that

$$\hat{R} = \int_{-\infty}^{+\infty} r \, dE_r. \tag{2.5}$$

The spectral family E_r, where $r \in \mathbb{R}$, is a one-parameter family of commuting orthogonal projection operators with the following properties:

1. The family of projections is monotonically increasing,

$$E_{r'} \geq E_r \quad \text{for} \quad r' > r. \tag{2.6}$$

2. The family is continuous from the right,

$$\lim_{\varepsilon \to +0} E_{r+\varepsilon} = E_r. \tag{2.7}$$

3. It has the following limits

$$\lim_{r \to -\infty} E_r = 0, \quad \lim_{r \to +\infty} E_r = I, \tag{2.8}$$

where I denotes the unit operator.

Equation (2.5) represents the spectral decomposition of self-adjoint operators on which the statistical interpretation of quantum mechanics is based.

The spectral theorem asserts that all projectors E_r of the spectral family commute with $\hat{R}$ and that the domain of $\hat{R}$ can be characterized as follows

$$
\mathcal{D}(\hat{R}) = \left\{ \psi \in \mathcal{H} \, \middle| \, \int_{-\infty}^{+\infty} r^2 \, d\langle \psi | E_r | \psi \rangle < \infty \right\}. \tag{2.9}
$$

Note that the integral on the right-hand side is an ordinary (Riemann–Stieltjes) integral which may be written in the following alternative way

$$
\int_{-\infty}^{+\infty} r^2 \, d\langle \psi | E_r | \psi \rangle = \int_{-\infty}^{+\infty} r^2 \, d||E_r \psi||^2 = ||\hat{R}\psi||^2. \tag{2.10}
$$

More generally, for any continuous function $f(r)$ one can define the operator function $f(\hat{R})$ in terms of the spectral decomposition

$$
f(\hat{R}) = \int_{-\infty}^{+\infty} f(r) dE_r. \tag{2.11}
$$

Furthermore, the spectral family gives rise to an orthogonal decomposition of the unit operator. To this end, consider any disjoint partition of the real axis into intervals $\Delta r_\alpha = (r_{\alpha-1}, r_\alpha]$, and define the corresponding projectors

$$
\Delta E_\alpha = E_{r_\alpha} - E_{r_{\alpha-1}}. \tag{2.12}
$$

With the help of the properties (2.6)–(2.8) of the spectral family we find the completeness relation

$$
\sum_\alpha \Delta E_\alpha = I \tag{2.13}
$$

and the orthogonality relation

$$
\Delta E_\alpha \, \Delta E_\beta = \delta_{\alpha\beta} \Delta E_\beta. \tag{2.14}
$$

The spectrum $\mathrm{spec}(\hat{R})$ of a self-adjoint operator $\hat{R}$ can be expressed in terms of the properties of its spectral family E_r. To this end, a point $r \in \mathbb{R}$ is said to be stationary if for some $\varepsilon > 0$

$$
E_{r+\varepsilon} - E_{r-\varepsilon} = 0. \tag{2.15}
$$

This means that the spectral family is constant in an ε-neighbourhood of r. The spectrum $\mathrm{spec}(\hat{R})$ is then defined to be the set of all non-stationary points.

The spectrum $\text{spec}(\hat{R})$ may be decomposed into the discrete spectrum and the continuous spectrum. The discrete spectrum consists of all so-called jump points. A point $r \in \mathbb{R}$ is called a jump point if for any $\varepsilon > 0$ one has

$$E_r - E_{r-\varepsilon} \neq 0, \tag{2.16}$$

which means that E_r is not continuous from the left in r. If, for example, the dimension of E_r increases by one in r, we have for some normalized vector φ

$$E_r - E_{r-\varepsilon} = |\varphi\rangle\langle\varphi|. \tag{2.17}$$

It follows that φ is an eigenvector of $\hat{R}$ with eigenvalue r, i.e. $\hat{R}\varphi = r\varphi$. On the other hand, a point $r \in \text{spec}(\hat{R})$ belongs to the continuous spectrum if the spectral family is continuous in r.

Example 2.1 The simplest example of a self-adjoint operator $\hat{R}$ is that of an operator with a purely discrete spectrum which is obtained from the eigenvalue equation

$$\hat{R}\psi_{n,k} = r_n\psi_{n,k}. \tag{2.18}$$

The index n labels the different eigenvalues r_n, whereas k distinguishes the eigenvectors belonging to the same eigenvalue. Thus, the projector onto the eigenspace with eigenvalue r_n may be written as

$$\Pi_n = \sum_k |\psi_{n,k}\rangle\langle\psi_{n,k}|. \tag{2.19}$$

The spectral family takes the form

$$E_r = \sum_{r_n \leq r} \Pi_n \tag{2.20}$$

and the spectral decomposition of $\hat{R}$ is given by

$$\hat{R} = \int_{-\infty}^{+\infty} r \, dE_r = \sum_n r_n \Pi_n. \tag{2.21}$$

Example 2.2 In the Schrödinger representation the position operator $\hat{Q}$ of a one-dimensional system with Hilbert space $\mathcal{H} = \mathbb{L}_2(\mathbb{R})$, the space of square-integrable functions on $\mathbb{R}$, is defined by

$$(\hat{Q}\psi)(x) = x\psi(x). \tag{2.22}$$

Its spectral family may be defined as follows,

$$(E_q\psi)(x) = \begin{cases} \psi(x), & \text{for } x \leq q \\ 0, & \text{for } x > q \end{cases} = \theta(q - x)\psi(x). \tag{2.23}$$

Obviously, $\hat{Q}$ has only a continuous spectrum which covers the whole real axis.

Example 2.3 The momentum operator $\hat{P}$ canonically conjugated to $\hat{Q}$ is defined by (setting Planck's constant $\hbar$ equal to 1)

$$(\hat{P}\psi)(x) = -i\frac{d}{dx}\psi(x). \tag{2.24}$$

The domain of $\hat{P}$ consists of all functions $\psi(x)$ which are absolutely continuous and whose derivative belongs to $\mathbb{L}_2(\mathbb{R})$. The spectral family of the momentum operator is most easily defined with the help of the Fourier transform

$$\tilde{\psi}(p) = \int_{-\infty}^{+\infty} dx\, e^{-ipx}\psi(x) \tag{2.25}$$

of $\psi(x)$ in terms of which we get

$$(E_p\psi)(x) = \int_{-\infty}^{p} \frac{dk}{2\pi} e^{ikx}\tilde{\psi}(k). \tag{2.26}$$

It can easily be verified that the momentum operator also has a purely continuous spectrum which covers the whole real axis and that (2.26) provides, in fact, the spectral decomposition of the momentum operator. Namely, if $\psi(x)$ is absolutely continuous we find with the help of (2.26)

$$\int_{-\infty}^{+\infty} p\, dE_p\psi(x) = \int_{-\infty}^{+\infty} \frac{dp}{2\pi}\, p\, e^{ipx}\tilde{\psi}(p)$$

$$= \int_{-\infty}^{+\infty} \frac{dp}{2\pi}\left(-i\frac{d}{dx}e^{ipx}\right)\tilde{\psi}(p) = -i\frac{d}{dx}\psi(x)$$

$$= (\hat{P}\psi)(x). \tag{2.27}$$

2.1.2 *Observables and random variables*

The statistical interpretation of quantum mechanics is closely connected to the spectral decomposition (2.5) of self-adjoint operators $\hat{R}$ introduced in the preceding subsection. It is based on the following postulates.

We consider a statistical ensemble $\mathcal{E}$ consisting of a large number of identically prepared quantum systems $S^{(1)}, S^{(2)}, \ldots, S^{(N)}$,

$$\mathcal{E} = \{S^{(1)}, S^{(2)}, \ldots, S^{(N)}\}. \tag{2.28}$$

As in the classical theory, the construction of such an ensemble requires the specification of a certain set of experimental conditions which may be realized, at least in principle, an infinite number of times. Each realization of this identical set of conditions leads to the preparation of a single quantum mechanical system $S^{(i)}$ which belongs to the ensemble $\mathcal{E}$. The first postulate is that, under certain conditions (more on this point later), a complete characterization of such a statistical ensemble is provided by a normalized state vector $|\psi\rangle$ in the Hilbert space $\mathcal{H}$ pertaining to the quantum mechanical system.

The second postulate is that the measurable quantities of the statistical ensemble $\mathcal{E}$ are represented by self-adjoint operators in the Hilbert space $\mathcal{H}$. The outcomes of the measurements of an observable $\hat{R}$, performed on the ensemble described by $|\psi\rangle$, represent a real-valued random variable R with a cumulative distribution function $F_R(r)$ which is given by

$$F_R(r) = \langle\psi|E_r|\psi\rangle, \tag{2.29}$$

where E_r is the spectral family of $\hat{R}$.

Equation (2.29) represents the basis of the statistical interpretation (von Neumann, 1955). It is easy to see that $F_R(r)$ has, in fact, the properties (1.18)–(1.20) of a cumulative distribution function of a real random variable R. Note that this is a direct consequence of the properties (2.6)–(2.8) of the spectral family E_r. In functional analysis $F_R(r)$ is referred to as the spectral measure. In view of eqn (2.29) it is also obvious that the possible outcomes of the measurement represented by $\hat{R}$ are given by the values of r which belong to the spectrum of $\hat{R}$. Namely, if r is not contained in $\mathrm{spec}(\hat{R})$ then the spectral family is constant in a neighbourhood of r and, thus, also $F_R(r)$ is constant there. The probability for the random variable R to fall into this neighbourhood is therefore equal to zero.

On the basis of eqn (2.29) one can also define the probabilities for more general events as follows. If B denotes some Borel set of $\mathbb{R}$ we can introduce a corresponding projection operator by means of

$$E(B) = \int_B dE_r, \tag{2.30}$$

such that the probability for the measurement outcome to fall into the set B is given by

$$P_R(B) = \langle\psi|E(B)|\psi\rangle. \tag{2.31}$$

For example, the probability for r to fall into the interval $(a, b]$ is found to be equal to $\langle\psi|E_b - E_a|\psi\rangle = F_R(b) - F_R(a)$. In view of eqn (2.30) we have a correspondence between the Borel sets B and the projection operators $E(B)$ defined in terms of the spectral family. For any sequence of disjoint Borel sets B_i we have

$$E\left(\cup_i B_i\right) = \sum_i E(B_i), \tag{2.32}$$

and a formally identical relation holds for the corresponding probabilities, namely

$$P_R\left(\cup_i B_i\right) = \sum_i P_R(B_i). \tag{2.33}$$

In mathematical terms, the map $B \mapsto E(B)$ with the property (2.32) is called a projection-valued measure.

Thus we see that an observable $\hat{R}$ leads, via its spectral family, to a real random variable R which describes the probabilities for all possible measurement outcomes. The sample space Ω is the real axis and the algebra of events $\mathcal{B}$ is given by the Borel sets of $\mathbb{R}$. In particular, on using the spectral decomposition one finds for the mean value of R the well-known expression

$$E(R) = \int_{-\infty}^{+\infty} r \, dF_R(r) = \int_{-\infty}^{+\infty} r \, d\langle\psi|E_r|\psi\rangle = \langle\psi|\hat{R}|\psi\rangle, \qquad (2.34)$$

whereas the variance of R reads

$$\text{Var}(R) = E(R^2) - [E(R)]^2 = \langle\psi|\hat{R}^2|\psi\rangle - \langle\psi|\hat{R}|\psi\rangle^2. \qquad (2.35)$$

2.1.3 Pure states and statistical mixtures

Our characterization of quantum statistical ensembles is not yet the most general one encountered in the applications of quantum theory. An obvious way to obtain more general ensembles is the following. Consider a number M of ensembles $\mathcal{E}_1, \mathcal{E}_2, \ldots, \mathcal{E}_M$ of the type introduced in the preceding subsection. Each of these ensembles is described by a normalized state vector ψ_α, $\alpha = 1, 2, \ldots, M$, in the underlying Hilbert space $\mathcal{H}$. It is then natural to study the statistics of the total ensemble $\mathcal{E}$ which is obtained by mixing all the $\mathcal{E}_\alpha$ with respective weights w_α satisfying

$$w_\alpha \geq 0, \quad \sum_{\alpha=1}^{M} w_\alpha = 1. \qquad (2.36)$$

The mixing is achieved by taking a large number N_α of systems from each $\mathcal{E}_\alpha$. The total number $N = \sum_\alpha N_\alpha$ of systems then constitutes the new ensemble $\mathcal{E}$ and the weights w_α are given by $w_\alpha = N_\alpha/N$.

2.1.3.1 The statistical formulae of quantum mechanics

According to the rules of classical probability theory any self-adjoint operator $\hat{R}$ now yields a random variable R with cumulative distribution function

$$F_R(r) = \sum_\alpha w_\alpha \langle\psi_\alpha|E_r|\psi_\alpha\rangle. \qquad (2.37)$$

This equation generalizes eqn (2.29) to the present case. Accordingly, the mean value of R is now given by

$$E(R) = \sum_\alpha w_\alpha \langle\psi_\alpha|\hat{R}|\psi_\alpha\rangle. \qquad (2.38)$$

These formulae can be cast into a compact form by introducing the *density matrix*, or *statistical operator*,

$$\rho = \sum_\alpha w_\alpha |\psi_\alpha\rangle\langle\psi_\alpha|, \qquad (2.39)$$

which enables one to write the distribution function of the random variable R as

$$F_R(r) = \text{tr}\{E_r\rho\}, \tag{2.40}$$

where tr denotes the trace of an operator. The trace of an operator A is defined by

$$\text{tr}A = \sum_i \langle\varphi_i|A|\varphi_i\rangle, \tag{2.41}$$

where $\{\varphi_i\}$ is an orthonormal basis of the Hilbert space $\mathcal{H}$. Provided the trace exists, it is easy to show that it does not depend on the particular choice of the basis $\{\varphi_i\}$.

By the same argument, the mean and the variance of R can now be written in the following form,

$$E(R) = \text{tr}\{\hat{R}\rho\}, \tag{2.42}$$

$$\text{Var}(R) = \text{tr}\{\hat{R}^2\rho\} - \left[\text{tr}\{\hat{R}\rho\}\right]^2. \tag{2.43}$$

Following the usual notation we also write the expectation value of R as

$$\langle\hat{R}\rangle = \text{tr}\{\hat{R}\rho\}, \tag{2.44}$$

and its variance as

$$\text{Var}(\hat{R}) = \langle\hat{R}^2\rangle - \langle\hat{R}\rangle^2. \tag{2.45}$$

Thus we have the result that the total ensemble $\mathcal{E}$ can, with regard to its statistical properties, be completely characterized by a density matrix ρ. With the help of eqn (2.39) one easily verifies that ρ is self-adjoint, positive[6] and has trace one,

$$\rho^\dagger = \rho, \quad \rho \geq 0, \quad \text{tr}\rho = 1. \tag{2.46}$$

The above reasoning can be confirmed further by an axiomatic approach (von Neumann, 1955). Namely, consider the set of bounded self-adjoint operators on the Hilbert space $\mathcal{H}$. We denote this set by $\mathcal{R}$. We seek a mathematical expression for the expectation value, which is regarded as a real functional $\tilde{E}(\hat{R})$ on $\mathcal{R}$, that is, which assigns a real number to every $\hat{R} \in \mathcal{R}$. A natural set of conditions any such functional must fulfil is the following:

$$\tilde{E}(I) = 1, \tag{2.47}$$

$$\tilde{E}(\Pi) \geq 0, \tag{2.48}$$

$$\tilde{E}\left(\sum_i c_i\hat{R}_i\right) = \sum_i c_i\tilde{E}(\hat{R}_i), \quad c_i \in \mathbb{R}. \tag{2.49}$$

The first condition means that the unit operator, corresponding to a deterministic (i.e. dispersion-free) variable $R = 1$, has expectation 1. The second condition expresses the natural requirement that the expectation of any projection

---

[6] An operator A is said to be positive if for all $\psi \in \mathcal{D}(A)$ the inequality $\langle\psi|A\psi\rangle \geq 0$ holds, which is often simply written as $A \geq 0$. Note that we include the equality sign in the definition of positivity.

operator Π must be positive. This requirement results from our former observation that the events of the theory correspond to the projection operators in the Hilbert space such that $\tilde{E}(\Pi)$ is the probability for the event represented by Π (see eqn (2.30) and (2.31)). Finally, the third condition states that $\tilde{E}$ should be a linear functional on $\mathcal{R}$. A fundamental theorem (Langerholc, 1965) then states that any functional $\tilde{E}(\hat{R})$ with these properties must necessarily be of the form (2.42), that is there exists a unique operator ρ with the properties (2.46) such that $\tilde{E}(\hat{R}) = E(R) = \text{tr}(\hat{R}\rho)$.

In our definition (2.39) of the density matrix we did not assume that the $|\psi_\alpha\rangle$ are orthogonal. However, given ρ we can, of course, diagonalize it. Since ρ is positive, its eigenvalues are greater than or equal to zero. The spectral theory of density operators asserts that ρ has only a countable number of strictly positive eigenvalues $p_i > 0$. The point 0 is the only possible accumulation point of the spectrum. Furthermore, the strictly positive eigenvalues are finitely degenerate, and 0 is the only possible infinitely degenerate eigenvalue. Hence, the spectral decomposition of ρ can be written as

$$\rho = \sum_i p_i |\varphi_i\rangle\langle\varphi_i|, \tag{2.50}$$

where the sum extends over a complete set of eigenstates $|\varphi_i\rangle$ with the eigenvalues p_i. The normalization condition therefore takes the form

$$\text{tr}\rho = \sum_i p_i = 1. \tag{2.51}$$

2.1.3.2 *Properties of the density matrix* We briefly discuss the most important general properties of density matrices. The first one is the inequality

$$\text{tr}\rho^2 \leq \text{tr}\rho = 1. \tag{2.52}$$

It may be shown that the equality sign holds if and only if ρ has the form $\rho = |\psi\rangle\langle\psi|$ for some unit state vector $|\psi\rangle$. This is the case considered in the previous subsection 2.1.2. An ensemble with this property is said to be in a pure state. It is clear that one can replace the above $|\psi\rangle$ by $\exp(i\chi)|\psi\rangle$, with an arbitrary phase $\chi \in [0, 2\pi)$, without changing the density matrix and the statistical formulae. Thus, a pure state is uniquely described by a ray in the underlying Hilbert space. In mathematical terms, one can therefore say that the set of pure states is isomorphic to the set of rays in $\mathcal{H}$, that is, to the projective Hilbert space.

If the left-hand side in (2.52) is strictly less than 1, the ensemble is called a statistical mixture. Alternatively, the difference between pure states and mixtures may be characterized as follows. The set of all density matrices will be denoted by $\mathcal{S}(\mathcal{H})$. This set is convex which means that for any two density matrices ρ_1 and ρ_2 the convex linear combination

$$\rho = \lambda\rho_1 + (1 - \lambda)\rho_2, \quad \lambda \in [0, 1], \tag{2.53}$$

is also a density matrix. Physically, this convex linear combination describes an ensemble which is obtained from the mixture of the ensembles $\mathcal{E}_1$ (corresponding to ρ_1) and $\mathcal{E}_2$ (corresponding to ρ_2) with weights λ and $1 - \lambda$, respectively. The pure states lie on the boundary of the set $\mathcal{S}(\mathcal{H})$ and are distinguished by the fact that they cannot be represented as a non-trivial convex linear combination of two different density matrices. More precisely, if ρ is a pure state and if (2.53) holds for some $\lambda \in (0,1)$, it follows that $\rho_1 = \rho_2 = \rho$. The physical implication is that any decomposition of a pure statistical ensemble $\mathcal{E}$ into two sub-ensembles $\mathcal{E}_1$ and $\mathcal{E}_2$ does not change in any way the statistical properties, that is, $\mathcal{E}_1$ and $\mathcal{E}_2$ have the same statistics as $\mathcal{E}$.

In eqn (2.39) the density matrix has been introduced as a convex linear combination of normalized, not necessarily orthogonal states $|\psi_\alpha\rangle$ with weights w_α satisfying (2.36). Given some density matrix ρ this decomposition into a convex linear combination of pure states is, in general, not unique (Hughston, Jozsa and Wootters, 1993; Nielsen and Chuang, 2000). In order to characterize the possible convex linear combinations which give rise to one and the same density matrix it is convenient to work with non-normalized states $|\tilde{\psi}_\alpha\rangle = \sqrt{w_\alpha}|\psi_\alpha\rangle$ and to write the density matrix as

$$\rho = \sum_\alpha |\tilde{\psi}_\alpha\rangle\langle\tilde{\psi}_\alpha|. \tag{2.54}$$

Such a set of states $\{|\tilde{\psi}_\alpha\rangle\}$ is said to generate the given density matrix ρ. Consider now two sets of states $\{|\tilde{\psi}_\alpha\rangle\}$ and $\{|\tilde{\chi}_\alpha\rangle\}$. These two sets generate the same density matrix ρ, that is

$$\rho = \sum_\alpha |\tilde{\psi}_\alpha\rangle\langle\tilde{\psi}_\alpha| = \sum_\alpha |\tilde{\chi}_\alpha\rangle\langle\tilde{\chi}_\alpha|, \tag{2.55}$$

if and only if there exists a unitary matrix $u = (u_{\alpha\beta})$ such that

$$|\tilde{\psi}_\alpha\rangle = \sum_\beta u_{\alpha\beta}|\tilde{\chi}_\beta\rangle. \tag{2.56}$$

Here, one appends zero vectors to the set with the smaller number of states in such a way that the number of states in both sets becomes equal.

This statement will be very useful later on, for example in the characterization of the freedom one has in the representation of generalized quantum measurements. To prove the statement we first assume that (2.56) holds. By direct substitution it is then easily seen that (2.55) holds, that is both sets generate the same density matrix. Conversely, suppose that (2.55) holds. We write the spectral decomposition (2.50) of ρ as $\rho = \sum_i |\tilde{\varphi}_i\rangle\langle\tilde{\varphi}_i|$, where $|\tilde{\varphi}_i\rangle = \sqrt{p_i}|\varphi_i\rangle$. Consider any state $|\psi\rangle$ which is orthogonal to the space spanned by the states $\{|\tilde{\varphi}_i\rangle\}$. Then we have

$$\langle\psi|\rho|\psi\rangle = 0 = \sum_\alpha |\langle\psi|\tilde{\psi}_\alpha\rangle|^2, \tag{2.57}$$

and we conclude that $\langle \psi | \tilde{\psi}_\alpha \rangle = 0$ for all α and for all $|\psi\rangle$ of the above form. This means that $|\tilde{\psi}_\alpha\rangle$ can be expressed as a linear combination of the $|\tilde{\varphi}_i\rangle$,

$$|\tilde{\psi}_\alpha\rangle = \sum_i c_{\alpha i} |\tilde{\varphi}_i\rangle. \tag{2.58}$$

This gives

$$\rho = \sum_i |\tilde{\varphi}_i\rangle\langle\tilde{\varphi}_i| = \sum_{ij}\left(\sum_\alpha c^*_{\alpha j}c_{\alpha i}\right)|\tilde{\varphi}_i\rangle\langle\tilde{\varphi}_j|, \tag{2.59}$$

from which we deduce that

$$\sum_\alpha c^\dagger_{j\alpha}c_{\alpha i} = \delta_{ji}. \tag{2.60}$$

This means that the matrix $c = (c_{\alpha i})$ can be supplemented to form a square and unitary matrix such that

$$|\tilde{\psi}_\alpha\rangle = \sum_\beta c_{\alpha\beta}|\tilde{\varphi}_\beta\rangle, \tag{2.61}$$

where, if appropriate, zero vectors must be appended to the set of the states $|\tilde{\varphi}_i\rangle$. In the same manner we get a relation of the form

$$|\tilde{\chi}_\alpha\rangle = \sum_\beta d_{\alpha\beta}|\tilde{\varphi}_\beta\rangle, \tag{2.62}$$

with a unitary matrix d. The last two equations yield a relation of the desired form (2.56), where the matrix u is defined by $u = cd^\dagger$ which is unitary since c and d are unitary.

2.1.3.3 *The non-existence of dispersion-free ensembles* It is a well-known fact that quantum mechanics cannot be formulated as a statistical theory on a classical probability space. There exist many mathematical theorems which express this fact in various ways (see, e.g. Gudder, 1979). Here and in the following subsection we shall elucidate this point from a probabilistic viewpoint.

Consider an arbitrary σ-algebra $\mathcal{A}$ of events on a sample space Ω (see Section 1.1). One can then always introduce a dispersion-free measure μ on $\mathcal{A}$ in the following way: Take some fixed point $\omega_0 \in \Omega$ and define for $A \in \mathcal{A}$,

$$\mu(A) \equiv \begin{cases} 1, & \text{for } \omega_0 \in A, \\ 0, & \text{for } \omega_0 \notin A. \end{cases} \tag{2.63}$$

The probability space thus defined has the property that *all* random variables X on it are dispersion-free. In fact, the cumulative distribution of X is given by (see eqn (1.17))

$$F_X(x) = \begin{cases} 0, & \text{for } x < x_0, \\ 1, & \text{for } x \geq x_0. \end{cases} \tag{2.64}$$

This shows that X is a sharp, deterministic variable which takes on the single value $x_0 = X(\omega_0)$ with unit probability.

By a well-known theorem of von Neumann (von Neumann, 1955) in quantum mechanics the situation is markedly different, namely there do not exist dispersion-free ensembles $\mathcal{E}$ of whatsoever type. This is easily demonstrated with the help of the statistical formulae derived above. Namely, if $\mathcal{E}$ were such an ensemble and ρ its statistical operator we must have $\mathrm{Var}(R) = 0$ for all observables $\hat{R}$, that is

$$\mathrm{tr}\left\{\hat{R}^2\rho\right\} = \left[\mathrm{tr}\left\{\hat{R}\rho\right\}\right]^2. \tag{2.65}$$

Taking for $\hat{R}$ the one-dimensional projections $\hat{R} = |\psi\rangle\langle\psi|$, $||\psi|| = 1$, one finds

$$\langle\psi|\rho|\psi\rangle = \langle\psi|\rho|\psi\rangle^2, \tag{2.66}$$

and, therefore, $\langle\psi|\rho|\psi\rangle$ must be identically 0 or 1 for all unit vectors in $\mathcal{H}$. It follows that $\rho = 0$ or $\rho = I$, which is a contradiction for both cases do not represent a statistical operator (they violate the normalization condition $\mathrm{tr}(\rho) = 1$; if one admits non-normalizable density matrices the case $\rho = 0$ is clearly excluded, but also the case $\rho = I$ is excluded since it does not represent a dispersion-free ensemble as is easily demonstrated directly).

2.1.4   Joint probabilities in quantum mechanics

As we have seen, a self-adjoint operator $\hat{R}$ leads via its spectral family E_r to a random variable R with cumulative distribution function $F_R(r)$ given by (2.29) (for simplicity we consider here only ensembles in pure states). Consider now two self-adjoint operators $\hat{R}_1$ and $\hat{R}_2$ and the corresponding random variables R_1, R_2 with distribution functions

$$F_i(r) = \langle\psi|E_r^i|\psi\rangle, \quad i = 1, 2, \tag{2.67}$$

where E_r^1 and E_r^2 denote the spectral families of $\hat{R}_1$ and $\hat{R}_2$, respectively. An important question is then, of what nature are the statistical correlations between the random variables R_1 and R_2 and can one describe these correlations in terms of a classical joint probability?

This question is answered by a theorem of Nelson (1967) which asserts, essentially, that two observables $\hat{R}_1$ and $\hat{R}_2$ can be represented in all states as random variables on a common, classical probability space if and only if they commute, that is, if and only if $[\hat{R}_1, \hat{R}_2] = 0$. Thus, the correlations between non-commuting observables do not in general admit the characterization with the help of a classical joint probability distribution.

Let us formulate this theorem more precisely. To this end, consider two random numbers R_1, R_2 on a common probability space with probability measure μ. We denote by

$$F(r_1, r_2) = \mu\left(R_1 \leq r_1, \ R_2 \leq r_2\right) \tag{2.68}$$

their cumulative joint probability distribution. If R_1 and R_2 are statistically independent (see eqn (1.26)) the joint probability distribution factorizes, $F(r_1, r_2) =$

$F_1(r_1) \cdot F_2(r_2)$. In that case, $F(r_1, r_2)$ can be reconstructed, of course, from knowledge of the marginal distributions $F_1(r_1)$ and $F_2(r_2)$. This is not possible if R_1 and R_2 are statistically dependent. However, if the distribution function of the random variable

$$k \cdot R \equiv k_1 R_1 + k_2 R_2, \tag{2.69}$$

namely the function

$$F_{k \cdot R}(r) = \mu \left(k_1 R_1 + k_2 R_2 \leq r \right), \tag{2.70}$$

is known for all real k_1, k_2, the joint distribution $F(r_1, r_2)$ is uniquely determined. This is a theorem of classical probability theory and can be shown as follows. Recall that $F(r_1, r_2)$ is uniquely determined by its characteristic function

$$G(k_1, k_2) = \int\limits_{\mathbb{R}^2} \exp\left[i(k_1 r_1 + k_2 r_2)\right] dF(r_1, r_2). \tag{2.71}$$

This equation can be written as

$$G(k_1, k_2) = \int\limits_{-\infty}^{+\infty} e^{ir} dF_{k \cdot R}(r), \tag{2.72}$$

which shows explicitly that the characteristic function of $F(r_1, r_2)$ and, therefore, $F(r_1, r_2)$ itself can be obtained from knowledge of $F_{k \cdot R}(r)$.

Turn now to quantum mechanics and consider the above self-adjoint operators $\hat{R}_1$, $\hat{R}_2$. We assume that also the linear combination

$$k \cdot \hat{R} \equiv k_1 \hat{R}_1 + k_2 \hat{R}_2 \tag{2.73}$$

is self-adjoint for all real k_1, k_2 and denote by $E_r^{k \cdot \hat{R}}$ the spectral family of the operator $k \cdot \hat{R}$. One should expect then that the self-adjoint operator (2.73) corresponds to the random variable (2.69) which means, according to the statistical formulae of quantum mechanics, that the distribution function of (2.69) is given by

$$F_{k \cdot R}(r) = \langle \psi | E_r^{k \cdot \hat{R}} | \psi \rangle. \tag{2.74}$$

If this is true the observables $\hat{R}_1$ and $\hat{R}_2$ have a joint probability distribution $F(r_1, r_2)$ which can be determined from $F_{k \cdot R}(r)$ with the help of eqn (2.72). Thus we are led to the following definition: Two observables $\hat{R}_1$ and $\hat{R}_2$ are said to have a joint probability distribution in the state ψ if there exist a classical probability space and two random numbers R_1, R_2 on it such that for all real k_1, k_2 eqn (2.74) holds.

Inserting eqn (2.74) into eqn (2.72) we get

$$G(k_1, k_2) = \int\limits_{-\infty}^{+\infty} e^{ir} dF_{k \cdot R}(r) = \left\langle \psi \left| \int\limits_{-\infty}^{+\infty} e^{ir} dE_r^{k \cdot \hat{R}} \right| \psi \right\rangle$$

$$= \left\langle \psi \left| \exp \left(ik \cdot \hat{R} \right) \right| \psi \right\rangle, \tag{2.75}$$

where we have applied eqn (2.11) to the function $f(r) = \exp(ir)$ and to the operator $k \cdot \hat{R}$. Thus we see that, if $\hat{R}_1$ and $\hat{R}_2$ have a joint probability distribution in the state ψ, then (2.75) is necessarily its characteristic function.

Let us investigate the expectation values generated by $G(k_1, k_2)$. The first terms of its Taylor expansion around $k = 0$ are given by

$$G(k_1, k_2) = 1 + ik_1 \langle \psi | \hat{R}_1 | \psi \rangle + ik_2 \langle \psi | \hat{R}_2 | \psi \rangle - \frac{k_1^2}{2} \langle \psi | \hat{R}_1^2 | \psi \rangle - \frac{k_2^2}{2} \langle \psi | \hat{R}_2^2 | \psi \rangle$$

$$- \frac{1}{2} k_1 k_2 \langle \psi | \hat{R}_1 \hat{R}_2 + \hat{R}_2 \hat{R}_1 | \psi \rangle + \mathcal{O}(k^3). \tag{2.76}$$

This shows that $G(k_1, k_2)$ generates the quantum expectation value of the symmetrically ordered product $(\hat{R}_1 \hat{R}_2 + \hat{R}_2 \hat{R}_1)/2$, but not that of the operator $i(\hat{R}_1 \hat{R}_2 - \hat{R}_2 \hat{R}_1)/2$. In other words, G does not contain any information on the difference between the expectations $\langle \psi | \hat{R}_1 \hat{R}_2 | \psi \rangle$ and $\langle \psi | \hat{R}_2 \hat{R}_1 | \psi \rangle$.

With the above definition of joint probability we can now give the precise formulation of Nelson's theorem: The observables $\hat{R}_1$ and $\hat{R}_2$ have a joint probability distribution in all states ψ if and only if they commute, that is if and only if $[\hat{R}_1, \hat{R}_2] = 0$.

It is clear that $\hat{R}_1$ and $\hat{R}_2$ have a joint probability distribution if they commute, namely

$$F(r_1, r_2) = \langle \psi | E_{r_1}^1 E_{r_2}^2 | \psi \rangle. \tag{2.77}$$

As is easily checked using the properties of the spectral families $E_{r_1}^1$ and $E_{r_2}^2$, this is a real, non-negative function with the properties of a true joint distribution function. In addition, it leads to the characteristic function (2.75) and the relation (2.74) is satisfied.

The general proof of the *only if* part of the theorem is more difficult. We refrain from giving the details of this proof, but merely discuss the important case $\hat{R}_1 = \hat{Q}$, the position operator, and $\hat{R}_2 = \hat{P}$, the momentum operator, satisfying the Heisenberg commutation relation ($\hbar = 1$),

$$\left[\hat{Q}, \hat{P} \right] = i. \tag{2.78}$$

It is then easily seen that the function (2.75) is *not* the characteristic function of a classical joint probability distribution. In fact, on Fourier transforming we find that the density of $F(q, p)$ is given by

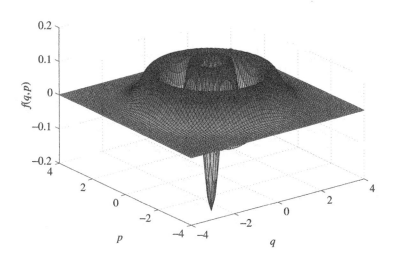

FIG. 2.1. The Wigner distribution (2.81) corresponding to the third excited eigenstate of the harmonic oscillator Hamiltonian $H = \frac{1}{2}(\hat{P}^2 + \hat{Q}^2)$.

$$f(q,p) \equiv \frac{\partial^2 F}{\partial q \partial p} = \int \frac{dk_1}{2\pi} \int \frac{dk_2}{2\pi} e^{-i(k_1 q + k_2 p)} \left\langle \psi \left| e^{i(k_1 \hat{Q} + k_2 \hat{P})} \right| \psi \right\rangle \qquad (2.79)$$

$$= \int \frac{dk_1}{2\pi} \int \frac{dk_2}{2\pi} \int dx e^{-i(k_1 q + k_2 p) + i k_1 k_2/2 + i k_1 x} \psi^*(x) \psi(x + k_2).$$

Here we have introduced the position representation of the state vector, $\psi(x) = \langle x|\psi\rangle$. Further we made use of

$$e^{i(k_1 \hat{Q} + k_2 \hat{P})} = e^{i k_1 \hat{Q}} e^{i k_2 \hat{P}} e^{i k_1 k_2/2}, \qquad (2.80)$$

and of the fact that $\exp(ik_2 \hat{P})$ induces the translation $x \mapsto x + k_2$. Carrying out the integrations over k_1 and x and substituting $s = -k_2$ we finally arrive at

$$f(q,p) = \int\limits_{-\infty}^{+\infty} \frac{ds}{2\pi} e^{isp} \psi^*(q + s/2) \psi(q - s/2). \qquad (2.81)$$

This is the famous *Wigner distribution* (see, e.g. Louisell, 1990). It is interesting to see that, in the present context, it arises as the only possible candidate for a joint probability density of position and momentum. However, as is well known it does not represent a true probability density since it is not positive for general ψ. An example is shown in Fig. 2.1.

In spite of the fact that the Wigner distribution is not a true probability distribution it is a useful quantity. Its Fourier transform is known as the *Wigner characteristic function.* For a system described by an arbitrary density matrix ρ it is given by

$$G(k_1, k_2) = \text{tr}\left\{\exp\left(i(k_1\hat{Q} + k_2\hat{P})\right)\rho\right\}. \qquad (2.82)$$

Introducing the creation and destruction operators $b^\dagger$ and b through the relations $\hat{Q} = (b+b^\dagger)/\sqrt{2}$, $\hat{P} = -i(b-b^\dagger)/\sqrt{2}$, and using the complex variable $\alpha = ik_1 - k_2$ one often writes the Wigner characteristic function as

$$\chi(\alpha, \alpha^*) = \text{tr}\left\{\exp\left(\alpha b^\dagger - \alpha^* b\right)\rho\right\}. \qquad (2.83)$$

Similarly to eqn (2.76), it may be shown by an expansion of the exponential that $\chi(\alpha, \alpha^*)$ generates the symmetrically ordered products of b and $b^\dagger$ through the formula

$$\text{tr}\left\{\left(b^n b^{\dagger m}\right)_s \rho\right\} = \left.\frac{\partial^{n+m}}{\partial(-\alpha^*)^n \partial\alpha^m}\chi(\alpha, \alpha^*)\right|_{\alpha=\alpha^*=0}. \qquad (2.84)$$

Here, the symmetrically ordered product $\left(b^n b^{\dagger m}\right)_s$ is defined to be the average over all possibilities of ordering the operators in the product $b^n b^{\dagger m}$.

2.2 Composite quantum systems

In the theory of open quantum systems the notion of composite quantum systems is fundamental. The Hilbert space of a composite quantum system is the tensor product space of the Hilbert spaces describing its subsystems. The present section serves to introduce some basic features of tensor products of Hilbert spaces and of the statistical properties of composite systems and their subsystems.

2.2.1 *Tensor product*

We consider two quantum systems $S^{(1)}$ and $S^{(2)}$ with respective Hilbert spaces $\mathcal{H}^{(1)}$ and $\mathcal{H}^{(2)}$. The two systems may represent two (distinguishable) particles, two different composite objects (e.g. two atoms or molecules), or two different degrees of freedom of the same object (e.g. rotational and translational degrees of freedom of a molecule). In general, $S^{(1)}$ and $S^{(2)}$ interact with each other. According to the postulates of quantum mechanics the state space $\mathcal{H}$ of the combined system $S = S^{(1)} + S^{(2)}$ is given by the tensor product of the Hilbert spaces pertaining to the subsystems $S^{(1)}$ and $S^{(2)}$,

$$\mathcal{H} = \mathcal{H}^{(1)} \otimes \mathcal{H}^{(2)}. \qquad (2.85)$$

If we take fixed orthonormal bases $\{|\varphi_i^{(1)}\rangle\}$ and $\{|\varphi_j^{(2)}\rangle\}$ in $\mathcal{H}^{(1)}$ and $\mathcal{H}^{(2)}$, respectively, a general state in the tensor product space $\mathcal{H}$ may be written as

$$|\Psi\rangle = \sum_{ij} \alpha_{ij} |\varphi_i^{(1)}\rangle \otimes |\varphi_j^{(2)}\rangle. \qquad (2.86)$$

This means that the elements $|\varphi_i^{(1)}\rangle \otimes |\varphi_j^{(2)}\rangle$ form a basis of the tensor product space and that the dimension of $\mathcal{H}$ is equal to the product of the dimensions of $\mathcal{H}^{(1)}$ and $\mathcal{H}^{(2)}$.

If $A^{(1)}$ is an operator acting in $\mathcal{H}^{(1)}$ and $A^{(2)}$ is an operator acting in $\mathcal{H}^{(2)}$, one defines their tensor product $A^{(1)} \otimes A^{(2)}$ by

$$(A^{(1)} \otimes A^{(2)})(|\varphi_i^{(1)}\rangle \otimes |\varphi_j^{(2)}\rangle) \equiv (A^{(1)}|\varphi_i^{(1)}\rangle) \otimes (A^{(2)}|\varphi_j^{(2)}\rangle), \qquad (2.87)$$

and by a linear extension of this formula for arbitrary states (2.86). Any operator A acting on $\mathcal{H}$ can be represented as a linear combination of tensor products,

$$A = \sum_{\alpha} A_{\alpha}^{(1)} \otimes A_{\alpha}^{(2)}. \qquad (2.88)$$

Specifically, the observables of system $S^{(1)}$ take the form $A^{(1)} \otimes I^{(2)}$, while observables of system $S^{(2)}$ are given by the expression $I^{(1)} \otimes A^{(2)}$. The identity operators in $\mathcal{H}^{(1)}$ and $\mathcal{H}^{(2)}$ are denoted, respectively, by $I^{(1)}$ and $I^{(2)}$.

The density matrix of the composite system S is an operator in the state space $\mathcal{S}(\mathcal{H})$. If the subsystems $S^{(1)}$ and $S^{(2)}$ are uncorrelated the total density matrix takes the form of a tensor product of the density matrices of the subsystems,

$$\rho = \rho^{(1)} \otimes \rho^{(2)}. \qquad (2.89)$$

This implies that the expectation value of any tensor product of operators pertaining to the subsystems factorizes,

$$\langle A^{(1)} \otimes A^{(2)} \rangle \equiv \mathrm{tr} \left\{ \left(A^{(1)} \otimes A^{(2)} \right) \rho \right\} = \mathrm{tr}^{(1)} \left\{ A^{(1)} \rho^{(1)} \right\} \cdot \mathrm{tr}^{(2)} \left\{ A^{(2)} \rho^{(2)} \right\}$$
$$= \langle A^{(1)} \rangle \cdot \langle A^{(2)} \rangle. \qquad (2.90)$$

Here, $\mathrm{tr}^{(1)}$ and $\mathrm{tr}^{(2)}$ denote the partial traces over the Hilbert spaces $\mathcal{H}^{(1)}$ and $\mathcal{H}^{(2)}$, respectively.

If one is only interested in observables of subsystem $S^{(1)}$, that is only in operators of the form

$$A = A^{(1)} \otimes I^{(2)}, \qquad (2.91)$$

it is convenient to introduce the reduced density matrix pertaining to the subsystem by means of

$$\rho^{(1)} \equiv \mathrm{tr}^{(2)} \rho. \qquad (2.92)$$

In this equation $\mathrm{tr}^{(2)}$ denotes the partial trace taken over the second Hilbert space $\mathcal{H}^{(2)}$. A density matrix which is obtained in this way through a partial trace of a density matrix in some larger space is sometimes called an improper mixture. It completely describes the statistical properties of all observables belonging to the subsystem under consideration since the expectation value of any observable of the form (2.91) can be determined with the help of the formula

$$\langle A \rangle = \mathrm{tr}^{(1)} \{ A^{(1)} \rho^{(1)} \}. \qquad (2.93)$$

We have seen that the reduced density matrix of subsystem $S^{(1)}$ is obtained by taking the partial trace over subsystem $S^{(2)}$. It might be interesting to note that this way of describing subsystem $S^{(1)}$ is the only possible. More precisely, if we seek a function $f : S(\mathcal{H}) \longrightarrow S(\mathcal{H}^{(1)})$ which maps the state space of the composite system to the state space of the subsystem, then the function given by $f(\rho) = \mathrm{tr}^{(2)}\rho$ is the only one with the property

$$\mathrm{tr}^{(1)}\{A^{(1)}f(\rho)\} = \mathrm{tr}\left\{\left(A^{(1)} \otimes I^{(2)}\right)\rho\right\}. \tag{2.94}$$

Thus, the definition of the reduced density matrix given above is the only possible one which is compatible with the statistical formulae.

To prove this statement it is convenient to introduce the *Liouville space*. Given some Hilbert space $\mathcal{H}$ the Liouville space is the space of Hilbert–Schmidt operators, that is the space of operators A in $\mathcal{H}$ for which $\mathrm{tr}A^\dagger A$ is finite. Equipped with the scalar product

$$(A, B) \equiv \mathrm{tr}\{A^\dagger B\} \tag{2.95}$$

the space of Hilbert–Schmidt operators becomes a Hilbert space. One can therefore introduce an orthonormal basis $\{B_i\}$ in this space satisfying the orthogonality and the completeness conditions,

$$(B_i, B_j) = \delta_{ij}, \tag{2.96}$$

$$A = \sum_i B_i(B_i, A). \tag{2.97}$$

Let us consider now the Liouville space associated with the state space $\mathcal{H}^{(1)}$ of subsystem $S^{(1)}$. We take an orthonormal basis $\{B_i\}$ of Hermitian operators B_i on $\mathcal{H}^{(1)}$. Using the completeness condition (2.97) and the requirement (2.94) we then get

$$\begin{aligned} f(\rho) &= \sum_i B_i(B_i, f(\rho)) \\ &= \sum_i B_i\mathrm{tr}^{(1)}\{B_i f(\rho)\} \\ &= \sum_i B_i\mathrm{tr}\{(B_i \otimes I^{(2)})\rho\} \\ &= \sum_i B_i\mathrm{tr}^{(1)}\{B_i \rho^{(1)}\} \\ &= \sum_i B_i(B_i, \rho^{(1)}) \\ &= \rho^{(1)} = \mathrm{tr}^{(2)}\rho, \end{aligned} \tag{2.98}$$

which proves that $f(\rho)$ is necessarily given by the partial trace of ρ taken over subsystem $S^{(2)}$.

2.2.2 Schmidt decomposition and entanglement

An important characterization of the states of a composite quantum system is obtained with the help of the Schmidt decomposition theorem. This theorem asserts that for any given state $|\Psi\rangle \in \mathcal{H}^{(1)} \otimes \mathcal{H}^{(2)}$ there exist orthonormal bases, the Schmidt bases $\{|\chi_i^{(1)}\rangle\}$ and $\{|\chi_j^{(2)}\rangle\}$ in $\mathcal{H}^{(1)}$ and $\mathcal{H}^{(2)}$, respectively, such that

$$|\Psi\rangle = \sum_i \alpha_i |\chi_i^{(1)}\rangle \otimes |\chi_i^{(2)}\rangle. \tag{2.99}$$

The α_i are complex numbers called Schmidt coefficients. For a normalized state we have obviously,

$$\langle \Psi | \Psi \rangle = \sum_i |\alpha_i|^2 = 1. \tag{2.100}$$

It must be noted that the Schmidt bases which allow a representation of the form (2.99) depend, in general, on the given state $|\Psi\rangle$.

The Schmidt decomposition theorem can be proven as follows. First, we may suppose without restriction that $\mathcal{H}^{(1)}$ and $\mathcal{H}^{(2)}$ have the same dimension. The matrix $\alpha = (\alpha_{ij})$ of coefficients in the decomposition (2.86) with respect to the fixed basis vectors $|\varphi_i^{(1)}\rangle$ and $|\varphi_j^{(2)}\rangle$ is then a square matrix. By use of the singular value decomposition this matrix can always be written as $\alpha = udv$, where u and v are unitary matrices and d is a diagonal matrix with non-negative diagonal elements $\alpha_i \geq 0$. Thus, the decomposition (2.86) takes the form

$$|\Psi\rangle = \sum_{ijk} u_{ij} \alpha_j v_{jk} |\varphi_i^{(1)}\rangle \otimes |\varphi_k^{(2)}\rangle. \tag{2.101}$$

By virtue of the unitarity of u and v the vectors defined by

$$|\chi_j^{(1)}\rangle \equiv \sum_i u_{ij} |\varphi_i^{(1)}\rangle, \tag{2.102}$$

$$|\chi_j^{(2)}\rangle \equiv \sum_k v_{jk} |\varphi_k^{(2)}\rangle, \tag{2.103}$$

form orthonormal bases in $\mathcal{H}^{(1)}$ and $\mathcal{H}^{(2)}$, respectively. Using these expressions in eqn (2.101) we immediately obtain the Schmidt decomposition (2.99). Note that the proof also demonstrates that the Schmidt bases can always be chosen such that the Schmidt coefficients are real and non-negative.

The number of non-zero Schmidt coefficients α_i is called the Schmidt number. This number is invariant with respect to unitary transformations $U^{(1)}$ and $U^{(2)}$ which act only in the respective spaces $\mathcal{H}^{(1)}$ and $\mathcal{H}^{(2)}$. For the same reason the Schmidt number does not depend on the particular Schmidt bases chosen and is thus uniquely defined for a given state $|\Psi\rangle$.

A state $|\Psi\rangle \in \mathcal{H}^{(1)} \otimes \mathcal{H}^{(2)}$ is said to be *entangled* if it cannot be written as a tensor product $|\varphi^{(1)}\rangle \otimes |\varphi^{(2)}\rangle$ of states of the subsystems. If $|\Psi\rangle$ *can* be written

as a tensor product we call it a product state. It follows from the Schmidt
decomposition theorem that $|\Psi\rangle$ is an entangled state if and only if the Schmidt
number is larger than 1. Equivalently, $|\Psi\rangle$ is a product state if and only if its
Schmidt number is equal to 1. Another notion will be important later on: If the
absolute values of all non-vanishing Schmidt coefficients for a given state $|\Psi\rangle$ are
equal to each other the state is said to be *maximally entangled*.

The Schmidt decomposition can be used to prove several interesting statements about the states of a composite system and its subsystems. For example,
if the combined system S is in a pure state $\rho = |\Psi\rangle\langle\Psi|$ the reduced density
matrices $\rho^{(1)} = \text{tr}^{(2)}\rho$ and $\rho^{(2)} = \text{tr}^{(1)}\rho$ have the same eigenvalues. In fact, the
Schmidt decomposition of $|\Psi\rangle$ immediately yields

$$\rho^{(1)} = \text{tr}^{(2)}\{|\Psi\rangle\langle\Psi|\} = \sum_i |\alpha_i|^2 |\chi_i^{(1)}\rangle\langle\chi_i^{(1)}|, \qquad (2.104)$$

$$\rho^{(2)} = \text{tr}^{(1)}\{|\Psi\rangle\langle\Psi|\} = \sum_i |\alpha_i|^2 |\chi_i^{(2)}\rangle\langle\chi_i^{(2)}|, \qquad (2.105)$$

from which the statement follows. We note that the subsystems are, in general,
described by mixed states. In particular, if $|\Psi\rangle$ is maximally entangled the re-
duced densities are proportional to the identities in the subspaces spanned by
the Schmidt basis vectors belonging to the non-vanishing Schmidt coefficients.
Moreover, these representations show that $|\Psi\rangle$ is a product state if and only if
the reduced density matrix $\rho^{(1)}$ and, therefore, also $\rho^{(2)}$ describe pure states.

2.3   Quantum entropies

Quantum entropies play a crucial rôle in quantum statistical mechanics and
quantum information theory. In this section we briefly review the basic definitions and some important properties of quantum entropies, which will be used
in later sections to characterize, for example, the information gained in a quantum measurement and the irreversible nature of the quantum dynamics of open
systems. More details about quantum entropies may be found in a review article
by Werl (1978) and in the book by Nielsen and Chuang (2000).

2.3.1   *Von Neumann entropy*

The von Neumann entropy provides an important entropy functional used in
quantum statistical mechanics and thermodynamics. Given the state of a quantum statistical ensemble in terms of a density matrix ρ it is defined by

$$S(\rho) \equiv -\text{tr}\{\rho\ln\rho\}. \qquad (2.106)$$

In the present section we set the Boltzmann constant k_B equal to 1, that is we
measure temperatures in units of energy. Using the spectral decomposition of
the density matrix,

$$\rho = \sum_i p_i |\varphi_i\rangle\langle\varphi_i|, \quad p_i \geq 0, \quad \sum_i p_i = 1, \qquad (2.107)$$

we can write

$$S(\rho) = -\sum_i p_i \ln p_i \equiv H(\{p_i\}). \qquad (2.108)$$

In these expressions it is understood that $0 \cdot \ln 0 \equiv 0$. For a mathematically precise definition one defines $S(\rho)$ by means of equation (2.106) if the trace is finite, and sets $S(\rho) = +\infty$ otherwise.

Equation (2.108) shows that the von Neumann entropy is equal to the *Shannon information entropy* $H(\{p_i\})$ of the distribution $i \mapsto p_i$, that is of the random number I with the distribution $p_i = \mu(I = i)$ given by the spectral decomposition of the density matrix ρ. A statistical mixture which is described by ρ can be obtained by mixing pure ensembles described by states $|\varphi_i\rangle$ with corresponding weights p_i. Then $S(\rho)$ expresses our uncertainty, or lack of knowledge about the realization of a particular state $|\varphi_i\rangle$ in the mixture.

We list without proof some important properties of the von Neumann entropy which make evident its importance for quantum statistical mechanics and quantum information theory.

1. For all density matrices one has

$$S(\rho) \geq 0, \qquad (2.109)$$

   where the equality sign holds if and only if ρ is a pure state.

2. If the dimension of the Hilbert space is finite, $\dim \mathcal{H} = D < \infty$, the von Neumann entropy is bounded from above $S(\rho) \leq \ln D$, where the equality sign holds if and only if ρ is the completely mixed or infinite temperature state $\rho = I/D$.

3. The von Neumann entropy is invariant with respect to unitary transformations U of the Hilbert space, that is $S(U\rho U^\dagger) = S(\rho)$.

4. The von Neumann entropy is a concave functional $\rho \mapsto S(\rho)$ on the space of density matrices. This means that for any collection of densities ρ_i and numbers $\lambda_i \geq 0$ satisfying $\sum_i \lambda_i = 1$ one has the inequality

$$S\left(\sum_i \lambda_i \rho_i\right) \geq \sum_i \lambda_i S(\rho_i). \qquad (2.110)$$

   The equality sign in this relation holds if and only if all ρ_i with non-vanishing λ_i are equal to each other. This property is called strict concavity of the entropy functional. In physical terms it means that our uncertainty about the state $\rho = \sum_i \lambda_i \rho_i$ is greater than or equal to the average uncertainty of the states ρ_i that constitute the total mixture.

5. Consider a composite system with Hilbert space $\mathcal{H} = \mathcal{H}^{(1)} \otimes \mathcal{H}^{(2)}$ and denote by ρ the density matrix of the total system and by $\rho^{(1)} = \text{tr}^{(2)}\rho$

and $\rho^{(2)} = \text{tr}^{(1)}\rho$ the densities of the subsystems. Then the von Neumann entropy obeys the so-called subadditivity condition,

$$S(\rho) \leq S(\rho^{(1)}) + S(\rho^{(2)}), \qquad (2.111)$$

where the equality sign holds if and only if the total density matrix describes an uncorrelated state, $\rho = \rho^{(1)} \otimes \rho^{(2)}$. Thus, our uncertainty about the product state $\rho^{(1)} \otimes \rho^{(2)}$ is, in general, greater than our uncertainty about the state ρ of the composite system. In other words, by tracing over the subsystems we lose information on correlations between the subsystems and, consequently, increase the entropy. If $\rho = |\Psi\rangle\langle\Psi|$ is a pure state we have $S(\rho) = 0$. It follows from the Schmidt decomposition that $\rho^{(1)}$ and $\rho^{(2)}$ have the same eigenvalues (see eqns (2.104) and (2.105)) and, thus the von Neumann entropies of the subsystems are equal to each other. This yields

$$S(\rho^{(1)}) = S(\rho^{(2)}) = -\sum_i |\alpha_i|^2 \ln|\alpha_i|^2 \geq 0, \qquad (2.112)$$

and the left-hand side of the inequality is strictly greater than zero if and only if $|\Psi\rangle$ is an entangled state.

2.3.2 *Relative entropy*

For a given pair of density matrices ρ and σ the relative entropy is defined by

$$S(\rho||\sigma) \equiv \text{tr}\{\rho \ln \rho\} - \text{tr}\{\rho \ln \sigma\}. \qquad (2.113)$$

To give a rigorous definition one introduces the kernel of a density matrix as the space spanned by the eigenstates with zero eigenvalue, and the image of a density matrix as the space spanned by the eigenstates belonging to the non-vanishing eigenvalues. The relative entropy is then defined to be equal to $+\infty$ if the kernel of σ has a non-trivial intersection with the image of ρ.

A physical interpretation for the relative entropy can be given if we consider again a composite quantum system as in point 5 of Subsection 2.3.1. Then we have

$$S(\rho||\rho^{(1)} \otimes \rho^{(2)}) = S(\rho^{(1)}) + S(\rho^{(2)}) - S(\rho). \qquad (2.114)$$

The entropy of the density matrix ρ of a combined system relative to the corresponding uncorrelated state $\rho^{(1)} \otimes \rho^{(2)}$ is thus a measure of the change of the von Neumann entropy resulting from the tracing over the subsystems, and provides a measure for the corresponding information loss.

Let us summarize some important properties of the relative entropy functional.

1. The relative entropy fulfils the inequality

$$S(\rho||\sigma) \geq 0 \qquad (2.115)$$

for all ρ and all σ, which is known as the Klein inequality. The equality sign holds if and only if $\rho = \sigma$. With the help of eqn (2.114) the Klein inequality leads to the subadditivity property (2.111) of the von Neumann entropy.

2. Similarly to the von Neumann entropy, the relative entropy is invariant with respect to unitary transformations U,

$$S(U\rho U^\dagger \| U\sigma U^\dagger) = S(\rho\|\sigma). \tag{2.116}$$

3. The relative entropy is jointly convex in its arguments. This means that for $0 \leq \lambda \leq 1$ the inequality

$$S(\rho\|\sigma) \leq \lambda S(\rho_1\|\sigma_1) + (1-\lambda)S(\rho_2\|\sigma_2) \tag{2.117}$$

is satisfied, where $\rho = \lambda\rho_1 + (1-\lambda)\rho_2$ and $\sigma = \lambda\sigma_1 + (1-\lambda)\sigma_2$.

4. Setting $\rho^{(1)} = \mathrm{tr}^{(2)}\rho$ and $\sigma^{(1)} = \mathrm{tr}^{(2)}\sigma$ we have

$$S(\rho^{(1)}\|\sigma^{(1)}) \leq S(\rho\|\sigma). \tag{2.118}$$

Thus, the tracing over a subsystem in both input arguments reduces the relative entropy. In particular, if ρ is an uncorrelated state we find

$$S(\rho^{(1)}\|\sigma^{(1)}) = S(\rho^{(1)} \otimes \rho^{(2)} \| \sigma^{(1)} \otimes \rho^{(2)}). \tag{2.119}$$

We finally comment briefly on a theorem which is of fundamental importance in many proofs of entropy inequalities. This is Lieb's theorem which states that the functional

$$f_t(A, B) = -\mathrm{tr}\left\{X^\dagger A^t X B^{1-t}\right\} \tag{2.120}$$

is jointly convex in its arguments A and B. Here, A and B are positive operators, while X is an arbitrary fixed operator and t is a fixed number in the interval $[0, 1]$. As an example, we note that Lieb's theorem plays a crucial rôle in the proof of the strong subadditivity of the von Neumann entropy (Lieb, 1973) which asserts that for any system which is composed of three subsystems $S^{(1)}$, $S^{(2)}$ and $S^{(3)}$ the inequality

$$S(\rho^{(1,2,3)}) + S(\rho^{(2)}) \leq S(\rho^{(1,2)}) + S(\rho^{(2,3)}) \tag{2.121}$$

holds, where $\rho^{(1,2,3)}$ is the density matrix of the total system and

$$\rho^{(1,2)} = \mathrm{tr}^{(3)}\rho^{(1,2,3)}, \tag{2.122}$$

$$\rho^{(2,3)} = \mathrm{tr}^{(1)}\rho^{(1,2,3)}, \tag{2.123}$$

$$\rho^{(2)} = \mathrm{tr}^{(1,3)}\rho^{(1,2,3)}, \tag{2.124}$$

are reduced density matrices pertaining to the subsystems $S^{(1)} + S^{(2)}$, $S^{(2)} + S^{(3)}$ and $S^{(2)}$, respectively. For classical systems the proof of strong subadditivity is relatively simple. By contrast, no easy proof of strong subadditivity is known in the quantum case (for details, see Werl, 1978; Nielsen and Chuang, 2000).

We shall use Lieb's theorem in Chapter 3 to prove that the entropy production rate for a quantum dynamical semigroup is a convex functional on the state space of an open quantum system. As a preparatory step we take $A = B = \rho$ in eqn (2.120) to get the convex functional

$$f_t(\rho) = -\mathrm{tr}\left\{X^\dagger \rho^t X \rho^{1-t}\right\}. \tag{2.125}$$

Since $f_0(\rho) = -\mathrm{tr}\{X^\dagger X \rho\}$ is a linear functional we conclude that the derivative of $f_t(\rho)$ with respect to t taken at $t = 0$ is also a convex functional. Thus, by Lieb's theorem the functional

$$\rho \mapsto -\mathrm{tr}\left\{\left(X\rho X^\dagger - X^\dagger X \rho\right) \ln \rho\right\} \tag{2.126}$$

is convex. To prove convexity of the entropy production rate we have to connect the expression under the trace to the generator of an irreversible quantum dynamical semigroup which will be derived in Chapter 3.

2.3.3   *Linear entropy*

As another measure for the purity of states it is sometimes useful to use, instead of the von Neumann entropy, the so-called linear entropy $S_l(\rho)$ which is defined by

$$S_l(\rho) = \mathrm{tr}\{\rho - \rho^2\} = 1 - \mathrm{tr}\rho^2. \tag{2.127}$$

We can immediately give an upper and a lower bound for this functional,

$$0 \le S_l(\rho) \le 1. \tag{2.128}$$

The first inequality is obvious from the inequality (2.52), which also shows that $S_l(\rho)$ is equal to zero if and only if ρ is a pure state. The second inequality follows from the fact that $\mathrm{tr}\rho^2$ is a positive operator.

For a D-dimensional space, $\dim \mathcal{H} = D$, we have the upper bound

$$S_l(\rho) \le 1 - \frac{1}{D}. \tag{2.129}$$

The functional $S_l(\rho)$ attains its maximal value for the infinite-temperature state $\rho = I/D$.

2.4   The theory of quantum measurement

In this section we discuss the quantum theory of measurement which is fundamental to the statistical interpretation of quantum mechanics. A general introduction may be found in the book by Gottfried (1974).

The measurement process in quantum mechanics plays a dual rôle. On the one hand, it describes the way in which the state of a quantum system changes if a measurement is performed on it, thereby influencing the predictions on the future behaviour of the system. On the other hand, it gives a unique prescription for the preparation of a quantum system in a definite state. Thus, the measurement process tells us how to choose the set of experimental conditions we have been talking about at the beginning of Section 2.1.2 and which led to the realization of quantum statistical ensembles.

2.4.1 *Ideal quantum measurements*

The fundamental measurement postulate of quantum mechanics may be formulated as follows. Suppose that some property B with corresponding projection operator $E(B)$ is measured on a quantum statistical ensemble $\mathcal{E}$ described by the density matrix ρ. Then, after the measurement, the density matrix

$$\rho' = \frac{E(B)\rho E(B)}{\text{tr}\left\{E(B)\rho E(B)\right\}} \qquad (2.130)$$

describes the sub-ensemble $\mathcal{E}'$ consisting of those systems for which property B has been found to be true. This is the well-known von Neumann–Lüders projection postulate (von Neumann, 1955; Lüders, 1951). Note that the denominator in (2.130) ensures that ρ' is normalized, $\text{tr}\rho' = 1$, and that it is just the probability for the measured property to occur, namely

$$P(B) = \text{tr}\left\{E(B)\rho E(B)\right\} = \text{tr}\left\{E(B)\rho\right\}. \qquad (2.131)$$

If we take some self-adjoint operator $\hat{R}$, then B can be, for example, a Borel set of $\mathbb{R}$, the corresponding event being that the random variable R takes a value in B. The projection operator $E(B)$ is defined in terms of the spectral family of $\hat{R}$ by eqn (2.30). For example, consider the orthogonal decomposition of unity which is given by the projections ΔE_α defined in eqn (2.12). The probability for the event $R \in \Delta r_\alpha$ then takes the form

$$P(\Delta r_\alpha) = \text{tr}\left\{\Delta E_\alpha \rho\right\}, \qquad (2.132)$$

whereas the sub-ensemble corresponding to this particular event is described by the density matrix

$$\rho'_\alpha = \frac{1}{P(\Delta r_\alpha)}\Delta E_\alpha \rho \Delta E_\alpha. \qquad (2.133)$$

The probabilities $P(\Delta r_\alpha)$ are normalized to 1 in view of the completeness relation (2.13),

$$\sum_\alpha P(\Delta r_\alpha) = 1. \qquad (2.134)$$

The density matrices ρ'_α of the sub-ensembles are orthogonal in the sense that $\rho'_\alpha \rho'_\beta = \delta_{\alpha\beta}\rho'_\beta$.

The measurement of an orthogonal decomposition ΔE_α of unity thus leads to a decomposition of the original ensemble $\mathcal{E}$ into the various sub-ensembles labelled by the index α. Such a splitting of the original ensemble into various sub-ensembles, each of which being conditioned on a specific measurement outcome, is called a *selective measurement*.

One could also imagine an experimental situation in which the various sub-ensembles are again mixed with the probabilities $P(\Delta r_\alpha)$ of their occurrence. The resulting ensemble is then described by the density matrix

$$\rho' = \sum_\alpha P(\Delta r_\alpha)\rho'_\alpha = \sum_\alpha \Delta E_\alpha \rho \Delta E_\alpha. \qquad (2.135)$$

This remixing of the sub-ensembles after the measurement is referred to as *non-selective measurement*.

As another example, take a self-adjoint operator $\hat{R}$ with a discrete, possibly degenerate spectrum (see Example 2.1) and the initial density matrix $\rho = |\psi\rangle\langle\psi|$. The probability for the measurement of the discrete eigenvalue r_n, that is, for the event $R = r_n$ corresponding to the projection operator Π_n, is given by

$$P(r_n) = \mathrm{tr}\{\Pi_n \rho\} = \langle\psi|\Pi_n|\psi\rangle. \qquad (2.136)$$

The corresponding sub-ensemble is described by the density matrix

$$\rho'_n = \frac{\Pi_n|\psi\rangle\langle\psi|\Pi_n}{P(r_n)} = |\phi\rangle\langle\phi|, \qquad (2.137)$$

which implies that the measured sub-ensemble may be represented by the normalized state vector

$$\phi = P(r_n)^{-1/2}\Pi_n\psi. \qquad (2.138)$$

Thus we see that the sub-ensemble is again a pure state. This is not true, in general, if the initial density matrix ρ is a true mixture. If however the eigenvalue r_n is non-degenerate, the projection being $\Pi_n = |\psi_n\rangle\langle\psi_n|$, then its measurement does lead to a sub-ensemble describable by a pure state ϕ, even if ρ is a true mixture, namely $\phi = \psi_n$ with probability $P(r_n) = \langle\psi_n|\rho|\psi_n\rangle$.

Finally, we study the position operator $\hat{Q}$ (see Example 2.2). Suppose that an ideal measurement of the event $Q \in \Delta \equiv (a, b]$, that is, of the event that the coordinate lies in the interval $(a, b]$ is performed on an ensemble $\mathcal{E}$ in the pure state $\rho = |\psi\rangle\langle\psi|$. The probability for this event is

$$P(\Delta) = \mathrm{tr}\{E(\Delta)\rho\} = \int_a^b dx\, |\psi(x)|^2, \qquad (2.139)$$

and the corresponding sub-ensemble is to be described by the pure state $\rho' = |\phi\rangle\langle\phi|$, where

$$\phi(x) = \begin{cases} P(\Delta)^{-1/2}\psi(x), & a < x \leq b \\ 0, & x \leq a, \quad x > b \end{cases} \qquad (2.140)$$

represents the new wave function after the measurement.

2.4.2 *Operations and effects*

It is important to emphasize that the last example of the preceding subsection describes the *ideal* quantum measurement of the projection $E(\Delta)$ which is derived from the continuous spectral family E_q of the position operator $\hat{Q}$. One must expect, however, that in practice it is not the exact spectral family E_q that is measured, but rather some kind of approximation which involves the finite resolution of the detector. Moreover, as we shall see, there are many other important variants of quantum measurement schemes that are encountered in practice. It turns out that in this context the generalized measurement theory based on the notions of operations and effects is extremely useful.

We consider some measurement scheme which yields a set $\mathcal{M}$ of possible outcomes $m \in \mathcal{M}$. One may consider $\mathcal{M}$ as a classical sample space, and the possible outcomes m as its elementary events. For simplicity we generally assume the set $\mathcal{M}$ to be discrete. The measurement is performed on some quantum statistical ensemble which is described by a density matrix ρ.

The generalized theory of quantum measurements (Kraus, 1983; Davies, 1976; Braginsky and Khalili, 1992) is based on the following concepts which are illustrated in Fig. 2.2 and which may be viewed as natural generalizations of the von Neumann–Lüders projection postulate (compare with eqns (2.130) and (2.131)):

1. The measurement outcome m represents a classical random number with probability distribution

$$P(m) = \text{tr}\left\{F_m \rho\right\}, \tag{2.141}$$

where F_m is a positive operator, called the *effect*, which satisfies the normalization condition

$$\sum_{m \in \mathcal{M}} F_m = I, \tag{2.142}$$

such that $P(m)$ is also normalized, that is,

$$\sum_{m \in \mathcal{M}} P(m) = 1. \tag{2.143}$$

2. For the case that the measurement is a selective one, the sub-ensemble of those systems for which the outcome m has been found is to be described by the density matrix

$$\rho'_m = P(m)^{-1} \Phi_m(\rho), \tag{2.144}$$

where $\Phi_m = \Phi_m(\rho)$ is a positive super-operator, called an *operation*, which maps positive operators to positive operators. The operation Φ_m is further assumed to obey the consistency condition

$$\text{tr}\, \Phi_m(\rho) = \text{tr}\left\{F_m \rho\right\}, \tag{2.145}$$

which yields, together with eqn (2.141), the normalization of the density matrix ρ'_m, namely

$$\text{tr}\rho'_m = P(m)^{-1}\text{tr}\, \Phi_m(\rho) = 1. \tag{2.146}$$

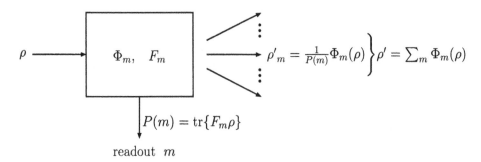

FIG. 2.2. Illustration of the operation Φ_m and the effect F_m for a generalized measurement scheme.

3. For the corresponding non-selective measurement we get the density matrix

$$\rho' = \sum_{m \in \mathcal{M}} P(m)\rho'_m = \sum_{m \in \mathcal{M}} \Phi_m(\rho), \qquad (2.147)$$

which is normalized according to eqns (2.145) and (2.142),

$$\mathrm{tr}\rho' = \sum_{m \in \mathcal{M}} \mathrm{tr}\,\Phi_m(\rho) = \sum_{m \in \mathcal{M}} \mathrm{tr}\,\{F_m\rho\} = \mathrm{tr}\rho = 1. \qquad (2.148)$$

The von Neumann–Lüders scheme for an ideal quantum measurement is obviously a special case of the above general setting. Namely, taking the orthogonal decomposition of unity ΔE_α and considering $\mathcal{M}$ to be the set of intervals Δr_α of that decomposition we find for the measurement outcome $m = \Delta r_\alpha$ the effect $F_m = \Delta E_\alpha$ and the operation $\Phi_m(\rho) = \Delta E_\alpha \rho \Delta E_\alpha$. Furthermore, the completeness relation (2.13) ensures that eqn (2.142) holds, whereas (2.145) follows from the fact that the ΔE_α are projection operators.

As a natural generalization of this example one may consider an operation given by

$$\Phi_m(\rho) = \Omega_m \rho \Omega_m^\dagger, \qquad (2.149)$$

and the corresponding effect

$$F_m = \Omega_m^\dagger \Omega_m, \qquad (2.150)$$

where the Ω_m are linear operators in the underlying Hilbert space satisfying the normalization condition

$$\sum_{m \in \mathcal{M}} F_m = \sum_{m \in \mathcal{M}} \Omega_m^\dagger \Omega_m = I. \qquad (2.151)$$

As we have seen, an ideal quantum measurement gives rise to a projection-valued measure which associates with each measurement outcome $m = \Delta r_\alpha$ a projection operator $F_m = \Delta E_\alpha$. The above concept of generalized measurement theory leads, in a natural way, to the more general idea of a *positive operator-valued measure* which associates with each measurement outcome m a positive operator $F_m = \Omega_m^\dagger \Omega_m$. A specific example of the above general concepts will be discussed in Subsections 2.4.5 and 2.4.6.

2.4.3 *Representation theorem for quantum operations*

For practical applications it is often very helpful to work with an explicit representation for the operations Φ_m. Such a representation is provided by an important theorem which is due to Kraus (1983). We first formulate the three conditions underlying this theorem and then briefly sketch the proof.

We consider a collection of positive maps $\Phi_m(\rho)$ labelled by the possible outcomes m of a generalized measurement. Each of these maps is required to fulfil the following three conditions.

1. Since we intend to interpret $P(m) = \operatorname{tr} \Phi_m(\rho)$ as the probability of the outcome m we first demand that

$$0 \le \operatorname{tr} \Phi_m(\rho) \le 1. \tag{2.152}$$

2. The map Φ_m is further required to be convex linear, that is for any collection of density matrices ρ_i and non-negative numbers $p_i \ge 0$ with $\sum_i p_i = 1$ we require

$$\Phi_m \left(\sum_i p_i \rho_i \right) = \sum_i p_i \Phi_m(\rho_i). \tag{2.153}$$

This condition may be motivated with the help of Bayes's theorem (1.11). To this end, we suppose that the state $\rho = \sum_i p_i \rho_i$ describes an ensemble which is obtained by preparing the states ρ_i with probabilities p_i. We then expect that the following relation holds,

$$\frac{\Phi_m(\rho)}{\operatorname{tr} \Phi_m(\rho)} = \sum_i p(i|m) \frac{\Phi_m(\rho_i)}{\operatorname{tr} \Phi_m(\rho_i)}. \tag{2.154}$$

The left-hand side represents the state of the total mixture under the condition that the outcome m occurred. The quantity $\Phi_m(\rho_i)/\operatorname{tr} \Phi_m(\rho_i)$ gives the state after the outcome m occurred under the condition that the state ρ_i has been prepared. This state is multiplied with the probability $p(i|m)$ which is the probability that the state ρ_i has been prepared under the condition that the outcome m occurred, that is under the condition that the operation Φ_m acts on the state. Applying Bayes's theorem (1.11) we obtain

$$p(i|m) = p(m|i) \frac{p_i}{\operatorname{tr} \Phi_m(\rho)} = \operatorname{tr} \Phi_m(\rho_i) \frac{p_i}{\operatorname{tr} \Phi_m(\rho)}, \tag{2.155}$$

where $p(m|i) = \operatorname{tr} \Phi_m(\rho_i)$ is the probability of the occurrence of the outcome m under the condition that ρ_i has been prepared. Note also that p_i

is the unconditioned probability of the preparation of ρ_i, while $\operatorname{tr} \Phi_m(\rho)$ is the unconditioned probability of the occurrence of the outcome m. Substituting (2.155) into (2.154) leads to the convex linearity (2.153) of the operation.

3. Finally, the map Φ_m is required to be *completely* positive. This condition is much stronger than positivity (for an example, see Gorini, Kossakowski and Sudarshan, 1976) and can be formulated as follows. Positivity of Φ_m means that $\Phi_m(A)$ is positive for any positive operator A acting on the Hilbert space $\mathcal{H}$. Consider now a second Hilbert space $\tilde{\mathcal{H}}$ of arbitrary dimension D. We can then define a combined operation $\Phi_m \otimes I$ which acts on the tensor product $\mathcal{H} \otimes \tilde{\mathcal{H}}$ as follows,

$$(\Phi_m \otimes I) \left(\sum_\alpha A_\alpha \otimes B_\alpha \right) = \sum_\alpha \Phi_m(A_\alpha) \otimes B_\alpha. \qquad (2.156)$$

Complete positivity means that not only is Φ_m positive, but also the combined operation $\Phi_m \otimes I$ for all dimensions D, that is $\Phi_m \otimes I$ maps positive operators of the composite system to positive operators. Physically, this is a reasonable condition since the combined operation $\Phi_m \otimes I$ may be viewed as an operation which operates locally on the first of two widely separated systems without influencing the second system.

The representation theorem for quantum operations now states that an operation Φ_m satisfies the above three conditions if and only if there exist a countable set of operators Ω_{mk} such that the operation can be written as

$$\Phi_m(\rho) = \sum_k \Omega_{mk} \rho \Omega_{mk}^\dagger, \qquad (2.157)$$

while the effects satisfy

$$F_m = \sum_k \Omega_{mk}^\dagger \Omega_{mk} \leq I. \qquad (2.158)$$

These are the most general forms for operations and effects.

We briefly sketch the proof of the representation theorem for finite-dimensional spaces. Suppose first that we have an operation Φ_m which is given by operators Ω_{mk} as in eqns (2.157) and (2.158). Then Φ_m obviously satisfies the three conditions given above. For example, complete positivity of Φ_m follows immediately since

$$\langle \varphi | (\Omega_{mk} \otimes I) A (\Omega_{mk}^\dagger \otimes I) | \varphi \rangle = \langle \varphi_{mk} | A | \varphi_{mk} \rangle \geq 0 \qquad (2.159)$$

for positive operators A acting on the product $\mathcal{H} \otimes \tilde{\mathcal{H}}$, where we have defined $|\varphi_{mk}\rangle \equiv (\Omega_{mk}^\dagger \otimes I)|\varphi\rangle$.

Conversely, suppose we have an operation Φ_m satisfying the above three conditions. Let us give an explicit construction for the operators Ω_{mk}. To this end, we take an auxiliary Hilbert space $\tilde{\mathcal{H}}$ of the same dimension as the original space $\mathcal{H}$, and choose fixed orthonormal bases $\{|\chi_i\rangle\}$ and $\{|\tilde{\chi}_i\rangle\}$ in $\mathcal{H}$ and $\tilde{\mathcal{H}}$, respectively. Next we introduce the maximally entangled state

$$|\varphi\rangle \equiv \sum_i |\chi_i\rangle \otimes |\tilde{\chi}_i\rangle, \tag{2.160}$$

and define the operator

$$\sigma_m \equiv (\Phi_m \otimes I)\left(|\varphi\rangle\langle\varphi|\right). \tag{2.161}$$

We note that σ_m is an operator in the tensor product space pertaining to the combined system. By virtue of the complete positivity of Φ_m this operator is positive since $|\varphi\rangle\langle\varphi|$ is positive. Therefore, σ_m has a spectral decomposition of the form

$$\sigma_m = \sum_k |\varphi_{mk}\rangle\langle\varphi_{mk}|. \tag{2.162}$$

The states $|\varphi_{mk}\rangle$ are states of the composite system and are, in general, not normalized. Finally, for any state

$$|\psi\rangle = \sum_i \alpha_i |\chi_i\rangle \tag{2.163}$$

in $\mathcal{H}$ we introduce a corresponding state in $\tilde{\mathcal{H}}$ by means of

$$|\tilde{\psi}\rangle = \sum_i \alpha_i^* |\tilde{\chi}_i\rangle. \tag{2.164}$$

For each index k the equation

$$\Omega_{mk}|\psi\rangle \equiv \langle\tilde{\psi}|\varphi_{mk}\rangle = \sum_i \alpha_i \langle\tilde{\chi}_i|\varphi_{mk}\rangle \tag{2.165}$$

then defines a linear operator on $\mathcal{H}$. We now prove that the collection of operators Ω_{mk} represents the operation Φ_m in accordance with eqn (2.157). To this end, we first consider the pure case. With the help of our definitions we obtain

$$\begin{aligned}
\sum_k \Omega_{mk}|\psi\rangle\langle\psi|\Omega_{mk}^\dagger &= \sum_k \langle\tilde{\psi}|\varphi_{mk}\rangle\langle\varphi_{mk}|\tilde{\psi}\rangle = \langle\tilde{\psi}|\sigma_m|\tilde{\psi}\rangle \\
&= \langle\tilde{\psi}|(\Phi_m \otimes I)(|\varphi\rangle\langle\varphi|)|\tilde{\psi}\rangle \\
&= \sum_{ij} \langle\tilde{\psi}|\Phi_m(|\chi_i\rangle\langle\chi_j|) \otimes (|\tilde{\chi}_i\rangle\langle\tilde{\chi}_j|)|\tilde{\psi}\rangle \\
&= \sum_{ij} \Phi_m(|\chi_i\rangle\langle\chi_j|)\alpha_j^*\alpha_i.
\end{aligned} \tag{2.166}$$

Hence, we have

$$\Phi_m(|\psi\rangle\langle\psi|) = \sum_k \Omega_{mk}|\psi\rangle\langle\psi|\Omega_{mk}^\dagger. \qquad (2.167)$$

This proves the representation (2.157) for pure states. The general case easily follows with the help of the convex linearity of Φ_m. Finally, the inequality in (2.158) is a direct consequence of the condition (2.152).

The representation theorem shows that a completely positive quantum operation Φ_m can be represented in terms of a set of operators Ω_{mk}. This gives rise to the interesting question concerning the freedom in the choice of the operators Ω_{mk}. Consider two sets $\{\Omega_{mk}\}$ and $\{\tilde{\Omega}_{mk}\}$ of operators which represent operations Φ_m and $\tilde{\Phi}_m$, respectively, in accordance with the representation theorem,

$$\Phi_m(\rho) = \sum_k \Omega_{mk}\rho\Omega_{mk}^\dagger, \qquad (2.168)$$

$$\tilde{\Phi}_m(\rho) = \sum_k \tilde{\Omega}_{mk}\rho\tilde{\Omega}_{mk}^\dagger. \qquad (2.169)$$

Adding zero operators to the smaller set we can always ensure that both sets contain the same number of elements. Then both operations are equal, that is $\Phi_m = \tilde{\Phi}_m$, if and only if there exists a unitary matrix $u = (u_{kl})$ such that

$$\Omega_{mk} = \sum_l u_{kl}\tilde{\Omega}_{ml}. \qquad (2.170)$$

Thus, two quantum operations are equal to each other if and only if their corresponding Ω-operators are related through a linear combination involving a unitary coefficient matrix.

To prove the above statement let us first suppose that (2.170) is satisfied. By use of the representations (2.168), (2.169) it follows immediately that $\Phi_m = \tilde{\Phi}_m$. Conversely, assuming $\Phi_m = \tilde{\Phi}_m$ we have

$$\sum_k \Omega_{mk}\rho\Omega_{mk}^\dagger = \sum_k \tilde{\Omega}_{mk}\rho\tilde{\Omega}_{mk}^\dagger \qquad (2.171)$$

for all densities ρ. Using the same construction and the same notation as in the proof of the representation theorem we define

$$|\varphi_{mk}\rangle = (\Omega_{mk} \otimes I)|\varphi\rangle, \quad |\tilde{\varphi}_{mk}\rangle = (\tilde{\Omega}_{mk} \otimes I)|\varphi\rangle, \qquad (2.172)$$

which yields

$$\sigma_m = \sum_k |\varphi_{mk}\rangle\langle\varphi_{mk}| = \sum_k |\tilde{\varphi}_{mk}\rangle\langle\tilde{\varphi}_{mk}|. \qquad (2.173)$$

This shows that the positive operator σ_m is generated by the two sets of states $\{|\varphi_{mk}\rangle\}$ and $\{|\tilde{\varphi}_{mk}\rangle\}$. These states are therefore related through

$$|\varphi_{mk}\rangle = \sum_l u_{kl}|\tilde{\varphi}_{ml}\rangle \qquad (2.174)$$

with some unitary matrix u (see eqn (2.56)). For arbitrary $|\psi\rangle$ we have

$$\Omega_{mk}|\psi\rangle = \langle\tilde{\psi}|\varphi_{mk}\rangle = \sum_l u_{kl}\langle\tilde{\psi}|\tilde{\varphi}_{ml}\rangle = \sum_l u_{kl}\tilde{\Omega}_{ml}|\psi\rangle, \qquad (2.175)$$

which gives eqn (2.170), as was to be shown.

An important implication of the above statement is that any operation Φ_m acting on the state space of a system with a D-dimensional Hilbert space can be represented with the help of at most D^2 operators Ω_{mk}.

2.4.4 Quantum measurement and entropy

In general, a quantum measurement induces a change of the von Neumann entropy. We define the entropy change in a non-selective quantum measurement described by operations Φ_m as

$$\Delta S = S(\rho') - S(\rho) = S\left(\sum_m \Phi_m(\rho)\right) - S(\rho). \qquad (2.176)$$

Consider first an ideal quantum measurement in which case $\Phi_m(\rho) = \Pi_m \rho \Pi_m$ with orthogonal projections Π_m, satisfying $\sum_m \Pi_m = I$. From the Klein inequality (2.115) we get

$$0 \le S(\rho||\rho') = -S(\rho) - \mathrm{tr}\{\rho\ln\rho'\}. \qquad (2.177)$$

With the help of the completeness relation and of the fact that $\rho' = \sum_m \Pi_m \rho \Pi_m$ commutes with the projections Π_m, the second term on the right-hand side of the equality sign is found to be

$$
\begin{aligned}
-\mathrm{tr}\{\rho\ln\rho'\} &= -\mathrm{tr}\left\{\sum_m \Pi_m \rho \left(\ln\rho'\right)\Pi_m\right\} \\
&= -\mathrm{tr}\left\{\sum_m \Pi_m \rho \Pi_m \ln\rho'\right\} \\
&= -\mathrm{tr}\{\rho'\ln\rho'\} \\
&= S(\rho'),
\end{aligned}
\qquad (2.178)
$$

and, hence,

$$\Delta S \ge 0. \qquad (2.179)$$

We conclude that non-selective, ideal quantum measurements never decrease the von Neumann entropy. We note also that $\Delta S = 0$ if and only if $\rho' = \rho$, that is if and only if the density matrix is left unchanged by the measurement.

The above conclusion is, in general, not true for a generalized measurement as can be seen from the following example. We consider a two-level system the Hilbert space of which is spanned by two states $|e\rangle$ and $|g\rangle$. We define the two operations $\Phi_m(\rho) = \Omega_m \rho \Omega_m^\dagger$, $m = 1, 2$, where $\Omega_1 = |g\rangle\langle g|$ and $\Omega_2 = |g\rangle\langle e|$. Thus we have the effects $\Omega_1^\dagger \Omega_1 = |g\rangle\langle g|$ and $\Omega_2^\dagger \Omega_2 = |e\rangle\langle e|$, and their sum adds to the identity I. However, for a normalized initial density matrix ρ we find the following state after the non-selective measurement,

$$\rho' = \Omega_1 \rho \Omega_1^\dagger + \Omega_2 \rho \Omega_2^\dagger = |g\rangle\langle g|. \tag{2.180}$$

The final state is therefore always a pure state and $S(\rho') = 0$. Thus, the entropy decreases to zero if we start from an initial density with positive entropy. This example shows that a generalized measurement can decrease the von Neumann entropy.

Another interesting question concerns the change of entropy involved in the transition from the selective to the non-selective level of a measurement. We define the quantity

$$\delta S = S\left(\sum_m P(m)\rho_m'\right) - \sum_m P(m)S(\rho_m'), \tag{2.181}$$

which may be referred to as the mixing entropy. It is the entropy of the final density matrix $\rho' = \sum_m P(m)\rho_m'$ on the non-selective level minus the average of the entropies of the sub-ensembles described by the states ρ_m'. This mixing entropy satisfies the inequalities

$$0 \leq \delta S \leq H(\{P(m)\}). \tag{2.182}$$

Thus, δS is always non-negative. This statement follows immediately from the concavity of the von Neumann entropy (see eqn (2.110)). The second inequality shows that δS is never larger than the Shannon entropy associated with the random variable m following the distribution $m \mapsto P(m)$. The proof of this statement may be found, for example, in Nielsen and Chuang (2000). We can interpret the Shannon entropy $H(\{P(m)\})$ as the information lost in the transition from the selective to the non-selective level, since during the mixing process we forget about the measurement results. In other words, the Shannon entropy of the distribution $P(m)$ is a measure of the information gained in the measurement by obtaining a specific realization of the random variable m. Thus we see that the mixing entropy is never larger than the information gained. It can be shown further that the mixing entropy is equal to the Shannon entropy if and only if the densities ρ_m' have orthogonal images. This occurs for an ideal measurement, in which case the mixing entropy becomes equal to the Shannon entropy.

2.4.5   *Approximate measurements*

The generalized framework for the description of quantum measurements introduced above is capable of describing, for example, approximate measurements, that is, measurement devices which measure the spectrum of some observable $\hat{R}$ only with finite resolution (Braginsky and Khalili, 1992).

We consider some physical situation in which an observable $\hat{R}$ with discrete, non-degenerate spectrum,

$$\hat{R} = \sum_m r_m |\psi_m\rangle\langle\psi_m|, \tag{2.183}$$

is measured. If the eigenvalues r_m are not too closely spaced one can measure such an observable exactly, provided the resolution of the experimental device is sufficiently large. A measurement which is only approximate arises, however, if the resolution of the apparatus is not high enough to distinguish, for example, neighbouring eigenvalues. For such a case we introduce a conditional probability distribution $W(m|m')$ which represents the probability for the measurement to yield the outcome m provided the measured system is known to be in the state $\psi_{m'}$ with eigenvalue m'. This conditional probability distribution serves to describe the finite resolution of the apparatus, possible disturbances by the environment, as well as possible uncertainties in the precise state of the apparatus before the measurement.

Assuming that the apparatus always yields a definite outcome we have

$$\sum_m W(m|m') = 1. \tag{2.184}$$

The probability distribution for the measurement outcomes r_m then takes the form

$$P(m) = \sum_{m'} W(m|m')\langle\psi_{m'}|\rho|\psi_{m'}\rangle, \tag{2.185}$$

which is obviously normalized to 1,

$$\sum_m P(m) = 1. \tag{2.186}$$

Defining the effect operator

$$F_m \equiv \sum_{m'} |\psi_{m'}\rangle W(m|m')\langle\psi_{m'}| \tag{2.187}$$

we can rewrite the probability distribution of the outcomes as

$$P(m) = \mathrm{tr}\{F_m\rho\}. \tag{2.188}$$

As is easily verified, (2.187) is a positive operator and satisfies the normalization condition (2.142).

To find an appropriate operation we invoke the ansatz (2.149) and conclude from the consistency condition (2.145) the relation (2.150). The most general solution of this relation takes the form

$$\Omega_m = U_m F_m^{1/2}, \tag{2.189}$$

where U_m is an (undetermined) unitary operator and

$$F_m^{1/2} = \sum_{m'} |\psi_{m'}\rangle \sqrt{W(m|m')} \langle \psi_{m'}| \qquad (2.190)$$

is the square root of F_m. Thus we have the operation

$$\Phi_m(\rho) = U_m F_m^{1/2} \rho F_m^{1/2} U_m^\dagger, \qquad (2.191)$$

showing that the transition from the initial density matrix ρ to that describing the sub-ensemble conditioned on the outcome r_m may be viewed formally as a two-step process:

$$\rho \longrightarrow \rho_m' = P(m)^{-1} F_m^{1/2} \rho F_m^{1/2} \longrightarrow \rho_m'' = U_m \rho_m' U_m^\dagger. \qquad (2.192)$$

The second step in (2.192) represents a unitary transformation of the density matrix. During this step the entropy thus remains constant and no information on the quantum system is gained. The precise form of the unitary operator U_m is determined by the details of the measuring device. The above analysis shows that one cannot determine the exact form of the operation from the probability distribution of the measurement outcomes alone. As we shall see, the unitary operator U_m describes the disturbance of the quantum object by the measurement apparatus (see Section 2.4.7).

The first step in (2.192) describes the gain of information during the measurement: It is this step which splits, in general, the initial ensemble into the various sub-ensembles. The effect F_m is uniquely determined by the spectral family of $\hat{R}$ and the conditional probability distribution $W(m|m')$ that describes the finite resolution of the measuring device. Since $\hat{R}$ and F_m have the same spectral family they commute, $[F_m, \hat{R}] = 0$. As we shall see this is a sufficient condition for the first step to be a quantum non-demolition measurement.

In the case of an infinite resolution we get $W(m|m') = \delta_{mm'}$ and, therefore, the effect reduces to that of an ideal measurement of $\hat{R}$, namely $F_m = |\psi_m\rangle\langle\psi_m|$. Accordingly, the operation takes a form which is similar to the one given in the von Neumann–Lüders postulate, namely

$$\rho_m'' = U_m |\psi_m\rangle \langle \psi_m| U_m^\dagger, \qquad (2.193)$$

where the projection is, in general, accompanied by an unitary transformation.

As another example we consider the approximate measurement of the position operator $\hat{Q}$ (Davies, 1976). The probability density $p(q)$ for the outcome q is taken to be

$$p(q) = \int dq' \, W(q - q') \langle q' | \rho | q' \rangle. \qquad (2.194)$$

Here, the distribution $W(q-q')$ yields the conditional probability density for the apparatus to respond with the value q provided the system is in the (generalized)

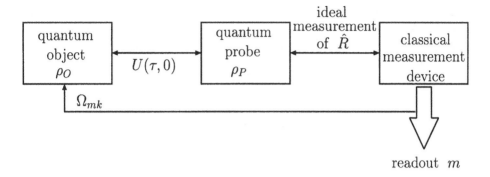

FIG. 2.3. Schematic picture of an indirect measurement.

eigenstate $|q'\rangle$ of $\hat{Q}$. Note that an exact position measurement would be obtained if $W(q)$ approaches the Dirac delta function. The corresponding effect reads

$$F_q = \int dq' \, |q'\rangle W(q - q')\langle q'|. \qquad (2.195)$$

It is interesting to determine the mean value $E(Q, W)$ and the variance $\text{Var}(Q, W)$ of the measurement outcomes in this approximate measurement. We find with the help of the distribution $p(q)$,

$$E(Q, W) = E(Q) + \int dq \, qW(q), \qquad (2.196)$$

$$\text{Var}(Q, W) = \text{Var}(Q) + \int dq \, q^2 W(q) - \left[\int dq \, qW(q)\right]^2, \qquad (2.197)$$

where, as usual, $E(Q)$ and $\text{Var}(Q)$ denote the mean and the variance as they would be obtained in an ideal measurement of $\hat{Q}$. The first equation shows that a non-vanishing mean value of $W(q)$ leads to a systematic bias of the measurement results. The second equation tells us that a finite variance of the distribution $W(q)$ yields an additional dispersion which is superimposed on the intrinsic quantum fluctuations given by $\text{Var}(Q)$.

2.4.6 Indirect quantum measurements

An important measurement scheme which can also be treated within the framework of the generalized theory of quantum measurements is provided by the concept of an indirect measurement (Braginsky and Khalili, 1992).

An indirect measurement can be viewed as consisting of three elements (see Fig. 2.3). The first element is the quantum system to be measured, that is the system from which information is to be extracted. This system is therefore called the *quantum object* and will be considered to live in the Hilbert space $\mathcal{H}_O$. The second element is the so-called *quantum probe* with Hilbert space $\mathcal{H}_P$. The quantum probe is again a certain quantum system which interacts with the quantum object. It is assumed that the probe system has been prepared, prior to this interaction, in a suitable state given by some density matrix ρ_P. As a result of the subsequent interaction, correlations between object and probe are built up. The third element of the scheme is a classical apparatus by which a measurement on the quantum probe is performed when the interaction between object and probe is over. The aim of this scheme is to obtain information on the state of the object by means of the measurement on the probe. This information is deduced from the correlations between object and probe built up during their interaction. Thus we can formulate three basic requirements for an ideal measurement:

1. Before the interaction between object and probe at time $t = 0$, the probe has been prepared in a well-defined quantum state ρ_P. At the same time the quantum object is in a state ρ_O.

2. The interaction between quantum object and probe starts at time $t = 0$ and is finished at some time $t = \tau > 0$ before the measurement by means of the apparatus sets in.

3. The third element of the scheme represents a classical apparatus and the measurement on the quantum probe can be described as an ideal measurement according to the von Neumann–Lüders projection postulate.

In the following we shall derive explicit expressions for the effect and the operation pertaining to such an ideal measurement scheme.

At time $t = 0$ the total quantum system consisting of object and probe is described by the density matrix $\rho_O \otimes \rho_P$ in the total Hilbert space given by the tensor product $\mathcal{H} = \mathcal{H}_O \otimes \mathcal{H}_P$. The Hamiltonian of the total system is taken to be

$$H(t) = H_O + H_P + H_I(t), \tag{2.198}$$

where H_O and H_P describe the free evolution of object and probe, respectively, and $H_I(t)$ their interaction. This time-dependent interaction is assumed to vanish outside the time interval $[0, \tau]$. The corresponding unitary operator describing the time evolution over this interval according to the Schrödinger equation will be denoted by U,

$$U \equiv U(\tau, 0) = \mathrm{T} \exp\left[-i \int_0^\tau dt H(t) \right], \tag{2.199}$$

where T denotes the chronological time ordering and we have set $\hbar = 1$.

The initial density matrix $\rho(0) = \rho_O \otimes \rho_P$ of the total system thus evolves after time τ into

$$\rho(\tau) = U \left(\rho_O \otimes \rho_P \right) U^\dagger. \tag{2.200}$$

We now assume that at time τ the classical apparatus measures the probe observable

$$\hat{R} = \sum_m r_m |\varphi_m\rangle\langle\varphi_m| \tag{2.201}$$

with a non-degenerate, discrete spectrum. Note that $\hat{R}$ is a self-adjoint operator which acts in the Hilbert space $\mathcal{H}_P$ of the quantum probe. The probability $P(m)$ for the outcome r_m of this measurement is found to be

$$P(m) = \operatorname{tr}\left\{|\varphi_m\rangle\langle\varphi_m|\rho(\tau)\right\} = \operatorname{tr}\left\{U^\dagger|\varphi_m\rangle\langle\varphi_m|U\left(\rho_O \otimes \rho_P\right)\right\}, \qquad (2.202)$$

where the trace is taken over the total Hilbert space $\mathcal{H}$. The partial traces over $\mathcal{H}_O$ and $\mathcal{H}_P$ will be denoted by tr_O and tr_P, respectively. The expression for $P(m)$ can then be rewritten as

$$P(m) = \operatorname{tr}_O\left\{F_m\rho_O\right\}, \qquad (2.203)$$

where the effect takes the form

$$F_m\rho_O = \operatorname{tr}_P\left\{U^\dagger|\varphi_m\rangle\langle\varphi_m|U\left(\rho_O \otimes \rho_P\right)\right\}. \qquad (2.204)$$

As is easily verified F_m is a positive operator which acts in the Hilbert space $\mathcal{H}_O$ of the quantum object and which satisfies the normalization condition, namely we have

$$\sum_m F_m\rho_O = \sum_m \operatorname{tr}_P\left\{U^\dagger|\varphi_m\rangle\langle\varphi_m|U\left(\rho_O \otimes \rho_P\right)\right\} = \operatorname{tr}_P\left\{\rho_O \otimes \rho_P\right\} = \rho_O,$$

$$(2.205)$$

showing that (2.142) is fulfilled.

The operation for the indirect measurement is obtained by applying the von Neumann–Lüders projection postulate to the measurement of the probe which immediately yields the following state of the quantum object:

$$\rho_m' = P(m)^{-1}\langle\varphi_m|U\left(\rho_O \otimes \rho_P\right)U^\dagger|\varphi_m\rangle. \qquad (2.206)$$

Introducing the spectral decomposition of the density matrix of the probe,

$$\rho_P = \sum_k p_k|\phi_k\rangle\langle\phi_k|, \qquad (2.207)$$

we can write the operation on the quantum object as

$$\Phi_m(\rho_O) = \sum_k \Omega_{mk}\rho_O\Omega_{mk}^\dagger, \qquad (2.208)$$

where

$$\Omega_{mk} = \sqrt{p_k}\langle\varphi_m|U|\phi_k\rangle. \qquad (2.209)$$

On using these operators the effect is given by

$$F_m = \sum_k \Omega_{mk}^\dagger\Omega_{mk}. \qquad (2.210)$$

Thus we observe that operation and effect take exactly the form given in the representation theorem, see eqns (2.157) and (2.158). If initially the probe is in a pure state $\rho_P = |\phi\rangle\langle\phi|$, we get

$$\Omega_m = \langle\varphi_m|U|\phi\rangle \tag{2.211}$$

which is proportional to the amplitude for a transition of the probe from the state ϕ to the eigenstate φ_m of the measured observable $\hat{R}$. Note that Ω_m is an operator in the Hilbert space of the quantum object which describes the change of the state of the object induced by the outcome r_m of the measurement on the probe. The additional sum over the index k in the general expression (2.208) results from the additional ignorance in the case that the initial state of the probe is a statistical mixture.

The above results are readily generalized to the case of an observable $\hat{R}$ with a continuous spectrum. Let us work out a specific example. We take $\hat{R}$ to be the momentum operator $\hat{P}$ in a specific direction, that is, the classical measurement apparatus measures a certain component of the momentum of some probe particle. Our aim is to measure by means of an indirect measurement the non-degenerate, discrete observable

$$\hat{A} = \sum_\nu a_\nu |\chi_\nu\rangle\langle\chi_\nu| \tag{2.212}$$

of the quantum object. To this end, we couple the position operator $\hat{Q}$ canonically conjugate to $\hat{P}$,

$$\left[\hat{Q}, \hat{P}\right] = i, \tag{2.213}$$

to the observable $\hat{A}$, that is, we consider the interaction Hamiltonian

$$H_I(t) = g(t)\hat{A}\hat{Q}, \tag{2.214}$$

where the time-dependent coupling $g(t)$ is assumed to vanish outside the time interval $[0, \tau]$. For simplicity we further assume that the interaction time τ is very small and that the free evolution generated by $H_O + H_P$ may be neglected over this time. The unitary time-evolution operator then takes the simple form

$$U = \exp\left[-iG\hat{A}\hat{Q}\right] \tag{2.215}$$

where

$$G \equiv \int_0^\tau dt\, g(t) \tag{2.216}$$

denotes the integrated coupling strength. The quantum probe is assumed to be in the initial state $\rho_P = |\phi\rangle\langle\phi|$ such that we obtain for the operator Ω_p pertaining to the measurement outcome p (see eqn (2.211))

$$\Omega_p = \langle p|U|\phi\rangle$$

$$= \langle p|\exp\left[-iG\hat{A}\hat{Q}\right]|\phi\rangle$$

$$= \sum_\nu |\chi_\nu\rangle\langle p|\exp\left[-iGa_\nu\hat{Q}\right]|\phi\rangle\langle\chi_\nu|. \tag{2.217}$$

The matrix element in the last expression is equal to $\phi(p + Ga_\nu)$, where $\phi(p) \equiv \langle p|\phi\rangle$ is the initial wave function of the probe in the momentum representation. Hence we have

$$\Omega_p = \sum_\nu |\chi_\nu\rangle\phi(p + Ga_\nu)\langle\chi_\nu| = \phi(p + G\hat{A}). \tag{2.218}$$

It follows that the probability density $f(p)$ for the momentum measurement (to avoid confusion with the momentum we denote here probability densities by f) is given by the expression

$$f(p) = \text{tr}\left\{\Omega_p^\dagger\Omega_p\rho_O\right\} = \sum_\nu |\phi(p + Ga_\nu)|^2\langle\chi_\nu|\rho_O|\chi_\nu\rangle. \tag{2.219}$$

The last two equations can now be interpreted as describing an approximate measurement of the observable $\hat{A}$. To see this we must answer the following question: What information on the observable $\hat{A}$ of the quantum object can be inferred from the measurement of $\hat{P}$? Suppose for a moment that $|\phi(p)|^2$ is sharply peaked around its mean value

$$\langle p\rangle \equiv \langle\phi|\hat{P}|\phi\rangle \equiv \int dp\, p|\phi(p)|^2. \tag{2.220}$$

Looking at eqn (2.218) we then see that the measurement of p effectively projects the state of the quantum object onto a group of eigenstates χ_ν of $\hat{A}$ whose eigenvalues a_ν satisfy $p + Ga_\nu \approx \langle p\rangle$. The precise number of states in the group is determined by the widths of the probe's wave function in the momentum representation. We thus define the a-value which is inferred from the measurement outcome p by means of the relation $p + Ga = \langle p\rangle$, that is,

$$a = \frac{1}{G}\left(\langle p\rangle - p\right). \tag{2.221}$$

We then get for the probability density of a

$$f(a) = \sum_\nu W(a - a_\nu)\langle\chi_\nu|\rho_O|\chi_\nu\rangle, \tag{2.222}$$

where

$$W(a - a_\nu) = |G| \cdot |\phi(\langle p\rangle - G[a - a_\nu])|^2. \tag{2.223}$$

We thus observe that the direct measurement of the momentum $\hat{P}$ of the probe particle yields an approximate measurement of the observable $\hat{A}$ of the quantum object. Expression (2.223) represents the conditional probability density of this approximate measurement of $\hat{A}$: It yields the density for obtaining the value a provided the object is in an eigenstate with eigenvalue a_ν.

Clearly, $W(a - a_\nu)$ is normalized to 1,

$$\int da\, W(a - a_\nu) = 1, \tag{2.224}$$

and satisfies

$$\int da\, a W(a - a_\nu) = a_\nu, \tag{2.225}$$

$$\int da\, a^2 W(a - a_\nu) = a_\nu^2 + \frac{1}{G^2} \text{Var}(P), \tag{2.226}$$

where $\text{Var}(P)$ denotes the variance of $\hat{P}$ in the initial state ϕ of the probe. It follows from eqn (2.225) that the expression (2.221) for the value of a inferred from the measurement outcome p has a vanishing bias, that is, its expectation value coincides with the exact mean value of $\hat{A}$,

$$\text{E}(a) \equiv \int da\, a f(a) = \text{tr}\left\{\hat{A}\rho_O\right\} = \text{E}(A). \tag{2.227}$$

Equation (2.226) on the other hand leads to the relation

$$\text{Var}(a) \equiv \int da\, a^2 f(a) - \left[\int da\, a f(a)\right]^2$$
$$= \text{Var}(A) + \frac{1}{G^2}\text{Var}(P). \tag{2.228}$$

This equation tells us that the variance of the inferred value of a equals the sum of the exact quantum variance $\text{Var}(A)$ of $\hat{A}$ and the variance of the probe momentum in the initial state divided by G^2. It follows that the measurement is the more accurate the lower the momentum uncertainty of the probe and the stronger the coupling between quantum object and probe.

Having obtained a particular value a, the corresponding operation is given by the operator

$$\Omega_a = \sum_\nu |\chi_\nu\rangle \sqrt{|G|}\phi(\langle p\rangle - G[a - a_\nu])\langle\chi_\nu| = \sqrt{|G|}\phi\left(\langle p\rangle - G(a - \hat{A})\right). \tag{2.229}$$

The density matrix of the resulting sub-ensemble then takes the form

$$\rho_a' = \frac{\sum_{\nu\nu'} |\chi_\nu\rangle\phi(\langle p\rangle - G[a - a_\nu])\langle\chi_\nu|\rho_O|\chi_{\nu'}\rangle\phi^*(\langle p\rangle - G[a - a_{\nu'}])\langle\chi_{\nu'}|}{\sum_\nu |\phi(\langle p\rangle - G[a - a_\nu])|^2 \langle\chi_\nu|\rho_O|\chi_\nu\rangle}$$
$$= \frac{\Omega_a \rho_O \Omega_a}{\text{tr}\{\Omega_a^\dagger \Omega_a \rho_O\}}. \tag{2.230}$$

In the ideal case of an infinitely sharp initial momentum p_0 of the probe we have $|\phi(p)|^2 \longrightarrow \delta(p - p_0)$ and, therefore,

$$f(a) = \sum_\nu \delta(a - a_\nu)\langle \chi_\nu|\rho_O|\chi_\nu\rangle, \qquad (2.231)$$

which shows that a takes on only a discrete set of values given by the eigenvalues of the operator $\hat{A}$. The corresponding probabilities

$$P(a_\nu) = \langle \chi_\nu|\rho_O|\chi_\nu\rangle \qquad (2.232)$$

as well as the new density matrix

$$\rho'_{a_\nu} = |\chi_\nu\rangle\langle\chi_\nu| \qquad (2.233)$$

are given by the correct quantum mechanical formulae corresponding to an ideal quantum measurement of $\hat{A}$.

2.4.7 Quantum non-demolition measurements

Suppose an ideal, non-selective measurement of some observable $\hat{A}$ is performed on a quantum system in the state ρ. As is well known if $\hat{A}$ commutes with ρ then the state of the system is not changed by the measurement. In fact, if

$$\hat{A} = \sum_\nu a_\nu \Pi_\nu \qquad (2.234)$$

represents the spectral decomposition of $\hat{A}$, the a_ν denoting the eigenvalues and Π_ν the corresponding projections, then we find for the non-selective density matrix ρ' after the measurement

$$\rho' = \sum_\nu \Pi_\nu \rho \Pi_\nu = \sum_\nu \Pi_\nu \rho = \rho, \qquad (2.235)$$

where we have used the completeness relation as well as the fact that ρ commutes with the projections Π_ν.

Consider now some generalized measurement scheme with operations Φ_m defined in terms of operators Ω_m (see eqn (2.149)). Such a scheme is defined to be a quantum non-demolition measurement (QND measurement) (Braginsky and Khalili, 1992) for some observable $\hat{A}$ if the probability distribution for an ideal measurement of $\hat{A}$ remains unchanged during the measurement process, that is if the distribution of the eigenvalues of $\hat{A}$ is the same in the initial and the final density matrix. Here, the generalized measurement is assumed to be a non-selective one. We remark that QND measurement devices are important for the design of high-precision quantum measurements (Bocko and Onofrio, 1996).

On using expression (2.149) for the operation of a generalized measurement we can cast the above definition into the following mathematical form,

$$\text{tr}\left\{\hat{A}^k \rho\right\} = \text{tr}\left\{\hat{A}^k \rho'\right\} = \sum_m \text{tr}\left\{\hat{A}^k \Omega_m \rho \Omega_m^\dagger\right\}. \qquad (2.236)$$

This means that all moments of $\hat{A}$ coincide in the initial and the final state and, therefore, also the distribution of its eigenvalues (assuming for simplicity that the moments exist). For the above example of an ideal measurement of $\hat{A}$ the Ω_m are given by the projections Π_ν and condition (2.236) is obviously satisfied provided ρ commutes with the Ω_m. This is a condition for the state of the quantum system on which the measurement is performed.

Alternatively, we can read eqn (2.236) as a condition for the operators Ω_m which is required to hold for all states ρ of the quantum system. Employing the normalization condition (2.151) we can cast eqn (2.236) into the form

$$\sum_m \text{tr}\left\{\Omega_m^\dagger \left[\hat{A}^k, \Omega_m\right] \rho\right\} = 0. \qquad (2.237)$$

Since this condition is assumed to hold for all ρ we conclude that a sufficient condition for a QND measurement is given by

$$\left[\hat{A}, \Omega_m\right] = 0. \qquad (2.238)$$

A measurement is therefore a QND measurement for the observable $\hat{A}$ if the operators Ω_m describing the change of the quantum system induced by the measurement commute with $\hat{A}$. Condition (2.238) is always fulfilled for an ideal measurement of the observable $\hat{A}$, that is an ideal measurement of $\hat{A}$ is always a QND measurement in the sense defined above. We also see immediately from eqn (2.229) that the example for an indirect measurement scheme discussed in the preceding subsection fulfils the QND condition (2.238). This is due to the fact that Ω_a is diagonal in the $\hat{A}$-representation which, in turn, results from the fact that the interaction Hamiltonian $H_I(t)$ commutes with the measured quantity. As a result, $\hat{A}$ is a constant of motion during the measurement process.

Let us study the general setting for an indirect measurement with a probe in a pure state ϕ. Substituting (2.211) into (2.238) we find

$$\langle \varphi_m | \left[\hat{A}, U(\tau, 0)\right] |\phi\rangle = 0. \qquad (2.239)$$

Since this must be true for all m and because the φ_m represent a basis in the Hilbert space $\mathcal{H}_P$ of the probe, we conclude that

$$\left[\hat{A}, U(\tau, 0)\right] |\phi\rangle = 0. \qquad (2.240)$$

This equation can possibly be achieved by preparing the probe in an appropriate state ϕ. Since this is obviously a difficult task in practice one requires that the commutator itself vanishes, which leads to the following sufficient QND condition

$$\left[\hat{A}, U(\tau, 0)\right] = 0. \tag{2.241}$$

According to this equation the Heisenberg picture operator

$$\hat{A}_H(t) = U^\dagger(t, 0)\hat{A}U(t, 0) \tag{2.242}$$

returns to its original value after time $t = \tau$, that is, $\hat{A}_H(\tau) = \hat{A}_H(0)$. The easiest way to achieve this is to require that $\hat{A}$ commutes with the total Hamiltonian $H(t) = H_O + H_P + H_I(t)$, that is

$$\left[\hat{A}, H_O + H_I(t)\right] = 0. \tag{2.243}$$

If $\hat{A}$ is a constant of motion under the free evolution of the quantum object, a general sufficient condition for a QND measurement is found to be

$$\left[\hat{A}, H_I(t)\right] = 0. \tag{2.244}$$

This condition is also known as the back-action evasion condition, for it guarantees that the interaction of the quantum object with the probe system does not disturb the measured quantity.

We remark finally that by the same reasoning one can show that a sufficient condition for the approximate measurement discussed in Subsection 2.4.5 to be a QND measurement is given by $\left[\hat{R}, U_m\right] = 0$, which means that the observable measured approximately by the scheme must commute with the unitary operators U_m that occur in the second step of (2.192). Thus, these unitary transformations disturb, in general, the probability distribution of the measured quantity.

References

Akhiezer, N. I. and Glazman, I. M. (1981). *Theory of Linear Operators in Hilbert Space*. Pitman, Boston.

Bocko, M. F. and Onofrio, R. (1996). On the measurement of a weak classical force coupled to a harmonic oscillator: Experimental progress. *Rev. Mod. Phys.*, **68**, 775–799.

Bohm, A. (1993). *Quantum Mechanics* (third edition). Springer-Verlag, New York.

Braginsky, V. B. and Khalili, F. Ya. (1992). *Quantum Measurement*. Cambridge University Press, Cambridge.

Cohen-Tannoudji, C., Diu, B. and Laloë, F. (1977). *Quantum Mechanics*, Volume I and II. John Wiley, New York.

Davies, E. B. (1976). *Quantum Theory of Open Systems*. Academic Press, London.

Gorini, V., Kossakowski, A. and Sudarshan, E. C. G. (1976). Completely positive dynamical semigroups of N-level systems. *J. Math. Phys.*, **17**, 821–825.

Gottfried, K. (1974). *Quantum Mechanics*. W. A. Benjamin, Reading, Massachusetts.

Gudder, S. P. (1979). *Stochastic Methods in Quantum Mechanics*. Elsevier, North Holland, New York.

Holevo, A. S. (1982). *Probabilistic and Statistical Aspects of Quantum Theory*, Volume I of *North-Holland Series in Statistics and Probability*. North-Holland, Amsterdam.

Hughston, L. P., Jozsa, R. and Wootters, W. K. (1993). A complete classification of quantum ensembles having a given density matrix. *Phys. Lett.*, **A183**, 14–18.

Kraus, K. (1983). *States, Effects, and Operations*, Volume 190 of *Lecture Notes in Physics*. Springer-Verlag, Berlin.

Landau, L. D. and Lifshitz, E. M. (1958). *Quantum Mechanics*. Pergamon Press, London.

Langerholc, J. (1965). Trace formalism for quantum mechanical expectation values. *J. Math. Phys.*, **6**, 1210–1218.

Lieb, E. H. (1973). Proof of the strong subadditivity of quantum-mechanical entropy. *J. Math. Phys.*, **14**, 1938–1941.

Louisell, W. H. (1990). *Quantum Statistical Properties of Radiation*. John Wiley, New York.

Lüders, G. (1951). Über die Zustandsänderung durch den Meßprozeß. *Ann. Phys. (Leipzig)*, **8**, 322–328.

Nelson, E. (1967). *Dynamical Theories of Brownian Motion*. Princeton University Press, Princeton.

Nielsen, M. A. and Chuang, I. L. (2000). *Quantum Computation and Quantum Information*. Cambridge University Press, Cambridge.

von Neumann, J. (1955). *Mathematical Foundations of Quantum Mechanics*. Princeton University Press, Princeton.

Werl, A. (1978). General properties of entropy. *Rev. Mod. Phys.*, **50**, 221–260.

Part II

Density matrix theory

3

QUANTUM MASTER EQUATIONS

In contrast to the case of a closed systems, the quantum dynamics of an open system cannot, in general, be represented in terms of a unitary time evolution. In many cases it turns out to be useful to formulate, instead, the dynamics of an open system by means of an appropriate equation of motion for its density matrix, a quantum master equation. In this chapter we shall give a survey of the various types of master equations which can be employed to analyse the quantum dynamics of open systems. The emphasis lies on the Markovian dynamics of open systems. A systematic treatment of non-Markovian quantum processes will be presented in Part IV.

To begin with, we give a general characterization of the dynamics of closed and open quantum systems. Quantum Markov processes are then introduced which represent the simplest case of the dynamics of open systems. They can be regarded, essentially, as a direct generalization of the classical probabilistic concept of a dynamical semigroup to quantum mechanics. In analogy to the differential Chapman–Kolmogorov equation of classical probability theory, a quantum dynamical semigroup gives rise to a first-order linear differential equation for the reduced density matrix, which is known as quantum Markovian master equation in Lindblad form.

A number of derivations of Markovian density matrix equations will be given which start from various microscopic models, such as the weak-coupling interaction of radiation with matter and the Caldeira–Leggett model. The resultant density matrix equations enable the investigation of general features like irreversibility, entropy production, and relaxation to equilibrium. In standard applications of the dynamics of open systems two limiting cases are of particular relevance. These are the quantum optical and the quantum Brownian motion master equation. We shall give a number of examples and discuss basic properties of the corresponding master equations. For later purposes the Caldeira–Leggett model will also be studied from a broader perspective using the exact Heisenberg equations of motion as well as influence functional and path integral techniques.

In the treatment of open many-body systems non-linear quantum master equations are also encountered. If one takes into account particle correlations on the mean-field level the reduced density matrix equation involves a non-linear time-evolution generator. Under certain conditions the solutions of such non-linear density matrix equations may be obtained with the help of a non-linear Schrödinger-type equation. As two prominent examples for mean-field master equations we treat the laser equations and the phenomenon of super-radiance.

3.1 Closed and open quantum systems

3.1.1 *The Liouville–von Neumann equation*

According to quantum mechanics the state vector $|\psi(t)\rangle$ evolves in time according to the Schrödinger equation,

$$i\frac{d}{dt}|\psi(t)\rangle = H(t)|\psi(t)\rangle, \tag{3.1}$$

where $H(t)$ is the Hamiltonian of the system and Planck's constant $\hbar$ has been set equal to 1. The solution of the Schrödinger equation may be represented in terms of the unitary time-evolution operator $U(t, t_0)$ which transforms the state $|\psi(t_0)\rangle$ at some initial time t_0 to the state $|\psi(t)\rangle$ at time t,

$$|\psi(t)\rangle = U(t, t_0)|\psi(t_0)\rangle. \tag{3.2}$$

If we substitute expression (3.2) into the Schrödinger equation (3.1) we get an operator equation for the time-evolution operator $U(t, t_0)$,

$$i\frac{\partial}{\partial t}U(t, t_0) = H(t)U(t, t_0), \tag{3.3}$$

which is subjected to the initial condition

$$U(t_0, t_0) = I. \tag{3.4}$$

It is easy to show with the help of (3.3) and (3.4) that $U^\dagger(t, t_0)U(t, t_0) = U(t, t_0)U^\dagger(t, t_0) \equiv I$ and, hence, that $U(t, t_0)$ is a unitary operator.

For a closed, isolated physical system the Hamiltonian H is time independent and eqn (3.3) is readily integrated to yield the well-known expression

$$U(t, t_0) = \exp\left[-iH(t - t_0)\right]. \tag{3.5}$$

In physical applications one often encounters the situation that the system under consideration is driven by external forces, an external electromagnetic field for example. If in such a case the dynamics of the system can still be formulated in terms of a possibly time-dependent Hamiltonian generator $H(t)$ the system will again be said to be *closed*, while we reserve the term *isolated* to mean that the Hamiltonian of the system is time independent. For a time-dependent Hamiltonian the solution of eqn (3.3) subjected to the initial condition (3.4) may be represented as a time-ordered exponential,

$$U(t, t_0) = \mathrm{T}_\leftarrow \exp\left[-i\int_{t_0}^{t} ds\, H(s)\right], \tag{3.6}$$

where $\mathrm{T}_\leftarrow$ denotes the chronological time-ordering operator which orders products of time-dependent operators such that their time-arguments increase from right to left as indicated by the arrow.

If the system under consideration is in a mixed state the corresponding quantum statistical ensemble may be characterized with the help of the statistical operator ρ. It is straightforward to derive an equation of motion for the density matrix starting from the Schrödinger equation (3.1). Let us assume that at some initial time t_0 the state of the system is characterized by the density matrix (see Section 2.1.3)

$$\rho(t_0) = \sum_\alpha w_\alpha |\psi_\alpha(t_0)\rangle\langle\psi_\alpha(t_0)|, \tag{3.7}$$

where the w_α are positive weights and the $|\psi_\alpha(t_0)\rangle$ are normalized state vectors which evolve in time according to the Schrödinger equation (3.2). The state of the system at time t will therefore be given by

$$\rho(t) = \sum_\alpha w_\alpha U(t, t_0)|\psi_\alpha(t_0)\rangle\langle\psi_\alpha(t_0)|U^\dagger(t, t_0), \tag{3.8}$$

which may be written more concisely as

$$\rho(t) = U(t, t_0)\rho(t_0)U^\dagger(t, t_0). \tag{3.9}$$

Differentiating this equation with respect to time we immediately get an equation of motion for the density matrix,

$$\frac{d}{dt}\rho(t) = -i[H(t), \rho(t)], \tag{3.10}$$

which is often referred to as the *von Neumann* or *Liouville–von Neumann equation*.

In order to stress the analogy of the von Neumann equation with the corresponding equation of motion for the probability density in classical statistical mechanics it is often written in a form analogous to the classical Liouville equation,

$$\frac{d}{dt}\rho(t) = \mathcal{L}(t)\rho(t). \tag{3.11}$$

Here, $\mathcal{L}$ is the *Liouville operator* which is defined through the condition that $\mathcal{L}(t)\rho$ is equal to $-i$ times the commutator of $H(t)$ with ρ. More precisely, $\mathcal{L}$ is often called a Liouville *super-operator* since it acts on an operator to yield another operator. In close analogy to eqn (3.6) the Liouville equation (3.11) leads to the formal expression

$$\rho(t) = \text{T}_\leftarrow \exp\left[\int_{t_0}^t ds\,\mathcal{L}(s)\right]\rho(t_0). \tag{3.12}$$

For the case of a time independent Hamiltonian the Liouville super-operator is also time independent and we obviously have

$$\rho(t) = \exp\left[\mathcal{L}(t - t_0)\right]\rho(t_0). \tag{3.13}$$

3.1.2 *Heisenberg and interaction picture*

In the Schrödinger picture the time dependence of the density matrix $\rho(t)$ is governed by the Liouville–von Neumann equation (3.11). An equivalent description of the quantum dynamics is obtained by transferring the time dependence from the density matrix to the operators in the Hilbert space $\mathcal{H}$. This leads to the Heisenberg picture. We assume that at some fixed initial time t_0 the quantum states in both pictures coincide, that is $\rho(t_0) = \rho_H(t_0)$. Here and in the following Heisenberg picture operators are indicated by an index H. Schrödinger picture and Heisenberg picture operators are related through the canonical transformation

$$A_H(t) = U^\dagger(t, t_0)A(t)U(t, t_0), \tag{3.14}$$

where we allow the Schrödinger picture operator $A(t)$ to depend explicitly on time. Obviously, at time t_0 Schrödinger and Heisenberg picture operators coincide, $A_H(t_0) = A(t_0)$. In the Heisenberg picture the quantum expectation values are determined through the fixed density matrix $\rho_H(t_0)$. The physical equivalence of the two pictures is made evident by the fact that the expectation value of an observable $A(t)$ is the same in both pictures,

$$\langle A(t) \rangle = \operatorname{tr} \left\{ A(t)\rho(t) \right\} = \operatorname{tr} \left\{ A_H(t)\rho_H(t_0) \right\}. \tag{3.15}$$

It is straightforward to derive an equation of motion for a Heisenberg picture operator $A_H(t)$ from the transformation law (3.14). Differentiating both sides of (3.14) with respect to time we obtain the Heisenberg equation of motion

$$\frac{d}{dt}A_H(t) = i[H_H(t), A_H(t)] + \frac{\partial A_H(t)}{\partial t}, \tag{3.16}$$

where $H_H(t)$ denotes the Hamiltonian in the Heisenberg picture,

$$H_H(t) = U^\dagger(t, t_0)H(t)U(t, t_0). \tag{3.17}$$

In eqn (3.16) d/dt denotes the total time derivative, while $\partial/\partial t$ is the partial derivative with respect to the explicit time dependence of the Schrödinger picture operator. Explicitly, we have

$$\frac{\partial A_H(t)}{\partial t} = U^\dagger(t, t_0)\frac{\partial A(t)}{\partial t}U(t, t_0). \tag{3.18}$$

If $dA_H(t)/dt = 0$, then A_H is a constant of motion. An important special case of the Heisenberg equation of motion is obtained by choosing A to be the Hamiltonian of the system, i.e. $A = H$. For an isolated system $\partial H/\partial t = 0$, and the time-evolution operator has the special form (3.5). Hence, the Hamiltonian commutes with the time-evolution operator $U(t, t_0)$ and it follows that the Heisenberg picture Hamiltonian $H_H(t)$ is a constant of motion, that is

$$\frac{d}{dt}H_H = 0. \tag{3.19}$$

Moreover, if the Schrödinger operator A has no explicit time dependence and if the system is isolated the Heisenberg equation of motion reduces to the form

$$\frac{d}{dt} A_H(t) = i[H, A_H(t)]. \tag{3.20}$$

Let us now look at the equation of motion for the expectation value of an arbitrary Schrödinger observable $A(t)$, which in general will depend explicitly on time. Differentiating eqn (3.15) with respect to time and invoking the Heisenberg equation of motion (3.16) one finds

$$\frac{d}{dt}\langle A(t)\rangle = \left\langle \frac{d}{dt} A_H(t) \right\rangle = \text{tr}\left\{ \left(i[H_H(t), A_H(t)] + \frac{\partial A_H(t)}{\partial t} \right) \rho_H(t_0) \right\}. \tag{3.21}$$

This equation is known as the Ehrenfest equation. It states that the time variation of the expectation value of an observable $A(t)$ is equal to the expectation value of the time derivative of the corresponding Heisenberg observable $A_H(t)$.

The Schrödinger and the Heisenberg pictures are the limiting cases of a more general picture, which is called the interaction picture. Let us write the Hamiltonian of the system as the sum of two parts

$$H(t) = H_0 + \hat{H}_I(t). \tag{3.22}$$

The precise form of the splitting of the Hamiltonian depends upon the particular physical situation under study. In general, H_0 represents the sum of the energies of two systems when the interaction between the systems is ignored and we shall assume that it is independent of time; $\hat{H}_I(t)$ is then the Hamiltonian describing the interaction between the systems. Again, the time evolution operator of the total system will be denoted by $U(t, t_0)$. According to eqn (3.15) the expectation value of a Schrödinger observable $A(t)$ (which may depend explicitly on time) at time t is given by

$$\langle A(t)\rangle = \text{tr}\left\{ A(t)U(t, t_0)\rho(t_0)U^\dagger(t, t_0) \right\}, \tag{3.23}$$

where $\rho(t_0)$ is the state of the system at time t_0.

We introduce the unitary time evolution operators

$$U_0(t, t_0) \equiv \exp[-iH_0(t - t_0)] \tag{3.24}$$

and

$$U_I(t, t_0) \equiv U_0^\dagger(t, t_0)U(t, t_0). \tag{3.25}$$

Then the expectation value (3.23) may be written as

$$\langle A(t)\rangle = \text{tr}\left\{ U_0^\dagger(t, t_0)A(t)U_0(t, t_0)U_I(t)\rho(t_0)U_I^\dagger(t, t_0) \right\}$$
$$\equiv \text{tr}\left\{ A_I(t)\rho_I(t) \right\}, \tag{3.26}$$

where we have introduced $A_I(t)$ as the interaction picture operator

$$A_I(t) \equiv U_0^\dagger(t, t_0) A(t) U_0(t, t_0), \tag{3.27}$$

and $\rho_I(t)$ as the interaction picture density matrix

$$\rho_I(t) \equiv U_I(t, t_0) \rho(t_0) U_I^\dagger(t, t_0). \tag{3.28}$$

It may be important to note that in contrast to the Heisenberg picture the time evolution of interaction picture operators is not generated by the full Hamiltonian H but only by the free part H_0. If $\hat{H}_I(t) = 0$ we have $U_0(t, t_0) = U(t, t_0)$ and $U_I(t, t_0) = I$, such that the interaction picture is identical to the Heisenberg picture. Conversely, in the other limiting case of a vanishing free Hamiltonian, $H_0 = 0$, we have $\hat{H}_I(t) = H(t)$ such that $U_0(t, t_0) = I$ and $U_I(t, t_0) = U(t, t_0)$, and we regain the Schrödinger picture.

It is clear that the interaction picture time-evolution operator $U_I(t, t_0)$ is the solution of the differential equation

$$i\frac{\partial}{\partial t} U_I(t, t_0) = H_I(t) U_I(t, t_0) \tag{3.29}$$

subject to the initial condition $U_I(t_0, t_0) = I$. Here, we have denoted the interaction Hamiltonian in the interaction picture by

$$H_I(t) \equiv U_0^\dagger(t, t_0) \hat{H}_I(t) U_0(t, t_0). \tag{3.30}$$

The corresponding von Neumann equation in the interaction picture thus takes the form

$$\frac{d}{dt} \rho_I(t) = -i \left[H_I(t), \rho_I(t) \right]. \tag{3.31}$$

Often, the interaction picture von Neumann equation is written in the equivalent integral form

$$\rho_I(t) = \rho_I(t_0) - i \int_{t_0}^{t} ds \left[H_I(s), \rho_I(s) \right]. \tag{3.32}$$

This form of the von Neumann equation may be used as a starting point of a perturbative approach to the construction of approximate solutions.

3.1.3 *Dynamics of open systems*

Having briefly sketched the fundamental equations describing the Hamiltonian dynamics of a closed quantum system let us now turn to the notion of an open system. In general terms, an open system is a quantum system S which is coupled to another quantum system B called the environment. It thus represents a subsystem of the combined total system $S + B$ (see Section 2.2), whereby in most cases it is assumed that the combined system is closed, following Hamiltonian

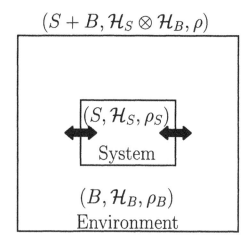

FIG. 3.1. Schematic picture of an open quantum system.

dynamics. The state of the subsystem S, however, will change as a consequence of its internal dynamics and of the interaction with the surroundings. The interaction leads to certain system–environment correlations such that the resulting state changes of S can no longer, in general, be represented in terms of unitary, Hamiltonian dynamics. The dynamics of the subsystem S induced by the Hamiltonian evolution of the total system is often referred to as *reduced system dynamics*, and S is also called the *reduced system*.

Let us denote by $\mathcal{H}_S$ the Hilbert space of the system and by $\mathcal{H}_B$ the Hilbert space of the environment. The Hilbert space of the total system $S + B$ is then given by the tensor product space $\mathcal{H} = \mathcal{H}_S \otimes \mathcal{H}_B$ (see Section 2.2.1). The total Hamiltonian $H(t)$ may be taken to be of the form

$$H(t) = H_S \otimes I_B + I_S \otimes H_B + \hat{H}_I(t), \tag{3.33}$$

where H_S is the self-Hamiltonian of the open system S, H_B is the free Hamiltonian of the environment B, and $\hat{H}_I(t)$ is the Hamiltonian describing the interaction between the system and the environment. Of course, one may also include time-dependent interactions into the model. A schematic picture of the typical situation under study is shown in Fig. 3.1.

If we speak of an open system S we shall use the general term *environment* for the system B coupled to it. The term *reservoir* refers to an environment with an infinite number of degrees of freedom such that the frequencies of the reservoir modes form a continuum. As will be seen, it is this property which generally leads to an irreversible behaviour of the open quantum system. Finally, the term *bath* or *heat bath* will be used for a reservoir which is in a thermal equilibrium state.

We are going to investigate several physical examples of open quantum systems in this book. Mainly, the motivation for the study of open systems is that in many physically important situations a complete mathematical model of the combined system's dynamics is much too complicated. The environment may represent, for example, a reservoir or a heat bath consisting of infinitely many degrees of freedom, in which case an exact treatment requires the solution of an infinite hierarchy of coupled equations of motion. Even if a solution is known one is confronted with the task of isolating and determining the interesting physical quantities through an average over the remaining, irrelevant degrees of freedom. Moreover, one often encounters the situation that the modes of the environment are neither known exactly nor controllable. One therefore tries to develop a simpler description in a reduced state space formed by a restricted set of physically relevant variables which is achieved by employing various analytical methods and approximation techniques.

We regard an open system S to be singled out by the fact that all observations of interest refer to this subsystem. According to Section 2.2.1 the observables referring to S are all of the form $A \otimes I_B$, where A is an operator acting on the Hilbert space $\mathcal{H}_S$ and I_B denotes the identity in the Hilbert space $\mathcal{H}_B$. If the state of the total system is described by some density matrix ρ, then the expectation values of all observables acting on the open system's Hilbert space are determined through the formula

$$\langle A \rangle = \mathrm{tr}_S \left\{ A \rho_S \right\}, \tag{3.34}$$

where

$$\rho_S = \mathrm{tr}_B \rho \tag{3.35}$$

is the reduced density matrix of the open quantum system S (see eqn (2.92)). Here and in the following tr_S denotes the partial trace over the open system's Hilbert space, while tr_B denotes the partial trace over the degrees of freedom of the environment B. It is clear that the reduced density operator ρ_S will be the quantity of central interest in the description of open quantum systems.

The reduced density matrix $\rho_S(t)$ at time t is obtained from the density matrix $\rho(t)$ of the total system by taking the partial trace over the degrees of freedom of the environment. Since the total density matrix evolves unitarily we have

$$\rho_S(t) = \mathrm{tr}_B \left\{ U(t, t_0) \rho(t_0) U^\dagger(t, t_0) \right\}, \tag{3.36}$$

where $U(t, t_0)$ is the time-evolution operator of the total system. In an analogous way the equation of motion for the reduced density matrix is obtained by taking the partial trace over the environment on both sides of the Liouville–von Neumann equation for the total system,

$$\frac{d}{dt} \rho_S(t) = -i \mathrm{tr}_B \left[H(t), \rho(t) \right]. \tag{3.37}$$

In the following sections we give a survey of the most important equations of motion for the reduced density matrix used to approximate the above exact dynamical equations.

3.2 Quantum Markov processes

The most important property of a classical, homogeneous Markov process is the semigroup property which is usually formulated in terms of the differential Chapman–Kolmogorov equation (1.68) involving a time-independent generator. The extension of this property to quantum mechanics leads to the concepts of a quantum dynamical semigroup and a quantum Markov process (Kraus, 1971; Davies, 1976; Alicki and Lendi, 1987; Alicki and Fannes, 2001). These concepts are introduced in this section and the general form of a quantum Markovian master equation is derived. We also discuss some general features of quantum Markov processes, such as the adjoint master equation, the quantum regression theorem, irreversibility and entropy production.

3.2.1 *Quantum dynamical semigroups*

In general the dynamics of the reduced system defined by the exact equations (3.36) and (3.37) will be quite involved. However, under the condition of short environmental correlation times one may neglect memory effects and formulate the reduced system dynamics in terms of a quantum dynamical semigroup.

We first introduce the concept of a dynamical map. To this end, let us suppose that we are able to prepare at the initial time $t = 0$ the state of the total system $S + B$ as an uncorrelated product state $\rho(0) = \rho_S(0) \otimes \rho_B$, where $\rho_S(0)$ is the initial state of the reduced system S and ρ_B represents some reference state of the environment, a thermal equilibrium state, for example. The transformation describing the change of the reduced system from the initial time $t = 0$ to some time $t > 0$ may then be written in the form

$$\rho_S(0) \mapsto \rho_S(t) = V(t)\rho_S(0) \equiv \mathrm{tr}_B \left\{ U(t,0) \left[\rho_S(0) \otimes \rho_B \right] U^\dagger(t,0) \right\}. \qquad (3.38)$$

If we regard the reference state ρ_B and the final time t to be fixed, this relation defines a map from the space $\mathcal{S}(\mathcal{H}_S)$ of density matrices of the reduced system into itself,

$$V(t) : \ \mathcal{S}(\mathcal{H}_S) \longrightarrow \mathcal{S}(\mathcal{H}_S). \qquad (3.39)$$

This map, describing the state change of the open system over time t, is called a *dynamical map* (see Fig. 3.2).

A dynamical map can be characterized completely in terms of operators pertaining to the open system's Hilbert space $\mathcal{H}_S$. To this end, we use the spectral decomposition of the density matrix ρ_B of the environment,

$$\rho_B = \sum_\alpha \lambda_\alpha |\varphi_\alpha\rangle\langle\varphi_\alpha|. \qquad (3.40)$$

$$\rho(0) = \rho_S(0) \otimes \rho_B \xrightarrow{\text{unitary evolution}} \rho(t) = U(t,0)[\rho_S(0) \otimes \rho_B]U^\dagger(t,0)$$

$$\text{tr}_B \downarrow \qquad\qquad\qquad\qquad\qquad\qquad\qquad \downarrow \text{tr}_B$$

$$\rho_S(0) \xrightarrow{\text{dynamical map}} \rho_S(t) = V(t)\rho_S(0)$$

FIG. 3.2. A commutative diagram showing the action of a dynamical map $V(t)$.

Here, the $|\varphi_\alpha\rangle$ form an orthonormal basis in $\mathcal{H}_B$ and the λ_α are non-negative real numbers satisfying $\sum_\alpha \lambda_\alpha = 1$. Definition (3.38) then immediately yields the following representation,

$$V(t)\rho_S = \sum_{\alpha,\beta} W_{\alpha\beta}(t)\rho_S W^\dagger_{\alpha\beta}(t), \qquad (3.41)$$

where the $W_{\alpha\beta}$, being operators in $\mathcal{H}_S$, are defined by

$$W_{\alpha\beta}(t) = \sqrt{\lambda_\beta}\langle\varphi_\alpha|U(t,0)|\varphi_\beta\rangle. \qquad (3.42)$$

According to eqn (3.41) the dynamical map $V(t)$ is of the form (2.157) of an operation Φ_m describing a generalized quantum measurement. Moreover, the operators $W_{\alpha\beta}(t)$ satisfy the condition

$$\sum_{\alpha,\beta} W^\dagger_{\alpha\beta}(t)W_{\alpha\beta}(t) = I_S, \qquad (3.43)$$

from which we deduce that

$$\text{tr}_S\{V(t)\rho_S\} = \text{tr}_S\rho_S = 1. \qquad (3.44)$$

Thus, we conclude that a dynamical map $V(t)$ represents a convex-linear, completely positive and trace-preserving quantum operation.

We have introduced above a dynamical map $V(t)$ for a fixed time $t \geq 0$. If we now allow t to vary we get a one-parameter family $\{V(t)|t \geq 0\}$ of dynamical maps, where $V(0)$ is the identity map. Such a family describes the whole future time evolution of the open system, which, in general, could be very involved. However, if the characteristic time scales over which the reservoir correlation functions decay are much smaller than the characteristic time scale of the systematic system evolution, it is justified to neglect memory effects in the reduced system dynamics. As in the classical theory one thus expects Markovian-type behaviour. For the homogeneous case the latter may be formalized with the help of the semigroup property (compare with eqn (1.70))

$$V(t_1)V(t_2) = V(t_1 + t_2), \quad t_1, t_2 \geq 0. \qquad (3.45)$$

The precise physical conditions underlying the Markovian approximation will be given in the microscopic derivations of Markovian quantum master equations (see

Sections 3.3 and 3.4). A systematic investigation and an expansion around the Markovian limit is postponed to Part IV. Summarizing, a quantum dynamical semigroup is a continuous, one-parameter family of dynamical maps satisfying the semigroup property (3.45).

3.2.2 *The Markovian quantum master equation*

Given a quantum dynamical semigroup there exists, under certain mathematical conditions (see below), a linear map $\mathcal{L}$, the generator of the semigroup, which allows us to represent the semigroup in exponential form,

$$V(t) = \exp(\mathcal{L}t). \tag{3.46}$$

This representation immediately yields a first-order differential equation for the reduced density matrix of the open system,

$$\frac{d}{dt}\rho_S(t) = \mathcal{L}\rho_S(t), \tag{3.47}$$

which is called the Markovian quantum master equation. The generator $\mathcal{L}$ of the semigroup represents a super-operator. It may be regarded as a generalization of the Liouville super-operator which has been introduced in eqn (3.11) and which is given by the commutator with some Hamiltonian.

Let us construct the most general form for the generator $\mathcal{L}$ of a quantum dynamical semigroup. To this end, we consider first the simple case of a finite-dimensional Hilbert space $\mathcal{H}_S$, $\dim\mathcal{H}_S = N$. The corresponding Liouville space is a complex space of dimension N^2 and we choose a complete basis (see eqn (2.97)) of orthonormal operators F_i, $i = 1, 2, \ldots, N^2$, in this space such that we have

$$(F_i, F_j) \equiv \mathrm{tr}_S\left\{F_i^\dagger F_j\right\} = \delta_{ij}. \tag{3.48}$$

For convenience one of the basis operators is chosen to be proportional to the identity, namely $F_{N^2} = (1/N)^{1/2}I_S$, such that the other basis operators are traceless, that is $\mathrm{tr}_S F_i = 0$ for $i = 1, 2, \ldots, N^2 - 1$. Applying the completeness relation to each of the operators $W_{\alpha\beta}(t)$ defined in eqn (3.42), we have

$$W_{\alpha\beta}(t) = \sum_{i=1}^{N^2} F_i(F_i, W_{\alpha\beta}(t)). \tag{3.49}$$

With the help of the representation (3.41) we can write the action of the dynamical map $V(t)$ as

$$V(t)\rho_S = \sum_{i,j=1}^{N^2} c_{ij}(t) F_i \rho_S F_j^\dagger, \tag{3.50}$$

where

$$c_{ij}(t) \equiv \sum_{\alpha\beta}(F_i, W_{\alpha\beta}(t))(F_j, W_{\alpha\beta}(t))^*. \tag{3.51}$$

The coefficient matrix $c = (c_{ij})$ is easily seen to be Hermitian and positive. In fact, for any N^2-dimensional complex vector v we have

$$\sum_{ij} c_{ij} v_i^* v_j = \sum_{\alpha\beta} \left| \left(\sum_i v_i F_i, W_{\alpha\beta}(t) \right) \right|^2 \geq 0, \qquad (3.52)$$

which proves that $c \geq 0$.

The definition (3.46) of the generator $\mathcal{L}$ now gives, by virtue of eqn (3.50),

$$\mathcal{L}\rho_S = \lim_{\varepsilon \to 0} \frac{1}{\varepsilon} \{ V(\varepsilon)\rho_S - \rho_S \}$$

$$= \lim_{\varepsilon \to 0} \left\{ \frac{1}{N} \frac{c_{N^2N^2}(\varepsilon) - N}{\varepsilon} \rho_S + \frac{1}{\sqrt{N}} \sum_{i=1}^{N^2-1} \left(\frac{c_{iN^2}(\varepsilon)}{\varepsilon} F_i \rho_S + \frac{c_{N^2i}(\varepsilon)}{\varepsilon} \rho_S F_i^\dagger \right) \right.$$

$$\left. + \sum_{i,j=1}^{N^2-1} \frac{c_{ij}(\varepsilon)}{\varepsilon} F_i \rho_S F_j^\dagger. \right\} \qquad (3.53)$$

Next, we define coefficients a_{ij} by

$$a_{N^2N^2} = \lim_{\varepsilon \to 0} \frac{c_{N^2N^2}(\varepsilon) - N}{\varepsilon}, \qquad (3.54)$$

$$a_{iN^2} = \lim_{\varepsilon \to 0} \frac{c_{iN^2}(\varepsilon)}{\varepsilon}, \quad i = 1, \ldots, N^2 - 1, \qquad (3.55)$$

$$a_{ij} = \lim_{\varepsilon \to 0} \frac{c_{ij}(\varepsilon)}{\varepsilon}, \quad i, j = 1, \ldots, N^2 - 1, \qquad (3.56)$$

and introduce the quantities

$$F = \frac{1}{\sqrt{N}} \sum_{i=1}^{N^2-1} a_{iN^2} F_i \qquad (3.57)$$

and

$$G = \frac{1}{2N} a_{N^2N^2} I_S + \frac{1}{2}(F^\dagger + F), \qquad (3.58)$$

as well as the Hermitian operator

$$H = \frac{1}{2i}(F^\dagger - F). \qquad (3.59)$$

We note that the matrix formed by the coefficients a_{ij}, $i, j = 1, 2, \ldots, N^2 - 1$, is Hermitian and positive. With the help of these definitions we can write the generator as

$$\mathcal{L}\rho_S = -i[H, \rho_S] + \{G, \rho_S\} + \sum_{i,j=1}^{N^2-1} a_{ij} F_i \rho_S F_j^\dagger. \qquad (3.60)$$

Since the semigroup is trace preserving we have for all density matrices ρ_S

$$0 = \mathrm{tr}_S\{\mathcal{L}\rho_S\} = \mathrm{tr}_S\left\{\left(2G + \sum_{i,j=1}^{N^2-1} a_{ij} F_j^\dagger F_i\right)\rho_S\right\}, \qquad (3.61)$$

from which we deduce that

$$G = -\frac{1}{2}\sum_{i,j=1}^{N^2-1} a_{ij} F_j^\dagger F_i. \qquad (3.62)$$

Hence, we get from (3.60) the *first standard form* of the generator,

$$\mathcal{L}\rho_S = -i\,[H,\rho_S] + \sum_{i,j=1}^{N^2-1} a_{ij}\left(F_i\rho_S F_j^\dagger - \frac{1}{2}\left\{F_j^\dagger F_i, \rho_S\right\}\right). \qquad (3.63)$$

Since the coefficient matrix $a = (a_{ij})$ is positive it may be diagonalized with the help of an appropriate unitary transformation u,

$$uau^\dagger = \begin{pmatrix} \gamma_1 & 0 & \cdots & 0 \\ 0 & \gamma_2 & \cdots & 0 \\ 0 & 0 & \ddots & 0 \\ 0 & 0 & \cdots & \gamma_{N^2-1} \end{pmatrix}, \qquad (3.64)$$

where the eigenvalues γ_i are non-negative. Introducing a new set of operators A_k through

$$F_i = \sum_{k=1}^{N^2-1} u_{ki} A_k, \qquad (3.65)$$

the following *diagonal* form of the generator is obtained

$$\mathcal{L}\rho_S = -i\,[H,\rho_S] + \sum_{k=1}^{N^2-1} \gamma_k \left(A_k\rho_S A_k^\dagger - \frac{1}{2}A_k^\dagger A_k\rho_S - \frac{1}{2}\rho_S A_k^\dagger A_k\right). \qquad (3.66)$$

This is the most general form for the generator of a quantum dynamical semigroup. The first term of the generator represents the unitary part of the dynamics generated by the Hamiltonian H. The operators A_k, introduced above as appropriate linear combinations of the basis operators F_i in Liouville space, are usually referred to as Lindblad operators and the corresponding density matrix equation (3.47) is called the Lindblad equation. We note that the non-negative quantities γ_k have the dimension of an inverse time provided the A_k are taken to be dimensionless. As will be seen later the γ_k are given in terms of certain correlation functions of the environment and play the rôle of relaxation rates for the different decay modes of the open system.

The detailed mathematical proof that (3.63) defines the most general generator of a quantum dynamical semigroup for the case of a finite-dimensional Hilbert space has been given by Gorini, Kossakowski, and Sudarshan (1976). At the same time Lindblad (1976) proved in a theorem that bears his name that (3.66) provides the most general form for a bounded generator in a separable Hilbert space if the index k is allowed to run over a countable set. In the physics literature quantum master equations of the form (3.63) or (3.66) appeared much earlier, for example in the context of spin relaxation dynamics and laser theory (see, e.g. Bausch, 1966; Haake, 1973; Haken, 1984). We emphasize that the Lindblad theorem presupposes that the generator $\mathcal{L}$ is bounded. In physical applications this is usually not the case: The self-Hamiltonian H_S of the reduced system as well as the Lindblad operators are, in general, unbounded. However, all known examples for generators of quantum dynamical semigroups are either of Lindblad form, or can be cast into it by slight modifications.

In the above presentation we have left out all mathematical details. Let us make, however, a few remarks on the mathematical theory (for a review see, e.g. Gorini, Frigerio, Verri, Kossakowski and Sudarshan, 1978; Spohn, 1980; Alicki and Lendi, 1987). First, we mention that the theory of quantum dynamical semigroups is usually formulated in terms of the Heisenberg, instead of the Schrödinger picture. To this end, one characterizes dynamical maps in the Heisenberg picture. The Heisenberg picture dynamical map will be denoted by $V^{\dagger}(t)$ and acts on operators A in the Hilbert space $\mathcal{H}_S$ of the open system such that we have

$$\mathrm{tr}_S\left\{A\left(V(t)\rho_S\right)\right\} = \mathrm{tr}_S\left\{\left(V^{\dagger}(t)A\right)\rho_S\right\} \qquad (3.67)$$

for all ρ_S. Since we know that $V(t)$ maps density matrices to density matrices, the Heisenberg picture map $V^{\dagger}(t)$ should also transform positive operators into positive operators and must preserve the identity operator, that is $V^{\dagger}(t)I_S = I_S$. If the mathematical theory is based on an axiomatic approach to dynamical maps one has to demand not only positivity but the stronger condition of complete positivity (see Section 2.4.3). However, from a physical point of view the additional requirement of complete positivity does not introduce a new condition if dynamical maps are introduced, as we have done above, through the reduced dynamics induced by the unitary dynamics in a larger system $S + B$.

We have also required the one-parameter family of dynamical maps $V(t)$ to be continuous. It can be shown that it suffices to demand ultraweak continuity, which means that we must have

$$\lim_{\varepsilon \to 0} \mathrm{tr}_S\left\{\left(V^{\dagger}(\varepsilon)A - A\right)\rho_S\right\} = 0 \qquad (3.68)$$

for all ρ_S and all bounded operators A. It can be shown further that quantum dynamical semigroups always have the property of being contracting, that is they fulfil the condition

$$\|V(t)^{\dagger}A\|_1 \leq \|A\|_1, \qquad (3.69)$$

where $||A||_1 = \text{tr}_S|A|$ is known as the trace norm. It is this property which allows the application of the theory of contracting semigroups and enables the introduction of the infinitesimal generator $\mathcal{L}$.

It is sometimes convenient to introduce the *dissipator*

$$\mathcal{D}(\rho_S) \equiv \sum_k \gamma_k \left(A_k \rho_S A_k^\dagger - \frac{1}{2} A_k^\dagger A_k \rho_S - \frac{1}{2} \rho_S A_k^\dagger A_k \right) \tag{3.70}$$

and to write the quantum master equation (3.47) in the form

$$\frac{d}{dt}\rho_S(t) = -i\,[H, \rho_S(t)] + \mathcal{D}(\rho_S(t)). \tag{3.71}$$

We remark that, in general, the operator H cannot be identified with the free Hamiltonian H_S of the reduced system S: The Hamiltonian H may contain additional terms which are due to the coupling of the system to its environment (see Section 3.3.1 for an example). We also note that the generator $\mathcal{L}$ does not uniquely fix the form of the Hamiltonian H and the Lindblad operators A_i. In fact, the generator is invariant under the following transformations:

1. unitary transformations of the set of Lindblad operators,

$$\sqrt{\gamma_i} A_i \longrightarrow \sqrt{\gamma_i'} A_i' = \sum_j u_{ij} \sqrt{\gamma_j} A_j, \tag{3.72}$$

   where u_{ij} is a unitary matrix,

2. inhomogeneous transformations

$$A_i \longrightarrow A_i' = A_i + a_i,$$

$$H \longrightarrow H' = H + \frac{1}{2i} \sum_j \gamma_j \left(a_j^* A_j - a_j A_j^\dagger \right) + b, \tag{3.73}$$

   where the a_i are complex numbers and b is real.

Because of the second invariance property it is always possible to choose traceless Lindblad operators.

As we already noticed the open system may be subjected to an external time-dependent field. The description of such an open system may require the help of time-dependent generators. An obvious time-dependent generalization of (3.47) may be written in the form

$$\frac{d}{dt}\rho_S(t) = \mathcal{L}(t)\rho_S(t), \tag{3.74}$$

where $\mathcal{L}(t)$ is the generator of a quantum dynamical semigroup for each fixed $t \geq 0$. Let us introduce the corresponding propagator by means of

$$V(t, t_0) = \text{T}_\leftarrow \exp\left(\int_{t_0}^{t} ds \mathcal{L}(s) \right), \tag{3.75}$$

which satisfies

$$\frac{\partial}{\partial t}V(t,t_0) = \mathcal{L}(t)V(t,t_0). \tag{3.76}$$

Instead of the semigroup property (3.45) we now have

$$V(t,t_1)V(t_1,t_0) = V(t,t_0). \tag{3.77}$$

We study Lindblad generators with time-dependent external fields in Section 8.4.

3.2.3 *The adjoint quantum master equation*

In full analogy to closed quantum systems, also for open quantum systems it is possible to define for each system operator A in the Schrödinger picture a corresponding operator $A_H(t)$ in the Heisenberg picture. This is achieved with the help of eqn (3.67) which we rewrite as

$$\mathrm{tr}_S\left\{AV(t,0)\rho_S\right\} = \mathrm{tr}_S\left\{\left(V^\dagger(t,0)A\right)\rho_S\right\} = \mathrm{tr}_S\left\{A_H(t)\rho_S\right\}, \tag{3.78}$$

where we allow the Lindblad generator to be time dependent. In analogy to eqn (3.75) we have introduced the adjoint propagator

$$V^\dagger(t,t_0) = \mathrm{T}_\rightarrow \exp\left(\int_{t_0}^t ds\mathcal{L}^\dagger(s)\right). \tag{3.79}$$

The symbol $\mathrm{T}_\rightarrow$ denotes the antichronological time-ordering operator, while the adjoint generator $\mathcal{L}^\dagger$ is defined by

$$\mathrm{tr}_S\left\{A\mathcal{L}(t)\rho_S\right\} = \mathrm{tr}_S\left\{\left(\mathcal{L}^\dagger(t)A\right)\rho_S\right\}. \tag{3.80}$$

The adjoint propagator $V^\dagger(t,t_0)$ satisfies the differential equation

$$\frac{\partial}{\partial t}V^\dagger(t,t_0) = V^\dagger(t,t_0)\mathcal{L}^\dagger(t), \tag{3.81}$$

and describes the time evolution of operators in the Heisenberg picture,

$$A_H(t) = V^\dagger(t,0)A. \tag{3.82}$$

This yields the following equation of motion for the Heisenberg operator $A_H(t)$,

$$\frac{d}{dt}A_H(t) = V^\dagger(t,0)\left\{\mathcal{L}^\dagger(t)A\right\}, \tag{3.83}$$

which is known as the *adjoint master equation*.

Notice that on the right-hand side of eqn (3.83) the adjoint Lindblad operator first acts on the operator A and then the result is propagated with the help of the adjoint propagator. In order to determine the form of the equation of motion at time t knowledge of the Heisenberg operator $A_H(t)$ is not sufficient; one also needs the adjoint propagator, in general.

An important special case is obtained if the Lindblad generator does not depend explicitly on time, as is the case in eqn (3.66). In this case the adjoint Lindblad generator $\mathcal{L}^\dagger$ commutes with $V^\dagger(t, 0)$ and the adjoint quantum master equation takes the simpler form

$$\frac{d}{dt}A_H(t) \tag{3.84}$$

$$= \mathcal{L}^\dagger A_H(t)$$

$$= i\,[H, A_H(t)] + \sum_k \gamma_k \left(A_k^\dagger A_H(t) A_k - \frac{1}{2} A_H(t) A_k^\dagger A_k - \frac{1}{2} A_k^\dagger A_k A_H(t) \right).$$

Now, the right-hand side of the adjoint master equation only depends on the Heisenberg operator $A_H(t)$ at time t. An example of the use of the adjoint master equation will be discussed in Section 3.4.6.

3.2.4 *Multi-time correlation functions*

The quantum master equation (3.47) and the adjoint master equation (3.83) describe the time evolution of the density matrix and of the Heisenberg picture operators. With their help we can determine expectation values and matrix elements of all system observables. However, it is important to be aware of the fact that for open quantum systems these are not the only observable quantities. A complete statistical description also requires knowledge of mean values of operators taken at different times, that is of multi-time correlations functions. In physical applications these are of great interest since they are often directly related to measurable quantities. A typical example is provided by a radiation mode $a(t)$ of the electromagnetic field. In this case one might be interested in the fluctuation spectrum which is defined in terms of the Fourier transform of the two-time correlation function $\langle a^\dagger(t)a(0)\rangle$.

To begin with, we first consider a two-time correlation function of the form

$$g(t, s) \equiv \langle B(t)C(s)\rangle = \mathrm{tr}_S\left\{B(t)C(s)\rho_S(0)\right\}, \tag{3.85}$$

where $t > s \geq 0$. Suppressing for simplicity the index H we write $B(t)$ and $C(t)$ for arbitrary Heisenberg operators, where we take the fixed initial time $t_0 = 0$. On using the propagator of the quantum master equation we can write eqn (3.85) as follows,

$$g(t, s) = \mathrm{tr}_S\left\{BV(t, s)CV(s, 0)\rho_S(0)\right\}, \tag{3.86}$$

where B and C are Schrödinger picture operators. We use here and in the following the convention that the super-operators $V(t, s)$ always act on anything standing to the right of them. Equation (3.86) thus means that, in order to calculate the two-time correlation function, one propagates the density matrix $\rho_S(0)$ from time 0 to time s, applies from the left the operator C, propagates the new 'density matrix' $C\rho_S(s)$ from time s to time t, and, finally, applies B from the left and takes the trace over $\mathcal{H}_S$.

This result is easily generalized to the most general form of a measurable multi-time correlation function which is given by (Gardiner and Zoller, 2000; Gardiner and Collett, 1985)

$$g(t_1, t_2, \ldots, t_n; s_1, s_2, \ldots, s_m) \qquad (3.87)$$
$$\equiv \langle B_1(s_1)B_2(s_2)\ldots B_m(s_m)C_n(t_n)\ldots C_2(t_2)C_1(t_1)\rangle$$
$$= \mathrm{tr}_S\left\{C_n(t_n)\ldots C_2(t_2)C_1(t_1)\rho_S(0)B_1(s_1)B_2(s_2)\ldots B_m(s_m)\right\}.$$

Here, the $B_j(s_j)$, $j = 1, \ldots, m$, and the $C_i(t_i)$, $i = 1, \ldots, n$, are arbitrary Heisenberg operators and the s_j and t_i are ordered in time such that

$$t_n > t_{n-1} > \ldots > t_1 \geq 0, \quad s_m > s_{m-1} > \ldots > s_1 \geq 0. \qquad (3.88)$$

In order to bring the multi-time correlation function into a form analogous to (3.86) we perform a time ordering of the set $\{t_1, t_2, \ldots, t_n, s_1, s_2, \ldots, s_m\}$. Note that the number q of elements of this set may be smaller than $n+m$ since we allow the possibility that $t_i = s_j$ for certain indices i, j. We denote the time-ordered elements by r_l, $l = 1, 2, \ldots, q$. Thus we have

$$r_q > r_{q-1} > \ldots > r_1 \geq 0. \qquad (3.89)$$

For each l we define a super-operator $\mathcal{F}_l$ through

$$\left.\begin{array}{l} \mathcal{F}_l\rho = C_i\rho \quad \text{if } r_l = t_i \neq s_j \text{ for some } i \text{ and all } j, \\ \mathcal{F}_l\rho = \rho B_j \quad \text{if } r_l = s_j \neq t_i \text{ for some } j \text{ and all } i, \\ \mathcal{F}_l\rho = C_i\rho B_j \quad \text{if } r_l = t_i = s_j \text{ for some } i \text{ and } j. \end{array}\right\} \qquad (3.90)$$

Then we have

$$g(t_1, t_2, \ldots, t_n; s_1, s_2, \ldots, s_m) \qquad (3.91)$$
$$= \mathrm{tr}_S\left\{\mathcal{F}_q V(r_q, r_{q-1})\mathcal{F}_{q-1}V(r_{q-1}, r_{q-2})\ldots\mathcal{F}_1 V(r_1, 0)\rho_S(0)\right\}.$$

This equation provides a general expression for all time-ordered multi-time correlation functions in terms of the propagator of the underlying quantum Markovian master equation.

In practical applications it is often useful to formulate the dynamics of correlation functions in terms of a system of differential equations. Suppose we have a set of system operators $\{B_i\}$ such that the Markovian master equation (3.47) gives rise to a closed linear system of first-order differential equations for their averages,

$$\frac{d}{dt}\langle B_i(t)\rangle = \sum_j G_{ij}\langle B_j(t)\rangle \qquad (3.92)$$

with some coefficient matrix G_{ij}. The two-point correlation function then satisfies the same system of differential equations,

$$\frac{d}{d\tau}\langle B_i(t+\tau)B_l(t)\rangle = \sum_j G_{ij}\langle B_j(t+\tau)B_l(t)\rangle. \tag{3.93}$$

This statement is known as the *quantum regression theorem* (Lax, 1963; Gardiner and Zoller, 2000). To prove it we first write

$$\frac{d}{dt}\langle B_i(t)\rangle = \mathrm{tr}_S\left\{\left(\mathcal{L}^\dagger B_i\right)\rho_S(t)\right\} = \mathrm{tr}_S\left\{\left(\sum_j G_{ij}B_j\right)\rho_S(t)\right\}. \tag{3.94}$$

Since this is supposed to hold, in particular, for any initial density $\rho_S(0)$ we conclude that

$$\mathcal{L}^\dagger B_i = \sum_j G_{ij}B_j. \tag{3.95}$$

Hence we obtain

$$\frac{d}{d\tau}\langle B_i(t+\tau)B_l(t)\rangle = \mathrm{tr}_S\left\{\left(\mathcal{L}^\dagger B_i\right)V(t+\tau,t)B_l\rho_S(t)\right\}$$
$$= \sum_j G_{ij}\langle B_j(t+\tau)B_l(t)\rangle, \tag{3.96}$$

which was to be shown. The significance of the quantum regression theorem derives from the fact that it allows us to determine explicit expressions for the correlation functions once a solution for the mean values is known. An example will be discussed in Section 3.4.5.

3.2.5 *Irreversibility and entropy production*

The quantum Markovian master equation (3.47) describes the irreversible evolution of an open quantum system. In order to discuss in some detail the irreversible character of quantum dynamical semigroups we shall derive in this section a general expression for the entropy production rate σ in an open quantum system which admits a stationary state and show that it is always non-negative.

To begin with, we consider an arbitrary dynamical map $V(t)$ and investigate the change of the relative entropy $S(\rho\|\rho_0)$ of two states ρ, ρ_0 of S which is induced by $V(t)$ (Lindblad, 1975). Representing $V(t)$ as in eqn (3.38) and employing eqns (2.116) and (2.119) and the inequality (2.118) we find[7]

$$S(V(t)\rho\|V(t)\rho_0) = S\left(\mathrm{tr}_B\{U(t,0)\rho\otimes\rho_B U^\dagger(t,0)\}\|\mathrm{tr}_B\{U(t,0)\rho_0\otimes\rho_B U^\dagger(t,0)\}\right)$$
$$\leq S\left(U(t,0)\rho\otimes\rho_B U^\dagger(t,0)\|U(t,0)\rho_0\otimes\rho_B U^\dagger(t,0)\right)$$
$$= S(\rho\otimes\rho_B\|\rho_0\otimes\rho_B)$$
$$= S(\rho\|\rho_0). \tag{3.97}$$

Let us suppose that ρ_0 is a stationary state of S, that is $V(t)\rho_0 = \rho_0$. We can then conclude that a dynamical map $V(t)$ decreases the relative entropy with respect to a stationary state,

---

[7]To simplify the notation we omit the index S on quantities of the reduced system.

$$S(V(t)\rho||\rho_0) \leq S(\rho||\rho_0).$$ (3.98)

If now $V(t) = \exp \mathcal{L}t$ is a dynamical semigroup it follows from inequality (3.98) that the negative time derivative of the relative entropy is positive, that is we have the inequality

$$\sigma(\rho(t)) \equiv -\frac{d}{dt} S(\rho(t)||\rho_0) \geq 0,$$ (3.99)

where $\rho(t) = V(t)\rho(0)$. Invoking arguments from non-equilibrium thermodynamics we may interpret the functional $\sigma(\rho)$ as the *entropy production rate* (see below). The definition of the relative entropy functional, eqn (2.113), leads to the explicit expression

$$\sigma(\rho) = -k_{\mathrm{B}} \mathrm{tr} \left\{ \mathcal{L}(\rho) \ln \rho \right\} + k_{\mathrm{B}} \mathrm{tr} \left\{ \mathcal{L}(\rho) \ln \rho_0 \right\} \geq 0.$$ (3.100)

Note that we have included here the Boltzmann constant k_{B} into the definition of the entropy.

Equation (3.100) shows that the entropy production is non-negative and vanishes in a stationary state. With the help of Lieb's theorem (see eqn (2.120)) we can prove even more, namely that the map $\rho \mapsto \sigma(\rho)$ is a convex functional on the state space of the open system (Spohn, 1978). To see this we first observe that the second term in expression (3.100) is a linear functional. Thus, it suffices to show that the map

$$\rho \mapsto -k_{\mathrm{B}} \mathrm{tr} \left\{ \mathcal{L}(\rho) \ln \rho \right\}$$ (3.101)

is a convex functional. Employing the general form (3.66) of the generator $\mathcal{L}$ this map takes the form

$$\rho \mapsto -k_{\mathrm{B}} \sum_k \gamma_k \mathrm{tr} \left\{ \left(A_k \rho A_k^\dagger - A_k^\dagger A_k \rho \right) \ln \rho \right\}.$$ (3.102)

The convexity of the entropy production functional $\sigma(\rho)$ now follows immediately from the convexity of the map (2.126).

Let us motivate the definition (3.100) for the entropy production rate by using arguments from non-equilibrium thermodynamics. To this end, we suppose that the canonical equilibrium distribution (Gibbs state)

$$\rho_{\mathrm{th}} = \frac{1}{Z} \exp\left(-\beta H\right)$$ (3.103)

is a stationary solution of the master equation. This means that $\mathcal{L}(\rho_{\mathrm{th}}) = \mathcal{D}(\rho_{\mathrm{th}}) = 0$, where $\mathcal{D}$ denotes the dissipator of the quantum master equation (see eqns (3.70) and (3.71)). The normalization factor $Z = \mathrm{tr} \exp(-\beta H)$ is the partition function and

$$\beta \equiv \frac{1}{k_{\mathrm{B}} T}$$ (3.104)

is the inverse temperature. The precise physical conditions underlying this assumption will be discussed later.

In non-equilibrium thermodynamics the entropy obeys a balance equation which can be written in the form

$$\sigma = \frac{dS}{dt} + J. \tag{3.105}$$

Here, S is the von Neumann entropy (see Section 2.3.1) of the open system. The quantity J denotes the entropy flux, that is the amount of entropy which is exchanged per unit of time between the open system and its environment, where we use the convention that for $J > 0$ entropy flows from the open system into the environment. Consequently, the quantity σ is the entropy production rate, that is the amount of entropy produced per unit of time as a result of irreversible processes.

The time derivative of the von Neumann entropy is easily evaluated to be

$$\frac{dS}{dt} = -k_{\mathrm{B}} \mathrm{tr}\{\mathcal{L}(\rho) \ln \rho\}. \tag{3.106}$$

On the other hand, the entropy flux J is due to those changes of the internal energy $E = \mathrm{tr}\{H\rho\}$ which result from dissipative effects. Thus we may define the entropy flux by means of

$$J = -\frac{1}{T} \frac{d}{dt}\bigg|_{\mathrm{diss}} E \equiv -\frac{1}{T} \mathrm{tr}\{H\mathcal{D}(\rho)\} = -\frac{1}{T} \mathrm{tr}\{H\mathcal{L}(\rho)\}. \tag{3.107}$$

Using the explicit expression (3.103) for the thermal distribution we find

$$-\frac{1}{T}H = k_{\mathrm{B}} \ln \rho_{\mathrm{th}} + k_{\mathrm{B}} \ln Z, \tag{3.108}$$

so that the entropy flow can be written as

$$J = k_{\mathrm{B}} \mathrm{tr}\left\{\mathcal{L}(\rho) \ln \rho_{\mathrm{th}}\right\}, \tag{3.109}$$

where we have made use of the fact that the generator is trace-preserving, i.e. $\mathrm{tr}\{\mathcal{L}(\rho)\} = 0$. Adding eqns (3.106) and (3.109) we see that the thermodynamic entropy production rate σ defined by the balance equation (3.105) coincides with expression (3.100) for the negative rate of change of the relative entropy with respect to the thermal equilibrium state. In this context the inequality $\sigma(\rho) \geq 0$ expresses the second law of thermodynamics. Note also that $\sigma(\rho_{\mathrm{th}}) = 0$, that is the entropy production vanishes in the thermal equilibrium state. Thus we conclude that the entropy production rate $\sigma(\rho)$ is a convex functional on the space of density matrices which vanishes in the thermal equilibrium state.

3.3 Microscopic derivations

From a fundamental viewpoint it is desirable to derive the generator of a quantum dynamical semigroup from the underlying Hamiltonian dynamics of the total system. The aim of this section is to show under which assumptions such derivations can be given on the grounds of various approximation schemes.

3.3.1 *Weak-coupling limit*

We begin by considering a quantum mechanical system S weakly coupled to a reservoir B. The Hamiltonian of the total system is assumed to be of the form

$$H = H_S + H_B + H_I, \tag{3.110}$$

where H_S and H_B denote respectively the free Hamiltonian of the system and of the reservoir and H_I describes the interaction between the system and the reservoir. The derivation of a quantum Markovian master equation is most easily performed in the interaction picture. Our starting point is thus the interaction picture von Neumann equation (see Section 3.1.2)

$$\frac{d}{dt}\rho(t) = -i[H_I(t), \rho(t)] \tag{3.111}$$

for the total density matrix $\rho(t)$ and its integral form

$$\rho(t) = \rho(0) - i \int_0^t ds[H_I(s), \rho(s)]. \tag{3.112}$$

Note that we omit here for ease of notation the index I which served to indicate the interaction picture in Section 3.1.2. Inserting the integral form into (3.111) and taking the trace over the reservoir we find

$$\frac{d}{dt}\rho_S(t) = -\int_0^t ds\,\mathrm{tr}_B\,[H_I(t), [H_I(s), \rho(s)]]. \tag{3.113}$$

Here, we have assumed that

$$\mathrm{tr}_B[H_I(t), \rho(0)] = 0. \tag{3.114}$$

Equation (3.113) still contains the density matrix of the total system $\rho(t)$ on its right-hand side. In order to eliminate $\rho(t)$ from the equation of motion we perform a first approximation, known as the *Born approximation*. This approximation assumes that the coupling between the system and the reservoir is weak, such that the influence of the system on the reservoir is small (weak-coupling approximation). Thus, the density matrix of the reservoir ρ_B is only negligibly affected by the interaction and the state of the total system at time t may be approximately characterized by a tensor product

$$\rho(t) \approx \rho_S(t) \otimes \rho_B. \tag{3.115}$$

We emphasize that this does not imply that there are no excitations in the reservoir caused by the reduced system. The Markovian approximation to be derived below provides a description on a coarse-grained time scale and the assumption is that environmental excitations decay over times which are not resolved. Inserting the tensor product into the exact equation of motion (3.113) we obtain a closed integro-differential equation for the reduced density matrix $\rho_S(t)$

$$\frac{d}{dt}\rho_S(t) = -\int_0^t ds \mathrm{tr}_B \left[H_I(t), [H_I(s), \rho_S(s) \otimes \rho_B] \right]. \qquad (3.116)$$

In order to simplify the above equation further we perform the *Markov approximation*, in which the integrand $\rho_S(s)$ is first replaced by $\rho_S(t)$. In this way we obtain an equation of motion for the reduced system's density matrix in which the time development of the state of the system at time t only depends on the present state $\rho_S(t)$,

$$\frac{d}{dt}\rho_S(t) = -\int_0^t ds \mathrm{tr}_B \left[H_I(t), [H_I(s), \rho_S(t) \otimes \rho_B] \right]. \qquad (3.117)$$

This equation is called the Redfield equation (Redfield, 1957; Blum, 1981).

The Redfield equation is local in time, but it is not yet a Markovian master equation since the time evolution of the reduced density matrix still depends upon an explicit choice for the initial preparation at time $t = 0$. This implies that the dynamics of the reduced system is not yet described by a dynamical semigroup. In order to achieve this we substitute s by $t - s$ in the integral in eqn (3.117) and let the upper limit of the integral go to infinity. This is permissible provided the integrand disappears sufficiently fast for $s \gg \tau_B$. The Markov approximation is therefore justified if the time scale τ_R over which the state of the system varies appreciably is large compared to the time scale τ_B over which the reservoir correlation functions decay. Thus, we finally obtain the Markovian quantum master equation

$$\frac{d}{dt}\rho_S(t) = -\int_0^\infty ds \mathrm{tr}_B \left[H_I(t), [H_I(t - s), \rho_S(t) \otimes \rho_B] \right]. \qquad (3.118)$$

It is important to realize that in a description of the reduced system dynamics on the basis of a Markovian quantum master equation the dynamical behaviour over times of the order of magnitude of the correlation time τ_B is not resolved. As mentioned before, the evolution is described in this sense on a coarse-grained time axis.

The approximations performed above are usually termed the *Born–Markov approximation*. In general they do not guarantee, however, that the resulting equation (3.118) defines the generator of a dynamical semigroup (Davies, 1974; Dümcke and Spohn, 1979). One therefore performs a further secular approximation which involves an averaging over the rapidly oscillating terms in the master equation and is known as the rotating wave approximation. To explain the procedure let us write the Schrödinger picture interaction Hamiltonian H_I in the form

$$H_I = \sum_\alpha A_\alpha \otimes B_\alpha, \tag{3.119}$$

where $A_\alpha^\dagger = A_\alpha$ and $B_\alpha^\dagger = B_\alpha$. This is the most general form of the interaction. The secular approximation is easily carried out if one decomposes the interaction Hamiltonian H_I into eigenoperators of the system Hamiltonian H_S. Supposing the spectrum of H_S to be discrete this may be achieved as follows. Let us denote the eigenvalues of H_S by ε and the projection onto the eigenspace belonging to the eigenvalue ε by $\Pi(\varepsilon)$. Then we can define the operators

$$A_\alpha(\omega) \equiv \sum_{\varepsilon' - \varepsilon = \omega} \Pi(\varepsilon) A_\alpha \Pi(\varepsilon'). \tag{3.120}$$

The sum in this expression is extended over all energy eigenvalues ε' and ε of H_S with a fixed energy difference of ω. An immediate consequence of this definition is that the following relations are satisfied,

$$[H_S, A_\alpha(\omega)] = -\omega A_\alpha(\omega), \tag{3.121}$$

$$[H_S, A_\alpha^\dagger(\omega)] = +\omega A_\alpha^\dagger(\omega). \tag{3.122}$$

The operators $A_\alpha(\omega)$ and $A_\alpha^\dagger(\omega)$ are therefore said to be eigenoperators of H_S belonging to the frequencies $\pm\omega$, respectively. It follows from relations (3.121) and (3.122) that the corresponding interaction picture operators take the form

$$e^{iH_S t} A_\alpha(\omega) e^{-iH_S t} = e^{-i\omega t} A_\alpha(\omega), \tag{3.123}$$

$$e^{iH_S t} A_\alpha^\dagger(\omega) e^{-iH_S t} = e^{+i\omega t} A_\alpha^\dagger(\omega). \tag{3.124}$$

Finally, we note that

$$[H_S, A_\alpha^\dagger(\omega) A_\beta(\omega)] = 0, \tag{3.125}$$

and

$$A_\alpha^\dagger(\omega) = A_\alpha(-\omega). \tag{3.126}$$

Summing (3.120) over all energy differences and employing the completeness relation we get

$$\sum_\omega A_\alpha(\omega) = \sum_\omega A_\alpha^\dagger(\omega) = A_\alpha. \tag{3.127}$$

This enables us to cast the interaction Hamiltonian into the following form

$$H_I = \sum_{\alpha,\omega} A_\alpha(\omega) \otimes B_\alpha = \sum_{\alpha,\omega} A_\alpha^\dagger(\omega) \otimes B_\alpha^\dagger. \tag{3.128}$$

This is the desired decomposition of the interaction into eigenoperators of the system Hamiltonian. Note that the frequency spectrum $\{\omega\}$ is, in general, degenerate: For a fixed ω the index α labels the different operators $A_\alpha(\omega)$ belonging to the same frequency. A specific example will be encountered in Section 3.4, where α labels the Cartesian components of the dipole operator.

The reason for introducing the eigenoperator decomposition (3.128) is that the interaction picture interaction Hamiltonian can now be written in the particularly simple form

$$H_I(t) = \sum_{\alpha,\omega} e^{-i\omega t} A_\alpha(\omega) \otimes B_\alpha(t) = \sum_{\alpha,\omega} e^{+i\omega t} A_\alpha^\dagger(\omega) \otimes B_\alpha^\dagger(t), \qquad (3.129)$$

where

$$B_\alpha(t) = e^{iH_B t} B_\alpha e^{-iH_B t} \qquad (3.130)$$

are interaction picture operators of the environment. We also note that condition (3.114) becomes

$$\langle B_\alpha(t) \rangle \equiv \operatorname{tr}\{B_\alpha(t)\rho_B\} = 0, \qquad (3.131)$$

which states that the reservoir averages of the $B_\alpha(t)$ vanish.

Inserting now the form (3.129) into the master equation (3.118) we obtain after some algebra

$$\frac{d}{dt}\rho_S(t) = \int_0^\infty ds \operatorname{tr}_B \left\{ H_I(t-s)\rho_S(t)\rho_B H_I(t) - H_I(t)H_I(t-s)\rho_S(t)\rho_B \right\} + \text{h.c.}$$

$$= \sum_{\omega,\omega'} \sum_{\alpha,\beta} e^{i(\omega'-\omega)t} \Gamma_{\alpha\beta}(\omega) \left(A_\beta(\omega)\rho_S(t)A_\alpha^\dagger(\omega') - A_\alpha^\dagger(\omega')A_\beta(\omega)\rho_S(t) \right)$$

$$+ \text{h.c.} \qquad (3.132)$$

Here h.c. means the Hermitian conjugated expression and we have introduced the one-sided Fourier transforms

$$\Gamma_{\alpha\beta}(\omega) \equiv \int_0^\infty ds\, e^{i\omega s} \langle B_\alpha^\dagger(t)B_\beta(t-s) \rangle \qquad (3.133)$$

of the *reservoir correlation functions*

$$\langle B_\alpha^\dagger(t)B_\beta(t-s) \rangle \equiv \operatorname{tr}_B \left\{ B_\alpha^\dagger(t)B_\beta(t-s)\rho_B \right\}. \qquad (3.134)$$

Let us suppose that ρ_B is a stationary state of the reservoir, that is $[H_B, \rho_B] = 0$. The reservoir correlation functions are then homogeneous in time which yields

$$\langle B_\alpha^\dagger(t)B_\beta(t-s) \rangle = \langle B_\alpha^\dagger(s)B_\beta(0) \rangle, \qquad (3.135)$$

showing that the quantities $\Gamma_{\alpha\beta}(\omega)$ do not depend on time. We remark that there are interesting cases in which the reservoir correlation functions do depend on the time argument t. This happens, for example, if the reservoir represents a squeezed vacuum state (see Section 3.4.3).

As mentioned before, the basic condition underlying the Markov approximation is that the reservoir correlation functions (3.135) decay sufficiently fast over
a time τ_B which is small compared to the relaxation time τ_R. Typical examples for the behaviour of these correlation functions will be discussed in Sections
3.6.2.1 and 12.1.1.3. It is important to note that a decay of the correlations can
only be strictly valid for an environment which is infinitely large and involves a
continuum of frequencies. In the typical situation the reservoir is provided by a
collection of harmonic oscillator modes b_n with frequencies ω_n and the B_α are
given by linear combinations of the modes b_n. If the frequency spectrum $\{\omega_n\}$
of the reservoir modes is discrete, it is easy to see that, in general, correlation
functions of the type (3.135) are quasi-periodic functions of s. A rapid decay of
the reservoir correlations therefore requires a continuum of frequencies: For an
infinitely small frequency spacing Poincaré recurrence times become infinite and
irreversible dynamics can emerge.

We denote by τ_S the typical time scale of the intrinsic evolution of the system
S. This time scale τ_S is defined by a typical value for $|\omega' - \omega|^{-1}$, $\omega' \neq \omega$, that
is by a typical value for the inverse of the frequency differences involved. If τ_S
is large compared to the relaxation time τ_R of the open system the non-secular
terms in (3.132), i.e. the terms for which $\omega' \neq \omega$, may be neglected, since they
oscillate very rapidly during the time τ_R over which ρ_S varies appreciably. This
condition is typically satisfied for quantum optical systems where it is known as
the rotating wave approximation. Thus we have

$$\frac{d}{dt}\rho_S(t) = \sum_\omega \sum_{\alpha,\beta} \Gamma_{\alpha\beta}(\omega)\left(A_\beta(\omega)\rho_S(t)A_\alpha^\dagger(\omega) - A_\alpha^\dagger(\omega)A_\beta(\omega)\rho_S(t)\right) + \text{h.c.}$$

(3.136)

It is convenient to decompose the Fourier transforms of the reservoir correlation
functions as follows

$$\Gamma_{\alpha\beta}(\omega) = \frac{1}{2}\gamma_{\alpha\beta}(\omega) + iS_{\alpha\beta}(\omega),$$ (3.137)

where for fixed ω the coefficients

$$S_{\alpha\beta}(\omega) = \frac{1}{2i}\left(\Gamma_{\alpha\beta}(\omega) - \Gamma_{\beta\alpha}^*(\omega)\right)$$ (3.138)

form a Hermitian matrix and the matrix defined by

$$\gamma_{\alpha\beta}(\omega) = \Gamma_{\alpha\beta}(\omega) + \Gamma_{\beta\alpha}^*(\omega) = \int\limits_{-\infty}^{+\infty} ds\, e^{i\omega s}\langle B_\alpha^\dagger(s)B_\beta(0)\rangle$$ (3.139)

is positive (see below). With these definitions we finally arrive at the interaction
picture master equation

$$\frac{d}{dt}\rho_S(t) = -i\left[H_{LS}, \rho_S(t)\right] + \mathcal{D}(\rho_S(t)).$$ (3.140)

The Hermitian operator

$$H_{LS} = \sum_{\omega} \sum_{\alpha,\beta} S_{\alpha\beta}(\omega) A_{\alpha}^{\dagger}(\omega) A_{\beta}(\omega) \tag{3.141}$$

provides a Hamiltonian contribution to the dynamics. This term is often called the *Lamb shift* Hamiltonian since it leads to a Lamb-type renormalization of the unperturbed energy levels induced by the system–reservoir coupling. Note that the Lamb shift Hamiltonian commutes with the unperturbed system Hamiltonian,

$$[H_S, H_{LS}] = 0, \tag{3.142}$$

by virtue of eqn (3.125). Finally, the dissipator of the master equation takes the form

$$\mathcal{D}(\rho_S) = \sum_{\omega} \sum_{\alpha,\beta} \gamma_{\alpha\beta}(\omega) \left(A_{\beta}(\omega)\rho_S A_{\alpha}^{\dagger}(\omega) - \frac{1}{2} \left\{ A_{\alpha}^{\dagger}(\omega) A_{\beta}(\omega), \rho_S \right\} \right). \tag{3.143}$$

We note that the master equation (3.140) is of the first standard form (3.63). It can be brought into Lindblad form (3.66) by diagonalization of the matrices $\gamma_{\alpha\beta}(\omega)$ defined in eqn (3.139). In order to prove that these matrices are positive one uses Bochner's theorem according to which the Fourier transform of a function $f(s)$ is positive provided $f(s)$ has the property of being of positive type. The latter property means that for arbitrary $t_1, t_2, \ldots, t_n$ and all n the $(n \times n)$ matrix $a_{kl} = f(t_k - t_l)$ must be positive. Since all homogeneous correlation functions $f(s) = \langle B^{\dagger}(s)B(0) \rangle$ are of positive type the positivity of the matrices $\gamma_{\alpha\beta}(\omega)$ follows immediately. Finally, we remark that the Schrödinger picture master equation is obtained from (3.140) simply by adding the free system Hamiltonian H_S to H_{LS}, as is easily verified with the help of the properties (3.121), (3.122) and (3.125) of the eigenoperators.

Let us summarize the different approximations used in the above derivation. The first approximation is a consequence of the weak-coupling assumption which allows us to expand the exact equation of motion for the density matrix to second order. Together with the condition $\rho(t) \approx \rho_S(t) \otimes \rho_B$ this leads to the Born approximation to the master equation. The second approximation is the Markov approximation in which the quantum master equation is made local in time by replacing the density matrix $\rho_S(s)$ at the retarded time s with that at the present time $\rho_S(t)$. Furthermore, the integration limit is pushed to infinity to get the Born–Markov approximation of the master equation. The relevant physical condition for the Born–Markov approximation is that the bath correlation time τ_B is small compared to the relaxation time of the system, that is $\tau_B \ll \tau_R$. Finally, in the rotating wave approximation rapidly oscillating terms proportional to $\exp[i(\omega' - \omega)t]$ for $\omega' \neq \omega$ are neglected, ensuring that the quantum master equation is in Lindblad form. The corresponding condition is that the inverse frequency differences involved in the problem are small compared to the relaxation time of the system, that is $\tau_S \sim |\omega' - \omega|^{-1} \ll \tau_R$.

3.3.2 *Relaxation to equilibrium*

In the previous section we have assumed that the environment is in a stationary state ρ_B which is invariant with respect to the dynamics of the reservoir. Now we want to consider a situation in which the environment is a heat bath at the inverse temperature β. In the absence of external time-dependent fields one expects the Gibbs state

$$\rho_{\text{th}} = \frac{\exp(-\beta H_S)}{\text{tr}_S \exp(-\beta H_S)} \tag{3.144}$$

to be a stationary solution of the quantum master equation (3.140). It can be shown then that for any initial state the system returns to equilibrium,

$$\rho_S(t) \longrightarrow \rho_{\text{th}}, \quad \text{for} \quad t \longrightarrow +\infty, \tag{3.145}$$

provided the quantum dynamical semigroup has the property of being ergodic. This means that the relations

$$[X, A_\alpha^\dagger(\omega)] = [X, A_\alpha(\omega)] = 0 \quad \text{for all} \quad \alpha, \omega \tag{3.146}$$

imply that X is proportional to the identity.

In order to show that (3.144) is indeed a stationary solution of the master equation (3.140) we make use of the KMS condition according to which the bath correlation functions are related through

$$\langle B_\alpha^\dagger(t) B_\beta(0) \rangle = \langle B_\beta(0) B_\alpha^\dagger(t + i\beta) \rangle. \tag{3.147}$$

The KMS condition can easily be verified if the reservoir is a heat bath with canonical equilibrium distribution

$$\rho_B = \frac{\exp(-\beta H_B)}{\text{tr}_B \exp(-\beta H_B)}. \tag{3.148}$$

It can also be shown to hold for thermal equilibrium systems in the thermodynamic limit. Equation (3.147) leads to the following relations between the Fourier transforms (3.139) of the bath correlation functions,

$$\gamma_{\alpha\beta}(-\omega) = \exp(-\beta\omega)\gamma_{\beta\alpha}(\omega). \tag{3.149}$$

We further have by virtue of eqns (3.121) and (3.122),

$$\rho_{\text{th}} A_\alpha(\omega) = e^{\beta\omega} A_\alpha(\omega)\rho_{\text{th}}, \tag{3.150}$$

$$\rho_{\text{th}} A_\alpha^\dagger(\omega) = e^{-\beta\omega} A_\alpha^\dagger(\omega)\rho_{\text{th}}. \tag{3.151}$$

The proof of the stationarity of ρ_{th} is now easily carried out with the help of the relations (3.142), (3.149), (3.150) and (3.151).

We mention a further important property of the quantum master equation (3.140). Namely, if the spectrum of the system Hamiltonian $H_S = \sum_n \varepsilon_n |n\rangle\langle n|$ is non-degenerate it gives rise to a closed equation of motion for the populations

$$P(n,t) = \langle n|\rho_S(t)|n\rangle \tag{3.152}$$

of the eigenstates $|n\rangle$. Thus, the equation for the diagonals of the density matrix in the eigenbasis of H_S decouples from the off-diagonal elements. As is easily checked using the quantum master equation the populations are governed by the equation

$$\frac{d}{dt}P(n,t) = \sum_m [W(n|m)P(m,t) - W(m|n)P(n,t)]. \tag{3.153}$$

This equation is of the form of the classical discrete master equation (1.89) with time-independent transition rates given by

$$W(n|m) = \sum_{\alpha,\beta} \gamma_{\alpha\beta}(\varepsilon_m - \varepsilon_n)\langle m|A_\alpha|n\rangle\langle n|A_\beta|m\rangle. \tag{3.154}$$

Equation (3.153) is also known as the Pauli master equation. The rates (3.154) are real and non-negative as a consequence of the positivity of the matrices $\gamma_{\alpha\beta}(\omega)$. They are just those obtained with the help of Fermi's golden rule.

The relations (3.149) give

$$W(m|n)\exp(-\beta\varepsilon_n) = W(n|m)\exp(-\beta\varepsilon_m) \tag{3.155}$$

which is known as the condition of detailed balance and which leads to the conclusion that the equilibrium populations $P_s(n)$ follow the Boltzmann distribution

$$P_s(n) = \text{const} \times \exp(-\beta\varepsilon_n) \tag{3.156}$$

over the energy eigenvalues ε_n.

3.3.3 *Singular-coupling limit*

In the weak-coupling limit the perturbation caused by the interaction between the system and the environment is assumed to be small. As a result the degrees of freedom of the environment are the fast variable and can be effectively eliminated. With an appropriate scaling of the time parameter it is possible to derive under certain conditions a linear quantum master equation also for the case of strong coupling. In this so-called singular-coupling limit one considers a total Hamiltonian of the form

$$H = H_S + \varepsilon^{-2}H_B + \varepsilon^{-1}H_I, \tag{3.157}$$

where the interaction Hamiltonian is again written as

$$H_I = \sum_\alpha A_\alpha \otimes B_\alpha \tag{3.158}$$

with $A_\alpha^\dagger = A_\alpha$ and $B_\alpha^\dagger = B_\alpha$. The aim is to derive an equation of motion for the reduced density matrix in the limit $\epsilon \to 0$. To motivate the form of the

Hamiltonian (3.157) we first observe that the decay time of reservoir correlation functions of the form (3.134) is decreased by a factor of ε^2 through the scaling $H_B \rightarrow \varepsilon^{-2} H_B$. The scaling $H_I \rightarrow \varepsilon^{-1} H_I$ of the interaction Hamiltonian ensures that the Fourier transforms of the reservoir correlation functions remain finite in the limit $\varepsilon \rightarrow 0$.

The derivation of the quantum master equation from this microscopic model is similar to the weak-coupling case and will not be presented here in detail. The only essential difference is that it is not necessary to perform the rotating wave approximation. The result of the derivation is the following Schrödinger picture master equation,

$$\frac{d}{dt}\rho_S(t) = -i[H_S + H_{LS}, \rho_S(t)] + \sum_{\alpha\beta} \gamma_{\alpha\beta} \left(A_\beta \rho_S(t) A_\alpha - \frac{1}{2}\{A_\alpha A_\beta, \rho_S(t)\} \right),$$

$$(3.159)$$

where the Lamb shift Hamiltonian reads

$$H_{LS} = \sum_{\alpha\beta} S_{\alpha\beta} A_\alpha A_\beta, \qquad (3.160)$$

with $S_{\alpha\beta}$ defined analogously to (3.138), and

$$\gamma_{\alpha\beta} = \int\limits_{-\infty}^{+\infty} ds \langle B_\alpha(s) B_\beta(0)\rangle. \qquad (3.161)$$

The matrix $\gamma_{\alpha\beta}$ is again Hermitian and positive. Hence, the generator of the quantum master equation (3.159) is of the first standard form (3.63) and may be written in Lindblad form by diagonalization of the matrix $\gamma_{\alpha\beta}$. Master equations of the general form (3.159) will be encountered, for example, in our study of the quantum Zeno effect in Section 3.5.2 and in the recoilless limit of quantum Brownian motion (see Section 3.6.4.5).

3.3.4 Low-density limit

We consider a gas of particles with low density n. The particles of the gas interact through collisions involving the excitation of internal degrees of freedom of the gas particles. The collisions of the gas particles are considered as statistically independent processes. The aim is to write down a master equation which effectively describes the dynamics of the internal degrees of freedom. To this end, the internal variables are taken to be the reduced system S, while the translational degrees of freedom of the gas particles provide the environment B.

The internal degrees of freedom of the particles may be described by a free Hamiltonian of the form

$$H_S = \sum_k \epsilon_k |k\rangle\langle k|. \qquad (3.162)$$

The translational degrees of freedom of a gas particle are described by the Hamiltonian

$$H_B = \int d^3p\, E(\vec{p}) |\vec{p}\rangle\langle\vec{p}|, \tag{3.163}$$

where $|\vec{p}\rangle$ is a momentum eigenstate and $E(\vec{p}) = \vec{p}^{\,2}/2m$ is the free particle energy. The one-particle states of the environment are given by the density matrix

$$\rho_B = \frac{1}{V} \int d^3p\, G(\vec{p}) |\vec{p}\rangle\langle\vec{p}|, \tag{3.164}$$

where $G(\vec{p})$ represents the momentum distribution of the gas particles. The density matrix ρ_B is normalized in some volume V according to

$$\mathrm{tr}\rho_B = \int d^3p\langle\vec{p}|\rho_B|\vec{p}\rangle = \int \frac{d^3p}{(2\pi)^3}\, G(\vec{p}) = 1. \tag{3.165}$$

The interaction of the internal degrees of freedom with the translational degrees of freedom is given by scattering reactions in which the incoming momentum changes from $\vec{p}$ to $\vec{q}$, while the internal level changes from $|l\rangle$ to $|k\rangle$. The scattering amplitude for this process can be written in the form

$$\langle k,\vec{q}|S|l,\vec{p}\rangle = \delta(\vec{q}-\vec{p})\delta_{kl} - 2\pi i\delta\left(\epsilon_k + E(\vec{q}) - \epsilon_l - E(\vec{p})\right)T(k,\vec{q}|l,\vec{p}), \tag{3.166}$$

with the S-matrix and the T-matrix known from scattering theory. In the low-density limit the reduced dynamics of the quantum systems formed by the internal degrees of freedom can be shown to yield the Schrödinger picture quantum master equation

$$\frac{d}{dt}\rho_S(t) = -i[H_S + H_{LS}, \rho_S(t)] + \mathcal{D}(\rho_S(t)). \tag{3.167}$$

The Hamiltonian H_{LS} yields a renormalization of the system Hamiltonian,

$$H_{LS} = \frac{n}{2} \sum_{\epsilon_k=\epsilon_l} \int d^3p\, G(\vec{p}) \left\{ T(k,\vec{p}|l,\vec{p}) + T^*(l,\vec{p}|k,\vec{p}) \right\} |k\rangle\langle l|, \tag{3.168}$$

and the dissipator is given by

$$\mathcal{D}(\rho_S) = 2\pi n \sum_\omega \int d^3p \int d^3q\, G(\vec{p})\delta\left(E(\vec{q}) - E(\vec{p}) + \omega\right) \tag{3.169}$$

$$\times \left\{ T_\omega(\vec{q},\vec{p})\rho_S T_\omega^\dagger(\vec{q},\vec{p}) - \frac{1}{2}\rho_S T_\omega^\dagger(\vec{q},\vec{p})T_\omega(\vec{q},\vec{p}) - \frac{1}{2}T_\omega^\dagger(\vec{q},\vec{p})T_\omega(\vec{q},\vec{p})\rho_S \right\},$$

where $T_\omega(\vec{q},\vec{p})$ is defined by

$$T_\omega(\vec{q},\vec{p}) = \sum_{\epsilon_k-\epsilon_l=\omega} T(k,\vec{q}|l,\vec{p})|k\rangle\langle l|. \tag{3.170}$$

Thus the Lindblad operator $T_\omega(\vec{q},\vec{p})$ is given by a coherent superposition of the scattering amplitudes for all processes which contribute to a given excitation energy ω of the internal degrees of freedom,

$$\omega = \epsilon_k - \epsilon_l = E(\vec{p}) - E(\vec{q}). \tag{3.171}$$

This is the low-density limit of the quantum master equation which can be shown to hold in the thermodynamic limit. A rigorous mathematical treatment may be found in Dümcke (1985). Equation (3.167) is obviously in Lindblad form, where we have, however, a continuous family of Lindblad operators. To derive the master equation an averaging procedure over times which are large compared to the mean collision time τ_C is performed, which is similar to the rotating wave approximation. Let d be the linear dimension of the interaction volume, v the mean particle velocity, and l the mean free path length. Then the low-density limit master equation is valid as long as

$$\tau_C \sim v^{-1}d \ll v^{-1}l \sim v^{-1}n^{-1/3}, \tag{3.172}$$

which means that the mean collision time τ_C must be small compared to the time between collisions. Hence, we get the low-density condition $n \ll d^{-3}$.

3.4 The quantum optical master equation

The interaction of matter with electromagnetic radiation in the quantum optical limit provides a typical field for the application of quantum dynamical semi-groups and quantum Markovian master equations (Cohen-Tannoudji, Dupont-Roc and Grynberg, 1998; Gardiner and Zoller, 2000; Walls and Milburn, 1994). The reason for this fact is that in a typical quantum optical situation the physical conditions underlying the Markovian approximation are very well satisfied. A number of examples will be discussed in this section.

3.4.1 *Matter in quantized radiation fields*

We consider a bound quantum system, e.g. an atom or a molecule, which interacts with a quantized radiation field. The radiation field represents a reservoir with an infinite number of degrees of freedom and the bound system is the reduced system we are interested in. The uncoupled atom or molecule is described by some Hamiltonian H_S, while the free quantized radiation field will be represented by the Hamiltonian (subtracting an infinite c-number for the vacuum energy)

$$H_B = \sum_{\vec{k}} \sum_{\lambda=1,2} \hbar\omega_k b_\lambda^\dagger(\vec{k}) b_\lambda(\vec{k}). \tag{3.173}$$

In this section we reintroduce all physical constants such as Planck's constant $\hbar$ and the speed of light c. For simplicity we decompose the radiation field into Fourier modes in a box of volume V, imposing periodic boundary conditions. These modes are labelled by the wave vector $\vec{k}$ and two corresponding, transverse unit polarization vectors $\vec{e}_\lambda(\vec{k})$, such that

$$\vec{k} \cdot \vec{e}_\lambda(\vec{k}) = 0, \tag{3.174}$$

$$\vec{e}_\lambda(\vec{k}) \cdot \vec{e}_{\lambda'}(\vec{k}) = \delta_{\lambda\lambda'}, \tag{3.175}$$

$$\sum_{\lambda=1,2} e_\lambda^i(\vec{k})e_\lambda^j(\vec{k}) = \delta_{ij} - \frac{k_i k_j}{|\vec{k}|^2}, \quad i,j = 1,2,3. \tag{3.176}$$

The dispersion relation is $\omega_k = ck = c|\vec{k}|$. The field operators $b_\lambda(\vec{k})$ and $b_\lambda^\dagger(\vec{k})$ describe the destruction and creation of photons with wave vector $\vec{k}$ and polarization $\vec{e}_\lambda(\vec{k})$. They obey the commutation relations

$$\left[b_\lambda(\vec{k}), b_{\lambda'}(\vec{k}') \right] = \left[b_\lambda^\dagger(\vec{k}), b_{\lambda'}^\dagger(\vec{k}') \right] = 0, \tag{3.177}$$

$$\left[b_\lambda(\vec{k}), b_{\lambda'}^\dagger(\vec{k}') \right] = \delta_{\vec{k}\vec{k}'}\delta_{\lambda\lambda'}. \tag{3.178}$$

Finally, we assume the interaction Hamiltonian to be given in the dipole approximation by

$$H_I = -\vec{D} \cdot \vec{E}, \tag{3.179}$$

where $\vec{D}$ is the dipole operator of the system under consideration and $\vec{E}$ is the electric field operator in the Schrödinger picture,

$$\vec{E} = i \sum_{\vec{k}} \sum_{\lambda=1,2} \sqrt{\frac{2\pi\hbar\omega_k}{V}} \vec{e}_\lambda(\vec{k}) \left(b_\lambda(\vec{k}) - b_\lambda^\dagger(\vec{k}) \right). \tag{3.180}$$

With these definitions the total Hamiltonian governing the coupled system of the matter degrees of freedom and the radiation degrees of freedom is given by

$$H = H_S + H_B + H_I. \tag{3.181}$$

3.4.1.1 *Performing the Born–Markov approximation* We proceed as in Section 3.3.1 and decompose the dipole operator $\vec{D}$ into eigenoperators of H_S. The latter take the form (see eqn (3.120))

$$\vec{A}(\omega) \equiv \sum_{\varepsilon'-\varepsilon=\hbar\omega} \Pi(\varepsilon)\vec{D}\Pi(\varepsilon'). \tag{3.182}$$

Note that the index α used in Section 3.3.1 labels here the Cartesian components D_i, $i = 1, 2, 3$, of the dipole operator. In accordance with eqns (3.121), (3.122) and (3.126) we now have

$$[H_S, \vec{A}(\omega)] = -\hbar\omega\vec{A}(\omega), \quad [H_S, \vec{A}^\dagger(\omega)] = +\hbar\omega\vec{A}^\dagger(\omega), \tag{3.183}$$

and

$$\vec{A}^\dagger(\omega) = \vec{A}(-\omega). \tag{3.184}$$

The decomposition of the interaction picture dipole operator $\vec{D}(t)$ into eigenoperators therefore reads

$$\vec{D}(t) = \sum_\omega e^{-i\omega t}\vec{A}(\omega) = \sum_\omega e^{+i\omega t}\vec{A}^\dagger(\omega), \tag{3.185}$$

and the interaction Hamiltonian can now be written in the interaction picture in a form analogous to eqn (3.129),

$$H_I(t) = -\sum_\omega e^{-i\omega t}\vec{A}(\omega) \cdot \vec{E}(t), \tag{3.186}$$

where $\vec{E}(t)$ denotes the electric field operator in the interaction picture.

Assuming as in eqn (3.131) that

$$\langle \vec{E}(t) \rangle \equiv \mathrm{tr} \left\{ \vec{E}(t) \rho_B \right\} = 0, \qquad (3.187)$$

we can immediately write down the equation of motion analogous to eqn (3.132),

$$\frac{d}{dt} \rho_S = \sum_{\omega, \omega'} \sum_{i,j} e^{i(\omega' - \omega)t} \Gamma_{ij}(\omega) \left(A_j(\omega) \rho_S A_i^\dagger(\omega') - A_i^\dagger(\omega') A_j(\omega) \rho_S \right) + \mathrm{h.c.}$$

$$(3.188)$$

The correlation functions of the electric field operator are defined through

$$\langle E_i(t) E_j(t - s) \rangle \equiv \mathrm{tr}_B \left\{ E_i(t) E_j(t - s) \rho_B \right\}, \qquad (3.189)$$

and their one-sided Fourier transforms are given by

$$\Gamma_{ij}(\omega) \equiv \frac{1}{\hbar^2} \int_0^\infty ds\, e^{i\omega s} \langle E_i(t) E_j(t - s) \rangle. \qquad (3.190)$$

The matrix $\Gamma_{ij}(\omega)$ will be referred to as the spectral correlation tensor. It depends, in general, on t. In fact, without further assumption on the state ρ_B of the reservoir we find

$$\Gamma_{ij}(\omega) = \frac{1}{\hbar^2} \sum_{\vec{k}, \vec{k}'} \sum_{\lambda, \lambda'} \sqrt{\frac{2\pi \hbar \omega_k}{V}} \sqrt{\frac{2\pi \hbar \omega_{k'}}{V}} e_\lambda^i(\vec{k}) e_{\lambda'}^j(\vec{k}') \qquad (3.191)$$

$$\times \int_0^\infty ds \left[\langle b_\lambda(\vec{k}) b_{\lambda'}^\dagger(\vec{k}') \rangle \exp\left[+i(\omega_{k'} - \omega_k)t - i(\omega_{k'} - \omega)s \right] \right.$$

$$+ \langle b_\lambda^\dagger(\vec{k}) b_{\lambda'}(\vec{k}') \rangle \exp\left[-i(\omega_{k'} - \omega_k)t + i(\omega_{k'} + \omega)s \right]$$

$$- \langle b_\lambda(\vec{k}) b_{\lambda'}(\vec{k}') \rangle \exp\left[-i(\omega_{k'} + \omega_k)t + i(\omega_{k'} + \omega)s \right]$$

$$\left. - \langle b_\lambda^\dagger(\vec{k}) b_{\lambda'}^\dagger(\vec{k}') \rangle \exp\left[+i(\omega_{k'} + \omega_k)t - i(\omega_{k'} - \omega)s \right] \right].$$

3.4.1.2 *Thermal reservoir* As our first example let us now take the reservoir of radiation modes to be in an equilibrium state at temperature T,

$$\rho_B = \frac{1}{Z_B} \exp\left[-\beta H_B \right] = \prod_{\vec{k}, \lambda} \left(1 - \exp\left[-\beta \hbar \omega_k \right] \right) \exp\left[-\beta \hbar \omega_k b_\lambda^\dagger(\vec{k}) b_\lambda(\vec{k}) \right].$$

$$(3.192)$$

As above, expectation values with respect to this equilibrium state are denoted by angular brackets. Note that $\langle \vec{E}(t) \rangle = 0$ since the field operator is linear in the creation and annihilation operators. Since the thermal equilibrium state is stationary with respect to the reservoir dynamics, the reservoir correlation functions are homogeneous in time and we have

$$\langle E_i(t)E_j(t-s)\rangle = \langle E_i(s)E_j(0)\rangle. \tag{3.193}$$

It follows that the spectral correlation tensor $\Gamma_{ij}(\omega)$ does not depend on t. Consequently, if we perform the rotating wave approximation in eqn (3.188) only the diagonal terms $\omega = \omega'$ of the double sum over the system frequencies survive and we are left with

$$\frac{d}{dt}\rho_S = \sum_\omega \sum_{i,j} \Gamma_{ij}(\omega)\left\{ A_j(\omega)\rho_S A_i^\dagger(\omega) - A_i^\dagger(\omega)A_j(\omega)\rho_S \right\} + \text{h.c.} \tag{3.194}$$

It remains to determine the spectral correlation tensor. To this end we use the relations

$$\langle b_\lambda(\vec{k})b_{\lambda'}(\vec{k}')\rangle = \langle b_\lambda^\dagger(\vec{k})b_{\lambda'}^\dagger(\vec{k}')\rangle = 0, \tag{3.195}$$

$$\langle b_\lambda(\vec{k})b_{\lambda'}^\dagger(\vec{k}')\rangle = \delta_{\vec{k}\vec{k}'}\delta_{\lambda\lambda'}(1 + N(\omega_k)), \tag{3.196}$$

$$\langle b_\lambda^\dagger(\vec{k})b_{\lambda'}(\vec{k}')\rangle = \delta_{\vec{k}\vec{k}'}\delta_{\lambda\lambda'}N(\omega_k), \tag{3.197}$$

where

$$N(\omega_k) = \frac{1}{\exp[\beta\hbar\omega_k] - 1} \tag{3.198}$$

denotes the Planck distribution, that is the average number of photons in a mode with frequency ω_k. We further perform the continuum limit,

$$\frac{1}{V}\sum_{\vec{k}} \longrightarrow \int \frac{d^3k}{(2\pi)^3} = \frac{1}{(2\pi)^3c^3}\int_0^\infty d\omega_k\omega_k^2 \int d\Omega. \tag{3.199}$$

The integration over the solid angle $d\Omega$ of the wave vector $\vec{k}$ is carried out with the help of

$$\int d\Omega \left(\delta_{ij} - \frac{k_ik_j}{k^2}\right) = \frac{8\pi}{3}\delta_{ij}. \tag{3.200}$$

Equation (3.191) then yields

$$\Gamma_{ij}(\omega) = \frac{2}{3\pi\hbar c^3}\delta_{ij}\int_0^\infty d\omega_k\omega_k^3 \left[(1 + N(\omega_k))\int_0^\infty ds\exp[-i(\omega_k - \omega)s] \right.$$

$$\left. + N(\omega_k)\int_0^\infty ds\exp[+i(\omega_k + \omega)s] \right]. \tag{3.201}$$

On making use of the formula

$$\int_0^\infty ds\, e^{-i\varepsilon s} = \pi\delta(\varepsilon) - i\mathrm{P}\frac{1}{\varepsilon}, \tag{3.202}$$

where P denotes the Cauchy principal value, we finally arrive at

$$\Gamma_{ij}(\omega) = \delta_{ij}\left(\frac{1}{2}\gamma(\omega) + iS(\omega)\right), \tag{3.203}$$

where we have introduced the quantities

$$\gamma(\omega) = \frac{4\omega^3}{3\hbar c^3}\left(1 + N(\omega)\right), \tag{3.204}$$

and

$$S(\omega) = \frac{2}{3\pi\hbar c^3}\mathrm{P}\int_0^{+\infty} d\omega_k\, \omega_k^3\left[\frac{1 + N(\omega_k)}{\omega - \omega_k} + \frac{N(\omega_k)}{\omega + \omega_k}\right]. \tag{3.205}$$

Note that the Planck distribution satisfies $N(-\omega) = -(1 + N(\omega))$ such that $\gamma(\omega) = 4\omega^3(1 + N(\omega))/3\hbar c^3$ for $\omega > 0$ and $\gamma(\omega) = 4|\omega|^3 N(|\omega|)/3\hbar c^3$ for $\omega < 0$.

Summing up our results we obtain a Markovian quantum master equation for the matter degrees of freedom, often called the *quantum optical master equation*, which can be written in Lindblad form,

$$\frac{d}{dt}\rho_S(t) = -\frac{i}{\hbar}[H_{LS}, \rho_S(t)] + \mathcal{D}(\rho_S(t)). \tag{3.206}$$

The Hamiltonian

$$H_{LS} = \sum_\omega \hbar S(\omega)\vec{A}^\dagger(\omega)\cdot\vec{A}(\omega) \tag{3.207}$$

leads to a renormalization of the system Hamiltonian H_S which is induced by the vacuum fluctuations of the radiation field (Lamb shift) and by thermally induced processes (Stark shift). A more detailed treatment of the Lamb shift is given in Section 12.2.3.2. The dissipator of the quantum master equation takes the form

$$\mathcal{D}(\rho_S) = \sum_{\omega>0}\frac{4\omega^3}{3\hbar c^3}(1 + N(\omega))\left(\vec{A}(\omega)\rho_S\vec{A}^\dagger(\omega) - \frac{1}{2}\left\{\vec{A}^\dagger(\omega)\vec{A}(\omega), \rho_S\right\}\right)$$

$$+ \sum_{\omega>0}\frac{4\omega^3}{3\hbar c^3}N(\omega)\left(\vec{A}^\dagger(\omega)\rho_S\vec{A}(\omega) - \frac{1}{2}\left\{\vec{A}(\omega)\vec{A}^\dagger(\omega), \rho_S\right\}\right). \tag{3.208}$$

To obtain this form we have transformed the sum over the negative frequencies into a sum over the positive frequencies and have made use of relation (3.184). We observe that the dissipator of the master equation describes spontaneous and thermally induced processes. By virtue of eqn (3.121) the Lindblad operators $A_i(\omega)$ lower the atomic energy by the amount $\hbar\omega$: If ψ is an eigenstate of H_S with energy ε, then $A_i(\omega)\psi$ is again an eigenstate of H_S belonging to the eigenvalue $\varepsilon - \hbar\omega$. Correspondingly, the operators $A_i^\dagger(\omega)$ raise the atomic energy by $\hbar\omega$. Thus, the Lindblad operators $A_i(\omega)$ describe spontaneous and thermally induced emission processes which occur with the rate $4\omega^3(1 + N(\omega))/3\hbar c^3$, while the $A_i^\dagger(\omega)$ describe thermally induced absorption processes taking place with the rate $4\omega^3 N(\omega)/3\hbar c^3$.

Let us finally comment on the validity of the quantum optical master equation. The relaxation time τ_R is given by the inverse of a typical relaxation rate γ_0. The latter is defined by a typical value for the transition rates $4\omega^3|\vec{d}|^2/3\hbar c^3$ of electric dipole transitions, where $\vec{d}$ is the corresponding matrix element of the dipole operator. On the other hand, the vacuum correlation time τ_B of the reservoir of radiation degrees of freedom is given by the inverse of a typical transition frequency ω_0. This will be seen explicitly in our discussion of a systematic perturbation expansion around the Markovian limit in Chapter 10. It can also be concluded from the well-known derivation of the golden rule in quantum mechanics. The condition $\tau_B \ll \tau_R$ for the Born–Markov approximation thus yields $\gamma_0 \ll \omega_0$, which is obviously a weak-coupling condition. This condition is usually very well satisfied in the quantum optical regime. For example, typical radiative inverse atomic lifetimes are of the order 10^7 s^{-1} to 10^9 s^{-1} or even much smaller, whereas optical frequencies are of the order 10^{15} s^{-1}. Furthermore, it must be noted that we need the condition $\tau_S \ll \tau_R$ if the master equation involves a number of different transition frequencies (see Section 3.3.1). This condition enables one to perform the rotating wave approximation and ensures that transitions belonging to different frequencies may be described as separate decay channels involving different Lindblad operators.

3.4.2 Decay of a two-level system

The simplest quantum system is a two-level system whose Hilbert space is spanned by just two states, an excited state $|e\rangle$ and a ground state $|g\rangle$. The Hilbert space of such a system is equivalent to that of a spin-$\frac{1}{2}$ system. In fact, the Pauli operators

$$\sigma_1 = |e\rangle\langle g| + |g\rangle\langle e|, \quad \sigma_2 = -i|e\rangle\langle g| + i|g\rangle\langle e|, \quad \sigma_3 = |e\rangle\langle e| - |g\rangle\langle g| \quad (3.209)$$

satisfy the commutation relations

$$[\sigma_i, \sigma_j] = 2i\varepsilon_{ijk}\sigma_k, \quad (3.210)$$

as well as the anticommutation relations

$$\{\sigma_i, \sigma_j\} = 2\delta_{ij}. \quad (3.211)$$

It will be convenient to define also the operators

$$\sigma_+ = |e\rangle\langle g| = \frac{1}{2}(\sigma_1 + i\sigma_2), \quad \sigma_- = |g\rangle\langle e| = \frac{1}{2}(\sigma_1 - i\sigma_2). \quad (3.212)$$

The corresponding matrix representations of these operators in the basis $|e\rangle$, $|g\rangle$ take the form

$$\sigma_1 = \begin{pmatrix} 0 & 1 \\ 1 & 0 \end{pmatrix}, \quad \sigma_2 = \begin{pmatrix} 0 & -i \\ i & 0 \end{pmatrix}, \quad \sigma_3 = \begin{pmatrix} 1 & 0 \\ 0 & -1 \end{pmatrix}, \tag{3.213}$$

and

$$\sigma_+ = \begin{pmatrix} 0 & 1 \\ 0 & 0 \end{pmatrix}, \quad \sigma_- = \begin{pmatrix} 0 & 0 \\ 1 & 0 \end{pmatrix}. \tag{3.214}$$

We take the free Hamiltonian H_S of the system to be diagonal in the basis $|e\rangle$, $|g\rangle$. With an appropriate choice for the ground state energy we then have

$$H_S = \frac{1}{2}\omega_0 \sigma_3, \tag{3.215}$$

where $\omega_0 > 0$ is the transition frequency and we set again $\hbar = 1$. In the following we work in the interaction picture.

Physically, such a two-level system arises whenever the dynamics of the system under study is effectively confined to a two-dimensional subspace, that is under the condition that transitions to other levels may be neglected. To be specific we regard here the model to describe the dynamics of a two-level atom with a transition frequency ω_0 in the optical range. We note that the Pauli operators $\sigma_\pm$ represent eigenoperators of the atomic Hamiltonian,

$$[H_S, \sigma_-] = -\omega_0 \sigma_-, \quad [H_S, \sigma_+] = +\omega_0 \sigma_+, \tag{3.216}$$

and, hence, $\sigma_\pm$ changes the atomic energy by the amount $\pm\omega_0$, corresponding to the absorption and emission processes, respectively. We thus have two Lindblad operators

$$\vec{A}(\omega_0) \equiv \vec{A} = \vec{d}\sigma_-, \quad \vec{A}(-\omega_0) \equiv \vec{A}^\dagger = \vec{d}^*\sigma_+, \tag{3.217}$$

where $\vec{d} = \langle g|\vec{D}|e\rangle$ is the transition matrix element of the dipole operator (it is assumed here that the diagonals of the dipole operator vanish). Within the two-level approximation the atomic dipole operator in the interaction picture can thus be written as

$$\vec{D}(t) = \vec{d}\sigma_- e^{-i\omega_0 t} + \vec{d}^*\sigma_+ e^{+i\omega_0 t}. \tag{3.218}$$

Neglecting the Lamb and Stark shift contribution we can now write the quantum optical master equation (3.206) in the form (writing ρ instead of ρ_S for simplicity)

$$\frac{d}{dt}\rho(t) = \gamma_0(N+1)\left(\sigma_-\rho(t)\sigma_+ - \frac{1}{2}\sigma_+\sigma_-\rho(t) - \frac{1}{2}\rho(t)\sigma_+\sigma_-\right)$$

$$+\gamma_0 N\left(\sigma_+\rho(t)\sigma_- - \frac{1}{2}\sigma_-\sigma_+\rho(t) - \frac{1}{2}\rho(t)\sigma_-\sigma_+\right), \tag{3.219}$$

with the spontaneous emission rate

$$\gamma_0 = \frac{4\omega_0^3|\vec{d}|^2}{3\hbar c^3}. \tag{3.220}$$

The dissipator of the master equation describes spontaneous emission processes (rate γ_0) as well as thermally induced emission and absorption processes (rates $\gamma_0 N$). The total transition rate will be denoted by

$$\gamma = \gamma_0(2N+1), \tag{3.221}$$

where $N = N(\omega_0)$ denotes the Planck distribution at the transition frequency.

To solve eqn (3.219) it is convenient to represent the density matrix as

$$\rho(t) = \frac{1}{2}\left(I + \langle\vec{\sigma}(t)\rangle \cdot \vec{\sigma}\right) = \begin{pmatrix} \frac{1}{2}\left(1 + \langle\sigma_3(t)\rangle\right) & \langle\sigma_-(t)\rangle \\ \langle\sigma_+(t)\rangle & \frac{1}{2}\left(1 - \langle\sigma_3(t)\rangle\right) \end{pmatrix}, \tag{3.222}$$

where the vector

$$\vec{v}(t) \equiv \langle\vec{\sigma}(t)\rangle = \mathrm{tr}\left\{\vec{\sigma}\rho(t)\right\} \tag{3.223}$$

is known as the Bloch vector. It represents a real 3-vector satisfying $|\vec{v}(t)| \leq 1$. This condition is equivalent to the requirement that $\rho(t)$ must be positive. For $|\vec{v}(t)| < 1$ the corresponding density matrix describes a true statistical mixture, while a Bloch vector satisfying $|\vec{v}(t)| = 1$ represents a pure state. Thus we see that the set of density matrices of the two-level system is isomorphic to the unit sphere, known as the Bloch sphere, the surface of which is equal to the set of pure states.

The matrix elements $\rho_{11}(t) = p_e(t)$ and $\rho_{22}(t) = p_g(t)$ are the populations of the excited and ground state levels, respectively. The off-diagonals $\rho_{12}(t) = \rho_{21}^*(t)$ are the coherences given by the expectation values of the atomic raising and lowering operators $\sigma_\pm$. With the help of the algebra of the Pauli matrices one easily derives the differential equations

$$\frac{d}{dt}\langle\sigma_1(t)\rangle = -\frac{\gamma}{2}\langle\sigma_1(t)\rangle, \tag{3.224}$$

$$\frac{d}{dt}\langle\sigma_2(t)\rangle = -\frac{\gamma}{2}\langle\sigma_2(t)\rangle, \tag{3.225}$$

$$\frac{d}{dt}\langle\sigma_3(t)\rangle = -\gamma\langle\sigma_3(t)\rangle - \gamma_0. \tag{3.226}$$

We observe that the 3-component of the Bloch vector decays exponentially with rate γ, while the coherences $\langle\sigma_\pm(t)\rangle$ decay with rate $\gamma/2$. The stationary solution is given by

$$\langle\sigma_1\rangle_s = \langle\sigma_2\rangle_s = 0, \quad \langle\sigma_3\rangle_s = -\frac{\gamma_0}{\gamma} = -\frac{1}{2N+1}, \tag{3.227}$$

and the stationary population of the upper level is found to be

$$p_e^s = \frac{1}{2}\left(1 + \langle\sigma_3\rangle_s\right) = \frac{N}{2N+1}. \tag{3.228}$$

Of course, the stationary solution of the master equation is equal to the thermal equilibrium state $\rho_s = \rho_{\text{th}}$ given in eqn (3.144).

If we choose, for example, the initial state $\rho(0) = |g\rangle\langle g|$ we find the corresponding time-dependent solution

$$\langle\sigma_3(t)\rangle = -e^{-\gamma t}\left(1 + \langle\sigma_3\rangle_s\right) + \langle\sigma_3\rangle_s, \tag{3.229}$$

$$p_e(t) = p_e^s\left(1 - e^{-\gamma t}\right), \tag{3.230}$$

showing explicitly the exponential approach to the thermal equilibrium values.

3.4.3 *Decay into a squeezed field vacuum*

In the preceding subsection we have considered a reservoir which is stationary with respect to the bath dynamics. In order to give an example for a non-stationary environmental state we consider here a two-level system interacting with a reservoir in a state of the form $\rho_B = |\phi\rangle\langle\phi|$, where $|\phi\rangle$ represents a squeezed vacuum

$$|\phi\rangle = \prod_{\vec{k},\lambda} S_{\vec{k},\lambda}(\xi_k)|0\rangle. \tag{3.231}$$

Here, we have introduced the unitary squeeze operator (Walls and Milburn, 1994)

$$S_{\vec{k},\lambda}(\xi) = \exp\left[\frac{1}{2}\xi_k^* b_\lambda(\vec{k})^2 - \frac{1}{2}\xi_k b_\lambda^\dagger(\vec{k})^2\right], \tag{3.232}$$

with $\xi_k = r_k e^{i\theta_k}$. With the help of the transformation properties

$$S_{\vec{k},\lambda}^\dagger(\xi_k)b_\lambda(\vec{k})S_{\vec{k},\lambda}(\xi_k) = b_\lambda(\vec{k})\cosh r_k - b_\lambda^\dagger(\vec{k})e^{i\theta_k}\sinh r_k, \tag{3.233}$$

$$S_{\vec{k},\lambda}^\dagger(\xi_k)b_\lambda^\dagger(\vec{k})S_{\vec{k},\lambda}(\xi_k) = b_\lambda^\dagger(\vec{k})\cosh r_k - b_\lambda(\vec{k})e^{-i\theta_k}\sinh r_k, \tag{3.234}$$

one obtains the expectation values (compare with eqns (3.195)–(3.197))

$$\langle b_\lambda(\vec{k})b_{\lambda'}^\dagger(\vec{k}')\rangle = \delta_{\vec{k}\vec{k}'}\delta_{\lambda\lambda'}(1 + N_k), \tag{3.235}$$

$$\langle b_\lambda^\dagger(\vec{k})b_{\lambda'}(\vec{k}')\rangle = \delta_{\vec{k}\vec{k}'}\delta_{\lambda\lambda'}N_k, \tag{3.236}$$

$$\langle b_\lambda(\vec{k})b_{\lambda'}(\vec{k}')\rangle = \delta_{\vec{k}\vec{k}'}\delta_{\lambda\lambda'}M_k, \tag{3.237}$$

$$\langle b_\lambda^\dagger(\vec{k})b_{\lambda'}^\dagger(\vec{k}')\rangle = \delta_{\vec{k}\vec{k}'}\delta_{\lambda\lambda'}M_k^*, \tag{3.238}$$

where we have defined the quantities

$$N_k \equiv \sinh^2 r_k, \quad M_k \equiv -\cosh r_k \sinh r_k e^{i\theta_k}. \tag{3.239}$$

We note that these relations are compatible with the commutation relations (3.177), (3.178) of the field operators and that

$$|M_k|^2 = N_k(N_k + 1). \tag{3.240}$$

As for a thermal reservoir we have $\langle \vec{E}(t) \rangle = \langle \phi | \vec{E}(t) | \phi \rangle = 0$. However, the correlation functions (3.189) of the electric field operator are no longer homogeneous in time since ρ_B is not an invariant state. As a consequence the spectral correlation tensor now depends explicitly on time t. With the help of eqn (3.191) the spectral correlation tensor for the squeezed vacuum can be written as follows,

$$\Gamma_{ij}(\omega) = \frac{2\delta_{ij}}{3\pi\hbar c^3} \int_0^\infty d\omega_k \omega_k^3 \left[(1 + N_k) \int_0^\infty ds e^{-i(\omega_k - \omega)s} + N_k \int_0^\infty ds e^{+i(\omega_k + \omega)s} \right.$$

$$\left. - M_k e^{-2i\omega_k t} \int_0^\infty ds e^{+i(\omega_k + \omega)s} - M_k^* e^{+2i\omega_k t} \int_0^\infty ds e^{-i(\omega_k - \omega)s} \right]$$

$$\equiv \Gamma_{ij}^{(1)}(\omega) + \Gamma_{ij}^{(2)}(\omega). \tag{3.241}$$

It has been assumed that the squeezing is uniform over the total 4π solid angle. We observe that the spectral correlation tensor consists of two parts: The first contribution $\Gamma_{ij}^{(1)}(\omega)$ is independent of t and is formally identical to the thermal correlation tensor given in eqn (3.201), while the second contribution $\Gamma_{ij}^{(2)}(\omega)$ involves the rapidly oscillating exponentials $\exp(\pm 2i\omega_k t)$.

Correspondingly, the dissipator of the master equation consists of two parts,

$$\frac{d}{dt}\rho(t) = \mathcal{D}^{(1)}(\rho(t)) + \mathcal{D}^{(2)}(\rho(t)). \tag{3.242}$$

The contribution $\mathcal{D}^{(1)}$ is determined through $\Gamma_{ij}^{(1)}(\omega)$. Formally it therefore has the same structure as the thermal dissipator given by the right-hand side of eqn (3.219), where N is the value of N_k at the resonance frequency $\omega_0 = ck_0$. It is given in terms of the resonant squeeze parameter r as

$$N = \sinh^2 r, \quad N + 1 = \cosh^2 r. \tag{3.243}$$

If we neglect the Lamb and Stark shift contributions we find for the time-dependent part of the spectral correlation tensor

$$\Gamma_{ij}^{(2)}(\omega) = -\frac{2\omega^3}{3\hbar c^3}\delta_{ij}M^* e^{2i\omega t}, \quad \text{for } \omega > 0, \tag{3.244}$$

and

$$\Gamma_{ij}^{(2)}(\omega) = -\frac{2|\omega|^3}{3\hbar c^3}\delta_{ij}M e^{2i\omega t}, \quad \text{for } \omega < 0, \tag{3.245}$$

where again M is the value of M_k taken at the resonance frequency, that is

$$M = -\cosh r \sinh r e^{i\theta}, \tag{3.246}$$

with the resonant squeeze parameter r and the resonant phase θ. In order to determine the contribution $\mathcal{D}^{(2)}$ of the dissipator of the master equation we substitute $\Gamma_{ij}^{(2)}(\omega)$ into the right-hand side of eqn (3.188) and perform the rotating wave approximation. Since $\Gamma_{ij}^{(2)}(\omega)$ oscillates with $\exp(2i\omega t)$ the rotating wave approximation selects exactly two resonant terms from the double frequency sum, namely the terms corresponding to the case $\omega = -\omega_0$, $\omega' = +\omega_0$ and to the case $\omega = +\omega_0$, $\omega' = -\omega_0$. The first case is seen to lead to the contribution

$$\vec{A}(\omega)\rho\vec{A}^{\dagger}(\omega') \equiv \vec{A}^{\dagger}\rho\vec{A}^{\dagger} = \vec{d}^{*2}\sigma_+\rho\sigma_+, \tag{3.247}$$

whereas the second case leads to the contribution

$$\vec{A}(\omega)\rho\vec{A}^{\dagger}(\omega') \equiv \vec{A}\rho\vec{A} = \vec{d}^{2}\sigma_-\rho\sigma_-. \tag{3.248}$$

Absorbing the phase of the dipole matrix element into the squeezing phase θ we therefore get

$$\mathcal{D}^{(2)}(\rho) = -\frac{\gamma_0}{2}\left[M\sigma_+\rho\sigma_+ + M^*\sigma_-\rho\sigma_-\right] + \text{h.c.} \tag{3.249}$$

Adding the first part $\mathcal{D}^{(1)}(\rho)$ of the dissipator we finally arrive at the density matrix equation

$$\frac{d}{dt}\rho(t) = \gamma_0(N+1)\left(\sigma_-\rho(t)\sigma_+ - \frac{1}{2}\sigma_+\sigma_-\rho(t) - \frac{1}{2}\rho(t)\sigma_+\sigma_-\right)$$
$$+ \gamma_0 N\left(\sigma_+\rho(t)\sigma_- - \frac{1}{2}\sigma_-\sigma_+\rho(t) - \frac{1}{2}\rho(t)\sigma_-\sigma_+\right)$$
$$- \gamma_0 M\sigma_+\rho(t)\sigma_+ - \gamma_0 M^*\sigma_-\rho(t)\sigma_-. \tag{3.250}$$

This is the master equation describing the decay of a two-level system into a squeezed vacuum. It is important to realize that it can be written in Lindblad form. In fact, introducing the Lindblad operator

$$C \equiv \sigma_-\cosh r + \sigma_+\sinh r e^{i\theta} \tag{3.251}$$

we find

$$\frac{d}{dt}\rho(t) = \gamma_0\left(C\rho(t)C^{\dagger} - \frac{1}{2}C^{\dagger}C\rho(t) - \frac{1}{2}\rho(t)C^{\dagger}C\right). \tag{3.252}$$

Invoking the algebra of the Pauli matrices we obtain the equations of motion

$$\frac{d}{dt}\langle\sigma_+(t)\rangle = \frac{d}{dt}\langle\sigma_-(t)\rangle^* \tag{3.253}$$
$$= -\frac{\gamma_0}{2}\left(\cosh^2 r + \sinh^2 r\right)\langle\sigma_+(t)\rangle + \gamma_0\sinh r\cosh r e^{-i\theta}\langle\sigma_-(t)\rangle,$$

$$\frac{d}{dt}\langle\sigma_3(t)\rangle = -\gamma_0\left(\cosh^2 r + \sinh^2 r\right)\langle\sigma_3(t)\rangle - \gamma_0. \tag{3.254}$$

These equations describe a phase-dependent relaxation of the coherences and a relaxation of the 3-component of the Bloch vector which is enhanced in comparison with the decay into the field vacuum. Setting $\theta = 0$ the equations of motion for the Bloch vector take the form

$$\frac{d}{dt}\langle\sigma_1(t)\rangle = -\frac{\gamma_0}{2}e^{-2r}\langle\sigma_1(t)\rangle, \tag{3.255}$$

$$\frac{d}{dt}\langle\sigma_2(t)\rangle = -\frac{\gamma_0}{2}e^{+2r}\langle\sigma_2(t)\rangle, \tag{3.256}$$

$$\frac{d}{dt}\langle\sigma_3(t)\rangle = -\gamma_0\left(2\sinh^2 r + 1\right)\langle\sigma_3(t)\rangle - \gamma_0. \tag{3.257}$$

The corresponding vacuum Bloch equations (eqns (3.224)–(3.226) for $\gamma = \gamma_0$) are obtained in the limit $r \longrightarrow 0$. We observe that for a squeezed vacuum the components $\langle\sigma_{1,2}\rangle$ of the Bloch vector decay with different relaxation rates: In comparison to the vacuum case the relaxation rate is enhanced in one direction and diminished in the other. The 3-component of the Bloch vector approaches the stationary state

$$\langle\sigma_3\rangle_s = -\frac{1}{2\sinh^2 r + 1}, \tag{3.258}$$

which has the same form as in eqn (3.227), where $N = \sinh^2 r$ here plays the rôle of the Planck distribution.

3.4.4 More general reservoirs

The form of the master equation (3.250) holds for a general environmental state ρ_B which can be represented by a product of independent modes satisfying $\langle b_\lambda(\vec{k})\rangle = \langle b_\lambda^\dagger(\vec{k})\rangle = 0$. For example, if we consider a squeezed thermal state, that is if we replace the vacuum state $|0\rangle\langle 0|$ used in the preceding subsection by a thermal equilibrium state and squeeze the latter, we get instead of eqns (3.243) and (3.246)

$$N = N_{\text{th}}\left(\cosh^2 r + \sinh^2 r\right) + \sinh^2 r, \tag{3.259}$$

$$M = -\cosh r \sinh r e^{i\theta}(2N_{\text{th}} + 1), \tag{3.260}$$

where we have written $N_{\text{th}} = 1/(\exp[\beta\omega_0] - 1)$ for the Planck distribution. Obviously, the quantities N and M are not independent. One easily verifies the relation

$$|M|^2 = N(N+1) - N_{\text{th}}(N_{\text{th}} + 1). \tag{3.261}$$

The last equation shows that the inequality

$$|M|^2 \leq N(N+1) \tag{3.262}$$

holds. This inequality is satisfied for general ρ_B of the above form and may be proven with the help of an appropriate uncertainty relation. To this end we take a fixed mode $b \equiv b_\lambda(\vec{k})$ and write

$$\langle b^\dagger b\rangle = N, \quad \langle bb^\dagger\rangle = N+1, \tag{3.263}$$

$$\langle bb\rangle = M = |M|e^{i\theta}, \quad \langle b^\dagger b^\dagger\rangle = M^* = |M|e^{-i\theta}. \tag{3.264}$$

This is the most general parametrization which is compatible with the commutation relations $[b, b^\dagger] = 1$. Next one defines the operators

$$X = \frac{1}{\sqrt{2}}\left(e^{-i\theta/2}b + e^{+i\theta/2}b^\dagger\right), \quad Y = \frac{i}{\sqrt{2}}\left(e^{-i\theta/2}b - e^{+i\theta/2}b^\dagger\right), \quad (3.265)$$

which are Hermitian and satisfy $\langle X \rangle = \langle Y \rangle = 0$ and $[X, Y] = -i$. The uncertainty relation $\langle X^2 \rangle \cdot \langle Y^2 \rangle \geq 1/4$ then immediately leads to the inequality (3.262).

The significance of the inequality (3.262) is that it guarantees the positivity of the generator of the master equation (3.250). In fact, the generator of eqn (3.250) is in the first standard form (3.63) with $F_1 = \sigma_-$, $F_2 = \sigma_+$ and the coefficient matrix

$$(a_{ij}) = \gamma_0 \begin{pmatrix} N+1 & -M^* \\ -M & N \end{pmatrix}. \quad (3.266)$$

The positivity of this matrix leads to the condition (3.262).

3.4.5 Resonance fluorescence

Let us consider the case that the atomic transition $|e\rangle \leftrightarrow |g\rangle$ is driven by an external, coherent single-mode field on resonance. Invoking the dipole approximation we obtain the following interaction picture Hamiltonian describing the interaction of the system with the driving mode,

$$H_L = -\vec{E}_L(t) \cdot \vec{D}(t). \quad (3.267)$$

Here, the quantity

$$\vec{E}_L(t) = \vec{\varepsilon}e^{-i\omega_0 t} + \vec{\varepsilon}^* e^{+i\omega_0 t} \quad (3.268)$$

is the electric field strength of the driving mode. The product

$$\Omega = 2\vec{\varepsilon} \cdot \vec{d}^* \quad (3.269)$$

is referred to as the Rabi frequency. In the following we choose the phase of the external field such that Ω is real and positive. With these definitions we can write the atom–field interaction in the rotating wave approximation as follows,

$$H_L = -\frac{\Omega}{2}(\sigma_+ + \sigma_-). \quad (3.270)$$

Now we couple the system to a thermal reservoir of radiation modes. According to Section 3.4.2 the quantum master equation becomes

$$\frac{d}{dt}\rho(t) = \frac{i\Omega}{2}[\sigma_+ + \sigma_-, \rho(t)]$$
$$+ \gamma_0(N+1)\left(\sigma_-\rho(t)\sigma_+ - \frac{1}{2}\sigma_+\sigma_-\rho(t) - \frac{1}{2}\rho(t)\sigma_+\sigma_-\right)$$
$$+ \gamma_0 N\left(\sigma_+\rho(t)\sigma_- - \frac{1}{2}\sigma_-\sigma_+\rho(t) - \frac{1}{2}\rho(t)\sigma_-\sigma_+\right). \quad (3.271)$$

3.4.5.1 *Equation of motion for the Bloch vector*   The master equation (3.271)
leads to the following system of differential equations,

$$\frac{d}{dt}\langle\vec{\sigma}(t)\rangle = G\langle\vec{\sigma}(t)\rangle + \vec{b}, \qquad (3.272)$$

where we have introduced the matrix

$$G = \begin{pmatrix} -\gamma/2 & 0 & 0 \\ 0 & -\gamma/2 & \Omega \\ 0 & -\Omega & -\gamma \end{pmatrix} \qquad (3.273)$$

and the 3-vector

$$\vec{b} = \begin{pmatrix} 0 \\ 0 \\ -\gamma_0 \end{pmatrix}. \qquad (3.274)$$

Equation (3.272) is known as the optical *Bloch equation.*
  The stationary solution of the differential equation (3.272) is found to be

$$\langle\sigma_3\rangle_s = -\frac{\gamma_0\gamma}{\gamma^2 + 2\Omega^2}, \qquad (3.275)$$

$$\langle\sigma_+\rangle_s = \langle\sigma_-\rangle_s^* = \frac{-i\Omega\gamma_0}{\gamma^2 + 2\Omega^2}. \qquad (3.276)$$

The stationary population of the upper level is therefore

$$p_e^s = \frac{1}{2}\frac{\gamma(\gamma - \gamma_0) + 2\Omega^2}{\gamma^2 + 2\Omega^2}. \qquad (3.277)$$

At zero temperature ($N = 0$, $\gamma = \gamma_0$) the stationary solutions take the form

$$\langle\sigma_3\rangle_s = -\frac{\gamma_0^2}{\gamma_0^2 + 2\Omega^2}, \qquad (3.278)$$

$$\langle\sigma_+\rangle_s = \langle\sigma_-\rangle_s^* = \frac{-i\Omega\gamma_0}{\gamma_0^2 + 2\Omega^2}, \qquad (3.279)$$

while the stationary population of the upper level is

$$p_e^s = \frac{\Omega^2}{\gamma_0^2 + 2\Omega^2}. \qquad (3.280)$$

In the strong driving limit $\Omega \gg \gamma_0$ we get the limiting forms $p_e^s = 1/2$ and
$\langle\sigma_+\rangle_s = -i\gamma_0/2\Omega$.

The time-dependent solution of the Bloch equation (3.272) is most easily found by introducing the vector

$$\langle\langle\vec{\sigma}(t)\rangle\rangle = \langle\vec{\sigma}(t)\rangle - \langle\vec{\sigma}\rangle_s. \tag{3.281}$$

Representing the difference to the stationary solution this vector satisfies the homogeneous equation

$$\frac{d}{dt}\langle\langle\vec{\sigma}(t)\rangle\rangle = G\langle\langle\vec{\sigma}(t)\rangle\rangle, \tag{3.282}$$

which can easily be solved by diagonalization of the matrix G. The eigenvalues of G are given by

$$\lambda_1 = -\frac{\gamma}{2}, \quad \lambda_{2,3} = -\frac{3\gamma}{4} \pm i\mu, \tag{3.283}$$

where we have introduced

$$\mu = \sqrt{\Omega^2 - \left(\frac{\gamma}{4}\right)^2}. \tag{3.284}$$

We note that all eigenvalues have negative real parts. It follows that any initial density matrix is driven to the unique stationary solution $\langle\vec{\sigma}\rangle_s$ of the Bloch equation determined above. The time-dependent solution can be represented by

$$\langle\langle\sigma_3(t)\rangle\rangle = e^{-3\gamma t/4}\left[\cos\mu t - \frac{\gamma}{4\mu}\sin\mu t\right]\langle\langle\sigma_3(0)\rangle\rangle$$
$$+\frac{i\Omega}{\mu}e^{-3\gamma t/4}\sin\mu t\left(\langle\langle\sigma_+(0)\rangle\rangle - \langle\langle\sigma_-(0)\rangle\rangle\right), \tag{3.285}$$

and

$$\langle\langle\sigma_+(t)\rangle\rangle = \frac{1}{2}e^{-\gamma t/2}\left(\langle\langle\sigma_+(0)\rangle\rangle + \langle\langle\sigma_-(0)\rangle\rangle\right)$$
$$+\frac{1}{2}e^{-3\gamma t/4}\left[\cos\mu t + \frac{\gamma}{4\mu}\sin\mu t\right]\left(\langle\langle\sigma_+(0)\rangle\rangle - \langle\langle\sigma_-(0)\rangle\rangle\right)$$
$$+\frac{i\Omega}{2\mu}e^{-3\gamma t/4}\sin\mu t\langle\langle\sigma_3(0)\rangle\rangle. \tag{3.286}$$

Let us consider here the case that the atom is initially in the ground state, $\rho(0) = |g\rangle\langle g|$, which gives

$$\langle\langle\sigma_\pm(0)\rangle\rangle = -\langle\sigma_\pm\rangle_s, \tag{3.287}$$
$$\langle\langle\sigma_3(0)\rangle\rangle = -1 - \langle\sigma_3\rangle_s. \tag{3.288}$$

At zero temperature ($\gamma = \gamma_0$) we find for arbitrary driving strength

$$p_e(t) = \frac{\Omega^2}{\gamma_0^2 + 2\Omega^2}\left[1 - e^{-3\gamma_0 t/4}\left\{\cos\mu t + \frac{3\gamma_0}{4\mu}\sin\mu t\right\}\right], \tag{3.289}$$
$$\langle\sigma_+(t)\rangle = \frac{-i\Omega\gamma_0}{\gamma_0^2 + 2\Omega^2}\left[1 - e^{-3\gamma_0 t/4}\left\{\cos\mu t + \left(\frac{\gamma_0}{4\mu} - \frac{\Omega^2}{\gamma_0\mu}\right)\sin\mu t\right\}\right]. \tag{3.290}$$

In the underdamped case, which is defined by $\Omega > \gamma_0/4$, the quantity μ is real and both the occupation of the upper level and the coherences exhibit exponentially

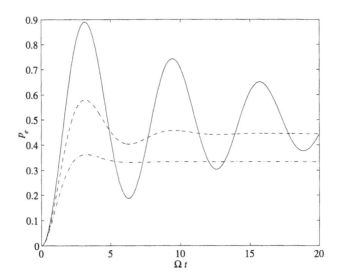

FIG. 3.3. The population of the upper level $p_e(t)$ for the driven two-level system
as a function of time. Parameters: $\gamma_0/\Omega = 0.1$ (solid line), $\gamma_0/\Omega = 0.5$ (dashed
line), and $\gamma_0/\Omega = 1$ (dashed-dotted line).

damped oscillations as shown in Figs. 3.3 and 3.4. By contrast, in the overdamped
case, $\Omega < \gamma_0/4$, the quantity μ becomes purely imaginary and we write

$$\mu = i\tilde{\mu} = i\sqrt{\left(\frac{\gamma_0}{4}\right)^2 - \Omega^2}. \tag{3.291}$$

The occupation probabilities and the coherences then show a monotonic approach to their stationary values.

In the limit of very strong driving, $\Omega \gg \gamma_0/4$, we obtain the asymptotic
expressions,

$$p_e(t) \approx \frac{1}{2}\left[1 - e^{-3\gamma_0 t/4}\cos\Omega t\right], \tag{3.292}$$

$$\langle\sigma_+(t)\rangle \approx -\frac{i}{2}e^{-3\gamma_0 t/4}\sin\Omega t. \tag{3.293}$$

The tip of the Bloch vector thus follows an exponential spiral in the $(2,3)$-plane
around the origin, the circular frequency being equal to the Rabi frequency Ω.

3.4.5.2 *Spectrum of resonance fluorescence* As an application of the quantum
regression theorem (see Section 3.2.4) we determine the fluorescence spectrum of
the emitted radiation in the stationary state. The radiation spectrum is determined by a certain two-time correlation function of the system. To see this we

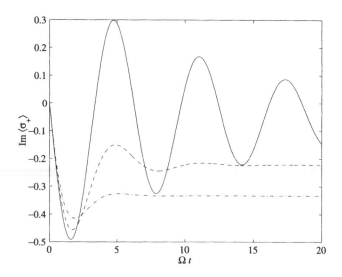

FIG. 3.4. The imaginary part of the coherence $\langle\sigma_+(t)\rangle$ for the driven two-level system as a function of time. Parameters: $\gamma_0/\Omega = 0.1$ (solid line), $\gamma_0/\Omega = 0.5$ (dashed line), and $\gamma_0/\Omega = 1$ (dashed-dotted line).

first note that the positive frequency component of the electric dipole field radiated by the source is found from the retarded solution of the Maxwell equations to be

$$\vec{E}^{(+)}(t, \vec{x}) = \frac{\omega_0^2}{c^2 r} \left(\left[\vec{n} \times \vec{d}\right] \times \vec{n} \right) \sigma_-(t - r/c). \qquad (3.294)$$

Here, the dipole is located at the origin of the coordinate system, $r = |\vec{x}|$ and $\vec{n} = \vec{x}/r$ is the unit vector pointing into the direction of $\vec{x}$. This equation relates the positive frequency component of the radiated field to the atomic lowering operator σ_- at the retarded time $t - r/c$. On the other hand, the spectral intensity radiated per unit solid angle by the oscillating dipole is given by

$$\frac{dI(\omega)}{d\Omega} = \frac{cr^2}{4\pi} \int\limits_{-\infty}^{+\infty} \frac{d\tau}{2\pi} e^{i\omega\tau} \langle \vec{E}^{(-)}(t, \vec{x}) \vec{E}^{(+)}(t + \tau, \vec{x}) \rangle_s. \qquad (3.295)$$

Substituting eqn (3.294) one finds

$$\frac{dI(\omega)}{d\Omega} = \frac{\omega_0^4}{8\pi^2 c^3} \left| \left(\vec{n} \times \vec{d} \right) \times \vec{n} \right|^2 S(\omega), \qquad (3.296)$$

which shows the typical angular distribution of dipole radiation, whereas the frequency dependence is embodied in the spectral function

$$S(\omega) = \int\limits_{-\infty}^{+\infty} d\tau e^{-i\omega\tau} \langle \sigma_+(\tau)\sigma_-(0) \rangle_s. \qquad (3.297)$$

Here, we have employed the homogeneity in time of the stationary correlation function.

Thus, in order to obtain the spectrum we need the atomic correlation function $\langle \sigma_+(\tau)\sigma_-(0)\rangle_s$. To this end, we consider the vector

$$\langle\langle \vec{\sigma}(\tau)\sigma_-(0)\rangle\rangle \equiv \langle \vec{\sigma}(\tau)\sigma_-(0)\rangle_s - \langle \vec{\sigma}\rangle_s\langle\sigma_-\rangle_s. \qquad (3.298)$$

It follows from the quantum regression theorem that the dynamics of this vector is determined through the homogeneous part of the optical Bloch equation, i.e.

$$\frac{d}{d\tau}\langle\langle \vec{\sigma}(\tau)\sigma_-(0)\rangle\rangle = G\langle\langle \vec{\sigma}(\tau)\sigma_-(0)\rangle\rangle. \qquad (3.299)$$

We conclude that the desired correlation function $\langle\langle \sigma_+(\tau)\sigma_-(0)\rangle\rangle$ is given by the right-hand side of eqn (3.286), where the initial values $\langle\langle \sigma_\pm(0)\rangle\rangle$ and $\langle\langle \sigma_3(0)\rangle\rangle$ must be replaced, respectively, by

$$\langle\langle \sigma_+(0)\sigma_-(0)\rangle\rangle = \langle \sigma_+\sigma_-\rangle_s - \langle\sigma_+\rangle_s\langle\sigma_-\rangle_s = p_e^s - |\langle\sigma_+\rangle_s|^2, \qquad (3.300)$$

$$\langle\langle \sigma_-(0)\sigma_-(0)\rangle\rangle = \langle \sigma_-\sigma_-\rangle_s - \langle\sigma_-\rangle_s\langle\sigma_-\rangle_s = -\langle\sigma_-\rangle_s^2, \qquad (3.301)$$

$$\langle\langle \sigma_3(0)\sigma_-(0)\rangle\rangle = \langle \sigma_3\sigma_-\rangle - \langle\sigma_3\rangle_s\langle\sigma_-\rangle_s = -\langle\sigma_-\rangle_s - \langle\sigma_3\rangle_s\langle\sigma_-\rangle_s. \qquad (3.302)$$

Finally, it must be taken into account that the Bloch equations have been solved in the interaction picture. To return to the Schrödinger picture we replace $\sigma_\pm(t)$ by $\exp[\pm i\omega_0 t]\sigma_\pm(t)$. This yields the two-time correlation function in the stationary state

$$\langle \sigma_+(\tau)\sigma_-(0)\rangle_s = \frac{\Omega^2}{\gamma_0^2 + 2\Omega^2}\left\{ \frac{\gamma_0^2}{\gamma_0^2 + 2\Omega^2}\exp\left[i\omega_0\tau\right] + \frac{1}{2}\exp\left[-(\gamma_0/2 - i\omega_0)\tau\right]\right.$$

$$+ A_+ \exp\left[-\left(\frac{3\gamma_0}{4} - \tilde{\mu} - i\omega_0\right)\tau\right]$$

$$\left. + A_- \exp\left[-\left(\frac{3\gamma_0}{4} + \tilde{\mu} - i\omega_0\right)\tau\right]\right\}, \qquad (3.303)$$

where we have defined the amplitudes

$$A_\pm = \frac{1}{4}\frac{1}{\gamma_0^2 + 2\Omega^2}\left[2\Omega^2 - \gamma_0^2 \pm \frac{\gamma_0}{4\tilde{\mu}}\left(10\Omega^2 - \gamma_0^2\right)\right]. \qquad (3.304)$$

Equation (3.303) gives the two-time correlation for $\tau \geq 0$; for negative times it is found with the help of the relation $\langle \sigma_+(-\tau)\sigma_-(0)\rangle_s = \langle\sigma_+(\tau)\sigma_-(0)\rangle_s^*$.

The correlation function is seen to consist of four terms. The first term represents the contribution from elastic Rayleigh scattering which gives rise to a δ-shape spectral function (see below). The other three terms describe inelastic scattering processes and yield Lorentzian line shapes. In the overdamped case, $\Omega < \gamma_0/4$, the quantity $\tilde{\mu}$ is real, such that the spectral function contains a sum of

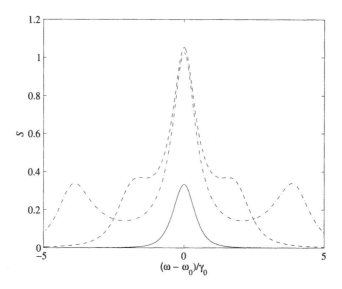

FIG. 3.5. The spectrum of resonance fluorescence according to (3.305) without the elastic contribution. The plot shows $S(\omega)$ for the three values $4\Omega/\gamma_0 = 2$ (solid line), $4\Omega/\gamma_0 = 8$ (dashed line), and $4\Omega/\gamma_0 = 16$ (dashed-dotted line).

three Lorentzian peaks centred at $\omega = \omega_0$. If one goes over to the underdamped case, $\Omega > \gamma_0/4$, the quantity $\tilde{\mu} = -i\mu$ becomes imaginary and the single inelastic peak splits into three peaks centred at $\omega = \omega_0$ and $\omega = \omega_0 \pm \mu$. In fact, on taking the Fourier transform of (3.303) we find the spectral function for the fluorescence radiation of an atomic dipole in the underdamped case (see Fig. 3.5):

$$S(\omega) = \frac{2\Omega^2}{\gamma_0^2 + 2\Omega^2} \left\{ \frac{\pi\gamma_0^2}{\gamma_0^2 + 2\Omega^2}\delta(\omega - \omega_0) + \frac{1}{2}\frac{\gamma_0/2}{(\gamma_0/2)^2 + (\omega - \omega_0)^2} \right. \tag{3.305}$$
$$\left. + \Re\frac{A_+(3\gamma_0/4 - i[\omega - \omega_0 + \mu])}{(3\gamma_0/4)^2 + (\omega - \omega_0 + \mu)^2} + \Re\frac{A_-(3\gamma_0/4 - i[\omega - \omega_0 - \mu])}{(3\gamma_0/4)^2 + (\omega - \omega_0 - \mu)^2} \right\}.$$

In the limit of strong driving, $\Omega \gg \gamma_0/4$, the amplitudes $A_\pm$ approach $1/4$ and the elastic scattering contribution is negligible. This yields the following spectrum in the strong driving limit,

$$S(\omega) = \frac{1}{2}\frac{\gamma_0/2}{(\gamma_0/2)^2 + (\omega - \omega_0)^2} \tag{3.306}$$
$$+ \frac{1}{4}\frac{3\gamma_0/4}{(3\gamma_0/4)^2 + (\omega - \omega_0 + \Omega)^2} + \frac{1}{4}\frac{3\gamma_0/4}{(3\gamma_0/4)^2 + (\omega - \omega_0 - \Omega)^2}.$$

The spectrum thus consists of three Lorentzian peaks: A central peak at $\omega = \omega_0$ with width $\gamma_0/2$, and two sideband peaks at $\omega = \omega_0 \pm \Omega$ with width $3\gamma_0/4$. The heights of the peaks are in the ratio $1 : 3 : 1$, while their integrated intensities are in the ratio $1 : 2 : 1$. The above spectrum was derived theoretically by Mollow (1969), and verified experimentally by Wu, Grove and Ezekiel (1975) and by Hartig *et al.* (1976). Further interesting features of the two-level system, such as the photon statistics and the phenomenon of photon antibunching, will be discussed in Section 6.3.2.

3.4.6 *The damped harmonic oscillator*

As a further example of a dynamical semigroup we study the master equation for a damped harmonic oscillator. The free evolution is generated by the system Hamiltonian $H_S = \omega_0 a^\dagger a$, describing a harmonic oscillator with frequency ω_0. The Schrödinger picture master equation reads

$$\frac{d}{dt}\rho_S(t) = -i\omega_0 \left[a^\dagger a, \rho_S(t)\right]$$
$$+\gamma_0(N+1)\left\{a\rho_S(t)a^\dagger - \frac{1}{2}a^\dagger a\rho_S(t) - \frac{1}{2}\rho_S(t)a^\dagger a\right\}$$
$$+\gamma_0 N\left\{a^\dagger\rho_S(t)a - \frac{1}{2}aa^\dagger\rho_S(t) - \frac{1}{2}\rho_S(t)aa^\dagger\right\}$$
$$\equiv \mathcal{L}\rho_S(t). \tag{3.307}$$

This equation may be used, for example, to describe the damping of an electromagnetic field mode inside a cavity, in which case $a^\dagger$ and a denote the creation and annihilation operators of the cavity mode. The environment is provided, for example, by the modes outside the cavity and leads to a damping of the cavity mode with a rate γ_0. The quantity

$$N = [\exp(\omega_0/k_B T) - 1]^{-1} \tag{3.308}$$

is the mean number of quanta in a mode with frequency ω_0 of the thermal reservoir.

3.4.6.1 *Pauli master equation and stationary solution* We denote by $|n\rangle$, $n = 0, 1, 2, \ldots$, the n-th oscillator eigenstate, i.e. $a^\dagger a|n\rangle = n|n\rangle$. The master equation (3.307) leads to a closed equation for the populations of the n-th level,

$$P(n, t) = \langle n|\rho_S(t)|n\rangle, \tag{3.309}$$

which takes on the form of a classical master equation for a one-step process (see Section 1.4.4.2), the *Pauli master equation*

$$\frac{d}{dt}P(n, t) = \gamma_0(N+1)\left[(n+1)P(n+1, t) - nP(n, t)\right]$$
$$+\gamma_0 N\left[nP(n-1, t) - (n+1)P(n, t)\right]. \tag{3.310}$$

The stationary solution of this equation is given by

$$P_s(n) = \frac{1}{1+N} \left(\frac{N}{1+N} \right)^n.$$

(3.311)

Invoking eqn (3.308) this can be rewritten as

$$P_s(n) = [1 - \exp(-\omega_0/k_B T)] \exp[-n\omega_0/k_B T],$$

(3.312)

which is recognized as the Boltzmann distribution over the eigenstates of the oscillator. The stationary value for the mean number of quanta in the mode is thus equal to the thermal average,

$$\langle a^\dagger a \rangle_s = \sum_{n=0}^{\infty} n P_s(n) = N.$$

(3.313)

3.4.6.2 Adjoint master equation To investigate the dynamics one may deduce equations of motion for expectation values of system operators directly from the master equation. Alternatively, one can use the adjoint master equation (3.84) which becomes

$$\frac{d}{dt} A_H(t) = +i\omega_0 \left[a^\dagger a, A_H(t) \right]$$
$$+ \gamma_0 (N+1) \left\{ a^\dagger A_H(t) a - \frac{1}{2} a^\dagger a A_H(t) - \frac{1}{2} A_H(t) a^\dagger a \right\}$$
$$+ \gamma_0 N \left\{ a A_H(t) a^\dagger - \frac{1}{2} a a^\dagger A_H(t) - \frac{1}{2} A_H(t) a a^\dagger \right\},$$

(3.314)

in order to determine the dynamics of the corresponding Heisenberg operators $A_H(t)$. Solving this equation with the initial condition that Heisenberg and Schrödinger operators coincide at $t = 0$ one finds, for example,

$$a_H(t) = e^{(-i\omega_0 - \gamma_0/2)t} a,$$

(3.315)

$$a_H^\dagger(t) = e^{(+i\omega_0 - \gamma_0/2)t} a^\dagger,$$

(3.316)

$$\left(a^\dagger a \right)_H (t) = e^{-\gamma_0 t} a^\dagger a + N \left(1 - e^{-\gamma_0 t} \right).$$

(3.317)

From these equations one immediately deduces the mean amplitude of the oscillator,

$$\langle a(t) \rangle = \text{tr} \{ a_H(t) \rho_S(0) \} = \langle a(0) \rangle e^{(-i\omega_0 - \gamma_0/2)t},$$

(3.318)

and the mean number of quanta,

$$\langle a^\dagger a(t) \rangle = \text{tr} \left\{ \left(a^\dagger a \right)_H (t) \rho_S(0) \right\} = \langle a^\dagger a(0) \rangle e^{-\gamma_0 t} + N \left(1 - e^{-\gamma_0 t} \right).$$

(3.319)

The mean oscillator amplitude thus follows an exponential spiral in the complex plane, converging for $\gamma_0 t \gg 1$ to the origin. The mean number of quanta is seen to start from the initial value $\langle a^\dagger a(0) \rangle$ and approaches, for times which are large compared to the inverse damping rate, the thermal average N which is independent of the initial value.

3.4.6.3 *Coherent state representation* In quantum optical applications one often studies the master equation with the help of phase space methods. As an example we consider here the *coherent state* or *P-representation* of the master equation. More details and various other related representations may be found in Gardiner and Zoller (2000).

For any complex number α the coherent state $|\alpha\rangle$ is defined in terms of the number states $|n\rangle$ of the oscillator by

$$|\alpha\rangle = \exp\left[-\frac{1}{2}|\alpha|^2\right] \sum_{n=0}^{\infty} \frac{\alpha^n}{\sqrt{n!}} |n\rangle = \exp\left[\alpha a^\dagger - \alpha^* a\right] |0\rangle. \tag{3.320}$$

The coherent states are normalized to 1 but are not orthogonal to each other. The overlap of two coherent states is given by

$$\langle\alpha|\beta\rangle = \exp\left[-\frac{1}{2}|\alpha|^2 - \frac{1}{2}|\beta|^2 + \alpha^*\beta\right]. \tag{3.321}$$

The coherent states are eigenstates of the non-Hermitian operator a, that is

$$a|\alpha\rangle = \alpha|\alpha\rangle. \tag{3.322}$$

They form an overcomplete set of states satisfying the completeness relation

$$\frac{1}{\pi} \int d^2\alpha |\alpha\rangle\langle\alpha| = I. \tag{3.323}$$

The coherent state representation of the system's density matrix goes back to Glauber (1963) and Sudarshan (1963) and is defined by

$$\rho_S(t) = \int d^2\alpha P(\alpha, \alpha^*, t)|\alpha\rangle\langle\alpha|. \tag{3.324}$$

Thus, $\rho_S(t)$ is represented as a mixture of coherent states $|\alpha\rangle$ with a corresponding *weight function* $P(\alpha, \alpha^*, t)$ which is uniquely defined by the given density matrix. It should be noted, however, that P is only some kind of quasi-probability for it is, in general, not a positive distribution function. It can been shown (Klauder, McKenna and Currie, 1965) that the distribution P exists for any density matrix provided one permits P to be a generalized, singular function. The normalization of the density matrix leads to the normalization condition

$$\mathrm{tr}_S\rho_S(t) = \int d^2\alpha P(\alpha, \alpha^*, t) = 1. \tag{3.325}$$

A product of (equal-time) creation and annihilation operators $a^\dagger(t)$, $a(t)$ is said to be normally ordered if all creation operators stand to the left of all annihilation operators. For such a normally ordered product we find

$$\langle (a^\dagger(t))^p a(t)^q \rangle = \int d^2\alpha (\alpha^*)^p \alpha^q P(\alpha, \alpha^*, t). \qquad (3.326)$$

This is an important property of the P-representation which means that the expectation value of any normally ordered product of creation and annihilation operators transforms into a corresponding moment of the distribution $P(\alpha, \alpha^*, t)$.

To derive an equation for $P(\alpha, \alpha^*, t)$ one substitutes the P-representation into the quantum master equation (3.307) and uses the properties

$$a|\alpha\rangle\langle\alpha| = \alpha|\alpha\rangle\langle\alpha|, \qquad (3.327)$$

$$|\alpha\rangle\langle\alpha|a^\dagger = \alpha^*|\alpha\rangle\langle\alpha|, \qquad (3.328)$$

$$a^\dagger|\alpha\rangle\langle\alpha| = \left(\frac{\partial}{\partial\alpha} + \alpha^*\right)|\alpha\rangle\langle\alpha|, \qquad (3.329)$$

$$|\alpha\rangle\langle\alpha|a = \left(\frac{\partial}{\partial\alpha^*} + \alpha\right)|\alpha\rangle\langle\alpha|, \qquad (3.330)$$

which may easily be deduced from the definition for the coherent states. Next, one performs an integration by parts (assuming zero boundary conditions at infinity) which introduces an additional minus sign for each differential operator. The above properties then show that we have the following correspondences,

$$a\rho_S \leftrightarrow \alpha P, \qquad (3.331)$$

$$\rho_S a^\dagger \leftrightarrow \alpha^* P, \qquad (3.332)$$

$$a^\dagger \rho_S \leftrightarrow \left(-\frac{\partial}{\partial\alpha} + \alpha^*\right) P, \qquad (3.333)$$

$$\rho_S a \leftrightarrow \left(-\frac{\partial}{\partial\alpha^*} + \alpha\right) P. \qquad (3.334)$$

With the help of these relations on easily derives the following Fokker–Planck type equation of motion in the P-representation, (Scully and Zubairy, 1997),

$$\frac{\partial}{\partial t} P(\alpha, \alpha^*, t) = -\left[\left(-i\omega_0 - \frac{\gamma_0}{2}\right)\frac{\partial}{\partial\alpha}\alpha + \left(+i\omega_0 - \frac{\gamma_0}{2}\right)\frac{\partial}{\partial\alpha^*}\alpha^*\right] P(\alpha, \alpha^*, t)$$

$$+ \gamma_0 N \frac{\partial^2}{\partial\alpha\partial\alpha^*} P(\alpha, \alpha^*, t). \qquad (3.335)$$

The structure of this equation is similar to that of the classical Fokker–Planck equation (1.129) for a diffusion process: The first term on the right-hand side represents the deterministic drift, while the second term has the structure of a diffusion term.

Let us solve eqn (3.335) for the initial value

$$P(\alpha, \alpha^*, 0) = \delta^2(\alpha - \alpha_0), \qquad (3.336)$$

which means that the initial state is the coherent state $|\alpha_0\rangle$. The corresponding solution can be obtained by substituting the Gaussian ansatz

$$P(\alpha, \alpha^*, t) = \frac{1}{\pi\sigma^2(t)} \exp\left[-\frac{|\alpha - \beta(t)|^2}{\sigma^2(t)}\right] \tag{3.337}$$

into (3.335) and deriving differential equations for the time-dependent functions $\sigma^2(t)$ and $\beta(t)$. However, a simpler way is to observe that $\beta(t)$ is equal to the mean amplitude since

$$\beta(t) = \int d^2\alpha\, \alpha P(\alpha, \alpha^*, t) = \langle\alpha_0|a_H(t)|\alpha_0\rangle = \alpha_0 e^{(-i\omega_0 - \gamma_0/2)t}. \tag{3.338}$$

Correspondingly, we find that $\sigma^2(t)$ is equal to the variance of $a(t)$ which can be obtained using (3.317),

$$\begin{aligned}\sigma^2(t) &= \mathrm{Var}\,(a(t)) \\ &\equiv \langle\alpha_0|\left(a^\dagger a\right)_H(t)|\alpha_0\rangle - \alpha_0^*\alpha_0 e^{-\gamma_0 t} \\ &= N\left(1 - e^{-\gamma_0 t}\right). \end{aligned} \tag{3.339}$$

The P-representation is therefore given by a Gaussian function whose mean value spirals around the origin of the phase plane. The width $\sigma^2(t)$ increases from zero to the thermal average N when the oscillator comes to equilibrium with the surrounding reservoir (see Fig. 3.6). We also note that at zero temperature ($N = 0$) the initial coherent state remains a coherent state under the time evolution.

The Fokker–Planck equation (3.335) describes the evolution of a complex Ornstein–Uhlenbeck process (see Section 1.4.5), the equivalent stochastic Itô differential equation (see Section 1.5.4.1) being given by

$$d\alpha(t) = \left(-i\omega_0 - \frac{\gamma_0}{2}\right)\alpha(t)dt + \sqrt{\gamma_0 N}dW(t). \tag{3.340}$$

Here, $dW(t)$ is the increment of a complex Wiener process which satisfies

$$\mathrm{E}[dW(t)] = \mathrm{E}[dW^*(t)] = 0, \tag{3.341}$$
$$dW(t)dW(t) = dW^*(t)dW^*(t) = 0, \tag{3.342}$$
$$dW(t)dW^*(t) = dt. \tag{3.343}$$

The complex increment can be represented in terms of two statistically independent real Wiener processes as follows

$$dW(t) = \frac{1}{\sqrt{2}}\left(dW_1(t) + idW_2(t)\right), \tag{3.344}$$

which is easily verified with the help of the relations (1.184)–(1.186).

Thus we observe that for appropriate initial conditions which give rise to a true probability distribution in the P-representation, the dynamics of the complex amplitude of the oscillator can be modelled in terms of the classical stochastic process $\alpha(t)$. It must be noted, however, that this does not imply, as

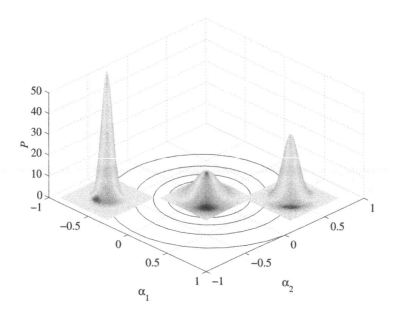

FIG. 3.6. Coherent state representation of the damped harmonic oscillator. The figure shows the distribution function $P(\alpha, \alpha^*, t)$ according to eqn (3.337) over the complex phase plane with coordinates $\alpha = \alpha_1 + i\alpha_2$ at three different times given by $\omega_0 t = 3\pi/4, 7\pi/4, 32\pi$. Parameters: $\gamma_0/\omega_0 = 0.1$, $\alpha_0 = 1$, and $N = 0.025$. The trajectory of the mean amplitude $\beta(t)$ is also shown.

we already know, that all features of the dynamics can be described with the help of a classical stochastic process on the phase plane. The quantum optical master equation (3.307) for the damped harmonic oscillator will be encountered at several occasions in the following chapters. For example, it will be used for a study of decoherence phenomena in Chapter 4, in the context of a stochastic analysis of continuous measurements in Section 6.7, and for a test of numerical simulation algorithms in Section 7.3.1.

3.5 Non-selective, continuous measurements

Density matrix equations can be used to describe the system evolution which is induced by a non-selective, continuous measurement of some observable of the system. An ideal and continuous measurement gives rise to the so-called quantum Zeno effect which will be described in the first subsection. We then turn to the description of a continuous, indirect measurement of a system observable and derive an equation of motion for the density matrix of the object system.

3.5.1 The quantum Zeno effect

We consider the measurement of an observable A whose spectrum is assumed, for simplicity, to be purely discrete and non-degenerate,

$$A = \sum_n a_n |\psi_n\rangle\langle\psi_n|. \tag{3.345}$$

A series of ideal and instantaneous measurements of the observable A is carried out in such a way that two successive measurements are separated by a fixed time interval θ. In between two successive measurements the state vector evolves according to the Schrödinger equation

$$i\frac{\partial}{\partial t}|\psi(t)\rangle = H|\psi(t)\rangle \tag{3.346}$$

with some Hamiltonian H. In the limit $\theta \longrightarrow 0$ one speaks of a *continuous measurement* of the observable A.

Let us assume that the system is initially in an eigenstate $|\psi_n\rangle$ belonging to the eigenvalue a_n of A,

$$|\psi(0)\rangle = |\psi_n\rangle. \tag{3.347}$$

For sufficiently small t we then have according to the Schrödinger equation,

$$|\psi(t)\rangle = \left[I - iHt - \frac{1}{2}H^2t^2 + \ldots\right]|\psi_n\rangle. \tag{3.348}$$

The first ideal measurement of A is carried out at time $t = \theta$. The probability of obtaining the eigenvalue a_n in this measurement is given by

$$w_{nn}(\theta) = |\langle\psi_n|\psi(\theta)\rangle|^2 = 1 - (\Delta E)_n^2\theta^2 + \ldots, \tag{3.349}$$

where the dots indicate terms of higher order and

$$(\Delta E)_n^2 = \langle\psi_n|H^2|\psi_n\rangle - \langle\psi_n|H|\psi_n\rangle^2 \tag{3.350}$$

is the energy uncertainty in the state $|\psi_n\rangle$. The quantity $w_{nn}(\theta)$ is just the probability that the system is still in the initial state $|\psi_n\rangle$ after time θ. After time $\tau = k\theta$, that is after k measurements this probability is thus

$$w_{nn}(\tau) \approx \left[1 - (\Delta E)_n^2\theta^2\right]^k. \tag{3.351}$$

For large k and fixed τ, that is in the limit $\theta = \tau/k \longrightarrow 0$, this leads to

$$w_{nn}(\tau) \approx \left[1 - (\Delta E)_n^2\frac{\tau\theta}{k}\right]^k \approx \exp\left[-(\Delta E)_n^2\tau\theta\right] \longrightarrow 1. \tag{3.352}$$

This equation tells us that the system remains with probability 1 in the initial state $|\psi_n\rangle$ if a continuous, ideal measurement of the observable A is carried out on the system. As a result of the continuous state reduction induced by the measurements the system cannot leave its initial state. Formally, the reason for this fact is that in the limit of small θ the probability of leaving the state $|\psi_n\rangle$ is proportional to the square of θ, that is we have $1 - w_{nn}(\theta) \propto \theta^2$, while the number k of measurements increases as θ^{-1}. The state reduction induced by the succession of measurements is thus faster than possible transitions into other states. In analogy to Zeno's paradox this phenomenon is known as the quantum Zeno effect.

3.5.2 Density matrix equation

The quantum Zeno effect described above results, of course, from an idealization which presupposes an ideal, continuous measurement. In order to investigate what happens for non-ideal measurements we consider the following model for an indirect continuous monitoring of the system.

As in the preceding subsection we first subdivide the time axis into intervals of length θ. Over each interval the object system evolves according to the Schrödinger equation (3.346). However, instead of performing ideal measurements at the beginning of each interval we carry out indirect measurements of A (see Section 2.4.6). At the beginning of each interval a single probe particle with coordinate Q interacts with the quantum object through an interaction Hamiltonian of the form $H_I(t) = g(t)AQ$. The probe particles are independent and supposed to be in identical initial states $|\phi_\theta\rangle$. After the object–probe interaction the momentum P canonically conjugated to Q is measured directly on each probe particle. This leads to a quantum operation which is described by the operators (compare eqn (2.217))

$$\langle p| \exp(-iGAQ)|\phi_\theta\rangle, \tag{3.353}$$

where $|p\rangle$ is a momentum eigenstate and $G = \int dtg(t) = \theta$ is the integrated coupling strength which is assumed to scale with θ.

Including the coherent evolution generated by H we thus have an operation described by the operators

$$\Omega_p = \exp\left(-iH\theta\right) \langle p| \exp\left(-iAQ\theta\right)|\phi_\theta\rangle. \tag{3.354}$$

The evolution of the object's density matrix over a time interval θ on the non-selective level can thus be written as

$$\rho(t + \theta) = \int dp\Omega_p\rho(t)\Omega_p^\dagger. \tag{3.355}$$

Substituting expression (3.354) and expanding the coherent part to first and the incoherent part to second order in θ one obtains

$$\rho(t + \theta) = \rho(t) - i\left[H, \rho(t)\right]\theta - i\left[A, \rho(t)\right]\langle Q\rangle_\theta\theta$$
$$+ \left[A\rho(t)A - \frac{1}{2}A^2\rho(t) - \frac{1}{2}\rho(t)A^2\right]\langle Q^2\rangle_\theta\theta^2, \tag{3.356}$$

where we have defined

$$\langle Q \rangle_\theta = \langle \phi_\theta | Q | \phi_\theta \rangle, \quad \langle Q^2 \rangle_\theta = \langle \phi_\theta | Q^2 | \phi_\theta \rangle. \tag{3.357}$$

We now assume that $\langle Q \rangle_\theta = 0$. This means that the drift contribution $-i\, [A, \rho(t)]\, \langle Q \rangle_\theta$ vanishes and, therefore, that the measuring device does not lead to a systematic back-action on the dynamics of the quantum object. Furthermore, it will be assumed that the limit

$$\sigma_Q^2 \equiv \lim_{\theta \to 0} \theta \langle Q^2 \rangle_\theta \tag{3.358}$$

exists and is finite. It follows that the stochastic back-action induced by the measurement device remains finite in the continuum limit $\theta \longrightarrow 0$.

In the continuum limit we therefore obtain the following equation of motion for the state of the object system,

$$\frac{d}{dt}\rho(t) = -i\, [H, \rho(t)] + \sigma_Q^2 \left[A\rho(t)A - \frac{1}{2}A^2 \rho(t) - \frac{1}{2}\rho(t)A^2 \right]$$

$$= -i\, [H, \rho(t)] - \frac{1}{2}\sigma_Q^2\, [A, [A, \rho(t)]] \,. \tag{3.359}$$

The right-hand side of this equation obviously takes on the form of the generator of a quantum dynamical semigroup. The generator is of the form obtained in Section 3.3.3 for the singular-coupling limit. It involves a single Hermitian Lindblad operator A given by the observable being measured indirectly.

The dissipator of the master equation describes the back-action on the object system induced by the measurement device. We observe that the strength of this back-action is related to the accuracy of the A-measurement. Namely, assuming $\langle P \rangle_\theta = 0$ we have by virtue of the uncertainty relation

$$\frac{\langle P^2 \rangle_\theta}{\theta} \cdot \theta \langle Q^2 \rangle_\theta \geq \frac{1}{4}. \tag{3.360}$$

Taking the limit $\theta \longrightarrow 0$ one obtains

$$\sigma_A^2 \cdot \sigma_Q^2 \geq \frac{1}{4}, \tag{3.361}$$

where we have defined

$$\sigma_A^2 = \lim_{\theta \to 0} \frac{\langle P^2 \rangle_\theta}{\theta}. \tag{3.362}$$

The quantity σ_A is a measure for the uncertainty of the A-measurement (see Section 2.4.6) and eqn (3.361) tells us that the back-action induced by the measurement is the stronger the more accurate is the A-measurement. Equation (3.361) thus provides a kind of uncertainty relation for the accuracy of the monitoring and the resulting fluctuating back-action on the measured system. If we take A to be dimensionless the quantity σ_Q^2 has the dimension of an inverse time,

$$\tau_0 \equiv [\sigma_Q^2]^{-1}, \tag{3.363}$$

which may be referred to as the Zeno time. In fact, the time τ_0 is proportional to the accuracy of the A-measurement. For vanishing Zeno time, $\tau_0 \longrightarrow 0$, that is for an arbitrarily accurate A-measurement, one is led back to the quantum Zeno effect described above.

The interplay between the characteristic time scale of the coherent motion of the object system and the Zeno time τ_0 can be nicely illustrated by considering again a two-level system. The free object Hamiltonian is taken to be $H = -\frac{\Omega}{2}\sigma_1$ describing coherent Rabi oscillations with frequency Ω. The measured quantity is chosen to be $A = \sigma_3$, that is the measurement device is designed to monitor whether the system is in the upper level $|e\rangle$ or in the ground state $|g\rangle$. The corresponding density matrix equation thus takes the form

$$\frac{d}{dt}\rho(t) = \frac{i\Omega}{2}[\sigma_1, \rho(t)] - \frac{1}{2\tau_0}[\sigma_3, [\sigma_3, \rho(t)]], \tag{3.364}$$

which leads to the following system of differential equations for the components of the Bloch vector,

$$\frac{d}{dt}\langle \sigma_1(t)\rangle = -\frac{2}{\tau_0}\langle \sigma_1(t)\rangle, \tag{3.365}$$

$$\frac{d}{dt}\langle \sigma_2(t)\rangle = -\frac{2}{\tau_0}\langle \sigma_2(t)\rangle + \Omega\langle \sigma_3(t)\rangle, \tag{3.366}$$

$$\frac{d}{dt}\langle \sigma_3(t)\rangle = -\Omega\langle \sigma_2(t)\rangle. \tag{3.367}$$

Let us consider the case that the object system is initially in the upper level, which means $\rho(0) = |e\rangle\langle e|$ such that $\langle \sigma_1(0)\rangle = \langle \sigma_2(0)\rangle = 0$ and $\langle \sigma_3(0)\rangle = 1$. The corresponding solution of the above differential equations leads to

$$\langle \sigma_3(t)\rangle = \frac{1}{\mu_2 - \mu_1}\left(\mu_2 e^{-\mu_1 t} - \mu_1 e^{-\mu_2 t}\right), \tag{3.368}$$

where the characteristic frequencies are given by

$$\mu_{1,2} = \frac{1}{\tau_0} \pm \sqrt{\frac{1}{\tau_0^2} - \Omega^2}. \tag{3.369}$$

With the help of these relations the population of the upper level is given by

$$p_e(t) = \frac{1}{2}\left(1 + \langle \sigma_3(t)\rangle\right). \tag{3.370}$$

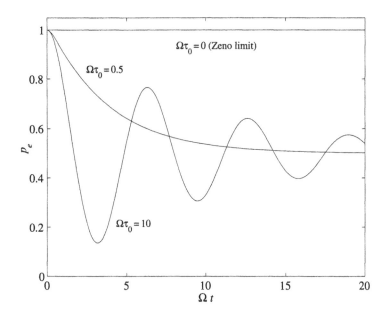

FIG. 3.7. The occupation probability $p_e(t)$ (eqn (3.370)) as a function of time for three different values of the Zeno time τ_0.

Since the real parts of $\mu_{1,2}$ are always positive the occupation probability $p_e(t)$ of the upper level approaches $1/2$ in the long time limit. This means that the back-action of the measurement device drives the quantum object into a stationary state with equal populations of upper and lower level. In the under-damped case $\Omega \tau_0 > 1$ the characteristic frequencies are complex. We write them as $\mu_{1,2} = 1/\tau_0 \pm i\nu$ with

$$\nu = \Omega \sqrt{1 - \frac{1}{\Omega^2 \tau_0^2}}. \tag{3.371}$$

Then we have

$$\langle \sigma_3(t) \rangle = \left(\cos \nu t + \frac{1}{\nu \tau_0} \sin \nu t \right) e^{-t/\tau_0}. \tag{3.372}$$

This shows that $p_e(t)$ follows an exponentially damped oscillation with frequency ν. For $\Omega \tau_0 \gg 1$, that is for a Zeno time which is much larger than the inverse of the Rabi frequency, the object system oscillates many times with a nearly unperturbed frequency $\nu \approx \Omega$. If the Zeno time is decreased the frequency ν becomes smaller and the coherent oscillations die out at $\Omega \tau_0 = 1$.

For the overdamped case $\Omega \tau_0 < 1$ one has a monotonic approach to the stationary state. Thus, for a sufficiently small Zeno time the coherent oscillations of the unperturbed motion are completely suppressed (see Fig. 3.7). In the case $\Omega \tau_0 \ll 1$ one finds the approximate solution

$$\langle \sigma_3(t) \rangle \approx e^{-t/\tau}, \tag{3.373}$$

where the time constant τ is given by

$$\tau = \frac{2}{\Omega^2 \tau_0}. \tag{3.374}$$

The decay time of the upper level thus becomes inversely proportional to the Zeno time τ_0. In the limit $\tau_0 \longrightarrow 0$ we have $\tau \longrightarrow \infty$. This limit corresponds to infinite accuracy of the σ_3-measurement and to the emergence of the quantum Zeno effect: The lifetime of the upper level becomes infinite, i.e. the object system remains frozen in its initial state. We remark that the occurrence of the quantum Zeno effect has been observed experimentally in a similar system (Itano, Heinzen, Bollinger and Wineland, 1990).

3.6   Quantum Brownian motion

In the preceding sections we were mainly concerned with the investigation of master equations which arise in the quantum optical limit. In this limit the rotating wave approximation enabled the derivation of Markovian density matrix equations defining a quantum dynamical semigroup. The physical condition underlying the rotating wave approximation is that the systematic evolution of the reduced system is fast, which means that the coherent dynamics goes through many cycles during a typical relaxation time. In several physical applications involving strong system–environment couplings and low temperatures this condition is violated. By contrast to the typical situation in quantum optics it may even happen that the systematic dynamics of the reduced system is slow compared to the correlation times of the environment. This is the characteristic situation of quantum Brownian motion. Under such circumstances another type of master equation arises which requires an approximation scheme that differs from the quantum optical one.

In this section we are going to introduce and discuss in detail a specific system–reservoir model which is appropriate for the study of several aspects of quantum Brownian motion. This is the famous Caldeira–Leggett model (Caldeira and Leggett, 1983) which is a prototype of a system–reservoir model for the description of dissipation phenomena in solid state physics (Weiss, 1999). In the high-temperature limit the model leads to a quantum master equation which is quite different from the ones obtained in the quantum optical case, but is still of Markovian nature. However, for low temperatures and/or strong couplings the reduced system dynamics exhibits a pronounced non-Markovian character and one must resort to different techniques in order to describe the reduced system dynamics. We shall discuss here the use of the exact Heisenberg equations of motion for the coupled system, the fluctuation–dissipation theorem and the prominent Feynman–Vernon influence functional technique (Feynman and Vernon, 1963). More details on the treatment of non-Markovian processes and a systematic perturbation expansion are postponed to Part IV.

3.6.1 The Caldeira–Leggett model

The model describes a Brownian particle of mass m with coordinate x which moves in a potential $V(x)$. The free Hamiltonian H_S of the particle is thus taken to be

$$H_S = \frac{1}{2m}p^2 + V(x), \qquad (3.375)$$

where p is the particle momentum. The particle is assumed to be coupled to a bath consisting of a large number of harmonic oscillators with masses m_n and frequencies ω_n described by the Hamiltonian

$$H_B = \sum_n \hbar\omega_n \left(b_n^\dagger b_n + \frac{1}{2} \right) = \sum_n \left(\frac{1}{2m_n}p_n^2 + \frac{1}{2}m_n\omega_n^2 x_n^2 \right). \qquad (3.376)$$

Here, the b_n, $b_n^\dagger$ denote the annihilation and creation operators of the bath modes, while the x_n and p_n are the corresponding coordinates and canonically conjugated momenta.

In the model under consideration the coordinate x of the Brownian particle is assumed to be coupled linearly to the coordinates x_n of the bath oscillators. The corresponding interaction Hamiltonian H_I takes the form

$$H_I = -x \sum_n \kappa_n x_n \equiv -xB, \qquad (3.377)$$

where the bath operator

$$B = \sum_n \kappa_n x_n = \sum_n \kappa_n \sqrt{\frac{\hbar}{2m_n\omega_n}}(b_n + b_n^\dagger) \qquad (3.378)$$

is a weighted sum over the coordinates x_n of the bath modes involving the corresponding coupling constants κ_n. This type of interaction will be seen later to yield a renormalization of the potential $V(x)$ of the Brownian particle. To compensate for this renormalization it is convenient to include a further term into the interaction Hamiltonian which is of the form

$$H_c = x^2 \sum_n \frac{\kappa_n^2}{2m_n\omega_n^2}. \qquad (3.379)$$

This term, known as the *counter-term*, acts only in the Hilbert space $\mathcal{H}_S$ of the Brownian particle. It ensures, as will be shown later, that the potential $V(x)$ involves the physical frequencies of the motion of the Brownian particle.

The total Hamiltonian of the combined system $S + B$ is thus given by

$$H = H_S + H_B + H_I + H_c \qquad (3.380)$$
$$= \frac{1}{2m}p^2 + V_c(x) + \sum_n \left(\frac{1}{2m_n}p_n^2 + \frac{1}{2}m_n\omega_n^2 x_n^2 \right) - x \sum_n \kappa_n x_n.$$

In the second expression we have included the counter-term H_c into the potential of the particle,

$$V_c(x) = V(x) + x^2 \sum_n \frac{\kappa_n^2}{2m_n\omega_n^2}. \qquad (3.381)$$

3.6.2 High-temperature master equation

We start our discussion with an investigation of the simplest case, namely of the motion of the Brownian particle in the weak-coupling and high-temperature limit. This limit allows the derivation of a Markovian master equation for the reduced density matrix of the particle, which is known as the Caldeira–Leggett master equation.

3.6.2.1 Derivation of the master equation The starting point is the Born–Markov approximation for the reduced density matrix of the Brownian particle which can be written in the Schrödinger picture as follows,

$$\frac{d}{dt}\rho_S(t) = -\frac{i}{\hbar}[H_S + H_c, \rho_S(t)] + \mathcal{K}\rho_S(t), \qquad (3.382)$$

where we have introduced the super-operator

$$\mathcal{K}\rho_S(t) = -\frac{1}{\hbar^2}\int_0^\infty d\tau \, \mathrm{tr}_B[H_I, [H_I(-\tau), \rho_S(t) \otimes \rho_B]]. \qquad (3.383)$$

This expression may be derived easily by transforming eqn (3.118) back to the Schrödinger picture. Note that the counter-term H_c must be treated as a term of second order in the coupling, while H_I is of first order, such that (3.382) provides a consistent expansion of the equation of motion to second order in the coupling. Here and in the following, operators with a time argument such as $H_I(t)$ denote interaction picture operators with respect to the unperturbed Hamiltonian $H_0 = H_S + H_B$. We shall assume factorizing initial conditions and that the bath is in a thermal equilibrium state ρ_B at temperature $T = 1/k_B\beta$,

$$\rho_B = \frac{\exp(-\beta H_B)}{\mathrm{tr}_B \exp(-\beta H_B)}. \qquad (3.384)$$

Again, averages with respect to ρ_B are denoted by angular brackets.

For the discussion of quantum Brownian motion it is convenient to introduce the following correlation functions,[8]

$$D(\tau) \equiv i\langle[B, B(-\tau)]\rangle = i[B, B(-\tau)], \qquad (3.385)$$
$$D_1(\tau) \equiv \langle\{B, B(-\tau)\}\rangle. \qquad (3.386)$$

For reasons which will become clear below, the functions $D(\tau)$ and $D_1(\tau)$ are often referred to as the dissipation and noise kernel, respectively. We note that

---

[8] We use a notation which is analogous to the usual notation for the corresponding correlation functions in QED, see Chapter 12.

the average over the bath may be omitted in the expression for the dissipation kernel $D(\tau)$, since the commutator $[B, B(-\tau)]$ is a c-number. Making use of the spectral density defined by

$$J(\omega) = \sum_n \frac{\kappa_n^2}{2m_n\omega_n}\delta(\omega - \omega_n) \tag{3.387}$$

we can express the bath correlation functions by

$$D(\tau) = 2\hbar \int_0^\infty d\omega\, J(\omega)\sin\omega\tau, \tag{3.388}$$

$$D_1(\tau) = 2\hbar \int_0^\infty d\omega\, J(\omega)\coth(\hbar\omega/2k_{\mathrm{B}}T)\cos\omega\tau. \tag{3.389}$$

After a little rearrangement the super-operator $\mathcal{K}$ can now be written as

$$\mathcal{K}\rho_S(t) = \frac{1}{\hbar^2}\int_0^\infty d\tau \left(\frac{i}{2}D(\tau)[x, \{x(-\tau), \rho_S(t)\}] - \frac{1}{2}D_1(\tau)[x, [x(-\tau), \rho_S(t)]]\right). \tag{3.390}$$

The properties of the generator $\mathcal{K}$ strongly depend on the behaviour of the dissipation and the noise kernel which, in turn, is determined by the spectral density $J(\omega)$. In order to obtain true irreversible dynamics one introduces a continuous distribution of bath modes and replaces the spectral density by a smooth function of the frequency ω of the bath modes. In phenomenological modelling one often introduces a frequency-independent damping constant γ and takes the spectral density to be proportional to the frequency for small ω,

$$J(\omega) = \frac{2m\gamma}{\pi}\omega, \quad \text{for } \omega \longrightarrow 0. \tag{3.391}$$

It will be seen below that this form for the spectral density, which is known as the Ohmic spectral density, gives rise to frequency-independent damping with the rate γ. On the other hand, the high-frequency modes of the environment lead to a renormalization of the physical parameters in the particle potential. To account for this renormalization of the particle Hamiltonian one introduces a high-frequency cutoff Ω into the spectral density. To give an example we take an Ohmic spectral density with a Lorentz–Drude cutoff function,

$$J(\omega) = \frac{2m\gamma}{\pi}\omega\frac{\Omega^2}{\Omega^2 + \omega^2}. \tag{3.392}$$

For this type of spectral density the bath correlations can be determined analytically as

Table 3.1 *Separation of the time scales in the quantum optical limit and in quantum Brownian motion.*

quantum optical limit	quantum Brownian motion
$\tau_B \ll \tau_R$	$\tau_B \ll \tau_R$
$\tau_S \ll \tau_R$	$\tau_B \ll \tau_S$

$$D(\tau) = 2m\gamma\hbar\Omega^2 e^{-\Omega|\tau|}\text{sign }\tau \tag{3.393}$$

$$D_1(\tau) = 4m\gamma k_B T\Omega^2 \sum_{n=-\infty}^{+\infty} \frac{\Omega e^{-\Omega|\tau|} - |\nu_n|e^{-|\nu_n|\cdot|\tau|}}{\Omega^2 - \nu_n^2}. \tag{3.394}$$

Here, the noise kernel $D_1(\tau)$ is calculated by using the formula

$$\coth(\hbar\omega/2k_B T) = \frac{2k_B T}{\hbar} \sum_{n=-\infty}^{+\infty} \frac{\omega}{\omega^2 + \nu_n^2}, \tag{3.395}$$

where the $\nu_n = 2\pi n k_B T/\hbar$ are known as the Matsubara frequencies. The remaining integrals are then determined with the help of the method of residues.

We observe that the correlation functions involve the correlation times Ω^{-1} and $|\nu_n|^{-1}$ for $n \neq 0$. The largest correlation time is therefore equal to $\tau_B = \text{Max}\{\Omega^{-1}, \hbar/2\pi k_B T\}$ and the condition for the applicability of the Born–Markov approximation becomes

$$\hbar\gamma \ll \text{Min}\{\hbar\Omega, 2\pi k_B T\}. \tag{3.396}$$

This condition corresponds to the condition $\tau_B \ll \tau_R$ which has already been used in the derivation of the weak-coupling and the quantum optical master equations in Sections 3.3.1 and 3.4. However, in the quantum optical case one proceeds by performing the rotating wave approximation which requires that the systematic evolution of the reduced system is fast compared to the typical relaxation time, that is $\tau_S \sim |\omega'-\omega|^{-1} \ll \tau_R$. By contrast, here we investigate the case that the systematic system's evolution is slow in comparison to the bath correlation time. If we denote by $\omega_0 = \tau_S^{-1}$ a typical frequency of the system evolution the latter condition becomes

$$\hbar\omega_0 \ll \text{Min}\{\hbar\Omega, 2\pi k_B T\}. \tag{3.397}$$

The fundamental difference between the quantum optical and the quantum Brownian motion case is summarized in Table 3.1.

In order to simplify the expression for the generator $\mathcal{K}$ we exploit condition (3.397) and approximate $x(-\tau)$ by the free dynamics,

$$x(-\tau) \equiv e^{-iH_S\tau/\hbar}xe^{iH_S\tau/\hbar} \approx x - \frac{i}{\hbar}[H_S, x]\tau = x - \frac{p}{m}\tau. \tag{3.398}$$

Substituting into eqn (3.390) the generator is seen to consist of four terms:

$$\mathcal{K}\rho_S = \frac{i}{2\hbar^2} \int_0^\infty d\tau D(\tau)[x, \{x, \rho_S\}] - \frac{i}{2\hbar^2 m} \int_0^\infty d\tau \tau D(\tau)[x, \{p, \rho_S\}]$$

$$-\frac{1}{2\hbar^2} \int_0^\infty d\tau D_1(\tau)[x, [x, \rho_S]] + \frac{1}{2\hbar^2 m} \int_0^\infty d\tau \tau D_1(\tau)[x, [p, \rho_S]]. \quad (3.399)$$

The first term on the right-hand side of this equation can be determined with the help of the relation

$$\int_0^\infty d\tau \sin \omega\tau = P\frac{1}{\omega}. \quad (3.400)$$

Hence, we have

$$\int_0^\infty d\tau D(\tau) = 2\hbar \int_0^\infty d\omega \frac{J(\omega)}{\omega} = 2\hbar \sum_n \frac{\kappa_n^2}{2m_n\omega_n^2}. \quad (3.401)$$

Since, in addition, $[x, \{x, \rho_S\}] = [x^2, \rho_S]$ the first term on the right-hand side of eqn (3.399) can be cast into the form

$$\frac{i}{\hbar} \sum_n \frac{\kappa_n^2}{2m_n\omega_n^2}[x^2, \rho_S] = \frac{i}{\hbar}[H_c, \rho_S]. \quad (3.402)$$

Evidently, this term compensates the contribution from the counter-term to the Hamiltonian part of eqn (3.382). Thus we indeed see, as claimed earlier, that the interaction with the bath induces a renormalization of the free-particle Hamiltonian H_S which is exactly cancelled by the counter-term H_c.

In order to determine the second term on the right-hand side of (3.399) we use

$$\int_0^\infty d\tau \tau \sin \omega\tau = -\frac{\partial}{\partial\omega} \int_0^\infty d\tau \cos \omega\tau = -\pi\delta'(\omega), \quad (3.403)$$

which yields

$$\int_0^\infty d\tau \tau D(\tau) = \hbar\pi J'(0) = 2m\gamma\hbar, \quad (3.404)$$

and allows the second term to be written as

$$-\frac{i\gamma}{\hbar}[x, \{p, \rho_S\}]. \quad (3.405)$$

Correspondingly, we find

$$\int_0^\infty d\tau D_1(\tau) = 2\hbar\pi \int_0^\infty d\omega J(\omega) \coth(\hbar\omega/2k_BT)\delta(\omega)$$

$$= \hbar\pi \lim_{\omega \to 0} J(\omega) \coth(\hbar\omega/2k_BT) = 4m\gamma k_BT. \quad (3.406)$$

The third term of (3.399) may thus be cast into the form

$$-\frac{2m\gamma k_B T}{\hbar^2}[x, [x, \rho_S]].$$ (3.407)

The fourth term of (3.399), finally, depends on the frequency cutoff Ω and will be determined by use of the explicit expression (3.394) for the Ohmic spectral density with Lorentz–Drude cutoff. In the limit of high temperatures such that $k_B T \gtrsim \hbar\Omega$ we get

$$\int_0^\infty d\tau \tau D_1(\tau) \approx \frac{4m\gamma k_B T}{\Omega}.$$ (3.408)

Hence, the fourth term reads

$$\frac{2\gamma k_B T}{\hbar^2 \Omega}[x, [p, \rho_S]].$$ (3.409)

To estimate the importance of this contribution we compare it with the third term (3.407). Since the momentum is of the order of $p = m\dot{x} \sim m\omega_0 x$, we see that the term (3.409) differs from the term (3.407) by a factor of ω_0/Ω, which, by assumption, is very small. Thus we may neglect the fourth term in the master equation.

Collecting our results we thus arrive finally at the Caldeira–Leggett master equation (Caldeira and Leggett, 1983)

$$\frac{d}{dt}\rho_S(t) = -\frac{i}{\hbar}[H_S, \rho_S(t)] - \frac{i\gamma}{\hbar}[x, \{p, \rho_S(t)\}] - \frac{2m\gamma k_B T}{\hbar^2}[x, [x, \rho_S(t)]].$$ (3.410)

The first term on the right-hand side of the master equation describes the free coherent dynamics of the system. The second term, which is proportional to the relaxation rate γ, is a dissipative term which stems from the contribution involving the dissipation kernel $D(\tau)$. The last term, which is proportional to the temperature, describes thermal fluctuations and will turn out, as we will see later on, to be of fundamental importance in the theoretical description of the phenomenon of decoherence.

One might ask the question of whether the generator of the Brownian motion master equation (3.410) can be brought into Lindblad form. The answer to this question is negative. However, it can be written in Lindblad form by just adding a term which is small in the high-temperature limit. To see this we observe that (3.410) can be written in the form

$$\frac{d}{dt}\rho_S = -\frac{i}{\hbar}[H_S, \rho_S] - \frac{i\gamma}{2\hbar}[xp + px, \rho_S] + \mathcal{D}(\rho_S).$$ (3.411)

The commutator $[xp + px, \rho_S]$ gives a Hamiltonian contribution to the generator, while $\mathcal{D}(\rho_S)$ is of the form of the dissipator in eqn (3.63) with $F_1 = x$, $F_2 = p$ and the coefficient matrix

$$(a_{ij}) = \begin{pmatrix} 4m\gamma k_B T/\hbar^2 & -i\gamma/\hbar \\ i\gamma/\hbar & 0 \end{pmatrix}. \tag{3.412}$$

Thus, the generator of the Caldeira–Leggett master equation is of the first standard form (3.63) under the condition that the matrix a_{ij} is positive. This condition is obviously violated for $\det a = -(\gamma/\hbar)^2 < 0$. However, we could introduce a non-vanishing coefficient a_{22} which corresponds to an additional term of the form $-a_{22}[p, [p, \rho_S]]/2$ in the generator of the master equation. With such a term the positivity condition reads

$$\det a = \frac{4m\gamma k_B T}{\hbar^2} a_{22} - (\gamma/\hbar)^2 \geq 0. \tag{3.413}$$

Let us employ a *minimally invasive* modification and require that $\det a = 0$, which yields $a_{22} = \gamma/4mk_B T$. This amount to adding the term

$$-\frac{\gamma}{8mk_B T}[p, [p, \rho_S]] \tag{3.414}$$

to the right-hand side of the master equation which makes the generator positive and allows the master equation to be written in Lindblad form (3.66) with a single relaxation rate γ and a single Lindblad operator

$$A = \sqrt{\frac{4mk_B T}{\hbar^2}} x + i\sqrt{\frac{1}{4mk_B T}} p. \tag{3.415}$$

The term (3.414) is small compared to the term with the double x-commutator provided $\hbar/k_B T \ll \omega_0^{-1}$. This shows that the term added above is in fact small under the conditions which were used to derive the master equation.

3.6.2.2 *Approximate stationary solution* In the position representation the master equation (3.410) takes the form

$$\frac{\partial}{\partial t}\rho_S(x, x't) = \left[\frac{i\hbar}{2m}\left(\frac{\partial^2}{\partial x^2} - \frac{\partial^2}{\partial x'^2} \right) - \frac{i}{\hbar}(V(x) - V(x')) \right. \tag{3.416}$$

$$\left. -\gamma(x - x')\left(\frac{\partial}{\partial x} - \frac{\partial}{\partial x'} \right) - \frac{2m\gamma k_B T}{\hbar^2}(x - x')^2 \right] \rho_S(x, x't).$$

Introducing new variables r and q defined through

$$x = r + \hbar q, \quad x' = r - \hbar q, \tag{3.417}$$

we obtain an equation for the function $f(r, q, t) = \rho_S(x, x', t)$,

$$\frac{\partial}{\partial t}f = \left[\frac{i}{2m}\frac{\partial^2}{\partial r \partial q} - \frac{i}{\hbar}(V(r + \hbar q) - V(r - \hbar q)) - 2\gamma q\frac{\partial}{\partial q} - 8\gamma mk_B T q^2 \right] f. \tag{3.418}$$

On approximating

$$V(r + \hbar q) - V(r - \hbar q) \approx 2V'(r)\hbar q \qquad (3.419)$$

the stationary solution can be written

$$f(r, q) = N \exp\left[-\frac{V(r)}{k_{\mathrm{B}}T} - 2mk_{\mathrm{B}}Tq^2\right], \qquad (3.420)$$

such that an approximate stationary solution of the master equation (3.410) is found to be

$$\rho_S(x, x') = N \exp\left[-\frac{V((x + x')/2)}{k_{\mathrm{B}}T} - \frac{mk_{\mathrm{B}}T(x - x')^2}{2\hbar^2}\right], \qquad (3.421)$$

with the normalization factor

$$N^{-1} = \int\limits_{-\infty}^{+\infty} dx \exp\left[-\frac{V(x)}{k_{\mathrm{B}}T}\right]. \qquad (3.422)$$

We see that the diagonal of the position space density matrix represents an equilibrium distribution proportional to $\exp[-V(x)/k_{\mathrm{B}}T]$ in accordance with the result of statistical mechanics. The off-diagonal elements decay exponentially with the distance $|x - x'|$ from the diagonal, where the relevant length scale is given by the thermal wavelength $\bar{\lambda}_{\mathrm{th}} = \hbar/\sqrt{2mk_{\mathrm{B}}T}$. Expanding the left-hand side of (3.419) to second order we thus see that (3.419) provides a good approximation as long as the thermal wavelength is small compared to $\sqrt{24|V'|/|V'''|}$.

We also note that (3.419) becomes an equality for a quadratic potential $V(x) = \frac{1}{2}m\omega_0^2 x^2$, in which case the stationary density is represented by a Gaussian function

$$\rho_S(x, x') = \frac{1}{\sqrt{2\pi\sigma_x^2}} \exp\left[-\frac{1}{2\sigma_x^2}\left(\frac{x + x'}{2}\right)^2 - \frac{\sigma_p^2}{2\hbar^2}(x - x')^2\right], \qquad (3.423)$$

with corresponding expressions for the variances of position and momentum coordinates,

$$\sigma_x^2 = \langle x^2 \rangle = \frac{k_{\mathrm{B}}T}{m\omega_0^2}, \quad \sigma_p^2 = \langle p^2 \rangle = mk_{\mathrm{B}}T. \qquad (3.424)$$

The product of position and momentum uncertainty is therefore

$$\sigma_x \cdot \sigma_p = \frac{k_{\mathrm{B}}T}{\omega_0} \gg \frac{\hbar}{2}. \qquad (3.425)$$

Thus, in the present high-temperature approximation the product of the uncertainties is large compared to the minimal possible value allowed by the uncertainty relation.

3.6.2.3 *Equations of motion for mean values and variances* With the help of
the master equation (3.410) one easily obtains the following equations for the
first and second moments of the coordinate and the momentum of the Brownian
particle,

$$\frac{d}{dt}\langle x \rangle = \frac{1}{m}\langle p \rangle, \tag{3.426}$$

$$\frac{d}{dt}\langle p \rangle = -\langle V'(x) \rangle - 2\gamma\langle p \rangle, \tag{3.427}$$

$$\frac{d}{dt}\langle x^2 \rangle = \frac{1}{m}\langle px + xp \rangle, \tag{3.428}$$

$$\frac{d}{dt}\langle px + xp \rangle = \frac{2}{m}\langle p^2 \rangle - 2\langle xV'(x) \rangle - 2\gamma\langle px + xp \rangle, \tag{3.429}$$

$$\frac{d}{dt}\langle p^2 \rangle = -\langle pV'(x) + V'(x)p \rangle - 4\gamma\langle p^2 \rangle + 4m\gamma k_{\mathrm{B}}T. \tag{3.430}$$

These are the Ehrenfest equations for the first and second moments of coordinate
and momentum of a damped particle. They involve the friction force $-2\gamma p$. The
corresponding classical equations of motion are given by the stochastic differential equations

$$dx(t) = \frac{1}{m}p(t)dt, \tag{3.431}$$

$$dp(t) = -V'(x(t))dt - 2\gamma p(t)dt + \sqrt{4m\gamma k_{\mathrm{B}}T}dW(t). \tag{3.432}$$

Let us solve the equations of motion (3.426)–(3.430) for a free Brownian
particle, that is for $V = 0$. The solutions for the first moments yield

$$\langle x(t) \rangle = \langle x(0) \rangle + \frac{1}{2m\gamma}\left(1 - e^{-2\gamma t}\right)\langle p(0) \rangle, \tag{3.433}$$

$$\langle p(t) \rangle = e^{-2\gamma t}\langle p(0) \rangle. \tag{3.434}$$

Thus, the initial momentum relaxes exponentially to zero over a time scale $1/2\gamma$,
while the average position is displaced asymptotically by the value $\langle \dot{x}(0) \rangle/2\gamma$.
Defining

$$\sigma_x^2(t) = \langle x^2(t) \rangle - \langle x(t) \rangle^2, \tag{3.435}$$

$$\sigma_p^2(t) = \langle p^2(t) \rangle - \langle p(t) \rangle^2, \tag{3.436}$$

$$\sigma_{px}(t) = \langle \{p(t), x(t)\} \rangle - 2\langle p(t) \rangle\langle x(t) \rangle, \tag{3.437}$$

we can represent the solutions for the second moments as follows,

$$\sigma_x^2(t) = \sigma_x^2(0) + \left(\frac{1 - e^{-2\gamma t}}{2\gamma}\right)^2 \frac{\sigma_p^2(0)}{m^2} + \frac{1 - e^{-2\gamma t}}{2m\gamma}\sigma_{px}(0)$$

$$+ \frac{k_B T}{m\gamma^2}\left[\gamma t - \left(1 - e^{-2\gamma t}\right) + \frac{1}{4}\left(1 - e^{-4\gamma t}\right)\right], \qquad (3.438)$$

$$\sigma_p^2(t) = e^{-4\gamma t}\sigma_p^2(0) + m k_B T\left(1 - e^{-4\gamma t}\right), \qquad (3.439)$$

$$\sigma_{px}(t) = e^{-2\gamma t}\sigma_{px}(0) + \frac{1}{m\gamma}\left(1 - e^{-2\gamma t}\right)e^{-2\gamma t}\sigma_p^2(0)$$

$$+ \frac{k_B T}{\gamma}\left(1 - e^{-2\gamma t}\right)^2. \qquad (3.440)$$

In the long time limit ($\gamma t \gg 1$) we thus get the asymptotic expressions

$$\sigma_x^2 \longrightarrow \frac{k_B T}{m\gamma}t, \quad \sigma_p^2 \longrightarrow m k_B T, \quad \sigma_{px} \longrightarrow \frac{k_B T}{\gamma}. \qquad (3.441)$$

Obviously, the process does not become stationary in the long time limit. The momentum uncertainty approaches a value which is equal to that given by the thermal equilibrium value (see eqn (3.424)). The position uncertainty, however, increases with the square root of time, exactly as in classical Brownian motion.

For a pure Gaussian (minimal uncertainty) initial state we have $\sigma_p^2(0) = \hbar^2/4\sigma_x^2(0)$ and $\sigma_{px}(0) = 0$, which leads to

$$\sigma_x^2(t) = \sigma_x^2(0) + \frac{\hbar^2}{4m^2\sigma_x^2(0)}\left(\frac{1 - e^{-2\gamma t}}{2\gamma}\right)^2$$

$$+ \frac{k_B T}{m\gamma^2}\left[\gamma t - \left(1 - e^{-2\gamma t}\right) + \frac{1}{4}\left(1 - e^{-4\gamma t}\right)\right]$$

$$\approx \sigma_x^2(0) + \frac{\hbar^2 t^2}{4m^2\sigma_x^2(0)} + \frac{4k_B T}{3m\gamma^2}(\gamma t)^3, \qquad (3.442)$$

where the second relation holds for $\gamma t \ll 1$. The second term of this second form describes the spreading of the Gaussian wave packet according to the free Schrödinger equation. For short times the influence of the environment is seen to yield an additional contribution to the spreading which is proportional to the third power of t.

3.6.3   The exact Heisenberg equations of motion

For a number of applications it is useful not to resort to an approximate master equation for the reduced density matrix, but to work instead directly with the Heisenberg equations of motion of the system. Since in the Caldeira–Leggett model the bath is a collection of harmonic oscillators and since the coupling to the reduced system in linear, one can eliminate the dynamics of the bath variables completely from the Heisenberg equations of motion for the variables pertaining to the reduced system. This will be done in the present subsection. The usage of the resulting Heisenberg equations of motion will be illustrated with the help of several examples for free Brownian motion and by means of the fluctuation–dissipation theorem.

3.6.3.1 *Derivation of the Heisenberg equations* The complete Caldeira–Leggett model leads to the following exact Heisenberg equations of motion for the Brownian particle and the environmental oscillators,

$$\dot{x}(t) = \frac{i}{\hbar}[H, x(t)] = \frac{1}{m}p(t), \tag{3.443}$$

$$\dot{x}_n(t) = \frac{i}{\hbar}[H, x_n(t)] = \frac{1}{m_n}p_n(t), \tag{3.444}$$

$$\dot{p}(t) = \frac{i}{\hbar}[H, p(t)] = -V_c'(x(t)) + \sum_n \kappa_n x_n(t), \tag{3.445}$$

$$\dot{p}_n(t) = \frac{i}{\hbar}[H, p_n(t)] = -m_n \omega_n^2 x_n(t) + \kappa_n x(t). \tag{3.446}$$

The corresponding equation for the coordinate of the Brownian particles is thus

$$m\ddot{x}(t) + V_c'(x(t)) - \sum_n \kappa_n x_n(t) = 0, \tag{3.447}$$

while the equations for the coordinates of the bath oscillators take the form

$$m_n \ddot{x}_n(t) + m_n \omega_n^2 x_n(t) - \kappa_n x(t) = 0. \tag{3.448}$$

The last equation shows that the n-th bath oscillator is driven by the force $\kappa_n x(t)$ which depends linearly on the coordinate of the Brownian particle. In order to get a closed equation of motion for $x(t)$ one solves eqn (3.448) in terms of $x(t)$ and of the initial conditions for the bath modes and substitutes the result into eqn (3.447). To this end, it is convenient to express the coordinates of the bath oscillators in terms of the creation and annihilation operators $b_n^\dagger$, b_n as

$$x_n(0) = \sqrt{\frac{\hbar}{2m_n\omega_n}}(b_n + b_n^\dagger), \quad p_n(0) = -i\sqrt{\frac{m_n\hbar\omega_n}{2}}(b_n - b_n^\dagger). \tag{3.449}$$

The solution of eqn (3.448) is then given by

$$x_n(t) = \sqrt{\frac{\hbar}{2m_n\omega_n}}\left(e^{-i\omega_n t}b_n + e^{i\omega_n t}b_n^\dagger\right) + \frac{\kappa_n}{m_n\omega_n}\int_0^t ds\,\sin[\omega_n(t-s)]x(s). \tag{3.450}$$

Substituting into eqn (3.447) yields

$$m\ddot{x}(t) + V_c'(x(t)) - \sum_n \frac{\kappa_n^2}{m_n\omega_n}\int_0^t ds\,\sin[\omega_n(t-s)]x(s) = B(t), \tag{3.451}$$

where we recall that the operator $B(t)$ which appears here on the right-hand side is the interaction picture operator

$$B(t) = \sum_n \kappa_n \sqrt{\frac{\hbar}{2m_n \omega_n}} \left(e^{-i\omega_n t} b_n + e^{i\omega_n t} b_n^\dagger \right) \qquad (3.452)$$

corresponding to the Schrödinger operator $B = \sum_n \kappa_n x_n(0)$. With the help of the dissipation kernel we can write the equation of motion for the coordinate of the Brownian particle as

$$\ddot{x}(t) + \frac{1}{m} V_c'(x(t)) - \frac{1}{\hbar m} \int_0^t ds D(t-s) x(s) = \frac{1}{m} B(t). \qquad (3.453)$$

In the theory of quantum Brownian motion it is useful to express the dissipation kernel in terms of another quantity which is known as the damping kernel

$$\gamma(t-s) = \frac{2}{m} \int_0^\infty d\omega \frac{J(\omega)}{\omega} \cos[\omega(t-s)], \qquad (3.454)$$

which satisfies

$$\frac{d}{dt} \gamma(t-s) = -\frac{1}{\hbar m} D(t-s), \qquad (3.455)$$

and

$$\gamma(0) = \frac{2}{m} \int_0^\infty d\omega \frac{J(\omega)}{\omega} = \sum_n \frac{\kappa_n^2}{m m_n \omega_n^2}. \qquad (3.456)$$

With the help of this damping kernel we can write the dissipative term of eqn (3.453) as follows

$$-\frac{1}{\hbar m} \int_0^t ds D(t-s) x(s) = \int_0^t ds \frac{d}{dt} \gamma(t-s) x(s) \qquad (3.457)$$

$$= \frac{d}{dt} \int_0^t ds \gamma(t-s) x(s) - \gamma(0) x(t).$$

In view of eqn (3.456) the last term $-\gamma(0)x(t)$ is seen to cancel the contribution from the counter-term contained in the potential $V_c(x)$. Thus we finally arrive at the following exact Heisenberg equation of motion,

$$\ddot{x}(t) + \frac{1}{m} V'(x(t)) + \frac{d}{dt} \int_0^t ds \gamma(t-s) x(s) = \frac{1}{m} B(t). \qquad (3.458)$$

Equation (3.458) is the desired equation of motion for the coordinate of the Brownian particle. It may be viewed as the quantum analogue of a classical stochastic differential equation, involving a damping kernel $\gamma(t-s)$ and a stochastic force $B(t)$ whose statistical properties depend on the initial distribution at $t = 0$. In the case of an environment with an Ohmic spectral density with an infinite cutoff, $\Omega \longrightarrow \infty$, we get the damping kernel (see eqns (3.392) and (3.454))

$$\gamma(t) = 4\gamma\delta(t) \tag{3.459}$$

such that the Heisenberg equations of motion take the form

$$\dot{x}(t) = \frac{1}{m}p(t), \tag{3.460}$$

$$\dot{p}(t) = -V'(x(t)) - 2m\gamma\dot{x}(t) + B(t). \tag{3.461}$$

As in the classical stochastic differential equations (3.431) and (3.432) the friction force is equal to $-2m\gamma\dot{x}$.

The statistics of the forcing term $B(t)$ in eqn (3.458) is described by the quantum correlation function $\langle\{B(t), B(t')\}\rangle$. We emphasize that the angular brackets denote here the average over the initial distribution $\rho(0)$ of the total system since we are working in the Heisenberg picture. If we use an uncorrelated initial state of the form $\rho(0) = \rho_S(0) \otimes \rho_B$ this correlation function is equal to the noise kernel $D_1(t - t')$ introduced in eqn (3.386). For an Ohmic spectral density it is given by

$$D_1(t - t') = \langle\{B(t), B(t')\}\rangle = \frac{4m\gamma\hbar}{\pi}\int\limits_0^\Omega d\omega\omega \coth(\hbar\omega/2k_BT)\cos[\omega(t - t')].$$
$$\tag{3.462}$$

In the high-temperature limit one may assume that $2k_BT \gg \hbar\omega$ for all relevant frequencies ω which gives

$$\frac{1}{2}\langle\{B(t), B(t')\}\rangle \approx \frac{2m\gamma\hbar}{\pi}\frac{2k_BT}{\hbar}\int\limits_0^\infty d\omega \cos[\omega(t - t')] = 4m\gamma k_BT\delta(t - t'). \tag{3.463}$$

This shows that $\hbar$ drops out of the correlation function and that we obtain precisely the correlation function of the noise term in the classical stochastic differential equations (3.431) and (3.432).

3.6.3.2 *Quadratic potentials* For a quadratic potential

$$V(x) = \frac{1}{2}m\omega_0^2 x^2 \tag{3.464}$$

the Heisenberg equation (3.458) becomes

$$\ddot{x}(t) + \omega_0^2 x(t) + \frac{d}{dt}\int\limits_0^t ds\gamma(t - s)x(s) = \frac{1}{m}B(t). \tag{3.465}$$

To solve this equation we introduce the fundamental solutions $G_1(t)$ and $G_2(t)$ of the homogeneous part of eqn (3.465), which is obtained by setting the right-hand side equal to zero. These solutions are defined through the initial conditions

$$G_1(0) = 1, \quad \dot{G}_1(0) = 0, \tag{3.466}$$
$$G_2(0) = 0, \quad \dot{G}_2(0) = 1. \tag{3.467}$$

Introducing the Laplace transformation

$$\hat{f}(z) = \int_0^\infty dt e^{-zt} f(t), \tag{3.468}$$

one can write the Laplace transforms of the fundamental solutions as follows,

$$\hat{G}_1(z) = \frac{z}{z^2 + \omega_0^2 + z\hat{\gamma}(z)},$$
$$\hat{G}_2(z) = \frac{1}{z^2 + \omega_0^2 + z\hat{\gamma}(z)}, \tag{3.469}$$

where $\hat{\gamma}(z)$ is the Laplace transform of the damping kernel.

In terms of the fundamental solutions one can write the general solution of the Heisenberg equation (3.465) as

$$x(t) = G_1(t)x(0) + G_2(t)\dot{x}(0) + \frac{1}{m}\int_0^t ds G_2(t-s)B(s). \tag{3.470}$$

With the help of this solution all desired mean values, variances and correlation functions of the Brownian particle may be expressed in terms of averages over the initial distribution $\rho(0)$ of the total system.

3.6.3.3 *Free Brownian motion*   As an example we study free Brownian motion, that is $V = 0$, for an Ohmic spectral density. For an infinite cutoff the fundamental solutions then take the form

$$G_1(t) = 1, \quad G_2(t) = \frac{1}{2\gamma}\left(1 - e^{-2\gamma t}\right), \tag{3.471}$$

which yields

$$x(t) = x(0) + \frac{1 - e^{-2\gamma t}}{2\gamma}\dot{x}(0) + \int_0^t ds \frac{1 - e^{-2\gamma(t-s)}}{2m\gamma}B(s), \tag{3.472}$$

$$p(t) = e^{-2\gamma t}p(0) + \int_0^t ds e^{-2\gamma(t-s)}B(s). \tag{3.473}$$

Let us consider a factorizing initial state $\rho(0) = \rho_S(0) \otimes \rho_B$. The initial coordinates of the Brownian particle and of the bath oscillators are then uncorrelated and averages of the form $\langle x(0)B(s) \rangle$ and $\langle p(0)B(s) \rangle$ vanish. With the help of (3.473) the mean kinetic energy of the Brownian particle, for example, is found to be

$$E(t) \equiv \frac{1}{2m} \langle p^2(t) \rangle \qquad (3.474)$$

$$= \frac{1}{2m} \langle p^2(0) \rangle e^{-4\gamma t} + \frac{1}{4m} \int_0^t ds \int_0^t ds' e^{-2\gamma(2t-s-s')} D_1(s-s')$$

$$= \frac{1}{2m} \langle p^2(0) \rangle e^{-4\gamma t} + \frac{\gamma\hbar}{\pi} \int_0^\Omega d\omega\, \omega \coth(\hbar\omega/2k_BT) \frac{\left| e^{i\omega t} - e^{-2\gamma t} \right|^2}{\omega^2 + (2\gamma)^2},$$

where we have used expression (3.462) for the noise kernel. If we perform here the limit $\gamma t \longrightarrow \infty$ we find that the mean kinetic energy approaches the value

$$E(\infty) = \frac{2\gamma}{\pi} \int_0^\Omega d\omega \frac{\hbar\omega \left(N(\omega) + \frac{1}{2} \right)}{\omega^2 + (2\gamma)^2} \equiv E_{\text{th}} + E_{\text{vac}}. \qquad (3.475)$$

This equation provides a spectral representation of the mean kinetic energy of the Brownian particle: Each frequency component is given by the mean energy of the environmental oscillators at that frequency times a Lorentzian-type spectral density.

The thermal contribution E_{th} to the kinetic energy is given by an integral which converges in the limit of an infinite cutoff,

$$E_{\text{th}} = \frac{2\gamma}{\pi} \int_0^\infty d\omega \frac{\hbar\omega N(\omega)}{\omega^2 + (2\gamma)^2}. \qquad (3.476)$$

In the high-temperature limit $k_BT \gg \hbar\gamma$ this expression leads to the well-known result of classical statistical mechanics, $E_{\text{th}} \approx \frac{1}{2}k_BT$. On the other hand, the vacuum contribution E_{vac} is seen to diverge logarithmically with the cutoff,

$$E_{\text{vac}} = \frac{2\gamma}{\pi} \int_0^\Omega d\omega \frac{\hbar\omega/2}{\omega^2 + (2\gamma)^2} = \frac{\gamma\hbar}{2\pi} \ln\left[1 + \left(\frac{\Omega}{2\gamma} \right)^2 \right]. \qquad (3.477)$$

The interaction of the Brownian particle with the vacuum fluctuations of the high-frequency modes of the environment thus yields a logarithmically divergent contribution to its mean kinetic energy.

As a further example we determine the mean square displacement

$$d^2(t, t') = \langle [x(t) - x(t')]^2 \rangle \qquad (3.478)$$

of the particle which is an important quantity in the theory of Brownian motion. On using eqns (3.472) and (3.462) one obtains

$$d^2(t, t') = \frac{1}{4m^2\gamma^2} \left(e^{-2\gamma t} - e^{-2\gamma t'} \right)^2 \langle p^2(0) \rangle \qquad (3.479)$$

$$+\frac{2\gamma\hbar}{m\pi} \int\limits_{0}^{\Omega} d\omega \frac{\omega \coth(\hbar\omega/2k_\mathrm{B}T)}{\omega^2 + (2\gamma)^2} \left| \frac{e^{i\omega t} - e^{i\omega t'}}{i\omega} + \frac{e^{-2\gamma t} - e^{-2\gamma t'}}{2\gamma} \right|^2 .$$

The second term within the squared modulus leads to an integral which diverges in the limit of an infinite cutoff. In fact, the cutoff-dependent contribution to the mean squared displacement is found to be

$$d^2(t, t')|_{\mathrm{singular}} = \frac{1}{4\gamma^2} \left(e^{-2\gamma t} - e^{-2\gamma t'} \right)^2 \frac{2E_\mathrm{vac}}{m}. \qquad (3.480)$$

This shows that the ultraviolet divergent vacuum part of the mean quadratic velocity $2E_\mathrm{vac}/m$ leads to a corresponding divergent contribution of the mean square displacement: The Brownian particle can absorb an arbitrary amount of energy from the high-frequency environmental modes and can travel an arbitrary distance within a finite time interval.

This singular behaviour, known as initial jolts, is clearly a result of the artificial assumption of an uncorrelated initial state. Namely, if we let the times t and t' go to infinity, keeping fixed their difference $\tau \equiv t - t'$, all transient terms vanish and the mean square displacement becomes a function of τ which is given by the ultraviolet convergent integral

$$d^2(\tau) = \frac{2\gamma\hbar}{m\pi} \int\limits_{0}^{\infty} d\omega \frac{\omega \coth(\hbar\omega/2k_\mathrm{B}T)}{\omega^2 + (2\gamma)^2} \frac{4\sin^2(\omega\tau/2)}{\omega^2}. \qquad (3.481)$$

For any finite temperature we can replace the second factor of the integrand by $2\pi|\tau|\delta(\omega)$ which leads to the asymptotic expression

$$d^2(\tau) \approx \frac{k_\mathrm{B}T}{m\gamma} |\tau|. \qquad (3.482)$$

This formula describes the well-known regime of classical diffusion which is valid for $\gamma|\tau| \gg 1$ and $|\tau| \gg \hbar/k_\mathrm{B}T$. The same result is found for a classical Brownian particle. On the other hand, for zero temperature the term $\omega \coth(\hbar\omega/k_\mathrm{B}T)$ in the integrand must be replaced by ω and one obtains the asymptotic expression

$$d^2(\tau) = \frac{8\gamma\hbar}{m\pi} \int\limits_{0}^{\infty} d\omega \frac{\sin^2(\omega\tau/2)}{\omega(\omega^2 + (2\gamma)^2)} \approx \frac{\hbar}{m\pi\gamma} \ln(\gamma|\tau|), \qquad (3.483)$$

which is valid for $\gamma|\tau| \gg 1$. We observe a strong subdiffusive behaviour of the particle in the quantum regime. The mean square displacement increases only weakly with the logarithm of time, the relevant length scale for the quantum diffusion being given by $\sqrt{\hbar/m\pi\gamma}$.

3.6.3.4 *Response function, equilibrium correlation function and the fluctuation–dissipation theorem* In order to determine the equilibrium fluctuations of the total system it is useful to invoke the fluctuation–dissipation theorem (FDT). Together with the exact solutions of the Heisenberg equations of motion the FDT allows us to determine all equilibrium correlations of the Brownian particle. To formulate the FDT we introduce the symmetrized autocorrelation function of some Heisenberg picture operator $z(t)$ of the total system,

$$S(\tau) = \frac{1}{2}\langle\{z(t+\tau), z(t)\}\rangle = \frac{1}{2}\langle\{z(\tau), z(0)\}\rangle. \qquad (3.484)$$

For simplicity we assume that $\langle z(t)\rangle = 0$ and $z^\dagger(t) = z(t)$. We emphasize that the angular brackets denote here the average over the equilibrium distribution of the total system given by

$$\rho = \frac{\exp(-\beta H)}{\operatorname{tr}\exp(-\beta H)}, \qquad (3.485)$$

where we recall that H is the Hamiltonian of the combined system $S + B$. Due to the homogeneity in time the autocorrelation only depends on τ and satisfies $S^*(\tau) = S(\tau) = S(-\tau)$. The spectrum of the equilibrium fluctuations is given by the Fourier transform of the autocorrelation function,

$$\tilde{S}(\omega) = \int\limits_{-\infty}^{+\infty} d\tau \cos\omega\tau S(\tau). \qquad (3.486)$$

We note that the spectrum has the property $\tilde{S}^*(\omega) = \tilde{S}(\omega) = \tilde{S}(-\omega)$.

The response function pertaining to the variable $z(t)$ is defined as

$$\chi(\tau) = \frac{i}{\hbar}\theta(\tau)\langle[z(\tau), z(0)]\rangle. \qquad (3.487)$$

It describes the linear response of the system to an external force $F(t)$ which is applied for times $t > 0$ and is represented by a time-dependent perturbation of the form $V(t) = -zF(t)$ in the Hamiltonian of the total system. First-order perturbation theory then yields the linear response of the system

$$\langle\delta z(t)\rangle = \int\limits_{0}^{\infty} ds\,\chi(t-s)F(s), \qquad (3.488)$$

where $\delta z(t) \equiv z_V(t) - z(t)$ and $z_V(t)$ denotes the Heisenberg operator corresponding to the perturbed Hamiltonian $H + V(t)$. The average first-order change of the variable $z(t)$ is thus given by a linear integral transform of the applied force where the integral kernel is provided by the response function. We note that the linear response at time t only depends on the values of the force at times prior to

t, showing the causal character of the system's response. The Fourier transform of the response function may be written as

$$\tilde{\chi}(\omega) = \int\limits_{-\infty}^{+\infty} d\tau e^{i\omega\tau}\chi(\tau) = \int\limits_{0}^{+\infty} d\tau e^{i\omega\tau}\chi(\tau) \equiv \tilde{\chi}'(\omega) + i\tilde{\chi}''(\omega), \qquad (3.489)$$

where we have decomposed $\tilde{\chi}(\omega)$ into real and imaginary parts.

The fluctuation–dissipation theorem provides a relation between the linear response of the system to an external force and the fluctuations in equilibrium. In frequency language this relation is given by (see, e.g. Landau and Lifshitz, 1958)

$$\tilde{S}(\omega) = \hbar\coth(\hbar\omega/2k_\mathrm{B}T)\tilde{\chi}''(\omega), \qquad (3.490)$$

which expresses the spectrum of the equilibrium fluctuations in terms of the imaginary part of the Fourier transform of the response function.

The fluctuation–dissipation theorem holds for a general system in thermal equilibrium and for any Heisenberg picture observable $z(t)$. Let us now apply the theorem to the coordinate $x(t)$ of a Brownian particle in a harmonic potential. Since the commutator $[x(\tau), x(0)]$ is a c-number we can immediately determine the response function with the help of the exact solution (3.470) of the Heisenberg equation of motion (3.465),

$$\chi(\tau) = \frac{i}{\hbar}\theta(\tau)[x(\tau), x(0)] = \frac{i}{\hbar}\theta(\tau)G_2(\tau)[\dot{x}(0), x(0)] = \theta(\tau)\frac{1}{m}G_2(\tau). \qquad (3.491)$$

This relation can also be inferred directly from (3.470). Namely, adding the perturbation $-xF(t)$ to the total Hamiltonian of the system amounts to replacing $B(t)$ by $B(t)+F(t)$ in the equation of motion (3.465). This replacement yields the additional term $\int_0^t dsG_2(t-s)F(s)/m$ on the right-hand side of (3.470), which shows that the response function is in fact related to the fundamental solution $G_2(\tau)$ by eqn (3.491).

Expressing the Fourier transform of the response function in terms of its Laplace transform we therefore obtain

$$\tilde{\chi}(\omega) = \frac{1}{m}\int\limits_{0}^{+\infty} d\tau e^{i\omega\tau}G_2(\tau) = \frac{1}{m}\hat{G}_2(-i\omega) = \frac{1}{m}\frac{1}{\omega_0^2 - \omega^2 - i\omega\hat{\gamma}(-i\omega)}. \qquad (3.492)$$

The imaginary part of the Fourier transform of the response function is thus

$$\tilde{\chi}''(\omega) = \frac{1}{m}\frac{\omega\Re[\hat{\gamma}(-i\omega)]}{(\omega_0^2 - \omega^2 + \omega\Im[\hat{\gamma}(-i\omega)])^2 + \omega^2(\Re[\hat{\gamma}(-i\omega)])^2}, \qquad (3.493)$$

and the FDT (3.490) immediately leads to the following expression for the autocorrelation function of the coordinate of the Brownian particle,

$$S(\tau) = \int\limits_{-\infty}^{+\infty} \frac{d\omega}{2\pi} \hbar \coth(\hbar\omega/2k_{\mathrm{B}}T) \tilde{\chi}''(\omega) \cos\omega\tau. \tag{3.494}$$

With the help of eqns (3.494) and (3.493) one can determine the exact equilibrium correlations in terms of the Laplace transform $\hat{\gamma}(z)$ of the damping kernel. For example, we have by definition of the autocorrelation function

$$\langle x^2 \rangle = S(0), \tag{3.495}$$

$$\langle px + xp \rangle = 2m\dot{S}(0) = 0, \tag{3.496}$$

$$\langle p^2 \rangle = -m^2 \ddot{S}(0), \tag{3.497}$$

and therefore

$$\langle x^2 \rangle = \int\limits_{-\infty}^{+\infty} \frac{d\omega}{2\pi} \hbar \coth(\hbar\omega/2k_{\mathrm{B}}T) \tilde{\chi}''(\omega), \tag{3.498}$$

$$\langle p^2 \rangle = m^2 \int\limits_{-\infty}^{+\infty} \frac{d\omega}{2\pi} \hbar\omega^2 \coth(\hbar\omega/2k_{\mathrm{B}}T) \tilde{\chi}''(\omega). \tag{3.499}$$

As an example, let us consider the Ohmic spectral density (3.392) with a Lorentz–Drude cutoff. According to definition (3.454) this leads to the damping kernel

$$\gamma(\tau) = 2\gamma\Omega e^{-\Omega|\tau|}, \tag{3.500}$$

with the Laplace transform

$$\hat{\gamma}(z) = \frac{2\gamma\Omega}{\Omega + z}. \tag{3.501}$$

Thus, we have

$$\Re[\hat{\gamma}(\varepsilon - i\omega)] = 2\gamma \frac{\Omega^2}{\Omega^2 + \omega^2} \longrightarrow 2\gamma, \tag{3.502}$$

$$\Im[\hat{\gamma}(\varepsilon - i\omega)] = 2\gamma \frac{\omega}{\Omega} \frac{\Omega^2}{\Omega^2 + \omega^2} \longrightarrow 0, \tag{3.503}$$

where we have indicated the limits of an infinite cutoff, $\Omega \longrightarrow \infty$. The Laplace transform of the damping kernel becomes real for an infinite cutoff and the spectrum of the fluctuations takes the form

$$\tilde{S}(\omega) = \frac{2\hbar\gamma}{m} \frac{\omega}{(\omega_0^2 - \omega^2)^2 + (2\gamma\omega)^2} \coth(\hbar\omega/2k_{\mathrm{B}}T). \tag{3.504}$$

For frequencies such that $k_{\mathrm{B}}T \gg \hbar\omega$ the Planck constant drops out and the spectrum becomes

$$\tilde{S}(\omega) = \frac{4\gamma k_{\mathrm{B}}T}{m} \frac{1}{(\omega_0^2 - \omega^2)^2 + (2\gamma\omega)^2}, \tag{3.505}$$

which exactly corresponds to the classical fluctuation spectrum of a damped harmonic oscillator in thermal equilibrium.

3.6.4   *The influence functional*

In the case of the Caldeira–Leggett model it is possible to eliminate the environmental variables exactly. This is due to the fact that the environment is supposed to be in a thermal state whose characteristic function is Gaussian, and because the reduced system couples linearly to the bath operators $B(t)$ whose commutator represents a c-number. In the present subsection we are going to construct an exact super-operator representation for the reduced density matrix and derive from it the corresponding path integral representation. The path integral involves a certain functional, known as the Feynman–Vernon influence functional, which represents exactly the influence of the heat bath on the reduced system.

3.6.4.1   *The influence super-operator*   To obtain an exact representation for the reduced density matrix, let us assume factorizing initial conditions and that the initial environmental state is a Gaussian (thermal) state. We emphasize that the assumption of an uncorrelated initial state is made here only for simplicity; it is possible to carry out the derivation also for correlated initial states (for a review, see Grabert, Schramm and Ingold, 1988).

An appropriate starting point of the derivation is the following exact representation of the reduced density matrix $\rho_S(t_f)$ at time t_f in terms of the total density $\rho(t_i)$ at some initial time t_i,

$$\rho_S(t_f) = \text{tr}_\text{B} \left\{ U(t_f, t_i)\rho(t_i)U^\dagger(t_f, t_i) \right\}. \tag{3.506}$$

Again, we are working in the interaction picture such that $U(t_f, t_i)$ is the interaction picture time-evolution operator of the total system. To eliminate the bath variables from this expression one can employ the following method of derivation. The interaction picture time-evolution operator involves a time-ordering of the reduced system as well as of the bath variables. First, one eliminates the time-ordering of the bath variables, which is easily done, for the commutator of the bath variables is a c-number. One is then left with a simple average taken over the state of the bath. In a second step this average is determined with the help of a cumulant expansion, which terminates at second order since the bath state is Gaussian. This procedure yields the final result

$$\rho_S(t_f) = \text{T}_{\leftarrow} \exp \left(\frac{i}{\hbar} \Phi[x_c, x_a] \right) \rho_S(t_i). \tag{3.507}$$

The details of the calculations leading to this result will be given in Section 12.2 (see, in particular, eqn (12.99)) in a more general context and will thus be omitted here. In eqn (3.507) the symbol $\text{T}_{\leftarrow}$ indicates, as usual, the chronological time ordering in the interaction picture. The phase of the time-ordered exponential represents a super-operator which is given by

$$\frac{i}{\hbar}\Phi[x_c, x_a] = \int\limits_{t_i}^{t_f} dt \mathcal{L}_c(t) \tag{3.508}$$

$$+ \int\limits_{t_i}^{t_f} dt \int\limits_{t_i}^{t} dt' \left\{ \frac{i}{2\hbar^2} D(t - t') x_c(t) x_a(t') - \frac{1}{2\hbar^2} D_1(t - t') x_c(t) x_c(t') \right\}.$$

The first term on the right-hand side stems from the counter-term H_c in the Hamiltonian of the total system and is defined in terms of the commutator super-operator

$$\mathcal{L}_c(t)\rho_S \equiv -\frac{i}{\hbar}[H_c(t), \rho_S] = -\frac{i}{\hbar}\frac{m}{2}\gamma(0)[x^2(t), \rho_S]. \tag{3.509}$$

The second term on the right-hand side of (3.508) involves the commutator and the anticommutator super-operators

$$x_c(t)\rho_S \equiv [x(t), \rho_S], \quad x_a(t)\rho_S \equiv \{x(t), \rho_S\}, \tag{3.510}$$

where all operators carrying a time argument are interaction picture operators with respect to the free Hamiltonian $H_0 = H_S + H_B$.

The phase $\Phi[x_c, x_a]$ subsumes completely the influence of the environment on the reduced system's dynamics and may thus be called the influence phase functional. Formally, $\Phi[x_c, x_a]$ is both a super-operator acting on the initial state $\rho_S(t_i)$ and a bilinear functional which involves the dissipation kernel $D(t - t')$ and the noise kernel $D_1(t - t')$. We note that the second-order generator (3.390) may be obtained from eqn (3.507) by expanding the exponential to first order in the influence phase and by taking into account that the time integrals in expression (3.508) for the influence phase are already time-ordered.

3.6.4.2 *Path integral representation* Equation (3.507) gives rise to an equivalent path integral representation for the reduced density matrix. It thus provides an operator formulation of the famous Feynman–Vernon influence functional. To construct the path integral representation we turn back to the Schrödinger picture and introduce the so-called propagator function J through the relation

$$\rho_S(x_f, x_f', t_f) = \int dx_i \int dx_i' J(x_f, x_f', t_f; x_i, x_i', t_i)\rho_S(x_i, x_i', t_i). \tag{3.511}$$

Thus, the propagator function is simply the Green function for the reduced density matrix in the position representation. In terms of the influence super-operator in eqn (3.507) it is given through the relation

$$J(x_f, x_f', t_f; x_i, x_i', t_i) = \langle x_f t_f | \left(\mathrm{T}_{\leftarrow} \exp\left(\frac{i}{\hbar}\Phi[x_c, x_a]\right) |x_i t_i\rangle\langle x_i' t_i| \right) |x_f' t_f\rangle, \tag{3.512}$$

where $|xt\rangle$ is the eigenstate of the interaction picture operator $x(t)$ belonging to the eigenvalue x, that is

$$|xt\rangle = \exp\left(+iH_S(t - t_i)/\hbar\right)|x\rangle. \tag{3.513}$$

We note that (3.512) leads to the following initial and normalization conditions,

$$\lim_{t_f \to t_i} J(x_f, x_f', t_f; x_i, x_i', t_i) = \langle x_f|x_i\rangle\langle x_i'|x_f'\rangle$$

$$= \delta(x_f - x_i)\delta(x_f' - x_i'), \tag{3.514}$$

$$\int dx_f J(x_f, x_f, t_f; x_i, x_i', t_i) = \langle x_i'|x_i\rangle = \delta(x_i - x_i'). \tag{3.515}$$

We also observe that for vanishing system–environment coupling ($\Phi = 0$) the propagator function reduces to a product of the Green function G_S for the Schrödinger equation with Hamiltonian H_S,

$$J(x_f, x_f', t_f; x_i, x_i', t_i) = \langle x_f t_f|x_i t_i\rangle\langle x_i' t_i|x_f' t_f\rangle$$

$$= G_S(x_f, t_f; x_i, t_i)G_S^*(x_f', t_f; x_i', t_i). \tag{3.516}$$

The path integral representation of the influence functional is obtained by writing G_S as a path integral:

$$G_S(x_f, t_f; x_i, t_i) = \langle x_f|\exp\left(-iH_S(t_f - t_i)/\hbar\right)|x_i\rangle = \int Dx \exp\left(\frac{i}{\hbar}S_0[x]\right), \tag{3.517}$$

where

$$S_0[x] = \int_{t_i}^{t_f} dt\left(\frac{1}{2}m\dot{x}^2 - V(x)\right) \tag{3.518}$$

is the classical action functional of the free Brownian particle. Here, the path integral is an integral over all paths $x(t)$ subjected to the boundary conditions $x(t_i) = x_i$, and $x(t_f) = x_f$.

Now, if $\mathrm{T}_{\leftarrow}\mathcal{O}[x(t)]$ is any time-ordered product, the corresponding matrix element can be obtained from the path integral as follows

$$\langle x_f t_f|\mathrm{T}_{\leftarrow}\mathcal{O}[x(t)]|x_i t_i\rangle = \int Dx \exp\left(\frac{i}{\hbar}S_0[x]\right)\mathcal{O}[x(t)]. \tag{3.519}$$

In other words, the path integral performs the time ordering automatically if the time-ordered operator product is replaced by the corresponding classical expression under the path integral. In a similar way the matrix elements of the commutator or the anticommutator super-operators $x_{c,a}(t)$ are given as double path integrals as follows,

$$\langle x_f t_f|x_{a,c}(t)\big(|x_i t_i\rangle\langle x_i' t_i|\big)|x_f' t_f\rangle$$

$$= \langle x_f t_f|x(t)|x_i t_i\rangle\langle x_i' t_i|x_f' t_f\rangle \pm \langle x_f t_f|x_i t_i\rangle\langle x_i' t_i|x(t)|x_f' t_f\rangle$$

$$= \int Dx \int Dx' \exp\left(\frac{i}{\hbar}(S_0[x] - S_0[x'])\right)(x(t) \pm x(t')). \tag{3.520}$$

Thus we see that the transition from the super-operator representation to the path integral representation may be achieved by performing the replacements

$$x_a(t) \longrightarrow x(t) + x'(t), \quad x_c(t) \longrightarrow x(t) - x'(t). \qquad (3.521)$$

A moment's thought shows that this prescription is also valid in the general case of an arbitrary time-ordered product of the super-operators $x_{c,a}(t)$. We are therefore led to the expression

$$J(x_f, x'_f, t_f; x_i, x'_i, t_i) = \int Dx \int Dx' \exp\left(\frac{i}{\hbar} A[x, x']\right). \qquad (3.522)$$

This is the double path integral representation for the propagator function which is well known from the Feynman–Vernon influence functional technique (Feynman and Vernon, 1963). It involves an integration over all paths $x(t)$ and $x'(t)$ satisfying the boundary conditions

$$x(t_i) = x_i, \qquad x(t_f) = x_f, \qquad x'(t_i) = x'_i, \qquad x'(t_f) = x'_f. \qquad (3.523)$$

The weight factor of the paths is given by the effective action functional

$$A[x, x'] = S_0[x] - S_0[x'] + \Phi[x, x'], \qquad (3.524)$$

where the path integral representation of the influence phase functional $\Phi[x, x']$ is obtained from (3.508) with the help of the replacements (3.521),

$$\Phi[x, x'] = -\int_{t_i}^{t_f} dt \frac{m}{2}\gamma(0)\left[x^2(t) - x'^2(t)\right] \qquad (3.525)$$

$$+ \int_{t_i}^{t_f} dt \int_{t_i}^{t} dt' \frac{1}{2\hbar} D(t - t')[x(t) - x'(t)][x(t') + x'(t')]$$

$$+ \int_{t_i}^{t_f} dt \int_{t_i}^{t} dt' \frac{i}{2\hbar} D_1(t - t')[x(t) - x'(t)][x(t') - x'(t')].$$

For the further discussion it is convenient to introduce new variables through

$$q = x - x', \qquad r = \frac{1}{2}(x + x'), \qquad (3.526)$$

and to consider the propagator function J as a function of $(r_f, q_f, t_f; r_i, q_i, t_i)$. The corresponding path integral representation reads

$$J(r_f, q_f, t_f; r_i, q_i, t_i) = \int Dr \int Dq \exp\left(\frac{i}{\hbar} A[r, q]\right), \qquad (3.527)$$

which is a double path integral over all paths $r(t)$, $q(t)$ satisfying the boundary conditions

$$r(t_i) = r_i, \qquad r(t_f) = r_f, \qquad q(t_i) = q_i, \qquad q(t_f) = q_f. \qquad (3.528)$$

The initial and normalization conditions for the propagator function read as follows,

$$\lim_{t_f \to t_i} J(r_f, q_f, t_f; r_i, q_i, t_i) = \delta(r_f - r_i)\delta(q_f - q_i), \qquad (3.529)$$

$$\int dr_f J(r_f, q_f = 0, t_f; r_i, q_i, t_i) = \delta(q_i). \qquad (3.530)$$

The effective action functional (3.524) is now given in the new variables by

$$\mathcal{A}[r,q] = \int_{t_i}^{t_f} dt \big(m\dot{r}\dot{q} - V(r+q/2) + V(r-q/2) - m\gamma(0)rq \big) \qquad (3.531)$$

$$+ \int_{t_i}^{t_f} dt \int_{t_i}^{t} dt' \left\{ \frac{1}{\hbar}D(t-t')q(t)r(t') + \frac{i}{2\hbar}D_1(t-t')q(t)q(t') \right\}.$$

3.6.4.3 Classical equation of motion The action functional (3.531) leads to classical equations of motion which are found by setting the first variation of $\mathcal{A}$ equal to zero. The variation with respect to $q(t)$ gives the equation of motion for $r(t)$,

$$\ddot{r}(t) + \frac{1}{2m}\frac{\partial}{\partial r}\big(V(r+q/2) + V(r-q/2)\big) + \frac{d}{dt}\int_{t_i}^{t} dt'\gamma(t-t')r(t')$$

$$= \frac{i}{2m\hbar}\int_{t_i}^{t_f} dt' D_1(t-t')q(t'), \qquad (3.532)$$

whereas the variation of $r(t)$ leads to the equation for $q(t)$,

$$\ddot{q}(t) + \frac{2}{m}\frac{\partial}{\partial q}\big(V(r+q/2) + V(r-q/2)\big) - \frac{d}{dt}\int_{t}^{t_f} dt'\gamma(t'-t)q(t') = 0. \quad (3.533)$$

Here we have expressed the equations of motion through the damping kernel $\gamma(t-t')$ (eqn (3.454)) and the noise kernel $D_1(t-t')$. We note that for $q=0$ the dynamical equation for the path $r(t)$ takes on the form of the homogeneous part of the Heisenberg equation of motion (3.458).

3.6.4.4 *Determining the path integral for harmonic potentials* To illustrate the path integral technique we consider a harmonic potential

$$V(x) = \frac{1}{2} m \omega_0^2 x^2, \tag{3.534}$$

which leads to the classical equations of motion

$$\ddot{r}(t) + \omega_0^2 r(t) + \frac{d}{dt} \int_{t_i}^{t} dt' \gamma(t - t') r(t') = \frac{i}{2m\hbar} \int_{t_i}^{t_f} dt' D_1(t - t') q(t'), \tag{3.535}$$

and

$$\ddot{q}(t) + \omega_0^2 q(t) - \frac{d}{dt} \int_{t}^{t_f} dt' \gamma(t' - t) q(t') = 0. \tag{3.536}$$

Note that eqn (3.536) represents the backward equation of the homogeneous part of eqn (3.535), that is, if $r(t)$ solves the homogeneous part of eqn (3.535), then $q(t) \equiv r(t_f + t_i - t)$ is a solution of eqn (3.536).

Since the action functional is quadratic the double path integral (3.527) can be determined exactly by evaluating the action along the classical solution and by taking into account the Gaussian fluctuations around the classical paths. Let $r(t)$, $q(t)$ be a solution of the classical equations of motion (3.535) and (3.536) satisfying the boundary conditions (3.528). The value of the action functional (3.531) along this solution can be written as

$$\mathcal{A}[r,q] = m \left(\dot{r}_f q_f - \dot{r}_i q_i \right)$$
$$- \int_{t_i}^{t_f} dt\, m q(t) \left(\ddot{r}(t) + \omega_0^2 r(t) + \frac{d}{dt} \int_{t_i}^{t} dt' \gamma(t - t') r(t') \right)$$
$$+ \frac{i}{4\hbar} \int_{t_i}^{t_f} dt \int_{t_i}^{t_f} dt'\, D_1(t - t') q(t) q(t')$$
$$= m \left(\dot{r}_f q_f - \dot{r}_i q_i \right) - \frac{i}{4\hbar} \int_{t_i}^{t_f} dt \int_{t_i}^{t_f} dt'\, D_1(t - t') q(t) q(t'), \tag{3.537}$$

where the equation of motion (3.535) has been used in the second step.

Due to the inhomogeneous term in eqn (3.535) $r(t)$ couples to $q(t)$ through the noise kernel $D_1(t - t')$. As a result the solution $r(t)$ is, in general, complex. We decompose $r(t)$ into real and imaginary parts,

$$r(t) = r^{(1)}(t) + i r^{(2)}(t). \tag{3.538}$$

The real part $r^{(1)}(t)$ solves the homogeneous part of eqn (3.535), while the imaginary part $r^{(2)}(t)$ is a solution of the inhomogeneous equation

$$\ddot{r}^{(2)}(t)+\omega_0^2 r^{(2)}(t)+\frac{d}{dt}\int_{t_i}^{t} dt'\gamma(t-t')r^{(2)}(t') = \frac{1}{2m\hbar}\int_{t_i}^{t_f} dt' D_1(t-t')q(t'). \quad (3.539)$$

We now show the following useful property: In order to find the classical action it suffices to determine the real solution $r^{(1)}(t)$ and to insert it into the action functional, which means that we have the relation

$$\mathcal{A}[r,q] = \mathcal{A}[r^{(1)},q] \quad (3.540)$$

$$= m\left(\dot{r}_f^{(1)}q_f - \dot{r}_i^{(1)}q_i\right) + \frac{i}{4\hbar}\int_{t_i}^{t_f} dt \int_{t_i}^{t_f} dt' D_1(t-t')q(t)q(t').$$

To prove this statement one first uses the equations of motion for the imaginary part $r^{(2)}(t)$ and for $q(t)$ to show that

$$\frac{i}{2\hbar}\int_{t_i}^{t_f} dt \int_{t_i}^{t_f} dt' D_1(t-t')q(t)q(t') = im\left(\dot{r}_f^{(2)}q_f - \dot{r}_i^{(2)}q_i\right). \quad (3.541)$$

To obtain this equation one performs two integrations by part and uses the condition that $r_i^{(2)} = r_f^{(2)} = 0$, since the boundary conditions for the paths are real. Combining the last equation with (3.537) immediately leads to eqn (3.540) which proves the statement.

The procedure to determine the propagator function can thus be summarized as follows. One first solves the homogeneous classical equations of motion,

$$\ddot{r}(t)+\omega_0^2 r(t)+\frac{d}{dt}\int_{t_i}^{t} dt'\gamma(t-t')r(t') = 0, \quad (3.542)$$

$$\ddot{q}(t)+\omega_0^2 q(t)-\frac{d}{dt}\int_{t}^{t_f} dt'\gamma(t'-t)q(t') = 0, \quad (3.543)$$

under the boundary conditions (3.528). Substituting these solutions into the action functional then yields the propagator function

$$J(r_f,q_f,t_f;r_i,q_i,t_i) = N(t_f,t_i)\exp\left(\frac{i}{\hbar}\mathcal{A}[r,q]\right) \quad (3.544)$$

$$= N(t_f,t_i)\exp\left(\frac{im}{\hbar}(\dot{r}_f q_f - \dot{r}_i q_i) + \Gamma(q_f,t_f;q_i,t_i)\right),$$

where

$$\Gamma(q_f, t_f; q_i, t_i) = -\frac{1}{4\hbar^2} \int_{t_i}^{t_f} dt \int_{t_i}^{t_f} dt' D_1(t - t') q(t) q(t'). \qquad (3.545)$$

In eqn (3.544) $N(t_f, t_i)$ is a time-dependent normalization factor. Let us set $t_i = 0$ for simplicity in the following. With the help of the fundamental solutions $G_1(t)$ and $G_2(t)$ (see eqns (3.466) and (3.467)) we can write the solutions of (3.542) and (3.543) as

$$r(t) = \left[G_1(t) - \frac{G_1(t_f)}{G_2(t_f)} G_2(t) \right] r_i + \frac{G_2(t)}{G_2(t_f)} r_f, \qquad (3.546)$$

$$q(t) = \frac{G_2(t_f - t)}{G_2(t_f)} q_i + \left[G_1(t_f - t) - \frac{G_1(t_f)}{G_2(t_f)} G_2(t_f - t) \right] q_f. \qquad (3.547)$$

Invoking the normalization condition (3.530) the normalization factor is found to be

$$N(t_f, 0) = \frac{m}{2\pi\hbar G_2(t_f)}. \qquad (3.548)$$

With the above expressions the problem of determining the exact propagator function is solved completely. Obviously, the propagator function is a Gaussian function whose exponent is a quadratic form in the coordinates (r_f, q_f, r_i, q_i), which is determined by substituting (3.546) and (3.547) into (3.544). We also note that the normalization factor is directly related to the response function through $N(t_f, 0) = 1/2\pi\hbar\chi(t_f)$.

3.6.4.5 *Solution of the Caldeira–Leggett master equation* To give an example we consider free Brownian motion ($\omega_0 = 0$) for an Ohmic environment. In the limit of an infinite cutoff we have the simple expressions $G_1(t) \equiv 1$ and $G_2(t) = [1 - \exp(-2\gamma t)]/2\gamma$ which yields

$$\dot{r}_i = \frac{1}{G_2(t_f)}(r_f - r_i), \quad \dot{r}_f = \frac{\dot{G}_2(t_f)}{G_2(t_f)}(r_f - r_i), \qquad (3.549)$$

showing the translational invariance of the porpagator function. Thus we have

$$J(r_f, q_f, t_f; r_i, q_i, t_i) \qquad (3.550)$$

$$= \frac{m}{2\pi\hbar G_2(t_f)} \exp\left(\frac{im}{\hbar G_2(t_f)}(r_f - r_i)[\dot{G}_2(t_f)q_f - q_i] + \Gamma(q_f, t_f; q_i, 0) \right).$$

This expression for the propagator function can be used to obtain the general solution of the Caldeira–Leggett master equation (3.410). To this end, we just have to evaluate the function $\Gamma(q_f, t_f; q_i, 0)$ in the high-temperature limit which is provided by the expression

$$\Gamma(q_f, t_f; q_i, 0) \approx -\frac{2m\gamma k_{\mathrm{B}} T}{\hbar^2} \int\limits_0^{t_f} dt\, q^2(t). \qquad (3.551)$$

Substituting the solution $q(t)$ (eqn (3.547)) and evaluating the time integral finally yields

$$J(r_f, q_f, t_f; r_i, q_i, 0) \qquad\qquad\qquad\qquad\qquad\qquad (3.552)$$

$$= \frac{2m\gamma}{2\pi\hbar(1 - e^{-2\gamma t_f})}$$

$$\times \exp\left(\frac{2im\gamma/\hbar}{1 - e^{-2\gamma t_f}} (r_f - r_i)(e^{-2\gamma t_f} q_f - q_i) \right.$$

$$-\frac{2mk_{\mathrm{B}}T}{\hbar^2} \left\{ \frac{\gamma t_f - (1 - e^{-2\gamma t_f}) + \frac{1}{4}(1 - e^{-4\gamma t_f})}{(1 - e^{-2\gamma t_f})^2} (q_f - q_i)^2 \right.$$

$$\left.\left. -\frac{\gamma t_f - \frac{1}{2}(1 - e^{-2\gamma t_f})}{1 - e^{-2\gamma t_f}} 2q_f(q_f - q_i) + \gamma t_f q_f^2 \right\} \right).$$

This is an explicit expression for the propagator function of the Caldeira–Leggett master equation for free Brownian motion. With the help of the relations

$$\gamma t_f - \frac{1}{2}(1 - e^{-2\gamma t_f}) = (\gamma t_f)^2 + \mathcal{O}((\gamma t_f)^3), \qquad (3.553)$$

$$\gamma t_f - (1 - e^{-2\gamma t_f}) + \frac{1}{4}(1 - e^{-4\gamma t_f}) = \frac{4}{3}(\gamma t_f)^3 + \mathcal{O}((\gamma t_f)^4), \qquad (3.554)$$

one easily shows that in the limit $\gamma t_f \longrightarrow 0$ the propagator function approaches

$$J(r_f, q_f, t_f; r_i, q_i, 0) \longrightarrow \frac{m}{2\pi\hbar t_f} \exp\left(\frac{im}{\hbar t_f}(r_f - r_i)(q_f - q_i) \right), \qquad (3.555)$$

which is recognized as the Green function corresponding to the von Neumann equation for a free Schrödinger particle.

Let us finally consider the so-called recoilless limit of the master equation. This is the limit of a heavy particle which is studied over times which are short in comparison to the relaxation time. It corresponds to the Caldeira–Leggett master equation (3.410) without the friction term $-i\gamma[x, \{p, \rho_S\}]/\hbar$. Thus we set $\Lambda = 2m\gamma k_{\mathrm{B}}T/\hbar^2$ and perform the limit $\gamma t_f \longrightarrow 0$, keeping Λ fixed. Making use of eqns (3.553), (3.554) the propagator function is found to be

$$J(x_f, x'_f, t_f; x_i, x'_i, 0) = \frac{m}{2\pi\hbar t_f} \exp\left(\frac{im}{2\hbar t_f}\{(x_f - x_i)^2 - (x'_f - x'_i)^2\} \right. \qquad (3.556)$$

$$\left. -\frac{\Lambda t_f}{3}\{(x_f - x'_f)^2 + (x_i - x'_i)^2 + (x_f - x'_f)(x_i - x'_i)\} \right),$$

where we have again used the original coordinates.

3.7 Non-linear quantum master equations

In classical statistical mechanics the dynamics of a many-particle system may often be approximated in terms of a non-linear kinetic equation for the time-dependent distribution function $f(\vec{p}, \vec{x}, t)$ on one-particle phase space (Reichl, 1998). The most prominent example for such a kinetic equation is the Boltzmann equation. For the corresponding approximation of a quantum mechanical many-body system one has to replace the classical distribution $f(\vec{p}, \vec{x}, t)$ by the one-particle density matrix $\rho(t)$. The latter is then expected to satisfy a non-linear master equation which provides a mean-field type description of the dynamics of the many-body system. Within this framework it is also possible to consider many-body systems interacting with external reservoirs. The coupling to an external reservoir may be performed either on the level of the mean-field approximation or else on the level of the many-particle system. A number of examples for the resulting non-linear master equations, such as the quantum Boltzmann equation, the time-dependent Hartree equation, and the non-linear Schrödinger equation, will be discussed below (Alicki and Messer, 1983).

The general structure of a non-linear quantum master equation may be postulated to be of the following form,

$$\frac{d}{dt}\rho(t) = \mathcal{L}[\rho(t)]\rho(t), \quad \rho(0) = \rho_0. \tag{3.557}$$

Here, $\mathcal{L}[\sigma]$ is a super-operator which is in Lindblad form for each fixed density matrix σ. Thus, eqn (3.557) is a Lindblad-type quantum master equation whose generator depends parametrically on the density matrix. Under certain conditions eqn (3.557) has a unique solution

$$\rho(t) = V_t(\rho_0), \quad t \geq 0, \tag{3.558}$$

where $\{V_t | t \geq 0\}$ is a one-parameter family of non-linear maps satisfying the semigroup property

$$V_t(V_s(\rho)) = V_{t+s}(\rho), \quad V_0 = I. \tag{3.559}$$

Equation (3.557) is therefore said to be in *parametric Lindblad form*.

3.7.1 *Quantum Boltzmann equation*

The discrete quantum Boltzmann equation is an example of a non-linear quantum master equation of the form (3.557). In a phenomenological way this equation can be obtained by exploiting the formal analogy to the classical Boltzmann equation for the one-particle density $f(\vec{p}, \vec{x}, t)$,

$$\frac{\partial}{\partial t} f(\vec{p}, \vec{x}, t) = -\{h, f(\vec{p}, \vec{x}, t)\} + C[f](\vec{p}, \vec{x}, t). \tag{3.560}$$

Here, $h = \vec{p}^2/2m$ is the one-particle Hamiltonian of a particle with momentum $\vec{p}$ and mass m, the curly brackets denote the Poisson bracket and $C[f]$ is the

Boltzmann collision operator, which is a bilinear functional of the one-particle density f.

To be specific we consider a system of N identical particles. The internal degrees of freedom of the particles are described by a one-particle Hamiltonian h with a discrete spectrum. The particles interact through collisions during which they may exchange internal energy, whereby the total energy is conserved. An appropriate master equation analogous to the classical Boltzmann equation (3.560) is given by

$$\frac{d}{dt}\rho(t) = -i[h, \rho(t)] + \mathcal{D}(\rho(t)), \tag{3.561}$$

which may be called the quantum Boltzmann equation. The dissipator $\mathcal{D}(\rho)$ which replaces the collision operator of the classical Boltzmann equation is postulated to be given by the bilinear expression

$$\mathcal{D}(\rho) = \text{tr}_2 \left\{ \mathcal{K}(\rho \otimes \rho) \right\}, \tag{3.562}$$

where $\mathcal{K}$ is a Lindblad generator which acts on the two-particle space $\mathcal{H}_1 \otimes \mathcal{H}_2$ as follows,

$$\mathcal{K}(R) = \sum_\alpha \left\{ T_\alpha R T_\alpha^\dagger - \frac{1}{2} T_\alpha^\dagger T_\alpha R - \frac{1}{2} R T_\alpha^\dagger T_\alpha \right\}. \tag{3.563}$$

We have further used the symbol tr_2 to denote the partial trace taken over the second Hilbert space $\mathcal{H}_2$.

The operators T_α introduced in eqn (3.563) act on the two-particle space $\mathcal{H}_1 \otimes \mathcal{H}_2$. They describe the exchange of energy between the particles during the collisions and have to satisfy certain conditions. First, to guarantee the conservation of energy one demands

$$[T_\alpha, h_1 + h_2] = 0. \tag{3.564}$$

Second, micro-reversibility requires that

$$T_\alpha^\dagger T_\alpha = T_\alpha T_\alpha^\dagger, \tag{3.565}$$

and, third, the T_α are supposed to be invariant under permutations $\pi(1, 2)$ of the particles,

$$[\pi(1, 2), T_\alpha] = 0. \tag{3.566}$$

It is easy to show that eqn (3.561) is of parametric Lindblad form (3.557). We note further that the quantum Boltzmann equation, in close analogy to the classical Boltzmann equation, has the following properties. It conserves energy

$$\frac{d}{dt}\text{tr}(h\rho(t)) = 0, \tag{3.567}$$

the H-theorem holds which states that the von Neumann entropy $S(\rho)$ is non-decreasing,

$$\frac{d}{dt}S(\rho(t)) \geq 0, \tag{3.568}$$

and the canonical distribution $\rho_{\text{th}} = \exp(-\beta h)/\text{tr}\exp(-\beta h)$ is a stationary solution, that is $\mathcal{D}(\rho_{\text{th}}) = 0$. These properties can be verified with the help of the conditions (3.564)–(3.566).

3.7.2 Mean field master equations

We consider N identical quantum systems associated with certain sites $j = 1, 2, \ldots, N$ of a lattice, a spin lattice for example. The Hilbert space of the j-th particle is denoted by $\mathcal{H}_j$ and its self-Hamiltonian by $h = h_j$ regarded as an operator acting on $\mathcal{H}_j$. The systems at two different sites i and j interact via a two-particle potential $V = V_{ij}$ which is considered to be an operator on $\mathcal{H}_i \otimes \mathcal{H}_j$. The Hamiltonian of the total system thus takes the form

$$H_N = \sum_{j=1}^{N} h_j + \frac{1}{N}\sum_{i \neq j}^{N} V_{ij} \tag{3.569}$$

and acts on the Hilbert space

$$\mathcal{H} = \mathcal{H}_1 \otimes \mathcal{H}_2 \otimes \ldots \otimes \mathcal{H}_N \tag{3.570}$$

of the total system. The interaction is scaled by a factor $1/N$ to ensure that the total Hamiltonian scales with the particle number $\langle H_N \rangle \sim N$.

Let us denote by tr_n the partial trace over the n-th Hilbert space $\mathcal{H}_n$ and let us introduce the shorthand notation

$$\text{tr}_{[n,N]} = \text{tr}_n \text{tr}_{n+1} \ldots \text{tr}_N \tag{3.571}$$

for the partial trace taken over the Hilbert spaces labelled by n, $n + 1$, ..., N. Then the following theorem on the *mean-field* or *Hartree approximation* holds (Spohn, 1980). Suppose that the initial state of the system is taken to be a product state

$$\rho_N = \rho \otimes \cdots \otimes \rho \quad (N \text{ factors}). \tag{3.572}$$

Then one has in the limit of an infinite particle number

$$\lim_{N \to \infty} \text{tr}_{[n+1,N]}\left\{e^{-iH_Nt}\rho_N e^{iH_Nt}\right\} = \rho(t) \otimes \cdots \otimes \rho(t) \quad (n \text{ factors}), \tag{3.573}$$

where $\rho(t)$ is the solution of the Hartree equation

$$\frac{d}{dt}\rho(t) = -i[h, \rho(t)] - i\text{tr}_2\left[V_{12} + V_{21}, \rho(t) \otimes \rho(t)\right] \tag{3.574}$$

corresponding to the initial condition $\rho(0) = \rho$. The relation (3.573) holds for all n and the convergence is to be understood as a convergence in the trace norm.

Essentially, the mean-field approximation consists in performing the approximation $\rho_2(t) \approx \rho(t) \otimes \rho(t)$ in the hierarchy of coupled equations of motion for the n-particle density matrices

$$\rho_n(t) = \mathrm{tr}_{[n+1,N]}\rho_N(t), \qquad (3.575)$$

where

$$\rho_N(t) = \exp(-iH_Nt)\rho_N \exp(+iH_Nt) \qquad (3.576)$$

represents the exact dynamics of the total system. The above theorem shows that this approximation holds in the limit $N \longrightarrow \infty$.

We will not present the detailed proof of eqn (3.573). Let us demonstrate, however, that the Hartree equation (3.574) is obtained as a consequence of (3.573). To this end, we first observe that for $n = 1$ eqn (3.573) yields

$$\rho(t) = \lim_{N\to\infty} \mathrm{tr}_{[2,N]}\rho_N(t), \qquad (3.577)$$

from which we get

$$\frac{d}{dt}\rho(t) = -i \lim_{N\to\infty} \mathrm{tr}_{[2,N]} \left\{ \sum_j [h_j, \rho_N(t)] + \frac{1}{N} \sum_{i\neq j} [V_{ij}, \rho_N(t)] \right\}. \qquad (3.578)$$

The first term on the right-hand side can be written as

$$-i \lim_{N\to\infty} \mathrm{tr}_{[2,N]} \left\{ [h_1, \rho_N(t)] + \sum_{j=2}^{N}[h_j, \rho_N(t)] \right\}$$

$$= -i \left[h, \lim_{N\to\infty} \mathrm{tr}_{[2,N]}\rho_N(t) \right] - i \lim_{N\to\infty} \sum_{j=2}^{N} \mathrm{tr}_{[2,N]}[h_j, \rho_N(t)]$$

$$= -i[h, \rho(t)], \qquad (3.579)$$

where we made use of the fact that the partial trace $\mathrm{tr}_{[2,N]}[h_j, \rho_N(t)]$ vanishes for $j = 2, 3, \ldots, N$. Hence, we can write the equation of motion for the density matrix $\rho(t)$ in the form

$$\frac{d}{dt}\rho(t) = -i[h, \rho(t)] - i \lim_{N\to\infty} \frac{1}{N} \sum_{i\neq j} \mathrm{tr}_{[2,N]}[V_{ij}, \rho_N(t)]. \qquad (3.580)$$

Due to the trace over $\mathcal{H}_2 \otimes \cdots \otimes \mathcal{H}_N$ only those terms in the second term on the right-hand side of (3.580) survive for which either $i = 1$ or $j = 1$. The second term in eqn (3.580) is thus equal to

$$-i \lim_{N\to\infty} \frac{1}{N} \sum_{j=2}^{N} \mathrm{tr}_{[2,N]}[V_{1j} + V_{j1}, \rho_N(t)]. \qquad (3.581)$$

Because of the symmetry of ρ_N and $\rho_N(t)$ all terms of the sum over j are equal to each other. Hence, expression (3.581) can be written as

$$-i \lim_{N \to \infty} \frac{N-1}{N} \text{tr}_{[2,N]}[V_{12} + V_{21}, \rho_N(t)] = -i\text{tr}_2[V_{12} + V_{21}, \rho(t) \otimes \rho(t)], \quad (3.582)$$

where we have used $\text{tr}_{[2,N]} = \text{tr}_2 \text{tr}_{[3,N]}$ and eqn (3.573) for $n = 2$, that is

$$\rho(t) \otimes \rho(t) = \lim_{N \to \infty} \text{tr}_{[3,N]} \rho_N(t). \quad (3.583)$$

This shows that the Hartree equation (3.574) follows from the mean-field approximation (3.573).

3.7.3 Mean field laser equations

As an example of the application of the Hartree equation with linear dissipation we consider a simple model for the laser, namely the model developed by Haken (1984), Lax (1966), Louisell (1990), and Gordon (1967). To this end, we first derive the mean-field equation governing the atom–laser mode dynamics, and then couple external reservoirs to the system to describe the pumping of the atoms and the field losses.

The atoms are described as two-level systems with transition frequency ω living in the Hilbert space $\mathcal{H}_A \cong \mathbb{C}^2$. The atoms are coupled to a laser mode described by creation and annihilation operators $a^\dagger$, a acting on the single-mode Fock space $\mathcal{H}_F$. Employing the rotating wave approximation and assuming exact resonance we can write the atom–laser mode Hamiltonian as

$$H = \sum_{j=1}^{N} \frac{\omega}{2} \sigma_j^3 + \omega a^\dagger a + i\tilde{g} \sum_{j=1}^{N} (a^\dagger \sigma_j^- - a\sigma_j^+) \equiv H_A + H_F + H_{AF}. \quad (3.584)$$

H_A and H_F are the free Hamiltonians for the atoms and for the laser mode, respectively, and $\sigma_j^3, \sigma_j^+, \sigma_j^-$ are the j-th atom's 3-component of the Pauli matrices and the corresponding raising and lowering operators in the usual notation. The atom–field interaction is given by H_{AF} with coupling constant $\tilde{g}$.

In order to bring the Hamiltonian of the system into the form (3.569) required to apply the mean-field theory we make use of the following trick. We introduce N identical copies a_j of the field mode, satisfying the commutation relations $[a_i, a_j^\dagger] = \delta_{ij}$, and perform the following replacements,

$$a \longrightarrow \frac{1}{\sqrt{N}} \sum_j a_j, \quad a^\dagger a \longrightarrow \sum_j a_j^\dagger a_j. \quad (3.585)$$

If we also scale the coupling constant as

$$\tilde{g} = \frac{g}{\sqrt{N}}, \quad (3.586)$$

we can write the atom–laser mode Hamiltonian as

$$H = \sum_{j=1}^{N} \omega a_j^\dagger a_j + \sum_{j=1}^{N} \frac{\omega}{2} \sigma_j^3 + \frac{ig}{N} \sum_{i \neq j} \left(a_i^\dagger \sigma_j^- - a_i \sigma_j^+ \right). \tag{3.587}$$

This representation allows us to exploit the following identification,

$$h_j = \omega a_j^\dagger a_j + \frac{\omega}{2} \sigma_j^3, \tag{3.588}$$

$$V_{ij} = ig \left(a_i^\dagger \sigma_j^- - a_i \sigma_j^+ \right). \tag{3.589}$$

Hence, the Hartree equation (3.574) becomes

$$\begin{aligned}
\frac{d}{dt} \rho(t) = &-i \left[\omega a^\dagger a + \frac{\omega}{2} \sigma^3, \rho(t) \right] \\
&+ g \left[\mathrm{tr}(\sigma^- \rho(t)) a^\dagger - \mathrm{tr}(\sigma^+ \rho(t)) a, \rho(t) \right] \\
&+ g \left[\mathrm{tr}(a^\dagger \rho(t)) \sigma^- - \mathrm{tr}(a\rho(t)) \sigma^+, \rho(t) \right].
\end{aligned} \tag{3.590}$$

This is the mean-field approximation of the atom–laser mode dynamics.

The next step is to consider the losses of the radiation field and the pumping of the two-level atoms. To this end, the laser mode and the atoms are coupled to two different reservoirs. These couplings are modelled within the weak-coupling approximation and lead to two Lindblad generators which must be added to the right-hand side of the mean-field equation:

$$\begin{aligned}
\frac{d}{dt} \rho(t) = &-i[\omega a^\dagger a + \frac{\omega}{2} \sigma^3, \rho(t)] \\
&+ g \left[\mathrm{tr}(\sigma^- \rho(t)) a^\dagger - \mathrm{tr}(\sigma^+ \rho(t)) a, \rho(t) \right] \\
&+ g \left[\mathrm{tr}(a^\dagger \rho(t)) \sigma^- - \mathrm{tr}(a\rho(t)) \sigma^+, \rho(t) \right] \\
&+ 2\kappa \left(a\rho(t) a^\dagger - \frac{1}{2} a^\dagger a\rho(t) - \frac{1}{2} \rho(t) a^\dagger a \right) \\
&+ W_{21} \left(\sigma^- \rho(t) \sigma^+ - \frac{1}{2} \sigma^+ \sigma^- \rho(t) - \frac{1}{2} \rho(t) \sigma^+ \sigma^- \right) \\
&+ W_{12} \left(\sigma^+ \rho(t) \sigma^- - \frac{1}{2} \sigma^- \sigma^+ \rho(t) - \frac{1}{2} \rho(t) \sigma^- \sigma^+ \right). \tag{3.591}
\end{aligned}$$

The damping of the laser mode through losses and output is described by the Lindblad operator a and by a damping constant 2κ. The reservoir for the atoms leads to the Lindblad operators σ^- and σ^+, where we have written W_{21} and W_{12} for the rates of downward and upward transitions, respectively. To account for the pumping the reservoir of the atoms is taken to be a reservoir with negative temperature, that is $W_{12} > W_{21}$.

It is now easy to verify that eqn (3.591) leads to the following closed system of equations for the mean values $\langle a(t) \rangle = \text{tr}(a\rho(t))$, $\langle \sigma^-(t) \rangle = \text{tr}(\sigma^-\rho(t))$ and $\langle \sigma^3(t) \rangle = \text{tr}(\sigma^3\rho(t))$,

$$\frac{d}{dt}\langle a(t) \rangle = (-i\omega - \kappa)\langle a(t) \rangle + g\langle \sigma^-(t) \rangle, \tag{3.592}$$

$$\frac{d}{dt}\langle \sigma^-(t) \rangle = (-i\omega - \gamma)\langle \sigma^-(t) \rangle + g\langle a(t) \rangle\langle \sigma^3(t) \rangle, \tag{3.593}$$

$$\frac{d}{dt}\langle \sigma^3(t) \rangle = -2g\left(\langle a^\dagger(t) \rangle\langle \sigma^-(t) \rangle + \langle a(t) \rangle\langle \sigma^+(t) \rangle\right)$$
$$-2\gamma\left(\langle \sigma^3(t) \rangle - d\right). \tag{3.594}$$

Here, we have introduced the abbreviations

$$\gamma \equiv \frac{1}{2}(W_{12} + W_{21}), \tag{3.595}$$

and

$$d \equiv \frac{W_{12} - W_{21}}{W_{12} + W_{21}}. \tag{3.596}$$

Introducing as new variables the negative frequency part of the laser field

$$A(t) = \langle a(t) \rangle \exp(+i\omega t), \tag{3.597}$$

the negative frequency part of the polarization

$$S(t) = \langle \sigma^-(t) \rangle \exp(+i\omega t), \tag{3.598}$$

and the inversion

$$D(t) = \langle \sigma^3(t) \rangle, \tag{3.599}$$

we may write the system of differential equations (3.592)–(3.594) as

$$\frac{d}{dt}A(t) = -\kappa A(t) + gS(t), \tag{3.600}$$

$$\frac{d}{dt}S(t) = -\gamma S(t) + gA(t)D(t), \tag{3.601}$$

$$\frac{d}{dt}D(t) = -2g(A^*(t)S(t) + A(t)S(t)^*) - 2\gamma(D(t) - d). \tag{3.602}$$

These are the mean-field laser equations. With the help of an appropriate transformation of the variables A, S, and D this non-linear system of differential equations can be shown to be equivalent to the famous Lorentz equations of fluid dynamics (Haken, 1975). We note that the non-linear terms are the terms which are proportional to g. These terms stem from the mean-field coupling between the field and the atoms. As is easily seen the system (3.600)–(3.602)

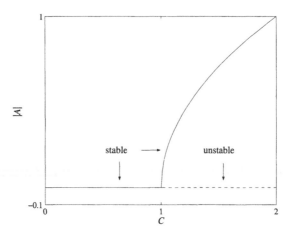

FIG. 3.8. Bifurcation diagram of the non-linear equation (3.604) for the laser field amplitude A.

has a non-trivial stationary solution $A \neq 0$ corresponding to the emergence of a mean coherent field (lasing action) provided the so-called pump parameter

$$C \equiv \frac{dg^2}{\gamma\kappa} \tag{3.603}$$

satisfies $C > 1$. This is obviously possible only for $g \neq 0$, i.e. lasing action may be viewed as a collective mean-field coupling.

The system (3.600)–(3.602) is determined through the three time constants κ^{-1}, g^{-1}, and γ^{-1}. Since γ^{-1} is the smallest time in the problem we may adiabatically eliminate the fast variables S and D. Setting $\dot{S} = \dot{D} = 0$ we find a differential equation for the field amplitude,

$$\frac{dA}{dt} = -\kappa A \left(1 - \frac{C}{1 + |A|^2/n_0} \right), \tag{3.604}$$

where $n_0 = \gamma^2/2g^2$. From this equation we read off the stationary solutions as well as their stability properties. The stationary solution $A = 0$ is unique and stable for $C < 1$. At $C = 1$ a Hopf bifurcation occurs: For $C > 1$ the state $A = 0$ becomes unstable, while a new one-parameter family of stationary states emerges which is determined by $|A|^2 = n_0(C - 1)$ (see Fig. 3.8).

3.7.4 Non-linear Schrödinger equation

In the previous example of the laser equations we first performed the mean-field approximation and then added linear Lindblad generators to describe the coupling to the reservoirs. This procedure corresponds to a weak-coupling approximation on the mean-field level. In the following we consider the case in which the weak-coupling approximation is performed first on the level of the N-particle system, and then the mean-field approximation is made. This approximation procedure leads to non-linear dissipative terms in the master equation and gives rise to a non-linear Schrödinger-type equation.

The starting point is again a total Hamiltonian of the form (3.569). The total system of N particles is coupled to some reservoir. In the weak-coupling approximation the interaction with the reservoir is supposed to lead to a Lindblad-type generator in the master equation for the total density matrix $\rho_N(t)$,

$$\frac{d}{dt}\rho_N(t) = -i\left[\sum_j h_j + \frac{1}{N}\sum_{i\neq j}V_{ij}, \rho_N(t)\right] \tag{3.605}$$

$$+\frac{1}{N}\sum_\alpha\left\{W^\alpha\rho_N(t)W^{\alpha\dagger} - \frac{1}{2}W^{\alpha\dagger}W^\alpha\rho_N(t) - \frac{1}{2}\rho_N(t)W^{\alpha\dagger}W^\alpha\right\}.$$

The operators W^α in this equation are supposed to be collective Lindblad operators of the form

$$W^\alpha = \sum_i V_i^\alpha, \tag{3.606}$$

where the V_i^α are single-particle operators acting on the Hilbert space of the i-th particle. An example for such a master equation will be given in the next subsection.

The mean-field approximation of the above master equation is obtained in a similar way as in Section 3.7.2. Namely, in the limit of large N one obtains the mean-field master equation

$$\frac{d}{dt}\rho(t) = -i[h, \rho(t)] - i\mathrm{tr}_2[V_{12} + V_{21}, \rho(t)\otimes\rho(t)] - i\left[V_D(\rho(t)), \rho(t)\right]. \tag{3.607}$$

The first two terms on the right-hand side provide the Hartree approximation, while the third term describes collective dissipative processes through a non-linear potential given by

$$V_D(\rho) = \frac{i}{2}\sum_\alpha\left\{\mathrm{tr}(V^{\alpha\dagger}\rho)V^\alpha - \mathrm{tr}(V^\alpha\rho)V^{\alpha\dagger}\right\}. \tag{3.608}$$

To demonstrate the emergence of this potential we determine the quantity

$$\lim_{N\to\infty}\mathrm{tr}_{[2,N]}\frac{1}{N}\sum_\alpha\sum_{ij}\left\{V_i^\alpha\rho_N V_j^{\alpha\dagger} - \frac{1}{2}V_j^{\alpha\dagger}V_i^\alpha\rho_N - \frac{1}{2}\rho_N V_j^{\alpha\dagger}V_i^\alpha\right\} \tag{3.609}$$

$$= \lim_{N\to\infty}\sum_\alpha\mathrm{tr}_{[2,N]}\left\{V_2^\alpha\rho_N V_1^{\alpha\dagger} - \frac{1}{2}V_1^{\alpha\dagger}V_2^\alpha\rho_N - \frac{1}{2}\rho_N V_1^{\alpha\dagger}V_2^\alpha\right\} + (1\leftrightarrow 2).$$

The symbol $(1\leftrightarrow 2)$ indicates that the terms with indices 1 and 2 interchanged must be added. Invoking the mean-field approximation (3.573) for $n=2$ we now get

$$\sum_\alpha \text{tr}_2 \left\{ V_2^\alpha \rho \otimes \rho V_1^{\alpha\dagger} - \frac{1}{2} V_1^{\alpha\dagger} V_2^\alpha \rho \otimes \rho - \frac{1}{2} \rho \otimes \rho V_1^{\alpha\dagger} V_2^\alpha \right\} + (1 \leftrightarrow 2)$$

$$= \sum_\alpha \left\{ \text{tr}(V^\alpha \rho)\rho V^{\alpha\dagger} - \frac{1}{2}\text{tr}(V^\alpha \rho)V^{\alpha\dagger}\rho - \frac{1}{2}\text{tr}(V^\alpha \rho)\rho V^{\alpha\dagger} \right.$$

$$\left. + \text{tr}(V^{\alpha\dagger}\rho)V^\alpha \rho - \frac{1}{2}\text{tr}(V^{\alpha\dagger}\rho)V^\alpha \rho - \frac{1}{2}\text{tr}(V^{\alpha\dagger}\rho)\rho V^\alpha \right\}$$

$$= \frac{1}{2}\sum_\alpha \left\{ \text{tr}(V^{\alpha\dagger}\rho)[V^\alpha, \rho] - \text{tr}(V^\alpha \rho)[V^{\alpha\dagger}, \rho] \right\}$$

$$= -i[V_D(\rho), \rho(t)]. \tag{3.610}$$

This shows that the master equation (3.605) gives rise to the non-linear potential (3.608) on the mean-field level.

The non-linear mean-field master equation (3.607) has the property that it can be written in the form of a *non-linear Schrödinger equation*. Namely, if the state vector $\psi(t)$ satisfies the non-linear Schrödinger-type equation

$$\frac{d}{dt}\psi(t) = -i\left[h + V_H(\psi) + V_D(\psi)\right]\psi(t) \tag{3.611}$$

then the pure state density matrix

$$\rho(t) = |\psi(t)\rangle\langle\psi(t)| \tag{3.612}$$

is a solution of eqn (3.607). In eqn (3.611) h is the one-particle Hamiltonian, $V_H(\psi) = \langle\psi|V_{12} + V_{21}|\psi\rangle$ represents the Hartree potential and $V_D(\psi)$ denotes the non-linear potential

$$V_D(\psi) = \frac{i}{2}\sum_\alpha \left\{ \langle\psi|V_\alpha^\dagger|\psi\rangle V_\alpha - \langle\psi|V_\alpha|\psi\rangle V_\alpha^\dagger \right\}. \tag{3.613}$$

Obviously, the non-linear Schrödinger equation (3.611) preserves the norm of the state vector since $\text{tr}\rho(t) = ||\psi(t)||^2 = 1$.

3.7.5 *Super-radiance*

As an example of the non-linear dissipative processes and the non-linear Schrödinger equation (3.611) we study here the phenomenon of super-radiance (Gross and Haroche, 1982). To this end, we consider N identical two-level atoms with transition frequency ω interacting with the vacuum of the electromagnetic field. The atoms are located at fixed positions $\vec{r}_j$, $j = 1, 2, \ldots, N$. The Pauli matrices associated with the j-th atom are again denoted by σ_j^3, $\sigma_j^\pm$. The Hamiltonian of the atoms is taken to be

$$H_A = \frac{\omega}{2}\sum_{j=1}^N \sigma_j^3, \tag{3.614}$$

while the interaction of the atoms with the radiation field will be described in the dipole approximation,

$$H_I = -\sum_{j=1}^{N} \left(\sigma_j^- \vec{d} \cdot \vec{E}(\vec{r}_j) + \sigma_j^+ \vec{d}^* \cdot \vec{E}(\vec{r}_j) \right). \tag{3.615}$$

The vector $\vec{D}_j = \vec{d}(\sigma_j^+ + \sigma_j^-)$ represents the dipole operator of the j-th atom, $\vec{d}$ being the dipole matrix element of the atomic transition. Finally, the Schrödinger picture operator for the electric field at position $\vec{r}$ is (compare eqn (3.180))

$$\vec{E}(\vec{r}) = i \sum_{\vec{k},\lambda} \sqrt{\frac{2\pi\omega_k}{V}} \vec{e}_\lambda(\vec{k}) \left(b_\lambda(\vec{k}) \exp(i\vec{k}\cdot\vec{r}) - b_\lambda^\dagger(\vec{k}) \exp(-i\vec{k}\cdot\vec{r}) \right). \tag{3.616}$$

From the general form (3.143) of the dissipator for the weak-coupling limit we immediately obtain the following master equation for the N-atom system,

$$\frac{d}{dt}\rho_N(t) = -i\omega \sum_{i=1}^{N} [\sigma_i^3, \rho_N(t)]$$
$$+ \sum_{ij} a_{ij} \left\{ \sigma_j^- \rho_N \sigma_i^+ - \frac{1}{2}\sigma_i^+ \sigma_j^- \rho_N - \frac{1}{2}\rho_N \sigma_i^+ \sigma_j^- \right\}. \tag{3.617}$$

The coefficients a_{ij} are given by

$$a_{ij} = \int_{-\infty}^{\infty} dt \exp(i\omega t) \langle \vec{d}^* \cdot \vec{E}(\vec{r}_i, t) \vec{d} \cdot \vec{E}(\vec{r}_j, 0) \rangle. \tag{3.618}$$

In eqn (3.617) Lamb shift contributions have been neglected and the temperature of the radiation field was supposed to be zero. Thus, the correlation functions in eqn (3.618) are vacuum expectation values.

The Fourier transforms of the reservoir correlation functions can be determined explicitly. Performing the continuum limit of the radiation modes and proceeding as in Section 3.4.1.2 one gets

$$a_{ij} = \frac{\omega^3}{2\pi} \int d\Omega \left(\vec{d}^* \cdot \vec{d} - \frac{(\vec{k}\cdot\vec{d}^*)(\vec{k}\cdot\vec{d})}{|k|^2} \right) \exp(i\vec{k}(\vec{r}_i - \vec{r}_j)). \tag{3.619}$$

The angular integration can be carried out to yield

$$a_{ij} = \frac{4}{3}\omega^3 |\vec{d}|^2 \left\{ j_0(x_{ij}) + P_2(\cos\theta_{ij}) j_2(x_{ij}) \right\}, \tag{3.620}$$

where we have introduced the Bessel functions

$$j_0(x) = \frac{\sin x}{x}, \quad j_2(x) = \left(\frac{3}{x^3} - \frac{1}{x} \right) \sin x - \frac{3}{x^2} \cos x, \tag{3.621}$$

and the Legendre polynomial

$$P_2(\cos\theta) = \frac{1}{2}\left(3\cos^2\theta - 1\right), \tag{3.622}$$

and defined

$$x_{ij} = \omega|\vec{r}_i - \vec{r}_j|, \quad \cos^2\theta_{ij} = \frac{|\vec{d}\cdot(\vec{r}_i - \vec{r}_j)|^2}{|\vec{d}|^2\cdot|\vec{r}_i - \vec{r}_j|^2}. \tag{3.623}$$

The eigenvalues of the $N \times N$ matrix a_{ij} determine the relaxation rates of the system. The diagonals of this matrix are equal to the spontaneous emission rate for a single atom,

$$a_{ii} = \frac{4}{3}\omega^3|\vec{d}|^2 \equiv \gamma_0. \tag{3.624}$$

Let us suppose that all atoms are initially in the excited state which yields the initial state

$$|\Psi_0\rangle = |e, e, \dots, e\rangle \quad (N \text{ factors}) \tag{3.625}$$

for the N-atom system. According to eqns (3.620) the matrix a_{ij} is approximately diagonal provided the mean distance between the atoms is large compared to the wavelength $\bar{\lambda} \equiv 1/\omega$ of the radiation. Hence, in this case we have $a_{ij} \approx \gamma_0\delta_{ij}$. The atoms therefore radiate essentially independently of each other and the radiated intensity follows an exponential decay law

$$I(t) = \gamma_0\omega N e^{-\gamma_0 t}. \tag{3.626}$$

The radiation emitted by the atoms adds incoherently which leads to a total intensity proportional to N.

We denote by r the linear dimension of the N-atom system. For $r \ll \bar{\lambda}$ the matrix elements are approximately equal to each other, i.e. $a_{ij} \approx \gamma_0$. One then finds a qualitatively different behaviour: After a certain delay time τ_D the emitted radiation occurs in a short burst whose maximal intensity is proportional to N^2, while its width scales with N^{-1}. This phenomenon is called super-radiance (see Fig. 3.9).

To understand this behaviour qualitatively we first note that due to the condition $r \ll \bar{\lambda}$ the coupling (3.615) is approximately symmetric under exchanges of the atoms. Therefore, we may assume that the wave function $|\Psi(t)\rangle$ of the total system stays symmetric under the time evolution. Identifying the two-level atoms with spin-$\frac{1}{2}$ particles we may introduce the total 'angular momentum' operator

$$\vec{J} = \sum_{j=1}^{N} \frac{1}{2}\vec{\sigma}_j. \tag{3.627}$$

Its 3-component is proportional to the atomic Hamiltonian H_A,

$$J^3 = \sum_{j=1}^{N} \frac{1}{2}\sigma_j^3 = \frac{1}{\omega}H_A. \tag{3.628}$$

The corresponding raising and lowering operators take the form

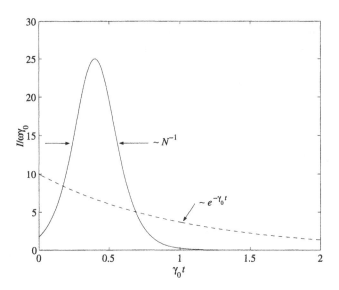

FIG. 3.9. Qualitative picture of the radiation intensity $I(t)$ of a system of N two-level atoms. Dashed line: Independent spontaneous emission processes lead to the exponential decay law (3.626). Solid line: The emission of super-radiance occurs in a short burst with a width proportional to N^{-1} and a height proportional to N^2. The super-radiance intensity profile was taken from the mean-field result (3.643) with $N = 10$ and $\gamma_0 t_D = 0.4$.

$$J^{\pm} = \sum_{j=1}^{N} \sigma_j^{\pm}. \tag{3.629}$$

With the help of these operators the energy eigenstates of H_A may be characterized by the simultaneous eigenstates $|J, M\rangle$ of $\vec{J}^2$ and J^3, where $J = 0, 1, 2, \ldots, N/2$ and $M = -J, \ldots, +J$. These states are known as Dicke states (see, e.g. Mandel and Wolf, 1995). As is well known the subspace which is totally symmetric under permutations of the atoms is spanned by the states with maximal angular momentum $J = N/2$. We can thus order the totally symmetric eigenstates of the system with decreasing energy as follows,

$$\begin{aligned}
|J, M = J\rangle &= |e, e, \ldots, e\rangle \\
|J, M = J - 1\rangle &= \mathcal{S}|g, e, \ldots, e\rangle \\
|J, M = J - 2\rangle &= \mathcal{S}|g, g, e, \ldots, e\rangle \\
&\cdots\cdots\cdots \\
|J, M = -J\rangle &= |g, \ldots, g\rangle,
\end{aligned}$$

where $\mathcal{S}$ is the symmetrization operator. In view of this scheme super-radiance can be considered as a cascade emission along a ladder of 'angular momentum' eigenstates. The emission rate $W(M)$ for the transition $|J, M\rangle \longrightarrow |J, M - 1\rangle$ is given by the square of a certain Clebsch–Gordan coefficient,

$$W(M) = \gamma_0 |\langle J, M - 1|J^-|J, M\rangle|^2 = \gamma_0 (J - M + 1)(J + M). \quad (3.630)$$

Thus we see that the emission rate increases from the initial value $W(M = J) = 2\gamma_0 J = \gamma_0 N$ to a maximal value

$$W(M = 0, 1) = \gamma_0 J(J + 1) = \frac{\gamma_0}{4} N(N + 2), \quad \text{for } N \text{ even}, \quad (3.631)$$

$$W(M = 1/2) = \gamma_0 (J + 1/2)^2 = \frac{\gamma_0}{4}(N + 1)^2, \quad \text{for } N \text{ odd}, \quad (3.632)$$

which scales with the square N^2 of the number of atoms. Finally, the rate decreases again to $W(M = -J) = 0$.

Let us now apply the mean-field theory. For $r \ll \bar{\lambda}$ we get the master equation

$$\frac{d}{dt}\rho_N(t) = -i\omega[J^3, \rho_N(t)]$$
$$+ \gamma_0 \left(J^- \rho_N(t)J^+ - \frac{1}{2}J^+J^- \rho_N(t) - \frac{1}{2}\rho_N(t)J^+J^- \right). \quad (3.633)$$

This equation is of the form of the mean-field master equation (3.605) with the collective Lindblad operator J^- and we can immediately write down the corresponding non-linear Schrödinger equation (3.611),

$$\frac{d}{dt}\psi(t) = -i\frac{\omega}{2}\sigma^3\psi(t) - iV_D(\psi(t))\psi(t), \quad (3.634)$$

where

$$V_D(\psi) = i\frac{N\gamma_0}{2} \left(\langle\psi|\sigma^+|\psi\rangle\sigma^- - \langle\psi|\sigma^-|\psi\rangle\sigma^+ \right) \quad (3.635)$$

is the non-linear potential. Introducing the parametrization

$$\psi(t) = \begin{pmatrix} \psi_1(t) \\ \psi_2(t) \end{pmatrix} = \begin{pmatrix} \sqrt{p(t)}\exp(i\theta(t)) \\ \sqrt{1 - p(t)}\exp(i\phi(t)) \end{pmatrix} \quad (3.636)$$

and using the representations (3.213) and (3.214) of the Pauli matrices one finds the system of differential equations

$$\dot{p} + 2ip\dot{\theta} = -i\omega p - N\gamma_0 p(1 - p), \quad (3.637)$$

$$\dot{p} - 2i(1 - p)\dot{\phi} = -i\omega(1 - p) - N\gamma_0 p(1 - p). \quad (3.638)$$

Separating real and imaginary parts we obtain $2\dot{\theta} = -\omega$ and $2\dot{\phi} = +\omega$. The phases θ and ϕ of the mean-field wave function thus evolve according to the free dynamics.

Since $\psi(t)$ represents the mean-field approximation of the N-particle system the quantity $Np(t) = N|\psi_1(t)|^2$ must be interpreted as the mean number of atoms in the excited state. Hence,

$$I(t) = -N\omega\frac{d}{dt}p(t) \tag{3.639}$$

is the mean intensity of the radiation. From the system of differential equations (3.637) and (3.638) we obtain a differential equation for $p(t)$,

$$\dot{p} = -N\gamma_0 p(1-p), \tag{3.640}$$

which has the solution

$$\frac{p(t)}{1-p(t)} = \frac{p_D}{1-p_D}\exp[-\gamma_0 N(t-t_D)]. \tag{3.641}$$

We identify t_D with the delay time of the super-radiance and $p(t = t_D) = p_D$ is the mean occupation probability of the excited state $|e\rangle$ at time $t = t_D$, i.e. at the time of maximal intensity. The above interpretation in terms of the cascade process (see eqns (3.631) and (3.632)) yields $p_D \approx 1/2$, such that we finally obtain

$$p(t) = \frac{1}{\exp[\gamma_0 N(t-t_D)]+1}. \tag{3.642}$$

The intensity of the radiation is therefore given by

$$I(t) = \frac{\gamma_0 \omega N^2}{4}\left[\cosh\left(\frac{\gamma_0 N}{2}(t-t_D)\right)\right]^{-2}. \tag{3.643}$$

We note that this formula already describes the characteristic features of super-radiance (Gross and Haroche, 1982) as shown in Fig. 3.9: The intensity is proportional to N^2, the maximal intensity is reached after a delay time t_D, and the width of the radiation pulse scales with N^{-1}. It must be noted, however, that the delay time exhibits strong fluctuations which are determined by the quantum fluctuations in the initial phase of the super-radiance process. The simple mean-field description given above cannot account for the statistics of the delay time.

References

Alicki, R. and Messer, J. (1983). Nonlinear quantum dynamical semigroups for many-body open systems. *J. Stat. Phys.*, **32**, 299–312.

Alicki, R. and Lendi, K. (1987). *Quantum Dynamical Semigroups and Applications*, Volume 286 of *Lecture Notes in Physics*. Springer-Verlag, Berlin.

Alicki, R. and Fannes, M. (2001). *Quantum Dynamical Systems*. Oxford University Press, Oxford.

Bausch, R. (1966). Bewegungsgesetze nicht abgeschlossener Quantensysteme. *Z. Phys.*, **193**, 246–265.

Blum, K. (1981). *Density Matrix Theory and Applications*. Plenum Press, New York, London.

Caldeira, A. O. and Leggett, A. J. (1983). Path integral approach to quantum Brownian motion. *Physica*, **121A**, 587–616.

Cohen-Tannoudji, C., Dupont-Roc, J. and Grynberg, G. (1998). *Atom–Photon Interactions*. John Wiley, New York.

Davies, E. B. (1974). Markovian master equations. *Commun. Math. Phys.*, **39**, 91–110.

Davies, E. B. (1976). *Quantum Theory of Open Systems*. Academic Press, London.

Dümcke, R. and Spohn, H. (1979). The proper form of the generator in the weak coupling limit. *Z. Phys.*, **B34**, 419–422.

Dümcke, R. (1985). The low density limit for an N-level system interacting with a free Bose or Fermi gas. *Commun. Math. Phys.*, **97**, 331–359.

Feynman, R. P. and Vernon, F. L. (1963). The theory of a general quantum system interacting with a linear dissipative system. *Ann. Phys. (N.Y.)*, **24**, 118–173.

Gardiner, C. W. and Collett, M. J. (1985). Input and output in damped quantum systems: Quantum stochastic differential equations and the quantum master equation. *Phys. Rev.*, **A31**, 3761–3774.

Gardiner, C. W. and Zoller, P. (2000). *Quantum Noise* (second edition). Springer-Verlag, Berlin.

Glauber, R. J. (1963). Coherent and incoherent states of the radiation field. *Phys. Rev.*, **131**, 2766–2788.

Gordon, J. P. (1967). Quantum theory of a simple maser oscillator. *Phys. Rev.*, **161**, 367–386.

Gorini, V., Kossakowski, A. and Sudarshan, E. C. G. (1976). Completely positive dynamical semigroups of N-level systems. *J. Math. Phys.*, **17**, 821–825.

Gorini, V., Frigerio, A., Verri, M., Kossakowski, A. and Sudarshan, E. C. G. (1978). Properties of quantum Markovian master equations. *Rep. Math. Phys.*, **13**, 149–173.

Grabert, H., Schramm, P. and Ingold, G.-L. (1988). Quantum Brownian motion: The functional integral approach. *Phys. Rep.*, **168**, 115–207.

Gross, M. and Haroche, S. (1982). Superradiance: An essay on the theory of collective spontaneous emission. *Phys. Rep.*, **93**, 301–396.

Haake, F. (1973). *Statistical treatment of open systems by generalized master equations*. Springer Tracts in Modern Physics, **66**, 98–168.

Haken, H. (1975). Analogy between higher instabilities in fluids and lasers. *Phys. Lett.*, **A53**, 77–78.

Haken, H. (1984). *Laser Theory*. Springer-Verlag, Berlin.

Hartig, W., Rasmussen, W., Schieder, R. and Walther, H. (1976). Study of the frequency distribution of the fluorescent light induced by monochromatic radiation. *Z. Phys.*, **A278**, 205–210.

Itano, W. M., Heinzen, D. J., Bollinger, J. J. and Wineland, D. J. (1990).

Quantum Zeno effect. *Phys. Rev.*, **A41**, 2295–2300.

Klauder, J. R., McKenna, J. and Currie, D. G. (1965). On "diagonal" coherent-state representations for quantum-mechanical density matrices. *J. Math. Phys.*, **6**, 734–739.

Kraus, K. (1971). General state changes in quantum theory. *Ann. Phys. (N.Y.)*, **64**, 311–335.

Landau, L. D. and Lifshitz, E. M. (1958). *Statistical Physics*. Pergamon Press, London.

Lax, M. (1963). Formal theory of quantum fluctuations from a driven state. *Phys. Rev.*, **129**, 2342–2348.

Lax, M. (1966). Quantum noise. IV. Quantum theory of noise sources. *Phys. Rev.*, **145**, 110–129.

Lindblad, G. (1975). Completely positive maps and entropy inequalities. *Commun. Math. Phys.*, **40**, 147–151.

Lindblad, G. (1976). On the generator of quantum dynamical semigroups. *Commun. Math. Phys.*, **48**, 119–130.

Louisell, W. (1990). *Quantum Statistical Properties of Radiation*. John Wiley, New York.

Mandel, L. and Wolf, E. (1995). *Optical Coherence and Quantum Optics*. Cambridge University Press, Cambridge.

Mollow, B. R. (1969). Power spectrum of light scattered by two-level systems. *Phys. Rev.*, **188**, 1969–1975.

Redfield, A. G. (1957). On the theory of relaxation processes. *IBM J. Res. Dev.*, **1**, 19–31.

Reichl, L. E. (1998). *A Modern Course in Statistical Physics* (second edition). John Wiley, New York.

Scully, M. O. and Zubairy, M. S. (1997). *Quantum Optics*. Cambridge University Press, Cambridge.

Spohn, H. (1978). Entropy production for quantum dynamical semigroups. *J. Math. Phys.*, **19**, 1227–1230.

Spohn, H. (1980). Kinetic equations from Hamiltonian dynamics: Markovian limits. *Rev. Mod. Phys.*, **53**, 569–615.

Sudarshan, E. C. G. (1963). Equivalence of semiclassical and quantum mechanical descriptions of statistical light beams. *Phys. Rev. Lett.*, **10**, 277–279.

Walls, D. F. and Milburn, G. J. (1994). *Quantum Optics*. Springer-Verlag, Berlin.

Weiss, U. (1999). *Quantum Dissipative Systems* (second edition). World Scientific, Singapore.

Wu, F. Y., Grove, R. E. and Ezekiel, S. (1975). Investigation of the spectrum of resonance fluorescence induced by a monochromatic field. *Phys. Rev. Lett.*, **35**, 1426–1429.

4

DECOHERENCE

The interaction of an open quantum system with its surroundings creates correlations between the states of the system and of the environment. The environment carries information on the open system in the form of these correlations. For certain system–environment interactions the environment behaves similarly to a quantum probe performing a kind of indirect measurement on the open system: After tracing over the environmental degrees of freedom a certain set of states of the open system's Hilbert space exhibits strong stability properties, while superpositions of these states are destroyed in the course of time, often very rapidly or even nearly instantaneously. This environment-induced, dynamical destruction of quantum coherence is called decoherence. It leads to a dynamical selection of a distinguished set of pure states of the open system and counteracts the superposition principle in the Hilbert space of the open system.

The theory of decoherence allows a number of interesting physical applications, ranging from fundamental questions of quantum mechanics to technological applications in quantum information processing. Moreover, modern experimental techniques enable the observation and control of decoherence phenomena and a quantitative comparison with the theoretical analysis. An important application of decoherence is the explanation of the extreme sensitivity of coherent superpositions of macroscopically distinguishable states to the influence of their surroundings.

In a variety of theoretical models it turns out that the environmental interaction leads to a decay of the coherences of such superpositions on extremely short time scales, much shorter than the corresponding relaxation time scales of the open system. Thus, the environment induces the emergence of effective superselection sectors. The latter give rise to a decomposition of the reduced system's Hilbert space into coherent subspaces in such a way that coherences between different subspaces are no longer observable locally, that is through measurements on the reduced system. This destruction of quantum coherences is of particular relevance for the formulation of dynamical models of quantum measurements. It provides a physical mechanism for the selection of a preferred pointer basis, that is for a definite set of apparatus states which designate classical alternative outcomes.

We start in Section 4.1 with a general discussion of the dynamical structure which leads to what may be called ideal environment-induced decoherence, that is to the destruction of quantum coherence without damping. The basic concepts developed, such as the dynamical selection of a preferred basis, the decoherence time, and the emergence of coherent subspaces, are illustrated with the help of a simple, analytically solvable model in Section 4.2. Section 4.3 is devoted to a discussion of various fundamental physical decoherence mechanisms that lead to the space localization of composite quantum objects. These are high-temperature quantum Brownian motion, decoherence through excitation and de-excitation of internal degrees of freedom and decoherence through the scattering of particles. The localization in space will be seen to occur almost instantaneously for macroscopic objects.

The interplay between decoherence and dissipation is investigated in Section 4.4 by means of the example of the damped harmonic oscillator. The central topic will be the determination of the time it takes for the destruction of quantum coherences through a reservoir in the vacuum state and in the presence of thermal noise. As an application of the theoretical analysis we study in Section 4.5 an experiment on the decoherence of electromagnetic field states which was performed by Haroche and coworkers. This experiment enabled them to observe the progression of the decoherence in a mesoscopic variant of Schrödinger's famous *gedanken* experiment involving the superposition of a dead and alive cat.

An exact treatment of decoherence in the Caldeira–Leggett model, including the full non-Markovian dynamics, is presented in Section 4.6. Finally, we discuss in Section 4.7 the rôle of environment-induced decoherence in the quantum theory of measurement.

4.1   The decoherence function

In its purest form decoherence arises for certain types of system–reservoir interactions. These are measurement-type interactions which are used to describe an ideal, indirect measurement (see Section 2.4.6) on the open system whereby the environment plays the rôle of the quantum probe.[9] The characteristic feature of this type of interaction is that the reduced system affects the environment in a way that leads to certain system–reservoir correlations whereby, however, the back-action of the reservoir on certain system states is negligibly small. As a result the damping of the populations of the reduced density matrix in a specific basis representation is small, while the coherences are often found to be strongly decaying on extremely short time scales.

The starting point of our discussion is thus a microscopic Hamiltonian of the form

$$H = H_S + H_B + H_I = H_0 + H_I, \tag{4.1}$$

where the interaction Hamiltonian is taken to be

$$H_I = \sum_n |n\rangle\langle n| \otimes B_n \equiv \sum_n A_n \otimes B_n. \tag{4.2}$$

[9]It must be emphasized that we are not talking here about a real quantum measurement involving a reduction of the state vector, but only about a certain type of system–reservoir interaction whose form is that of an indirect measurement.

This interaction Hamiltonian singles out a specific set of orthonormal basis vectors $|n\rangle$ of the reduced system, while the $B_n = B_n^\dagger$ are arbitrary reservoir operators. We assume further that the system Hamiltonian H_S commutes with the projections $A_n = |n\rangle\langle n|$ which yields

$$[H_0 + H_I, A_n] = [H_0, A_n] = 0, \qquad (4.3)$$

such that the system operators A_n are conserved quantities. As a consequence the mean energy is constant in time, that is

$$\frac{d}{dt}\langle H_S(t)\rangle = 0. \qquad (4.4)$$

The interaction picture interaction Hamiltonian therefore takes the form

$$H_I(t) = e^{iH_0 t} H_I e^{-iH_0 t} = \sum_n |n\rangle\langle n| \otimes B_n(t), \quad B_n(t) = e^{iH_0 t} B_n e^{-iH_0 t}, \qquad (4.5)$$

while the interaction picture time-evolution operator for the combined system can be written as

$$U(t) = \mathrm{T}_\leftarrow \exp\left[-i\int_0^t ds \sum_n |n\rangle\langle n| \otimes B_n(s)\right]. \qquad (4.6)$$

This expressions shows that, as an immediate consequence of the commutation relation (4.3), the basis states $|n\rangle$ are not affected by the coupled dynamics and that the initial state

$$|\Psi(0)\rangle = \sum_n c_n |n\rangle \otimes |\phi\rangle, \qquad (4.7)$$

where $|\phi\rangle$ is an arbitrary reservoir state, evolves into

$$|\Psi(t)\rangle = \sum_n c_n |n\rangle \otimes |\phi_n(t)\rangle, \qquad (4.8)$$

where

$$|\phi_n(t)\rangle = \mathrm{T}_\leftarrow \exp\left[-i\int_0^t ds B_n(s)\right]|\phi\rangle \equiv V_n(t)|\phi\rangle. \qquad (4.9)$$

The state (4.8) is an entangled system–reservoir state given by a superposition of the states $|n\rangle \otimes |\phi_n(t)\rangle$. The latter represent perfect correlations between the various system states $|n\rangle$ and corresponding reservoir states $|\phi_n(t)\rangle$. Due to the measurement-type interaction the reservoir carries information on the system state. However, $|\Psi(t)\rangle$ is still a superposition involving all system states $|n\rangle$

already present in the initial state $|\Psi(0)\rangle$. As a consequence the coherences are still present in the reduced system's density matrix which is given by

$$\rho_S(t) = \text{tr}_B\left\{|\Psi(t)\rangle\langle\Psi(t)|\right\} = \sum_{n,m} c_n c_m^* |n\rangle\langle m| \langle\phi_m(t)|\phi_n(t)\rangle. \tag{4.10}$$

It follows from unitarity that $\langle\phi_n(t)|\phi_n(t)\rangle = 1$, and, thus, the diagonal elements of $\rho_S(t)$ are constant in time. The off-diagonal elements of $\rho_S(t)$, however, do change with time, in general. The time dependence of the matrix element $\langle n|\rho_S(t)|m\rangle$ is given by the overlap of the corresponding reservoir states $|\phi_n(t)\rangle$ and $|\phi_m(t)\rangle$ which will be written as

$$|\langle\phi_n(t)|\phi_m(t)\rangle| = \exp\left[\Gamma_{nm}(t)\right], \quad \Gamma_{nm}(t) \le 0. \tag{4.11}$$

For $n \ne m$ the quantity $\Gamma_{nm}(t)$ describes the behaviour of the off-diagonals of the reduced density matrix and will be called the *decoherence function* in the following.

The time dependence of the decoherence function $\Gamma_{nm}(t)$ strongly depends, in general, on the specific form of the system–reservoir coupling, on the various parameters of the underlying microscopic model, and also on the properties of the initial state. For many physical systems it turns out that the irreversible dynamics induced by the system–reservoir interaction leads to a rapid decrease of the overlap $\langle\phi_n(t)|\phi_m(t)\rangle$ when $n \ne m$. Several examples of this behaviour will be discussed in the following sections. Let us consider here the extreme case that for $n \ne m$ the overlap of the states $|\phi_n(t)\rangle$ and $|\phi_m(t)\rangle$ decreases to zero after times which are large compared to a typical time scale τ_D, the decoherence time,

$$\langle\phi_n(t)|\phi_m(t)\rangle \longrightarrow \delta_{nm} \quad \text{for} \quad t \gg \tau_D. \tag{4.12}$$

This leads to the reduced system's density matrix

$$\rho_S(t) \longrightarrow \sum_n |c_n|^2 |n\rangle\langle n|. \tag{4.13}$$

The coherences of the density matrix in the basis $|n\rangle$ have disappeared as a result of the interaction with the environment: After times $t \gg \tau_D$ the state $\rho_S(t)$ of the reduced system behaves as an incoherent mixture of the states $|n\rangle$ in the sense that interference terms of the form $\langle m|A|n\rangle$, $n \ne m$, no longer appear in the expectation value of any system observable A. Superpositions of the states $|n\rangle$ are therefore effectively destroyed locally which means that they are unobservable for all measurements performed solely on the system S.

The dynamical transition expressed by (4.13) is called decoherence. According to this relation the reduced density matrix becomes diagonal in a particular set of basis states $|n\rangle$ which is sometimes called the preferred basis. It is clear that these basis states are distinguished by the form of the interaction (4.2) between system and environment and by the behaviour of the decoherence function embodied in

(4.12). Moreover, by virtue of the condition (4.3) the preferred basis consists of those states which are not affected during the time evolution. These states are thus stable with respect to the system–environment interaction.

More generally, one can study an initial state of the combined total system of the form

$$\rho(0) = |\psi(0)\rangle\langle\psi(0)| \otimes \rho_B(0), \tag{4.14}$$

where

$$|\psi(0)\rangle = \sum_n c_n |n\rangle \tag{4.15}$$

is the system's initial state and $\rho_B(0)$ may be any reservoir density matrix, a thermal equilibrium state, for example. The reduced system's density at time t can then be written as

$$\rho_S(t) = \sum_{nm} c_n c_m^* |n\rangle\langle m| \operatorname{tr}_B \left\{ V_m^{-1}(t) V_n(t) \rho_B(0) \right\}, \tag{4.16}$$

such that the decoherence function takes the form

$$\Gamma_{nm}(t) = \ln |\langle V_m^{-1}(t) V_n(t) \rangle |. \tag{4.17}$$

Here, the angular brackets denote the expectation value taken over the initial density $\rho_B(0)$ of the reservoir.

As we have seen in eqn (4.4) the mean system energy is constant in time for the simple class of models studied here. By contrast, the entropy of the reduced system does depend on time, in general, since the initial pure state is transformed into a statistical mixture in the course of time. Let us determine the behaviour of the linear entropy (2.127),

$$S_l(\rho_S(t)) = 1 - \sum_{nm} |c_n|^2 |c_m|^2 \exp\left[2\Gamma_{nm}(t)\right]. \tag{4.18}$$

This shows that the linear entropy can be expressed in terms of the initial populations $|c_n|^2$ and of the decoherence function $\Gamma_{nm}(t)$. Note that $S_l(\rho_S(0)) = 0$ since we started out from a pure initial state. For complete decoherence in the long time limit we get

$$S_l(\rho_S(+\infty)) = 1 - \sum_n |c_n|^4. \tag{4.19}$$

For example, if the initial state (4.15) is maximally entangled, that is if all non-vanishing amplitudes c_n are equal in magnitude, $|c_n| = 1/\sqrt{D}$ for $n = 1, 2, \ldots, D$, we find that the linear entropy attains its maximal possible value in a D-dimensional space,

$$S_l(\rho_S(+\infty)) = 1 - \frac{1}{D}. \tag{4.20}$$

The interaction Hamiltonian (4.2) distinguishes a specific set of basis states $|n\rangle$. It is precisely this basis of states which is not influenced by the environmental interaction since the projections $A_n = |n\rangle\langle n|$ are conserved quantities.

By contrast, superpositions of these basis states react, in general, extremely sensitively to this interaction. Let us consider again an interaction of the form $H_I = \sum_n A_n \otimes B_n$, where now the operators

$$A_n = \sum_{j=1}^{d_n} |nj\rangle\langle nj| \tag{4.21}$$

single out an orthogonal decomposition of the system's state space $\mathcal{H}_S$ into linear subspaces $A_n\mathcal{H}_S$ of dimensions $d_n \geq 1$. The solution of the time-dependent Schrödinger equation corresponding to the initial condition

$$|\psi(0)\rangle = \sum_{n,j} c_{nj}|nj\rangle \otimes |\phi\rangle, \tag{4.22}$$

then immediately yields the reduced density matrix

$$\rho_S(t) = \sum_{nm}\sum_{jj'} c_{nj}c_{mj'}^*|nj\rangle\langle mj'|\langle\phi_m(t)|\phi_n(t)\rangle. \tag{4.23}$$

Under the condition of complete decoherence expressed by (4.12) this becomes

$$\rho_S(t) \longrightarrow \sum_n\sum_{jj'} c_{nj}c_{nj'}^*|nj\rangle\langle nj'| \quad \text{for} \quad t \gg \tau_D. \tag{4.24}$$

Thus we see that the coherences between the states $|nj\rangle$ for different j and a fixed n are still visible in the reduced system's density matrix. In other words, coherences referring to one and the same subspace $A_n\mathcal{H}_S$ are not affected by the environmental interaction. The latter thus gives rise to an orthogonal decomposition of the systems Hilbert space into *coherent subspaces* $A_n\mathcal{H}_S$, or *superselection sectors*,

$$\mathcal{H}_S = \sum_n A_n\mathcal{H}_S, \tag{4.25}$$

such that coherences are only observable locally within one and the same subspace. The measurement of any system observable of the form $A = \sum_n a_n A_n$, that is of any system observable which commutes with the projections A_n, constitutes a QND measurement on the state (4.24). This is the general framework for an environment-induced superselection rule, which has been called *einselection* by Zurek (1998).

We close this section with an investigation of the short-time behaviour of the decoherence function. To this end, we consider for simplicity the important case that the average of the reservoir operators B_n taken over the initial state $\rho_B(0)$ are equal to zero, that is $\langle B_n(t)\rangle = 0$. Using then the short-time expansion of the unitary evolution operator $V(t)$ introduced in eqn (4.9) one easily deduces

$$\Gamma_{nm}(t) \approx -\frac{t^2}{2}\left\langle (B_n - B_m)^2 \right\rangle, \tag{4.26}$$

showing that $\Gamma_{nm}(t)$ is proportional to the square of t for short times. Correspondingly, the short-time behaviour of the linear entropy (4.18) is found to be

$$S_l(\rho_S(t)) \approx t^2 \sum_{n \neq m} |c_n|^2 |c_m|^2 \left\langle (B_n - B_m)^2 \right\rangle, \tag{4.27}$$

which grows with the square of the time.

4.2 An exactly solvable model

The general discussion of the preceding section can be illustrated with the help of a specific system–reservoir model (Unruh, 1995; Palma, Suominen and Ekert, 1996). The advantage of this model is that, on the one hand, it shows several important features of decoherence and that, on the other hand, it is simple enough to allow for an exact analytic solution. As will be seen the model exhibits decoherence in its purest form, namely the destruction of quantum coherence without decay of populations.

4.2.1 *Time evolution of the total system*

We consider a two-state system which is coupled to a reservoir of harmonic oscillators. The total Hamiltonian in the Schrödinger picture is taken to be

$$
\begin{aligned}
H &= H_S + H_B + H_I \\
&= H_0 + H_I \\
&= \frac{\omega_0}{2}\sigma_3 + \sum_k \omega_k b_k^\dagger b_k + \sum_k \sigma_3 \left(g_k b_k^\dagger + g_k^* b_k \right),
\end{aligned} \tag{4.28}
$$

where ω_0 is the level spacing of the two-state system and k labels the reservoir modes with frequencies ω_k and bosonic creation and annihilation operators $b_k^\dagger$ and b_k, satisfying

$$\left[b_k, b_{k'}^\dagger \right] = \delta_{kk'}. \tag{4.29}$$

The g_k are coupling constants which describe the coupling of the two-state system to the reservoir modes b_k through the Pauli matrix σ_3. The Hamiltonian (4.28) provides a particularly simple example of a spin-boson model. Originally, the model was introduced to study the influence of decoherence in quantum computers (DiVincenzo, 1995). In this context the two-state system is referred to as a *qubit*, the elementary building block of a quantum computer (Steane, 1998; Bouwmeester, Ekert and Zeilinger, 2000).

Let us introduce a basis of states vectors of the qubit through $\sigma_3|0\rangle = -|0\rangle$ and $\sigma_3|1\rangle = +|1\rangle$. In the usual representation we thus have

$$\sigma_3 = \begin{pmatrix} 1 & 0 \\ 0 & -1 \end{pmatrix}, \quad |0\rangle = \begin{pmatrix} 0 \\ 1 \end{pmatrix}, \quad |1\rangle = \begin{pmatrix} 1 \\ 0 \end{pmatrix}. \tag{4.30}$$

The Hamiltonian (4.28) is of the form (4.1) and (4.2) used in our general discussion. In particular, the Pauli matrix σ_3 is a conserved quantity for this model, since it commutes with the total Hamiltonian, $[H, \sigma_3] = 0$. It follows that the populations

$$\rho_{11} = \mathrm{tr}_{S+B} \{|1\rangle\langle 1|\rho(t)\} = \langle 1|\rho_S(t)|1\rangle, \tag{4.31}$$

$$\rho_{00} = \mathrm{tr}_{S+B} \{|0\rangle\langle 0|\rho(t)\} = \langle 0|\rho_S(t)|0\rangle, \tag{4.32}$$

are constant in time, where $\rho(t)$ is the density matrix of the total system. To determine the decoherence function for this model we first observe that the interaction picture interaction Hamiltonian is given by

$$H_I(t) = e^{iH_0 t} H_I e^{-iH_0 t} = \sum_k \sigma_3 \left(g_k b_k^\dagger e^{i\omega_k t} + g_k^* b_k e^{-i\omega_k t} \right), \tag{4.33}$$

and the unitary time-evolution operator in the interaction picture can be written

$$U(t) = \mathrm{T}_{\leftarrow} \exp\left[-i \int_0^t ds H_I(s) \right]. \tag{4.34}$$

Since the commutator of the interaction Hamiltonian at two different times is a c-number function,

$$[H_I(t), H_I(t')] = -2i \sum_k |g_k|^2 \sin \omega_k (t - t') \equiv -2i\varphi(t - t'), \tag{4.35}$$

we obtain,

$$U(t) = \exp\left[-\frac{1}{2} \int_0^t ds \int_0^t ds' \, [H_I(s), H_I(s')] \, \theta(s - s') \right] \exp\left[-i \int_0^t ds H_I(s) \right]$$

$$= \exp\left[i \int_0^t ds \int_0^t ds' \varphi(s - s') \theta(s - s') \right] V(t), \tag{4.36}$$

where the unitary operator $V(t)$ is defined by,

$$V(t) = \exp\left[\frac{1}{2}\sigma_3 \sum_k \left(\alpha_k b_k^\dagger - \alpha_k^* b_k \right) \right], \tag{4.37}$$

with the amplitudes

$$\alpha_k = 2g_k \frac{1 - e^{i\omega_k t}}{\omega_k}. \tag{4.38}$$

Thus, we see that apart from an overall, time-dependent phase factor, the time evolution of the total system is governed by the operator $V(t)$ defined above. The time development is exactly of the form given in the preceding section. Namely, we find for an arbitrary reservoir state $|\phi\rangle$,

$$V(t)\left(|0\rangle \otimes |\phi\rangle\right) = |0\rangle \otimes \prod_k D\left(-\alpha_k/2\right)|\phi\rangle \equiv |0\rangle \otimes |\phi_0(t)\rangle, \qquad (4.39)$$

$$V(t)\left(|1\rangle \otimes |\phi\rangle\right) = |1\rangle \otimes \prod_k D\left(+\alpha_k/2\right)|\phi\rangle \equiv |1\rangle \otimes |\phi_1(t)\rangle, \qquad (4.40)$$

where

$$D(\alpha_k) = \exp\left[\alpha_k b_k^\dagger - \alpha_k^* b_k\right] \qquad (4.41)$$

is the coherent state generator. Thus, the interaction of the system with its environment creates correlations between the system states $|0\rangle$ and $|1\rangle$ and certain reservoir states $|\phi_0(t)\rangle$ and $|\phi_1(t)\rangle$, respectively. If the reservoir is in the vacuum state initially, we find that the reservoir states

$$|\phi_0(t)\rangle = \prod_k |-\alpha_k/2\rangle, \quad |\phi_1(t)\rangle = \prod_k |+\alpha_k/2\rangle \qquad (4.42)$$

are products of coherent states with amplitudes $\pm\alpha_k/2$, where the sign of the shift generated by $D(\alpha_k)$ is triggered by the quantum number of the system state.

4.2.2   Decay of coherences and the decoherence factor

Let us suppose that the initial state of the total system is given by

$$\rho(0) = \rho_S(0) \otimes \rho_B, \quad \rho_B = \frac{1}{Z_B} e^{-\beta H_B}. \qquad (4.43)$$

Here, the reservoir is in a thermal equilibrium state at temperature T, where $\beta = 1/k_B T$ and Z_B is the reservoir partition function. The matrix elements of the system's density matrix can be determined from the relation

$$\rho_{ij}(t) = \langle i|\rho_S(t)|j\rangle = \langle i|\mathrm{tr}_B\left\{V(t)\rho(0)V^{-1}(t)\right\}|j\rangle, \qquad (4.44)$$

where $i, j = 0, 1$. As is easily verified, the populations stay constant in time, $\rho_{11}(t) = \rho_{11}(0)$, $\rho_{00}(t) = \rho_{00}(0)$, while the coherences behave as

$$\rho_{10}(t) = \rho_{01}(t)^* = \rho_{10}(0)e^{\Gamma(t)}. \qquad (4.45)$$

With the help of eqns (4.37) and (4.44) the decoherence function is found to be

$$\Gamma(t) = \ln \mathrm{tr}_B \left\{ \exp\left[\sum_k \left(\alpha_k b_k^\dagger - \alpha_k^* b_k\right)\right] \rho_B \right\} = \sum_k \ln \left\langle \exp\left[\alpha_k b_k^\dagger - \alpha_k^* b_k\right] \right\rangle. \qquad (4.46)$$

The angular brackets denote the expectation value with respect to the thermal distribution ρ_B. The expectation value

$$\chi(\alpha_k, \alpha_k^*) \equiv \left\langle \exp\left[\alpha_k b_k^\dagger - \alpha_k^* b_k\right]\right\rangle \qquad (4.47)$$

is the Wigner characteristic function of the bath mode k. It can easily be determined by noting that it represents a Gaussian function, which immediately leads to the expression

$$\chi(\alpha_k, \alpha_k^*) = \exp\left[-\frac{1}{2}|\alpha_k|^2 \left\langle \{b_k, b_k^\dagger\}\right\rangle\right]. \qquad (4.48)$$

Thus, we find

$$\Gamma(t) = -\sum_k \frac{1}{2}|\alpha_k|^2 \left\langle \{b_k, b_k^\dagger\}\right\rangle = -\sum_k \frac{4|g_k|^2}{\omega_k^2} \coth\left(\omega_k/2k_\mathrm{B}T\right)\left(1 - \cos\omega_k t\right). \qquad (4.49)$$

We now perform the continuum limit of the bath modes. Introducing the density $f(\omega)$ of the modes of frequency ω and defining the spectral density as

$$J(\omega) = 4f(\omega)|g(\omega)|^2, \qquad (4.50)$$

we can write the decoherence function as

$$\Gamma(t) = -\int_0^\infty d\omega\, J(\omega) \coth\left(\omega/2k_\mathrm{B}T\right)\frac{1 - \cos\omega t}{\omega^2}. \qquad (4.51)$$

Thus, we have obtained an explicit expression for the decoherence function for the present model. Obviously, $\Gamma(t)$ depends on the temperature T of the environment and on the form of the spectral density $J(\omega)$. To illustrate the dynamical behaviour of the decoherence function let us take a spectral density of the form

$$J(\omega) = A\omega e^{-\omega/\Omega}. \qquad (4.52)$$

We assume a linear increase of $J(\omega)$ for small frequencies and an exponential frequency cutoff at the cutoff frequency Ω. Such a form for the spectral density is typically obtained in the quantum optical regime, where $g(\omega) \sim \sqrt{\omega}$ and by assuming a one-dimensional field of bath modes with a constant mode density, $f(\omega) = $ constant. Note also that for this case we have a dimensionless coupling constant, that is, the constant prefactor A of the spectral density is dimensionless and will be taken to be equal to 1.

In order to determine the decoherence function it is helpful to split it into a vacuum part $\Gamma_\mathrm{vac}(t)$ and a thermal part $\Gamma_\mathrm{th}(t)$ as follows,

$$\Gamma(t) = \Gamma_\mathrm{vac}(t) + \Gamma_\mathrm{th}(t). \qquad (4.53)$$

The vacuum contribution can be determined explicitly,

$$\Gamma_\mathrm{vac}(t) \equiv -\int_0^\infty d\omega\, e^{-\omega/\Omega}\frac{1 - \cos\omega t}{\omega} = -\frac{1}{2}\ln\left(1 + \Omega^2 t^2\right). \qquad (4.54)$$

It is independent of the temperature and describes how the fluctuations of the field vacuum affect the coherence of the open system. This part depends on the cutoff frequency Ω. The thermal contribution to the decoherence function is given by

$$\Gamma_{\text{th}}(t) \equiv -\int_0^\infty d\omega e^{-\omega/\Omega}\left[\coth\left(\omega/2k_BT\right) - 1\right]\frac{1 - \cos\omega t}{\omega}$$

$$= -\frac{1}{\beta}\int_0^t ds \int_0^\infty dx\, e^{-k_BTx/\Omega}[\coth(x/2) - 1]\sin(sx/\beta). \quad (4.55)$$

If we assume that $k_BT \ll \Omega$ the thermal contribution is found to be

$$\Gamma_{\text{th}}(t) \approx -\frac{1}{\beta}\int_0^t ds \int_0^\infty dx\,[\coth(x/2) - 1]\sin(sx/\beta)$$

$$= -\ln\left[\frac{\sinh\left(t/\tau_B\right)}{t/\tau_B}\right]. \quad (4.56)$$

Here we have carried out the x-integration with the help of the formula

$$\int_0^\infty dx[\coth(x/2) - 1]\sin(\alpha x) = \pi\coth(\pi\alpha) - \frac{1}{\alpha} \quad (4.57)$$

and we have introduced the thermal correlation time

$$\tau_B = \frac{\beta}{\pi} = \frac{1}{\pi k_BT} \approx 2.43 \cdot 10^{-12}\frac{\text{s}}{T[\text{K}]}. \quad (4.58)$$

Summarizing these results, we can write the decoherence function as

$$\Gamma(t) = -\frac{1}{2}\ln\left(1 + \Omega^2 t^2\right) - \ln\left[\frac{\sinh\left(t/\tau_B\right)}{t/\tau_B}\right]. \quad (4.59)$$

One clearly recognizes three different regimes of time:

1. The short-time regime $t \ll \Omega^{-1}$: In this regime the magnitude of Γ increases with the square of t,

$$\Gamma(t) \approx -\frac{1}{2}\Omega^2 t^2, \quad (4.60)$$

which also follows directly from the short-time expansion of the time-evolution operator (see eqn 4.26). Note that the short-time behaviour in this regime is fully determined by the vacuum contribution Γ_{vac}.

2. The vacuum regime $\Omega^{-1} \ll t \ll \tau_B$: Here, we may approximate the decoherence function as

$$\Gamma(t) \approx -\ln\Omega t. \quad (4.61)$$

In this range of time, decoherence effects are mainly due to the vacuum fluctuations of the field.

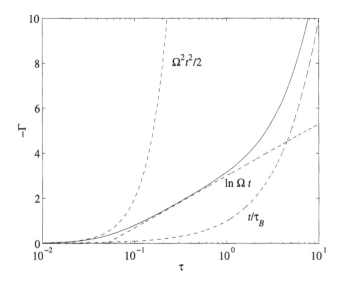

FIG. 4.1. Semilogarithmic plot of the decoherence function $\Gamma(t)$ according to eqn (4.59) as a function of the dimensionless time variable $\tau = t/\tau_B$, where $\Omega\tau_B = 20$ (solid line). The three approximations given in eqns (4.60), (4.61) and (4.62) are also indicated (dashed lines).

3. The thermal regime $\tau_B \ll t$: This regime may also be called the Markovian regime since the magnitude of the decoherence function increases linearly with time,

$$\Gamma(t) \approx -\frac{t}{\tau_B}, \tag{4.62}$$

which means that the off-diagonals of the reduced density matrix decay exponentially with a rate given by τ_B^{-1}.

A plot of the decoherence function $\Gamma(t)$ is shown in Fig. 4.1 together with these approximations.

4.2.3 Coherent subspaces and system-size dependence

The above simple model can also be used to illustrate the emergence of coherent subspaces and the dependence of the decoherence function on the system size (Palma, Suominen and Ekert, 1996). To this end, we consider N qubits, labelled by an index j, which interact with the reservoir through an interaction of the form used in eqn (4.28). Thus, the total Hamiltonian reads

$$H = \frac{\omega_0}{2} \sum_{j=1}^{N} \sigma_3^{(j)} + \sum_k \omega_k b_k^\dagger b_k + \sum_{j=1}^{N} \sum_k \sigma_3^{(j)} \left(g_{kj} b_k^\dagger + g_{kj}^* b_k \right). \tag{4.63}$$

We assume here that the qubits do not interact directly with each other. The coupling constant g_{kj} describes the coupling of the j-th qubit with mode k. We further suppose that the qubits take on fixed positions $\vec{r}^{(j)}$ in space and that the reservoir may be characterized through a certain correlation length r_c. We can then distinguish two extreme situations.

First, we consider the case that the minimal distance between the qubits is large compared to the correlation length r_c of the reservoir. The qubits may then be supposed to interact with independent reservoirs, one for each qubit. Denoting a specific qubit configuration by $\{m^{(j)}\}$, where the $m^{(j)}$ take on the values 0 or 1, we have for the matrix elements of the N-qubit density matrix

$$\left\langle \{\widetilde{m}^{(j)}\} \left| \rho_S(t) \right| \{m^{(j)}\} \right\rangle = \left\langle \{\widetilde{m}^{(j)}\} \left| \rho_S(0) \right| \{m^{(j)}\} \right\rangle \tag{4.64}$$

$$\times \prod_{j=1}^{N} \left\langle \exp\left[\left(\widetilde{m}^{(j)} - m^{(j)}\right) \sum_k \left(\alpha_k b_k^\dagger - \alpha_k^* b_k\right) \right] \right\rangle.$$

The decoherence function can therefore be written as

$$\Gamma(\widetilde{m}^{(j)}, m^{(j)}, t) = \sum_{j=1}^{N} \left| \widetilde{m}^{(j)} - m^{(j)} \right| \Gamma(t), \tag{4.65}$$

where $\Gamma(t)$ is the decoherence function for a single qubit, given by eqn (4.46). Thus, the decoherence function is an integer multiple of the decoherence function of a single qubit. It vanishes if and only if $\widetilde{m}^{(j)} = m^{(j)}$ for all qubits, that is on the diagonal of the N-qubit density matrix. If exactly two single-qubit quantum numbers are different from each other it is given by the expression for a single qubit, while it is proportional to the total number of qubits if the quantum numbers of all qubits are different. In the latter case we have

$$\Gamma(\widetilde{m}^{(j)}, m^{(j)}, t) = N \cdot \Gamma(t), \tag{4.66}$$

corresponding to maximal decoherence.

Let us now consider the case that the linear dimension of the total N-qubit system is small in comparison to the correlation length r_c. The qubits then interact collectively with the reservoir, that is we may assume that for a fixed mode k all g_{kj} are equal to each other. The system of qubits therefore effectively interacts with the reservoir through the collective operator $\sum_j \sigma_3^{(j)}$. The decoherence function is thus found to be

$$\Gamma(\widetilde{m}^{(j)}, m^{(j)}, t) = \left(\widetilde{M} - M\right)^2 \Gamma(t), \tag{4.67}$$

where $M = \sum_j m^{(j)}$ and $\widetilde{M} = \sum_j \widetilde{m}^{(j)}$. This shows that the decoherence function for the N-qubit system is proportional to the square of the difference of the sums M and $\widetilde{M}$ of the quantum numbers $\{m^{(j)}\}$ and $\{\widetilde{m}^{(j)}\}$.

The coherent subspaces are defined by the condition $M = \widetilde{M}$, that is a specific coherent subspace is spanned by the basis states $|\{m^{(j)}\}\rangle$ with a fixed value for the sum $M = \sum_j m^{(j)}$. For example, if $N = 2$ the two-qubit states $|10\rangle$ and $|01\rangle$ form a non-trivial, two-dimensional coherent subspace.

The formula (4.67) demonstrates that collective interactions may lead to a strong amplification of decoherence. The case of maximal decoherence is obtained for $\widetilde{M} = N$, $M = 0$, or for $\widetilde{M} = 0$, $M = N$, which yields

$$\Gamma(\widetilde{m}^{(j)}, m^{(j)}, t) = N^2 \cdot \Gamma(t). \tag{4.68}$$

Summarizing, in the case of independent reservoir interactions the maximal decoherence increases linearly with the system size N, while it grows with the square N^2 of the system size in the case of collective interaction.

4.3 Markovian mechanisms of decoherence

A number of important physical processes leading to decoherence can already be analysed with the help of simple Markovian models. Here we discuss some basic physical mechanisms which cause the decoherence of the centre of mass coordinate of a composite quantum object and, thus, lead to a localization of the object in position space. To make the discussion as simple as possible we restrict ourselves here to the recoilless limit, that is we neglect damping effects. The generalizations to include these and, in addition, non-Markovian effects will be developed in the following sections.

4.3.1 The decoherence rate

We consider a master equation of the following general form,

$$\frac{d}{dt}\rho_S(t) = -i\left[H_S, \rho_S(t)\right] - \Lambda\left[\vec{x}, [\vec{x}, \rho_S(t)]\right], \tag{4.69}$$

where

$$H_S = \frac{1}{2m}\vec{p}^{\,2} \tag{4.70}$$

generates the free evolution of the centre-of-mass coordinate $\vec{x}$ of the quantum object with total mass m. As will be shown in the next sections, the decay of the off-diagonals in the position representation of the reduced density matrix ρ_S often occurs on an extremely short time scale, much shorter than those corresponding to the damping of the diagonals and those of the free evolution of the reduced system. To a first approximation we may thus neglect the free evolution altogether and solve the master equation by disregarding the Hamiltonian part. This yields

$$\rho_S(t, \vec{x}, \vec{x}') \approx \exp\left[-\Lambda\left(\vec{x} - \vec{x}'\right)^2 t\right]\rho_S(0, \vec{x}, \vec{x}'), \tag{4.71}$$

showing that the off-diagonals are damped by a factor $\exp\left[-\Lambda\Delta x^2 t\right]$, where

$$\Delta x \equiv |\vec{x} - \vec{x}'| \tag{4.72}$$

measures the distance to the diagonal of the density matrix. The quantity Λ is referred to as the decoherence rate with dimension $(\text{time})^{-1} \times (\text{length})^{-2}$.

The decoherence function introduced in the preceding section therefore takes the form[10]

$$\Gamma(t) = -\Lambda \Delta x^2 t. \tag{4.73}$$

This allows us to define a corresponding decoherence time as

$$\tau_D = \frac{1}{\Lambda \Delta x^2}. \tag{4.74}$$

In the following we discuss three basic physical mechanisms which may be described by a Markovian master equation of the form (4.69). These are quantum Brownian motion in the high-temperature limit, decoherence through spontaneous and thermally induced transitions of internal degrees of freedom, and decoherence by scattering of an incoming particle flux. It will be seen that in all cases the decoherence rate typically takes the form of a product of a characteristic rate and a squared wavenumber,

$$\Lambda \sim (\text{rate}) \times (\text{wavenumber})^2. \tag{4.75}$$

The rate characterizes the physical process causing the decoherence, i.e. it is equal to the relaxation, to the transition, or to the scattering rate. The wavenumber is determined by the thermal wavelength of the decohering object, by the wavelength of the emitted radiation, or by the de Broglie wavelength of the scattered particles.

4.3.2 *Quantum Brownian motion*

As our first case we discuss the emergence of decoherence in those situations in which the motion of the collective degree of freedom can be described by the Caldeira–Leggett quantum master equation (3.410). In the recoilless limit (see the end of Section 3.6.4.5) the master equation for the collective degree of freedom $\vec{x}$ reduces to the form (4.69), where the decoherence rate is given by (reintroducing factors of $\hbar$)

$$\Lambda = \frac{2m\gamma k_B T}{\hbar^2}. \tag{4.76}$$

Introducing the thermal wavelength

$$\bar{\lambda}_{\text{th}} = \frac{\hbar}{\sqrt{2mk_B T}} \tag{4.77}$$

and the corresponding wavenumber $k_{\text{th}} = 1/\bar{\lambda}_{\text{th}}$ we can cast the decoherence rate into the general form (4.75),

$$\Lambda = \gamma k_{\text{th}}^2. \tag{4.78}$$

As expected, the relevant rate here is the relaxation rate γ while the characteristic wavenumber is given by the thermal wavelength of the object.

---

[10]The above argument of neglecting the free evolution will be made more rigorous in the next sections, see in particular eqn (4.234).

Due to the neglect of the free evolution and of damping effects, the diagonal elements of the density matrix are not affected in the course of the time evolution whereas the off-diagonal elements are strongly decaying. We generally define the relaxation time τ_R to be the decay time of the square of the momentum of the particle. According to eqn (3.427) the momentum of a free Brownian particle relaxes as $\exp(-2\gamma t)$ which gives $\tau_R = 1/4\gamma$. From eqns (4.74) and (4.78) we thus infer that the ratio of decoherence time to relaxation time may be written as

$$\frac{\tau_D}{\tau_R} = 4 \left(\frac{\bar{\lambda}_{\text{th}}}{\Delta x} \right)^2. \tag{4.79}$$

Since for macroscopic objects the thermal wavelength is extremely small this estimate shows that the decoherence time can differ from the relaxation time by many orders of magnitude (Caldeira and Leggett, 1985; Unruh and Zurek, 1989). For example, if we consider a particle of mass $m = 1$ g at room temperature, $T = 300$ K, and assume that $\Delta x = 1$ cm, we find that the ratio of the two time scales τ_D and τ_R is of the order $\tau_D/\tau_R \sim 10^{-40}$ (Zurek, 1991). Thus, even if we choose τ_R to be of the order of the age of the universe ($\tau_R \sim 10^{17}$ s) the decoherence time scale is found to be extremely small, namely $\tau_D \sim 10^{-23}$ s.

4.3.3 Internal degrees of freedom

The destruction of coherences of the centre-of-mass density matrix can be caused by spontaneous or induced transitions involving internal degrees of freedom of the composite object. Again we denote the centre-of-mass coordinate of the object by $\vec{x}$ and concentrate on two internal energy levels $|e\rangle$ (excited state) and $|g\rangle$ (ground state) separated by a transition frequency ω_0. The environment is taken to be a thermal radiation field at temperature T. Thus, we have spontaneous emissions with a rate γ_0 and thermally induced emission and absorption processes. The Markovian master equation for the density matrix $\rho(t)$ of the object, including the centre-of-mass as well as the internal degree of freedom, takes the form of a Lindblad equation

$$\frac{d}{dt}\rho(t) = -i \left[H_S + \omega_0 |e\rangle\langle e|, \rho(t) \right] \tag{4.80}$$

$$+ \frac{\gamma_0(N+1)}{4\pi} \int d\Omega \left[A(\vec{k})\rho(t)A^\dagger(\vec{k}) - \frac{1}{2}\left\{ A^\dagger(\vec{k})A(\vec{k}), \rho(t) \right\} \right]$$

$$+ \frac{\gamma_0 N}{4\pi} \int d\Omega \left[A^\dagger(\vec{k})\rho(t)A(\vec{k}) - \frac{1}{2}\left\{ A(\vec{k})A^\dagger(\vec{k}), \rho(t) \right\} \right].$$

Here, $N = N(\omega_0)$ is the Planck distribution and the integration is performed over the solid angle element $d\Omega$ into the directions $\vec{k}/|\vec{k}|$ of the photon emission, which, for simplicity, is taken to be isotropic.

The Lindblad operator $A(\vec{k})$ provides both the transition $e \longrightarrow g$ of the internal state and the momentum recoil caused by the emission of a photon with wave vector $\vec{k}$. The momentum recoil is described by the operator $\exp(-i\vec{k}\vec{x})$ which changes the momentum of the object by $-\vec{k}$. Hence we have

$$A(\vec{k}) = e^{-i\vec{k}\cdot\vec{x}}\sigma_-, \quad A^\dagger(\vec{k}) = e^{i\vec{k}\cdot\vec{x}}\sigma_+, \tag{4.81}$$

where $\sigma_- = |g\rangle\langle e|$ and $\sigma_+ = |e\rangle\langle g|$. We remark that in the master equation (4.80) the frequency shifts induced by Doppler effect and recoil are neglected, which is justified in the limit of a large total mass m of the object.

Our aim is now to derive an effective master equation for the reduced density matrix of the centre-of-mass coordinate,

$$\rho_S(t) = \mathrm{tr}_{\mathrm{int}}\{\rho(t)\}, \tag{4.82}$$

where the trace is taken over the internal degrees of freedom. On using $\rho = \rho_S \otimes \rho_{\mathrm{int}}$ we immediately find with the help of eqn (4.80),

$$
\frac{d}{dt}\rho_S(t) = -i\,[H_S,\rho_S(t)] \tag{4.83}
$$

$$
+ \frac{\gamma_0(N+1)}{4\pi}p_e(t)\int d\Omega\,\left[e^{-i\vec{k}\cdot\vec{x}}\rho_S(t)e^{i\vec{k}\cdot\vec{x}} - \rho_S(t)\right]
$$

$$
+ \frac{\gamma_0 N}{4\pi}p_g(t)\int d\Omega\,\left[e^{i\vec{k}\cdot\vec{x}}\rho_S(t)e^{-i\vec{k}\cdot\vec{x}} - \rho_S(t)\right],
$$

where $p_e(t) = \langle e|\rho_{\mathrm{int}}(t)|e\rangle$ and $p_g(t) = 1 - p_e(t)$ are the populations of the excited and ground state levels, respectively. These populations are to be determined from the internal density matrix $\rho_{\mathrm{int}}(t)$, which is obtained from $\rho(t)$ by taking the trace over the centre-of-mass coordinate. Since the dynamics governing the internal degrees of freedom decouples from the centre-of-mass motion one immediately finds

$$\frac{d}{dt}p_e(t) = -\gamma_0(2N+1)p_e(t) + \gamma_0 N. \tag{4.84}$$

If the object is initially in the excited state, $p_e(0) = 1$, for example, we get the solution

$$p_e(t) = \frac{N+1}{2N+1}\exp[-\gamma_0(2N+1)t] + \frac{N}{2N+1}. \tag{4.85}$$

Performing the angular integration in (4.83) now yields the following position space representation for the master equation governing the centre-of-mass density matrix,

$$\frac{d}{dt}\rho_S(t,\vec{x},\vec{x}') = -i\langle\vec{x}|\,[H_S,\rho_S(t)]\,|\vec{x}'\rangle \tag{4.86}$$

$$-\gamma_0\,(N+p_e(t))\left[1 - \frac{\sin(k|\vec{x}-\vec{x}'|)}{k|\vec{x}-\vec{x}'|}\right]\rho_S(t,\vec{x},\vec{x}').$$

This is the master equation which describes the motion of the centre-of-mass coordinate $\vec{x}$ of the composite object, where $k = |\vec{k}| = \omega_0/c = 1/\bar{\lambda}$ is the wavenumber of the emitted radiation.

We can distinguish two important limiting cases. First, we suppose that $k\Delta x \gg 1$, that is $\Delta x \gg \bar{\lambda}$. In this limit the second term in (4.86) approaches

$$-\gamma_0 \left(N + p_e(t)\right) \rho_S(t, \vec{x}, \vec{x}'), \tag{4.87}$$

showing that the decay rate of the off-diagonals becomes independent of Δx. The decoherence thus saturates for distances from the diagonal which are large compared to the wavelength of the radiation. Moreover, we note that the decoherence rate is approximately equal to the sum of the rates for spontaneous emission (contribution $\gamma_0 p_e$), for induced emission (contribution $\gamma_0 N p_e$), and for induced absorption (contribution $\gamma_0 N p_g$). The rate given in eqn (4.87) depends on time through the internal dynamics given by eqn (4.84). However, for small times, $\gamma_0(2N+1)t \ll 1$, and with the initial condition $p_e(0) = 1$ the decoherence time is found to be

$$\tau_D \approx \frac{1}{\gamma_0 \left(N + 1\right)}. \tag{4.88}$$

For example, at zero temperature, $N = 0$, we simply get $\tau_D = 1/\gamma_0$. This is an obvious result: If the wavelength of the radiation is small compared to the distance $\Delta x = |\vec{x} - \vec{x}'|$ between two superimposed localized wave packets describing the centre-of-mass coordinate, the emission of a single photon already enables an approximate localization of the object and thus leads to partial destruction of the coherence of the superposition. At zero temperature the decoherence time must therefore be equal to the average time for the emission of a photon, that is to the inverse of the spontaneous emission rate.

Let us now consider the other extreme case, namely $k\Delta x \ll 1$ which means $\Delta x \ll \bar{\lambda}$. An expansion of the term $1 - \sin(k\Delta x)/k\Delta x$ in eqn (4.86), yields a master equation of the general form (4.69) with the following expression for the decoherence rate,

$$\Lambda \approx \frac{1}{6}\gamma_0(N + p_e(t))k^2. \tag{4.89}$$

Again, the decoherence rate depends on time. However, for high temperatures, $N \gg 1$, we get a time-independent rate,

$$\Lambda \approx \frac{1}{6}\gamma_0 N k^2 \approx \frac{1}{6}\gamma_0 \frac{k_B T}{\hbar \omega_0} k^2. \tag{4.90}$$

This decoherence rate is again of the general form (4.75): The characteristic wavenumber k is that of the transition radiation, while the characteristic rate is provided by the rate of thermally induced processes.

On the other hand, at low temperatures and for times t satisfying $\gamma_0 t \ll 1$ the decoherence rate is found to be

$$\Lambda \approx \frac{1}{6}\gamma_0 k^2 \tag{4.91}$$

if the system is initially in the excited state. To give an example, let us consider the $2p \longrightarrow 1s$ transition of hydrogen. For a hydrogen atom we have $\gamma_0 \sim \alpha^3 \omega_0$, where $\alpha = e^2/\hbar c$ is the fine structure constant. This leads to the simple estimate

$$\Lambda \sim \frac{1}{6}c(\alpha k)^3 \sim 10^{20}\frac{1}{\text{cm}^2\text{s}} \tag{4.92}$$

for the decoherence rate of the atomic transition.

4.3.4 Scattering of particles

We consider finally the destruction of coherence through the scattering of an incoming flux of particles off the composite object. In general, many types of scattering reactions contribute to the decoherence. In the case of an incoming photon flux, for example, we could have Thomson scattering, elastic Rayleigh or inelastic Raman scattering. To derive a master equation for the reduced density matrix of the centre-of-mass motion (Gallis and Fleming, 1990; Joos and Zeh, 1985; Joos, 1996) we shall use here the assumptions that the scattering is elastic and that recoil may again be neglected. This means that only the state of the scattered particles changes during the scattering reaction, while the state of a localized target object is left unchanged.

If we take the centre-of-mass coordinate of the object to be a position eigenstate $|\vec{x}\rangle$ we may thus write for the scattering reaction

$$|\vec{x}\rangle|\phi\rangle \longrightarrow S\left(|\vec{x}\rangle|\phi\rangle\right) = |\vec{x}\rangle|\phi_{\vec{x}}\rangle, \tag{4.93}$$

where $|\phi\rangle$ is the incoming wave function and S the scattering matrix. The outgoing wave function is denoted by $|\phi_{\vec{x}}\rangle$. It represents the scattered wave for a scattering centre located at $\vec{x}$. Note that it is assumed here that the scattering time τ_{scatt} is small compared to the typical time scale for the systematic system evolution. As a result of a single scattering reaction the density matrix of the centre-of-mass coordinate of the object then undergoes the transition:

$$\rho_S(\vec{x}, \vec{x}') \longrightarrow \rho_S(\vec{x}, \vec{x}')\langle\phi_{\vec{x}'}|\phi_{\vec{x}}\rangle. \tag{4.94}$$

Thus, the matrix element $\rho_S(\vec{x}, \vec{x}')$ is multiplied by the overlap of the waves scattered by $\vec{x}$ and $\vec{x}'$.

To determine the overlap of the scattered waves we invoke the assumption that the S-matrix commutes with the total momentum, that is with the sum of the object's momentum $\vec{p}$ and of the momentum $\vec{q}$ of the scattered particle,

$$[S, \vec{p} + \vec{q}] = 0. \tag{4.95}$$

On writing the initial state as

$$|\vec{x}\rangle|\phi\rangle = \exp(-i\vec{p}\cdot\vec{x})|\vec{x}=0\rangle|\phi\rangle = \exp\left[-i(\vec{p}+\vec{q})\cdot\vec{x}\right]|\vec{x}=0\rangle\exp(i\vec{q}\cdot\vec{x})|\phi\rangle \tag{4.96}$$

the conservation of total momentum thus gives

$$\begin{aligned}
S\left(|\vec{x}\rangle|\phi\rangle\right) &= \exp\left[-i(\vec{p}+\vec{q})\cdot\vec{x}\right]S\left(|\vec{x}=0\rangle\exp(i\vec{q}\cdot\vec{x})|\phi\rangle\right)\\
&= \exp\left[-i(\vec{p}+\vec{q})\cdot\vec{x}\right]|\vec{x}=0\rangle S_0\left(\exp(i\vec{q}\cdot\vec{x})|\phi\rangle\right)\\
&= |\vec{x}\rangle\exp\left[-i\vec{q}\cdot\vec{x}\right]S_0\left(\exp(i\vec{q}\cdot\vec{x})|\phi\rangle\right)\\
&\equiv |\vec{x}\rangle|\phi_{\vec{x}}\rangle,
\end{aligned} \tag{4.97}$$

where S_0 denotes the S-matrix for the scattering at $\vec{x} = 0$. The overlap of the scattered waves can therefore be written as

$$\langle \phi_{\vec{x}'} | \phi_{\vec{x}} \rangle = \left\langle \phi \left| \exp(-i\vec{q} \cdot \vec{x}') S_0^\dagger \exp\left[-i\vec{q} \cdot (\vec{x} - \vec{x}') \right] S_0 \exp(i\vec{q} \cdot \vec{x}) \right| \phi \right\rangle. \quad (4.98)$$

Next we suppose that the incoming particle state $|\phi\rangle$ represents a momentum eigenstate $|\vec{k}\rangle$ which is normalized to 1 in a quantization volume L^3. Introducing the T-matrix through

$$S_0 = I + iT_0 \quad (4.99)$$

and employing the unitarity of the S-matrix one easily deduces

$$\langle \phi_{\vec{x}'} | \phi_{\vec{x}} \rangle = 1 + \sum_{\vec{k}'} \left(\exp\left[i(\vec{k} - \vec{k}')(\vec{x} - \vec{x}') \right] - 1 \right) |\langle \vec{k}' | T_0 | \vec{k} \rangle|^2. \quad (4.100)$$

We perform the continuum limit, replacing

$$\sum_{\vec{k}'} \longrightarrow \left(\frac{2\pi}{L} \right)^3 \int d^3 k' = \left(\frac{2\pi}{L} \right)^3 \int dk' k'^2 \int d\Omega', \quad (4.101)$$

and

$$|\delta(k' - k)|^2 \longrightarrow \frac{L}{2\pi} \delta(k' - k). \quad (4.102)$$

The scattering amplitude $f(\vec{k}', \vec{k})$ is then defined in terms of the T-matrix as

$$\langle \vec{k}' | T_0 | \vec{k} \rangle = \frac{i}{2\pi k} f(\vec{k}', \vec{k}) \delta(k' - k), \quad (4.103)$$

such that the overlap of the scattered waves can be cast into the form,

$$\langle \phi_{\vec{x}'} | \phi_{\vec{x}} \rangle = 1 + \frac{1}{L^2} \int d\Omega' \left(\exp\left[i(\vec{k} - \vec{k}')(\vec{x} - \vec{x}') \right] - 1 \right) |f(\vec{k}', \vec{k})|^2. \quad (4.104)$$

With this expression we can write the change $\Delta \rho_S$ of the density matrix during a time interval Δt as a result of a single scattering event as follows,

$$\frac{\Delta \rho_S}{\Delta t} = \frac{\langle \phi_{\vec{x}'} | \phi_{\vec{x}} \rangle - 1}{\Delta t} \rho_S. \quad (4.105)$$

Let us now suppose that many scattering reactions take place during the time interval Δt, which implies that the interval Δt must be chosen in such a way that it is much larger than τ_{scatt}, and much smaller than the characteristic time scale of the free evolution of the system. In addition, we also assume that the incoming state may be described by an incoherent mixture of momentum eigenstates $|\vec{k}\rangle$ and that the corresponding distribution of incoming momenta is isotropic. To describe the incoming state we define $dk\, I(k)$ to be the flux of incoming momenta in the interval $[k, k + dk]$, that is the quantity $dk\, I(k) L^2 \Delta t$

is the number of incoming particles during the time interval Δt and having momenta lying in the interval $[k, k + dk]$. The total rate of change of the density matrix is obtained with the help of an integral over all momenta and by an average over all directions. Hence, we have

$$\frac{d\rho_S}{dt} \approx \int dk \int \frac{d\Omega}{4\pi} I(k) L^2 \Delta t \frac{\Delta \rho_S}{\Delta t}. \tag{4.106}$$

Substituting expression (4.105) and adding the free evolution, we thus obtain the density matrix equation (Gallis and Fleming, 1990)

$$\frac{d}{dt}\rho_S(t, \vec{x}, \vec{x}') = i\langle \vec{x}| [H_S, \rho_S(t)] |\vec{x}'\rangle - F(\vec{x} - \vec{x}')\rho_S(t, \vec{x}, \vec{x}'), \tag{4.107}$$

where

$$F(\vec{x} - \vec{x}') = \int dk I(k) \int \frac{d\Omega d\Omega'}{4\pi} \left(1 - \exp\left[i(\vec{k} - \vec{k}')(\vec{x} - \vec{x}') \right] \right) |f(\vec{k}', \vec{k})|^2. \tag{4.108}$$

As in the preceding subsection we may distinguish two important limiting cases. Suppose first that $k_0 \Delta x \gg 1$, where k_0 is a typical wavenumber of the scattered particles. The exponential in expression (4.108) then averages to zero and we get

$$F \approx \int dk I(k) \int \frac{d\Omega d\Omega'}{4\pi} |f(\vec{k}', \vec{k})|^2 = \int dk I(k)\sigma(k) = \gamma_{\text{scatt}}, \tag{4.109}$$

where $\sigma(k)$ is the total cross-section and γ_{scatt} the total scattering rate. Similarly to the decoherence through transitions between internal levels, the decoherence induced by scattering thus saturates for large distances Δx and occurs at a rate which is equal to the total scattering rate. The saturation of the decoherence function is easily understood: For $k_0 \Delta x \gg 1$ the wavelength of the scattered particles is small compared to the distance Δx. A single scattering reaction thus provides sufficient information to localize the object. Increasing Δx any further cannot enhance this information.

On the other hand, for small distances, $k_0 \Delta x \ll 1$, we find

$$F(\vec{x} - \vec{x}') \approx \int dk I(k) \int \frac{d\Omega d\Omega'}{8\pi} \left[(\vec{k} - \vec{k}')(\vec{x} - \vec{x}') \right]^2 |f(\vec{k}', \vec{k})|^2. \tag{4.110}$$

Introducing spherical coordinates such that $d\Omega = d\cos\theta d\varphi$, $d\Omega' = d\cos\theta' d\varphi'$ we find

$$F(\vec{x} - \vec{x}') \approx \int dk I(k)\sigma_{\text{eff}}(k) k^2 (\vec{x} - \vec{x}')^2, \tag{4.111}$$

where

$$\sigma_{\text{eff}}(k) = \int \frac{d\Omega d\Omega'}{8\pi} (\cos\theta - \cos\theta')^2 |f(\vec{k}', \vec{k})|^2 \tag{4.112}$$

may be viewed as an effective cross-section. For example, if the differential cross-section is isotropic one obtains

$$\sigma_{\text{eff}}(k) = \frac{1}{3}\sigma(k). \tag{4.113}$$

These expressions reveal that the decoherence rate describing the decoherence through scattering off the object may be written

$$\Lambda = \int dk I(k)\sigma_{\text{eff}}(k)k^2 \equiv \gamma_{\text{scatt}}k_{\text{av}}^2. \tag{4.114}$$

This is an expression which is again of the general form (4.75). Obviously the relevant wavenumber is here equal to an appropriate average k_{av} of the de Broglie wavenumber of the scattered particles and the characteristic rate is given by the total scattering rate γ_{scatt}.

As a specific example we consider the decoherence through scattering in a photon gas at temperature T (Joos and Zeh, 1985). If the thermal wavelength of the photons in the gas is large compared to the radius a of the object we may assume that the scattering cross-section is given by the Rayleigh cross-section. The evaluation of the formula (4.114), employing an average over the Planck distribution of the photon gas, yields the following estimate for the decoherence rate,

$$\Lambda \sim 10^{20} \left(\frac{T}{K}\right)^9 \left(\frac{a}{\text{cm}}\right)^6 \frac{1}{\text{cm}^2\text{s}}. \tag{4.115}$$

Note the extremely strong dependence on the size a of the object and on the temperature of the gas. The decoherence rate increases with the sixth power of a which is due to the a-dependence of the Rayleigh cross-section in the limit of large wavelengths. The increase of Λ with the nineth power of the temperature T can be understood as follows. First, the photon flux I is proportional to the third power of T, which is just the Stefan–Boltzmann law of black-body radiation. In addition, the average of the product $\sigma(k)k^2$ increases with the sixth power of the thermal wavenumber $k_{\text{th}} = k_BT/\hbar c$ since the Rayleigh cross-section increases with the fourth power of the wavenumber. For example, for an object of size $a = 10^{-6}$ cm corresponding to a large molecule we get $\Lambda = 10^{-12}$ cm^{-2} s^{-1} at $T = 3$ K (cosmic background radiation), and $\Lambda = 10^6$ cm^{-2} s^{-1} at $T = 300$ K (room temperature). For a small dust particle, say $a = 10^{-5}$ cm, the corresponding decoherence rates increase to $\Lambda = 10^{-6}$ cm^{-2} s^{-1} ($T = 3$ K) and $\Lambda = 10^{12}$ cm^{-2} s^{-1} ($T = 300$ K).

For objects whose size is large compared to the wavelength the cross-section is approximately equal to the geometric cross-section. The decoherence rate may then be estimated from

$$\Lambda \sim 10^{14} \left(\frac{T}{K}\right)^5 \left(\frac{a}{\text{cm}}\right)^2 \frac{1}{\text{cm}^2\text{ s}}. \tag{4.116}$$

In this range Λ increases with the second power of a and with the fifth power of T. For an object of size $a = 10^{-1}$ cm this yields $\Lambda \sim 10^{24}$ cm^{-2} s^{-1} at $T = 300$

K. Further examples of the decoherence caused by scattering have been discussed by Tegmark (1993).

4.4 The damped harmonic oscillator

In the foregoing section we have treated the simplest case, namely the destruction of quantum coherence without damping. In more realistic physical models decoherence occurs together with a decay of populations. To study the effects of the combination of both phenomena we first investigate the damped harmonic oscillator in the quantum optical limit given by the master equation (3.307).

Let us consider as initial state a superposition of two coherent states,

$$|\psi(0)\rangle = \mathcal{N}\left(|\alpha\rangle + |\beta\rangle\right). \qquad (4.117)$$

In reference to Schrödinger's famous *gedanken* experiment (Schrödinger, 1935) such a superposition is sometimes called a Schrödinger cat state. The superposed states $|\alpha\rangle$ and $|\beta\rangle$ are considered to be macroscopically distinct with an extremely small overlap, representing the dead and the alive cat, respectively. The normalization factor $\mathcal{N}$ reads

$$\mathcal{N} = (2 + 2\Re\langle\alpha|\beta\rangle)^{-1/2}, \qquad (4.118)$$

where the overlap $\langle\alpha|\beta\rangle$ is given by eqn (3.321). The corresponding initial density matrix of the oscillator is thus

$$\rho_S(0) = \mathcal{N}^2 \left(|\alpha\rangle\langle\alpha| + |\beta\rangle\langle\beta| + |\alpha\rangle\langle\beta| + |\beta\rangle\langle\alpha|\right). \qquad (4.119)$$

Our aim is to determine the time evolution of this initial state and to define the decoherence function through the behaviour of the interference term $|\alpha\rangle\langle\beta| + |\beta\rangle\langle\alpha|$ in this equation. In the following Section 4.4.1 we present the solution to this problem for zero temperature. The general finite temperature case will be discussed in Section 4.4.2.

4.4.1 Vacuum decoherence

At zero temperature the master equation (3.307) can be written as

$$\frac{d}{dt}\rho_S(t) = \left(-i\omega_0 - \frac{\gamma_0}{2}\right) a^\dagger a \rho_S(t) + \left(+i\omega_0 - \frac{\gamma_0}{2}\right) \rho_S(t) a^\dagger a + \gamma_0 a \rho_S(t) a^\dagger$$

$$\equiv \mathcal{L}\rho_S(t). \qquad (4.120)$$

As we know already, a coherent state remains a coherent state under time evolution, which makes the solution of the problem particularly simple for the vacuum optical case. We therefore try the ansatz

$$\sigma(t) \equiv \exp\left[\mathcal{L}t\right] |\alpha\rangle\langle\beta| = f(t)|\alpha(t)\rangle\langle\beta(t)|, \qquad (4.121)$$

where $\alpha(0) \equiv \alpha$ and $\beta(0) \equiv \beta$ and $f(t)$ is a c-number function with $f(0) = 1$, such that $\sigma(0) = |\alpha\rangle\langle\beta|$. On using

$$\frac{d}{dt}|\alpha(t)\rangle = \left(\frac{\dot{\alpha}(t)}{\alpha(t)}a^{\dagger}a - \frac{1}{2}\frac{d}{dt}|\alpha(t)|^2\right)|\alpha(t)\rangle \tag{4.122}$$

as well as the fact that the coherent states are eigenstates of the annihilation operator, one verifies that $\sigma(t)$ given by (4.121) indeed solves the master equation (4.120) provided the differential equations

$$\frac{\dot{\alpha}(t)}{\alpha(t)} = \frac{\dot{\beta}(t)}{\beta(t)} = -i\omega_0 - \frac{\gamma_0}{2}, \tag{4.123}$$

$$\frac{\dot{f}(t)}{f(t)} = \frac{1}{2}\frac{d}{dt}\left(|\alpha(t)|^2 + |\beta(t)|^2\right) + \gamma_0\beta^*(t)\alpha(t) \tag{4.124}$$

are satisfied. These equations are easily solved to yield

$$\alpha(t) = \alpha\exp\left(-i\omega_0 - \frac{\gamma_0 t}{2}\right), \tag{4.125}$$

$$\beta(t) = \beta\exp\left(-i\omega_0 - \frac{\gamma_0 t}{2}\right), \tag{4.126}$$

$$f(t) = \exp\left[\left(1 - e^{-\gamma_0 t}\right)\left(-\frac{1}{2}|\alpha|^2 - \frac{1}{2}|\beta|^2 + \beta^*\alpha\right)\right]$$
$$= \langle\beta|\alpha\rangle^{[1-\exp(-\gamma_0 t)]}. \tag{4.127}$$

Summarizing, the solution of the master equation (4.120) corresponding to the initial state (4.119) takes the following form,

$$\rho_S(t) = \mathcal{N}^2\left(|\alpha(t)\rangle\langle\alpha(t)| + |\beta(t)\rangle\langle\beta(t)| + f(t)|\alpha(t)\rangle\langle\beta(t)| + f^*(t)|\beta(t)\rangle\langle\alpha(t)|\right). \tag{4.128}$$

The decoherence function $\Gamma(t)$ can now be defined as the logarithm of the modulus of the factor $f(t)$ multiplying the off-diagonals in the coherent state representation,

$$\Gamma(t) = \ln|f(t)| = -\frac{1}{2}|\alpha - \beta|^2\left(1 - e^{-\gamma_0 t}\right). \tag{4.129}$$

One observes that $\Gamma(t)$ is proportional to the square of the distance of the initial coherent state amplitudes α and β in the complex plane. For $\gamma_0 t \gg 1$ the decoherence function approaches the value $-|\alpha - \beta|^2/2$, that is $\exp(\Gamma)$ approaches the absolute value $|\langle\alpha|\beta\rangle|$ of the overlap of the initial states. For widely separated coherent states this overlap is extremely small which means that the coherences practically vanish in the long time limit.

On the other hand, for $\gamma_0 t \ll 1$ the decoherence function is proportional to time,

$$\Gamma(t) \approx -\frac{1}{2}|\alpha - \beta|^2\gamma_0 t. \tag{4.130}$$

This allows us to define a decoherence time τ_D through

$$\frac{\tau_D}{\tau_R} = \frac{2}{|\alpha - \beta|^2}, \tag{4.131}$$

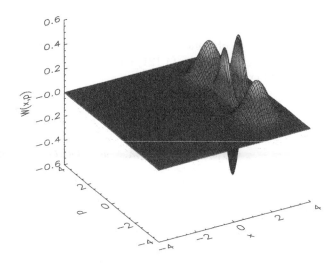

FIG. 4.2. Wigner distribution corresponding to the superposition (4.117) of two coherent states with $\alpha = \alpha_0 \exp[i\phi]$ and $\beta = \alpha_0 \exp[-i\phi]$, where $\alpha_0 = \sqrt{10}$ and $\phi = 0.7$.

where we make use of the relaxation time $\tau_R = \gamma_0^{-1}$. One observes that the ratio of the decoherence time to the relaxation time is inversely proportional to the square of the distance of the initial states in the complex plane. Equation (4.131) is an important relation which is encountered in many microscopic models of decoherence. It tells us that for widely separated initial states the decoherence time, that is the time over which coherences are destroyed through the interaction with the environment, is much smaller than the relaxation time, which characterizes the time after which the system loses its energy through dissipative effects.

The difference of relaxation and decoherence time is illustrated in Figs. 4.2 and 4.3, where we have taken the initial superposition (4.117) with $\alpha = \alpha_0 \exp[i\phi]$, $\beta = \alpha_0 \exp[-i\phi]$ and α_0 real. The initial phase components are thus separated by an angle 2ϕ in the complex plane, such that $|\alpha - \beta|^2 = 4\alpha_0^2 \sin^2 \phi$. Figure 4.2 shows the interaction picture Wigner distribution (2.81) of this initial state while Fig. 4.3 shows the corresponding state at time $t = 0.15\tau_R$. The time τ_R characterizes the time scale it takes for the two phase components to merge at the origin of the phase plane. We observe that the coherences of the initial state rapidly decay over a time which is much shorter than the relaxation time.

To interpret the result (4.131) physically (Caldeira and Leggett, 1985) we assume for simplicity that $\beta = 0$, which means that one of the superposed states is the ground state of the oscillator. Equation (4.131) then yields

$$\frac{\tau_D}{\tau_R} = \frac{2}{n}, \qquad (4.132)$$

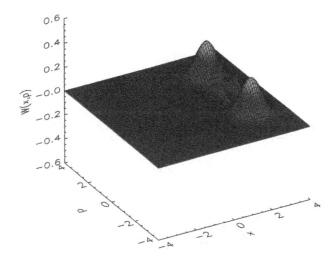

FIG. 4.3. Wigner distribution of the state which evolves from the initial state (4.117) shown in Fig. 4.2 after time $t = 0.15\gamma_0^{-1}$ according to the master equation (4.120).

where $|\alpha|^2 = \langle \alpha | a^\dagger a | \alpha \rangle \equiv n \gg 1$. The question is then, why is the decoherence time obtained above smaller than the relaxation time by the factor n, that is by the average number of quanta in the initial state? First, we note that at zero temperature the environment is in the ground state vacuum which will be denoted by $|0\rangle_B$. Thus we have the following initial state of the combined system (neglecting the overlap $\langle \alpha | 0 \rangle$),

$$|\Psi_i\rangle \approx \frac{1}{\sqrt{2}} \left(|\alpha\rangle + |0\rangle \right) \otimes |0\rangle_B. \qquad (4.133)$$

The rate for emission of a quantum of energy from this initial state is approximately equal to $W = \gamma_0 n/2$. Accordingly, the time it takes for the emission of a quantum is of the order $1/W = 2\tau_R/n$. After this time the initial state evolves approximately into the state

$$|\Psi_f\rangle \approx \frac{1}{\sqrt{2}} \left(|\alpha'\rangle \otimes |1\rangle_B + |0\rangle \otimes |0\rangle_B \right). \qquad (4.134)$$

Here, $|\alpha'\rangle$ is the state $|\alpha\rangle$ after the emission of one quantum, while $|1\rangle_B$ denotes a state of the reservoir containing one quantum of energy. The reduced density matrix of the oscillator is then given by

$$\rho_S = \mathrm{tr}_B\{|\Psi_f\rangle\langle\Psi_f|\} \approx \frac{1}{2} \left(|\alpha'\rangle\langle\alpha'| + |0\rangle\langle 0| \right), \qquad (4.135)$$

since the reservoir states $|0\rangle_B$ and $|1\rangle_B$ are orthogonal. This shows that coherences are already destroyed after the emission of one quantum, that is already after a time of order $2\tau_R/n$, whereas the dissipation of energy takes the emission of n quanta, that is a time of order τ_R. This is what is expressed by eqn (4.132).

4.4.2 Thermal noise

It must be expected that thermal noise leads to an enhancement of decoherence. To determine the decoherence time in the case of finite temperatures we write the solution of the master equation (3.307) corresponding to the initial state (4.119) as

$$\rho_S(t) = \mathcal{N}^2 \left(\rho_{\alpha\alpha}(t) + \rho_{\beta\beta}(t) + \rho_{\alpha\beta}(t) + \rho_{\beta\alpha}(t) \right), \tag{4.136}$$

where we have introduced

$$\rho_{\alpha\alpha}(t) = \exp\left[\mathcal{L}t \right] |\alpha\rangle\langle\alpha|, \tag{4.137}$$

$$\rho_{\beta\beta}(t) = \exp\left[\mathcal{L}t \right] |\beta\rangle\langle\beta|, \tag{4.138}$$

$$\rho_{\alpha\beta}(t) = \rho_{\beta\alpha}^\dagger(t) = \exp\left[\mathcal{L}t \right] |\alpha\rangle\langle\beta|. \tag{4.139}$$

Our strategy is to investigate the probability density in position space

$$\begin{aligned} p(x,t) &\equiv \langle x|\rho_S(t)|x\rangle \\ &= \mathcal{N}^2 \left[\rho_{\alpha\alpha}(x,t) + \rho_{\beta\beta}(x,t) + \rho_{\alpha\beta}(x,t) + \rho_{\beta\alpha}(x,t) \right], \end{aligned} \tag{4.140}$$

involving the matrix elements

$$\rho_{\alpha\alpha}(x,t) = \langle x|\rho_{\alpha\alpha}(t)|x\rangle, \quad \rho_{\beta\beta}(x,t) = \langle x|\rho_{\beta\beta}(t)|x\rangle, \tag{4.141}$$

$$\rho_{\alpha\beta}(x,t) = \langle x|\rho_{\alpha\beta}(t)|x\rangle = \rho_{\beta\alpha}^*(x,t), \tag{4.142}$$

and to determine the decoherence function from the reduction of the interference terms occurring in the expression for $p(x,t)$. In fact, it will be shown that $p(x,t)$ can be written as

$$\begin{aligned} p(x,t) = & \tag{4.143} \\ \mathcal{N}^2 & \left[\rho_{\alpha\alpha}(x,t) + \rho_{\beta\beta}(x,t) + 2\sqrt{\rho_{\alpha\alpha}(x,t)\rho_{\beta\beta}(x,t)} \exp[\Gamma(t)] \cos\varphi(x,t) \right]. \end{aligned}$$

The density $p(x,t)$ exhibits the typical structure of an interference pattern shown in Fig. 4.4. The first two contributions in eqn (4.143) represent an incoherent sum of the superposed wave packets, while the third term describes the interference pattern. As will be seen, the coupling to the environment yields a time and temperature dependent modulation of the pattern given by the phase $\varphi(x,t)$, as well as a reduction of the interference contrast (or the visibility of the pattern) which is determined by the factor $\exp\Gamma(t)$ in eqn (4.143). Figure 4.4 shows a plot of the function $p(x,t)$, where we use the analytical expressions for $\Gamma(t)$ and $\varphi(x,t)$ which will be derived below. The figure shows the decrease of the interference

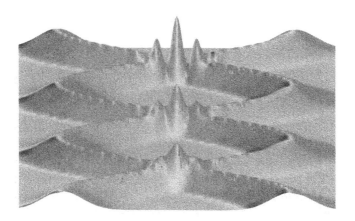

FIG. 4.4. The probability density $p(x,t)$ according to eqns (4.143), (4.160) and
(4.161). The x-coordinate increases from left to right and time increases from
back to front. The picture shows the time evolution over 1.5 periods of the
oscillator. The initial distance between the centres of the superposed wave
packets is equal to $12\sigma_0$. Parameters: $2\pi\gamma_0/\omega_0 = 0.05$ and $N = 0.5$.

contrast with increasing time, while damping effects are still negligible over the
time interval shown in the figure.

The task is now to determine the quantities defined in (4.141) and (4.142).
This can be done, of course, by determining the solution of the master equation
(3.307) corresponding to the initial condition (4.119). However, we are going to
employ a more direct method which turns out to be useful for the treatment of
other models as well. The key point is that $\rho_{\alpha\alpha}(x,0)$, $\rho_{\beta\beta}(x,0)$ and $\rho_{\alpha\beta}(x,0)$
are Gaussian functions. Since the system considered here is linear this Gaussian
property is preserved under time evolution. We thus conclude that the quantities
defined in (4.141) and (4.142) must take on the form of Gaussian functions of x
and all we have to do is to evaluate their mean and their variance. For example,
the function $\rho_{\alpha\alpha}(x,t)$ may be written

$$\rho_{\alpha\alpha}(x,t) = \frac{1}{\sqrt{2\pi\sigma_\alpha^2(t)}} \exp\left[-\frac{(x - x_\alpha(t))^2}{2\sigma_\alpha^2(t)}\right], \tag{4.144}$$

where

$$x_\alpha(t) = \int dx\, x \rho_{\alpha\alpha}(x,t) = \mathrm{tr}_S\left\{x\rho_{\alpha\alpha}(t)\right\} = \langle\alpha|x(t)|\alpha\rangle, \tag{4.145}$$

and

$$\sigma_\alpha^2(t) = \langle\alpha|x^2(t)|\alpha\rangle - \langle\alpha|x(t)|\alpha\rangle^2. \tag{4.146}$$

Here, $x(t)$ and $x^2(t)$ are Heisenberg picture operators which have to be determined from the adjoint master equation (see below). Of course, analogous ex-
pressions hold for $\rho_{\beta\beta}(x,t)$. Note that both $\rho_{\alpha\alpha}(x,t)$ and $\rho_{\beta\beta}(x,t)$ are correctly
normalized to 1. Finally, the function $\rho_{\alpha\beta}(x,t)$ must be written as

$$\rho_{\alpha\beta}(x,t) = \frac{\langle\beta|\alpha\rangle}{\sqrt{2\pi\sigma_{\alpha\beta}^2(t)}} \exp\left[-\frac{(x - x_{\alpha\beta}(t))^2}{2\sigma_{\alpha\beta}^2(t)}\right], \tag{4.147}$$

with

$$x_{\alpha\beta}(t) = \frac{1}{\langle\beta|\alpha\rangle}\int dx\, x\rho_{\alpha\beta}(x,t) = \frac{\langle\beta|x(t)|\alpha\rangle}{\langle\beta|\alpha\rangle}, \tag{4.148}$$

and

$$\sigma_{\alpha\beta}^2(t) = \frac{\langle\beta|x^2(t)|\alpha\rangle}{\langle\beta|\alpha\rangle} - \frac{\langle\beta|x(t)|\alpha\rangle^2}{\langle\beta|\alpha\rangle^2}. \tag{4.149}$$

Note that (4.147) is a complex-valued Gaussian function and that it is correctly normalized to the overlap of the initial coherent states,

$$\int dx\, \rho_{\alpha\beta}(x,t) = \text{tr}_S\left\{\rho_{\alpha\beta}(t)\right\} = \langle\beta|\alpha\rangle. \tag{4.150}$$

Our next step is to calculate the mean values and variances introduced in the above formulae. To this end we solve the adjoint master equation with the initial condition that at $t = 0$ the Heisenberg and Schrödinger operators coincide. This yields

$$x(t) = \frac{1}{\sqrt{2m\omega_0}}\left(e^{-i\omega_0 t}a + e^{i\omega_0 t}a^\dagger\right)e^{-\gamma_0 t/2} \tag{4.151}$$

$$x^2(t) = \frac{1}{2m\omega_0}\left(e^{-2i\omega_0 t}a^2 + e^{2i\omega_0 t}a^{\dagger 2} + 2a^\dagger a\right)e^{-\gamma_0 t}$$

$$+\frac{1}{2m\omega_0}\left[2N\left(1 - e^{-\gamma_0 t}\right) + 1\right], \tag{4.152}$$

where $a^\dagger$, a are Schrödinger operators and $N = [\exp(\omega_0/k_B T) - 1]^{-1}$. With the help of these relations the desired quantities are easily found to be

$$x_\alpha(t) = \frac{1}{\sqrt{2m\omega_0}}\left(e^{-i\omega_0 t}\alpha + e^{i\omega_0 t}\alpha^*\right)e^{-\gamma_0 t/2}, \tag{4.153}$$

$$x_\beta(t) = \frac{1}{\sqrt{2m\omega_0}}\left(e^{-i\omega_0 t}\beta + e^{i\omega_0 t}\beta^*\right)e^{-\gamma_0 t/2}, \tag{4.154}$$

$$x_{\alpha\beta}(t) = \frac{1}{\sqrt{2m\omega_0}}\left(e^{-i\omega_0 t}\alpha + e^{i\omega_0 t}\beta^*\right)e^{-\gamma_0 t/2}, \tag{4.155}$$

$$\sigma^2(t) \equiv \sigma_\alpha^2(t) = \sigma_\beta^2(t) = \sigma_{\alpha\beta}^2(t) = \sigma_0^2\left[2N(1 - e^{-\gamma_0 t}) + 1\right]. \tag{4.156}$$

We observe that the variances introduced are equal to each other and that $\sigma^2(0)$ equals the initial width of the wave packets,

$$\sigma^2(0) = \sigma_0^2 \equiv \frac{1}{2m\omega_0}. \tag{4.157}$$

As our final step we substitute the expressions found for $\rho_{\alpha\alpha}(x,t)$, $\rho_{\beta\beta}(x,t)$, and $\rho_{\alpha\beta}(x,t)$ into eqn (4.140). After a little rearrangement one obtains the decoherence function

$$\Gamma(t) = -\frac{1}{2}|\alpha - \beta|^2 - \frac{1}{2\sigma^2(t)}\Re\left[(x - x_{\alpha\beta}(t))^2 - \frac{1}{2}(x - x_\alpha(t))^2 - \frac{1}{2}(x - x_\beta(t))^2\right]$$

(4.158)

and the phase

$$\varphi(x,t) = \Im\left[\beta^*\alpha - \frac{1}{2\sigma^2(t)}\left\{(x - x_{\alpha\beta}(t))^2 - \frac{1}{2}(x - x_\alpha(t))^2 - \frac{1}{2}(x - x_\beta(t))^2\right\}\right].$$

(4.159)

If we insert here the relations (4.153)–(4.156) we finally get

$$\varphi(x,t) = \frac{\Delta p(t) \cdot x}{2N\left(1 - e^{-\gamma_0 t}\right) + 1} + \varphi_0(t),$$

(4.160)

and

$$\Gamma(t) = -\frac{1}{2}|\alpha - \beta|^2 \frac{2N + 1}{2N\left(1 - e^{-\gamma_0 t}\right) + 1}\left(1 - e^{-\gamma_0 t}\right).$$

(4.161)

The quantity $\varphi_0(t)$ in eqn (4.160) is a space-independent phase given by

$$\varphi_0(t) = \Im\left[\beta^*\alpha - \frac{1}{2}\frac{e^{-\gamma_0 t}}{2N\left(1 - e^{-\gamma_0 t}\right) + 1}\left(e^{-i\omega_0 t}\alpha + e^{i\omega_0 t}\beta^*\right)^2\right],$$

(4.162)

while

$$\Delta p(t) \equiv \langle\alpha|p(t)|\alpha\rangle - \langle\beta|p(t)|\beta\rangle$$

(4.163)

$$= -i\sqrt{\frac{m\omega_0}{2}}e^{-\gamma_0 t/2}\left[\left(e^{-i\omega_0 t}\alpha - e^{i\omega_0 t}\alpha^*\right) - \left(e^{-i\omega_0 t}\beta - e^{i\omega_0 t}\beta^*\right)\right]$$

is the difference of the average momenta of the superposed wave packets.

These equations describe the modulation of the phase and the reduction of the interference contrast through the interaction with the environment (Savage and Walls, 1985a). For $\gamma_0 t \ll 1$ we have from eqns (4.160) and (4.162)

$$\varphi(x,t) \approx \Delta p(t) \cdot x - \frac{1}{2}\Im\left[e^{-2i\omega_0 t}\alpha^2 + e^{2i\omega_0 t}\beta^{*2}\right].$$

(4.164)

This expression describes the usual interference pattern as it occurs for the case of a vanishing system–environment coupling. Equation (4.161) shows that $\Gamma(t)$ reduces to expression (4.129) in the limit of zero temperature ($N = 0$). In the long time limit expression (4.161) approaches the temperature-independent value given by the overlap of the initial states of the superposition,

$$\Gamma(t) \approx -\frac{1}{2}|\alpha - \beta|^2, \quad \gamma_0 t \gg 1.$$

(4.165)

For small times, $\gamma_0 t \ll 1$, we find

$$\Gamma(t) \approx -\frac{1}{2}|\alpha - \beta|^2(2N + 1)\gamma_0 t = -\frac{1}{2}|\alpha - \beta|^2 \coth\left(\frac{\omega_0}{2k_{\mathrm{B}}T}\right)\gamma_0 t.$$

(4.166)

Thus, the ratio of decoherence time to relaxation time becomes

$$\frac{\tau_D}{\tau_R} = \frac{2}{|\alpha - \beta|^2 (2N+1)}. \tag{4.167}$$

Comparing this expression with eqn (4.131) we see that in the presence of thermal noise the decoherence time τ_D is reduced by an additional factor of $1/(2N+1)$.

Let us assume that α and β are real. This means that the initial momenta of the superposed states vanish and that

$$(\alpha - \beta)^2 = \frac{\Delta x^2}{4\sigma_0^2}, \tag{4.168}$$

where

$$\Delta x \equiv x_\alpha(0) - x_\beta(0) \tag{4.169}$$

represents the initial space distance between the centres of the superposed states. We then find with the help of eqn (4.167) in the high-temperature limit, $k_B T \gg \hbar\omega_0$,

$$\frac{\tau_D}{\tau_R} \approx \frac{2}{|\alpha - \beta|^2} \frac{\hbar\omega_0}{2k_B T} = 4 \left(\frac{\bar{\lambda}_{\text{th}}}{\Delta x} \right)^2. \tag{4.170}$$

Here, we have included factors of $\hbar$ and in the second equation we have introduced the thermal wavelength $\bar{\lambda}_{\text{th}} = \hbar/\sqrt{2mk_B T}$. The same result has been obtained previously from the high-temperature Brownian motion master equation (compare with eqn (4.79)).

4.5   Electromagnetic field states

Having discussed some basic features of the theory, let us take a look at experiments on decoherence. Present day experimental technology is not only capable of investigating and controlling decoherence phenomena, but also to resolve them in time as dynamical processes and to verify results of the theory in quantitative terms. For example, it has been shown to be possible to prepare a superposition of two motional states of the centre-of-mass coordinate of a Be^+ ion stored in a Paul-trap and to measure its decoherence in controllable environments (Myatt *et al.*, 2000; Turchette *et al.*, 2000). Moreover, experimental evidences for a coherent superposition of two macroscopically distinct magnetic-flux states in a superconducting quantum interference device have been reported by Friedman *et al.* (2000).

In another famous experiment, which was performed by Brune *et al.* (1996), the progression of the decoherence of a mesoscopic superposition of two coherent field states in a high-Q cavity has been observed for the first time. The decohering field state observed in the experiment was of the form of the Schrödinger cat state (4.117). In the present section we wish to discuss this experiment in some detail. For our discussion the most important feature of the experiment is that it enabled a direct measurement of the decoherence function $\Gamma(t)$ and, thus, a direct comparison with the theoretical analysis (Maître *et al.*, 1997).

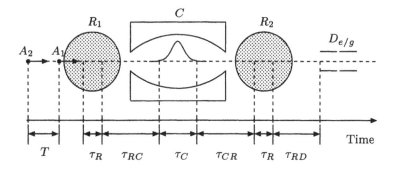

FIG. 4.5. Schematic setup of the experiment performed by Brune *et al.* The atoms A_1 and A_2 move along the dashed horizontal line with a time delay T and cross the resonator R_1, the cavity C, the resonator R_2 and detectors D_e and D_g. The times of flight between and within the different components are also indicated.

4.5.1   *Atoms interacting with a cavity field mode*

The experimental setup is sketched in Fig. 4.5. With a variable time delay T, two atoms A_1 and A_2 traverse the setup which consists of two microwave resonators R_1 and R_2 and a superconducting microwave cavity C with cavity frequency ν. The high cavity quality factor of $Q = 5 \cdot 10^7$ leads to a cavity relaxation rate of $\gamma_0 = (160 \ \mu s)^{-1}$. The atoms are described by two circular Rydberg levels (with principal quantum numbers $n = 51$ and $n = 50$) which will be denoted by $|e\rangle$ and $|g\rangle$, respectively. The corresponding transition frequency ω is detuned from the cavity frequency by an amount Δ. The resonators R_1 and R_2 are fed by the same classical microwave source of frequency ω_R. Finally, the state of the atoms is analysed in field ionization detectors D_e and D_g. The atomic level scheme is depicted in Fig. 4.6.

To begin with let us briefly describe the basic idea underlying the experiment. The first resonator R_1 serves to prepare the atoms in a certain superposition of the states $|e\rangle$ and $|g\rangle$. Initially, the cavity C contains a small coherent field $|\alpha\rangle$ as shown in Fig. 4.7. As will be demonstrated below, the interaction of the atoms with the field in C effectively induces a phase shift of $\pm\phi$ on the field state whose sign is triggered by the atomic state. Thus, the interaction of the first atom A_1 with the field in C leads to an entanglement between the two atomic states and the two phase components $|\alpha \exp(\pm i\phi)\rangle$. The second resonator R_2 induces a further mixing of the states of A_1 such that the final measurement of the atomic states in the field emission detector D_e and D_g does not give any information on the state in which A_1 has passed cavity C. As a result, the state measurement on A_1 projects the field in C onto a Schrödinger cat-type superposition of two phase components which are separated by an angle 2ϕ in the complex plane (see Fig. 4.8). Ignoring for a moment field damping, normalization factors, and further phase factors this state essentially takes the form (for details see the following

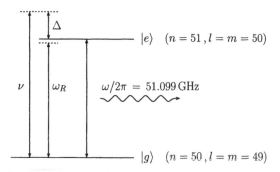

FIG. 4.6. Two-level scheme displaying the atomic transition frequency ω, the frequency ν of the field in the cavity C and the frequency ω_R of the fields in the resonators R_1 and R_2.

subsection)

$$|\alpha e^{-i\phi}\rangle + e^{i(\chi_1+\phi)}|\alpha e^{i\phi}\rangle, \tag{4.171}$$

where $\chi_1 = 0$ if A_1 was found to be in the state $|g\rangle$ and $\chi_1 = \pi$ if it was found to be in the state $|e\rangle$.

A similar transformation is then induced by the interaction of the second atom with the cavity field. The final state of the cavity field is thus

$$|\alpha e^{-2i\phi}\rangle + e^{i(\chi_1+\chi_2+2\phi)}|\alpha e^{2i\phi}\rangle + \left(e^{i\chi_1} + e^{i\chi_2}\right)e^{i\phi}|\alpha\rangle, \tag{4.172}$$

where χ_2 again takes on the values 0 or π, depending on the outcome of the measurement on A_2. Thus, if both atoms are found to be in the same state ($\chi_1 = \chi_2$) the field state (4.172) becomes (see Fig. 4.9)

$$|\alpha e^{-2i\phi}\rangle + e^{2i\phi}|\alpha e^{2i\phi}\rangle \pm 2e^{i\phi}|\alpha\rangle, \tag{4.173}$$

while it takes the form

$$|\alpha e^{-2i\phi}\rangle - e^{2i\phi}|\alpha e^{2i\phi}\rangle \tag{4.174}$$

if the atoms are found to be in different states ($\chi_1 \neq \chi_2$). We see that the interaction of the field with both atoms leads to some kind of constructive interference of the phase component $|\alpha\rangle$ if both atoms are detected in the same state. If the atoms are detected in different states the phase component $|\alpha\rangle$ interferes destructively and disappears from the final state.

In the experiment the conditional probabilities $W_{\varepsilon\varepsilon'}$ are measured. These are defined to be the probabilities of finding atom A_2 in state $|\varepsilon'\rangle$ under the condition that atom A_1 has been found to be in state $|\varepsilon\rangle$, where $\varepsilon, \varepsilon' = e, g$. The detailed analysis reveals that in the absence of any field damping the probability of finding both atoms in the same state (i.e. to have constructive interference) is larger than that of finding them in different states (destructive interference). If field damping and, therefore, the decoherence of the Schrödinger cat state

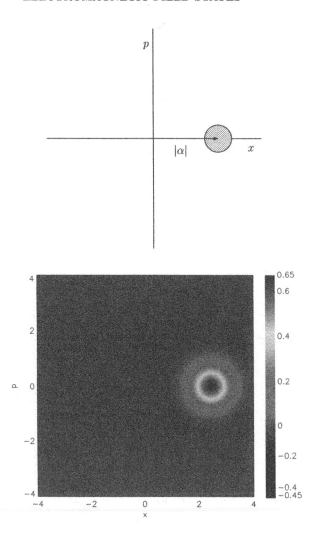

FIG. 4.7. Schematic representation and Wigner distribution of the initial coherent field in the central cavity C for α real and $\alpha^2 = 6$. The quadratures x
and p have been scaled such that $x = \operatorname{Re} \alpha$ and $p = \operatorname{Im} \alpha$.

(4.171) is taken into account this difference of conditional probabilities decays
to zero for increasing delay time T. Thus, by measuring a certain difference of
conditional probabilities as a function of the time delay T between the two atoms,
it is possible to measure the decoherence of the Schrödinger cat state.

Before presenting the detailed theoretical analysis of the experiment in the
next subsection, we investigate the dynamics induced by the interaction of the
atoms in the central cavity C. The atom–field interaction in C may be described
by the Jaynes–Cummings Hamiltonian,

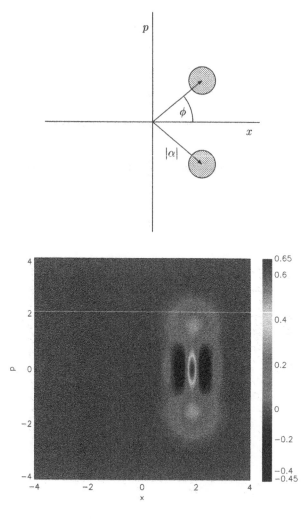

FIG. 4.8. Schematic representation and Wigner distribution of the superposition of coherent fields in the cavity C after the first atom has crossed C and was detected in the ground state. Parameters: $\alpha^2 = 6$, $\phi = 0.7$ and $\varphi_0 = 0$.

$$H_{JC}(t) = \frac{1}{2}\omega\sigma_3 + \nu a^\dagger a + \Omega(t)\left[a\sigma_+ + a^\dagger\sigma_-\right]. \qquad (4.175)$$

Note that the Rabi frequency $\Omega(t)$, providing the coupling of the field mode a to the atomic raising and lowering operators $\sigma_\pm$, depends on time. This is due to the space-dependence of the mode function, given by a Gaussian envelope, that describes the field mode in C. When traversing the cavity the atoms thus feel a time-varying coupling to the field mode and the Hamiltonian $H_{JC}(t)$ depends parametrically on time.

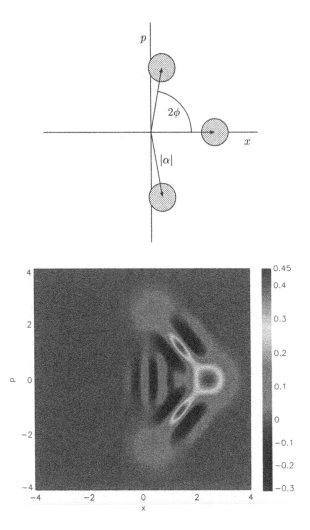

FIG. 4.9. Schematic representation and Wigner distribution of the superposition of coherent fields in the cavity C after detection of the second atom in the ground state. Parameters: $\alpha^2 = 6$, $\phi = 0.7$ and $\varphi_0 = 0$.

For a fixed time, that is for a fixed position of the atom within the cavity, the dressed energy eigenvalues of the Jaynes–Cummings Hamiltonian are given in linear approximation by

$$E_{\pm,n} = \nu \left(n + \frac{1}{2}\right) \pm \frac{\Delta}{2} \pm \frac{\Omega^2(t)(n+1)}{\Delta}, \qquad (4.176)$$

with the corresponding dressed states (we consider without restriction the case of positive detuning, $\Delta \equiv \omega - \nu > 0$)

$$|+, n\rangle = |e, n\rangle, \quad |-, n\rangle = |g, n+1\rangle. \tag{4.177}$$

Under the given experimental circumstances the off-resonant atom–field interaction in C is predominantly adiabatic, which means that real transitions between the dressed atomic states may be neglected and that only virtual processes must be taken into account. Applying the adiabatic theorem of quantum mechanics we then see that the dressed states of an atom that traverses the cavity take up a phase factor which is given by the integrated dynamical phase $\exp[-i \int dt E_{\pm,n}(t)]$. Thus, in addition to the contribution from the unperturbed energies, the interaction of the atom with the field mode a leads to the phase shifts given by

$$|e, n\rangle \longrightarrow \exp(i\varphi_e)|e, n\rangle, \quad |g, n\rangle \longrightarrow \exp(i\varphi_g)|g, n\rangle, \tag{4.178}$$

where

$$\exp(i\varphi_e) = \exp\left[-i \int\limits_{-\infty}^{+\infty} dt \frac{\Omega^2(t)(n+1)}{\Delta}\right] = \exp\left[i\phi(n+1)\right],$$

$$\exp(i\varphi_g) = \exp\left[-i \int\limits_{-\infty}^{+\infty} dt \frac{-\Omega^2(t)n}{\Delta}\right] = \exp\left[-i\phi n\right], \tag{4.179}$$

and

$$\phi \equiv - \int\limits_{-\infty}^{+\infty} dt \frac{\Omega^2(t)}{\Delta}. \tag{4.180}$$

Accordingly, we may write the transformations of the states $|e\rangle|\alpha\rangle$ and $|g\rangle|\alpha\rangle$ in the form

$$|e\rangle|\alpha\rangle \longrightarrow \exp\left[i\phi\left(a^\dagger a + 1\right)\right]|e\rangle|\alpha\rangle = e^{i\phi}|e\rangle|\alpha e^{i\phi}\rangle, \tag{4.181}$$

$$|g\rangle|\alpha\rangle \longrightarrow \exp\left[-i\phi a^\dagger a\right]|g\rangle|\alpha\rangle = |g\rangle|\alpha e^{-i\phi}\rangle. \tag{4.182}$$

This result will be used in the next subsection to describe the preparation of the superposition of field states whose decoherence is studied in the experiment.

4.5.2 Schrödinger cat states

We now turn to a detailed analysis of the experiment. Our man intention is to derive a formula for the difference $W_{ee} - W_{ge}$ of the conditional probabilities measured in the experiment and to relate it to the decoherence function pertaining to the Schrödinger cat. The experiment may be described as a succession of seven steps as follows.

4.5.2.1 *Initial state of A_1 and interaction in R_1*   First, atom A_1 is prepared in state $|e\rangle$ and undergoes a $\pi/2$-pulse in resonator R_1, yielding the following state of the atom-field system prior to the interaction in C,

$$\frac{1}{\sqrt{2}} \left(|e\rangle + |g\rangle\right) |\alpha\rangle. \tag{4.183}$$

4.5.2.2 *Atom–field interaction in C*   Atom A_1 enters the central cavity C. The field mode in C induces a phase shift which can be described by the transformations (4.181) and (4.182) such that the state (4.183) is transformed into the state

$$\frac{1}{\sqrt{2}} \left(e^{i\phi}|e\rangle|\alpha e^{i\phi}\rangle + |g\rangle|\alpha e^{-i\phi}\rangle\right). \tag{4.184}$$

4.5.2.3 *Interaction in R_2 and measurement on A_1*   The second $\pi/2$-pulse in R_2 induces the transformations

$$|e\rangle \longrightarrow \frac{1}{\sqrt{2}} \left(|e\rangle + e^{i\varphi_0}|g\rangle\right), \quad |g\rangle \longrightarrow \frac{1}{\sqrt{2}} \left(-e^{-i\varphi_0}|e\rangle + |g\rangle\right) \tag{4.185}$$

of the states of atom A_1, where $\varphi_0 = (\omega_R - \omega)\tau$ is the dynamical phase difference acquired during the time of flight τ between R_1 and R_2. Thus, the atom–field state takes the following form after atom A_1 has left resonator R_2,

$$\frac{1}{2} \left[|g\rangle \left(|\alpha e^{-i\phi}\rangle + e^{i(\phi+\varphi_0)}|\alpha e^{i\phi}\rangle\right) - e^{-i\varphi_0}|e\rangle \left(|\alpha e^{-i\phi}\rangle - e^{i(\phi+\varphi_0)}|\alpha e^{i\phi}\rangle\right) \right]. \tag{4.186}$$

The measurement of the state of atom A_1 projects the field state onto the states

$$\frac{1}{\mathcal{N}(\chi)} \left(|\alpha e^{-i\phi}\rangle + e^{i(\chi+\phi+\varphi_0)}|\alpha e^{i\phi}\rangle\right), \tag{4.187}$$

where the measurement of $|g\rangle$ implies $\chi = 0$, while the measurement of $|e\rangle$ means $\chi = \pi$. The normalization factor is given by

$$\mathcal{N}(\chi) = \sqrt{2 \left(1 + \exp\left[-2|\alpha|^2 \sin^2 \phi\right] \cos\left[\chi + \varphi_0 + \phi + |\alpha|^2 \sin 2\phi\right]\right)}. \tag{4.188}$$

4.5.2.4 *Field damping in C*   The state (4.187) corresponds to the density matrix

$$\rho_F(0, \chi) = \frac{1}{\mathcal{N}^2(\chi)} \left[|\alpha e^{-i\phi}\rangle\langle\alpha e^{-i\phi}| + |\alpha e^{i\phi}\rangle\langle\alpha e^{i\phi}| + e^{i\xi}|\alpha e^{i\phi}\rangle\langle\alpha e^{-i\phi}| + \text{h.c.} \right] \tag{4.189}$$

with $\xi = \chi + \varphi_0 + \phi$. Here and in the following h.c. means that the Hermitian conjugate of the last term must be added to the expression. We denote the time delay between the first and the second atom by T. In the experiment the temperature of the environment corresponds to a mean number $N = 0.05$ of thermal photons in the field mode, such that we may use the vacuum optical master equation (4.120). During the time interval T the field mode thus evolves into the density matrix

$$\rho_F(T,\chi) = \frac{1}{\mathcal{N}^2(\chi)} \left[|\alpha(T)\rangle\langle\alpha(T)| + |\beta(T)\rangle\langle\beta(T)| + f(T)|\beta(T)\rangle\langle\alpha(T)| + \text{h.c.} \right],$$
(4.190)

where (compare with eqns (4.125)–(4.127))

$$\alpha(T) = \alpha \exp(-\gamma_0 T/2 - i\phi),$$
(4.191)

$$\beta(T) = \alpha \exp(-\gamma_0 T/2 + i\phi),$$
(4.192)

$$f(T) = e^{i\xi} \exp\left[-|\alpha|^2 \left(1 - e^{2i\phi}\right)\left(1 - e^{-\gamma_0 T}\right)\right]$$

$$= e^{i\xi} \langle \alpha e^{-i\phi} | \alpha e^{i\phi} \rangle^{(1-e^{-\gamma_0 T})}.$$
(4.193)

Note that dynamical phase factors are absent since we are working in the interaction picture. We see that the decoherence function is given by

$$\Gamma(T) = \ln|f(T)| = -2|\alpha|^2 \sin^2 \phi \left(1 - e^{-\gamma_0 T}\right),$$
(4.194)

which leads to a decoherence time of the form

$$\tau_D = \frac{\tau_R}{2|\alpha|^2 \sin^2 \phi}.$$
(4.195)

4.5.2.5 *Initial state of A_2 and interaction in R_1* Atom A_2 is prepared into the initial state $|e\rangle$ and subjected to a $\pi/2$-pulse in R_1 such that the state of atom A_2 and the field, just before the atom enters the cavity, is given by

$$\frac{1}{2} \left[|e\rangle\langle e| + |g\rangle\langle g| + |e\rangle\langle g| + |g\rangle\langle e| \right] \rho_F(T,\chi).$$
(4.196)

4.5.2.6 *Atom–field interaction in C* The interaction of atom A_2 with the field in C yields the state

$$\frac{1}{2} \left[|e\rangle\langle e| e^{i\phi(a^\dagger a + 1)} \rho_F(T,\chi) e^{-i\phi(a^\dagger a + 1)} + |g\rangle\langle g| e^{-i\phi a^\dagger a} \rho_F(T,\chi) e^{i\phi a^\dagger a} \right.$$

$$\left. + |e\rangle\langle g| e^{i\phi(a^\dagger a + 1)} \rho_F(T,\chi) e^{i\phi a^\dagger a} + \text{h.c} \right].$$
(4.197)

4.5.2.7 *Interaction in R_2 and measurement on A_2* Atom A_2 traverses resonantor R_2, which transforms the state (4.197) into the state

$$\rho_{AF}(T,\chi) = \frac{1}{4} \left(|e\rangle + e^{i\varphi_0}|g\rangle\right) \left(\langle e| + e^{-i\varphi_0}\langle g|\right) e^{i\phi(a^\dagger a + 1)} \rho_F(T,\chi) e^{-i\phi(a^\dagger a + 1)}$$

$$+ \frac{1}{4} \left(|g\rangle - e^{-i\varphi_0}|e\rangle\right) \left(\langle g| - e^{i\varphi_0}\langle e|\right) e^{-i\phi a^\dagger a} \rho_F(T,\chi) e^{i\phi a^\dagger a}$$

$$+ \frac{1}{4} \left(|e\rangle + e^{i\varphi_0}|g\rangle\right) \left(\langle g| - e^{i\varphi_0}\langle e|\right) e^{i\phi(a^\dagger a + 1)} \rho_F(T,\chi) e^{i\phi a^\dagger a}$$

$$+ \frac{1}{4} \left(|g\rangle - e^{-i\varphi_0}|e\rangle\right) \left(\langle e| + e^{-i\varphi_0}\langle g|\right) e^{-i\phi a^\dagger a} \rho_F(T,\chi) e^{-i\phi(a^\dagger a + 1)}.$$

Finally, the state of atom A_2 is measured and the conditional probabilities W_{ee} and W_{ge} are determined. Here, W_{ee} is defined to be the probability of finding atom A_2 in the state $|e\rangle$ under the condition that atom A_1 was detected in the state $|\varepsilon\rangle$, where $\varepsilon = e, g$, depending on the outcome of the measurement on atom A_1.

According to the obtained expression for ρ_{AF} we now have

$$W_{ee} = \mathrm{tr}_{AF} \left\{ |e\rangle\langle e|\rho_{AF}(T, \chi) \right\} \tag{4.198}$$
$$= \frac{1}{2} \left[1 - \Re\, e^{i(\varphi_0 + \phi)}\, \mathrm{tr}_F \left\{ e^{2i\phi a^\dagger a} \rho_F(T, \chi) \right\} \right].$$

In the first expression the trace is taken over atom A_2 and the field mode, while in the second expression the trace is taken over the field mode only. Furthermore, we have to set $\chi = 0$ if $\varepsilon = g$ and $\chi = \pi$ if $\varepsilon = e$. With the help of the field density matrix (4.190) we find

$$\mathrm{tr}_F \left\{ e^{2i\phi a^\dagger a} \rho_F(T, \chi) \right\} = \frac{1}{\mathcal{N}^2(\chi)} \left[\langle \alpha(T)|\alpha(T) \exp 2i\phi \rangle + \langle \beta(T)|\beta(T) \exp 2i\phi \rangle \right.$$
$$+ e^{-i\xi} f^*(T) \langle \beta(T)|\alpha(T) \exp 2i\phi \rangle$$
$$\left. + e^{i\xi} f(T) \langle \alpha(T)|\beta(T) \exp 2i\phi \rangle \right], \tag{4.199}$$

which yields

$$W_{\varepsilon e} = \frac{1}{2} \left[1 - \frac{1}{2} \frac{2B + C\cos\chi + D\cos\chi}{1 + A\cos\chi} \right], \tag{4.200}$$

where

$$A = \exp\left[-2|\alpha|^2 \sin^2\phi\right] \cos\left[\varphi_0 + \phi + |\alpha|^2 \sin 2\phi\right], \tag{4.201}$$
$$B = \exp\left[-2|\alpha|^2 e^{-\gamma_0 T} \sin^2\phi\right] \cos\left[\varphi_0 + \phi + |\alpha|^2 e^{-\gamma_0 T} \sin 2\phi\right], \tag{4.202}$$
$$C = \exp\left[-2|\alpha|^2(1 - e^{-\gamma_0 T}) \sin^2\phi\right] \cos\left[|\alpha|^2(1 - e^{-\gamma_0 T}) \sin 2\phi\right], \tag{4.203}$$
$$D = \exp\left[-2|\alpha|^2 e^{-\gamma_0 T} \sin^2 2\phi - 2|\alpha|^2(1 - e^{-\gamma_0 T}) \sin^2\phi\right]$$
$$\times \cos\left[2\varphi_0 + 2\phi + |\alpha|^2 e^{-\gamma_0 T} \sin 4\phi + |\alpha|^2(1 - e^{-\gamma_0 T}) \sin 2\phi\right]. \tag{4.204}$$

The term $1 + A\cos\chi$ in eqn (4.200) stems from the normalization factor of the field density matrix, while the quantities B, C, and D arise from the four scalar products that contribute to the trace in eqn (4.199). Namely, the term B is due to the product

$$\langle \alpha(T)|\alpha(T) \exp 2i\phi \rangle = \langle \beta(T)|\beta(T) \exp 2i\phi \rangle = \langle \alpha(T)|\beta(T) \rangle$$
$$= \exp\left[-|\alpha|^2 e^{-\gamma_0 T} \left(1 - e^{2i\phi}\right)\right]. \tag{4.205}$$

Thus, B is determined by the overlap of the two phase components of the original superposition created by the first atom. These phase components are separated by an angle 2ϕ (see Fig. 4.8). The contribution C stems from the product

$$\langle \beta(T)|\alpha(T) \exp 2i\phi \rangle = \langle \beta(T)|\beta(T) \rangle = 1, \tag{4.206}$$

whereas D is determined by the product

$$\langle\alpha(T)|\beta(T)\exp 2i\phi\rangle = \exp\left[-|\alpha|^2 e^{-\gamma_0 T}\left(1 - e^{4i\phi}\right)\right]. \qquad (4.207)$$

The term D is therefore determined by the overlap of two phase components which are separated by an angle of 4ϕ, while C is obtained from the overlap of two phase components corresponding to the same angle. The term C thus describes the contribution from the recombination of the phase components at the location of the original coherent field state $|\alpha\rangle$ (see Fig. 4.9).

It may be seen from the relations (4.201)–(4.204) that A, B and D are exponentially small in $|\alpha|^2$ for $\gamma_0 T \ll 1$, provided the angle ϕ is not too close to 0, $\pi/2$, or π. Thus we have

$$W_{ee} \approx \frac{1}{2}\left[1 + \frac{1}{2}C\right], \quad W_{ge} \approx \frac{1}{2}\left[1 - \frac{1}{2}C\right]. \qquad (4.208)$$

In the experiment the following difference of the conditional probabilities is determined,

$$\eta(T, \varphi_0) = W_{ee} - W_{ge}, \qquad (4.209)$$

or rather the angle-averaged one

$$\bar{\eta}(T) = \frac{1}{\pi}\int\limits_0^\pi d\varphi_0 \eta(T, \varphi_0). \qquad (4.210)$$

With the help of eqn (4.200) one finds

$$\eta(T, \varphi_0) = \frac{1}{2}\frac{C + D - 2AB}{1 - A^2}. \qquad (4.211)$$

Since C represents the dominant contribution in $\eta(T, \varphi_0)$ and since it is independent of φ_0 we find that $\eta(T, \varphi_0) \approx \bar{\eta}(T) \approx C/2$. The expression (4.203) for C shows that it is equal to the real part of the quantity $f(T)$ (see eqn (4.193)) which multiplies the interference term of the Schrödinger cat state (4.190). Hence we get

$$\begin{aligned}
\eta(T, \varphi_0) &\approx \bar{\eta}(T) \\
&\approx \frac{1}{2}\mathrm{Re} f(T) \\
&= \frac{1}{2}\exp\left[\Gamma(T)\right]\cos\left[|\alpha|^2(1 - e^{-\gamma_0 T})\sin 2\phi\right]. \qquad (4.212)
\end{aligned}$$

This is the desired expression which relates the experimentally observed quantity $\bar{\eta}(T)$ to the decoherence function $\Gamma(T)$.

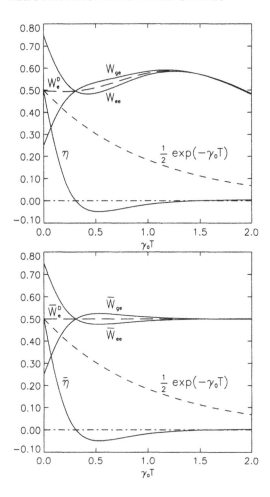

FIG. 4.10. Top: The conditional probabilities W_{ge} and W_{ee}, their difference η, and the probability W_e^D as a function of the delay time T for $\varphi_0 = 0$, $\phi = 0.7$ and $|\alpha|^2 = 6$. To make visible the difference between decoherence and relaxation of the photon number we also show the function $0.5 \exp[-\gamma_0 T]$. Bottom: The same quantities averaged over the angle φ_0.

We show in Fig. 4.10 the conditional probabilities W_{ge} and W_{ee} and their difference η as a function of the delay time T. For comparison we also depict the corresponding angle-averaged quantities, the function $\exp[-\gamma_0 T]/2$, and

$$W_e^D = \frac{1}{2}(1 - B), \qquad (4.213)$$

which is defined to be the probability of detecting A_2 in the excited state under the assumption of complete decoherence, that is under the condition that the interference terms of the field density matrix are put equal to zero.

According to eqn (4.212) the difference η of the conditional probabilities is directly related to the relevant off-diagonal element of the field density matrix and thus also to the decoherence function $\Gamma(t)$. As was demonstrated in the experiment performed by Haroche and coworkers this relation enabled the direct observation of the decoherence of the Schrödinger cat and a nice quantitative verification of the theory. A more detailed theoretical analysis of the experiment was given by Breuer, Dorner and Petruccione (2001).

4.6   Caldeira–Leggett model

Up to now we have discussed decoherence and dissipation for Markovian dynamics defined by the quantum optical or by the quantum Brownian motion master equation. The use of these equations presupposes the weak coupling and/or the high-temperature limit to be valid. In order to investigate effects from non-Markovian dynamics and from strong system–environment coupling we study here the destruction of quantum coherence arising in the full Caldeira–Leggett model for the damped harmonic oscillator discussed in Section 3.6 (Caldeira and Leggett, 1985; Unruh and Zurek, 1989).

4.6.1   General decoherence formula

To determine the decoherence function $\Gamma(t)$ corresponding to the superposition of two Gaussian wave packets $|\alpha\rangle$ and $|\beta\rangle$ we apply the same technique as in Section 4.4. A simple consideration reveals that eqn (4.158) can also be used for an exact treatment of the Caldeira–Leggett model for arbitrary spectral densities, temperatures and coupling strengths. The reason is that the derivation for this equation relies solely on the Gaussian property of the initial states and of the propagator function. Thus, eqn (4.158) holds also in the general case with the only modification that the quantities $x_\alpha(t)$, $x_\beta(t)$, $x_{\alpha\beta}(t)$ and $\sigma^2(t)$ are now defined in terms of expectation values of the exact Heisenberg picture operator $x(t)$ through

$$x_\alpha(t) = \text{tr}\left\{x(t)|\alpha\rangle\langle\alpha|\rho_B\right\}, \tag{4.214}$$

$$x_\beta(t) = \text{tr}\left\{x(t)|\beta\rangle\langle\beta|\rho_B\right\}, \tag{4.215}$$

$$x_{\alpha\beta}(t) = \frac{\text{tr}\left\{x(t)|\alpha\rangle\langle\beta|\rho_B\right\}}{\langle\beta|\alpha\rangle}, \tag{4.216}$$

$$\sigma^2(t) = \text{tr}\left\{x^2(t)|\alpha\rangle\langle\alpha|\rho_B\right\} - \left[\text{tr}\left\{x(t)|\alpha\rangle\langle\alpha|\rho_B\right\}\right]^2. \tag{4.217}$$

Here the trace is taken over the total system. We also note that $\sigma_\alpha^2(t) = \sigma_\beta^2(t) = \sigma_{\alpha\beta}^2(t) \equiv \sigma^2(t)$, as in the case of the quantum optical master equation. We may therefore determine the variance $\sigma^2(t)$ with the help of the state $|\alpha\rangle$, for example. It is assumed for simplicity that initially there are no system–bath correlations. A similar technique may be used, however, if correlations are present in the initial state.

To determine the decoherence function we first express the quantity $|\alpha - \beta|^2$ used in eqn (4.158) in terms of the initial separation of the superposed wave packets in position space,

$$\Delta x \equiv x_\alpha(0) - x_\beta(0), \qquad (4.218)$$

and in momentum space,

$$\Delta p \equiv m\dot{x}_\alpha(0) - m\dot{x}_\beta(0). \qquad (4.219)$$

This yields the identity

$$|\alpha - \beta|^2 \equiv \frac{\Delta x^2 + 4\sigma_0^4 \Delta p^2}{4\sigma_0^2}. \qquad (4.220)$$

We know already that the Heisenberg picture position operator $x(t)$ and the corresponding momentum operator $p(t)$ obey the equations of motion $p(t) = m\dot{x}(t)$ and (3.465). As in Section 3.6.3.2 we introduce the fundamental solutions $G_1(t)$ and $G_2(t)$ of the homogeneous part of eqn (3.465) satisfying $G_1(0) = \dot{G}_2(0) = 1$ and $\dot{G}_1(0) = G_2(0) = 0$ and write the Heisenberg operator $x(t)$ as

$$x(t) = G_1(t)x(0) + G_2(t)\dot{x}(0) + \frac{1}{m}\int_0^t ds\, G_2(t-s)B(s)$$

$$\equiv x_h(t) + I(t), \qquad (4.221)$$

where $x_h(t)$ denotes the solution of the homogeneous equation, while $I(t)$ is a solution of the inhomogeneous equation satisfying $I(0) = \dot{I}(0) = 0$.

With the help of eqn (4.221) it is now an easy task to evaluate the required quantities. We find that

$$\Re\left[(x - x_{\alpha\beta}(t))^2 - \frac{1}{2}(x - x_\alpha(t))^2 - \frac{1}{2}(x - x_\beta(t))^2\right]$$

$$= -|\alpha - \beta|^2 \left(\sigma_0^2 G_1^2(t) + \frac{G_2^2(t)}{4m^2\sigma_0^2}\right), \qquad (4.222)$$

and that the variance is given by

$$\sigma^2(t) = \sigma_0^2 G_1^2(t) + \frac{G_2^2(t)}{4m^2\sigma_0^2} + \langle I^2(t)\rangle, \qquad (4.223)$$

where

$$\langle I^2(t)\rangle = \frac{1}{2m^2}\int_0^t ds \int_0^t ds'\, G_2(t-s)G_2(t-s')D_1(s-s'), \qquad (4.224)$$

$$= \frac{1}{m^2}\int_0^\infty d\omega\, J(\omega) \coth\left(\frac{\omega}{2k_BT}\right)\left|\int_0^t ds\, G_2(s)e^{i\omega s}\right|^2.$$

We recall that

$$D_1(s - s') = 2 \int_0^\infty d\omega J(\omega) \coth\left(\frac{\omega}{2k_\mathrm{B}T}\right) \cos\omega(s - s') \tag{4.225}$$

is the noise kernel expressed here in terms of the spectral density $J(\omega)$ of the underlying model.

Substituting eqns (4.220), (4.222) and (4.223) into (4.158) we arrive at

$$\Gamma(t) = -\frac{\Delta x^2 + 4\sigma_0^4 \Delta p^2}{8\sigma_0^2} \frac{\langle I^2(t)\rangle}{\langle I^2(t)\rangle + \sigma_0^2 G_1^2(t) + G_2^2(t)/4m^2\sigma_0^2}. \tag{4.226}$$

This equation provides a general expression for the decoherence function $\Gamma(t)$. It may be used for all linear models with coordinate–coordinate coupling, involving Gaussian initial states, arbitrary coupling strengths and spectral densities. In eqn (4.226) the decoherence function has been expressed in terms of the initial width σ_0^2 of the superposed wave packets. If σ_0^2 is taken to be related to the oscillator frequency through eqn (4.157) the superposed wave packets represent coherent states of the oscillator. However, since σ_0^2 may be chosen arbitrarily eqn (4.226) is also valid for squeezed initial states.

Obviously, $\Gamma(0) = 0$ and $\Gamma(t)$ tends to the value given by the initial overlap in the long time limit provided $\langle I^2(t)\rangle \gg \sigma_0^2 G_1^2(t) + G_2^2(t)/4m^2\sigma_0^2$ in this limit. Moreover, we have the following limits

$$\Gamma(t) \approx -\Delta x^2 \frac{\langle I^2(t)\rangle}{2G_2^2(t)/m^2}, \qquad \sigma_0^2 \longrightarrow 0, \tag{4.227}$$

$$\Gamma(t) \approx -\Delta p^2 \frac{\langle I^2(t)\rangle}{2G_1^2(t)}, \qquad \sigma_0^2 \longrightarrow \infty. \tag{4.228}$$

The problem of determining the decoherence function $\Gamma(t)$ is thus reduced to the determination of the fundamental solutions $G_1(t)$ and $G_2(t)$ of the homogeneous part of the Heisenberg equation of motion, and of the reservoir average $\langle I^2(t)\rangle$ of the square of the inhomogeneous part.

4.6.2 Ohmic environments

Let us discuss in some detail the case of an Ohmic spectral density $J(\omega) = 2m\gamma\omega\theta(\Omega - \omega)/\pi$ with some cutoff frequency Ω.

4.6.2.1 High-temperature limit
In the high-temperature limit, that is in the case $2k_\mathrm{B}T \gg \Omega \gg \omega_0, \gamma$, we get (see eqns (3.462) and (3.463))

$$D_1(s - s') \approx 8m\gamma k_\mathrm{B}T\delta(s - s'), \tag{4.229}$$

which gives, by virtue of eqn (4.224),

$$\langle I^2(t) \rangle \approx \frac{4\gamma k_B T}{m} \int\limits_0^t ds G_2^2(s). \tag{4.230}$$

To give an example we study free Brownian motion for which we have the fundamental solutions $G_1(t) = 1$ and $G_2(t) = (1 - \exp(-2\gamma t))/2\gamma$. This leads to

$$\langle I^2(t) \rangle \approx \frac{k_B T}{m\gamma^2} \left(\gamma t - [1 - \exp(-2\gamma t)] + \frac{1}{4}[1 - \exp(-4\gamma t)] \right). \tag{4.231}$$

Substitution of these relations into eqn (4.226) yields the decoherence function for high-temperature free Brownian motion. Let us investigate times such that $2\gamma t \ll 1$. Introducing $\Lambda = 2m\gamma k_B T$ we find

$$\Gamma(t) = -\frac{\Delta x^2 + 4\sigma_0^4 \Delta p^2}{8\sigma_0^2} \frac{\frac{2\Lambda}{3m^2}t^3}{\frac{2\Lambda}{3m^2}t^3 + \sigma_0^2 + \frac{t^2}{4m^2\sigma_0^2}}. \tag{4.232}$$

This shows that the behaviour of the decoherence function depends crucially on the initial width of the superposed wave packets. For example, if the noise contribution to the variance $\sigma^2(t)$ is small compared to the initial width σ_0^2 of the wave packets and if the free spreading $t^2/4m^2\sigma_0^2$ may be neglected, we get

$$\Gamma(t) \approx -\frac{\Delta x^2 + 4\sigma_0^4 \Delta p^2}{8\sigma_0^2} \frac{2\Lambda}{3m^2\sigma_0^2}t^3, \tag{4.233}$$

showing that the magnitude of $\Gamma(t)$ increases as the third power of t.

In the limit of vanishing initial width (see eqn (4.227)) eqn (4.232) leads to

$$\Gamma(t) \approx -\frac{1}{3}\Lambda\Delta x^2 t. \tag{4.234}$$

Apart from the factor $\frac{1}{3}$ this relation was already used in the estimation of the decoherence rate in Section 4.3.1. On the other hand, if we let the initial width tend to infinity (see eqn (4.228)) we get

$$\Gamma(t) \approx -\frac{1}{3}\frac{\Lambda}{m^2}\Delta p^2 t^3. \tag{4.235}$$

Again, the magnitude of the decoherence function grows with the third power of time. This result corresponds to the case of an interference device involving plane waves which has been discussed by Savage and Walls (1985b).

4.6.2.2 *Harmonic oscillator* The homogeneous part of the Heisenberg equation of motion,

$$\ddot{x} + \omega_0^2 x + 2\gamma\dot{x} = 0, \tag{4.236}$$

is easily solved to yield the fundamental solutions

$$G_1(t) = \left[\frac{\gamma}{\nu}\sin\nu t + \cos\nu t\right]e^{-\gamma t}, \qquad (4.237)$$

$$G_2(t) = \frac{1}{\nu}\sin\nu t e^{-\gamma t}, \qquad (4.238)$$

where

$$\nu = \sqrt{\omega_0^2 - \gamma^2}, \quad \gamma < \omega_0, \qquad (4.239)$$

is the characteristic frequency in the underdamped case. In the overdamped case ν becomes imaginary and we write

$$\nu = -i\tilde{\nu} = -i\sqrt{\gamma^2 - \omega_0^2}, \quad \gamma > \omega_0, \qquad (4.240)$$

such that the fundamental solutions take the form

$$G_1(t) = \left[\frac{\gamma}{\tilde{\nu}}\sinh\tilde{\nu}t + \cosh\tilde{\nu}t\right]e^{-\gamma t}, \qquad (4.241)$$

$$G_2(t) = \frac{1}{\tilde{\nu}}\sinh\tilde{\nu}t e^{-\gamma t}. \qquad (4.242)$$

We use the spectral representation of the noise kernel (4.225) and carry out the s-integration in eqn (4.224) to get

$$\langle I^2(t)\rangle = \frac{\gamma}{2\pi m|\nu|^2}\int_{-\Omega}^{+\Omega}d\omega\,\omega\coth\left(\frac{\omega}{2k_BT}\right)h(\omega), \qquad (4.243)$$

with the function

$$h(\omega) \equiv \frac{1}{2}\left|\frac{e^{-[\gamma+i(\omega-\nu)]t}-1}{\gamma+i(\omega-\nu)} - \frac{e^{-[\gamma+i(\omega+\nu)]t}-1}{\gamma+i(\omega+\nu)}\right|^2. \qquad (4.244)$$

These expressions are valid for both the overdamped and the underdamped case. Inserting eqn (4.243) and eqns (4.237), (4.238) or (4.241), (4.242) into (4.226) yields the decoherence function for the harmonic oscillator (Caldeira and Leggett, 1985). The obtained expression is valid for arbitrary coupling strengths and temperatures. It turns out, however, that a general analysis of the frequency integral in eqn (4.243) is extremely difficult. In particular, the integral depends logarithmically on Ω for large cutoff frequencies. However, simple statements may be obtained for certain limiting cases. For the following discussion we set $\sigma_0^2 = 1/2m\omega_0$ corresponding to an initial superposition of coherent states.

In the weak damping limit we have $\gamma \ll \omega_0$ and $\nu \approx \omega_0$. The function $h(\omega)$ has thus two sharp peaks at $\omega \approx \pm\omega_0$ and in the limit $\gamma \longrightarrow 0$ we may approximate

$$\langle I^2(t)\rangle \approx \frac{\gamma(2N+1)}{2\pi m\omega_0}\int_{-\infty}^{+\infty}d\omega\,h(\omega), \qquad (4.245)$$

where we have introduced the Planck distribution $N = N(\omega_0)$ through the relation $2N(\omega_0) + 1 = \coth(\omega_0/2k_\mathrm{B}T)$. The remaining frequency integral may be determined with the help of the method of residues which leads to

$$\langle I^2(t)\rangle \approx \frac{2N+1}{2m\omega_0}\left[1 - \exp(-2\gamma t)\right]. \tag{4.246}$$

Substituting into the general decoherence formula (4.226) and noting that

$$\sigma_0^2 G_1^2(t) + \frac{1}{4m^2\sigma_0^2}G_2^2(t) = \frac{1}{2m\omega_0}\exp(-2\gamma t) \tag{4.247}$$

in the weak damping limit we see, as expected, that the expression for the decoherence function reduces to (4.161) which was obtained in the quantum optical limit. Note that the relaxation constants are related through $\gamma_0 \equiv 2\gamma$.

Let us also discuss the high-temperature limit for arbitrary couplings. Employing eqn (4.230) yields

$$\langle I^2(t)\rangle \approx \frac{k_\mathrm{B}T}{m\omega_0^2}\left[1 - e^{-2\gamma t}\left(1 + \frac{\gamma}{\nu}\sin 2\nu t + 2\frac{\gamma^2}{\nu^2}\sin^2\nu t\right)\right]. \tag{4.248}$$

The corresponding high-temperature decoherence function $\Gamma(t)$ is provided by inserting this expression into the general formula (4.226). The result is valid for both the underdamped and the overdamped case. As far as the underdamped case is concerned we observe, as expected, that the exact decoherence function differs from the quantum optical one by terms of order γ/ω_0. Figures 4.11 and 4.12 show $\Gamma(t)$ for two different values of γ/ω_0 and a comparison with the corresponding quantum optical result (4.161) in the high-temperature limit. The exact decoherence function oscillates around the quantum optical one for short times. With growing time these oscillations die out and $\Gamma(t)$ converges to the quantum optical limit.

The weak damping limit leads to

$$\Gamma(t) \approx -\frac{\Delta x^2 + 4\sigma_0^4\Delta p^2}{8\sigma_0^2}\frac{2k_\mathrm{B}T}{\omega_0}2\gamma t \tag{4.249}$$

for times satisfying $2\gamma t \ll 1$ and $(2k_\mathrm{B}T)(2\gamma t)/\omega_0 \ll 1$. Accordingly, the ratio of the decoherence time τ_D to the relaxation time $\tau_R = 1/2\gamma$ is

$$\frac{\tau_D}{\tau_R} = \frac{8\sigma_0^2}{\Delta x^2 + 4\sigma_0^4\Delta p^2}\frac{\omega_0}{2k_\mathrm{B}T}. \tag{4.250}$$

Let us compare this result with the strongly overdamped case. This case is defined by the limit $\gamma \gg \omega_0$ such that $\tilde{\nu} \approx \gamma$. The decoherence function is determined by substituting $\nu = -i\tilde{\nu}$ in eqn (4.248). Let us investigate times t satisfying

$$2(\gamma - \tilde{\nu})t \approx \frac{\omega_0^2}{\gamma^2}\gamma t \ll 1, \quad 2(\gamma + \tilde{\nu})t \approx 4\gamma t \gg 1. \tag{4.251}$$

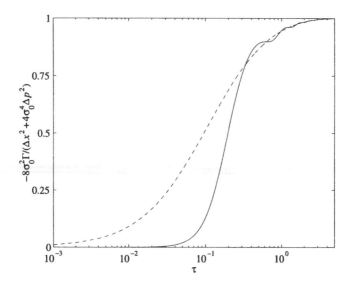

FIG. 4.11. Semilogarithmic plot of the underdamped high-temperature decoherence function $\Gamma(t)$ as given by eqns (4.226) and (4.248) (solid line), and of
the corresponding quantum optical result (4.161) (dashed line) as a function
of $\tau = 2\gamma t$. Parameters: $2k_B T/\hbar\omega_0 = 10$, $\gamma/\omega_0 = 0.1$.

The decoherence function for the strongly overdamped particle then becomes

$$\Gamma(t) \approx -\frac{\Delta x^2 + 4\sigma_0^4 \Delta p^2}{8\sigma_0^2}\frac{2k_B T}{\omega_0}\frac{\omega_0^2}{2\gamma^2}2\gamma t \qquad (4.252)$$

We note that this expression differs from the weak damping result (4.249) by the
factor $\omega_0^2/2\gamma^2$. It must be noted, however, that the relaxation rate is $\tau_R = 1/2\gamma$
in the weak damping case, while it takes the form

$$\tilde{\tau}_R = \frac{1}{2(\gamma - \tilde{\nu})} \approx \frac{\gamma}{\omega_0^2} = \frac{2\gamma^2}{\omega_0^2}\tau_R \qquad (4.253)$$

in the strong damping limit. It follows that the strong damping decoherence time
$\tilde{\tau}_D$ satisfies

$$\frac{\tilde{\tau}_D}{\tilde{\tau}_R} = \frac{\tau_D}{\tau_R}. \qquad (4.254)$$

Thus we arrive at the remarkable result that the quotient of decoherence time
and relaxation time is the same for the weak and for the strong damping case,
and that it coincides with the result (4.170) found in the quantum optical limit.

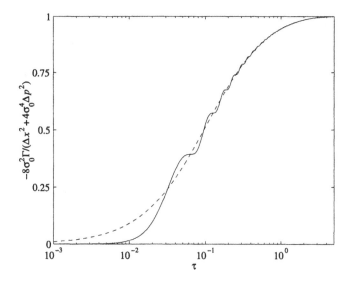

FIG. 4.12. The same as Fig. 4.11 but for $\gamma/\omega_0 = 0.01$.

4.7   Decoherence and quantum measurement

The quantum theory of measurement provides an important application of environment-induced decoherence. The destruction of quantum coherence through the influence of an environment singles out a specific basis set, known as the *pointer basis*, in the Hilbert space of some quantum apparatus M which is employed to measure a quantum object S. Since decoherence, as it is understood here, is ultimately linked to the tracing over degrees of freedom of the environment, it cannot, of course, solve the measurement problem. This means that decoherence cannot be used to deduce the reduction of the state vector and the statistical interpretation of quantum mechanics from the unitary evolution given by the Schrödinger equation. However, in any realistic measurement scheme the coupling of a macroscopic apparatus to its environment can be shown to lead, under quite general physical circumstances, to a dynamical selection of a specific pointer basis and, thus, to a unique definition of what is being measured by the apparatus (Zurek, 1981, 1982).

4.7.1   Dynamical selection of a pointer basis

We investigate a quantum system S which is coupled to an apparatus or *meter* M through an interaction Hamiltonian $H_{SM}(t)$. The meter degree of freedom will be described fully quantum mechanically such that the Hilbert space of the combined system is the tensor product space $\mathcal{H} = \mathcal{H}_S \otimes \mathcal{H}_M$. We further denote the interaction Hamiltonian in the interaction picture by

$$H_{SM}^I(t) = e^{iH_0t} H_{SM}(t) e^{-iH_0t}, \tag{4.255}$$

where $H_0 = H_S + H_M$ is the sum of the self-Hamiltonians of system and meter. As in Section 4.1 the system–meter interaction is supposed to lead, in the ideal case, to perfect correlations between certain orthogonal basis states $|S_n\rangle$ of S and meter states $|M_n\rangle$ such that the time evolution over a time interval τ takes the form (see eqns (4.7) and (4.8))

$$|\psi_{SM}(0)\rangle = \sum_n c_n|S_n\rangle \otimes |M\rangle \longrightarrow |\psi_{SM}(\tau)\rangle = \sum_n c_n|S_n\rangle \otimes |M_n\rangle. \quad (4.256)$$

This is precisely the type of dynamics underlying an indirect, QND measurement scheme: The system S represents the quantum object, while the meter M provides the quantum probe. The initial probe state is $|M\rangle$ and the quantity being measured is

$$\widehat{S} = \sum_n a_n|S_n\rangle\langle S_n|, \quad (4.257)$$

or some function $f(\widehat{S})$ thereof. In eqn (4.256) it is assumed that the back-action evasion condition holds,

$$\left[\widehat{S}, H_{SM}^I(t)\right] = 0. \quad (4.258)$$

This ensures, as we know, that the basis states $|S_n\rangle$ are unaffected by the system–meter interaction. Let us suppose that the meter states are, at least approximately, orthogonal (see also the example treated in Section 4.7.2), such that we have[11]

$$\langle S_n|S_m\rangle = \delta_{nm}, \quad \langle M_n|M_m\rangle = \delta_{nm}. \quad (4.259)$$

The application of the projection postulate to the meter quantity

$$\widehat{M} = \sum_n b_n|M_n\rangle\langle M_n|, \quad (4.260)$$

then yields that the readout b_n is found with probability $|c_n|^2$ and that the system's wave function conditioned on this event is subsequently given by $|S_n\rangle$. Within the framework of an indirect measurement scheme we may thus say that a measurement carried out on the meter system leads to a measurement of a system's quantity $f(\widehat{S})$ and induces the reduction of the system's state vector in accordance with the projection postulate. After the measurement the reduced density matrix of S takes the form,

$$\rho_S(\tau) = \sum_n |c_n|^2|S_n\rangle\langle S_n|, \quad (4.261)$$

describing the measurement on the non-selective level.

Although being strongly suggested by the decomposition of $|\psi_{SM}(\tau)\rangle$ in (4.256), the above interpretation is incomplete for the following reason. After

[11] To facilitate the following discussion we assume here that the interaction is non-degenerate, i.e. that it leads to an indirect measurement of a non-degenerate system observable.

the system–meter interaction the combined system ends up in an entangled state $|\psi_{SM}(\tau)\rangle$, describing perfect correlations between the system states $|S_n\rangle$ and the meter states $|M_n\rangle$. However, $|\psi_{SM}(\tau)\rangle$ is still a superposition of these correlated states. The latter *coexist* in $|\psi_{SM}(\tau)\rangle$ and without application of the reduction postulate to the meter system M there is no a priori reason of accepting only one of the states $|S_n\rangle \otimes |M_n\rangle$ as physically real. In fact, without making a definite decision on an observable being measured on M we could consider another set of basis states,

$$|\widetilde{M}_n\rangle = \sum_{n'} |M_{n'}\rangle\langle M_{n'}|\widetilde{M}_n\rangle, \tag{4.262}$$

and write the state vector of the combined system after time τ alternatively as

$$|\psi_{SM}(\tau)\rangle = \sum_n \widetilde{c}_n |\widetilde{S}_n\rangle \otimes |\widetilde{M}_n\rangle, \tag{4.263}$$

where we have introduced new normalized system states through

$$\widetilde{c}_n |\widetilde{S}_n\rangle \equiv \sum_{n'} c_{n'} |S_{n'}\rangle \langle \widetilde{M}_n | M_{n'}\rangle. \tag{4.264}$$

With the same right as before we could now claim that the measurement of the meter observable

$$\widehat{\widetilde{M}} = \sum_n \widetilde{b}_n |\widetilde{M}_n\rangle\langle \widetilde{M}_n| \tag{4.265}$$

leads to the measurement of the system observable

$$\widehat{\widetilde{S}} = \sum_n \widetilde{a}_n |\widetilde{S}_n\rangle\langle \widetilde{S}_n|, \tag{4.266}$$

or of some function $f(\widehat{\widetilde{S}})$. The problem is that, although we did not change the system–meter interaction and although the reduced system's density matrices obviously coincide,

$$\rho_S(\tau) = \sum_n |c_n|^2 |S_n\rangle\langle S_n| = \sum_n |\widetilde{c}_n|^2 |\widetilde{S}_n\rangle\langle \widetilde{S}_n|, \tag{4.267}$$

the to-be-measured system observable is ambiguous since, in general, the quantities $\widehat{S}$ and $\widehat{\widetilde{S}}$ do not commute,

$$\left[\widehat{S}, \widehat{\widetilde{S}}\right] \neq 0. \tag{4.268}$$

The question is thus: Do we have a measurement device which measures $\widehat{S}$ or $\widehat{\widetilde{S}}$?

Once a basis of meter states $|\widetilde{M}_n\rangle$ is given, the corresponding states $|\widetilde{S}_n\rangle$ in a decomposition of the form (4.263) are denoted as *relative states*. They are, in general, not orthogonal to each other but may be taken to be normalized in which case they are defined uniquely up to a phase factor. The concept of relative states has been used by Everett (1957) in his relative state formulation of quantum mechanics. We may assume the $\tilde{c}_n$ in eqn (4.264) to be real and non-negative which yields

$$\tilde{c}_n = \sqrt{\sum_{n'} |c_{n'}|^2 |\langle \widetilde{M}_n | M_{n'} \rangle|^2}, \tag{4.269}$$

and

$$|\tilde{S}_n\rangle = \frac{1}{\tilde{c}_n} \sum_{n'} c_{n'} |S_{n'}\rangle \langle \widetilde{M}_n | M_{n'} \rangle. \tag{4.270}$$

Hence, the scalar product of the relative states $|\tilde{S}_n\rangle$ is given by

$$\langle \tilde{S}_m | \tilde{S}_n \rangle = \frac{1}{\tilde{c}_m \tilde{c}_n} \sum_{n'} |c_{n'}|^2 \langle \widetilde{M}_n | M_{n'} \rangle \langle M_{n'} | \widetilde{M}_m \rangle. \tag{4.271}$$

We note that for non-orthogonal relative states eqn (4.266) still defines a self-adjoint operator, where, however, the $\tilde{a}_n$ are not its eigenvalues. Moreover, eqn (4.267) represents the system's density matrix after the measurement in two alternative ways as a mixture of pure states, where in the second case these pure states are not orthogonal.

The above problem becomes particularly acute if all non-zero coefficients c_n, $n = 1, 2, \ldots, D$, in the superposition of the initial state $|\psi_{SM}(0)\rangle$ have the same absolute value, that is if $|\psi_{SM}(0)\rangle$ is maximally entangled. In this case $\rho_S(\tau)$ becomes proportional to the identity in the subspace spanned by the corresponding states $|S_n\rangle$,

$$\rho_S(\tau) = \frac{1}{D} \sum_{n=1}^{D} |S_n\rangle \langle S_n|. \tag{4.272}$$

Assuming further that the corresponding basis states $|M_n\rangle$ and $|\widetilde{M}_n\rangle$ span the same subspace we immediately find with the help of eqn (4.271) that the relative states are orthogonal, that is

$$\langle \tilde{S}_m | \tilde{S}_n \rangle = \delta_{mn} \tag{4.273}$$

for any choice of meter basis states (with the above restriction, of course). By varying the meter basis we can measure any system basis. Thus we are led to the surprising conclusion that our measurement device is capable of measuring any system's observable.

It is important to realize that the problem discussed above does not provide a contradiction to the orthodox interpretation of quantum mechanics, for it arises only if one refuses to make a definite decision on the meter observable and to apply the reduction postulate. However, the situation *is* somehow dissatisfying

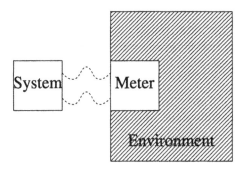

FIG. 4.13. Schematic picture of the measurement scheme. In addition to the system–meter interaction the meter is coupled to a large environment which dynamically selects a specific pointer basis through decoherence.

since according to experience a certain measuring device which has been designed to measure, for example, the momentum of a quantum system, *does* measure momentum, and *not* position. The ambiguity in the measured system observable is obviously due to the fact that the system–meter interaction does not fix a unique basis of states $|M_n\rangle$ in the meter's Hilbert space $\mathcal{H}_M$ (or in a certain subspace thereof). This ambiguity can only be avoided if for some reason a specific basis is singled out, that is if only a specific physical quantity $\widehat{M} = \sum_n b_n |M_n\rangle\langle M_n|$ (or some function of it) can be measured on M. The system observable $\widehat{S}$ measured by the device is then determined by the corresponding set of relative states $|S_n\rangle$.

At this point it must be taken into account that the meter M is usually assumed to represent a macroscopic degree of freedom, which, in turn, is coupled to a huge (usually infinite) number of further degrees of freedom. Environment-induced decoherence then leads to a dynamical selection of a specific basis. This means that the meter's Hilbert space $\mathcal{H}_M$ is decomposed into coherent subspaces (which are supposed to be one-dimensional here) in such a way that a local observer cannot observe coherences between different subspaces. Superpositions of the form (4.262) are thus effectively destroyed and a unique basis $\{|M_n\rangle\}$ emerges. This basis is often referred to as the *pointer basis*. The environment E acts like a measuring apparatus: While the total system $S + M + E$ evolves unitarily, of course, the reduced system $S+M$ behaves as if its state continuously collapses into one of the correlated states $|S_n\rangle\otimes|M_n\rangle$. The reduced state of $S+M$ then behaves, with regard to local measurements, as an incoherent statistical mixture of the states $|S_n\rangle \otimes |M_n\rangle$.

Thus, we consider a measurement scheme of the type depicted in Fig. 4.13. The total interaction Hamiltonian in the interaction picture will be supposed to be of the form

$$H_I(t) = H_{SM}^I(t) + H_{ME}^I(t). \qquad (4.274)$$

Here, we assume for simplicity that $H_{SME} = 0$, that is there are no triple

interaction terms in the total Hamiltonian, and that $H_{SE} = 0$, which means
that the direct interaction between system and environment is zero. The whole
measurement process can then be decomposed into two phases as follows (Zurek,
1981).

1. As before $H_{SM}^I(t)$ acts over a time interval $[0, \tau]$ in which $H_{SM}^I(t) \gg$
   $H_{ME}^I(t)$. During this time interval the desired system–meter correlations
   are built up under the influence of $H_{SM}^I(t)$,

$$\sum_n c_n |S_n\rangle \otimes |M\rangle \otimes |E\rangle \longrightarrow \sum_n c_n |S_n\rangle \otimes |M_n\rangle \otimes |E\rangle, \qquad (4.275)$$

   where $|E\rangle$ is the initial state of the environment. As before, the back-
   action evasion condition (4.258) is assumed to hold, which guarantees that
   the system states $|S_n\rangle$ are not affected by the system–meter interaction.
   This first phase of the process may be called pre-measurement.

2. For $t > \tau$ the meter–environment interaction dominates, that is $H_{ME}^I(t) \gg$
   $H_{SM}^I(t)$. This yields the dynamical selection of the pointer basis through
   environment-induced decoherence. Assuming also the back-action evasion
   condition for the meter–environment interaction,

$$\left[\widehat{M}, H_{ME}^I(t) \right] = 0, \qquad (4.276)$$

we may write the second phase of the measurement process as

$$\sum_n c_n |S_n\rangle \otimes |M_n\rangle \otimes |E\rangle \longrightarrow \sum_n c_n |S_n\rangle \otimes |M_n\rangle \otimes |E_n\rangle. \qquad (4.277)$$

We note that the most general solution of condition (4.276) reads

$$H_{ME}^I(t) = \sum_n |M_n\rangle\langle M_n| \otimes B_n(t), \qquad (4.278)$$

where the $B_n(t)$ are arbitrary operators of the environment. Condition
(4.276) guarantees that the eigenbasis of $\widehat{M}$ is not affected during the second phase of the process. The specific system–meter correlations created
during the pre-measurement are thus not destroyed through the environment; they are still present in the reduced system–meter density matrix,

$$\rho_{SM}(t) = \sum_n |c_n|^2 |S_n\rangle\langle S_n| \otimes |M_n\rangle\langle M_n|. \qquad (4.279)$$

This is, of course, only a schematic picture of what is going on in a realistic
measurement scheme. In particular, it turns out that in a more realistic description the ideal back-action evasion condition (4.276) cannot, usually, be fulfilled
exactly. The pointer basis can, however, be still approximately orthogonal, which
is fully sufficient for the measuring device to work. This point is illustrated with
the help of a specific model in the next subsection.

4.7.2 Dynamical model for a quantum measurement

Let us discuss here a dynamical model for a quantum measurement process which has been introduced and investigated by Walls, Collett and Milburn (1985). It consists of two coupled oscillator modes a and b, where a is the to-be-measured quantum object and b represents the meter. The system–meter interaction is taken to be of the form (the entire discussion will be performed in the interaction picture and we omit for ease of notation the index I)

$$H_{SM} = -\frac{i}{2} a^\dagger a \left(b \varepsilon^* - b^\dagger \varepsilon \right). \tag{4.280}$$

The physical background for this type of coupling is a four wave mixing interaction (Walls and Milburn, 1994) which leads to a back-action evasion interaction, satisfying $\left[f(a^\dagger a), H_{SM} \right] = 0$. Disregarding the coupling of mode b to the environment we get the following pre-measurement time evolution:

$$|\psi_{SM}(0)\rangle = \sum_n c_n |n\rangle_S \otimes |0\rangle_M \longrightarrow |\psi_{SM}(t)\rangle = \sum_n c_n |n\rangle_S \otimes |n\varepsilon t/2\rangle_M. \tag{4.281}$$

Thus, if the meter mode b is initially in the vacuum state $|0\rangle_M$, the number states $|n\rangle_S$ of mode a become correlated with the coherent states

$$|M_n\rangle = |n\varepsilon t/2\rangle_M, \tag{4.282}$$

which will be seen to play the rôle of the pointer basis states. We note that these states are approximately orthogonal since their overlap is exponentially small for $|\varepsilon| t \gg 1$,

$$\langle M_m | M_n \rangle = \exp\left[-\frac{|\varepsilon|^2 t^2}{8} (m-n)^2 \right]. \tag{4.283}$$

The environment is represented by a collection of modes c_j which are linearly coupled to the meter mode b,

$$H_{ME} = b \sum_j \kappa_j^* c_j^\dagger + b^\dagger \sum_j \kappa_j c_j. \tag{4.284}$$

Tracing over the environment and performing the Born–Markov approximation in the quantum optical limit we get the following master equation for the system–meter density matrix,

$$\frac{d}{dt} \rho_{SM}(t) = -\frac{1}{2} \left[a^\dagger a \left(b\varepsilon^* - b^\dagger \varepsilon \right), \rho_{SM}(t) \right]$$

$$+ \gamma_0 \left(b \rho_{SM}(t) b^\dagger - \frac{1}{2} b^\dagger b \rho_{SM}(t) - \frac{1}{2} \rho_{SM}(t) b^\dagger b \right), \tag{4.285}$$

with the initial condition

$$\rho_{SM}(0) = |\psi_{SM}(0)\rangle\langle\psi_{SM}(0)| = \sum_{nm} c_m^* c_n \left(|n\rangle\langle m|\right)_S \otimes \left(|0\rangle\langle 0|\right)_M. \quad (4.286)$$

We note that H_{ME} violates the back-action evasion condition (4.276). We know, however, that the master equation (4.285) leads to a rapid destruction of coherences between coherent states. We thus expect an approximately orthogonal pointer basis to emerge, consisting of the coherent states (4.282).

To solve the master equation we employ the ansatz

$$\rho_{SM}(t) = \sum_{nm} f_{nm}(t) \left(|n\rangle\langle m|\right)_S \otimes \left(|\beta_n(t)\rangle\langle\beta_m(t)|\right)_M, \quad (4.287)$$

where $f_{nm}(t)$ and $\beta_n(t)$ are c-number functions with initial conditions $f_{nm}(0) = c_m^* c_n$ and $\beta_n(0) = 0$. Inserting this ansatz into the master equation (4.285) one is led to the differential equations

$$\dot{\beta}_n = -\frac{\gamma_0}{2}\beta_n + \frac{n\varepsilon}{2}, \quad (4.288)$$

$$\frac{\dot{f}_{nm}}{f_{nm}} = \gamma_0 \beta_m^* \beta_n + \frac{1}{2}\frac{d}{dt}\left(|\beta_n|^2 + |\beta_m|^2\right) - \frac{n\varepsilon^*}{2}\beta_n - \frac{m\varepsilon}{2}\beta_m^*, \quad (4.289)$$

which are easily solved to yield

$$\beta_n(t) = \frac{n\varepsilon}{\gamma_0}\left(1 - e^{-\gamma_0 t/2}\right), \quad (4.290)$$

$$f_{nm}(t) = c_m^* c_n \exp\left[\frac{|\varepsilon|^2}{2\gamma_0^2}(n-m)^2\left\{3 - \gamma_0 t - 4e^{-\gamma_0 t/2} + e^{-\gamma_0 t}\right\}\right]. \quad (4.291)$$

The solution of the master equation can thus be written in the following two alternatives forms,

$$\begin{aligned}\rho_{SM}(t) &= \sum_{nm} c_m^* c_n \exp\left[\frac{|\varepsilon|^2}{2\gamma_0^2}(n-m)^2\left\{3 - \gamma_0 t - 4e^{-\gamma_0 t/2} + e^{-\gamma_0 t}\right\}\right] \\ &\quad \times \left(|n\rangle\langle m|\right)_S \otimes \left(|\beta_n(t)\rangle\langle\beta_m(t)|\right)_M \\ &= \sum_{nm} c_m^* c_n \exp\left[\frac{|\varepsilon|^2}{\gamma_0^2}(n-m)^2\left\{1 - \frac{\gamma_0 t}{2} - e^{-\gamma_0 t/2}\right\}\right] \\ &\quad \times \left(|n\rangle\langle m|\right)_S \otimes \left(\frac{|\beta_n(t)\rangle\langle\beta_m(t)|}{\langle\beta_m(t)|\beta_n(t)\rangle}\right)_M. \end{aligned} \quad (4.292)$$

Let us first discuss the limit $\gamma_0 t \ll 1$, that is the limit of times which are short compared to the relaxation time γ_0^{-1}. Then we have to lowest order in $\gamma_0 t$,

$$\beta_n(t) \approx \frac{n\varepsilon t}{2}, \quad (4.293)$$

$$3 - \gamma_0 t - 4e^{-\gamma_0 t/2} + e^{-\gamma_0 t} \approx -\frac{1}{12}\left(\gamma_0 t\right)^3, \quad (4.294)$$

and, hence,

$$\rho_{SM}(t) \approx \sum_{nm} c_m^* c_n \exp\left[-\frac{1}{24}\left(|\varepsilon|t\right)^2 \gamma_0 t \left(n-m\right)^2\right]$$
$$\times \left(|n\rangle\langle m|\right)_S \otimes \left(|n\varepsilon t/2\rangle\langle m\varepsilon t/2|\right)_M. \quad (4.295)$$

The measurement will be the more accurate the smaller the overlap (4.283) of the pointer states (4.282). Thus we consider the limiting case,

$$|\varepsilon|t \longrightarrow \infty, \quad \gamma_0 t \longrightarrow 0, \quad \left(|\varepsilon|t\right) \cdot \left(\gamma_0 t\right) = \text{fixed}. \quad (4.296)$$

Equation (4.295) shows that in this limit $\rho_{SM}(t)$ becomes diagonal in the pointer basis $|M_n\rangle = |n\varepsilon t/2\rangle$,

$$\rho_{SM}(t) \approx \sum_n |c_n|^2 \left(|n\rangle\langle n|\right)_S \otimes |M_n\rangle\langle M_n|. \quad (4.297)$$

The decohering influence of the environment has thus singled out an approximately orthogonal basis of pointer states. This happens on a time scale which, under the given conditions, is small compared to the relaxation time. Therefore, damping effects play a negligible rôle and the pointer basis is independent of the relaxation rate.

We finally remark on another interesting property of the model. The second equation in (4.292) gives the following reduced density matrix for mode a,

$$\rho_S(t) = \sum_{nm} c_m^* c_n \exp\left[\frac{|\varepsilon|^2}{\gamma_0^2}\left(n-m\right)^2 \left\{1 - \frac{\gamma_0 t}{2} - e^{-\gamma_0 t/2}\right\}\right] \left(|n\rangle\langle m|\right)_S. \quad (4.298)$$

For $\gamma_0 t \ll 1$ we have to lowest order

$$1 - \frac{\gamma_0 t}{2} - e^{-\gamma_0 t/2} \approx -\frac{1}{8}(\gamma_0 t)^2. \quad (4.299)$$

It follows that the off-diagonals of $\rho_S(t)$ in the number state basis of mode a decay exponentially with an exponent which is proportional to t^2, indicating the non-Markovian character of the model for short times. For large times, $\gamma_0 t \gg 1$, we find

$$\rho_S(t) = \sum_{nm} c_m^* c_n \exp\left[-\frac{|\varepsilon|^2 t}{2\gamma_0}\left(n-m\right)^2\right] \left(|n\rangle\langle m|\right)_S. \quad (4.300)$$

On differentiating this equation with respect to time we thus obtain

$$\frac{d}{dt}\rho_S(t) = -\frac{|\varepsilon|^2}{2\gamma_0}\left[a^\dagger a, \left[a^\dagger a, \rho_S(t)\right]\right]. \quad (4.301)$$

This is a Markovian master equation for the reduced system S and its form is that of a master equation in the singular coupling limit. The model thus shows a transition from non-Markovian behaviour for short times to Markovian dynamics for long times.

References

Bouwmeester, D., Ekert, A. and Zeilinger, A. (eds.) (2000). *The Physics of Quantum Information*. Springer-Verlag, Berlin.

Breuer, H. P., Dorner, U. and Petruccione, F. (2001). Preparation and decoherence of superpositions of electromagnetic field states. *Eur. Phys. J.*, **D14**, 377–386.

Brune, M., Hagley, E., Dreyer, J., Maître, X., Maali, A., Wunderlich, C., Raimond, J. M. and Haroche, S. (1996). Observing the progressive decoherence of the "meter" in a quantum measurement. *Phys. Rev. Lett.*, **77**, 4887–4890.

Caldeira, A. O. and Leggett, A. J. (1985). Influence of damping on quantum interference: An exactly soluble model. *Phys. Rev.*, **A31**, 1059–1066.

DiVincenzo, D. P. (1995). Two-bit gates are universal for quantum computation. *Phys. Rev.*, **A51**, 1015–1022.

Everett III, H. (1957). "Relative state" formulation of quantum mechanics. *Rev. Mod. Phys.*, **29**, 454–462.

Friedman, J. R., Patel, V., Chen, W., Tolpygo, S. K. and Lukens, J. E. (2000). Quantum superposition of distinct macroscopic states. *Nature*, **406**, 43–46.

Gallis, M. R. and Fleming, G. N. (1990). Environmental and spontaneous localization. *Phys. Rev.*, **A42**, 38–42.

Joos, E. and Zeh, H. D. (1985). The emergence of classical properties through interaction with the environment. *Z. Phys.*, **B59**, 223–243.

Joos, E. (1996). Decoherence through interaction with the environment. In *Decoherence and the Appearence of a Classical World in Quantum Theory* (eds. Giulini, D., Joos, E., Kiefer, C., Kupsch, J., Stamatescu, I.-O. and Zeh, H. D.), Springer-Verlag, Berlin, pp. 35–136.

Maître, X., Hagley, E., Dreyer, J., Maali, A., Wunderlich, C., Brune, M., Raimond, J. M. and Haroche, S. (1997). An experimental study of a Schrödinger cat decoherence with atoms and cavities. *J. Mod. Opt.*, **44**, 2023–2032.

Myatt, C. J., King, B. E., Turchette, Q. A., Sackett, C. A., Kielpinski, D., Itano, W. M., Monroe, C. and Wineland, D. J. (2000). Decoherence of quantum superpositions through coupling to engineered reservoirs. *Nature*, **403**, 269–273.

Palma, G. M., Suominen, K.-A. and Ekert, A. K. (1996). Quantum computers and dissipation. *Proc. R. Soc. Lond.*, **A452**, 567–584.

Savage, C. M. and Walls, D. F. (1985a). Damping of quantum coherence: The master-equation approach. *Phys. Rev.*, **A32**, 2316–2323.

Savage, C. M. and Walls, D. F. (1985b). Quantum coherence and interference of damped free particles. *Phys. Rev.*, **A32**, 3487–3492.

Schrödinger, E. (1935). Die gegenwärtige Situation in der Quantenmechanik. *Naturwissenschaften*, **23**, 807–812, 823–828, 844–849.

Steane, A. (1998). Quantum computing. *Rep. Prog. Phys.*, **61**, 117–173.

Tegmark, M. (1993). Apparent wave function collapse caused by scattering. *Found. Phys. Lett.*, **6**, 571–590.

Turchette, Q. A., Myatt, C. J., King, B. E., Sackett, C. A., Kielpinski, D., Itano,

W. M., Monroe, C. and Wineland, D. J. (2000). Decoherence and decay of motional quantum states of a trapped atom coupled to engineered reservoirs. *Phys. Rev.*, **A62**, 053807, 1–22.

Unruh, W. G. and Zurek, W. H. (1989). Reduction of a wave packet in quantum Brownian motion. *Phys. Rev.*, **D40**, 1071–1094.

Unruh, W. G. (1995). Maintaining coherence in quantum computers. *Phys. Rev. A*, **51**, 992–997.

Walls, D. F., Collet, M. J. and Milburn, G. J. (1985). Analysis of a quantum measurement. *Phys. Rev.*, **D32**, 3208–3215.

Walls, D. F. and Milburn, G. J. (1994). *Quantum Optics*. Springer-Verlag, Berlin.

Zurek, W. H. (1981). Pointer basis of quantum apparatus: Into what mixture does the wave packet collapse? *Phys. Rev.*, **D24**, 1516–1525.

Zurek, W. H. (1982). Environment-induced superselection rules. *Phys. Rev.*, **D26**, 1862–1880.

Zurek, W. H. (1991). Decoherence and the transition from quantum to classical. *Phys. Today*, **44**, 36–44.

Zurek, W. H. (1998). Decoherence, einselection and the existential interpretation (the rough guide). *Phil. Trans. R. Soc. Lond.*, **A356**, 1793–1821.

Part III

Stochastic processes in Hilbert space

5

PROBABILITY DISTRIBUTIONS ON HILBERT SPACE

We saw in Section 2.1.3 that the statistical properties of a quantum mechanical ensemble are fully characterized in terms of a density matrix. However, if selective measurements of one or several observables are carried on the ensemble, it will split into a number of sub-ensembles, each sub-ensemble being conditioned on a particular outcome of the measurements. The theoretical description of the collection of sub-ensembles thus created leads to a new kind of quantum statistical ensemble which differs from the one encountered in Section 2.1.3. The central subject of the present chapter will be to introduce these new ensembles. They will enable us in Chapter 6 to look at the dynamics of open quantum systems from a new perspective, namely from the viewpoint of stochastic processes in Hilbert space.

The considerations of Section 5.1 serve to give a precise physical interpretation of the new type of ensembles. The required mathematical framework of functional integration in Hilbert space will be developed in Section 5.2. There we introduce probability density functionals on Hilbert space which allow a general characterization of the new kind of ensembles and enable the construction of appropriate stochastic time-evolution equations for the state vector. As a further generalization required in later chapters we are going to study in Section 5.3 probability distributions on the space of density matrices.

5.1 The state vector as a random variable in Hilbert space

We introduce in this section a new type of quantum statistical ensemble which may be characterized by means of a random state vector in Hilbert space. The general concept will also be illustrated with the help of a simple example.

5.1.1 A new type of quantum mechanical ensemble

In Section 2.1.3 we studied a certain type of ensembles $\mathcal{E}$ the statistical properties of which are completely characterized by means of a density matrix ρ. In order to distinguish such ensembles from the new ensembles introduced below they will be denoted by $\mathcal{E} = \mathcal{E}_\rho$ in the following.

Recall that an ensemble $\mathcal{E}_\rho$ was constructed by mixing M ensembles $\mathcal{E}_\alpha$, $\alpha = 1, 2, \ldots, M$, of a certain quantum mechanical system with the weights $w_\alpha = N_\alpha/N$. Each $\mathcal{E}_\alpha$ represents a pure ensemble describable by a normalized state vector ψ_α. The number N_α denotes the number of systems in $\mathcal{E}_\alpha$ and

$$N = \sum_{\alpha=1}^{M} N_\alpha \qquad (5.1)$$

is the total number of systems in $\mathcal{E}_\rho$. Denoting by $S_\alpha^{(1)}, S_\alpha^{(2)}, \ldots, S_\alpha^{(N_\alpha)}$ the individual systems of $\mathcal{E}_\alpha$ we may write the result of the mixing process symbolically as

$$\mathcal{E}_\rho = \left\{ S_1^{(1)}, \ldots, S_1^{(N_1)}, \ldots, S_\alpha^{(1)}, \ldots, S_\alpha^{(N_\alpha)}, \ldots, S_M^{(1)}, \ldots, S_M^{(N_M)} \right\}. \qquad (5.2)$$

This notation serves to emphasize that the ensemble $\mathcal{E}_\rho$ is simply a collection of N quantum systems.

The important point to note is the following: Saying that the statistical properties of $\mathcal{E}_\rho$ are completely described by the density matrix

$$\rho = \sum_{\alpha=1}^{M} w_\alpha |\psi_\alpha\rangle\langle\psi_\alpha| \qquad (5.3)$$

one presupposes that the N systems $S_\alpha^{(i)}$ making up the ensemble $\mathcal{E}_\rho$ are indistinguishable in the sense that no information on the original grouping into the subsets $\mathcal{E}_\alpha$ is available. An experimenter performing measurements on $\mathcal{E}_\rho$ only knows that, taking at random a member from the ensemble, this particular system is to be described with probability w_α with the help of the state vector ψ_α. It is certainly true that an experimenter with only this information to hand can by no means predict anything other than the probabilities and expectation values determined with the help of the density matrix ρ. It is for this reason that one calls $\mathcal{E}_\rho$ a mixture, that is, a totally disordered collection of quantum systems.

As discussed in Section 2.1.3.2 a given density matrix ρ can be expressed in an infinite number of ways in the form (5.3) as a convex linear combination of pure states $|\psi_\alpha\rangle\langle\psi_\alpha|$ which need not be orthogonal. There is thus always an infinite number of ensembles of the type $\mathcal{E}_\rho$ which all lead to the same density matrix. Consider two such ensembles $\mathcal{E}_\rho$ and $\mathcal{E}_\rho'$ such that

$$\rho = \sum_{\alpha=1}^{M} w_\alpha |\psi_\alpha\rangle\langle\psi_\alpha| = \sum_{\beta=1}^{M'} w_\beta' |\psi_\beta'\rangle\langle\psi_\beta'|, \qquad (5.4)$$

where w_α, ψ_α are the weights and states of ensemble $\mathcal{E}_\rho$, and w_β', ψ_β' are those of $\mathcal{E}_\rho'$. Although both ensembles could have been prepared in entirely different ways, they are described by the same density matrix ρ. Physically, the meaning of this statement is that there is no way for an observer to distinguish these ensembles by means of any experimental setup. Of course, the observer could decompose both ensembles into a number of pure sub-ensembles by measuring some discrete, non-degenerate observable R (or else, by measuring a complete set of commuting observables). However, not only are the probabilities for the

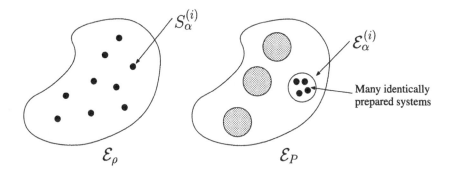

FIG. 5.1. Picture of an ensemble of type $\mathcal{E}_\rho$ (left) and of an ensemble of type $\mathcal{E}_P$ (right).

measurement outcomes the same for $\mathcal{E}_\rho$ and $\mathcal{E}'_\rho$, but also the collections of sub-ensembles thus created. In this sense it is justified to say that an ensemble of type $\mathcal{E}_\rho$ is completely characterized by a density matrix ρ.

As will now be demonstrated, one can, however, design an entirely different type of ensemble the statistical properties of which are not fully characterized by a density matrix. We denote such ensembles by the symbol $\mathcal{E}_P$. To construct them we again start from the ensembles $\mathcal{E}_\alpha$ describable by pure states ψ_α. One may imagine that each of these states has been prepared by the measurement of some complete set of commuting observables. The various ψ_α need not, however, to be orthogonal. We again combine the $\mathcal{E}_\alpha$ with respective weights w_α. However, this time we would like to keep the information that a particular quantum system belongs to a particular ensemble $\mathcal{E}_\alpha$. For this purpose we consider N_α identically prepared copies $\mathcal{E}_\alpha^{(1)}, \mathcal{E}_\alpha^{(2)}, \ldots, \mathcal{E}_\alpha^{(N_\alpha)}$ of $\mathcal{E}_\alpha$ for each α. The new ensemble $\mathcal{E}_P$ is then the collection of these ensembles, that is, an *ensemble of ensembles*:

$$\mathcal{E}_P = \left\{ \mathcal{E}_1^{(1)}, \ldots, \mathcal{E}_1^{(N_1)}, \ldots, \mathcal{E}_\alpha^{(1)}, \ldots, \mathcal{E}_\alpha^{(N_\alpha)}, \ldots, \mathcal{E}_M^{(1)}, \ldots, \mathcal{E}_M^{(N_M)} \right\}. \tag{5.5}$$

Note the decisive difference between (5.2) and (5.5): $\mathcal{E}_\rho$ is a disordered set of N elements each of which represents an individual quantum system prepared in one of the states ψ_α. On the other hand, $\mathcal{E}_P$ is a set whose elements are again sets, namely the ensembles $\mathcal{E}_\alpha^{(i)}$. The distinction between both types of ensembles is illustrated in Fig. 5.1.

The advantage of this construction is that $\mathcal{E}_P$ can now be regarded as a sample which has been drawn from the sample space

$$\Omega = \{\mathcal{E}_1, \ldots, \mathcal{E}_\alpha, \ldots, \mathcal{E}_M\} \tag{5.6}$$

furnished with the probability measure

$$\mu(\mathcal{E}_\alpha) = w_\alpha. \tag{5.7}$$

Thus we have constructed a well-defined classical probability space: The sample space Ω is a set of ensembles $\mathcal{E}_\alpha$, the algebra $\mathcal{A}$ of events is provided by the subsets of Ω, and the probability measure μ assigns the probabilities w_α to the elements of Ω.

The physical significance of the sample space Ω, or else, of the new type of ensembles $\mathcal{E}_P$ stems from the following fact: An experimenter confronted with an ensemble of the type $\mathcal{E}_P$ can always find out by an appropriate measurement which particular ensemble $\mathcal{E}_\alpha$ is realized, that is, he can determine to any desired precision the state ψ_α describing $\mathcal{E}_\alpha$. In addition, he can do so in such a way that the ensemble $\mathcal{E}_\alpha$ is only negligibly disturbed by the measurement. To achieve this he merely measures the state on a sub-ensemble of $\mathcal{E}_\alpha$, leaving the remaining part of $\mathcal{E}_\alpha$ completely unchanged. Thus we see also that the elements of Ω can be considered as classical objects such that the rules for combing the various $\mathcal{E}_\alpha$ are just the rules of ordinary, classical logic.

The probability measure (5.7) assigns to each $\mathcal{E}_\alpha$ a probability w_α. Since $\mathcal{E}_\alpha$ is represented by a state vector ψ_α, the measure (5.7) gives rise to a probability distribution on the space of state vectors. The conclusion is that the state vector becomes a random variable in Hilbert space which is given by the assignment

$$\mathcal{E}_\alpha \mapsto \psi_\alpha. \tag{5.8}$$

The corresponding probability distribution is given by

$$P(\psi_\alpha) = w_\alpha. \tag{5.9}$$

This equation yields a probability distribution on the underlying Hilbert space $\mathcal{H}$ or, more precisely, on the space of rays in $\mathcal{H}$.

Having defined the random state vector (5.8) characterized by the distribution (5.9) we can interpret the corresponding density matrix (5.3) as the covariance matrix of the random variable, that is, we may set

$$\rho = \mathrm{E}\left(|\psi\rangle\langle\psi|\right), \tag{5.10}$$

where E denotes the expectation value defined through the probability distribution (5.9). A more general definition will be given in the next section.

It is very important to realize that the concept of a random state vector is by no means a hidden-variable theory. On the contrary, the above construction is fully consistent with the statistical interpretation of quantum mechanics. The reason for this fact is that the stochastic state vector is defined as a map on the sample space Ω whose elements are again ensembles. In complete agreement with the statistical interpretation of quantum mechanics, a particular realization ψ_α of the random state vector thus represents a pure statistical ensemble $\mathcal{E}_\alpha$ of quantum systems, all typical quantum correlations and interference effects being embodied in this particular ψ_α.

It is obvious that an $\mathcal{E}_P$-ensemble contains more information than a corresponding $\mathcal{E}_\rho$-ensemble. Thus, there exist observable quantities for $\mathcal{E}_P$ which are not observables for $\mathcal{E}_\rho$ and which cannot be expressed in terms of the density matrix. To give an example we consider some self-adjoint operator R. On $\mathcal{E}_\rho$ we can measure, of course, the variance

$$\text{Var}(R) = \text{tr}\left\{R^2\rho\right\} - \left[\text{tr}\left\{R\rho\right\}\right]^2$$

$$= \sum_\alpha w_\alpha \langle\psi_\alpha|R^2|\psi_\alpha\rangle - \left[\sum_\alpha w_\alpha\langle\psi_\alpha|R|\psi_\alpha\rangle\right]^2 \qquad (5.11)$$

of R in the usual way: Take a sufficiently large sample from $\mathcal{E}_\rho$, and estimate from it the dispersion of R. The same quantity can be measured also on $\mathcal{E}_P$, of course. Consider now the following decomposition of $\text{Var}(R)$,

$$\text{Var}(R) = \text{Var}_1(R) + \text{Var}_2(R), \qquad (5.12)$$

where

$$\text{Var}_1(R) = \sum_\alpha w_\alpha\left[\langle\psi_\alpha|R^2|\psi_\alpha\rangle - \langle\psi_\alpha|R|\psi_\alpha\rangle^2\right], \qquad (5.13)$$

and

$$\text{Var}_2(R) = \sum_\alpha w_\alpha\langle\psi_\alpha|R|\psi_\alpha\rangle^2 - \left[\sum_\alpha w_\alpha\langle\psi_\alpha|R|\psi_\alpha\rangle\right]^2. \qquad (5.14)$$

Both $\text{Var}_1(R)$ and $\text{Var}_2(R)$ represent non-negative quantities and their sum equals the usual quantum variance (5.11) which is determined by the density matrix ρ. Note, however, that $\text{Var}_1(R)$ and $\text{Var}_2(R)$ cannot, in general, be expressed as density-matrix expectation values of some self-adjoint operator, which means that these quantities are not observables for $\mathcal{E}_\rho$. But they do represent measurable quantities for $\mathcal{E}_P$-ensembles.

A definite prescription for the measurement of both $\text{Var}_1(R)$ and $\text{Var}_2(R)$ on $\mathcal{E}_P$ can be given. For this purpose we take a sample of ensembles $\mathcal{E}_\alpha^{(i)}$ from $\mathcal{E}_P$. To measure $\text{Var}_1(R)$ we first determine the dispersion $\langle\psi_\alpha|R^2|\psi_\alpha\rangle - \langle\psi_\alpha|R|\psi_\alpha\rangle^2$ of R in each of the individual ensembles $\mathcal{E}_\alpha^{(i)}$ by carrying out a sufficiently large number of measurements on the systems making up $\mathcal{E}_\alpha^{(i)}$. The obtained values for the dispersion of R will, in general, be different for the various $\mathcal{E}_\alpha^{(i)}$. The dispersion is thus a real random variable X whose classical statistical average $\langle X\rangle$ equals $\text{Var}_1(R)$. To obtain $\text{Var}_2(R)$ one could, of course, simply subtract $\text{Var}_1(R)$ from $\text{Var}(R)$. But $\text{Var}_2(R)$ can also be measured directly as follows. One determines the expectation value $\langle\psi_\alpha|R|\psi_\alpha\rangle$ of R in each individual ensemble $\mathcal{E}_\alpha^{(i)}$. In general, this expectation value takes on different values for the different $\mathcal{E}_\alpha^{(i)}$. Thus, the quantity $Y = \langle\psi_\alpha|R|\psi_\alpha\rangle$ represents a real random variable whose classical statistical variance $\text{Var}(Y) = \langle Y^2\rangle - \langle Y\rangle^2$ is equal to $\text{Var}_2(R)$.

We infer from eqn (5.13) that $\text{Var}_1(R)$ vanishes if and only if the dispersion of R vanishes for all ψ_α which occur with non-zero probability, that is if and only if all these ψ_α are eigenstates of R. Thus, if $\text{Var}_1(R)$ is found to be zero we can immediately conclude that $\mathcal{E}_P$ is an ensemble which consists of pure ensembles in eigenstates of R. The variance $\mathrm{Var}_1(R)$ can therefore be considered as a measure for the distance of the $\mathcal{E}_P$-ensemble to an ensemble made up of eigenstates of R. On the other hand, definition (5.14) shows that $\mathrm{Var}_2(R)$ vanishes if and only if the random variable $\langle \psi_\alpha | R | \psi_\alpha \rangle$ is sharp, i.e. takes on a single value. In the general case $\mathrm{Var}_2(R)$ is the dispersion of the pure state quantum expectation value and represents a measure of the statistical fluctuations of $\langle \psi_\alpha | R | \psi_\alpha \rangle$ over the ensemble.

A partition of the quantum statistical variance $\mathrm{Var}(R)$ similar to (5.12) has been considered by Wiseman in the context of the examination of a continuously monitored laser (1993), while the variance $\mathrm{Var}_2(R)$ has been introduced by Mølmer, Castin and Dalibard (1993) as a measure for the statistical fluctuations of the stochastic wave function. Examples of the physical significance of these variances for the dynamics of quantum stochastic processes will be discussed in Sections 6.7.2 and 8.2.

5.1.2 Stern–Gerlach experiment

Two $\mathcal{E}_\rho$-ensembles belonging to the same density matrix ρ cannot be distinguished by any experimental procedure. Two $\mathcal{E}_P$-ensembles, however, can be distinguished even if they lead to one and the same density matrix. Let us illustrate this point with the help of a simple example. We consider two observers, called A and B, each with a Stern–Gerlach apparatus. A uses her apparatus to prepare a large sequence of atomic beams consisting of spin-$\frac{1}{2}$ atoms, whereas B analyses the beams with his own apparatus.

The Hilbert space belonging to the internal degree of freedom of the spin-$\frac{1}{2}$ atoms is $\mathcal{H} = \mathbb{C}^2$. As we know from Section 3.4.2, any density matrix ρ in this space can be written as a unique linear combination of the unit matrix I and the Pauli spin matrices σ_1, σ_2, and σ_3,

$$\rho(\vec{v}) = \frac{1}{2} \left(I + \vec{v} \cdot \vec{\sigma} \right), \tag{5.15}$$

where $\vec{v}$ denotes the Bloch vector, satisfying $|\vec{v}| = 1$ for a pure state. Thus, a pure state is uniquely represented by a point $\vec{v}$ on the surface of the Bloch sphere. Consequently, the probability distribution (5.9) describing an ensemble of the type $\mathcal{E}_P$ can be represented by means of a probability density $P(\vec{v})$ which is concentrated on the surface of the Bloch sphere, that is which vanishes outside this surface and satisfies the normalization condition

$$\int d^3 v \, P(\vec{v}) = 1. \tag{5.16}$$

Thus, the Bloch vector $\vec{v}$ becomes a random unit vector following the distribution $P(\vec{v})$. Note that $P(\vec{v})$ is, in fact, a distribution on the space of rays in $\mathcal{H}$. Of course, one could characterize the distribution more explicitly as a density $P = P(\theta, \varphi)$ introducing spherical coordinates (θ, φ) on the surface of the Bloch sphere

(Wiseman and Milburn, 1993a, 1993b). We prefer, however, to work here with the above general representation.

Let us illustrate how observer A can generate an $\mathcal{E}_P$-ensemble. We suppose that A chooses a fixed unit vector $\vec{n}$, $|\vec{n}| = 1$. Employing her Stern–Gerlach apparatus, she then prepares by the measurement of $\vec{n} \cdot \vec{\sigma}$ a large sequence of atomic beams. Each individual beam, consisting of a large number of atoms, represents an ensemble in a definite pure state which is an eigenstate of $\vec{n} \cdot \vec{\sigma}$ with eigenvalue $+1$ or -1 and which is therefore given by one of the antipodal points $\pm\vec{n}$ on the Bloch sphere. Furthermore A ensures that in the sequence of beams she produces both eigenstates occur with an equal weight of $\frac{1}{2}$. The resulting ensemble of the type $\mathcal{E}_P$ is therefore described by the following probability density on the surface of the Bloch sphere,

$$P(\vec{v}) = \frac{1}{2}\delta(\vec{v} - \vec{n}) + \frac{1}{2}\delta(\vec{v} + \vec{n}). \tag{5.17}$$

This ensemble is now passed to observer B who performs measurements on it.

Let us first note that the $\mathcal{E}_P$-ensemble given by (5.17) leads to the density matrix

$$\rho = \int d^3v\, \rho(\vec{v})P(\vec{v}) = \begin{pmatrix} \frac{1}{2} & 0 \\ 0 & \frac{1}{2} \end{pmatrix}, \tag{5.18}$$

as is easily verified with the help of (5.15). This density matrix corresponds to an unpolarized atomic beam and is independent of $\vec{n}$. Thus, different choices for $\vec{n}$ obviously lead to different $\mathcal{E}_P$-ensembles but to one and the same density matrix ρ. The question is therefore, can B find out with the help of his own Stern–Gerlach apparatus which particular direction has been chosen by A, that is, can he determine the direction of the unit vector $\vec{n}$ which is obviously not contained in the density matrix ρ? The answer to this question is affirmative, of course.

Picking up a particular beam from the ensemble, B does not know by which state it should be described. But he does know that each beam is in some definite pure state and since each particular beam consists of many atoms he can do statistics with it. To determine the direction of $\vec{n}$ observer B can proceed as follows. First he picks a sample $\{\mathcal{E}_\alpha^{(i)}\}$ of atomic beams. Each element $\mathcal{E}_\alpha^{(i)}$ is an atomic beam describable by a pure state of the form (5.15). To determine the Bloch vector $\vec{v}$ observer B rotates his Stern–Gerlach magnet until there is no splitting of the atomic beam into two parts. The direction defined by this condition is the direction of the Bloch vector $\vec{v}$. Observer B then knows that the beam is in a pure state given by an eigenstate of $\vec{v} \cdot \vec{\sigma}$, and he can choose the orientation of $\vec{v}$ in such a way that the eigenvalue is $+1$. The beam is thus determined to be in the state (5.15) with this Bloch vector $\vec{v}$. Repeating this procedure for his sample $\{\mathcal{E}_\alpha^{(i)}\}$ of beams B will find out that only two Bloch vectors $\vec{v} = \pm\vec{n}$ occur in the sample with equal probabilities of $\frac{1}{2}$. This information enables B to reconstruct the probability distribution (5.17) created by A and to fix the direction of the unit vector $\vec{n}$.

It should be clear from these considerations that B can also reconstruct, to any desired degree of accuracy, any distribution $P(\vec{v})$ in a similar manner. For example, A could have created, at least in principle, the $\mathcal{E}_P$-ensemble given by the distribution

$$P(\vec{v}) = \frac{1}{4\pi}\delta\left(|\vec{v}| - 1\right). \tag{5.19}$$

To generate this distribution A prepares atomic beams in the pure states $\rho(\vec{v})$, where the Bloch vector $\vec{v}$ is uniformly distributed over the surface of the Bloch sphere. The preparation of this $\mathcal{E}_P$-ensemble requires that A performs preparation measurements of all observables $\vec{v} \cdot \vec{\sigma}$. We note that for different $\vec{v}$ these observables do not commute, in general, and that they are measured on different atomic beams. We also note that the normalization condition (5.16) is satisfied and that (5.19) again yields the density matrix (5.18) of an unpolarized beam. Following the procedure described above, B will now find, within the usual statistical uncertainties caused by the finiteness of his sample, that the Bloch vector is uniformly distributed over the whole surface of the Bloch sphere and concludes that the distribution is given by (5.19).

The fact that an $\mathcal{E}_P$-ensemble contains more information than a corresponding $\mathcal{E}_\rho$-ensemble may also be seen by determining the variances $\text{Var}_1(R)$ or $\text{Var}_2(R)$. Choosing a fixed unit vector $\vec{m}$ observer B may take R to be $R = \vec{m} \cdot \vec{\sigma}$, for example. The expression for $\text{Var}_1(R)$ then becomes

$$\text{Var}_1(\vec{m} \cdot \vec{\sigma}) = \int d^3v P(\vec{v}) \left(\text{tr}\left\{(\vec{m} \cdot \vec{\sigma})^2 \rho(\vec{v})\right\} - [\text{tr}\left\{(\vec{m} \cdot \vec{\sigma})\rho(\vec{v})\right\}]^2\right)$$

$$= \int d^3v P(\vec{v}) \left(1 - (\vec{m} \cdot \vec{v})^2\right), \tag{5.20}$$

which is valid for all distributions $P(\vec{v})$. For the distribution (5.17) this expression reduces to

$$\text{Var}_1(\vec{m} \cdot \vec{\sigma}) = \sin^2 \chi, \tag{5.21}$$

where $\chi \in [0, \pi)$ denotes the angle between the directions of $\vec{m}$ and $\vec{n}$. Correspondingly, we find

$$\text{Var}_2(\vec{m} \cdot \vec{\sigma}) = \cos^2 \chi, \tag{5.22}$$

such that the decomposition (5.12) of the variance $\text{Var}(\vec{m} \cdot \vec{\sigma})$ takes the form

$$\text{Var}(\vec{m} \cdot \vec{\sigma}) = \sin^2 \chi + \cos^2 \chi = 1. \tag{5.23}$$

We observe that the variances $\text{Var}_{1,2}(\vec{m} \cdot \vec{\sigma})$, which are measurable on $\mathcal{E}_P$, depend on the angle between the directions chosen by A and by B. By contrast, since ρ describes an unpolarized beam, all quantities which are measurable on a corresponding $\mathcal{E}_\rho$-ensemble, for example the variance $\text{Var}(\vec{m} \cdot \vec{\sigma})$, are independent of χ.

According to eqn (5.21) the variance $\text{Var}_1(\vec{m} \cdot \vec{\sigma})$ vanishes if $\vec{m}$ is parallel or antiparallel to $\vec{n}$. In this case B concludes that $\mathcal{E}_P$ is an ensemble which consists of ensembles in eigenstates of $\vec{m} \cdot \vec{\sigma}$. This is an example of the general property of $\mathrm{Var}_1(R)$ discussed at the end of the preceding subsection, namely that $\mathrm{Var}_1(R) = 0$ if and only if the $\mathcal{E}_P$-ensemble consists of eigenstates of R.

5.2 Probability density functionals on Hilbert space

The example discussed in the preceding section clearly demonstrates the physical significance of ensembles of type $\mathcal{E}_P$. They result from selective preparation measurements of a set of (not necessarily commuting) observables and give rise to certain probability distributions on projective Hilbert space. In this section we develop an appropriate mathematical framework which enables the general description of such a distribution. A rigorous mathematical treatment of probability measures and functional integration in Hilbert space may be found in Skorohod (1974) and Gihman and Skorohod (1980). The rôle of probability distributions on Hilbert space in the formulation of quantum mechanics has been discussed by Bach (1979, 1980, 1981) (see also Bach and Wenning, 1982; Cyranski, 1982). We emphasize, however, that the physical interpretation developed above is entirely different from that suggested by Bach.

5.2.1 *Probability measures on Hilbert space*

Following the construction of a probability space given in Section 1.1 we consider a system $\mathcal{A}$ of subsets of the Hilbert space $\mathcal{H}$ which plays the rôle of the sample space.[12] The system $\mathcal{A}$ is assumed to form a σ-algebra. Thus, each element $A \in \mathcal{A}$ is a subset of $\mathcal{H}$ and the system $\mathcal{A}$ of subsets satisfies the conditions of a σ-algebra formulated in Section 1.1.1. According to Section 1.1.2 a set function $\mu = \mu(A)$ defined for all $A \in \mathcal{A}$ is called a probability measure if it satisfies the Kolmogorov axioms (1.2), (1.3) and (1.5).

To give an example, we consider the simplest measures, i.e. the Dirac measures $\mu = \delta_{\psi_0}$. Taking an arbitrary fixed state vector ψ_0 the corresponding Dirac measure δ_{ψ_0} is defined by

$$\delta_{\psi_0}(A) = \begin{cases} 1, \text{ if } & \psi_0 \in A, \\ 0, \text{ if } & \psi_0 \notin A, \end{cases} \tag{5.24}$$

for all $A \in \mathcal{A}$. As is easily verified this definition yields a probability measure satisfying the Kolmogorov axioms.

In physical applications it is customary to work with densities. We thus introduce a probability density functional $P = P[\psi]$ of the state vector ψ and write

$$\mu(A) = \int_A D\psi D\psi^* P[\psi] \tag{5.25}$$

for the probability of A. Accordingly, the probability density functional of the Dirac measure δ_{ψ_0} will be denoted by $P[\psi] = \delta[\psi - \psi_0]$ such that we have the relation

---

[12]More precisely speaking, the sample space is the projective Hilbert space, see Section 5.2.2.

$$\delta_{\psi_0}(A) = \int_A D\psi D\psi^* \delta[\psi - \psi_0]. \tag{5.26}$$

The integration in these formulae is carried out over a subset A of the Hilbert space. They thus represent a multidimensional or even an infinite-dimensional integral, that is a functional integral. Our next task will be to give an explicit construction for the corresponding functional volume element $D\psi D\psi^*$ in Hilbert space.

We take a fixed orthonormal basis $\{\phi_n\}$ in $\mathcal{H}$ and write the decomposition of $\psi \in \mathcal{H}$ with respect to the basis vectors as follows,

$$\psi = \sum_n z_n \phi_n. \tag{5.27}$$

The probability density functional $P = P[\psi]$ can thus be considered as a function $P = P[z_n, z_n^*]$ of the complex variables z_n, z_n^*. Alternatively, we can regard P as a function $P = P[a_n, b_n]$ of the real variables a_n, b_n defined by

$$z_n = a_n + ib_n. \tag{5.28}$$

An appropriate expression for the volume element in Hilbert space is the usual Euclidean volume element in a real space with coordinates a_n, b_n, that is we set

$$D\psi D\psi^* = \prod_n da_n db_n. \tag{5.29}$$

Writing the differentials da_n and db_n as

$$da_n = \frac{1}{2}(dz_n + dz_n^*), \tag{5.30}$$

$$db_n = \frac{1}{2i}(dz_n - dz_n^*), \tag{5.31}$$

we can write the volume element (5.29) in terms of the coordinates z_n and z_n^*,

$$D\psi D\psi^* = \prod_n \frac{i}{2} dz_n dz_n^*. \tag{5.32}$$

Summarizing we have the following explicit formulae for functional integration in Hilbert space

$$\int_A D\psi D\psi^* P[\psi] = \int_A \prod_n da_n db_n P[a_n, b_n] = \int_A \prod_n \frac{i}{2} dz_n dz_n^* P[z_n, z_n^*]. \tag{5.33}$$

The variables z_n and z_n^* must be treated as independent integration variables corresponding to the two independent real integration variables a_n and b_n.

In accordance with definition (5.29) the density of the Dirac measure δ_{ψ_0} belonging to the state

$$\psi_0 = \sum_n z_n^0 \phi_n, \quad z_n^0 = a_n^0 + ib_n^0, \tag{5.34}$$

is defined to be

$$\delta[\psi - \psi_0] = \prod_n \delta(a_n - a_n^0)\delta(b_n - b_n^0). \tag{5.35}$$

This is a functional Dirac function given by a product of ordinary δ-functions.

If one works in the position representation, for example, one may proceed as follows. We consider a finite volume V and discretize it into cells of equal volumes ΔV centred at x_l, $l = 1, 2, \ldots, M$. The continuous wave function $\psi(x)$ is thus approximated by the finite-dimensional vector

$$|\psi\rangle = \sum_l \psi_l |x_l\rangle, \quad \psi_l = \psi(x_l). \tag{5.36}$$

In this case the volume element takes the form

$$D\psi D\psi^* = \prod_l \frac{i}{2} d\psi_l d\psi_l^*. \tag{5.37}$$

In the continuum limit this will be written formally as

$$D\psi D\psi^* = \prod_x \frac{i}{2} d\psi(x) d\psi^*(x), \tag{5.38}$$

where the product is to be extended over all space coordinates.

In the following we will often make use of an important property of the functional volume element on Hilbert space constructed above, namely its invariance under linear unitary transformations

$$\psi' = U\psi \tag{5.39}$$

of the Hilbert space. This invariance may be expressed by the formula,

$$D\psi' D\psi'^* = D\psi D\psi^*. \tag{5.40}$$

To prove this formula we regard U as a unitary matrix in the basis introduced in eqn (5.27) and decompose it into real and imaginary parts,

$$U = \Re(U) + i\Im(U). \tag{5.41}$$

The unitarity of U leads to the relations

$$\Re(U)\Re(U)^{\mathrm{T}} + \Im(U)\Im(U)^{\mathrm{T}} = I, \tag{5.42}$$

$$\Im(U)\Re(U)^{\mathrm{T}} - \Re(U)\Im(U)^{\mathrm{T}} = 0, \tag{5.43}$$

where T denotes the transposed matrix. In the chosen representation the matrix U describes a unitary transformation $z_n \mapsto z_n'$ from the coefficients z_n in the basis decomposition of ψ to the coefficients z_n' in the basis decomposition of $\psi' = U\psi$. The corresponding transformation of the real coefficients a_n, b_n defined by $z_n = a_n + ib_n$ is provided by the real matrix

$$\tilde{U} = \begin{pmatrix} \Re(U) & -\Im(U) \\ \Im(U) & \Re(U) \end{pmatrix}. \tag{5.44}$$

It is now easy to check with the help of eqns (5.42) and (5.43) that $\tilde{U}$ is an orthogonal matrix satisfying $|\det \tilde{U}| = 1$. Thus, as was to be expected, the unitary transformation U of the Hilbert space $\mathcal{H}$ induces an orthogonal transformation $\tilde{U}$ of the real variables a_n, b_n which were introduced to define the volume element in Hilbert space. We can now apply the transformation formula for multidimensional integrals to conclude that

$$\prod_n da_n' db_n' = |\det \tilde{U}| \prod_n da_n db_n = \prod_n da_n db_n, \tag{5.45}$$

which proves the unitary invariance (5.40) of the volume element.

An important consequence of this result is that our definition (5.29) for the volume element $D\psi D\psi^*$ is independent of the choice of the basis $\{\phi_n\}$. Moreover, we immediately deduce from the transformation formula (5.40) that the Dirac density functional $\delta[\psi]$ is invariant under unitary transformations,

$$\delta[U\psi] = \delta[\psi], \tag{5.46}$$

and, hence,

$$\delta[\psi - U\psi_0] = \delta[U^{-1}\psi - \psi_0]. \tag{5.47}$$

These relations are useful for calculations with probability density functionals.

5.2.2 Distributions on projective Hilbert space

We have already met the simplest examples of probability measures on Hilbert space, namely the Dirac measures δ_{ψ_0}. The corresponding density functionals $\delta[\psi - \psi_0]$ satisfy

$$\int D\psi D\psi^* \delta[\psi - \psi_0] = 1. \tag{5.48}$$

The general normalization condition for the probability density functional $P[\psi]$ reads

$$\int D\psi D\psi^* P[\psi] = 1, \tag{5.49}$$

where the integration is extended over the whole Hilbert space.

According to the general principles of quantum mechanics the physical state of a pure ensemble is uniquely described by a ray in projective Hilbert space (see Section 2.1.3.2). In other words, pure states are characterized by normalized wave functions and wave functions which differ by a phase factor are equivalent. Since we want to characterize quantum statistical ensembles of the type $\mathcal{E}_P$ with the help of a probability density functional $P[\psi]$ we postulate that the probability distribution is concentrated on the surface of the unit sphere in Hilbert space. That is we demand the existence of a functional $Q[\psi]$ such that

$$P[\psi] = \delta(||\psi|| - 1)Q[\psi]. \tag{5.50}$$

This means that only normalized wave functions occur with non-zero probability. The equivalence of wave functions which differ by a phase factor leads to the requirement that the density functional does not depend on the phase of the wave function, i.e. for all $\chi \in [0, 2\pi)$ we must have

$$P[\exp(i\chi)\psi] = P[\psi]. \tag{5.51}$$

By virtue of the conditions (5.50) and (5.51) the functional $P[\psi]$ can be regarded as a probability density on projective Hilbert space (Breuer and Petruccione, 1995). In fact, a representation of projective Hilbert space is obtained by taking the surface of the unit sphere in $\mathcal{H}$ defined by $||\psi|| = 1$, and by identifying those points on it that differ by a phase factor. Since $P[\psi]$ is constant along the rays in $\mathcal{H}$ it can be considered as a functional on projective Hilbert space. Summarizing, an $\mathcal{E}_P$-ensemble is characterized by a probability density functional $P = P[\psi]$ satisfying the normalization condition (5.49) and the requirements (5.50) and (5.51) of a functional on projective Hilbert space.

The simplest example for a density functional on projective Hilbert space is provided by the expression

$$P[\psi] = \int\limits_0^{2\pi} \frac{d\chi}{2\pi} \, \delta[\psi - e^{i\chi}\psi_0], \tag{5.52}$$

where ψ_0 is a normalized state. As is easily checked this functional satisfies the normalization condition (5.49), as well as the postulates (5.50) and (5.51). In fact, since ψ_0 is supposed to be normalized the functional (5.52) is concentrated on the unit sphere in $\mathcal{H}$. Moreover, $P[\psi]$ is invariant under changes of the phase of ψ by virtue of the χ-integration in eqn (5.52).

Another simple situation occurs, if the experimental preparation yields an ensemble which consists of a discrete set of ensembles $\mathcal{E}_\alpha$ labelled by an index α, each $\mathcal{E}_\alpha$ being describable by a normalized state vector ψ_α. This is the situation that was investigated at the beginning of this chapter, and also encountered in the example of the Stern–Gerlach experiment. Again we denote by w_α the statistical weight of the ensemble $\mathcal{E}_\alpha$, such that $w_\alpha \geq 0$, and $\sum_\alpha w_\alpha = 1$. The probability density $P[\psi]$ which describes the corresponding ensemble of the type $\mathcal{E}_P$ consists of a sum of Dirac densities which are concentrated around the various ψ_α, each Dirac density being weighted with the corresponding factor w_α. Thus, in this case we have

$$P[\psi] = \sum_\alpha w_\alpha \int_0^{2\pi} \frac{d\chi}{2\pi} \, \delta[\psi - e^{i\chi}\psi_\alpha]. \qquad (5.53)$$

Again, it can be immediately verified that our basic requirements (5.49), (5.50) and (5.51) are fulfilled for this probability density functional.

The densities $P[\psi]$ encountered so far are all representable by sums of Dirac measures. Of course, one can also easily construct continuous densities. An example of such a density has already been given in Section 5.1.2. The density $P(\vec{v})$ given in eqn (5.19) is constant on the surface of the Bloch sphere $|\vec{v}| = 1$ which represents the projective Hilbert space of a spin-$\frac{1}{2}$ particle. To generalize this idea we consider an arbitrary s-dimensional linear subspace V of $\mathcal{H}$ which is spanned by an orthonormal set ψ_i, $i = 1, 2, \ldots, s$,

$$V = \mathrm{span}\{\psi_1, \psi_2, \ldots, \psi_s\}. \qquad (5.54)$$

Let us assume that the preparation process yields a mixture of all normalized states in V with equal weights (von Neumann, 1955). The correct probability density describing the corresponding $\mathcal{E}_P$-ensemble takes the form

$$P[\psi] = \frac{1}{|K|} \int_K d\omega(\lambda) \, \delta[\psi - \phi(\lambda)]. \qquad (5.55)$$

Here we have introduced the surface K of the unit sphere in V. This surface represents a manifold of real dimension $(2s - 1)$ given by

$$K \equiv \left\{ \phi(\lambda) \in \mathcal{H} \;\middle|\; \phi(\lambda) = \sum_{i=1}^s \lambda_i \psi_i, \; \lambda_i \in \mathbb{C}, \; \sum_{i=1}^s |\lambda_i|^2 = 1 \right\}. \qquad (5.56)$$

Furthermore, $d\omega(\lambda)$ denotes the usual Euclidean surface element of K and

$$|K| = \int_K d\omega(\lambda) \qquad (5.57)$$

is the total volume of K. Hence we see that eqn (5.55) provides a uniform probability density which is concentrated on the unit sphere $K \subset V$. Note that the integration over the sphere K in eqn (5.55) ensures that the probability density functional $P[\psi]$ does not depend upon the special choice of the basis functions ψ_i. Equation (5.55) makes sense also for the case $s = 1$. In that case K is isomorphic to the unit circle and the $d\omega(\lambda)$-integration reduces to the integration over the phase χ in eqn (5.52).

More generally, we may consider an ensemble $\mathcal{E}_P$ which results from the combination of ensembles of the above type with weights $w_\alpha \geq 0$ satisfying

$\sum_\alpha w_\alpha = 1$. Using an obvious generalization of the above notation this leads to the density functional

$$P[\psi] = \sum_\alpha w_\alpha \frac{1}{|K_\alpha|} \int\limits_{K_\alpha} d\omega_\alpha(\lambda)\, \delta[\psi - \phi_\alpha(\lambda)]. \tag{5.58}$$

It should again be clear that this density functional (5.58) satisfies the basic properties (5.49), (5.50), and (5.51).

5.2.3 Expectation values

With the help of the probability density functional $P[\psi]$ corresponding to an $\mathcal{E}_P$-ensemble we can now define the expectation value of a functional $F[\psi]$ of the random state vector ψ through the relation

$$\mathrm{E}\left(F[\psi]\right) \equiv \int D\psi D\psi^*\, P[\psi] F[\psi]. \tag{5.59}$$

Taking a self-adjoint operator R we may consider the functional $F[\psi] = \langle\psi|R|\psi\rangle$ and its expectation value

$$\mathrm{E}\left(\langle\psi|R|\psi\rangle\right) = \int D\psi D\psi^*\, P[\psi]\langle\psi|R|\psi\rangle. \tag{5.60}$$

The quantity

$$\langle R\rangle = \mathrm{E}\left(\langle\psi|R|\psi\rangle\right) = \mathrm{tr}\{R\rho\} \tag{5.61}$$

is the expectation value of R in an $\mathcal{E}_\rho$-ensemble corresponding to the $\mathcal{E}_P$-ensemble described by $P[\psi]$. This expectation value for the measurement of R on $\mathcal{E}_\rho$ is obtained by averaging $\langle\psi|R|\psi\rangle$ over the probability density functional $P[\psi]$. Hence, it is evident that the density matrix ρ describing $\mathcal{E}_\rho$ is equal to the covariance matrix

$$\rho = \mathrm{E}\left(|\psi\rangle\langle\psi|\right) \equiv \int D\psi D\psi^*\, P[\psi]\, |\psi\rangle\langle\psi| \tag{5.62}$$

of the random state vector ψ defined in terms of the density functional $P[\psi]$ (compare with eqn (5.10)). This shows again that the statistics of an $\mathcal{E}_\rho$-ensemble is completely determined by a probability density functional $P[\psi]$. The converse is of course not true, in general, since a probability density functional is not fixed uniquely by giving only its covariance matrix.

Let us evaluate expression (5.62) for the density functional given in eqn (5.53). Substituting the expression for $P[\psi]$ we get

$$\rho = \sum_\alpha w_\alpha \int D\psi D\psi^* \int\limits_0^{2\pi} \frac{d\chi}{2\pi}\, \delta[\psi - e^{i\chi}\psi_\alpha]|\psi\rangle\langle\psi|, \tag{5.63}$$

and, carrying out the functional integration with the help of the properties of the Dirac density, we arrive at

$$\rho = \sum_\alpha w_\alpha \int_0^{2\pi} \frac{d\chi}{2\pi} |\psi_\alpha\rangle\langle\psi_\alpha| = \sum_\alpha w_\alpha |\psi_\alpha\rangle\langle\psi_\alpha|, \qquad (5.64)$$

which is recognized as the density matrix of the corresponding $\mathcal{E}_\rho$-ensemble. As another example we substitute the density functional (5.55) into (5.62) which yields

$$\rho = \int D\psi D\psi^* \frac{1}{|K|} \int_K d\omega(\lambda)\, \delta[\psi - \phi(\lambda)]\, |\psi\rangle\langle\psi| = \frac{1}{|K|} \int_K d\omega(\lambda)\, |\phi(\lambda)\rangle\langle\phi(\lambda)|$$

$$= \frac{1}{|K|} \int_K d\omega(\lambda) \sum_{i,j=1}^s \lambda_i \lambda_j^* |\psi_i\rangle\langle\psi_j|, \qquad (5.65)$$

where we have again carried out the functional integration over the Hilbert space and used the representation (5.56) for the surface K of the unit sphere. Performing finally the $d\omega(\lambda)$-integration one finds

$$\rho = \frac{1}{|K|} \int_K d\omega(\lambda) \sum_i^s |\lambda_i|^2 |\psi_i\rangle\langle\psi_i| = \frac{1}{s} \sum_i^s |\psi_i\rangle\langle\psi_i| = \frac{1}{s} I_V. \qquad (5.66)$$

As expected, the corresponding $\mathcal{E}_\rho$-ensemble is described by a density matrix ρ which is proportional to the identity I_V in the subspace V spanned by the ψ_i.

All quantities measurable on an $\mathcal{E}_\rho$-ensemble are expressible in terms of the covariance matrix ρ of $P[\psi]$ which involves moments of ψ of second order. Higher-order moments of the random state vector ψ represent quantities which are only measurable on $\mathcal{E}_P$. Examples are the variances $\mathrm{Var}_1(R)$ and $\mathrm{Var}_2(R)$ introduced already in eqns (5.13) and (5.14). With the help of the probability density functional $P[\psi]$ these equations can now be written as (Breuer and Petruccione, 1996)

$$\mathrm{Var}_1(R) = \int D\psi D\psi^* P[\psi] \left(\langle\psi|R^2|\psi\rangle - \langle\psi|R|\psi\rangle^2 \right), \qquad (5.67)$$

and

$$\mathrm{Var}_2(R) = \int D\psi D\psi^* P[\psi]\, \langle\psi|R|\psi\rangle^2 - \left[\int D\psi D\psi^* P[\psi]\, \langle\psi|R|\psi\rangle \right]^2. \qquad (5.68)$$

These expressions involve fourth-order moments of the state vector. $\mathrm{Var}_1(R)$ represents the dispersion of R averaged over the pure states contained in the ensemble and is thus a measure of the average intrinsic quantum fluctuations. On the other hand, $\mathrm{Var}_2(R)$ is the variance of the real random variable $\langle\psi|R|\psi\rangle$. It is thus a measure of the classical statistical fluctuations of this quantity.

5.3 Ensembles of mixtures

In the foregoing discussion we have considered $\mathcal{E}_P$ as an ensemble of various ensembles $\mathcal{E}_\alpha$ which are describable by pure quantum mechanical states ψ_α.

This idea led us to the introduction of probability density functionals $P[\psi]$ on projective Hilbert space. An even more general type of ensembles arises, if one takes into account that, as a result of the preparation process for example, the various $\mathcal{E}_\alpha$ might be in true statistical mixtures described by density matrices ρ_α. This idea then forces us to consider probability density functionals on the space $\mathcal{S}(\mathcal{H})$ of density matrices.

5.3.1 Probability density functionals on state space

We denote probability density functionals on the state space of density matrices by $P = P[\sigma]$, where $\sigma \in \mathcal{S}(\mathcal{H})$ is a density matrix. They characterize the statistical properties of the random density matrix σ. To construct the corresponding volume element we proceed in close analogy to the case of distributions $P[\psi]$ on the Hilbert space $\mathcal{H}$.

Taking an orthonormal basis $\{\phi_n\}$ in $\mathcal{H}$ we write the matrix representation of an operator σ as

$$\sigma = \sum_{nm} |\phi_n\rangle \sigma_{nm} \langle \phi_m|. \tag{5.69}$$

Next we decompose the matrix elements σ_{nm} into real and imaginary parts,

$$\sigma_{nm} = u_{nm} + iv_{nm}. \tag{5.70}$$

An appropriate volume element may now be defined by

$$D\sigma \equiv \prod_{n,m} du_{nm} dv_{nm}. \tag{5.71}$$

A functional $F = F[\sigma]$ can be considered as a multidimensional function $F[\sigma] = F[u_{nm}, v_{nm}]$ of the variables u_{nm} and v_{nm} and we define the functional integration by means of

$$\int D\sigma F[\sigma] = \int \prod_{n,m} du_{nm} dv_{nm} \, F[u_{nm}, v_{nm}]. \tag{5.72}$$

The corresponding Dirac density is given by

$$\delta[\sigma] = \prod_{n,m} \delta(u_{nm}) \delta(v_{nm}). \tag{5.73}$$

A probability density functional on the space of density matrices is a non-negative functional $P = P[\sigma]$ which vanishes if any of the following conditions are violated,

$$u_{mn} = u_{nm}, \quad v_{mn} = -v_{nm}, \tag{5.74}$$

$$\mathrm{tr}\,\sigma = \sum_n u_{nn} = 1, \tag{5.75}$$

$$\sigma \geq 0. \tag{5.76}$$

The first condition ensures that only Hermitian matrices occur with non-zero probability and implies, in particular, that $v_{nn} = 0$. The second condition is the normalization condition for the random density matrix σ, while the third condition requires that $P[\sigma]$ must be identically zero outside the convex region defined by the positivity of σ. Finally, the normalization condition for the probability density functional takes the form

$$\int D\sigma P[\sigma] = 1. \tag{5.77}$$

As for the volume element $D\psi D\psi^*$ one can easily prove that the volume element $D\sigma$ is invariant under unitary transformations given by

$$\sigma' = U\sigma U^\dagger, \tag{5.78}$$

where U is an unitary operator in $\mathcal{H}$, i.e. that we have

$$D\sigma' = D\sigma. \tag{5.79}$$

In particular, this invariance implies that our definition (5.71) of $D\sigma$ does not depend on the basis $\{\phi_n\}$ chosen.

The expectation value of some self-adjoint operator R which is obtained by measuring on a corresponding ensemble of the type $\mathcal{E}_\rho$ is given by the first moment of $P[\sigma]$,

$$\langle R \rangle = \int D\sigma P[\sigma] \operatorname{tr}\{R\sigma\}. \tag{5.80}$$

Consequently, the density matrix describing $\mathcal{E}_\rho$ takes the form,

$$\rho = \mathrm{E}(\sigma) \equiv \int D\sigma P[\sigma]\, \sigma. \tag{5.81}$$

Of course, the different types of variances defined in the preceding section can be introduced in an analogous way for density functionals $P[\sigma]$ on the space of density matrices. More precisely, we have again the decomposition (5.12) of the variance $\mathrm{Var}(R)$ into two variances $\mathrm{Var}_1(R)$ and $\mathrm{Var}_2(R)$, where now

$$\mathrm{Var}_1(R) = \int D\sigma P[\sigma] \left(\operatorname{tr}\{R^2\sigma\} - [\operatorname{tr}\{R\sigma\}]^2 \right), \tag{5.82}$$

and

$$\mathrm{Var}_2(R) = \int D\sigma P[\sigma]\, [\operatorname{tr}\{R\sigma\}]^2 - \left[\int D\sigma P[\sigma]\operatorname{tr}\{R\sigma\} \right]^2. \tag{5.83}$$

The physical interpretation and the measurement prescription for these quantities are analogous to those given in Section 5.1.1.

5.3.2 *Description of selective quantum measurements*

As an example of the use of the probability density functionals $P[\sigma]$ we consider the selective measurement described by effects F_m and operations Φ_m (see Section 2.4.2). Performing such a measurement on an ensemble $\mathcal{E}_P$ described by the density $P[\sigma]$ yields a new ensemble $\mathcal{E}'_P$ which is described by the density $P'[\sigma']$ given by

$$P'[\sigma'] = \sum_m \int D\sigma P[\sigma] \operatorname{tr} \{F_m \sigma\} \delta \left[\frac{\Phi_m(\sigma)}{\operatorname{tr}\{\Phi_m(\sigma)\}} - \sigma' \right]. \qquad (5.84)$$

Obviously, the new density functional is normalized in view of condition (2.148),

$$\int D\sigma' P[\sigma'] = \sum_m \int D\sigma P[\sigma] \operatorname{tr} \{F_m \sigma\} = \int D\sigma P[\sigma] = 1. \qquad (5.85)$$

On using the first representation theorem of quantum measurement theory (see eqns (2.157) and (2.158)) we can rewrite (5.84) as

$$P'[\sigma'] = \sum_m \int D\sigma P[\sigma] \operatorname{tr} \left\{ \sum_k \Omega_{mk}^\dagger \Omega_{mk} \sigma \right\} \delta \left[\frac{\sum_k \Omega_{mk} \sigma \Omega_{mk}^\dagger}{\operatorname{tr}\left\{ \sum_k \Omega_{mk}^\dagger \Omega_{mk} \sigma \right\}} - \sigma' \right].$$

$$(5.86)$$

In the simplest case the index k in the last equation takes on a single value only. This situation occurs, for example, if one considers an indirect measurement with a quantum probe which is in a pure state initially. The last equation then simplifies to

$$P'[\sigma'] = \sum_m \int D\sigma P[\sigma] \operatorname{tr} \{\Omega_m^\dagger \Omega_m \sigma\} \delta \left[\frac{\Omega_m \sigma \Omega_m^\dagger}{\operatorname{tr}\left\{ \Omega_m^\dagger \Omega_m \sigma \right\}} - \sigma' \right]. \qquad (5.87)$$

It follows from eqn (5.87) that a distribution $P[\psi]$ of pure states transforms again into a distribution $P'[\psi']$ of pure states: If we perform the measurement on an ensemble in the state ψ, the new state conditioned on the outcome m is given by $\psi' = \Omega_m \psi / \|\Omega_m \psi\|$. Thus we get the following relation between the initial and the final density functional on Hilbert space,

$$P'[\psi'] = \sum_m \int D\psi D\psi^* \, \|\Omega_m \psi\|^2 \delta \left[\frac{\Omega_m \psi}{\|\Omega_m \psi\|} - \psi' \right] P[\psi]. \qquad (5.88)$$

The above relations describe the change of a probability density functional induced by the selective measurement on ensembles of the type $\mathcal{E}_P$. They constitute the starting point for our derivation of quantum stochastic processes which will be presented in the next chapter.

References

Bach, A. (1979). Quantum mechanics and integration in Hilbert space. *Phys. Lett.*, **A73**, 287–288.

Bach, A. (1980). A probabilistic formulation of quantum theory. *J. Math. Phys.*, **21**, 789–793.

Bach, A. (1981). Aspects of a representation of quantum theory in terms of classical probability theory by means of integration in Hilbert space. *J. Phys. A: Math. Gen.*, **14**, 125–132.

Bach, A. and Wenning, T. (1982). A probabilistic formulation of quantum theory. II. *J. Math. Phys.*, **23**, 1078–1081.

Breuer, H. P. and Petruccione, F. (1995). Stochastic dynamics of open quantum systems: Derivation of the differential Chapman–Kolmogorov equation. *Phys. Rev.*, **E51**, 4041–4054.

Breuer, H. P. and Petruccione, F. (1996). Quantum measurement and the transformation from quantum to classical probabilities. *Phys. Rev.*, **A54**, 1146–1153.

Cyranski, J. F. (1982). Quantum theory as a probability theory on Hilbert space. *J. Math. Phys.*, **23**, 1074–1077.

Gihman, I. I. and Skorohod, A. V. (1980). *The Theory of Stochastic Processes I*. Springer-Verlag, Berlin.

Mølmer, K., Castin, Y., and Dalibard, J. (1993). Monte-Carlo wave function method in quantum optics. *J. Opt. Soc. Am.*, **B10**, 524–538.

von Neumann, J. (1955). *Mathematical Foundations of Quantum Mechanics*. Princeton University Press, Princeton.

Skorohod, A. V. (1974). *Integration in Hilbert Space*. Springer-Verlag, New York.

Wiseman, H. M. (1993). Stochastic quantum dynamics of a continuously monitored laser. *Phys. Rev.*, **A47**, 5180–5192.

Wiseman, H. M. and Milburn, G. (1993a). Quantum theory of field-quadrature measurements. *Phys. Rev.*, **A47**, 642–662.

Wiseman, H. M. and Milburn, G. (1993b). Interpretation of quantum jump and diffusion processes illustrated on the Bloch sphere. *Phys. Rev.*, **A47**, 1652–1666.

6

STOCHASTIC DYNAMICS IN HILBERT SPACE

A quantum master equation describes the dynamics of an open system by an equation of motion for its reduced density matrix $\rho_S(t)$. In the language of the previous chapter this provides the evolution of an $\mathcal{E}_\rho$-ensemble. On the other hand, $\mathcal{E}_P$-ensembles are characterized through density functionals $P[\psi]$ on the underlying Hilbert space. Drawing on the analogy to the theory of classical stochastic processes developed in Chapter 1, one could ask the question as to whether the unfolding in time of an $\mathcal{E}_P$-ensemble may be formulated in terms of an appropriate dynamics for a time-dependent density functional $P[\psi, t]$. It will be shown in the present chapter that such a formulation is indeed possible. It leads to a stochastic process $\psi(t)$ in the open system's Hilbert space which reproduces the density matrix through its covariance matrix, that is through the expectation value $\rho_S(t) = \mathrm{E}(|\psi(t)\rangle\langle\psi(t)|)$.

This is the idea underlying the so-called *unravelling* of the master equation. Instead of representing the dynamics of an open system by a quantum master equation for its density matrix, it is formulated in terms of a stochastic process for the open system's wave function. The structure of the Lindblad generator for a quantum Markov process leads to a close relation between dynamical semigroups and piecewise deterministic processes in Hilbert space. This chapter gives a detailed account of this relation and of its physical interpretation and significance. The theory will also be illustrated with a number of applications.

The stochastic representation of quantum Markov processes already appeared in a fundamental paper by Davies (1969), and was applied by Srinivas and Davies (1981) to a derivation of the photocounting formula. While the theory was originally formulated in terms of a stochastic process for the reduced density matrix, it was recognized by Barchielli and Belavkin (1991), Dalibard, Castin and Mølmer (1992) and by Dum, Zoller and Ritsch (1992) that it can be formulated also as a stochastic process for the state vector in the reduced system's Hilbert space and that it leads to efficient numerical simulation algorithms. At the same time, Carmichael (1993) developed the idea of the unravelling of the master equation in terms of an ensemble of quantum trajectories which are the realizations of the underlying stochastic process.

The physical basis for the stochastic state vector evolution is provided by continuous measurement theory: The stochastic dynamics of an $\mathcal{E}_P$-ensemble pertaining to an open system results from a continuous monitoring of certain observables of its environment. The process $\psi(t)$ thus embodies the random changes of the state vector conditioned on the outcomes of this monitoring.

Accordingly, different detection schemes yield different stochastic processes for the conditioned state vector, as was demonstrated by Wiseman and Milburn (1993a, 1993b). Microscopic derivations of various stochastic processes will be given in this chapter.

6.1 Dynamical semigroups and PDPs in Hilbert space

This section serves to introduce, on a formal mathematical level, the relation between dynamical semigroups and a certain class of piecewise deterministic processes in Hilbert space. We further discuss some important mathematical properties of this class of PDPs. The physical meaning of the stochastic representation as well as its derivation from microscopic models are shown in the next section. Specific physical examples are investigated in Sections 6.3–6.7.

6.1.1 Reduced system dynamics as a PDP

6.1.1.1 *Closed systems* Let us first consider a closed system whose dynamics over the time interval from time t_0 to time t is described by a unitary time-evolution operator

$$U(t, t_0) = \exp\left[-iH(t - t_0)\right], \tag{6.1}$$

where H denotes the Hamiltonian of the system. In order to construct the probability density $P[\psi, t]$ for the process $\psi(t)$ describing the evolution of a corresponding $\mathcal{E}_P$-ensemble (see Chapter 5) we have to introduce an initial probability distribution $P_0[\psi]$ which represents the state of the ensemble at time t_0. The probability density for $\psi(t)$ to take the value ψ at time t is then given by

$$P[\psi, t] = \int D\psi_0 D\psi_0^* P_0[\psi_0] \delta\left[\psi - U(t, t_0)\psi_0\right] = P_0[U^{-1}(t, t_0)\psi]. \tag{6.2}$$

In the second equality we have made use of the definition (5.35) of the Dirac density in Hilbert space and of eqn (5.47). The above equation simply expresses the fact that any ψ_0 drawn from the initial distribution P_0 evolves according to the Schrödinger equation, that is the density at ψ at time t is equal to the initial density at the corresponding initial value $U^{-1}(t, t_0)\psi$. In the language of the theory of stochastic processes ψ is a deterministic Markov process (see Section 2.4.3) and the δ-function in (6.2) is nothing but the transition probability of the process.

Equation (6.2) can be written in differential form. To this end we introduce the functional derivatives $\delta/\delta\psi(x)$ and $\delta/\delta\psi^*(x)$ with the properties

$$\frac{\delta}{\delta\psi(x)}\psi(y) = \frac{\delta}{\delta\psi^*(x)}\psi^*(y) = \delta(x - y), \tag{6.3}$$

$$\frac{\delta}{\delta\psi(x)}\psi^*(y) = \frac{\delta}{\delta\psi^*(x)}\psi(y) = 0. \tag{6.4}$$

If $F[\psi(t)]$ is any functional of a time-dependent wave function the chain rule may be expressed in terms of the functional derivatives as follows

$$\frac{d}{dt}F[\psi(t)] = \int dx \left\{ \frac{\delta F}{\delta \psi(x)} \frac{d\psi(x,t)}{dt} + \frac{\delta F}{\delta \psi^*(x)} \frac{d\psi^*(x,t)}{dt} \right\}. \qquad (6.5)$$

With the help of these relations we immediately obtain from eqn (6.2) the differential form

$$\frac{\partial}{\partial t} P[\psi, t] = i \int dx \left\{ \frac{\delta}{\delta \psi(x)} (H\psi)(x) - \frac{\delta}{\delta \psi^*(x)} (H\psi)^*(x) \right\} P[\psi, t]. \qquad (6.6)$$

This is just the Liouville equation corresponding to the flow induced by the Schrödinger equation $\dot{\psi}(t) = -iH\psi(t)$. It is easy to check that (6.6) preserves the basic conditions (5.49), (5.50) and (5.51) required for a probability density functional on projective Hilbert space.

6.1.1.2 *Unravelling the quantum master equation* Let us now turn to an open system S. The dynamics of the density matrix $\rho_S(t)$ of S is assumed to be describable by a quantum dynamical semigroup satisfying a Markovian master equation in Lindblad form (see Section 3.2.2),

$$\frac{d}{dt}\rho_S(t) = -i[H, \rho_S(t)] + \sum_i \gamma_i \left(A_i \rho_S(t) A_i^\dagger - \frac{1}{2} A_i^\dagger A_i \rho_S(t) - \frac{1}{2} \rho_S(t) A_i^\dagger A_i \right). \qquad (6.7)$$

Our aim is to demonstrate that the dynamics given by this equation can be represented as a piecewise deterministic process $\psi(t)$ in the Hilbert space of the open system S in the following sense. The process will be defined through an appropriate time-evolution equation for its probability density functional $P[\psi, t]$. The density functional at time t is related to the density functional at time t_0 by

$$P[\psi, t] = \int D\tilde{\psi} D\tilde{\psi}^* \, T[\psi, t | \tilde{\psi}, t_0] P[\tilde{\psi}, t_0], \qquad (6.8)$$

where $T[\psi, t | \tilde{\psi}, t_0]$ is the conditional transition probability of the process. The process $\psi(t)$ gives rise to a certain equation of motion for the covariance matrix which was defined in eqn (5.62) and will be written explicitly as follows,

$$\rho_S(x, x', t) \equiv \mathrm{E}\left[\psi(x, t)\psi^*(x', t)\right] = \int D\psi D\psi^* P[\psi, t]\psi(x)\psi^*(x'). \qquad (6.9)$$

The covariance matrix is identified with the density matrix $\rho_S(x, x', t)$ of S. The basic requirement is then that the expectation value (6.9) satisfies the Lindblad equation (6.7). Thus, the process $\psi(t)$ reproduces the reduced system's density matrix through its covariance matrix, as is illustrated in Fig. 6.1. A process which satisfies this requirement is sometimes called *unravelling* of the master equation (Carmichael, 1993).

6.1.1.3 *The Liouville master equation* As discussed in Section 1.5.1 a PDP may be defined in terms of a Liouville master equation for its probability density

$$\rho_S(t_0) \xrightarrow{\ V(t,t_0)\ } \rho_S(t)$$

$$\mathrm{E}[|\tilde{\psi}\rangle\langle\tilde{\psi}|] \Big\uparrow \qquad\qquad \Big\uparrow \mathrm{E}[|\psi(t)\rangle\langle\psi(t)|]$$

$$P[\tilde{\psi}, t_0] \xrightarrow{\ T[\psi,t|\tilde{\psi},t_0]\ } P[\psi, t]$$

FIG. 6.1. The unravelling of the master equation leads to a commutative diagram: Starting from $P[\tilde{\psi}, t_0]$ one can either form the covariance matrix to get $\rho_S(t_0)$ and propagate with the help of the super-operator $V(t, t_0)$ which represents the time evolution according to the master equation (6.7), or else one can first propagate the stochastic process to get $P[\psi, t]$ and form the covariance matrix. Both ways lead to the same density matrix $\rho_S(t)$.

(Breuer and Petruccione, 1995a, 1995b). We consider here a Liouville master equation for a PDP in Hilbert space which takes the following form analogous to eqn (1.150),

$$\frac{\partial}{\partial t} P[\psi, t] = i \int dx \left\{ \frac{\delta}{\delta\psi(x)} G(\psi)(x) - \frac{\delta}{\delta\psi^*(x)} G(\psi)^*(x) \right\} P[\psi, t]$$
$$+ \int D\tilde{\psi} D\tilde{\psi}^* \left\{ W[\psi|\tilde{\psi}] P[\tilde{\psi}, t] - W[\tilde{\psi}|\psi] P[\psi, t] \right\}. \qquad (6.10)$$

The first term on the right-hand side provides the Liouville part. It represents the generator corresponding to the deterministic time-evolution equation

$$\frac{d}{dt}\psi(t) = -iG(\psi(t)), \qquad (6.11)$$

where $G(\psi)$ is a non-linear operator defined by

$$G(\psi) = \hat{H}\psi + \frac{i}{2} \sum_i \gamma_i ||A_i\psi||^2 \psi. \qquad (6.12)$$

The linear operator $\hat{H}$ is given by

$$\hat{H} = H - \frac{i}{2} \sum_i \gamma_i A_i^\dagger A_i. \qquad (6.13)$$

Thus, $\hat{H}$ consists of two parts, namely of the Hamiltonian H which appears in the Lindblad equation (6.7), and a non-Hermitian part which is defined in terms of the Lindblad operators A_i.

The jump part of eqn (6.10) (given by the second term on the right-hand side) describes the rate of change of $P[\psi, t]$ due to discontinuous jumps of the wave function. The gain term represents the total rate for all transitions from any state $\tilde{\psi}$ into the state ψ, whereas the loss term gives the total rate for all transition leaving the state ψ. The corresponding transition rate is defined to be

$$W[\psi|\tilde{\psi}] = \sum_i \gamma_i ||A_i\tilde{\psi}||^2 \, \delta \left[\frac{A_i\tilde{\psi}}{||A_i\tilde{\psi}||} - \psi \right]. \tag{6.14}$$

6.1.1.4 *Deterministic evolution*   The periods of deterministic evolution of the process are given by eqn (6.11) which has the form of a non-linear Schrödinger equation. The formal solution corresponding to the normalized initial value $\psi(0) = \tilde{\psi}$ can be written in terms of the non-Hermitian Hamiltonian $\hat{H}$ as follows,

$$\psi(t) \equiv g_t(\tilde{\psi}) = \frac{\exp(-i\hat{H}t)\tilde{\psi}}{||\exp(-i\hat{H}t)\tilde{\psi}||}. \tag{6.15}$$

To prove this equation we differentiate it with respect to time,

$$\frac{d}{dt}\psi(t) = -i\hat{H}\psi(t) - \frac{1}{2}\psi(t)||e^{-i\hat{H}t}\tilde{\psi}||^{-2}\frac{d}{dt}||e^{-i\hat{H}t}\tilde{\psi}||^2. \tag{6.16}$$

The time derivative of the norm is found to be

$$\frac{d}{dt}||e^{-i\hat{H}t}\tilde{\psi}||^2 = i\langle\tilde{\psi}|e^{i\hat{H}^\dagger t}\left(\hat{H}^\dagger - \hat{H}\right)e^{-i\hat{H}t}|\tilde{\psi}\rangle$$

$$= -\sum_i \gamma_i\langle\tilde{\psi}|e^{i\hat{H}^\dagger t}A_i^\dagger A_i e^{-i\hat{H}t}|\tilde{\psi}\rangle, \tag{6.17}$$

and, hence,

$$||e^{-i\hat{H}t}\tilde{\psi}||^{-2}\frac{d}{dt}||e^{-i\hat{H}t}\tilde{\psi}||^2 = -\sum_i \gamma_i\langle\psi(t)|A_i^\dagger A_i|\psi(t)\rangle. \tag{6.18}$$

Substituting this into eqn (6.16) immediately yields eqn (6.11).

Equation (6.18) shows that the time evolution generated by $\hat{H}$ leads to a monotonic decrease of the norm of the wave function. The non-linear part of eqn (6.11) compensates this decrease, such that the full non-linear Schrödinger equation preserves the norm of the wave function.

6.1.1.5 *Jump process and waiting time distribution*   According to eqn (6.14) the total rate for transitions from a given state $\tilde{\psi}$ to some other state is given by

$$\Gamma[\tilde{\psi}] = \int D\psi D\psi^* W[\psi|\tilde{\psi}] = \sum_i \gamma_i ||A_i\tilde{\psi}||^2. \tag{6.19}$$

Let us assume that the normalized state $\tilde{\psi}$ was reached through a jump at time t. Due to the continuous time evolution between the jumps, the total rate for the next jump depends on the time τ elapsed since time t. With the help of the flow $\psi(\tau) = g_\tau(\tilde{\psi})$ defined in eqn (6.15) the time-dependent total transition rate may be written

$$\Gamma[g_\tau(\tilde{\psi})] = \sum_i \gamma_i \langle \psi(\tau) | A_i^\dagger A_i | \psi(\tau) \rangle = -\frac{d}{d\tau} \ln \| \exp(-i\hat{H}\tau)\tilde{\psi} \|^2, \qquad (6.20)$$

where we have used eqn (6.18) in the second step. According to the general theory of PDPs (see Section 1.5) the distribution function for the random waiting time τ is thus given by

$$F[\tilde{\psi}, \tau] = 1 - \exp\left(-\int\limits_0^\tau ds\, \Gamma[g_s(\tilde{\psi})] \right) = 1 - \| \exp(-i\hat{H}\tau)\tilde{\psi} \|^2. \qquad (6.21)$$

This quantity yields the probability for the next jump to occur somewhere in the time interval $[t, t + \tau]$. We see that the waiting time distribution is simply determined by the decrease of the norm of $\exp(-i\hat{H}\tau)\tilde{\psi}$, that is by the decrease of the norm of the state vector that obeys the linear part of the non-linear Schrödinger equation.

Following our general discussion of the waiting time distribution function of PDPs in Section 1.5.2, we first note that the limit

$$\lim_{\tau \to \infty} \| \exp(-i\hat{H}\tau)\tilde{\psi} \|^2 \equiv q \qquad (6.22)$$

exists and thus $F[\tilde{\psi}, \infty] = 1 - q$. The number q is the defect and satisfies $0 \leq q \leq 1$. For vanishing defect it follows that $F[\tilde{\psi}, \infty] = 1$. This means that the next jump occurs with probability 1 in some finite time. However, if $\hat{H}$ has a zero-mode it is possible that $q > 0$. In this case, the defect q is the probability that after time t no further jumps occur. A physical example will be given in Section 8.2.

The quantity $W[\psi|\tilde{\psi}]$ (eqn 6.14) denotes the probability density per unit of time for a jump from $\tilde{\psi}$ to ψ. Since it is given by a discrete sum of functional δ-functions we have a discrete set of possible transitions. Under the condition that the state just before the jump is given by $\tilde{\psi}$, the particular jump

$$\tilde{\psi} \longrightarrow \psi = \frac{A_i \tilde{\psi}}{\|A_i \tilde{\psi}\|} \qquad (6.23)$$

takes place with probability

$$p_i = \frac{\gamma_i \|A_i \tilde{\psi}\|^2}{\Gamma[\tilde{\psi}]}. \qquad (6.24)$$

Note that these probabilities sum up to 1, that is $\sum_i p_i = 1$ by virtue of eqn (6.19). The transitions (6.23) will be referred to as quantum jumps.

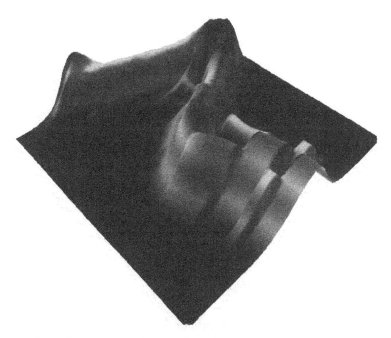

FIG. 6.2. A single realization of the PDP for a one-dimensional harmonic oscillator. The oscillator is coupled to a low-temperature heat bath and is driven by a time-dependent force. The picture shows $|\psi(x,t)|^2$ as a function of x and t. The initial state is the ground state of the oscillator. In addition to smooth evolution periods, one observes the wave function performing sudden, discontinuous jumps.

Again, one easily verifies that the Liouville master equation (6.10) preserves the basic conditions (5.49), (5.50) and (5.51). In particular, the conservation of the norm with probability one follows from the fact that both the deterministic evolution and the jumps preserve the norm of the wave function. The Liouville master equation thus defines a stochastic process in projective Hilbert space.

As an example of a PDP in Hilbert space we show in Fig. 6.2 the square $|\psi(x,t)|^2$ of a single realization of an unravelling of the master equation (3.307) for the damped harmonic oscillator discussed in Section 3.4.6. The jump operators are just the creation and annihilation operators a, $a^\dagger$, and we have added a time-dependent force to the Hamiltonian part of the dynamics.

6.1.1.6 *Stochastic differential equation* As we know from Section 1.5, a PDP defined by a Liouville master equation of the form (6.10) can also be formulated in terms of an equivalent stochastic differential equation. In the present case the latter is given by (compare with eqn (1.197))

$$d\psi(t) = -iG(\psi(t))dt + \sum_i \left(\frac{A_i\psi(t)}{||A_i\psi(t)||} - \psi(t) \right) dN_i(t), \qquad (6.25)$$

where the Poisson increments $dN_i(t)$ satisfy

$$dN_i(t)dN_j(t) = \delta_{ij}dN_j(t), \qquad (6.26)$$

$$\mathrm{E}\left[dN_i(t)\right] = \gamma_i ||A_i\psi(t)||^2 dt. \qquad (6.27)$$

This shows that the process $N_i(t)$ counts the number of jumps of type i, i.e. the number of jumps (6.23) with Lindblad operator A_i. The processes $N_i(t)$ are inhomogeneous since the expectation values of the increments $dN_i(t)$ depend on time through the time-dependence of the state vector.

6.1.1.7 *Quantum master equation* To prove that the PDP defined by the Liouville master equation (6.10) indeed provides an unravelling of the quantum master equation (6.7) we have to derive the equation of motion governing the covariance matrix (6.9). On differentiating (6.9) with respect to time we find

$$\frac{\partial}{\partial t}\rho_S(x,x',t) = \int D\psi D\psi^* \frac{\partial}{\partial t}P[\psi,t]\psi(x)\psi^*(x') \equiv \frac{\partial}{\partial t}\rho_S\bigg|_L + \frac{\partial}{\partial t}\rho_S\bigg|_J. \qquad (6.28)$$

Here, we have decomposed the total rate of change of ρ_S into the rate of change induced by the deterministic Liouvillean part and the rate of change induced by the jump part of (6.10). The first contribution is obtained from the first term on the right-hand side of (6.10),

$$\frac{\partial}{\partial t}\rho_S\bigg|_L = i \int D\psi D\psi^* \int dy\, \psi(x)\psi^*(x')$$

$$\times \left\{ \frac{\delta}{\delta\psi(y)}G(\psi)(y) - \frac{\delta}{\delta\psi^*(y)}G(\psi)^*(y) \right\} P[\psi,t]$$

$$= -i \int D\psi D\psi^* \left\{ G(\psi)(x)\psi^*(x') - \psi(x)G^*(\psi)(x') \right\} P[\psi,t]. \qquad (6.29)$$

In the second step we have performed a functional integration by parts and used the properties (6.3), (6.4) of the functional derivatives. On substituting (6.12) we thus get

$$\frac{\partial}{\partial t}\rho_S\bigg|_L = -i\left(\hat{H}\rho_S - \rho_S\hat{H}^\dagger \right) + \sum_i \gamma_i \mathrm{E}\left[||A_i\psi||^2\psi(x)\psi^*(x') \right]. \qquad (6.30)$$

Note that the expectation values on the right-hand side represent certain fourth-order correlation functions of ψ.

With the help of the second term on the right-hand side of (6.10) the jump contribution to the rate of change of ρ_S is found to be

$$\frac{\partial}{\partial t}\rho_S\bigg|_J = \int D\psi D\psi^* \int D\tilde{\psi}D\tilde{\psi}^*\psi(x)\psi^*(x') \left\{ W[\psi|\tilde{\psi}]P[\tilde{\psi},t] - W[\tilde{\psi}|\psi]P[\psi,t] \right\}. \qquad (6.31)$$

We insert (6.14) and carry out one of the two functional integrations in each term. This yields

$$\left.\frac{\partial}{\partial t}\rho s\right|_J = \sum_i \gamma_i A_i \rho s A_i^\dagger - \sum_i \gamma_i \mathrm{E}\left[||A_i\psi||^2\psi(x)\psi^*(x')\right]. \qquad (6.32)$$

If we now add eqns (6.30) and (6.32) we see that the contributions from the fourth-order correlation functions cancel each other and we are left with a closed equation for the covariance matrix. The latter is easily seen to take the form of the Lindblad equation (6.7), which concludes the proof.

6.1.2 The Hilbert space path integral

As for any PDP the propagator $T[\psi, t|\tilde\psi, t_0]$ can be represented in terms of a sum over all possible realizations of the process. An immediate adaptation of the formalism developed in Section 1.5.3 shows that this yields a path integral taken over all possible paths in the Hilbert space which connect the state $\tilde\psi$ at time t_0 with the state ψ at time t. Since the stochastic process is homogeneous in time we may set $t_0 \equiv 0$ without restriction in the following.

To begin with, we write the Kolmogorov forward equation for the process as follows (compare with eqn (1.167))

$$T[\psi, t|\tilde\psi, 0] = T^{(0)}[\psi, t|\tilde\psi, 0] \qquad (6.33)$$

$$+ \int_0^t ds \int D\psi_1 D\psi_1^* \int D\psi_2 D\psi_2^* T_0[\psi, t|\psi_1, s] W[\psi_1|\psi_2] T[\psi_2, s|\tilde\psi, 0].$$

The physical interpretation of this equation is the same as the one given in Section 1.5.3. In particular, the quantity

$$T^{(0)}[\psi, t|\tilde\psi, 0] = \left(1 - F[\tilde\psi, t]\right) \delta[\psi - g_t(\tilde\psi)] \qquad (6.34)$$

is the contribution to the propagator from the path without jump, that is the contribution that stems from pure deterministic evolution. More generally, we write $T^{(N)}[\psi, t|\tilde\psi, 0]$ for the contribution to the propagator which involves the paths with exactly N jumps. The full propagator can then be written as an expansion in the number of jumps as follows,

$$T[\psi, t|\tilde\psi, 0] = \sum_{N=0}^\infty T^{(N)}[\psi, t|\tilde\psi, 0]. \qquad (6.35)$$

The N-jump contributions satisfy the recursion relation

$$T^{(N)}[\psi, t|\tilde\psi, 0] = \int_0^t ds \int D\psi_1 D\psi_1^* \int D\psi_2 D\psi_2^* \qquad (6.36)$$

$$\times T^{(0)}[\psi, t|\psi_1, s] W[\psi_1|\psi_2] T^{(N-1)}[\psi_2, s|\tilde\psi, 0]$$

which holds for $N \geq 1$. It is convenient to introduce the non-unitary time-evolution operator

$$\hat{U}(t) = \exp(-i\hat{H}t). \tag{6.37}$$

On using eqn (6.14) and

$$T^{(0)}[\psi, t | \psi_1, s] = ||\hat{U}(t - s)\psi_1||^2 \delta\left[\psi - \frac{\hat{U}(t - s)\psi_1}{||\hat{U}(t - s)\psi_1||}\right], \tag{6.38}$$

and carrying out the integration over ψ_1 in eqn (6.36), we find that the recursion relation can be cast into the form

$$T^{(N)}[\psi, t | \tilde{\psi}, 0] = \int\limits_0^t ds \int D\phi D\phi^* \sum_i \gamma_i ||\hat{U}(t - s)A_i\phi||^2 \tag{6.39}$$

$$\times \delta\left[\psi - \frac{\hat{U}(t - s)A_i\phi}{||\hat{U}(t - s)A_i\phi||}\right] T^{(N-1)}[\phi, s | \tilde{\psi}, 0].$$

N-fold iteration of the recursion relation (6.39) finally leads to

$$T^{(N)}[\psi, t | \tilde{\psi}, 0] = \int\limits_0^t dt_N \int\limits_0^{t_N} dt_{N-1} \cdots \int\limits_0^{t_2} dt_1 \sum_{i_N} \sum_{i_{N-1}} \cdots \sum_{i_1} \tag{6.40}$$

$$\times p_N^t(t_1, i_1; \ldots; t_N, i_N)\delta\left[\psi - \frac{\psi_N^t(t_1, i_1; \ldots; t_N, i_N)}{||\psi_N^t(t_1, i_1; \ldots; t_N, i_N)||}\right].$$

This relation can easily be verified by induction over N with the help of the recursion relation (6.36). The quantity

$$\psi_N^t(t_1, i_1; \ldots; t_N, i_N) \tag{6.41}$$
$$= \hat{U}(t - t_N)A_{i_N}\hat{U}(t_N - t_{N-1})A_{i_{N-1}} \cdots \hat{U}(t_2 - t_1)A_{i_1}\hat{U}(t_1)\tilde{\psi}$$

is the state vector at time t conditioned on the following event: A quantum jump with jump operator A_{i_1} occurs at time t_1, a jump with operator A_{i_2} at time t_2, ..., a jump with operator A_{i_N} at time t_N, and no further jumps take place in the time interval $[0, t]$. The corresponding probability density for this event, known as the multi-time exclusive probability, is given by

$$p_N^t(t_1, i_1; \ldots; t_N, i_N) = \gamma_{i_N} \cdots \gamma_{i_1} ||\psi_N^t(t_1, i_1; \ldots; t_N, i_N)||^2. \tag{6.42}$$

According to eqn (6.40) the N-jump propagator is found by integrating the contributions from the ψ_N^t over all jump times $t_1, t_2, \ldots, t_N$, and by a summation over all jump types $i_1, i_2, \ldots, i_N$.

Substituting (6.40) into (6.35) yields the Hilbert space path integral (Breuer and Petruccione, 1996a). It must be emphasized that this path integral is entirely different from the Feynman–Vernon path integral that was introduced in Section 3.6.4.2. The latter is a sum over paths in configuration space, each path being weighted by a complex factor given through an effective action functional. By contrast, the Hilbert space path integral is a sum over paths $\psi(t)$ in Hilbert space, where each path contributes a real and positive weight factor which is just the probability of the path. It thus provides a mathematical formulation of Carmichael's idea of *quantum trajectories* (Carmichael, 1993) and of the *Monte Carlo wave function method* (Mølmer, Castin and Dalibard, 1993; Dum *et al.*, 1992; Gardiner, Parkins and Zoller, 1992; Castin and Mølmer, 1995; Mølmer and Castin, 1996)

If we have only a single jump operator A with corresponding rate γ_0 eqn (6.40) reduces to the simpler form

$$T^{(N)}[\psi, t|\tilde{\psi}, 0] = \int\limits_0^t dt_N \int\limits_0^{t_N} dt_{N-1} \ldots \int\limits_0^{t_2} dt_1 \qquad (6.43)$$

$$\times \, p_N^t(t_1, \ldots, t_N) \delta \left[\psi - \frac{\psi_N^t(t_1, \ldots, t_N)}{||\psi_N^t(t_1, \ldots, t_N)||} \right],$$

where

$$\psi_N^t(t_1, \ldots, t_N) = \hat{U}(t - t_N) A \hat{U}(t_N - t_{N-1}) A \ldots \hat{U}(t_2 - t_1) A \hat{U}(t_1) \tilde{\psi} \qquad (6.44)$$

and

$$p_N^t(t_1, \ldots, t_N) = \gamma_0^N ||\psi_N^t(t_1, \ldots, t_N)||^2. \qquad (6.45)$$

If we integrate $T^{(N)}[\psi, t|\tilde{\psi}, 0]$ over ψ we get the probability $p_N(t)$ for exactly N quantum jumps in the time interval $[0, t]$,

$$p_N(t) = \int\limits_0^t dt_N \int\limits_0^{t_N} dt_{N-1} \ldots \int\limits_0^{t_2} dt_1 p_N^t(t_1, \ldots, t_N). \qquad (6.46)$$

These expressions will be used in the examples of the following sections to determine the statistics of jump events.

6.1.3 Diffusion approximation

Under certain conditions the diffusion limit of the Liouville master equation (6.10) exists and yields a Fokker–Planck equation for the probability density functional which, in turn, is equivalent to a stochastic Schrödinger-type equation. The diffusion limit of the PDP thus gives rise to an unravelling of the quantum master equation in terms of a diffusion process in Hilbert space.

For the sake of simple notation we assume in the following that we have only one jump operator A with corresponding rate γ_0. According to the general theory of stochastic processes a diffusion expansion of a given master equation can be performed if the size of the transitions among the states becomes arbitrarily small and if, at the same time, the number of transitions in any finite time interval becomes arbitrarily large. In order to formulate these conditions we introduce a small (dimensionless) parameter ε and write the Lindblad operator as

$$A = I + \varepsilon C, \tag{6.47}$$

where I denotes the identity operator and the operator C is independent of ε. Our aim is to investigate the behaviour of the Liouville master equation (6.10) in the limit $\varepsilon \longrightarrow 0$. On using eqn (6.47) we obtain to second order in ε

$$G(\psi) = H\psi - \frac{i}{2}\gamma_0 \left\{ I + \varepsilon(C^\dagger + C) + \varepsilon^2 C^\dagger C \right\} \psi$$
$$+ \frac{i}{2}\gamma_0 \left\{ 1 + \varepsilon\langle C^\dagger + C\rangle_\psi + \varepsilon^2 \langle C^\dagger C\rangle_\psi \right\} \psi \tag{6.48}$$

and

$$W[\psi|\tilde{\psi}] = \gamma_0 \left(1 + \varepsilon\langle C^\dagger + C\rangle_{\tilde{\psi}} + \varepsilon^2 \langle C^\dagger C\rangle_{\tilde{\psi}} \right) \delta[\tilde{\psi} - \psi + \varepsilon M(\tilde{\psi}) + \varepsilon^2 N(\tilde{\psi})], \tag{6.49}$$

where we have introduced the abbreviation

$$\langle O\rangle_\psi \equiv \langle\psi|O|\psi\rangle, \tag{6.50}$$

and defined the non-linear operators

$$M(\psi) \equiv \left\{ C - \frac{1}{2}\langle C^\dagger + C\rangle_\psi \right\} \psi, \tag{6.51}$$

$$N(\psi) \equiv -\frac{1}{2} \left\{ \langle C^\dagger C\rangle_\psi - \frac{3}{4}\langle C^\dagger + C\rangle_\psi^2 + \langle C^\dagger + C\rangle_\psi C \right\} \psi. \tag{6.52}$$

Inserting these expressions into the Liouville master equation we obtain to second order in ε

$$\frac{\partial}{\partial t}P[\psi,t] = i \int dx \left\{ \frac{\delta}{\delta\psi(x)}K(\psi)(x) - \frac{\delta}{\delta\psi^*(x)}K(\psi)^*(x) \right\} P[\psi,t] \tag{6.53}$$
$$+ \frac{1}{2}\gamma_0\varepsilon^2 \int dx \int dx' \left\{ \frac{\delta^2}{\delta\psi(x)\delta\psi(x')}M(\psi)(x)M(\psi)(x') \right.$$
$$+ \frac{\delta^2}{\delta\psi^*(x)\delta\psi^*(x')}M(\psi)^*(x)M(\psi)^*(x')$$
$$\left. + 2\frac{\delta^2}{\delta\psi(x)\delta\psi^*(x')}M(\psi)(x)M(\psi)^*(x') \right\} P[\psi,t].$$

This is a Fokker–Planck equation for the probability density functional. The non-linear drift operator $K(\psi)$ takes the form

$$K(\psi) = H\psi + \frac{i}{2}\gamma_0\varepsilon \left\{ C - C^\dagger \right\} \psi$$
$$+ i\gamma_0\varepsilon^2 \left\{ \frac{1}{2}\langle C^\dagger + C\rangle_\psi C - \frac{1}{8}\langle C^\dagger + C\rangle_\psi^2 - \frac{1}{2}C^\dagger C \right\} \psi. \tag{6.54}$$

As can be seen from eqn (6.53) the diffusion part of the Fokker–Planck equation, involving the second-order functional derivatives, scales as $\gamma_0 \varepsilon^2$. Thus, in order to obtain a non-vanishing and finite diffusive contribution in the limit $\varepsilon \longrightarrow 0$, we assume that the ε-dependence of the relaxation time γ_0 reads

$$\gamma_0 = \varepsilon^{-2} \bar{\gamma}_0, \tag{6.55}$$

where $\bar{\gamma}_0$ is independent of ε. On the other hand, the drift operator (6.54) contains a term which is proportional to $\gamma_0 \varepsilon = \bar{\gamma}_0 \varepsilon^{-1}$. This term diverges in the limit $\varepsilon \longrightarrow 0$ unless we impose the condition that the operator C is self-adjoint, that is $C = C^\dagger$. Using this condition we obtain for the drift operator

$$K(\psi) = H\psi + i\bar{\gamma}_0 \left\{ \langle C \rangle_\psi C - \frac{1}{2} \langle C \rangle_\psi^2 - \frac{1}{2} C^2 \right\} \psi, \tag{6.56}$$

while the operator $M(\psi)$ takes the form

$$M(\psi) = (C - \langle C \rangle_\psi) \psi. \tag{6.57}$$

With these expressions the Fokker–Planck equation (6.53) is equivalent to the following stochastic Schrödinger equation in Itô form,

$$d\psi(t) = -iK(\psi(t))dt + \sqrt{\bar{\gamma}_0} M(\psi(t))dW(t), \tag{6.58}$$

where $dW(t)$ is the increment of a real Wiener process.

To get a well-defined diffusion limit we have assumed above that the jump operator is self-adjoint. Of course, it is possible to formulate other conditions and to perform a different expansion of the Liouville master equation. In particular, if we have several Lindblad operators the conditions leading to a diffusion limit may be different. A physical example will be discussed in Section 6.4.2.

6.1.4 Multi-time correlation functions

The representation of the dynamics of a Markovian quantum master equation in terms of a stochastic process in the underlying Hilbert space $\mathcal{H}$ can be generalized to lead also to a stochastic formulation of multi-time correlation functions (see Section 3.2.4). Here we show that this can be achieved by the construction of stochastic processes in a suitably enlarged Hilbert space, namely in the doubled Hilbert space $\mathcal{H} \oplus \mathcal{H}$ (Breuer, Kappler and Petruccione, 1997; Breuer, Kappler and Petruccione, 1998).

6.1.4.1 Reduced Heisenberg picture operators

According to Section 3.2.3 we can write the matrix elements of an arbitrary reduced Heisenberg picture operator $A(t)$ (we suppress the index H) in the following way

$$\langle \phi_0 | A(t) | \psi_0 \rangle = \langle \phi_0 | V^\dagger(t, 0) A | \psi_0 \rangle = \text{tr} \left\{ A V(t, 0) | \psi_0 \rangle \langle \phi_0 | \right\}, \tag{6.59}$$

where A is the corresponding Schrödinger picture operator at time $t_0 = 0$ and $V(t, 0)$ is the super-operator describing the time evolution according to the master equation (6.7). Equation (6.59) can be interpreted in the following way. For

(a) Unitary time evolution

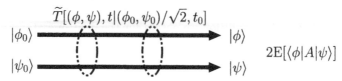

(b) Stochastic time evolution

FIG. 6.3. Determination of matrix elements of Heisenberg picture operators for (a) a closed system, and (b) for an open system.

the calculation of the matrix element $\langle\phi_0|A(t)|\psi_0\rangle$ start with the initial 'density matrix' $|\psi_0\rangle\langle\phi_0|$ and propagate it up to the time t. Then calculate the expectation value of A with respect to the propagated 'density matrix'. However, $|\psi_0\rangle\langle\phi_0|$ is, in general, not a true density matrix for it is neither Hermitian nor positive. Thus, it cannot be characterized as the covariance matrix $\mathrm{E}[|\psi\rangle\langle\psi|]$ of a distribution in the Hilbert space $\mathcal{H}$ of the open system. Hence a direct application of the stochastic unravelling developed in the preceding sections to the calculation of Heisenberg picture operators is not possible.

In a closed system where the time evolution of states is given through a unitary evolution operator $U(t,0)$ we can calculate the matrix element $\langle\phi_0|A(t)|\psi_0\rangle$ in the following way (see Fig. 6.3): Propagate ϕ_0 and ψ_0 independently to obtain $\phi = U(t,0)\phi_0$ and $\psi = U(t,0)\psi_0$, and evaluate the scalar product $\langle\phi|A|\psi\rangle$. This suggests the following method for the determination of the matrix elements (6.59) for an open system: Instead of propagating independently the state vectors $\phi_0 \in \mathcal{H}$ and $\psi_0 \in \mathcal{H}$, we design a stochastic process $\theta(t)$ in the *doubled* Hilbert space, that is the direct sum

$$\widetilde{\mathcal{H}} = \mathcal{H} \oplus \mathcal{H},\tag{6.60}$$

and use this process to propagate the normalized *pair* of state vectors

$$\theta_0 = \frac{1}{\sqrt{2}}\begin{pmatrix}\phi_0\\\psi_0\end{pmatrix} \in \widetilde{\mathcal{H}}.\tag{6.61}$$

This is illustrated in Fig. 6.3.

To formalize the idea we introduce the propagator $\widetilde{T}[\theta,t|\theta_0,t_0]$ of the process $\theta(t)$ in the doubled Hilbert space, where we write

$$\theta(t) \equiv \begin{pmatrix}\phi(t)\\\psi(t)\end{pmatrix} \in \widetilde{\mathcal{H}}.\tag{6.62}$$

Again, it is assumed that the process is normalized,

$$\|\theta(t)\|^2 = \|\phi(t)\|^2 + \|\psi(t)\|^2 \equiv 1. \tag{6.63}$$

We can now formulate our basic definition as follows. The process $\theta(t)$ is said to be a stochastic representation of the reduced Heisenberg picture if the condition

$$\langle \phi_0 | A(t) | \psi_0 \rangle = 2 \int D\theta D\theta^* \langle \phi | A | \psi \rangle \widetilde{T}[\theta, t | \theta_0, 0] \equiv 2\mathrm{E}\left[\langle \phi(t) | A | \psi(t) \rangle\right] \tag{6.64}$$

holds for all ϕ_0, ψ_0, and for all operators A. Thus, once we have constructed such a process $\theta(t)$ we can determine any matrix element of the Heisenberg operator $A(t)$ with the help of the expectation value of the quantity $\langle \phi(t) | A | \psi(t) \rangle$.

The question is now, how can we construct a process $\theta(t)$ which satisfies eqn (6.64)? To answer this question we consider the following master equation in the doubled Hilbert space,

$$\dot{\widetilde{\rho}}(t) = -i\left[\widetilde{H}, \widetilde{\rho}(t)\right] + \sum_i \gamma_i \left\{ \widetilde{A}_i \widetilde{\rho}(t) \widetilde{A}_i^\dagger - \frac{1}{2}\widetilde{A}_i^\dagger \widetilde{A}_i \widetilde{\rho}(t) - \frac{1}{2}\widetilde{\rho}(t)\widetilde{A}_i^\dagger \widetilde{A}_i \right\}, \tag{6.65}$$

where the Hamiltonian and the Lindblad operators in the extended space are defined as

$$\widetilde{H} = \begin{pmatrix} H & 0 \\ 0 & H \end{pmatrix}, \quad \widetilde{A}_i = \begin{pmatrix} A_i & 0 \\ 0 & A_i \end{pmatrix}. \tag{6.66}$$

We take $\theta(t)$ to be an arbitrary unravelling of the master equation (6.65) in the doubled Hilbert space with the initial condition $\theta(0) = \theta_0$. Thus, $\theta(t)$ may be, for example, a PDP of the type discussed in Section 6.1.1, or else a diffusion process of the type given in Section 6.1.3. We now claim that the process $\theta(t)$ is a stochastic representation of the reduced Heisenberg picture, that is it fulfils condition (6.64). Thus, *any* unravelling of the master equation in the doubled Hilbert space gives rise to a stochastic representation for the matrix elements of Heisenberg operators in the reduced space.

To prove this claim we write the density matrix in $\widetilde{\mathcal{H}}$ as follows,

$$\widetilde{\rho}(t) = \begin{pmatrix} \widetilde{\rho}_{11}(t) & \widetilde{\rho}_{12}(t) \\ \widetilde{\rho}_{21}(t) & \widetilde{\rho}_{22}(t) \end{pmatrix}. \tag{6.67}$$

By assumption this is a solution of the extended master equation (6.65) corresponding to the initial condition

$$\widetilde{\rho}(0) = |\theta_0\rangle\langle\theta_0| \equiv \frac{1}{2}\begin{pmatrix} |\phi_0\rangle\langle\phi_0| & |\phi_0\rangle\langle\psi_0| \\ |\psi_0\rangle\langle\phi_0| & |\psi_0\rangle\langle\psi_0| \end{pmatrix}. \tag{6.68}$$

By virtue of the block-diagonal structure of the operators (6.66) the master equation (6.65) yields four independent equations such that all components $\widetilde{\rho}_{ij}(t)$

separately solve the original master equation (6.7). In particular, $\tilde{\rho}_{21}(t)$ solves this equation with the initial value

$$\tilde{\rho}_{21}(0) = \frac{1}{2}|\psi_0\rangle\langle\phi_0|. \tag{6.69}$$

Thus, we have

$$\tilde{\rho}_{21}(t) = V(t,0)\left(\frac{1}{2}|\psi_0\rangle\langle\phi_0|\right), \tag{6.70}$$

and, hence, by virtue of (6.59)

$$\langle\phi_0|A(t)|\psi_0\rangle = 2\mathrm{tr}\Big\{A\tilde{\rho}_{21}(t)\Big\}. \tag{6.71}$$

On the other hand, since $\theta(t)$ is an unravelling of the master equation (6.65) with initial condition $\theta(0) = \theta_0$ we have

$$\tilde{\rho}(t) = \int D\theta D\theta^* \; |\theta\rangle\langle\theta| \; \tilde{T}[\theta, t|\theta_0, 0], \tag{6.72}$$

and, consequently,

$$\tilde{\rho}_{21}(t) = \int D\theta D\theta^* \; |\psi\rangle\langle\phi| \; \tilde{T}[\theta, t|\theta_0, 0]. \tag{6.73}$$

By inserting eqn (6.73) into eqn (6.71) we arrive at

$$\langle\phi_0|A(t)|\psi_0\rangle = 2 \int D\theta D\theta^* \; \mathrm{tr}\Big\{A|\psi\rangle\langle\phi|\Big\}\tilde{T}[\theta, t|\theta_0, 0]$$
$$= 2 \int D\theta D\theta^* \; \langle\phi|A|\psi\rangle\tilde{T}[\theta, t|\theta_0, 0], \tag{6.74}$$

which concludes the proof.

Thus we have shown that matrix elements of reduced Heisenberg picture operators are calculated correctly if the stochastic process in the doubled Hilbert space unravels the extended quantum master equation (6.65). It is important to note that the above proof does not rely on a specific unravelling of the quantum master equation (6.65). On the contrary, it is valid for any stochastic process the covariance matrix of which is governed by eqn (6.65).

6.1.4.2 *Multi-time correlations* The process $\theta(t)$ in the doubled Hilbert space $\tilde{\mathcal{H}}$ can now be used to construct a stochastic representation of multi-time correlation functions. Consider for example the two-time correlation function (see eqn (3.86))

$$g(t, s) = \langle\phi_0|B(t)C(s)|\phi_0\rangle, \tag{6.75}$$

where $t > s \geq 0$ and we may assume, without restriction, that the initial state is a pure state. The stochastic process which represents this correlation function may then be defined by the following algorithm.

1. Start in the state ϕ_0 at time $t = 0$ and use the stochastic time evolution in the Hilbert space $\mathcal{H}$ to obtain the stochastic wave function $\phi(s)$.

2. Propagate the normalized state

$$\theta(s) = \frac{1}{\|(\phi(s), C\phi(s))\|} \begin{pmatrix} \phi(s) \\ C\phi(s) \end{pmatrix} \in \tilde{\mathcal{H}} \tag{6.76}$$

using the stochastic time evolution in the doubled Hilbert space $\tilde{\mathcal{H}}$ to obtain the normalized state vector

$$\theta(t) = \begin{pmatrix} \phi(t) \\ \psi(t) \end{pmatrix} \in \tilde{\mathcal{H}}. \tag{6.77}$$

3. The above correlation function is then obtained by computing the expectation value

$$g(t, s) = \mathrm{E}\left[\|(\phi(s), C\phi(s))\|^2 \langle \phi(t)|B|\psi(t)\rangle\right]. \tag{6.78}$$

This scheme may be generalized to the treatment of arbitrary time-ordered multi-time correlation functions of the form (see eqn (3.87))

$$g(t_1, ..., t_n; s_1, ..., s_m) = \langle \phi_0|B_1(s_1)...B_m(s_m)C_n(t_n)...C_1(t_1)|\phi_0\rangle. \tag{6.79}$$

We use the same notation as in Section 3.2.4, and define the Schrödinger operators (compare this definition with the one given in eqn (3.90))

$$\left.\begin{array}{l} F_l = I, \ G_l = C_i, \ \text{if } r_l = t_i \neq s_j \text{ for some } i \text{ and all } j, \\ F_l = B_j^\dagger, \ G_l = I, \ \text{if } r_l = s_j \neq t_i \text{ for some } j \text{ and all } i, \\ F_l = B_j^\dagger, \ G_l = C_i, \ \text{if } r_l = t_i = s_j \text{ for some } i \text{ and } j. \end{array}\right\} \tag{6.80}$$

The multi-time correlation function is then obtained in the following way:

1. Start with the state ϕ_0 at time $t = 0$ and propagate it to time r_1 to obtain $\phi(r_1)$ with the help of the process in $\mathcal{H}$.

2. Employing the process in the doubled Hilbert space, propagate the state

$$\theta(r_1) = \frac{1}{\|(F_1\phi(r_1), G_1\phi(r_1))\|} \begin{pmatrix} F_1\phi(r_1) \\ G_1\phi(r_1) \end{pmatrix} \tag{6.81}$$

to obtain the state

$$\theta(r_2) = \begin{pmatrix} \phi(r_2) \\ \psi(r_2) \end{pmatrix}. \tag{6.82}$$

Then, jump to the state

$$\theta(r_2) = \frac{1}{\|(F_2\phi(r_2), G_2\psi(r_2))\|} \begin{pmatrix} F_2\phi(r_2) \\ G_2\psi(r_2) \end{pmatrix} \tag{6.83}$$

and propagate it to time r_3, and so on.

3. Finally, the multi-time correlation is found from

$$
\begin{aligned}
g(t_1, &..., t_n; s_1, ..., s_m) \\
&= \mathrm{E}\big[\big\|\big(F_1\phi(r_1), G_1\phi(r_1)\big)\big\|^2 \\
&\quad \times \big\|\big(F_2\phi(r_2), G_2\psi(r_2)\big)\big\|^2 \cdots \big\|\big(F_{q-1}\phi(r_{q-1}), G_{q-1}\psi(r_{q-1})\big)\big\|^2 \\
&\quad \times \langle\phi(r_q)|F_q^\dagger G_q|\psi(r_q)\rangle\big].
\end{aligned}
\tag{6.84}
$$

It should be remarked that this algorithm enables the determination of correlation functions of arbitrary order with the help of a stochastic process in the doubled Hilbert space. An example is shown in Fig. 6.5.

We finally note that the algorithm developed above is related to the method proposed by Dum, Zoller and Ritsch (1992) (see also the discussion by Marte et al., 1993a, 1993b; Mølmer and Castin, 1996). A further method has been proposed by Dalibard, Castin and Mølmer (1992).

6.2 Stochastic representation of continuous measurements

The stochastic unravelling of the quantum master equation introduced in the preceding section yields an intuitive physical picture which may be very helpful for the identification of the basic mechanism underlying the reduced system dynamics. This will be illustrated by means of a number of examples in this and the next two chapters. The method also leads to an efficient tool for the numerical simulation of the open system's dynamics (see Chapter 7).

However, it must be realized that the stochastic representation of a given master equation in terms of a PDP in Hilbert space is not unique. Suppose we are given a Lindblad equation with a certain Hamiltonian H and Lindblad operators A_i with corresponding rates γ_i. We may then transform to a new H', and new A_i', γ_i' by any of the transformations given by eqns (3.72) and (3.73), without changing the Lindblad generator. However, the corresponding PDP does change, in general, under these transformations. For a precise physical interpretation and for a microscopic derivation of a unique process one therefore needs an additional input which does not enter the derivation of the corresponding quantum master equation.

Thus, in this section we investigate the following question: What is the physical basis for the description of the reduced system dynamics in terms of a stochastic wave function $\psi(t)$, and can one give a microscopic derivation of such dynamics which is in agreement with the basic principles of quantum mechanics? It will be demonstrated that the theory of continuous measurements provides an appropriate framework to answer these questions, as was demonstrated by several authors (Barchielli and Belavkin, 1991; Hegerfeldt and Wilser, 1991; Wiseman and Milburn, 1993a, 1993b; Breuer and Petruccione, 1997) (see also the review article by Plenio and Knight (1998) and the references cited therein).

6.2.1 Stochastic time evolution of $\mathcal{E}_P$-ensembles

The physical situation studied here is similar to that of an indirect measurement scheme as already investigated in Section 2.4.6. Here, the quantum object, living in some Hilbert space $\mathcal{H}_S$, represents an open system whose evolution we wish to describe in terms of a stochastic wave function $\psi(t)$. The quantum object is coupled to another quantum mechanical system, the environment, whose Hilbert space is denoted by $\mathcal{H}_B$. The environment is continuously monitored by some measuring device. The environmental state thus acts as a quantum probe. The continuous monitoring of the latter yields certain information on the object system which can be deduced from the correlations between the object and probe system.

In Chapter 3 the dynamics of the reduced density matrix ρ_S of the object system S was obtained from the density matrix ρ of the combined system by taking the partial trace over the degrees of freedom of the environment B. Fixing an orthonormal basis of state vectors φ_α in $\mathcal{H}_B$ we may write the reduced density matrix as

$$\rho_S = \mathrm{tr}_B \rho = \sum_\alpha \langle \varphi_\alpha | \rho | \varphi_\alpha \rangle. \tag{6.85}$$

The trace over the degrees of freedom of the environment can thus be viewed as a non-selective measurement of an environmental observable with eigenvectors φ_α. In a non-selective measurement the information on the measurement outcomes is thrown away and the sub-ensembles conditioned on the measurement outcomes are re-mixed during the evolution. As a result the reduced density matrix ρ_S does not depend on the basis φ_α, that is, it does not depend in any way on the measured observable of the environment. Therefore, the time development of ρ_S describes the unfolding of an ensemble of type $\mathcal{E}_\rho$.

However, a time evolution for an ensemble of type $\mathcal{E}_P$ is obtained if we describe the process on the selective level, that is, if we keep the information obtained by the measurement record and if we take into account the splitting into the various sub-ensembles conditioned on it. This leads to a stochastic dynamics for the object system's wave function $\psi(t)$. Thus, the stochastic wave function dynamics describes the evolution on the level of an indirect, selective measurement which is performed on the environment and which leads to a stochastic back-action on the state vector of the object system. This back-action has already been discussed in connection with the general scheme of an indirect measurement. In view of this picture it must be expected that an average over the realizations of the process $\psi(t)$ leads to the dynamics on the non-selective level, that is to the corresponding equation of motion for $\rho_S(t)$.

6.2.2 Short-time behaviour of the propagator

To formulate these ideas in mathematical terms we take some initial time t_0 and some later time $t \equiv t_0 + \tau$, where $\tau > 0$. The conditional transition probability $T[\psi, t | \tilde{\psi}, t_0]$ of the process $\psi(t)$ is defined through eqn (6.8). Suppose that at time t an ideal quantum measurement of a non-degenerate environmental observable with eigenvectors φ_α is carried out. The corresponding operation is then given by the operators (compare with eqn (2.211))

$$\Omega_\alpha \equiv \langle \varphi_\alpha | U(t, t_0) | \phi_0 \rangle, \qquad (6.86)$$

where the environment is assumed to be in a pure state given by ϕ_0, and $U(t, t_0)$ is the time-evolution operator of the combined system.

The situation under study is the same as the one considered in the preceding chapter. In fact, on comparing eqn (5.88) with the definition (6.8) we immediately see that the conditional transition probability is given by the following exact expression,

$$T[\psi, t | \tilde{\psi}, t_0] = \sum_\alpha ||\Omega_\alpha \tilde{\psi}||^2 \delta \left[\frac{\Omega_\alpha \tilde{\psi}}{||\Omega_\alpha \tilde{\psi}||} - \psi \right]. \qquad (6.87)$$

This is the probability of finding after a complete orthogonal measurement of the environment at time t the reduced system in state ψ under the condition that at time t_0 the state $\tilde{\psi}$ was given. According to eqn (6.8) the probability density functional $P[\psi, t]$ describes the ensemble of type $\mathcal{E}_P$ that results from the complete measurement on the environment at time t. The rôle of the operators Ω_α is to describe the operation corresponding to the back-action on the object system S induced by the measurement. Note that T is considered to be a conditional transition probability in projective Hilbert space. Namely, if $P[\tilde{\psi}, t_0]$ is a phase-invariant density functional, then the density functional $P[\psi, t]$ according to eqn (6.8) is automatically phase invariant for a propagator of the form (6.87). This can be easily checked with the help of the invariance properties of the Dirac measure. It also follows from eqns (6.8) and (6.87) that the total probability for being in any state at time t is equal to 1, namely

$$\int D\psi D\psi^* T[\psi, t | \tilde{\psi}, t_0] = 1, \qquad (6.88)$$

and that for $t \longrightarrow t_0$ only the probability of being in the initial state is different from zero, that is,

$$\lim_{t \to t_0} T[\psi, t | \tilde{\psi}, t_0] = \delta[\tilde{\psi} - \psi]. \qquad (6.89)$$

In order to derive a PDP describing the continuous measurement one now has to invoke the Markov approximation. For a sufficiently small time increment τ (but still $\tau \gg \tau_B$) the short-time behaviour for the conditional transition probability takes the general form pertaining to a PDP, namely

$$T[\psi, t | \tilde{\psi}, t_0] = \left(1 - \tau \Gamma[\tilde{\psi}]\right) \delta \left[\tilde{\psi} - i\tau G(\tilde{\psi}) - \psi \right] + \tau W[\psi | \tilde{\psi}]. \qquad (6.90)$$

The strategy to derive the PDP is thus simply this: One first derives the short-time behaviour of the conditional transition probability given by eqn (6.87) employing the Markov approximation. A comparison with the general structure (6.90) then yields the transition rate $W[\psi|\tilde{\psi}]$ for the quantum jumps, as well as the generator $G(\psi)$ for the deterministic pieces of the process.

We close this section with a few remarks. Involving the variables of the environment and the exact time-evolution operator $U(t, t_0)$, eqn (6.87) for the conditional transition probability is an exact expression. However, in most physical applications one is interested in, the environment constitutes a system with a large (practically infinite) number of degrees of freedom with a quasi-continuous spectrum of frequencies, e.g. the continuum of modes of the electromagnetic field. In such cases, it is of course impossible in practice to design a measurement scheme that corresponds to a complete and orthogonal decomposition of unity. However, as will be seen in the following examples, the Markov approximation enables one to decompose the conditional transition probability into a small number of terms that can be interpreted as the different alternative outcomes of an *incomplete* measurement scheme. The latter is easily realized physically since it corresponds to a resolution of the identity in terms of projection operators that project, in general, onto high-dimensional subspaces of $\mathcal{H}_B$.

The process defined by the short-time behaviour (6.90) of the conditional transition probability has to be interpreted as resulting from a *continuous* measurement of the environment according to the measurement scheme that is defined by the operators Ω_α. This interpretation is necessary because each application of the infinitesimal generator of the process implies a state reduction which is fixed by the measurement scheme. Furthermore, it is clear that the term *continuous* has to be understood in the sense of a coarse graining in time which enables the Markovian approximation of the dynamics (compare the discussion of the quantum Zeno effect in Section 3.5.1). It should also be clear that one uses at each time the assumption of (approximate) statistical independence of the system and environment. In other words, it is assumed that the probe is only weakly disturbed by the object system and that after each time interval of the order of τ one can apply anew the short-time expression (6.90) for the propagator. It should be clear that this assumption of weak disturbance can only be valid if the environmental state ϕ_0 is only weakly influenced by the measurement process. This condition is satisfied, for example, if the bath state is contained in the reduction basis φ_α that fixes the measurement scheme. The examples treated below are precisely of this type.

Above, we have assumed the Markov process to be homogeneous in time since this is sufficient to treat the examples that will be given below. A non-homogeneous Markov process can result if, for example, the reduced system is coupled to time-dependent external fields, or if the state of the environment depends on time through the time evolution generated by the bath Hamiltonian H_B. For such a non-homogeneous Markov process the generator G as well as the transition rate W may depend explicitly on time. In the examples below we shall treat the case of time-dependent external fields. However, since the Hamiltonian will still be time independent in the interaction picture, the Markov processes are time homogeneous in that picture. Essential modifications occur, however, in the case of *strong* external fields (see Section 8.4).

6.3   Direct photodetection

As our first example we study in this section the driven two-level atom considered in Section 3.4.5. At zero temperature eqn (3.271) yields the vacuum optical master equation

$$\frac{d}{dt}\rho_S(t) = \frac{i\Omega}{2}\left[\sigma_+ + \sigma_-, \rho_S(t)\right] \tag{6.91}$$

$$+\gamma_0\left(\sigma_-\rho_S(t)\sigma_+ - \frac{1}{2}\sigma_+\sigma_-\rho_S(t) - \frac{1}{2}\rho_S(t)\sigma_+\sigma_-\right)$$

for the atomic density matrix $\rho_S(t)$.

The measurement scheme underlying the derivation of the PDP is the direct, continuous detection of the fluorenscence photons emitted by the two-level source. Thus, the environment is provided by the radiation modes and the vacuum of the electromagnetic field acts as the probe state. The continuous monitoring of the photons radiated gives rise to a stochastic process for the atomic state vector.

6.3.1   *Derivation of the PDP*

We write the Hamiltonian of the radiation field as follows,

$$H_B = \sum_j \omega_j b_j^\dagger b_j, \tag{6.92}$$

where $b_j^\dagger$, b_j are the creation and annihilation operators of the field modes labelled by j. The frequency of the mode j is denoted by ω_j. The interaction picture Hamiltonian $H_I(t)$ takes the form

$$H_I(t) = e^{i\omega_0 t}\sigma_+ B(t) + e^{-i\omega_0 t}\sigma_- B^\dagger(t) + H_L. \tag{6.93}$$

The first two terms describe the coupling of the atomic dipole moment to the radiation field, where

$$B(t) = \sum_j \kappa_j e^{-i\omega_j t} b_j, \tag{6.94}$$

and the κ_j are coupling constants. The third term yields the interaction with the resonant driving field,

$$H_L = -\frac{\Omega}{2}\left(\sigma_- + \sigma_+\right). \tag{6.95}$$

Suppose that the photons of the modes b_j emitted by the atomic source are observed through direct detection by a photocounter. This means that the basis vectors φ_α must be taken to be the Fock states created by the operators $b_j^\dagger$ out of the field vacuum,

$$\varphi_\alpha \equiv |\{N_j\}\rangle. \tag{6.96}$$

Here, the index α stands for a complete set $\{N_j\}$ of occupation numbers of the modes b_j. In particular, we define the vacuum of the electromagnetic field

$$\varphi_0 \equiv |0\rangle, \tag{6.97}$$

and the one-photon states

$$\varphi_j \equiv |j\rangle \equiv b_j^\dagger |0\rangle. \tag{6.98}$$

The two-level atom constitutes the quantum object, while the electromagnetic field plays the rôle of the quantum probe. The pure probe state will be taken to be the vacuum state, that is, we set $\phi_0 = \varphi_0 \equiv |0\rangle$. This choice satisfies the requirement of weak disturbance, since the demolition measurement of the field quanta puts the electromagnetic field back into the vacuum state.

Our task is now to derive the operators Ω_α which describe the operation for the detection of the photons. Employing second-order perturbation theory we have

$$U_I(t, t_0) \approx I - i \int_{t_0}^t dt'\, H_I(t') - \int_{t_0}^t dt' \int_{t_0}^{t'} dt''\, H_I(t') H_I(t''). \tag{6.99}$$

and, hence, eqn (6.86) leads to

$$\Omega_\alpha = \delta_{\alpha,0} + f_\alpha + g_\alpha, \tag{6.100}$$

$$||\Omega_\alpha \tilde{\psi}||^2 = \delta_{\alpha,0} \left\{ 1 - \sum_{\alpha'} \langle f_{\alpha'}^\dagger f_{\alpha'} \rangle_{\tilde{\psi}} \right\} + \langle f_\alpha^\dagger f_\alpha \rangle_{\tilde{\psi}}, \tag{6.101}$$

where we have introduced the operators

$$f_\alpha \equiv -i \int_{t_0}^t dt'\, \langle \varphi_\alpha | H_I(t') | 0 \rangle, \tag{6.102}$$

$$g_\alpha \equiv - \int_{t_0}^t dt' \int_{t_0}^{t'} dt''\, \langle \varphi_\alpha | H_I(t') H_I(t'') | 0 \rangle. \tag{6.103}$$

For $\varphi_\alpha = |0\rangle$ (vacuum state) this yields

$$f_0 = -i\tau H_L, \tag{6.104}$$

and for $\varphi_\alpha = |j\rangle$ (one-photon state) we get

$$f_j = -i\kappa_j^* \int_{t_0}^t dt'\, e^{i(\omega_j - \omega_0)t'} \sigma_- \equiv \tilde{f}_j \sigma_-. \tag{6.105}$$

Note that $f_\alpha = 0$ if φ_α is any N-photon state with $N \geq 2$. Thus, to second order we have to keep only the first-order term in Ω_j which corresponds to the projection onto the one-photon sector, and the second-order term in Ω_0 which corresponds to the projection onto the field vacuum. This yields, on using eqns (6.100), (6.101), (6.104), and (6.105),

$$\Omega_j = \tilde{f}_j \sigma_-, \tag{6.106}$$

$$||\Omega_j \tilde{\psi}||^2 = |\tilde{f}_j|^2 \langle \sigma_+ \sigma_- \rangle_{\tilde{\psi}}, \tag{6.107}$$

and

$$\Omega_0 = I - i\tau H_L + g_0, \tag{6.108}$$

$$||\Omega_0 \tilde{\psi}||^2 = 1 - \sum_j |\tilde{f}_j|^2 \langle \sigma_+ \sigma_- \rangle_{\tilde{\psi}}. \tag{6.109}$$

Equation (6.106) shows that Ω_j is the product of two factors: The first one is $\tilde{f}_j$ which is simply the probability amplitude for the observation of a photon in the mode j; the second factor is the atomic lowering operator σ_- which represents the back-action on the object system conditioned on this observation.

Collecting these results we see that the conditional transition probability may be split into two parts

$$T[\psi, t | \tilde{\psi}, t_0] = T_0[\psi, t | \tilde{\psi}, t_0] + T_1[\psi, t | \tilde{\psi}, t_0], \tag{6.110}$$

where

$$T_0[\psi, t | \tilde{\psi}, t_0] = ||\Omega_0 \tilde{\psi}||^2 \delta \left[\frac{\Omega_0 \tilde{\psi}}{||\Omega_0 \tilde{\psi}||} - \psi \right], \tag{6.111}$$

and

$$T_1[\psi, t | \tilde{\psi}, t_0] = \sum_j ||\Omega_j \tilde{\psi}||^2 \delta \left[\frac{\tilde{f}_j}{|\tilde{f}_j|} \frac{\sigma_- \tilde{\psi}}{||\sigma_- \tilde{\psi}||} - \psi \right]. \tag{6.112}$$

The factor $\tilde{f}_j / |\tilde{f}_j|$ under the argument of the δ-function is a pure phase factor. Remembering that the propagator represents a transition probability in projective Hilbert space, we see that this factor drops out if the propagator is convoluted with any phase-invariant density functional. Thus we can write

$$T_1[\psi, t | \tilde{\psi}, t_0] = \left(\sum_j ||\Omega_j \tilde{\psi}||^2 \right) \delta \left[\frac{\sigma_- \tilde{\psi}}{||\sigma_- \tilde{\psi}||} - \psi \right]. \tag{6.113}$$

This expression shows that the operation conditioned on the measurement of a photon does not depend on the mode j. The reason for this fact is, obviously, that Ω_j depends on j only through a scalar factor. Physically, this means that the back-action on the quantum object is the same for all field modes j.

The decomposition (6.110) of the conditional transition probability corresponds to two disjoint classical alternatives. T_1 describes the detection of precisely one photon. The factor in front of the δ-functional in (6.113) is the probability for the detection of a photon, whereas the argument of the δ-functional shows that the state

$$\psi = \frac{\sigma_- \tilde{\psi}}{||\sigma_- \tilde{\psi}||} = (\text{phase factor}) \times |g\rangle \qquad (6.114)$$

is the new state of the atom conditioned on that measurement result. Note that this state is just the ground state $|g\rangle$ of the two-level system, that is the detection of a photon puts the atom into the ground state.

On the other hand, T_0 gives the contribution from no photodetection. The probability for this event is given by eqn (6.109) and the state of the reduced system conditioned on that result is given by

$$\psi = \frac{\Omega_0 \tilde{\psi}}{||\Omega_0 \tilde{\psi}||}. \qquad (6.115)$$

Next we perform the Markov approximation. Under the assumption that the electromagnetic field may be approximated by a continuum of field modes we obtain for $\tau \gg 1/\omega_0$ from eqns (6.105) and (6.103)

$$\sum_j |\tilde{f}_j|^2 \approx \tau \gamma_0, \qquad (6.116)$$

$$g_0 \approx -\frac{1}{2} \tau \gamma_0 \sigma_+ \sigma_-, \qquad (6.117)$$

where we have neglected the Lamb shift, and

$$\gamma_0 = \int\limits_{-\infty}^{+\infty} dt\, e^{i\omega_0 t} \langle 0|B(t)B^\dagger(0)|0\rangle = 2\pi D(\omega_0)|\kappa(\omega_0)|^2 \qquad (6.118)$$

is the atomic damping rate, $D(\omega)$ being the density of field modes.

Thus we have from eqn (6.116) and eqn (6.107)

$$\sum_j ||\Omega_j \tilde{\psi}||^2 \approx \tau \gamma_0 \langle \sigma_+ \sigma_- \rangle_{\tilde{\psi}}, \qquad (6.119)$$

and from eqn (6.116) and eqn (6.109)

$$||\Omega_0 \tilde{\psi}||^2 \approx 1 - \tau \gamma_0 \langle \sigma_+ \sigma_- \rangle_{\tilde{\psi}}, \qquad (6.120)$$

and, hence, inserting (6.119) into (6.113)

$$T_1[\psi, t|\tilde{\psi}, t_0] \approx \tau \gamma_0 \langle \sigma_+ \sigma_- \rangle_{\tilde{\psi}}\, \delta \left[\frac{\sigma_- \tilde{\psi}}{||\sigma_- \tilde{\psi}||} - \psi \right]. \qquad (6.121)$$

From eqn (6.117) and eqn (6.100) we get

$$\Omega_0 = I + f_0 + g_0 \approx I - i\tau H_L - \frac{1}{2}\tau\gamma_0\sigma_+\sigma_-, \tag{6.122}$$

and therefore it follows with eqn (6.120) that

$$\frac{\Omega_0\tilde{\psi}}{||\Omega_0\tilde{\psi}||} \approx \left\{I - i\tau H_L - \frac{1}{2}\tau\gamma_0\left(\sigma_+\sigma_- - \langle\sigma_+\sigma_-\rangle_{\tilde{\psi}}\right)\right\}\tilde{\psi}. \tag{6.123}$$

Thus, we find the contribution T_0 of no photodetection by inserting (6.120) and (6.123) into (6.111),

$$T_0[\psi, t|\tilde{\psi}, t_0] \approx \left\{1 - \tau\gamma_0\langle\sigma_+\sigma_-\rangle_{\tilde{\psi}}\right\} \tag{6.124}$$

$$\times\ \delta\left[\left\{I - i\tau H_L - \frac{1}{2}\tau\gamma_0\left(\sigma_+\sigma_- - \langle\sigma_+\sigma_-\rangle_{\tilde{\psi}}\right)\right\}\tilde{\psi} - \psi\right].$$

If we finally add eqns (6.124) and (6.121) we obtain the short-time behaviour of the conditional transition probability $T[\psi, t|\tilde{\psi}, t_0]$. The latter is seen to be precisely of the general form (6.90) characteristic of a PDP in Hilbert space. In fact, from a comparison with (6.90) we infer that the generator of the deterministic evolution periods reads

$$G(\psi) = \hat{H}\psi + \frac{i\gamma_0}{2}||\sigma_-\psi||^2\psi, \tag{6.125}$$

with the non-Hermitian Hamiltonian

$$\hat{H} = H_L - \frac{i\gamma_0}{2}\sigma_+\sigma_-. \tag{6.126}$$

The transition functional takes the form

$$W[\psi|\tilde{\psi}] = \gamma_0||\sigma_-\tilde{\psi}||^2\delta\left[\frac{\sigma_-\tilde{\psi}}{||\sigma_-\tilde{\psi}||} - \psi\right]. \tag{6.127}$$

Accordingly, the corresponding Liouville master equation (6.10) describes a PDP with only one type of quantum jump

$$\tilde{\psi} \longrightarrow \psi = \frac{\sigma_-\tilde{\psi}}{||\sigma_-\tilde{\psi}||} \tag{6.128}$$

which occur with a rate, i.e. probability per unit of time, which is given by $\gamma_0||\sigma_-\tilde{\psi}||^2$. These jumps represent the back-action on the object system: Conditioned on the detection of a photon, they put the atom into its ground state.

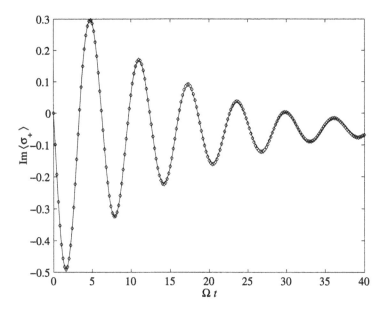

FIG. 6.4. Simulation of the PDP for direct photodetection of a driven two-level
system interacting with the field vacuum. The solid curve represents the
imaginary part of the density matrix element $\langle g|\rho_S(t)|e\rangle = \langle\sigma_+(t)\rangle$ obtained
from the analytical solution of the quantum master equation (see eqn (3.289))
for $\gamma_0/\Omega = 0.1$. The symbols give the average taken over an ensemble of 10^4
realizations of the PDP.

Finally, the equivalent stochastic differential equation (see eqn (6.25)) now
reads explicitly,

$$d\psi(t) = -i\left(H_L - \frac{i\gamma_0}{2}\left[\sigma_+\sigma_- - ||\sigma_-\psi(t)||^2\right]\right)\psi(t)dt$$
$$+ \left(\frac{\sigma_-\psi(t)}{||\sigma_-\psi(t)||} - \psi(t)\right)dN(t), \tag{6.129}$$

where, as usual, $dN(t)^2 = dN(t)$ and $\mathrm{E}\left[dN(t)\right] = \gamma_0||\sigma_-\psi(t)||^2dt$.

Following the reasoning of Section 6.1.1 we conclude that the PDP constructed above leads to a covariance matrix

$$\rho_S(t) = \mathrm{E}\left(|\psi(t)\rangle\langle\psi(t)|\right) \tag{6.130}$$

which satisfies the Bloch equation (6.91) for the density matrix. As an example we depict in Fig. 6.4 the simulation results for the stochastic quantity
$\langle g|\psi(t)\rangle\langle\psi(t)|e\rangle$. The figure shows the average of this quantity taken over a sample of many realizations of the PDP. The average provides a statistical estimate

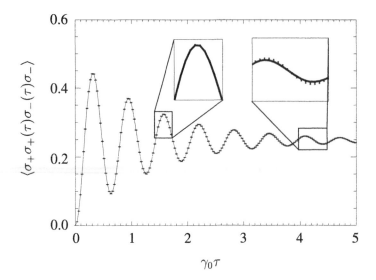

FIG. 6.5. The correlation function $\langle\sigma_+\sigma_+(\tau)\sigma_-(\tau)\sigma_-\rangle$ for a driven two-level atom on resonance with Rabi frequency $\Omega = 10\gamma_0$. The figure shows the simulation results for a sample of 10^5 realizations of the PDP in the doubled Hilbert space (thin line and error bars), and the analytical solution (thick line).

for the off-diagonal element $\langle g|\rho_S(t)|e\rangle = \langle\sigma_+(t)\rangle$ of the density matrix, as is demonstrated by a comparison with the solution of the Bloch equation.

As a further example we show in Fig. 6.5 the results of a numerical simulation of the stationary four-point correlation function $\langle\sigma_+\sigma_+(\tau)\sigma_-(\tau)\sigma_-\rangle$. This correlation has been estimated from a simulation of the corresponding stochastic process $\theta(t)$ in the doubled Hilbert space $\widetilde{\mathcal{H}}$, as explained in Section 6.1.4.

6.3.2 Path integral solution

The Liouville master equation for the PDP of direct photodetection may be solved by means of the path integral technique of Section 6.1.2.

Let us concentrate in the following on the underdamped case $\Omega > \gamma_0/2$, and consider the special initial condition $\psi(0) = |g\rangle$. Using eqn (6.126) in the definition (6.37) we find

$$\hat{U}(t) = e^{-\gamma_0 t/4}\begin{pmatrix} \cos\mu t - \frac{\gamma_0}{4\mu}\sin\mu t & i\frac{\Omega}{2\mu}\sin\mu t \\ i\frac{\Omega}{2\mu}\sin\mu t & \cos\mu t + \frac{\gamma_0}{4\mu}\sin\mu t \end{pmatrix}, \qquad (6.131)$$

where we have introduced the parameter μ which is defined by

$$\mu = \frac{1}{2}\sqrt{\Omega^2 - \left(\frac{\gamma_0}{2}\right)^2}. \qquad (6.132)$$

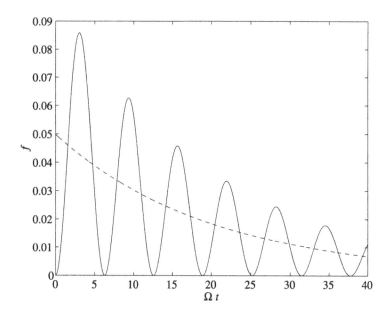

FIG. 6.6. Plots of the density $f(t)$ of the waiting time distribution according to eqn (6.134) with $\gamma_0/\Omega = 0.1$ (solid curve), and of the corresponding density for an exponential waiting time distribution with the same mean value (dashed curve).

With the help of eqns (6.21) and (6.131) the waiting time distribution function is easily found to be

$$F[|g\rangle, t] = 1 - ||\hat{U}|g\rangle||^2 \qquad (6.133)$$

$$= 1 - e^{-\gamma_0 t/2}\left(1 + \frac{\gamma_0^2}{8\mu^2}\sin^2\mu t + \frac{\gamma_0}{2\mu}\sin\mu t \cos\mu t\right).$$

The corresponding density $f(t)$ is given by

$$f(t) \equiv \frac{d}{dt}F[|g\rangle, t] = \gamma_0||\sigma_-\hat{U}(t)|g\rangle||^2 = \frac{\gamma_0\Omega^2}{4\mu^2}e^{-\gamma_0 t/2}\sin^2\mu t. \qquad (6.134)$$

These are the expressions for the waiting time distribution for direct photodetection of resonance fluorescence (Carmichael *et al.*, 1989). Note the oscillating behaviour of the density $f(t)$, as shown in Figs. 6.6 and 6.7. The comparison with the corresponding exponential distribution shows that the probability of small waiting times is strongly suppressed. This is obviously due to the fact that the atom is in the ground state immediately after a quantum jump. It therefore takes some time to get excited again with appreciable probability and to be able to emit another photon.

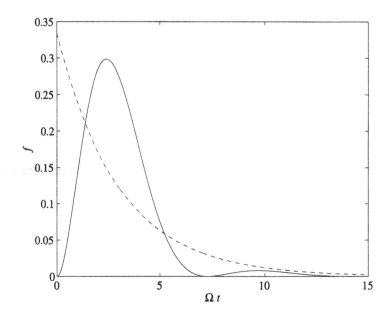

FIG. 6.7. The same as Fig. 6.6 but for $\gamma_0/\Omega = 1$.

Let us investigate the path integral representation (6.43) for the N-jump contribution to the propagator. First, we infer from eqn (6.44) that

$$\frac{\psi_N^t(t_1, \ldots, t_N)}{||\psi_N^t(t_1, \ldots, t_N)||} = \frac{\hat{U}(t - t_N)|g\rangle}{||\hat{U}(t - t_N)|g\rangle||} \qquad (6.135)$$

apart from an irrelevant phase factor. Moreover, eqn (6.45) gives the multi-time exclusive probabilities

$$p_N^t(t_1, t_2, \ldots, t_N) = p_0(t - t_N)f(t_N - t_{N-1}) \cdots f(t_2 - t_1)f(t_1), \qquad (6.136)$$

where $f(t)$ is given by eqn (6.134) and

$$p_0(t) = 1 - F[|g\rangle, t] = ||\hat{U}(t)|g\rangle||^2 \qquad (6.137)$$

denotes the no-jump probability, that is the probability that no count is observed in the time interval $[0, t]$. Thus, with the help of eqn (6.43) we find the following compact form for the full propagator

$$T[\psi, t||g\rangle, 0] = p_0(t)\delta \left[\psi - \frac{\hat{U}(t)|g\rangle}{||\hat{U}(t)|g\rangle||} \right] \qquad (6.138)$$

$$+ \sum_{N=1}^{\infty} \int_0^t dt_N \int_0^{t_N} dt_{N-1} \cdots \int_0^{t_2} dt_1 \, p_N^t(t_1, \ldots, t_N)\delta \left[\psi - \frac{\hat{U}(t - t_N)|g\rangle}{||\hat{U}(t - t_N)|g\rangle||} \right].$$

It is evident from these formulae that the complete statistics of the counting events can by reconstructed from the exclusive probabilities $p^t(t_1, t_2, \ldots, t_N)$. In particular the probability of N counts in the interval $[0, t]$ may be found from eqn (6.46). If we introduce the Laplace transformed quantities

$$\hat{p}_N(\lambda) = \int_0^\infty dt \exp(-\lambda t) p_N(t), \tag{6.139}$$

and

$$\hat{f}(\lambda) = \int_0^\infty dt \exp(-\lambda t) f(t), \tag{6.140}$$

eqn (6.46) together with (6.136) leads to

$$\hat{p}_N(\lambda) = \frac{1 - \hat{f}(\lambda)}{\lambda} \left[\hat{f}(\lambda) \right]^N, \tag{6.141}$$

where we have used the fact that the Laplace transform of the no-counting probability is given by (see eqn (6.137))

$$\hat{p}_0(\lambda) = \frac{1}{\lambda} \left(1 - \hat{f}(\lambda) \right). \tag{6.142}$$

The Laplace transform of the density f may be found with the help of (6.134),

$$\hat{f}(\lambda) = \frac{\frac{1}{2}\gamma_0 \Omega^2}{\bar{\lambda} \left(\bar{\lambda}^2 + (2\mu)^2 \right)}, \tag{6.143}$$

where we have introduced the abbreviation $\bar{\lambda} = \lambda + \gamma_0/2$. Thus, the Laplace transform of the probability $p_N(t)$ can be written as

$$\hat{p}_N(\lambda) = \frac{\left(\frac{1}{2}\gamma_0 \Omega^2 \right)^N \left[\bar{\lambda}^2 + \frac{1}{2}\gamma_0 \bar{\lambda} + \Omega^2 \right]}{\left[\bar{\lambda} \left(\bar{\lambda}^2 + (2\mu)^2 \right) \right]^{N+1}}. \tag{6.144}$$

This is the photocounting formula for direct photodetection of flourescence radiation derived by Mollow (1968, 1969, 1975).

Figure 6.8 shows the photocounting probability $p_N(t)$ for a fixed time t as a function of N. The picture has been obtained from a numerical simulation of the PDP by counting the number of quantum jumps in a sample of realizations of the process. The picture also shows the corresponding Poisson distribution with the same mean value. The sub-Poissonian character of $p_N(t)$ is clearly visible. It may be characterized with the help of the Mandel Q parameter defined by

$$Q = \frac{\text{Var}(N(t))}{\langle N(t) \rangle} - 1. \tag{6.145}$$

For a Poisson distribution we have $Q = 0$, while $Q < 0$ characterizes a sub-Poissonian distribution. For the simulation shown in the figure the Mandel Q

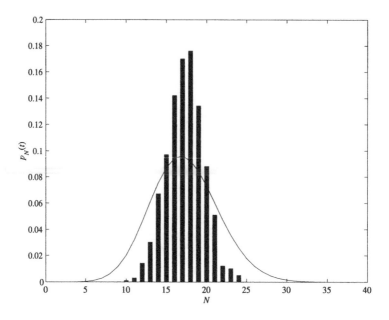

FIG. 6.8. The probability $p_N(t)$ for N photon counts in the time interval $[0, t]$ for the driven two-level system. The figure shows the simulation results obtained from an ensemble of 10^3 realizations of the PDP for direct photodetection (histogram). The solid line represents the Poisson distribution with the same mean value as $p_N(t)$. Parameter: $\gamma_0/\Omega = 1$ and $\Omega t = 50$.

parameter was found to be $Q \approx -0.69$ which is in good agreement with the analytical result (Mandel and Wolf, 1995).

The sub-Poissonian character may be understood from the waiting time distribution shown in Fig. 6.7 for the same parameters. The density $f(t)$ for the PDP exhibits a pronounced bump such that the variance of the random waiting time is less than that of the corresponding exponential distribution. In particular, the photon emission events exhibit a strong antibunching effect. The distribution $p_N(t)$ of the photon counts is therefore sharper than the corresponding Poisson distribution which would be obtained from an exponential waiting time distribution.

6.4 Homodyne photodetection

As a second example we consider in this section a driven two-level atom whose emitted light is detected by homodyne photodetection (Walls and Milburn, 1994; Wiseman and Milburn, 1993a, 1993b). A schematic picture of the setup is shown in Fig. 6.9.

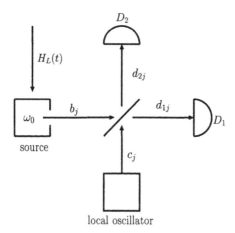

FIG. 6.9. Experimental setup of homodyne photodetection. The source is a two-level atom with frequency ω_0 which is driven by an external interaction $H_L(t)$. The light emitted by the source (modes b_j) traverses a beam splitter and is detected at detectors D_1 and D_2. The emitted light interferes with the light field from the local oscillator (modes c_j). The detectors register the quanta of the modes d_{1j} and d_{2j} defined in the text.

6.4.1 Derivation of the PDP for homodyne detection

The Hamiltonian H_B of the environment is taken to be

$$H_B = \sum_j \omega_j b_j^\dagger b_j + \sum_j \omega_j c_j^\dagger c_j. \tag{6.146}$$

Again, the b_j-modes are the radiation modes that couple directly to the atom. Furthermore, we have added a term that contains the modes c_j of the local oscillator whose light interferes with the radiation emitted by the atom. We take into account the presence of the local oscillator by a coherent driving of the c_j-modes. The interaction picture Hamiltonian now takes the following form,

$$H_I(t) = e^{-i\omega_0 t}\sigma_- B^\dagger(t) + e^{i\omega_0 t}\sigma_+ B(t) + \beta e^{-i\omega_0 t}C^\dagger(t) + \beta^* e^{i\omega_0 t}C(t) + H_L, \tag{6.147}$$

where H_L is defined in (6.95) and $B(t)$ is given by (6.94). The quantity β represents the amplitude of the local oscillator field and

$$C(t) = \sum_j \kappa_j e^{-i\omega_j t} c_j. \tag{6.148}$$

As for direct photodetection, the electromagnetic field serves as the quantum probe, the pure probe state being the field vacuum $\phi_0 = |0\rangle$.

The crucial point of the derivation of the stochastic process for homodyne photodetection is the correct choice of the basis φ_α in the reduction formula.

What is measured at the detectors D_1 and D_2 (see Fig. 6.9) is the superposition of the light emitted from the atom and the light of the local oscillator. Under the assumption that the beam splitter has transmittivity $\frac{1}{2}$ and including a phase shift of $\pi/2$, the field quanta that are detected at D_1 and D_2 are defined, respectively, by the annihilation operators

$$d_{1,j} = \frac{1}{\sqrt{2}}\left(b_j + ic_j\right),\tag{6.149}$$

$$d_{2,j} = \frac{1}{\sqrt{2}}\left(b_j - ic_j\right).\tag{6.150}$$

Note that these operators satisfy the usual boson commutation relations

$$[d_{k,j}, d_{k',j'}^\dagger] = \delta_{kk'}\delta_{jj'},\tag{6.151}$$

where the index $k = 1,2$ labels the two detectors D_1 and D_2. Thus, $d_{k,j}^\dagger$ can be interpreted as the creation operator for a photon in mode j at detector D_k. Consequently, we take as the basis φ_α the Fock basis which is generated by the creation operators $d_{k,j}^\dagger$ out of the field vacuum, that is, we write

$$\varphi_\alpha = |\{N_j^{(1)}\}; \{N_j^{(2)}\}\rangle.\tag{6.152}$$

The index α now represents two sets of occupation numbers: $\{N_j^{(k)}\}$ is the set of occupation numbers of the $d_{k,j}$-modes, where $k = 1,2$. In particular, we have two types of one-photon states given by

$$\varphi_{k,j} \equiv |k,j\rangle = d_{k,j}^\dagger|0\rangle, \quad k = 1,2.\tag{6.153}$$

We can now proceed in close analogy to Section 6.3. The first step is to determine the operators Ω_α which yield the operation corresponding to the measurement scheme. These operators are given by eqn (6.86), where for the states φ_α expression (6.152) must be used. We shall use again the definitions (6.102) and (6.103) for the operators f_α and g_α. We then find that for the vacuum state $f_0 = -i\tau H_L$. For a one-photon state $\varphi_\alpha = |k,j\rangle$, however, we have

$$f_{k,j} = -i\int_{t_0}^{t} dt'\, e^{-i\omega_0 t'}\left\{\langle k,j|B^\dagger(t')|0\rangle\sigma_- + \langle k,j|C^\dagger(t')|0\rangle\beta\right\}.\tag{6.154}$$

The matrix elements in the above equation are easily calculated to be

$$\langle k,j|B^\dagger(t')|0\rangle = \frac{1}{\sqrt{2}}\kappa_j^* e^{i\omega_j t'},\tag{6.155}$$

and

$$\langle k,j|C^\dagger(t')|0\rangle = \pm\frac{1}{\sqrt{2}}\kappa_j^* e^{i\omega_j t'},\tag{6.156}$$

where we use the convention that the upper sign corresponds to $k = 1$ and the lower sign to $k = 2$. Hence we obtain

$$f_{k,j} = \tilde{f}_j(\sigma_- \pm i\beta), \tag{6.157}$$

where

$$\tilde{f}_j \equiv \frac{-i}{\sqrt{2}}\kappa_j^* \int_{t_0}^{t} dt' e^{i(\omega_j - \omega_0)t'}. \tag{6.158}$$

On using these results we can decompose the conditional transition probability into three parts as follows

$$T[\psi, t|\tilde{\psi}, t_0] = T_0[\psi, t|\tilde{\psi}, t_0] + T_1[\psi, t|\tilde{\psi}, t_0] + T_2[\psi, t|\tilde{\psi}, t_0], \tag{6.159}$$

where

$$T_0[\psi, t|\tilde{\psi}, t_0] = ||\Omega_0\tilde{\psi}||^2 \delta\left[\frac{\Omega_0\tilde{\psi}}{||\Omega_0\tilde{\psi}||} - \psi\right], \tag{6.160}$$

$$T_1[\psi, t|\tilde{\psi}, t_0] = \left(\sum_j ||\Omega_{1,j}\tilde{\psi}||^2\right) \delta\left[\frac{(\sigma_- + i\beta)\tilde{\psi}}{||(\sigma_- + i\beta)\tilde{\psi}||} - \psi\right], \tag{6.161}$$

$$T_2[\psi, t|\tilde{\psi}, t_0] = \left(\sum_j ||\Omega_{2,j}\tilde{\psi}||^2\right) \delta\left[\frac{(\sigma_- - i\beta)\tilde{\psi}}{||(\sigma_- - i\beta)\tilde{\psi}||} - \psi\right]. \tag{6.162}$$

The factors in front of the δ-functions are given by

$$||\Omega_0\tilde{\psi}||^2 = 1 - \sum_{k,j} ||\Omega_{k,j}\tilde{\psi}||^2 \tag{6.163}$$

$$\sum_j ||\Omega_{1,j}\tilde{\psi}||^2 = \sum_j \langle f_{1,j}^\dagger f_{1,j}\rangle_{\tilde{\psi}} = \sum_j |\tilde{f}_j|^2 \langle(\sigma_- + i\beta)^\dagger(\sigma_- + i\beta)\rangle_{\tilde{\psi}}, \tag{6.164}$$

$$\sum_j ||\Omega_{2,j}\tilde{\psi}||^2 = \sum_j \langle f_{2,j}^\dagger f_{2,j}\rangle_{\tilde{\psi}} = \sum_j |\tilde{f}_j|^2 \langle(\sigma_- - i\beta)^\dagger(\sigma_- - i\beta)\rangle_{\tilde{\psi}}. \tag{6.165}$$

The above decomposition of T into three terms expresses three classically disjoint alternatives. Either no photon is detected (contribution T_0), or a photon is detected at D_1 (contribution T_1), or a photon is detected at D_2 (contribution T_2).

In the Markov approximation we find in the same way as in the previous section that

$$\sum_j |\tilde{f}_j|^2 \approx \frac{1}{2}\tau\gamma_0 \tag{6.166}$$

and, therefore,

$$\sum_j ||\Omega_{k,j}\tilde{\psi}||^2 \approx \frac{1}{2}\tau\gamma_0\langle(\sigma_- \pm i\beta)^\dagger(\sigma_- \pm i\beta)\rangle_{\tilde{\psi}}, \tag{6.167}$$

$$||\Omega_0\tilde{\psi}||^2 \approx 1 - \tau\gamma_0\left\{\langle\sigma_+\sigma_-\rangle_{\tilde{\psi}} + |\beta|^2\right\}. \tag{6.168}$$

Moreover, we have

$$g_0 \approx -\frac{1}{2}\tau\gamma_0\left\{\sigma_+\sigma_- + |\beta|^2\right\}. \tag{6.169}$$

Thus we get

$$\frac{\Omega_0\tilde{\psi}}{||\Omega_0\tilde{\psi}||} \approx \left\{I - i\tau H_L - \frac{1}{2}\tau\gamma_0\left(\sigma_+\sigma_- - \langle\sigma_+\sigma_-\rangle_{\tilde{\psi}}\right)\right\}\tilde{\psi}. \tag{6.170}$$

Note that the terms involving $|\beta|^2$ cancel each other and that we therefore find the same expression (6.123) as in the previous section.

Summarizing these results, we finally obtain

$$T_0[\psi, t|\tilde{\psi}, t_0] = \left\{1 - \tau\gamma_0\left[||\sigma_-\tilde{\psi}||^2 + |\beta|^2\right]\right\} \tag{6.171}$$

$$\times \delta\left[\left\{I - i\tau H_L - \frac{1}{2}\tau\gamma_0\left(\sigma_+\sigma_- - \langle\sigma_+\sigma_-\rangle_{\tilde{\psi}}\right)\right\}\tilde{\psi} - \psi\right],$$

$$T_1[\psi, t|\tilde{\psi}, t_0] = \frac{1}{2}\tau\gamma_0||(\sigma_- + i\beta)\tilde{\psi}||^2\delta\left[\frac{(\sigma_- + i\beta)\tilde{\psi}}{||(\sigma_- + i\beta)\tilde{\psi}||} - \psi\right], \tag{6.172}$$

$$T_2[\psi, t|\tilde{\psi}, t_0] = \frac{1}{2}\tau\gamma_0||(\sigma_- - i\beta)\tilde{\psi}||^2\delta\left[\frac{(\sigma_- - i\beta)\tilde{\psi}}{||(\sigma_- - i\beta)\tilde{\psi}||} - \psi\right]. \tag{6.173}$$

We add these three contributions to get the short-time behaviour of the propagator and compare with the general form (6.90). This shows that the generator $G(\psi)$ for the deterministic evolution periods of the PDP is of the same form as in eqn (6.125). However, the transition functional is now given by

$$W[\psi|\tilde{\psi}] = \frac{1}{2}\gamma_0||(\sigma_- + i\beta)\tilde{\psi}||^2\delta\left[\frac{(\sigma_- + i\beta)\tilde{\psi}}{||(\sigma_- + i\beta)\tilde{\psi}||} - \psi\right]$$

$$+ \frac{1}{2}\gamma_0||(\sigma_- - i\beta)\tilde{\psi}||^2\delta\left[\frac{(\sigma_- - i\beta)\tilde{\psi}}{||(\sigma_- - i\beta)\tilde{\psi}||} - \psi\right]. \tag{6.174}$$

The Liouville master equation thus describes a piecewise deterministic process with two types of quantum jump, corresponding to the detection of photons at D_1 or D_2,

$$\tilde{\psi} \longrightarrow \psi = \frac{(\sigma_- \pm i\beta)\tilde{\psi}}{||(\sigma_- \pm i\beta)\tilde{\psi}||}, \quad \text{rate} = \frac{1}{2}\gamma_0||(\sigma_- \pm i\beta)\tilde{\psi}||^2. \tag{6.175}$$

The deterministic pieces of the PDP are obtained from the same differential equation as for direct photodetection.

The equivalent stochastic differential equation for the PDP describing homodyne detection takes the form

$$d\psi(t) = -i\left(\hat{H} + \frac{i\gamma_0}{2}||\sigma_-\psi(t)||^2\right)\psi(t)dt \tag{6.176}$$

$$+ \left(\frac{(\sigma_- + i\beta)\psi(t)}{||(\sigma_- + i\beta)\psi(t)||} - \psi(t)\right)dN_1(t) + \left(\frac{(\sigma_- - i\beta)\psi(t)}{||(\sigma_- - i\beta)\psi(t)||} - \psi(t)\right)dN_2(t).$$

The processes $N_k(t)$ count the events at detector D_k and satisfy

$$dN_k(t)dN_{k'}(t) = \delta_{kk'}dN_k(t), \tag{6.177}$$

and

$$E\left[dN_1(t)\right] = \frac{\gamma_0}{2}||(\sigma_- + i\beta)\psi(t)||^2 dt, \quad E\left[dN_2(t)\right] = \frac{\gamma_0}{2}||(\sigma_- - i\beta)\psi(t)||^2 dt. \tag{6.178}$$

6.4.2 *Stochastic Schrödinger equation*

The PDP for homodyne photodetection yields a certain stochastic Schrödinger equation in the diffusion limit. To derive it we write

$$\beta = i|\beta|e^{i\theta} \tag{6.179}$$

and investigate the limit $\varepsilon \equiv 1/|\beta| \longrightarrow 0$ of a strongly excited local oscillator. This can easily be done by invoking the results of Section 6.1.3 on the diffusion limit of the Liouville master equation. We first observe that, instead of $\sigma_- \pm i\beta$, we may use the two jump operators

$$I \pm \varepsilon C \equiv I \pm \varepsilon e^{-i\theta}\sigma_-, \tag{6.180}$$

if we scale the damping constant by a factor of $|\beta|^2$, that is if we replace $\gamma_0 \longrightarrow \gamma_0|\beta|^2$. Then we have a situation which is similar to the one discussed in Section 6.1.3, where, however, we now have two jump operators. Since $C \equiv \exp(-i\theta)\sigma_-$ appears with a positive and with a negative sign, the linear part (in ε) of the drift generator (6.54) of the Fokker–Planck equation drops out and a well-defined diffusion limit is obtained without restriction on C.

The resulting stochastic Schrödinger equation for homodyning, first derived by Wiseman and Milburn (1993a, 1993b) and by Carmichael (1993), may be written as follows,

$$d\psi(t) = -iK(\psi(t))dt + \sqrt{\gamma_0}M(\psi(t))dW(t), \tag{6.181}$$

where $dW(t)$ is a real Wiener increment. The drift operator $K(\psi)$ reads

$$K(\psi) = H_L\psi + \frac{i\gamma_0}{2}\left\{\langle C + C^\dagger\rangle_\psi C - C^\dagger C - \frac{1}{4}\langle C + C^\dagger\rangle_\psi^2\right\}\psi, \tag{6.182}$$

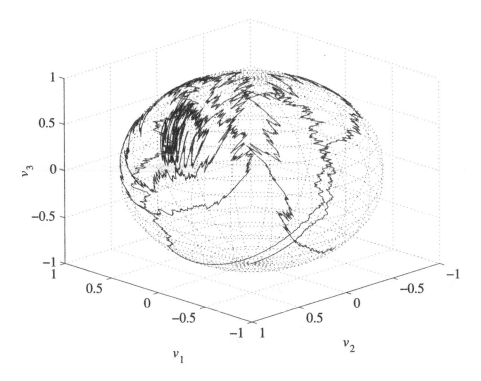

FIG. 6.10. Diffusion on the surface of the Bloch sphere. The picture shows a single realization of the random Bloch vector $\vec{v}(t) = \langle\psi(t)|\vec{\sigma}|\psi(t)\rangle$ for the process $\psi(t)$ given by the stochastic Schrödinger equation (6.181). Parameters: $\gamma_0/\Omega = 0.1$ and $\theta = 0$.

while the noise term is given by

$$M(\psi) = \left\{ C - \frac{1}{2}\langle C + C^\dagger\rangle_\psi \right\}\psi. \tag{6.183}$$

By construction, the stochastic Schrödinger equation (6.181) exactly conserves the norm of the wave function. It leads to a stochastic differential equation for the random Bloch vector

$$\vec{v}(t) = \langle\psi(t)|\vec{\sigma}|\psi(t)\rangle, \tag{6.184}$$

describing a diffusion process on the surface of the Bloch sphere. A single realization of this process is depicted in Fig. 6.10. We observe that the noise induced by the homodyne measurement scheme drives the Bloch vector out of the plane defined by $v_1 = 0$. Satisfying the Bloch equation, the expectation of $\vec{v}(t)$, however, remains in this plane as illustrated in Fig. 6.11 which shows the evolution of an ensemble of realizations.

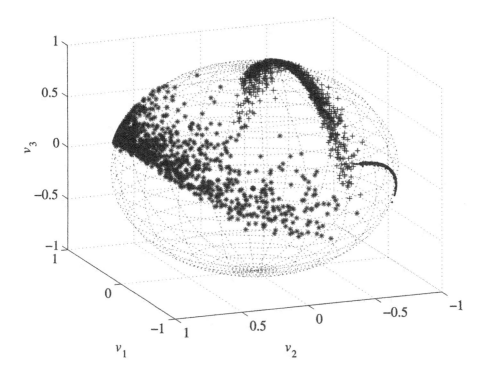

FIG. 6.11. 10^3 realizations of the random Bloch vector $\vec{v}(t) = \langle\psi(t)|\vec{\sigma}|\psi(t)\rangle$ plotted at three fixed times given by $\Omega t = 1.5$ (dots), $\Omega t = 3.0$ (crosses), and $\Omega t = 4.5$ (stars). All trajectories started out at the south pole of the sphere. Parameters: $\gamma_0/\Omega = 0.1$ and $\theta = 0$.

6.5 Heterodyne photodetection

The measurement scheme for heterodyne photodetection is the same as that for homodyne detection, the only difference being that the frequency of the local oscillator is detuned from the system's frequency ω_0. In the notation of the previous section this means that we have to replace the amplitude of the local oscillator β by

$$\beta_t = \beta \exp(-i\Omega t), \tag{6.185}$$

where Ω is the finite detuning of the local oscillator (not to be confused with the Rabi frequency).

6.5.1 Stochastic Schrödinger equation

We perform this replacement directly in the stochastic Schrödinger equation (6.181) for homodyne detection, which yields the drift operator

$$K(\psi) = H_L\psi - i\frac{\gamma_0}{2}\left\{\sigma_-\sigma_+ - \langle e^{-2i(\theta-\Omega t)}\sigma_- + \sigma_+\rangle_\psi\sigma_- \right. \tag{6.186}$$

$$\left. + \frac{1}{4}[\langle\sigma_-\rangle_\psi^2 e^{-2i(\theta-\Omega t)} + \langle\sigma_+\rangle_\psi^2 e^{2i(\theta-\Omega t)} + 2\langle\sigma_-\rangle_\psi\langle\sigma_+\rangle_\psi]\right\}\psi,$$

while the noise term becomes

$$M(\psi) = \left\{e^{-i(\theta-\Omega t)}\sigma_- - \frac{1}{2}\langle e^{-i(\theta-\Omega t)}\sigma_- + e^{i(\theta-\Omega t)}\sigma_+\rangle_\psi\right\}\psi. \tag{6.187}$$

Following Wiseman and Milburn we assume that the detuning is much higher than the system's characteristic damping rate which is of the same order as the interaction strength with the driving field, i.e. we assume

$$\Omega \gg \gamma_0 \sim H_L. \tag{6.188}$$

Hence, we may consider a time interval Δt such that $\Omega\Delta t \gg 1$ but $\gamma_0\Delta t \ll 1$. Now we integrate eqn (6.181) with (6.186) and (6.187) over the interval Δt, and neglect terms of second order in $\gamma_0\Delta t$. We consider the drift and the diffusion terms separately. The integration of the drift term leads to

$$\int_t^{t+\Delta t} dsK(\psi) \tag{6.189}$$

$$= H_L\psi\Delta t - i\frac{\gamma_0}{2}\left\{\sigma_+\sigma_-\Delta t - \int_t^{t+\Delta t} ds\langle e^{-2i(\theta-\Omega s)}\sigma_-\rangle_\psi\sigma_- - \int_t^{t+\Delta t} ds\langle\sigma_+\rangle_\psi\sigma_- \right.$$

$$\left. + \frac{1}{4}\int_t^{t+\Delta t} ds\left[\langle\sigma_-\rangle_\psi^2 e^{-2i(\theta-\Omega s)} + \langle\sigma_+\rangle_\psi^2 e^{2i(\theta-\Omega s)}\right] + \frac{1}{2}\int_t^{t+\Delta t} ds\langle\sigma_-\rangle_\psi\langle\sigma_+\rangle_\psi\right\}\psi.$$

Since the first and the third time integral in this expression are of order γ_0/Ω they can be omitted since by assumption $\Omega \gg \gamma_0$. So we are left with

$$\int_t^{t+\Delta t} dsK(\psi) = \left[H_L\psi - i\frac{\gamma_0}{2}\left(\sigma_+\sigma_- - \langle\sigma_+\rangle_\psi\sigma_- + \frac{1}{2}\langle\sigma_-\rangle_\psi\langle\sigma_+\rangle_\psi\right)\psi\right]\Delta t. \tag{6.190}$$

Accordingly, the integration of the noise term leads to

$$\int_t^{t+\Delta t} M(\psi)dW(s) = \int_t^{t+\Delta t} dW(s)e^{-i(\theta-\Omega s)}\sigma_-\psi \tag{6.191}$$

$$- \frac{1}{2}\int_t^{t+\Delta t} dW(s)\langle e^{-i(\theta-\Omega s)}\sigma_- + e^{i(\theta-\Omega s)}\sigma_+\rangle_\psi\psi.$$

Introducing the two quadratures

$$X_\theta \equiv \frac{1}{2} \left(e^{-i\theta} \sigma_- + e^{i\theta} \sigma_+ \right), \tag{6.192}$$

$$Y_\theta \equiv \frac{i}{2} \left(e^{i\theta} \sigma_+ - e^{-i\theta} \sigma_- \right), \tag{6.193}$$

we can write

$$\int_t^{t+\Delta t} M(\psi) dW(s) = \int_t^{t+\Delta t} dW(s) \cos(\Omega s) \left\{ e^{-i\theta} \sigma_- - \langle X_\theta \rangle_\psi \right\} \psi \tag{6.194}$$

$$- \int_t^{t+\Delta t} dW(s) \sin(\Omega s) \left\{ -ie^{-i\theta} \sigma_- - \langle Y_\theta \rangle_\psi \right\} \psi.$$

The integrals occurring in the above equation suggest we define two new Gaussian variables

$$\delta W_x(t) \equiv \sqrt{2} \int_t^{t+\Delta t} dW(s) \cos(\Omega s), \tag{6.195}$$

$$\delta W_y(t) \equiv -\sqrt{2} \int_t^{t+\Delta t} dW(s) \sin(\Omega s). \tag{6.196}$$

It is easy to show that to zeroth order in Ω^{-1}, δW_x and δW_y have zero mean and satisfy (Wiseman, 1994)

$$E\left[\delta W_q(t) \delta W_{q'}(t')\right] = \delta_{q,q'}(\Delta t - |t - t'|)\theta(\Delta t - |t - t'|), \tag{6.197}$$

where q and q' stand for x and y and θ denotes the Heaviside step function. When taking the continuum limit $\gamma_0/\Omega \longrightarrow 0$ the quantity Δt is infinitesimal on the system's time scale, i.e. $\Delta t \longrightarrow dt$, so $\delta W_q(t)$ can be regarded as an infinitesimal Wiener increment $dW_q(t)$ satisfying

$$dW_q(t) dW_{q'}(t') = \delta_{q,q'} dt. \tag{6.198}$$

For the averaged noise contribution (6.194) we may therefore make the replacement

$$\int_t^{t+\Delta t} M(\psi) dW(s) \longrightarrow \frac{1}{\sqrt{2}} \left\{ e^{-i\theta} \sigma_- - \langle X_\theta \rangle_\psi \right\} \psi dW_x(t) \tag{6.199}$$

$$+ \frac{1}{\sqrt{2}} \left\{ -ie^{-i\theta} \sigma_- - \langle Y_\theta \rangle_\psi \right\} \psi dW_y(t).$$

Summing up the results obtained so far the stochastic Schrödinger equation pertaining to heterodyne detection reads

$$d\psi(t) = -iH_L\psi dt - \frac{\gamma_0}{2}\left(\sigma_+\sigma_- - \langle\sigma_+\rangle_\psi\sigma_- + \frac{1}{2}\langle\sigma_-\rangle_\psi\langle\sigma_+\rangle_\psi\right)\psi dt \qquad (6.200)$$

$$+\sqrt{\frac{\gamma_0}{2}}\left(e^{-i\theta}\sigma_- - \langle X_\theta\rangle_\psi\right)\psi dW_x(t) - \sqrt{\frac{\gamma_0}{2}}\left(ie^{-i\theta}\sigma_- + \langle Y_\theta\rangle_\psi\right)\psi dW_y(t).$$

As is easy to verify this equation is norm preserving. It is also straightforward to check that it is equivalent to the following stochastic equation for the unnormalized wave function $\tilde{\psi}$,

$$d\tilde{\psi} = \left\{-iH_L - \frac{\gamma_0}{2}\sigma_+\sigma_- + \gamma_0\langle\sigma_+\rangle_\psi\sigma_-\right\}\psi dt + \sqrt{\gamma_0}\sigma_-\psi dW(t), \qquad (6.201)$$

where $d\tilde{\psi}$ is the increment of the unnormalized wave function and $dW(t)$ is the differential of a complex Wiener process

$$dW(t) \equiv \frac{1}{\sqrt{2}}\left(dW_x(t) - idW_y(t)\right). \qquad (6.202)$$

Note, that the right-hand side of eqn (6.201) involves the normalized wave function $\psi(t)$.

The examples considered in the present and the previous sections clearly illustrate the general concept behind the stochastic representation of continuous measurements. As we have discussed already the stochastic state vector describes the dynamics of an ensemble of type $\mathcal{E}_P$ for the object system, namely the atomic source. The output light field is observed by different detection schemes which, consequently, yield different stochastic processes for the atomic wave function: Information on the quantum object is extracted in different ways by monitoring the environment and, thus, also the back-action on the object turns out to be different. On the other hand, the evolution of the corresponding ensemble of type $\mathcal{E}_\rho$ is given, in all cases considered above, by one and the same density matrix equation, namely by the vacuum optical master equation (6.91). This reflects the fact that in all cases we have the same local coupling between atom and radiation field.

6.5.2 Stochastic collapse models

It might be important to mention at this point that stochastic Schrödinger equations of the type of eqns (6.181) and (6.200) have been proposed, in an entirely different physical context, by Pearle (1976, 1989), Ghirardi, Rimini and Weber (1986), Ghirardi, Pearle and Rimini (1990), Gisin (1984, 1989), and by Gisin and Percival (1992, 1993). Namely, these authors suggest modifying the Schrödinger equation on a fundamental level by the addition of non-linear and stochastic terms. Various models have been proposed, such as the continuous spontaneous localization model (eqn 6.181) and the quantum state diffusion model. The aim is to explain the non-existence of the superposition of certain macroscopically distinct states. The random terms in the Schrödinger equation lead to a dynamical destruction of such superpositions in a way that macroscopic objects are practically always in definite localized states. In these theories the destruction of superpositions is a real process going on in physical space and time. It is not, as in environment-induced decoherence, a result of the tracing over an external world that became entangled with an open system. Thus, the fundamental modifications of quantum mechanics based on random Schrödinger-type equations provide dynamical collapse models in which the stochastic state vector describes an individual quantum system and enables some kind of macro-realistic interpretation.

Stochastic collapse theories of this kind are not the interpretation followed here. As discussed in Chapter 5 stochastic Schrödinger equations of the form derived above admit a clear interpretation as the evolution of an $\mathcal{E}_P$-ensemble. Nevertheless, it might be instructive to show that a stochastic Schrödinger equation of the type suggested in the quantum state diffusion model, which was developed by Gisin and Percival (Percival, 1998), can be derived from eqn (6.200) by an appropriate phase transformation.

We begin by noting that eqn (6.200) can be rewritten with the help of the complex Wiener differential as

$$
\begin{aligned}
d\psi(t) = &-iH_L\psi dt - \frac{\gamma_0}{2}\left(C^\dagger C - \langle C^\dagger\rangle_\psi C + \frac{1}{2}\langle C\rangle_\psi\langle C^\dagger\rangle_\psi\right)\psi dt \\
&+\sqrt{\gamma_0}\left(C - \langle C\rangle_\psi\right)\psi dW(t) \\
&+\frac{\sqrt{\gamma_0}}{2}\left(\langle C\rangle_\psi dW(t) - \langle C^\dagger\rangle_\psi dW^*(t)\right)\psi,
\end{aligned} \tag{6.203}
$$

where we again use $C \equiv \exp(-i\theta)\sigma_-$. The first noise term in this equation is already of the form of the noise term in the stochastic Schrödinger equation proposed in quantum state diffusion (see below). Therefore, we will now try to eliminate the second noise term with the help of an appropriate phase transformation. To this end, we try the ansatz

$$
\hat{\psi}(t) = \exp[i\varphi(t)]\psi(t) \tag{6.204}
$$

with the stochastic phase

$$
\varphi(t) = \frac{i}{2}\sqrt{\gamma_0}\int_0^t \left[\langle C\rangle_{\psi(t')}dW(t') - \langle C^\dagger\rangle_{\psi(t')}dW^*(t')\right]. \tag{6.205}
$$

Since the phase is a real random number the phase transformation preserves the norm of the wave function and leaves unchanged the expression for the density matrix. We now determine the stochastic Schrödinger equation governing the dynamics of $\hat{\psi}$ using the rules of the Itô calculus.

Following our previous discussion the differential of the phase is

$$d\varphi(t) = \frac{i}{2}\sqrt{\gamma_0}\left(\langle C\rangle_{\psi(t)}dW(t) - \langle C^\dagger\rangle_{\psi(t)}dW^*(t)\right).\qquad(6.206)$$

Accordingly, the differential of the transformed wave function $\hat\psi$ is given by

$$\begin{aligned}d\hat\psi(t) &= \exp[i\varphi(t+dt)]\psi(t+dt) - \exp[i\varphi(t)]\psi(t)\\ &= \exp[i(\varphi(t)+d\varphi(t))](\psi(t)+d\psi(t)) - \exp[i\varphi(t)]\psi(t)\\ &= \exp[i\varphi(t)]\left\{\exp[id\varphi(t)](\psi(t)+d\psi(t)) - \psi(t)\right\}\qquad(6.207)\end{aligned}$$

and, hence,

$$d\hat\psi(t) = \exp[i\varphi(t)]\left\{(\exp[id\varphi(t)] - 1)\,\psi(t) + \exp[id\varphi(t)]d\psi(t)\right\}.\qquad(6.208)$$

Since the differential $d\varphi$ is of order $\sqrt{dt}$ we have to expand $(\exp[id\varphi]-1)$ to second order. Conversely, since $d\psi$ already contains terms of order $\sqrt{dt}$ it is sufficient to expand the exponential function in the term $\exp[id\varphi]d\psi$ to first order. To the desired order eqn (6.208) therefore reads

$$d\hat\psi = \left(id\varphi - \frac{1}{2}(d\varphi)^2\right)\hat\psi + \exp[i\varphi]\,(1 + id\varphi)\,d\psi.\qquad(6.209)$$

Upon inserting eqn (6.203) into the above equation we get

$$\begin{aligned}d\hat\psi =& \left(id\varphi - \frac{1}{2}(d\varphi)^2\right)\hat\psi - iH_L\hat\psi dt\\ &- \frac{\gamma_0}{2}\left(C^\dagger C - \langle C^\dagger\rangle_{\hat\psi}C + \frac{1}{2}\langle C\rangle_{\hat\psi}\langle C^\dagger\rangle_{\hat\psi}\right)\hat\psi dt\\ &+ \sqrt{\gamma_0}\left(C - \langle C\rangle_{\hat\psi}\right)\hat\psi dW(t)\\ &+ \frac{\sqrt{\gamma_0}}{2}\left(\langle C\rangle_{\hat\psi}dW - \langle C^\dagger\rangle_{\hat\psi}dW^*\right)\hat\psi + e^{i\varphi}id\varphi d\psi.\qquad(6.210)\end{aligned}$$

The terms appearing in the above equation are easily evaluated. It follows from (6.206) that

$$id\varphi\hat\psi = -\frac{\sqrt{\gamma_0}}{2}\left(\langle C\rangle_{\hat\psi}dW - \langle C^\dagger\rangle_{\hat\psi}dW^*\right)\hat\psi,\qquad(6.211)$$

and this term is seen to compensate the second and the third noise term in eqn (6.210). Equation (6.206) leads to

$$-\frac{1}{2}(d\varphi)^2\hat\psi = -\frac{\gamma_0}{4}\langle C^\dagger\rangle_{\hat\psi}\langle C\rangle_{\hat\psi}\hat\psi dt,\qquad(6.212)$$

because of the properties of the complex Wiener increment. Furthermore, we find

$$\exp[i\varphi]id\varphi d\psi = \frac{\gamma_0}{2}\langle C^\dagger\rangle_{\hat{\psi}} C\hat{\psi}dt. \tag{6.213}$$

Inserting eqns (6.211), (6.212) and (6.213) into eqn (6.210) we finally obtain

$$d\hat{\psi} = -iH_L\hat{\psi}dt + \gamma_0 \left(\langle C^\dagger\rangle_{\hat{\psi}} C - \frac{1}{2}C^\dagger C - \frac{1}{2}\langle C^\dagger\rangle_{\hat{\psi}}\langle C\rangle_{\hat{\psi}} \right) \hat{\psi}dt$$

$$+\sqrt{\gamma_0}\left(C - \langle C\rangle_\psi \right)\hat{\psi}dW, \tag{6.214}$$

which is just the stochastic differential equation of the quantum state diffusion model involving a single Lindblad operator C. We thus conclude that the stochastic Schrödinger equation of quantum state diffusion, which has been proposed originally as a model describing dynamic state vector localization, appears here as an equation which represents the stochastic dynamics of heterodyne photodetection in the diffusion limit.

6.6 Stochastic density matrix equations

The stochastic processes considered so far describe the dynamics of $\mathcal{E}_P$-ensembles of pure states $\psi(t)$. In a completely analogous manner we can also derive the processes for the corresponding ensembles of mixtures. This will lead us to equations of motion for the time-dependent distributions $P[\sigma,t]$ on the space of density matrices σ or, equivalently, to stochastic density matrix equations.

Probability distributions on the space of density matrices have already been introduced in Section 5.3. We have also discussed there the description of selective measurements. Our starting point will be eqn (5.84) which, applied to the present case, yields the new distribution $P[\sigma, t_0+\tau]$ of the object system at time $t = t_0+\tau$ as it results from the measurement on the probe system after the interaction between object and probe during the time interval τ. Thus, eqn (5.84) leads to

$$P[\sigma, t_0 + \tau] = \int D\tilde{\sigma} \sum_m \text{tr}\left\{\Phi_m(\tilde{\sigma})\right\}\delta\left[\frac{\Phi_m(\tilde{\sigma})}{\text{tr}\left\{\Phi_m(\tilde{\sigma})\right\}} - \sigma\right]P[\tilde{\sigma}, t_0]. \tag{6.215}$$

In view of this relation the conditional probability for the density matrix at time $t_0 + \tau$ to be σ under the condition that the density matrix at time t_0 is $\tilde{\sigma}$ takes the form

$$T[\sigma, t_0 + \tau|\tilde{\sigma}, t_0] = \sum_m \text{tr}\left\{\Phi_m(\tilde{\sigma})\right\}\delta\left[\frac{\Phi_m(\tilde{\sigma})}{\text{tr}\left\{\Phi_m(\tilde{\sigma})\right\}} - \sigma\right]. \tag{6.216}$$

The next step is again the determination of the operation Φ_m pertaining to the measurement scheme under consideration. As an example, let us consider direct photodetection. As in Section 6.3 we have to distinguish two cases, that is the index m takes on two values $m = 0, 1$. The operation

$$\Phi_0(\tilde{\sigma}) = \text{tr}_B\left\{\pi_0 U_I(t, t_0)\left(\tilde{\sigma}\otimes|0\rangle\langle 0|\right)U_I^\dagger(t, t_0)\pi_0\right\} = \Omega_0\tilde{\sigma}\Omega_0^\dagger \tag{6.217}$$

corresponds to the case that no photon is detected, where $\pi_0 = |0\rangle\langle 0|$ is the projection onto the vacuum state which serves as the probe state. The other case is that a photon is detected (note that in second-order perturbation theory there are no contributions from sectors with a higher number of photons). The corresponding operation is

$$\Phi_1(\tilde{\sigma}) = \mathrm{tr}_B \left\{ \pi_1 U_I(t, t_0) \left(\tilde{\sigma} \otimes |0\rangle\langle 0| \right) U_I^\dagger(t, t_0)\pi_1 \right\} = \sum_j \Omega_j \tilde{\sigma}\Omega_j^\dagger, \qquad (6.218)$$

where $\pi_1 = \sum_j |j\rangle\langle j|$ denotes the projector onto the one-photon subspace. In the same manner as before (see eqns (6.106), (6.116) and (6.122)) we find that in the Born–Markov approximation[13]

$$\Omega_0 \approx I - i\tau H_L - \frac{1}{2}\tau\gamma_0 A^\dagger A, \qquad (6.219)$$

$$\Omega_j \approx \tilde{f}_j A, \quad \text{where} \quad \sum_j |\tilde{f}_j|^2 \approx \tau\gamma_0. \qquad (6.220)$$

The second equation shows that the operation for the measurement of a photon can be written

$$\Phi_1(\tilde{\sigma}) \approx \sum_j |\tilde{f}_j|^2 A\tilde{\sigma}A^\dagger \approx \tau\gamma_0 A\tilde{\sigma}A^\dagger. \qquad (6.221)$$

Hence we find that the propagator (6.216) is again a sum of two terms, $T = T_0 + T_1$, expressing disjoint classical alternatives: T_0 corresponds to the event of no photon detection, and T_1 to the event that a photon is detected. These contributions are given by

$$T_0[\sigma, t_0 + \tau | \tilde{\sigma}, t_0] = \mathrm{tr}\left\{ \Omega_0^\dagger \Omega_0 \tilde{\sigma} \right\} \delta \left[\frac{\Omega_0 \tilde{\sigma}\Omega_0^\dagger}{\mathrm{tr}\left\{\Omega_0^\dagger \Omega_0 \tilde{\sigma}\right\}} - \sigma \right]$$

$$\approx \left(1 - \tau\gamma_0 \mathrm{tr}\left\{ A^\dagger A\tilde{\sigma} \right\} \right) \delta \left[\tilde{\sigma} - i\tau\mathcal{G}(\tilde{\sigma}) - \sigma \right], \qquad (6.222)$$

and

$$T_1[\sigma, t_0 + \tau | \tilde{\sigma}, t_0] = \mathrm{tr}\left\{ \Phi_1(\tilde{\sigma}) \right\} \delta \left[\frac{\Phi_1(\tilde{\sigma})}{\mathrm{tr}\left\{\Phi_1(\tilde{\sigma})\right\}} - \sigma \right]$$

$$\approx \tau\gamma_0 \mathrm{tr}\left\{ A^\dagger A\tilde{\sigma} \right\} \delta \left[\frac{A\tilde{\sigma}A^\dagger}{\mathrm{tr}\left\{A^\dagger A\tilde{\sigma}\right\}} - \sigma \right]. \qquad (6.223)$$

Here, we have introduced the non-linear super-operator

$$\mathcal{G}(\sigma) \equiv \hat{H}\sigma - \sigma\hat{H}^\dagger + i\gamma_0 \mathrm{tr}\left\{ A^\dagger A\sigma \right\} \sigma, \qquad (6.224)$$

where (see eqn (6.126))

$$\hat{H} = H_L - i\frac{\gamma_0}{2} A^\dagger A. \qquad (6.225)$$

It is clear that the above form for the propagator yields a PDP for the stochastic density matrix $\sigma(t)$, where $\mathcal{G}(\sigma)$ generates the deterministic parts

---

[13]To avoid confusion with the random density matrix σ we write here A for σ_- and $A^\dagger$ for σ_+.

of the evolution. The equivalent stochastic differential equation for the PDP is given by

$$d\sigma(t) = -i\mathcal{G}(\sigma(t))dt + \left[\frac{A\sigma(t)A^\dagger}{\mathrm{tr}\left\{A^\dagger A\sigma(t)\right\}} - \sigma(t)\right]dN(t), \qquad (6.226)$$

where the Poisson increments have the properties,

$$dN(t)^2 = dN(t), \quad \mathrm{E}\left[dN(t)\right] = \gamma_0 \mathrm{tr}\left\{A^\dagger A\sigma(t)\right\}dt. \qquad (6.227)$$

The stochastic process $\sigma(t)$ is easily seen to have the following property. If we suppose that we have a pure state $\sigma(t) = |\psi(t)\rangle\langle\psi(t)|$ at some time t, then purity is preserved under the stochastic time evolution. For such a case, the PDP thus describes a stochastic state vector evolution which is precisely the one derived in Section 6.3, namely the one given by eqn (6.129). This property of the stochastic density matrix equation is due to the fact that we have just one Lindblad operator and, thus, the operation Φ_1 corresponding to the detection of a photon transforms pure states into pure states, as can be seen directly from eqn (6.221).

The above conservation of purity must be carefully distinguished from the irreversible nature of the process on the level of the corresponding ensemble of type $\mathcal{E}_\rho$. Namely, the mean density matrix

$$\rho_S(t) \equiv \int D\sigma P[\sigma, t]\,\sigma \qquad (6.228)$$

obeys the irreversible dynamics given by the vacuum optical master equation (6.91), as is easily verified by taking the average over eqn (6.226).

6.7   Photodetection on a field mode

6.7.1   *The photocounting formula*

As our last example, we consider in this section the quantum measurement of the photons in a mode a of the electromagnetic field by a photon detector. It will be assumed that the operation describing the back-action on the field mode is proportional to the annihilation operator of the mode, that is we have a single jump operator given by $A = a$. If we denote the damping constant of the mode by γ_0 we thus have a PDP which is defined by the non-Hermitian generator

$$\hat{H} = -\frac{i\gamma_0}{2}a^\dagger a, \qquad (6.229)$$

with the corresponding evolution operator

$$\hat{U}(t) = \exp(-i\hat{H}t) = \exp\left[-\frac{\gamma_0 t}{2}a^\dagger a\right], \qquad (6.230)$$

and by the transition functional

$$W[\psi|\tilde{\psi}] = \gamma_0 ||a\tilde{\psi}||^2 \delta \left[\frac{a\tilde{\psi}}{||a\tilde{\psi}||} - \psi \right].$$ (6.231)

For this process the path integral representation developed in Section 6.1.2 can immediately be written down. First, we note that

$$\hat{U}(t)a = \exp(\gamma_0 t/2)a\hat{U}(t).$$ (6.232)

This relation can be used in eqn (6.44) to bring all jump operators to the left which gives

$$\psi_N^t(t_1, \ldots, t_N) = \exp\left[N\gamma_0 t/2 - \gamma_0(t_1 + \ldots + t_N)/2 \right] a^N \hat{U}(t)\tilde{\psi}.$$ (6.233)

The full propagator $T[\psi, t|\tilde{\psi}, 0]$ can therefore be written as follows,

$$T[\psi, t|\tilde{\psi}, 0] = \sum_{N=0}^{\infty} p_N(t)\delta \left[\psi - \frac{a^N \hat{U}(t)\tilde{\psi}}{||a^N \hat{U}(t)\tilde{\psi}||} \right].$$ (6.234)

The probability $p_N(t)$ for N counts in $[0, t]$ is found from eqn (6.46) with the help of the multi-time exclusive probabilities

$$\begin{aligned} p_N^t(t_1, \ldots, t_N) &= \gamma_0^N ||\psi_N^t(t_1, \ldots, t_N)||^2 \\ &= \gamma_0^N \exp\left[N\gamma_0 t - \gamma_0(t_1 + \ldots + t_N) \right] ||a^N \hat{U}(t)\tilde{\psi}||^2. \end{aligned}$$ (6.235)

Carrying out the time integrations one arrives at

$$p_N(t) = \frac{[\mu(t)]^N}{N!} \exp(N\gamma_0 t)||a^N \hat{U}(t)\tilde{\psi}||^2.$$ (6.236)

Here we have introduced the quantum efficiency

$$\mu(t) \equiv 1 - \exp(-\gamma_0 t).$$ (6.237)

This quantity is the probability of counting a photon in the time interval $[0, t]$ from a one-photon state $\tilde{\psi} = a^\dagger |0\rangle$. In fact, for a one-photon state we have obviously $p_1(t) = \mu(t)$, $p_0(t) = 1 - \mu(t)$ and $p_N(t) = 0$ for $N \geq 2$.

The above representation of the process has been obtained for an initial pure state $\tilde{\psi}$. If the field mode is initially in a mixed state given by $\tilde{\rho}$ we have to average (6.236) over a corresponding initial distribution $P[\tilde{\psi}, t = 0]$ which leads to

$$p_N(t) = \frac{[\mu(t)]^N}{N!} \exp(N\gamma_0 t)\mathrm{tr}\left\{ \hat{U}^\dagger(t)(a^\dagger)^N a^N \hat{U}(t)\tilde{\rho} \right\}.$$ (6.238)

As is easily demonstrated the above result is equivalent to the photon counting formula derived by Mollow (1968), Scully and Lamb (1969), and Selloni et al. (1978), which is usually written in the form

$$p_N(t) = \mathrm{tr}\left\{ : \frac{[\mu(t)a^\dagger a]^N}{N!} \exp\left[-\mu(t)a^\dagger a \right] : \tilde{\rho} \right\},$$ (6.239)

where $: f(a^\dagger, a) :$ denotes the normal ordered expression of some function f of the annihilation and creation operators. In the remainder of this subsection we are going to analyse the above expansion of the propagator (6.234) for different initial state vectors $\tilde{\psi}$.

6.7.1.1 *Fock state* Let us first take an initial Fock state, i.e. $\tilde{\psi} = |n\rangle$. Then we have the trivial relations

$$\frac{a^N \hat{U}(t)|n\rangle}{||a^N \hat{U}(t)|n\rangle||} = |n - N\rangle, \tag{6.240}$$

and

$$||a^N \hat{U}(t)|n\rangle||^2 = \exp(-n\gamma_0 t)\frac{n!}{(n - N)!}, \tag{6.241}$$

which yield

$$T[\psi, t|\tilde{\psi}, 0] = \sum_{N=0}^{\infty} p_N(t)\delta[\psi - |n - N\rangle]. \tag{6.242}$$

By virtue of (6.236) the counting probability is found to represent a binomial distribution,

$$p_N(t) = \binom{n}{N}[\mu(t)]^N[1 - \mu(t)]^{n-N}. \tag{6.243}$$

If the initial state is a mixed state $\tilde{\rho}$ we have

$$p_N(t) = \sum_{n=N}^{\infty} \langle n|\tilde{\rho}|n\rangle\binom{n}{N}[\mu(t)]^N[1 - \mu(t)]^{n-N}, \tag{6.244}$$

which corresponds to the expression given by Srinivas and Davies (1981).

6.7.1.2 *Coherent state* For an initial coherent state, $\tilde{\psi} = |\alpha\rangle$, we have (omitting an irrelevant phase factor)

$$\frac{a^N \hat{U}(t)|\alpha\rangle}{||a^N \hat{U}(t)|\alpha\rangle||} = |\alpha \exp(-\gamma_0 t/2)\rangle, \tag{6.245}$$

and

$$||a^N \hat{U}(t)|\alpha\rangle||^2 = |\alpha|^{2N} \exp(-N\gamma_0 t) \exp\left[-|\alpha|^2\mu(t)\right]. \tag{6.246}$$

This yields the propagator

$$T[\psi, t|\tilde{\psi}, 0] = \sum_{N=0}^{\infty} p_N(t)\delta\left[\psi - |\alpha \exp(-\gamma_0 t/2)\rangle\right] = \delta\left[\psi - |\alpha \exp(-\gamma_0 t/2)\rangle\right], \tag{6.247}$$

showing that a pure coherent state remains a pure coherent state under time evolution which, as a consequence of damping, shrinks continuously to the vacuum state. The counting probability now represents a Poisson distribution,

$$p_N(t) = \frac{\langle N(t)\rangle^N}{N!} \exp\left[-\langle N(t)\rangle\right], \tag{6.248}$$

with mean value $\langle N(t)\rangle = |\alpha|^2\mu(t)$.

6.7.1.3 *Schrödinger cat initial states* We now consider a Schrödinger cat type initial state (Garraway and Knight, 1994; Goetsch, Graham and Haake, 1995) given by a symmetric superposition of two coherent states,

$$\tilde{\psi} = |\alpha, +\rangle \equiv \frac{1}{N_+(\alpha)} \left(|\alpha\rangle + | - \alpha\rangle \right). \tag{6.249}$$

The corresponding antisymmetric superposition is written as

$$|\alpha, -\rangle \equiv \frac{1}{N_-(\alpha)} \left(|\alpha\rangle - | - \alpha\rangle \right). \tag{6.250}$$

The normalization constants are

$$N_+(\alpha) = 2e^{-|\alpha|^2/2} \sqrt{\cosh |\alpha|^2}, \tag{6.251}$$

$$N_-(\alpha) = 2e^{-|\alpha|^2/2} \sqrt{\sinh |\alpha|^2}. \tag{6.252}$$

With the help of

$$a^N \hat{U}(t)|\alpha, +\rangle = \frac{1}{N_+(\alpha)} \alpha^N \exp(-N\gamma_0 t/2) \exp(-|\alpha|^2 \mu(t)/2$$
$$\times \left\{ |\alpha \exp(-\gamma_0 t/2)\rangle + (-1)^N | - \alpha \exp(-\gamma_0 t/2)\rangle \right\} \tag{6.253}$$

it is easy to show that for N even we have

$$||a^N \hat{U}(t)|\alpha, +\rangle||^2 = |\alpha|^{2N} \exp(-N\gamma_0 t) \frac{\cosh \left(|\alpha|^2 e^{-\gamma_0 t} \right)}{\cosh |\alpha|^2}, \tag{6.254}$$

and

$$\frac{a^N \hat{U}(t)|\alpha, +\rangle}{||a^N \hat{U}(t)|\alpha, +\rangle||} = |\alpha \exp(-\gamma_0 t/2), +\rangle, \tag{6.255}$$

while for N odd we have

$$||a^N \hat{U}(t)|\alpha, +\rangle||^2 = |\alpha|^{2N} \exp(-N\gamma_0 t) \frac{\sinh \left(|\alpha|^2 e^{-\gamma_0 t} \right)}{\cosh |\alpha|^2}, \tag{6.256}$$

and

$$\frac{a^N \hat{U}(t)|\alpha, +\rangle}{||a^N \hat{U}(t)|\alpha, +\rangle||} = |\alpha \exp(-\gamma_0 t/2), -\rangle. \tag{6.257}$$

This yields the counting formulae

$$p_N(t) = \frac{[|\alpha|^2 \mu(t)]^N}{N!} \frac{\cosh \left(|\alpha|^2 e^{-\gamma_0 t} \right)}{\cosh |\alpha|^2} \tag{6.258}$$

for even N, and

$$p_N(t) = \frac{[|\alpha|^2 \mu(t)]^N}{N!} \frac{\sinh \left(|\alpha|^2 e^{-\gamma_0 t} \right)}{\cosh |\alpha|^2} \tag{6.259}$$

for odd N. Summing separately over the even and the odd values of N we immediately get the following expression for the propagator,

$$T[\psi, t | \tilde{\psi}, 0] = p^{\text{even}}(t) \delta \left[\psi - |\alpha \exp(-\gamma_0 t/2), +\rangle \right]$$
$$+ p^{\text{odd}}(t) \delta \left[\psi - |\alpha \exp(-\gamma_0 t/2), -\rangle \right], \qquad (6.260)$$

where

$$p^{\text{even}}(t) = \sum_{N \text{even}} p_N(t) = \frac{\cosh \left(|\alpha|^2 \mu(t) \right) \cosh \left(|\alpha|^2 [1 - \mu(t)] \right)}{\cosh |\alpha|^2}, \qquad (6.261)$$

and

$$p^{\text{odd}}(t) = \sum_{N \text{odd}} p_N(t) = \frac{\sinh \left(|\alpha|^2 \mu(t) \right) \sinh \left(|\alpha|^2 [1 - \mu(t)] \right)}{\cosh |\alpha|^2}. \qquad (6.262)$$

It is clear from eqn (6.260) that the initial Schrödinger cat remains a Schrödinger cat. The effect of the quantum jumps is to switch from the even superposition to the odd superposition. As a consequence of damping both cats move towards the vacuum state, while the coherence of the superposition is completely maintained. This requires, however, that the photon detection provides a complete record. As we saw in Section 4.4.1 the coherence is effectively destroyed already by a single photon that escapes undetected.

6.7.2 QND measurement of a field mode

In Section 4.7 we have analysed a specific model for the quantum measurement process developed by Walls, Collet, and Milburn (1985). This model consists of a field mode a (the to-be-measured system), and a field mode b (the meter), which are coupled by a four-wave mixing interaction. The meter mode b is coupled via amplitude coupling to a zero-temperature environment.

Assuming that the environment acts as a perfect photoelectron counter, we investigate here this model from the viewpoint of continuous monitoring of the photons in the meter mode b. The corresponding unravelling of the master equation (4.285) gives rise to a PDP for the state vector ψ_{SM} of system-plus-meter. We determine the photon statistics for this measurement scheme and demonstrate that the quantum demolition measurement of the photons of mode b leads to a QND measurement of a certain system observable, namely of the square $(a^\dagger a)^2$ of the photon number in mode a.

For the present model, the PDP is defined through the jump operator b, with corresponding rate γ_0, describing the demolition measurement of the quanta of the meter mode, and by the non-unitary time-evolution operator

$$\hat{U}(t) = \exp(-i\hat{H}_{SM} t) = \exp \left[-\frac{t}{2} a^\dagger a \left(b\varepsilon^* - b^\dagger \varepsilon \right) - \frac{\gamma_0 t}{2} b^\dagger b \right], \qquad (6.263)$$

corresponding to the generator

$$\hat{H}_{SM} = H_{SM} - \frac{i\gamma_0}{2}b^\dagger b, \qquad (6.264)$$

where H_{SM} is given by (4.280). As in eqn (4.281) we investigate an initial state of the form

$$\tilde{\psi}_{SM} = \sum_n c_n |n\rangle_S \otimes |0\rangle_M. \qquad (6.265)$$

Let us construct the propagator of the process. To this end we note that

$$b\hat{U}(t) = \hat{U}(t)\left[e^{-\gamma_0 t/2}b - \frac{\varepsilon}{\gamma_0}\left(e^{-\gamma_0 t/2} - 1\right)a^\dagger a\right]. \qquad (6.266)$$

This relation can be used in eqn (6.44) to bring all the $\hat{U}$'s to the left. Since mode b is initially in the ground state we are left with the expression

$$\psi_N^t(t_1,\ldots,t_N) = \left(\frac{-\varepsilon}{\gamma_0}\right)^N \left(e^{-\gamma_0 t_N/2} - 1\right)\left(e^{-\gamma_0 t_{N-1}/2} - 1\right)\ldots\left(e^{-\gamma_0 t_1/2} - 1\right)$$
$$\times \hat{U}(t)(a^\dagger a)^N \tilde{\psi}_{SM}. \qquad (6.267)$$

From this we conclude that the full propagator of the process reads

$$T[\psi_{SM}, t | \tilde{\psi}_{SM}, 0] = \sum_{N=0}^{\infty} p_N(t)\delta\left[\psi_{SM} - \frac{\hat{U}(t)(a^\dagger a)^N \tilde{\psi}_{SM}}{||\hat{U}(t)(a^\dagger a)^N \tilde{\psi}_{SM}||}\right], \qquad (6.268)$$

where the probability $p_N(t)$ for N counts is found from eqns (6.45) and (6.46). Performing the time integrations we get

$$p_N(t) = \frac{1}{N!}\lambda(t)^N ||\hat{U}(t)(a^\dagger a)^N \tilde{\psi}_{SM}||^2, \qquad (6.269)$$

with

$$\lambda(t) \equiv \frac{|\varepsilon|^2}{\gamma_0^2}\left(\gamma_0 t + 4e^{-\gamma_0 t/2} - e^{-\gamma_0 t} - 3\right). \qquad (6.270)$$

With the help of the relation

$$\hat{U}(t)(a^\dagger a)^N \tilde{\psi}_{SM} = \sum_n c_n e^{-\lambda(t)n^2/2} n^N |n\rangle_S \otimes |\beta_n(t)\rangle_M, \qquad (6.271)$$

where $\beta_n(t)$ is defined in eqn (4.290), we can determine the norm in the expression for $p_N(t)$, which finally yields

$$p_N(t) = \sum_n |c_n|^2 \frac{1}{N!}[\mu_n(t)]^N e^{-\mu_n(t)}. \qquad (6.272)$$

We observe that the distribution $p_N(t)$ for the counting events represents a weighted sum of Poisson distributions with mean values

$$\mu_n(t) \equiv \lambda(t)n^2. \qquad (6.273)$$

In the limit $\gamma_0 t \longrightarrow \infty$, $|\varepsilon|$ fixed, as well as in the limit (4.296) the quantity $\lambda(t)$ becomes arbitrarily large. The Poisson distributions in the sum (6.272) are thus strongly peaked around the average values $\mu_n(t)$, the relative widths being $1/\sqrt{\mu_n(t)} \longrightarrow 0$. Since further the Poisson distribution with mean value $\mu_n(t)$ occurs with the relative weight $|c_n|^2$ we conclude that the measurement of the number N of photon counts practically always yields one of the values in the vicinity of $N = \mu_n(t)$ with the respective probabilities $|c_n|^2$. For a given outcome N we conclude from eqn (6.273) that

$$n^2 = \frac{N}{\lambda(t)}, \tag{6.274}$$

that is we infer a measurement value for the observable

$$A = (a^\dagger a)^2. \tag{6.275}$$

This measurement of the square of the photon number in mode a through the measurement of the quanta in mode b provides a QND measurement as may be seen from eqn (6.271). The function

$$f(n) = e^{-\lambda(t)n^2/2} n^N \tag{6.276}$$

is sharply peaked around the value n given by eqn (6.274). The record N thus yields the conditional state vector

$$\frac{\hat{U}(t)(a^\dagger a)^N \tilde{\psi}_{SM}}{\|\hat{U}(t)(a^\dagger a)^N \tilde{\psi}_{SM}\|} \approx |n\rangle_S \otimes |\beta_n(t)\rangle_M \tag{6.277}$$

where n is related to the outcome N by eqn (6.274). Summarizing, in the limits given above the propagator of the process becomes

$$T[\psi_{SM}, t|\tilde{\psi}_{SM}, 0] \longrightarrow \sum_n |c_n|^2 \delta \left[\psi_{SM} - |n\rangle_S \otimes |\beta_n\rangle_M\right]. \tag{6.278}$$

Equation (6.278) expresses the transformation from quantum mechanical amplitudes to classical probabilities during the measurement process. The initial pure state $\tilde{\psi}_{SM}$ is transformed into a statistical mixture consisting of definite states $|n\rangle_S \otimes |\beta_n\rangle_M$. This mixture is an ensemble of type $\mathcal{E}_P$, where the various pointer states $|n\rangle_S \otimes |\beta_n\rangle_M$ represent classical alternatives.

The transition (6.278) may be illustrated further with the help of the variances $\text{Var}_1(A)$ and $\text{Var}_2(A)$ introduced in eqns (5.67) and (5.68). Since A commutes with H_{SM} the quantum mechanical variance

$$\text{Var}(A) = \text{Var}_1(A) + \text{Var}_2(A) \tag{6.279}$$

stays constant. Since further the initial state is pure, it follows that

$$\text{Var}_2(A(0)) = 0, \quad \text{Var}_1(A(0)) = \text{Var}(A). \tag{6.280}$$

In the long time limit $\gamma_0 t \longrightarrow \infty$, $|\varepsilon|$ fixed, $\text{Var}_1(A(t))$ decreases to zero, that is

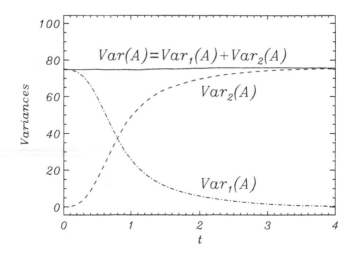

FIG. 6.12. The variances $\text{Var}_1(A)$, $\text{Var}_2(A)$, and $\text{Var}(A)$ for the observable $A = (a^\dagger a)^2$. We depict the simulation results for 10^4 realizations of the PDP describing the monitoring of the meter mode b. The simulation was performed with $\gamma_0 = 1$ and $\epsilon = 2$ and the initial state $\tilde{\psi}_{SM}$ was taken to be a superposition of the states $n = 1, 2, 3, 4, 5$ with equal amplitudes.

$$\lim_{t\to\infty} \text{Var}_1(A(t)) = 0, \qquad (6.281)$$

showing that the final ensemble is an ensemble of eigenstates of A. This may be shown directly with the help of the Liouville master equation pertaining to the process $\psi_{SM}(t)$. In fact, one finds

$$\frac{\partial}{\partial t} \text{Var}_1(A(t)) = -\gamma_0 \text{E}\left(Q[\psi_{SM}]^2\right) \leq 0, \qquad (6.282)$$

where we have introduced the functional

$$Q[\psi_{SM}] = \frac{1}{||b\psi_{SM}||}\left(\langle b\psi_{SM}|A|b\psi_{SM}\rangle - ||b\psi_{SM}||^2\langle\psi_{SM}|A|\psi_{SM}\rangle\right). \qquad (6.283)$$

Since the expectation on the right-hand side in (6.282) is taken over a non-negative functional Q^2 we see that $\text{Var}_1(A(t))$ decreases monotonically and, therefore, approaches a definite value in the limit $t \longrightarrow \infty$. On the other hand, it is easy to show that the functional $Q[\psi_{SM}]$ vanishes if and only if ψ_{SM} is one of the correlated states $|n\rangle_S \otimes |\beta_n\rangle_M$. These states are eigenstates of A and, thus, eqn (6.281) follows.

Thus we see that the states $|n\rangle_S \otimes |\beta_n\rangle_M$ are the stable states of the process $\psi_{SM}(t)$. The stochastic dynamics drives the initial state (6.265) with probability

$|c_n|^2$ into the state $|n\rangle_S \otimes |\beta_n\rangle_M$. On the other hand, the variance $\text{Var}_2(A(t))$ approaches $\text{Var}(A)$, demonstrating that the statistical fluctuations of the random variable $\langle \psi_{SM}|A|\psi_{SM}\rangle$, being zero initially, increase monotonically until they become equal to the quantum fluctuations given by $\text{Var}(A)$. This dynamical behaviour of the variances expresses the fact that the potential outcomes of the measurement contained in the amplitudes of the initial state are made objective during the process of measurement. These features are illustrated in Fig. 6.12, where we show the time dependence of the variances obtained from a numerical simulation of the PDP (Breuer and Petruccione, 1996b).

References

Barchielli, A. and Belavkin, V. P. (1991). Measurements continuous in time and *a posteriori* states in quantum mechanics. *J. Phys. A: Math. Gen.*, **24**, 1495–1514.

Breuer, H. P. and Petruccione, F. (1995a). Reduced system dynamics as a stochastic process in Hilbert space. *Phys. Rev. Lett.*, **74**, 3788–3791.

Breuer, H. P. and Petruccione, F. (1995b). Stochastic dynamics of quantum jumps. *Phys. Rev.*, **E52**, 428–441.

Breuer, H. P. and Petruccione, F. (1996a). Hilbert space path integral representation for the reduced dynamics of matter in thermal radiation fields. *J. Phys. A: Math. Gen.*, **29**, 7837–7853.

Breuer, H. P. and Petruccione, F. (1996b). Quantum measurement and the transformation from quantum to classical probabilities. *Phys. Rev.*, **A54**, 1146–1153.

Breuer, H. P. and Petruccione, F. (1997). Stochastic dynamics of reduced wave functions and continuous measurement in quantum optics. *Fortschr. Phys.*, **45**, 39–78.

Breuer, H. P., Kappler, B. and Petruccione, F. (1997). Stochastic wave-function approach to the calculation of multitime correlation functions of open quantum systems. *Phys. Rev.*, **A56**, 2334–2351.

Breuer, H. P., Kappler, B. and Petruccione, F. (1998). Heisenberg picture operators in the stochastic wave function approach to open quantum systems. *Eur. Phys. J.*, **D1**, 9–13.

Carmichael, H. (1993). *An Open Systems Approach to Quantum Optics*, Volume m18 of *Lecture Notes in Physics*. Springer-Verlag, Berlin.

Carmichael, H. J., Singh, S., Vyas, R. and Rice, P. R. (1989). Photoelectron waiting times and atomic state reduction in resonance fluorescence. *Phys. Rev.*, **A39**, 1200–1218.

Castin, Y. and Mølmer, K. (1995). Monte Carlo wave-function analysis of 3D optical molasses. *Phys. Rev. Lett.*, **74**, 3772–3775.

Dalibard, J., Castin, Y. and Mølmer, K. (1992). Wave-function approach to dissipative processes in quantum optics. *Phys. Rev. Lett.*, **68**, 580–583.

Davies, E. B. (1969). Quantum stochastic processes. *Commun. Math. Phys.*, **15**, 277–304.

Dum, R., Parkins, A. S., Zoller, P. and Gardiner, C. W. (1992). Monte Carlo simulation of master equations in quantum optics for vacuum, thermal, and squeezed reservoirs. *Phys. Rev.*, **A46**, 4382–4396.

Dum, R., Zoller, P. and Ritsch, H. (1992). Monte Carlo simulation of the atomic master equation for spontaneous emission. *Phys. Rev.*, **A45**, 4879–4887.

Gardiner, C. W., Parkins, A. S. and Zoller, P. (1992). Wave-function quantum stochastic differential equations and quantum-jump simulation methods. *Phys. Rev.*, **A46**, 4363–4381.

Garraway, B. M. and Knight, P. L. (1994). Evolution of quantum superpositions in open environments: Quantum trajectories, jumps, and localization in phase space. *Phys. Rev.*, **A50**, 2548–2563.

Ghirardi, G. C., Rimini, A. and Weber, T. (1986). Unified dynamics for microscopic and macroscopic systems. *Phys. Rev.*, **D34**, 470–491.

Ghirardi, G. C., Pearle, P. and Rimini, A. (1990). Markov processes in Hilbert space and continuous spontaneous localization of systems of identical particles. *Phys. Rev.*, **A42**, 78–89.

Gisin, N. (1984). Quantum measurements and stochastic processes. *Phys. Rev. Lett.*, **52**, 1657–1660.

Gisin, N. (1989). Stochastic quantum dynamics and relativity. *Helv. Phys. Acta*, **62**, 363–371.

Gisin, N. and Percival, I. C. (1992). The quantum-state diffusion model applied to open systems. *J. Phys. A: Math. Gen.*, **25**, 5677–5691.

Gisin, N. and Percival, I. C. (1993). Quantum state diffusion, localization and quantum dispersion entropy. *J. Phys. A: Math. Gen.*, **26**, 2233–2243.

Goetsch, P., Graham, R. and Haake, F. (1995) Schrödinger cat states and single runs for the damped harmonic oscillator. *Phys. Rev.*, **A51**, 136–142.

Hegerfeldt, G. C. and Wilser, T. S. (1991). Ensemble or individual system, collapse or no collapse: A description of a single radiating atom. In *Classical and Quantum Systems: Foundations and Symmetries*, (eds. Doebner, H. D., Scherer, W. and Schroeck, F., Jr.), pp. 104–115, World Scientific, Singapore.

Mandel, L. and Wolf, E. (1995). *Optical Coherence and Quantum Optics*. Cambridge University Press, Cambridge.

Marte, P., Dum, R., Taïeb, R., Lett, P. D. and Zoller, P. (1993a). Quantum wave function simulation of the resonance fluorescence spectrum from one-dimensional optical molasses. *Phys. Rev. Lett.*, **71**, 1335–1338.

Marte, P., Dum, R., Taïeb, R. and Zoller, P. (1993b). Resonance fluorescence from quantized one-dimensional molasses. *Phys. Rev.*, **A47**, 1378–1390.

Mollow, B. R. (1968). Quantum theory of field attenuation. *Phys. Rev.*, **168**, 1896–1919.

Mollow, B. R. (1969). Power spectrum of light scattered by two-level systems. *Phys. Rev.*, **188**, 1969–1975.

Mollow, B. R. (1975). Pure-state analysis of resonant light scattering: Radiative damping, saturation, and multiphoton effects. *Phys. Rev.*, **A12**, 1919–1943.

Mølmer, K. and Castin, Y. (1996). Monte Carlo wavefunctions in quantum optics. *Quantum Semiclass. Opt.*, **8**, 49–72.

Mølmer, K., Castin, Y. and Dalibard, J. (1993). Monte Carlo wave function method in quantum optics. *J. Opt. Soc. Am.*, **B10**, 524–538.

Pearle, P. (1976). Reduction of the state vector by a nonlinear Schrödinger equation. *Phys. Rev.*, **D13**, 857–868.

Pearle, P. (1989). Combining stochastic dynamical state-vector reduction with spontaneous localization. *Phys. Rev.*, **A39**, 2277–2289.

Percival, I.C. (1998). *Quantum State Diffusion.* Cambridge University Press, Cambridge.

Plenio, M. B. and Knight, P. L. (1998). The quantum-jump approach to dissipative dynamics in quantum optics. *Rev. Mod. Phys.*, **70**, 101–144.

Scully, M. O. and Lamb, W. E., Jr. (1969). Quantum theory of an optical maser. III. Theory of photoelectron counting statistics. *Phys. Rev*, **179**, 368–374.

Selloni, A., Schwendimann, P., Quattropani, A. and Baltes, H. P. (1978). Open-system theory of photodetection: Dynamics of field and atomic moments. *J. Phys. A: Math. Gen.*, **11**, 1427–1438.

Srinivas, M. D. and Davies, E. B. (1981). Photon counting probabilities in quantum optics. *Opt. Acta*, **28**, 981–996.

Walls, D. F., Collet, M. J., and Milburn, G. J. (1985). Analysis of a quantum measurement. *Phys. Rev.*, **D32**, 3208–3215.

Walls, D. F. and Milburn, G. J. (1994). *Quantum Optics.* Springer-Verlag, Berlin.

Wiseman, H. M. (1994). *Quantum Trajectories and Feedback.* PhD Thesis, Department of Physics, University of Queensland.

Wiseman, H. M. and Milburn, G. J. (1993a). Quantum theory of field-quadrature measurements. *Phys. Rev.*, **A47**, 642–662.

Wiseman, H. M. and Milburn, G. J. (1993b). Interpretation of quantum jump and diffusion processes illustrated on the Bloch sphere. *Phys. Rev.*, **A47**, 1652–1666.

7

THE STOCHASTIC SIMULATION METHOD

Most stochastic processes encountered in science cannot be solved by analytical methods. One therefore has to rely on numerical methods to derive predictions from a given stochastic dynamics. Numerical algorithms can be employed to solve the equation of motion governing the probability distribution for the stochastic variable itself, the master equation or the Fokker–Planck equation, for example. Another approach is to derive a system of deterministic differential equations for the moments of the process and to solve it by numerical integration. However, these methods are often not feasible in practice due to the high degree of complexity of the underlying problem: The probability distribution may be represented by a density in a very high (or even infinite) dimensional space, while the dynamics of the moments may lead to an infinite hierarchy of coupled, non-linear equations. In both cases the numerical integration by deterministic algorithms is extremely difficult.

An alternative and very powerful tool in the numerical analysis of stochastic processes is provided by the so-called Monte Carlo or stochastic simulation methods. The basic idea of these methods is to generate by a numerical algorithm independent realizations of the underlying stochastic process and to estimate with the help of statistical means all desired expectation values from a sample of such realizations. A stochastic simulation thus amounts to performing an experiment on a computer. It yields the outcomes of single runs with their correct probabilities and provides, in addition to the mean values, estimates for the statistical errors of the quantities of interest.

In the present chapter we discuss in some detail the application of Monte Carlo methods to the stochastic processes in Hilbert space that were derived in Chapter 6. Depending on whether the stochastic process is a piecewise deterministic process or a diffusion process in Hilbert space the corresponding individual realizations consist of intervals of deterministic evolution periods interrupted by sudden jumps, or of continuous, nowhere differentiable paths. Appropriate simulation techniques for both types of processes will be described in Sections 7.1 and 7.2 and illustrated by specific examples in Section 7.3.

To determine numerically the density matrix of an open quantum system one can either integrate the density matrix equation directly or else simulate the process for the stochastic wave function and estimate the covariance matrix. In general, the density matrix equation leads to a system of linear equations involving N^2 complex variables, N denoting the effective dimension of the Hilbert space. By contrast, stochastic simulation only requires the treatment of N complex variables characterizing the state vector. For large N, that is for high-dimensional Hilbert spaces, one thus expects Monte Carlo simulations to be numerically more efficient than the integration of the density matrix equation, provided the size of the required sample of realizations does not increase too strongly with N. This point will be illustrated in Section 7.4 which deals with a detailed comparison of the numerical performance of the stochastic wave function method and of the integration of the corresponding density matrix equation.

7.1   Numerical simulation algorithms for PDPs

At several places in Chapter 6 we have already shown results obtained from numerical simulations of PDPs in Hilbert space. Here we discuss in some detail the simulation technique and special features of appropriate numerical algorithms. A general account of the use of Monte Carlo methods in statistical and condensed matter physics may be found in Binder (1995) and Landau and Binder (2000).

7.1.1   *Estimation of expectation values*

A stochastic simulation algorithm serves to generate a sample of independent realizations of the stochastic process $\psi(t)$ for the wave function. Let us denote these realizations by $\psi^r(t)$, $r = 1, 2, \ldots, R$, where R is the number of realizations in the sample. The aim of the simulation is, in general, to estimate the expectation values

$$M_t \equiv \mathrm{E}\left(F[\psi, t]\right) = \int D\psi D\psi^* P[\psi, t] F[\psi, t] \tag{7.1}$$

of real functionals $F[\psi, t]$ of the random state vector which may depend explicitly on time. An unbiased and consistent estimator for the expectation value M_t is provided by the sample average

$$\hat{M}_t = \frac{1}{R} \sum_{r=1}^{R} F[\psi^r(t), t]. \tag{7.2}$$

Here and in the following a hat is used to indicate an estimator. Of particular interest are the expectation values

$$M_t \equiv \mathrm{E}\left(\langle \psi(t)|B|\psi(t)\rangle\right) = \int D\psi D\psi^* P[\psi, t] \langle \psi|B|\psi\rangle \tag{7.3}$$

of the observables B of the open system, for which the estimator takes the form

$$\hat{M}_t = \frac{1}{R} \sum_{r=1}^{R} \langle \psi^r(t)|B|\psi^r(t)\rangle. \tag{7.4}$$

It is clear that the estimated expectation value is subjected to statistical errors. A natural measure for the statistical fluctuations of the random variable $F[\psi, t]$ is the variance

$$\sigma_t^2 = \mathrm{E}\left(F[\psi, t]^2\right) - \{\mathrm{E}\left(F[\psi, t]\right)\}^2. \tag{7.5}$$

In the case $F = \langle \psi | B | \psi \rangle$ this is just the variance

$$\sigma_t^2 = \mathrm{Var}_2(B), \tag{7.6}$$

which was already introduced in eqns (5.14) and (5.68). An appropriate estimator for the statistical errors in the determination of M_t from a finite sample of size R is given by

$$\hat{\sigma}_t^2 = \frac{1}{R(R-1)} \sum_{r=1}^{R} \left(F[\psi^r(t), t] - \hat{M}_t\right)^2. \tag{7.7}$$

The quantity $\hat{\sigma}_t$ is known as the standard error of the mean value M_t. For the estimation of the mean value of an observable B it is given by

$$\hat{\sigma}_t^2 = \frac{1}{R(R-1)} \sum_{r=1}^{R} \left(\langle \psi^r(t) | B | \psi^r(t) \rangle - \hat{M}_t\right)^2. \tag{7.8}$$

If the realizations in the sample are statistically independent the standard error $\hat{\sigma}_t$ decreases with the square root of the sample size R,

$$\hat{\sigma}_t \sim \frac{1}{\sqrt{R}}. \tag{7.9}$$

This provides a relation between the standard error of the mean value and the number of realizations of the sample.

7.1.2 Generation of realizations of the process

Let us consider a PDP in Hilbert space of the type studied in Section 6.1.1 which can be represented through the stochastic differential equation (see eqn (6.25))

$$d\psi(t) = -i\left(\hat{H} + \frac{i}{2}\sum_i \gamma_i ||A_i \psi(t)||^2\right)\psi(t)dt + \sum_i \left(\frac{A_i \psi(t)}{||A_i \psi(t)||} - \psi(t)\right) dN_i(t). \tag{7.10}$$

The Poisson increments $dN_i(t)$ satisfy eqns (6.26) and (6.27), while the non-Hermitian Hamiltonian $\hat{H}$ is given by (see eqn (6.13))

$$\hat{H} = H - \frac{i}{2}\sum_i \gamma_i A_i^\dagger A_i. \tag{7.11}$$

We also recall from Section 6.1.1 that the cumulative waiting time distribution (6.21) for the quantum jumps is given by

$$F[\tilde{\psi}, \tau] = 1 - ||\exp(-i\hat{H}\tau)\tilde{\psi}||^2, \tag{7.12}$$

where τ represents the random waiting time between successive jumps.

From the discussion of Sections 1.5.2 and 6.1.1 we immediately see that a sample of realizations $\psi^r(t)$ of the process defined by eqn (7.10) in the time interval $[0, t_f]$ can be generated by means of the following algorithm:

1. Assume that the normalized state $\psi^r(t)$ was reached through a jump at time t and set $\psi^r(t) = \tilde{\psi}$. If t is the initial time $t = 0$, $\tilde{\psi}$ is the initial state of the process which must be drawn from the initial distribution $P[\tilde{\psi}, t = 0]$.

2. Determine a random waiting time τ according to the distribution function (7.12). This can be done, for example, by drawing a random number η which is uniformly distributed over the interval $[0, 1]$ and by determining τ from the equation

$$\eta = 1 - F[\tilde{\psi}, \tau] = \| \exp(-i\hat{H}\tau)\tilde{\psi} \|^2. \tag{7.13}$$

For $\eta > q$ there exists a unique solution, q being the defect of the waiting time distribution defined in eqn (6.22). For $\eta \leq q$ we set $\tau = \infty$ in which case there will be no further jumps. Within the time interval $[t, t + \tau]$ the realization follows the deterministic time evolution given in (6.15),

$$\psi^r(t + s) = \frac{\exp(-i\hat{H}s)\tilde{\psi}}{\| \exp(-i\hat{H}s)\tilde{\psi} \|}, \quad 0 \leq s \leq \tau. \tag{7.14}$$

3. At time $t+\tau$ (if τ is finite and $t+\tau < t_f$) one of the possible jumps labelled by the index i in eqn (7.10) occurs. Select a specific jump of type i with probability

$$p_i = \frac{\gamma_i\|A_i\psi^r(t + \tau)\|^2}{\sum_i \gamma_i\|A_i\psi^r(t + \tau)\|^2} \tag{7.15}$$

and replace

$$\psi^r(t + \tau) \longrightarrow \frac{A_i\psi^r(t + \tau)}{\|A_i\psi^r(t + \tau)\|}. \tag{7.16}$$

4. Repeat steps 1 to 3 until the desired final time t_f is reached, which yields the realization $\psi^r(t)$ over the whole time interval $[0, t_f]$.

5. Once a sample of realizations $\psi^r(t)$, $r = 1, 2, \ldots, R$, has been generated according to this algorithm any statistical quantity can be estimated through an appropriate ensemble average, as described in the previous subsection.

Let us now discuss the various parts of the algorithm in more detail.

7.1.3 *Determination of the waiting time*

An essential part of the simulation algorithm is the determination of the random waiting time.

7.1.3.1 *Exponential waiting time distributions* The easiest case is provided by an exponentially distributed waiting time,

$$F[\tilde{\psi}, \tau] = 1 - \exp\left(-\Gamma\tau\right). \tag{7.17}$$

Such a waiting time distribution always arises when the total jump rate does not depend on time, i.e. $\Gamma[\tilde{\psi}, \tau] = \Gamma = $ constant > 0. If η is an uniformly distributed random number in the interval $[0, 1]$, the waiting time τ is easily determined with the help of (7.13) which yields

$$\tau = -\frac{1}{\Gamma} \ln \eta. \tag{7.18}$$

The development of appropriate random number generators has been discussed, for example, by Press, Flannery, Teukolsky and Vetterling (1992) and by Knuth (1981).

7.1.3.2 *Multi-exponential waiting time distributions* A waiting time distribution which is a sum of exponential functions arises if the jump operators A_i are eigenoperators of the Hamiltonian H, which leads to (see eqn (3.125))

$$\left[H, \sum_i \gamma_i A_i^\dagger A_i \right] = 0. \tag{7.19}$$

This was the case, for example, in the weak-coupling master equation (3.140) without external driving fields. It follows that H and the positive operator $\sum_i \gamma_i A_i^\dagger A_i$ have a common eigenbasis $\{|\alpha\rangle\}$, i.e.

$$H|\alpha\rangle = E_\alpha |\alpha\rangle \tag{7.20}$$

and

$$\left(\sum_i \gamma_i A_i^\dagger A_i \right) |\alpha\rangle = \Gamma_\alpha |\alpha\rangle, \tag{7.21}$$

with real eigenvalues E_α and $\Gamma_\alpha \geq 0$. Hence, also $\hat{H}$ is diagonal in this basis and we may write

$$\hat{H}|\alpha\rangle = \left(E_\alpha - \frac{i}{2}\Gamma_\alpha \right) |\alpha\rangle. \tag{7.22}$$

The waiting time distribution is therefore a sum of exponential functions whose exponents are given by the imaginary parts of the eigenvalues of $\hat{H}$. Accordingly, the waiting time τ is determined by the equation

$$\eta = || \exp(-i\hat{H}\tau)\tilde{\psi} ||^2 = \sum_\alpha |\langle \alpha | \tilde{\psi} \rangle|^2 \exp(-\Gamma_\alpha \tau). \tag{7.23}$$

If $\tilde{\psi}$ is proportional to one of the $|\alpha\rangle$ we recover the exponential waiting time distribution (7.17). In general, eqn (7.23) can no longer be solved analytically for τ. However, it may be solved numerically provided the basis $|\alpha\rangle$ and the corresponding eigenvalues Γ_α are known. Equation (7.23) also shows that we will have a defect if $\Gamma_\alpha = 0$ for some α with $\langle \alpha | \tilde{\psi} \rangle \neq 0$.

7.1.3.3 General waiting time distributions

Step 2 of the simulation algorithm described in Section 7.1.2 rests on the assumption that the waiting time distribution $F[\tilde{\psi}, t]$ is known explicitly such that one can solve eqn (7.13) for τ, either analytically or numerically. If $F[\tilde{\psi}, \tau]$ is known the algorithm described is usually the most efficient one, in particular, if the random time steps τ become large and if an analytical expression for $\exp(-i\hat{H}\tau)$ is also known.

However, in the general case neither the deterministic evolution nor the waiting time distribution are known explicitly. The waiting time τ must then be determined along with the numerical solution of the deterministic dynamics. To achieve this one starts from the normalized initial state $\tilde{\psi}$ and solves numerically the equation

$$\frac{d}{dt}\psi(t) = -i\hat{H}\psi(t), \qquad (7.24)$$

which yields the non-normalized state vector $\psi(t) = \exp(-i\hat{H}t)\tilde{\psi}$ the norm of which decreases monotonically. According to eqn (7.13) the waiting time τ can be determined by checking after each integration step whether the square of the norm has decreased to the value η. Thus, step 2 of the algorithm given in Section 7.1.2 may be replaced by the following procedure:

2. Draw a random number η uniformly distributed in $[0, 1]$ and determine the non-normalized solution $\psi(t)$ of the linear Schrödinger equation (7.24) corresponding to the initial value $\psi(0) = \tilde{\psi}$. The waiting time τ is then determined by the condition

$$||\psi(\tau)||^2 = \eta. \qquad (7.25)$$

The deterministic evolution between the jumps takes the form

$$\psi^r(t+s) = \frac{\psi(s)}{||\psi(s)||}, \quad 0 \le s \le \tau. \qquad (7.26)$$

This form of the algorithm should be preferred when performing simulations in the general case. For the numerical solution of eqn (7.24) between the jumps it is often convenient to expand the wave function with respect to an eigenbasis of the Hamiltonian H and to bring (7.24) into a linear system of ordinary differential equations for the components of ψ. The integration can then be performed by employing standard numerical routines appropriate for deterministic differential equations. We remark finally that the above form of the algorithm, in particular condition (7.25) and eqn (7.26), also applies, of course, if the generator in eqn (7.24) depends on time, that is if $\hat{H} = \hat{H}(t)$.

7.1.4 Selection of the jumps

According to step 3 of the algorithm described in section 7.1.2 we have to select a specific quantum jump of type i with probability p_i given by eqn (7.15) and to carry out the corresponding transition (7.16). The index i specifying the jump is a random number with values in some index set $I = \{1, 2, \ldots, K\}$ following the distribution p_i.

Essentially, there are two ways to implement numerically the selection of the jumps. If the index set I is small, then i may be drawn with the help of the method of linear search (Gillespie, 1976, 1992). To this end, one subdivides the interval $[0, 1]$ into K sub-intervals J_i of lengths p_i and draws a uniform random number $x \in [0, 1]$. The index i is then found by searching for the interval J_i which contains x. When applying this algorithm, it is important to check that the probabilities p_i are larger than the resolution of the random number generator (Hanusse and Blanché, 1981).

If the values of the probabilities p_i are approximately equal to each other it may be more efficient to use the following rejection method (Press, Flannery, Teukolsky and Vetterling, 1992): Draw two independent uniform random numbers x and y in $[0, 1]$. The first random number is used to select an index $i = \text{int}(Kx) + 1$ from the index set I. If the second random number satisfies $y < p_i/p_{\max}$ this index i is accepted, where $p_{\max}$ denotes the maximum of the p_i. Otherwise i is rejected and the procedure is repeated.

The advantage of the rejection method over the linear search is that the former is numerically efficient even if the number K of elements in the index set becomes large. A condition is that the product of $p_{\max}$ and K should not be much larger then 1.

The linear search and the rejection method get inefficient if the index set I is very large and if the distribution p_i is very inhomogeneous. In these cases it might be helpful to cluster possible jumps into logarithmic classes (Fricke and Schnakenberg, 1991) or to make use of search trees (Maksym, 1988; Blue, Beichl and Sullivan, 1995). However, the time which is saved due to improved search algorithms might be lost again in the actualization of the data structures (Breuer, Huber and Petruccione, 1996).

7.2 Algorithms for stochastic Schrödinger equations

In this section we develop numerical algorithms for the simulation of stochastic Schrödinger equations (SSEs) of diffusion type. Such equations were encountered in Section 6.1.3 as diffusion approximations of a corresponding PDP in Hilbert space. They also arise in the context of fundamental dynamical collapse models (see Section 6.5.2).

Below, we are going to discuss four different algorithms to solve SSEs, namely the well-known Euler scheme, an adaptation of the Heun and the Runge–Kutta scheme to SSEs and a second-order weak scheme for stochastic differential equations proposed by Platen (Kloeden and Platen, 1992). Special emphasis is laid on the order of convergence of these schemes using as an example a stochastic Schrödinger equation of the type encountered in Section 6.4.2 (Breuer, Dorner and Petruccione, 2000). The methods discussed here can of course be applied to other kinds of SSEs. In particular, they may easily be generalized to simulate SSEs involving complex instead of real Wiener increments.

7.2.1   General remarks on convergence

Let us consider a stochastic Schrödinger equation of the following general Itô form,

$$d\psi(t) = D_1\left(\psi(t)\right)dt + D_2\left(\psi(t)\right)dW(t), \qquad (7.27)$$

with the drift term

$$D_1\left(\psi\right) = -iH\psi + \frac{\gamma}{2}\left(\langle A + A^\dagger\rangle_\psi A - A^\dagger A - \frac{1}{4}\langle A + A^\dagger\rangle_\psi^2\right)\psi, \qquad (7.28)$$

and the diffusion term

$$D_2\left(\psi\right) = \sqrt{\gamma}\left(A - \frac{1}{2}\langle A + A^\dagger\rangle_\psi\right)\psi, \qquad (7.29)$$

where we have made use of the shorthand notation

$$\langle A\rangle_\psi \equiv \langle\psi|A|\psi\rangle, \qquad (7.30)$$

and $dW(t)$ is a real Wiener increment. An equation of the form (7.27) was obtained in Section 6.4.2 by performing the diffusion limit of homodyne photodetection (see eqn (6.181)). For simplicity we consider the case of a single Lindblad operator A. The corresponding density matrix equation is given by

$$\frac{d}{dt}\rho(t) = -i[H, \rho(t)] + \gamma\left(A\rho(t)A^\dagger - \frac{1}{2}A^\dagger A\rho - \frac{1}{2}\rho A^\dagger A\right) \equiv \mathcal{L}\rho(t). \qquad (7.31)$$

We note that eqn (7.27) is a stochastic differential equation with multiplicative noise which means that the diffusion term $D_2(\psi(t))$ which multiplies the Wiener increment depends on the stochastic variable $\psi(t)$. The noise is called additive if the diffusion term does not depend on the stochastic variable.

It must be emphasized that $\psi(t)$ is a process in the Hilbert space $\mathcal{H}_S$ of an open quantum system. It thus represents, in general, a process in an infinite-dimensional vector space. However, in many cases of interest the dissipative character of the equations of motion ensures that the dynamics is confined to a finite-dimensional subspace of $\mathcal{H}_S$. It is therefore justified physically to restrict the discussion to finite-dimensional spaces, that is, we may regard eqn (7.27) as defining a process in an effectively finite-dimensional space.

Consider now a numerical scheme to integrate eqn (7.27) over the time interval $[t_0, t_0 + T]$. Such a scheme leads to a discrete time approximation ψ_k for the exact process $\psi(t_k)$ at times $t_k \equiv t_0 + k\Delta t$, where $k = 0, \dots, n = T/\Delta t$. The discretization does not have to be equidistant but to simplify the notation this will be supposed here. In this section, ψ_k will always denote a numerically generated approximation, while $\psi(t)$ denotes the exact process defined by eqn (7.27). Analogously, we denote by

$$\rho_k = \mathrm{E}[|\psi_k\rangle\langle\psi_k|] \qquad (7.32)$$

the discrete time approximation of the density matrix, whereas

$$\rho(t) = \mathrm{E}[|\psi(t)\rangle\langle\psi(t)|] \tag{7.33}$$

is the exact density matrix satisfying the Lindblad equation (7.31).

In order to characterize the convergence behaviour of the numerical schemes to be discussed below we will compare the discrete approximation ρ_k with the Taylor expansion of the exact density matrix $\rho(t)$ which is given by

$$\rho(t + \Delta t) = \rho(t) + \mathcal{L}\rho(t)\Delta t + \frac{1}{2}\mathcal{L}^2\rho(t)\Delta t^2 + \mathcal{O}(\Delta t^3). \tag{7.34}$$

Without loss of generality, we will always assume that a deterministic (i.e. sharp) initial state $\psi_0 \equiv \psi(t_0)$ is given. The single-step error of a certain numerical scheme may then be expressed through the difference

$$\rho_1 - \rho(t_1) = \mathcal{O}(\Delta t^{\beta'}), \tag{7.35}$$

which means that the scheme reproduces the Taylor expansion of $\rho(t)$ including terms of order $\beta' - 1$ in Δt.

It is straightforward to prove that the integration over a finite time interval $[t_0, t_0 + T]$ decreases the order of convergence by one since we need $n = T/\Delta t$ time steps to calculate the density at time $t_0 + T = t_n$, i.e.

$$\rho_n - \rho(t_n) = \mathcal{O}(\Delta t^{\beta}) \tag{7.36}$$

with $\beta = \beta' - 1$. If this equation is satisfied the numerical scheme will be said to be a scheme of order β in the following.

Equation (7.36) is a special case of *weak* convergence of order β. The degree of an approximation of a stochastic differential equation can also be characterized by the notions of *strong* convergence (Kloeden and Platen, 1992). If the numerical algorithm leads to an approximation ψ_n for $\psi(t_n)$ satisfying

$$\mathrm{E}\left[\|\psi_n - \psi(t_n)\|\right] = \mathcal{O}(\Delta t^{\beta}) \tag{7.37}$$

for sufficiently small Δt, it is said that the scheme converges strongly with order β. This definition imposes a condition on the closeness of the random variables ψ_n and $\psi(t_n)$ at the end of the integration interval. By contrast, weak convergence only requires that the probability distributions of ψ_n and $\psi(t_n)$ are close to each other which is a much weaker criterion. In practice, one is often only interested in this weaker form of convergence, for example, when considering the approximation of functionals of the stochastic variable.

7.2.2 The Euler scheme

The Euler scheme is probably the simplest recursive algorithm to get an approximation for the state vector. It takes the form

$$\psi_{k+1} = \psi_k + D_1(\psi_k)\Delta t + D_2(\psi_k)\Delta W_k. \tag{7.38}$$

The Wiener increments ΔW_k, $k = 0, 1, 2, \ldots, n-1$, are independent Gaussian random variables with zero mean and variance Δt, the index k indicating different realizations at each time step. Thus, we can write

$$\Delta W_k = \sqrt{\Delta t}\xi_k, \tag{7.39}$$

where ξ_k is a Gaussian distributed random variable with mean zero and *unit* variance, that is a random variable following a standard normal distribution. It should be mentioned that it is not necessary to use Gaussian distributed random variables if one requires convergence in the sense of eqn (7.36). One may use other random variables instead which coincide, for example, only in the first and the second moment with those of a standard normal distribution.

It is easy to demonstrate that eqn (7.36) is satisfied with $\beta = 1$. The Euler scheme thus provides a weak scheme of order 1. In spite of the low order of convergence the Euler algorithm is useful because it is very easy to implement numerically and often yields a reasonable degree of approximation. In general, the scheme does not conserve the norm of the state vector, by contrast to the SSE (7.27). The state vector should therefore be normalized after every iteration.

7.2.3   The Heun scheme

The following method, which we shall call the Heun scheme, is a generalization of the Heun method known from the numerical integration of deterministic differential equations. It is defined by the recursion relation

$$\psi_{k+1} = \psi_k + \frac{1}{2}\left\{D_1(\psi_k) + D_1(\tilde{\psi}_k)\right\}\Delta t + D_2(\psi_k)\Delta W_k, \tag{7.40}$$

with

$$\tilde{\psi}_k = \psi_k + D_1(\psi_k)\Delta t + D_2(\psi_k)\Delta W_k. \tag{7.41}$$

ΔW_k again takes the form of eqn (7.39). For vanishing diffusion one recovers the Heun method for deterministic differential equations. It turns out that, in general, the Heun algorithm again leads to relation (7.36) with $\beta = 1$. It is therefore a scheme of order 1, similar to the Euler method. This is in contrast to SSEs with additive noise, where the Heun method is actually of order two.

7.2.4   The fourth-order Runge–Kutta scheme

For deterministic differential equations there is a great variety of numerical integration methods and one is easily misled when heuristically modifying these methods in order to integrate SSEs. The following example shows that it is indeed wise to proceed cautiously.

A well-known numerical scheme for deterministic differential equations is the fourth-order Runge–Kutta method. A heuristic modification to approximate solutions of SSEs is obtained if one integrates in each time step the drift with the

Runge–Kutta method and the diffusion term of the SSE with the Euler scheme. This leads to the following scheme,

$$\psi_{k+1} = \psi_k + \frac{1}{6}\left\{\psi_k^1 + 2\psi_k^2 + 2\psi_k^3 + \psi_k^4\right\}\Delta t + D_2(\psi_k)\Delta W_k, \qquad (7.42)$$

with

$$\psi_k^1 = D_1(\psi_k), \qquad (7.43)$$

$$\psi_k^2 = D_1(\psi_k + \frac{1}{2}\Delta t\psi_k^1), \qquad (7.44)$$

$$\psi_k^3 = D_1(\psi_k + \frac{1}{2}\Delta t\psi_k^2), \qquad (7.45)$$

$$\psi_k^4 = D_1(\psi_k + \Delta t\psi_k^3). \qquad (7.46)$$

The order of convergence can again be determined by calculating the difference in eqn (7.36). The surprising result is that the scheme is of order $\beta = 1$. The stochastic generalization of the Runge–Kutta method converges with order 1, by contrast to the Runge–Kutta method used for deterministic differential equations which is of order 4.

From this example one can see that in general simple heuristic generalizations of higher-order Runge–Kutta schemes do not necessarily lead to higher-order methods for SSEs. The above method converges of course but, in general, there will be no advantage over the Euler method, for example.

7.2.5 A second-order weak scheme

The final algorithm we are going to consider has been proposed by Platen (Kloeden and Platen, 1992). In our notation this scheme is provided by the recursion relation

$$\psi_{k+1} = \psi_k + \frac{1}{2}\left(D_1(\tilde{\psi}_k) + D_1(\psi_k)\right)\Delta t$$
$$+ \frac{1}{4}\left(D_2(\psi_k^+) + D_2(\psi_k^-) + 2D_2(\psi_k)\right)\Delta W_k$$
$$+ \frac{1}{4}\left(D_2(\psi_k^+) - D_2(\psi_k^-)\right)\left\{(\Delta W_k)^2 - \Delta t\right\}\Delta t^{-1/2}, \qquad (7.47)$$

where

$$\tilde{\psi}_k = \psi_k + D_1(\psi_k)\Delta t + D_2(\psi_k)\Delta W_k, \qquad (7.48)$$

$$\psi_k^{\pm} = \psi_k + D_1(\psi_k)\Delta t \pm D_2(\psi_k)\sqrt{\Delta t}. \qquad (7.49)$$

Here, the increments ΔW_k must fulfil certain conditions. These are satisfied if eqn (7.39) is assumed. For vanishing noise the scheme reduces to the Heun scheme for deterministic differential equations.

Performing again the error analysis one finds that Platen's scheme converges with order $\beta = 2$. Thus, contrary to the algorithms studied before, Platen's scheme is really a higher-order scheme in the sense defined above. This point will be illustrated in the examples below.

7.3 Examples

In this section we apply the simulation algorithms for PDPs and stochastic
Schrödinger equations to some specific examples.

7.3.1 The damped harmonic oscillator

7.3.1.1 *Simulating the PDP* We study the following process for the damped
harmonic oscillator (see Section 6.7),

$$d\psi(t) = -\frac{\gamma}{2}\left(a^\dagger a - \langle\psi(t)|a^\dagger a|\psi(t)\rangle\right)\psi(t)dt + \left(\frac{a\psi(t)}{||a\psi(t)||} - \psi(t)\right)dN(t). \quad (7.50)$$

Let us simulate this equation for the simplest initial condition, i.e. for a number
state $\psi(0) = |n\rangle$. In this case the deterministic evolution periods of the process
are trivial since

$$\exp(-i\hat{H}t)|n\rangle = \exp(-\gamma n t/2)|n\rangle. \quad (7.51)$$

Also the jumps are trivially given by

$$|n\rangle \longrightarrow \frac{a|n\rangle}{||a|n\rangle||} = |n-1\rangle. \quad (7.52)$$

Since the waiting time distribution is

$$F[|n\rangle, \tau] = 1 - ||\exp(-i\hat{H}\tau)|n\rangle||^2 = 1 - \exp(-\gamma n \tau), \quad (7.53)$$

we find that the random waiting time may be obtained from the equation

$$\tau = -\frac{1}{\gamma n}\ln\eta, \quad (7.54)$$

where η is a uniformly distributed random number in the interval $[0, 1]$.

Figure 7.1 shows one realization of the process for the initial condition $\psi(0) = |n_0\rangle = |9\rangle$. As an example of the calculation of expectation values we also show in
Fig. 7.2 the mean of the number operator determined from a sample of $R = 100$
realizations through (see eqn (7.3))

$$\hat{M}_t = \frac{1}{R}\sum_{r=1}^{R}\langle\psi^r(t)|a^\dagger a|\psi^r(t)\rangle. \quad (7.55)$$

The analytical expression for this quantity is, of course, given by

$$\langle a^\dagger a(t)\rangle = n_0\exp(-\gamma t). \quad (7.56)$$

According to eqn (7.4) the standard error for this mean value is provided by the
expression

$$\hat{\sigma}_t^2 = \frac{1}{R(R-1)}\sum_{r=1}^{R}\left(\langle\psi^r(t)|a^\dagger a|\psi^r(t)\rangle - \hat{M}_t\right)^2. \quad (7.57)$$

The values for the standard error are indicated in Fig. 7.2 by error bars.

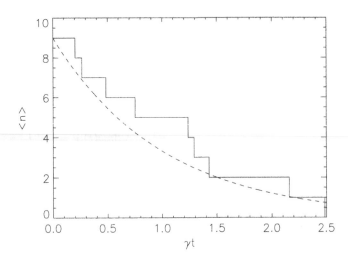

FIG. 7.1. One realization of the process (7.50) for the damped harmonic oscillator with initial condition $|n_0\rangle = |9\rangle$ (continuous line). The dashed line represents the analytical expression (7.56) for the expectation value of the number operator.

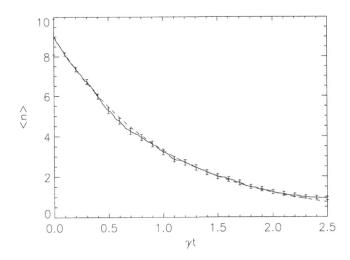

FIG. 7.2. The average over 100 realizations of the damped harmonic oscillator. The initial condition is $|n_0\rangle = |9\rangle$. The dotted line gives the analytical result.

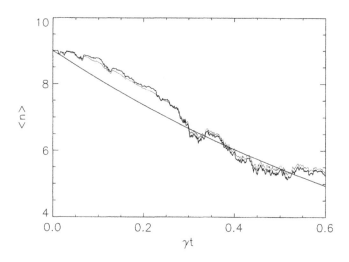

FIG. 7.3. A single realization of the mean number operator of the damped harmonic oscillator. The thin line corresponds to the second-order weak scheme and the bold line to the Euler, the Heun and the Runge–Kutta schemes (not distinguishable on the scale of the figure). The smooth line represents the analytical result (7.56).

7.3.1.2 *Simulating the SSE* The stochastic Schrödinger equation (7.27) for the damped harmonic oscillator reads

$$d\psi(t) = \frac{\gamma}{2}\left(\langle a + a^\dagger\rangle_{\psi(t)} a - a^\dagger a - \frac{1}{4}\langle a + a^\dagger\rangle^2_{\psi(t)}\right)\psi(t)dt$$

$$+\sqrt{\gamma}\left(a - \frac{1}{2}\langle a + a^\dagger\rangle_{\psi(t)}\right)\psi(t)dW(t). \qquad (7.58)$$

In the numerical simulations the state vector has been represented in the basis of number states $|n\rangle$. Again, the initial state is taken to be $\psi(0) = |n_0 = 9\rangle$ and the Hilbert space has been truncated at $n_{\max} = 12$, that is the simulation was performed in a subspace of dimension $N = 13$.

A single realization of the process is depicted in Fig. 7.3. The solid line shows the analytical curve (7.56). The thin line represents the solution calculated with the second-order weak scheme of Platen and the bold line the solutions calculated with the other methods, which are not distinguishable in this example. The size of the time steps is $\Delta t = 0.01$. Note that the same sequence of pseudo-random numbers was used for each run to stress the differences or similarities between the methods.

In order to study the convergence behaviour, we compute the mean value (7.55) of the number operator at a fixed time T for $R = 10^5$ realizations and for different step sizes Δt. The results are displayed in Fig. 7.4. Error bars are not

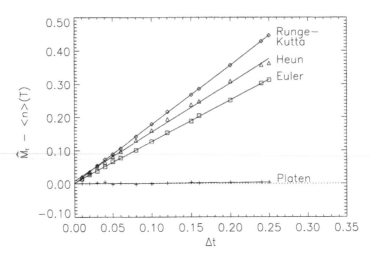

FIG. 7.4. A plot of the mean photon number $\hat{M}_T$ relative to the analytically calculated value $\langle n(T) \rangle$ at $\gamma T = 1.2$ versus the step size Δt for different methods. The solid lines correspond to linear and quadratic fits.

drawn in this figure because they would be roughly of the size of the symbols. Obviously Platen's second-order weak scheme converges very well already for quite large step sizes compared to the other schemes. The heuristic Runge–Kutta method shows the worst convergence behaviour.

Based on the knowledge of the order of convergence it is possible to fit linear functions (in the case of the Euler, Heun and Runge–Kutta methods) or quadratic functions (in the case of Platen's methods) to the data points. These fits (also shown in Fig. 7.4) are in good agreement with the data points and confirm the results derived in Section 7.2.1.

Another criterion for the assessment of numerical algorithms is, of course, the CPU time which is required to achieve a given accuracy of the results. Figure 7.5 displays the CPU time (normalized to one) versus the error for the data shown in Fig. 7.4. We see that Platen's scheme shows the best performance, followed by the Euler scheme.

7.3.2 The driven two-level system

As a second example we consider the PDP given by the equation

$$d\psi(t) = -i\left(\hat{H} + \frac{i\gamma}{2}||\sigma_-\psi(t)||^2\right)\psi(t)dt + \left(\frac{\sigma_-\psi(t)}{||\sigma_-\psi(t)||} - \psi(t)\right)dN(t). \quad (7.59)$$

As shown in Section 6.3 this equation describes the direct photodetection of a driven two-level system (see eqn (6.129)), while the stochastic Schrödinger equation

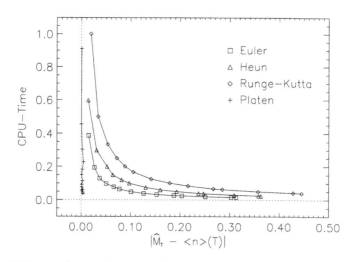

FIG. 7.5. CPU time (normalized to one) versus absolute error for the data points of Fig. 7.4.

$$
d\psi(t) = - iH_L\psi(t)dt
$$
$$
+ \frac{\gamma}{2}\left(\langle\sigma_- + \sigma_+\rangle_\psi\sigma_- - \sigma_+\sigma_- - \frac{1}{4}\langle\sigma_- + \sigma_+\rangle_\psi^2\right)\psi(t)dt
$$
$$
+ \sqrt{\gamma}\left(\sigma_- - \frac{1}{2}\langle\sigma_- + \sigma_+\rangle_\psi\right)\psi(t)dW(t) \tag{7.60}
$$

corresponds to homodyne photodetection (see Section 6.4). In the numerical simulations we start with the atom in its ground state $|g\rangle$ and compute the probability

$$
\rho_{11}(t) = \langle e|\rho(t)|e\rangle, \tag{7.61}
$$

of finding the atom in the excited state $|e\rangle$. From a sample of realizations this probability is estimated by determining the average

$$
\hat{M}_t = \frac{1}{R}\sum_{r=1}^{R}|\langle e|\psi^r(t)\rangle|^2, \tag{7.62}
$$

while the corresponding standard error $\hat{\sigma}_t$ is given by

$$
\hat{\sigma}_t^2 = \frac{1}{R(R-1)}\sum_{r=1}^{R}\left(|\langle e|\psi^r(t)\rangle|^2 - \hat{M}_t\right)^2. \tag{7.63}
$$

The discussion of the algorithm for the generation of realizations of the piecewise deterministic process may be kept brief because we have already collected all relevant formulas for the waiting time distribution and the jumps in Section

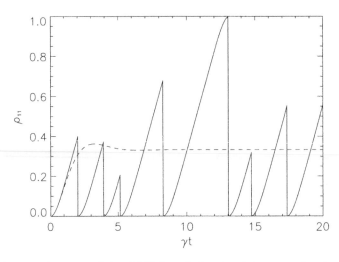

FIG. 7.6. One realization of the PDP (7.59) for the driven two-level atom. The continuous line shows ρ_{11} computed from the realization. The dashed line represents the analytical solution for ρ_{11} according to eqn (3.289). Parameters: $\Omega = \gamma = 0.4$.

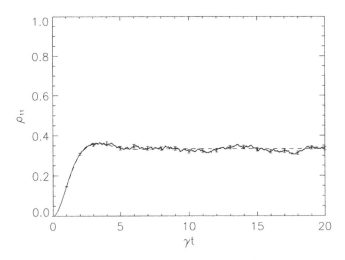

FIG. 7.7. The average over 1000 realizations of the PDP (7.59) for the driven two-level atom. The continuous line shows ρ_{11} computed from the average over the realizations. The dashed line represents the analytical solution for ρ_{11} according to eqn (3.289). The standard error of the mean is indicated by error bars. The parameters are the same as in Fig. 7.6.

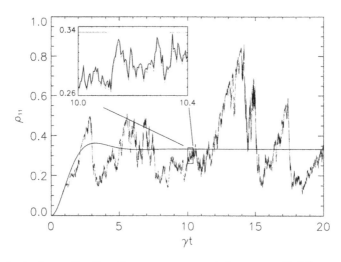

FIG. 7.8. A single realization for ρ_{11} computed from eqn (7.60) according to the different numerical schemes of Section 7.2 with step size $\Delta t = 0.01$. The results are not distinguishable on the scale of the figure. The smooth line gives the analytically solution (3.289). The inset shows a magnification of a small part (thin line: Platen's scheme; bold line: Euler, Heun and Runge–Kutta schemes). Parameters: $\Omega = \gamma = 0.4$.

6.3. According to eqns (7.13) and (6.133) the random waiting time τ can be determined by solving the equation

$$\eta = \exp(-\gamma\tau/2)\left(1 + \frac{\gamma^2}{8\mu^2}\sin^2\mu\tau + \frac{\gamma}{2\mu}\sin\mu\tau\cos\mu\tau\right) \qquad (7.64)$$

for τ. We have used a numerical routine to find the root of this equation. In Fig. 7.6 we show $|\langle e|\psi(t)\rangle|^2$ for one realization of the PDP. The corresponding average over an ensemble of 1000 relizations can be seen in Fig. 7.7 which clearly reveals the convergence of the algorithm.

Let us now turn our attention to the simulation of the stochastic Schrödinger equation (7.60). Single realizations calculated with the different numerical methods introduced in Section 7.2 are shown in Fig. 7.8. In the inset, showing an enlarged part of the image, it is seen that the weak scheme of order two differs from the other schemes which are still hardly distinguishable on the enlarged scale.

The results of simulations with different step sizes and $R = 5 \times 10^5$ realizations per point are shown in Fig. 7.9. For the sake of clarity we display only error bars at points which are not too close to other points. They have been calculated according to eqn (7.63). The error bars omitted are of about the same size. In any of the four cases the numerical results are in very good agreement with

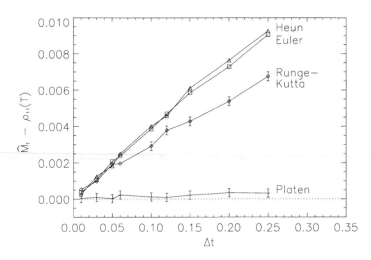

FIG. 7.9. Estimated means of $|\langle e|\psi^r(T)\rangle|^2$ minus the exact values for different step sizes and methods. The parameters are $\Omega = 0.4$, $\gamma = 0.4$ and $\gamma T = 8.4$.

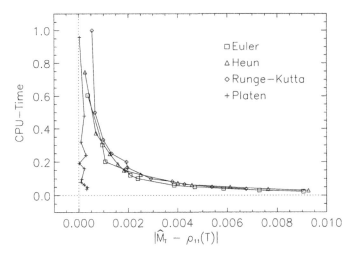

FIG. 7.10. The CPU time (normalized to one) versus the absolute error for the data points of Fig. 7.9.

the analytical predictions. The Runge–Kutta method seems to converge better than the Euler and the Heun scheme for bigger step sizes. However, if one takes into account the corresponding CPU times, as is done in Fig. 7.10, it becomes obvious that these three methods are more or less equal with respect to the ratio of accuracy and speed. Anyhow, Platen's scheme shows again the best performance. This changes if one chooses parameters such that $\gamma \ll \Omega$, i.e. if the deterministic part of the SSE prevails over the noise part. The Runge–Kutta method then shows the best performance, while Platen's scheme behaves like the Heun method. This is, of course, to be expected since for $\gamma \ll \Omega$ the deterministic part of the equation of motion dominates and, therefore, the fourth-order Runge–Kutta method becomes the most efficient one.

7.4 A case study on numerical performance

Monte Carlo algorithms are often the only way to study high-dimensional systems, for which deterministic calculations based on the numerical integration of the equations of motion are beyond the capacity of any computer. On the other hand, deterministic methods may be preferred for low-dimensional systems. In this section we want to present a systematic analysis of the numerical efficiency of the Monte Carlo wave function method and compare it with that of the integration of the corresponding density matrix equation. The main interest lies in the dependence of the time-consumption on the system size N. It will be shown that the CPU time consumption for the two approaches may be expressed in terms of simple power laws.

7.4.1 Numerical efficiency and scaling laws

The results of a Monte Carlo simulation are always subject to statistical errors. These errors in turn are related to the number of realizations that are generated in the simulation. Hence the total CPU time required by a Monte Carlo simulation depends crucially on the desired accuracy. A typical task of the numerical investigation could be, for instance, 'estimate the expectation value of the energy with a relative error of 1 %', or: 'calculate the density matrix with a precision of better than 10^{-4} in each of its elements'. The question to be answered is then, of course: Under which circumstances should a stochastic simulation be preferred?

The crucial quantity for the relative performance is therefore the number of realizations R. Hence, the dependence of the errors of the simulation with the sample size deserves a more detailed discussion. To this end let us denote by N the number of complex variables which are used for the numerical representation of the wave function, that is, the number of basis states. Accordingly, the number of complex variables which are necessary to represent the density matrix is then $N^2/2$. The relation (7.9) allows us to write down the following equation

$$\hat{\sigma}_t^2 = \frac{\lambda_B(N)}{R}, \tag{7.65}$$

in which the factor $\lambda_B(N)$ takes into account the dependence of the statistical error on the observable B and on the system size N, but does *not* depend on the sample size R. Using a sufficiently large sample of realizations, $\lambda_B(N)$ can be determined by fitting eqn (7.65) to the simulated data. Then, eqn (7.65) can be solved for R,

372 THE STOCHASTIC SIMULATION METHOD

$$R \equiv R(N) = \frac{\lambda_B(N)}{\hat{\sigma}_t^2}. \qquad (7.66)$$

This is the number of realizations that is necessary to achieve a given accuracy $\hat{\sigma}_t$ for observable B and system size N. If $\lambda_B(N)$ varies as a power of the system size, $\lambda_B(N) \sim N^{-x}$, the following classification can be made (Ferrenberg, Landau and Binder, 1991):

1. If $x = 1$, the observable B is strongly self-averaging.
2. If $0 < x < 1$, the observable B is self-averaging.
3. If $x = 0$, the observable B is not self-averaging.

With these considerations we can now investigate the scaling behaviour of the CPU time required by the deterministic and by the stochastic approach to the numerical estimation of expectation values with a prescribed accuracy.

In the case of the density matrix equation approach, the part of the numerical integration routine that dominates the CPU time consumption is the calculation of the generator $\mathcal{L}\rho$ of the Lindblad equation. One such calculation requires, for large enough N, an amount of CPU time proportional to a power of N, and the CPU time needed to integrate the density matrix over a given physical time interval is, to leading order in N,

$$T_{\text{DME}} = k_1 \, s_1(N) \, N^\beta. \qquad (7.67)$$

Here, $s_1(N)$ is the number of times the quantity $\mathcal{L}\rho$ has to be evaluated. k_1 and β depend on the type of the specific problem, but not on N. Besides, k_1 depends on the particular implementation on a computer.

Analogously, in many examples the time-critical part of the stochastic simulation is the calculation of the generator $\hat{H}\psi$, and the CPU time required for the simulation is

$$T_{\text{StS}} = k_2 \, R(N) \, s_2(N) \, N^\alpha. \qquad (7.68)$$

$R(N)$ is the number of realizations of the process that are generated to treat the system of size N, $s_2(N)$ is the number of evaluations of $\hat{H}\psi$ for one realization, and k_2 is analogous to k_1.

In many situations, $s_1(N)$ and $s_2(N)$ will be roughly equal. Provided that similar numerical integration routines are used, this is the case if the smallest time scale of the dynamics of the stochastic wave function is about equal to that of the density matrix. Since we want to separate the effects of system size from dynamical phenomena, this case is the one of interest for us, and the example presented in Section 7.4.2 illustrates that case.

Let us briefly note that there are also situations where $s_1(N)$ and $s_2(N)$ are quite different. In general their ratio might depend on N and they do not necessarily grow in the same way with the physical time over which the system is studied. Consider the case where the time scale of the dynamics of a *single* realization of the stochastic process varies, during its temporal evolution, in a wide range, as for example, in laser cooling (see Section 8.3). The simulation of one realization will then contain stretches with very long time steps, interrupted by phases of more rapid development and short time steps. The integrator of a density matrix equation, on the other hand, which describes the dynamics of the whole ensemble, must always adapt to the short time scale. Clearly, in such cases the stochastic wave function method is the preferred choice.

The number of floating point operations to calculate $\mathcal{L}\rho$ and to calculate $\hat{H}\psi$ differ by about a factor N, and one expects

$$\beta \approx \alpha + 1. \tag{7.69}$$

This relation will be verified in the example below. Concluding these general considerations, we can write eqns (7.67) and (7.68) in a more succinct form

$$T_{\text{DME}} = k_1 \, N^{\alpha+1} \tag{7.70}$$
$$T_{\text{StS}} = k_2 \, N^{\alpha-x}. \tag{7.71}$$

Here we have assumed that the numbers of steps s_1 and s_2 are roughly equal and can be absorbed into the constants α, k_1 and k_2. The performance of the stochastic wave function method versus that of the density matrix integration can be measured by the difference between the exponents, which is 1 in the non-self-averaging ($x = 0$) and 2 in the strongly self-averaging case ($x = 1$).

7.4.2 The damped driven Morse oscillator

Let us illustrate the scaling laws (7.70) and (7.71) for the CPU time consumption. To this end, we consider a non-trivial example, namely a damped Morse oscillator which is driven by a time-dependent force. It will be seen that the scaling laws are very well satisfied for this non-linear problem.

7.4.2.1 Description of the model
The Hamiltonian describing the coherent part of the dynamics is taken to be

$$H_S(t) = H_M + H_L(t), \tag{7.72}$$

where

$$H_M = \frac{1}{2m} p^2 + V(q) \tag{7.73}$$

and

$$V(q) = D[1 - \exp(-bq)]^2 \tag{7.74}$$

is the Morse potential. The Morse Hamiltonian (7.73) may be used to model, e.g. a molecular degree of freedom within a single electronic potential energy surface. With an appropriate choice of the parameters it yields a fairly realistic description of, for example, the vibrational dynamics of the local O-H bond in the water molecule. The external driving is described by the interaction term

$$H_L(t) = \mu q F_0 s(t) \sin \omega_L t. \tag{7.75}$$

Here, μq is the relevant component of the molecular dipole moment, F_0 is the maximum field strength and $s(t)$ is the envelope of the driving pulse. To be specific, we take

$$s(t) = \sin^2 \left(\frac{\pi t}{t_p} \right) \tag{7.76}$$

in the following, where t_p represents the pulse length. For the physical background of this model see Manz and Wöste (1995).

The Morse Hamiltonian H_M has a finite number N of bound states which depends on the parameters of the Morse potential $V(q)$. The following investigation will be restricted to the case where the dynamics is well confined to the bound state sector of the Morse oscillator. The simulation will therefore be performed in the energy eigenbasis $|j\rangle$, $j = 0, 1, 2, \ldots, N - 1$, of H_M with the corresponding energy eigenvalues E_j which are known analytically.

The jump operators are taken to be

$$A_{jk} = |j\rangle\langle k|, \qquad j, k = 0, 1, \ldots, N - 1, \tag{7.77}$$

with corresponding rates γ_{jk}. These rates are assumed to be given by the expressions for the weak-coupling master equation describing the dissipative dynamics in a thermal reservoir. The operators A_{jk} are eigenoperators of H_M belonging to the eigenvalues $(E_j - E_k)$. The effect of the jump operator A_{jk} on the system wave function may thus be interpreted as a transition describing the emission or absorption of a vibration quantum of energy $|E_j - E_k|$. In the energy representation the Markovian master equation for the matrix elements $\rho_{jk} = \langle j|\rho_S|k\rangle$ therefore reads

$$\frac{d\rho_{jk}}{dt} = -i(E_j - E_k)\,\rho_{jk} - if(t) \sum_l (Q_{jl}\,\rho_{lk} - Q_{lk}\,\rho_{jl})$$

$$+\delta_{jk} \left(\sum_l \gamma_{jl}\,\rho_{ll} \right) - \frac{1}{2}(\Gamma_j + \Gamma_k)\,\rho_{jk}. \tag{7.78}$$

The corresponding PDP is governed by the following equation for the components $\psi_j = \langle j|\psi\rangle$ of the normalized state vector ψ,

$$\frac{d}{dt}\psi_j = -i \left(E_j\psi_j + f(t) \sum_k Q_{jk}\psi_k \right) - \frac{1}{2} \left(\sum_k \gamma_{kj} - \sum_{kl} \gamma_{kl}|\psi_l|^2 \right) \psi_j$$

$$+ \sum_{kl} (\delta_{kj} - \psi_j)\, dN_{kl}(t). \tag{7.79}$$

In these equations $f(t)$ is the time-dependent external force (cf. eqn 7.75)

$$f(t) = \mu F_0 s(t) \sin(\omega_L t), \tag{7.80}$$

and the $Q_{jk} = \langle j|q|k\rangle$ are the matrix elements of the dipole operator. The matrix (Q_{jk}) is real and symmetric. Γ_j is the total rate of all jumps away from $|j\rangle$,

$$\Gamma_j = \sum_{m=0}^{N-1} \gamma_{mj}. \tag{7.81}$$

Finally, the Poisson increments satisfy

$$dN_{kl}(t)dN_{k'l'}(t) = \delta_{kk'}\delta_{ll'}dN_{kl}(t), \tag{7.82}$$

$$\mathrm{E}\left[dN_{kl}(t)\right] = \gamma_{kl}|\psi_l|^2 dt. \tag{7.83}$$

7.4.2.2 *Simulation results* In the following we solve eqns (7.78) and (7.79) and compare the numerical performance for both equations. The initial state is always taken to be the ground state of the Morse oscillator. In order to study the effect of the system size on the time consumption of the numerical routines, a series of similar oscillators with varying number N of bound states was investigated. The system size N was varied in the range $N = 12, \ldots, 78$. In order to not just blow up the number of states, with the actual dynamics always staying in the same number of low-lying states, it is necessary to appropriately scale the driving field as well. For the following study two parameter combinations are employed, one corresponding to weak damping, the other to strong damping. We also note that the same Runge–Kutta routine is used for the integration of the density matrix equation and for the integration of the deterministic pieces of the PDP. The details of the physical parameters used and of the numerical implementation are described by Breuer, Huber and Petruccione (1997).

Let us first look at the exponents α and β which were introduced in eqns (7.67) and (7.68). Figure 7.11 shows the CPU time per time step of the numerical integrator as a function of the system size N. Measuring the slope of the lines one finds

$$\beta = 3.0 \pm 0.1, \qquad \alpha = 2.0 \pm 0.1, \tag{7.84}$$

which confirms eqn (7.69). These values can be easily understood: In the case of the stochastic simulation, the most time-consuming part is the multiplication of ψ with the dipole matrix Q (cf. eqn (7.79)), which requires $\mathcal{O}(N^2)$ floating point operations. Analogously, for the density matrix integration the calculation of the right-hand side of eqn (7.78) for all j and k involves $\mathcal{O}(N^3)$ floating point operations.

Figure 7.12 displays the number of integrator steps s_1 and s_2 that are necessary to calculate a whole pulse. As we can see s_1 and s_2 increase with N, but remain roughly equal. Their increase is due to particular properties of the presented example. The systems of differential equations which have to be solved for the density matrix calculation and for the stochastic simulation both become stiffer with increasing N. In particular, the ratio between the highest and the lowest eigenenergy of the oscillator grows about linearly with increasing N.

In order to investigate the behaviour of $R(N)$, the number of Monte Carlo realizations that have to be generated to treat a system of size N, it is necessary to look at the standard error of the observables of the system. Let us consider,

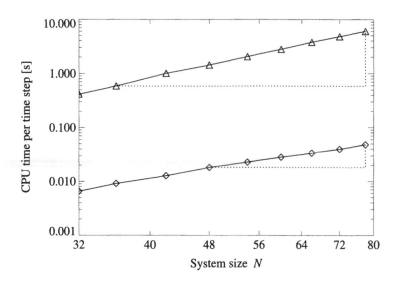

FIG. 7.11. The CPU time per step for the integration of the density matrix equation ($\triangle$) and for the propagation of the wave vector ψ ($\diamond$). The dotted lines indicate how the exponents α and β were calculated from the slope; the continuous lines simply connect the data points.

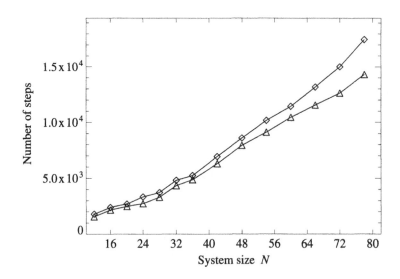

FIG. 7.12. The triangles ($\triangle$) display s_1, the number of integrator steps to calculate one pulse using the density matrix equation. The diamonds ($\diamond$) show s_2, the average number of steps to calculate one realization of the process.

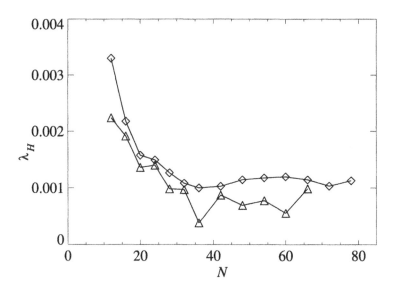

FIG. 7.13. The function $\lambda_H(N)$ measures the self-averaging property of the mean energy of the Morse oscillator H_M, cf. eqn (7.65). The triangles $\triangle$ correspond to weak dissipation, the diamonds $\diamond$ to strong dissipation.

for example, the energy H_M of the Morse oscillator, whose standard error will be denoted by $\hat{\sigma}_H$. According to eqn (7.66) we have

$$R(N) = \frac{\lambda_H(N)}{\hat{\sigma}_H^2}. \qquad (7.85)$$

Figure 7.13 shows the behaviour of the function $\lambda_H(N)$ for two different environment parameter sets. Up to statistical fluctuations, $\lambda(N)$ and, therefore, $R(N)$ is a non-increasing function of N. Similar results are obtained for other observables and for the matrix elements of ρ. This means that, in order to achieve a constant statistical error in the simulation results when the system size N is increased, the number of realizations need not be increased. It follows that the stochastic simulation will eventually, for large system sizes, be always faster than solving the density matrix equation.

This result is exemplified in Fig. 7.14. The plots show the CPU times needed to integrate the density matrix equation and to generate as many realizations of the stochastic process as are necessary to obtain the standard error $\hat{\sigma}_H = 4 \cdot 10^{-3}$ of the mean oscillator energy. The number R of realizations was calculated with the help of eqn (7.85). According to Fig. 7.13, we chose $\lambda_H(N) = 10^{-3}$ independently of N. The curves follow different power laws and at some point N_0 they intersect. In the present example, $N_0 \approx 35$ for the weak dissipation case and $N_0 \approx 55$ for the case of strong dissipation. Above N_0, the stochastic simulation is faster.

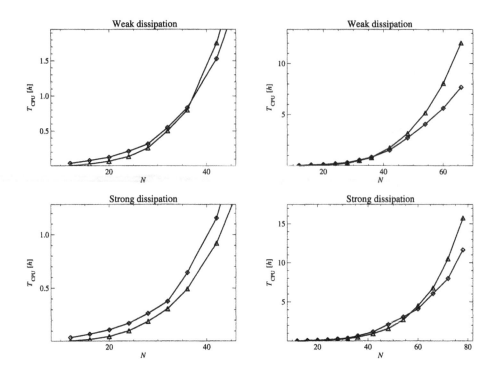

FIG. 7.14. CPU times needed to integrate the density matrix equation ($\triangle$) and to generate as many realizations of the stochastic process ($\diamond$) as are necessary to obtain the standard error $\hat{\sigma}_H = 4 \cdot 10^{-3}$ for the mean oscillator energy. The plots on the right-hand side cover the full variation of the system size while the plots on the left side zoom in at lower N.

The main result of the numerical study is that the CPU time for the density matrix integration T_{DME} and the time for the stochastic wave function simulation T_{StS} scale with the system size N as

$$T_{\text{DME}} \sim N^3, \tag{7.86}$$
$$T_{\text{StS}} \sim N^2. \tag{7.87}$$

Although the numerical study was performed on a specific example, the considerations made in Section 7.4.1 are far more general. In particular, whereas the absolute values of the exponents in eqns (7.86) and (7.87) depend on specific properties of the system under study, their difference is of more general significance. Under general conditions, the exponent in the expression for T_{DME} is expected to be larger by 1 to 2 than the exponent of T_{StS}.

The quantities T_{DME} and T_{StS} that have been investigated above stand for the time that the programs run on a single processor. When comparing the efficiency of numerical codes that are designed for processing on a parallel computer with many processors, other important criteria are speed-up and scalability. The speed-up is defined as the ratio between the wall-clock times needed to do the job on a single processor and on the parallel computer. Scalability means that the speed-up is close to the number of processors of the parallel machine, for a wide range of numbers of processors. This implies that little time is spent on the communication between processors and that the synchronization overhead is small. Whereas an efficient, scalable parallelization of the density matrix integration might be a complicated task, the stochastic wave function method is an intrinsically parallel and very well scalable algorithm. Since the individual realizations of the sample are generated independently, the only communication needed is for the final averaging or archiving, and for parameter control. Monte Carlo methods are therefore ideally suited for parallel processing.

References

Binder, K. (ed.) (1995). *The Monte Carlo Method in Condensed Matter Physics.* Springer-Verlag, Berlin.

Blue, J. L., Beichl, I. and Sullivan, F. (1995). Faster Monte Carlo simulations. *Phys. Rev.*, **E51**, R867–R868.

Breuer, H. P., Huber, W. and Petruccione, F. (1996). Fast Monte Carlo algorithm for nonequilibrium systems. *Phys. Rev.*, **E53**, 4232–4235.

Breuer, H. P., Huber, W. and Petruccione, F. (1997). Stochastic wave-function method versus density matrix: a numerical comparison. *Comp. Phys. Commun.*, **104**, 46–58.

Breuer, H. P., Dorner, U. and Petruccione, F. (2000). Numerical integration methods for stochastic wave function equations. *Comp. Phys. Commun.*, **132**, 30–43.

Ferrenberg, A. M., Landau, D. P. and Binder, K. (1991). Statistical and systematic errors in Monte Carlo sampling. *J. Stat. Phys.*, **63**, 867–882.

Fricke, T. and Schnakenberg, J. (1991). Monte Carlo simulation of an inhomogeneous reaction-diffusion system in the biophysics of receptor cells. *Z. Phys.*, **83**, 277–284.

Gillespie, D. T. (1976). A general method for numerically simulating the stochastic time evolution of coupled chemical reactions. *J. Comp. Phys.*, **22**, 403–434.

Gillespie, D. T. (1992). *Markov Processes.* Academic Press, Boston.

Hanusse, P. and Blanché, A. (1981). A Monte Carlo method for large reaction-diffusion systems. *J. Chem. Phys.*, **74**, 6148–6153.

Kloeden, P. E. and Platen, E. (1992). *Numerical Solutions of Stochastic Differential Equations.* Springer-Verlag, Berlin.

Knuth, D. E. (1981). *Seminumerical Algorithms.* Addison-Wesley, Reading, Mass.

Landau, D. P. and Binder, K. (2000). *A Guide to Monte Carlo Simulations in Statistical Physics.* Cambridge University Press, Cambridge.

Maksym, P. A. (1988). Fast Monte Carlo simulation of mbe growth. *Semiconductor Science and Technology*, **3**, 594–596.

Manz, J. and Wöste, L. (eds.) (1995). *Femtosecond Chemistry*. VCH, Weinheim, New York.

Press, W. H., Flannery, B. P., Teukolsky, S. A. and Vetterling, W. T. (1992). *Numerical Recipes*. Cambridge University Press, Cambridge.

8

APPLICATIONS TO QUANTUM OPTICAL SYSTEMS

Chapter 6 provided a number of simple examples for stochastic processes in Hilbert space and their derivations in the framework of continuous measurement theory. The present chapter deals with a presentation of the theory on the basis of the microscopic equations of quantum electrodynamics (QED). The emphasis lies on the derivation of the quantum operations and the corresponding stochastic processes for various detection schemes directly from the Hamiltonian which describes the interaction of the matter degrees of freedom with the quantized electromagnetic field. Throughout this chapter we will treat the matter degrees of freedom non-relativistically, having in mind applications to quantum optics.

After a brief survey of the quantization of the electromagnetic field we start with the derivation of the general expression for the QED operation which describes the back-action on the matter degrees of freedom induced by the measurement on the field variables. From the general expression we deduce the stochastic representation for the multipole radiation of the matter current. The stochastic dynamics will be formulated in terms of a stochastic equation of motion for the reduced density matrix of the source. For the case of a complete measurement of the quantum numbers of the radiated photons a stochastic evolution equation for the state vector of the source is obtained. We shall also deal with incomplete measurements. They lead to an unravelling of the dynamics in the form of a stochastic density matrix equation which does not preserve the purity of quantum states.

The chapter includes a detailed discussion of some important physical problems for which the interplay of coherent quantum evolution and dissipative processes plays an important rôle. First, we investigate the emergence of dark states in the interaction of atoms with coherent laser fields. In addition, we study the mechanism of coherent population trapping in the sub-recoil cooling of atoms in laser fields. The corresponding piecewise deterministic process for the quantized motion of the atoms is derived. It provides an interesting example of the emergence of long-range Lévy-type distributions resulting from a quantum interference effect. Finally, we examine dissipative phenomena in the dynamics of systems in strong periodic driving fields. Appropriate stochastic wave function dynamics can be derived by employing the representation of the reduced system's state vector in a basis consisting of Floquet states.

8.1   Continuous measurements in QED

8.1.1   *Constructing the microscopic Hamiltonian*

To begin with, we briefly recall the quantization of the free radiation field (see, e.g. Bjorken and Drell, 1965) which may be represented by the free field operator $\vec{A}(\vec{x}, t)$ for the vector potential. Imposing the Coulomb gauge condition

$$\vec{\nabla} \cdot \vec{A}(\vec{x}, t) = 0, \tag{8.1}$$

electric and magnetic field operators are given by[14]

$$\vec{E}(\vec{x}, t) = -\frac{1}{c}\frac{\partial}{\partial t}\vec{A}(\vec{x}, t), \tag{8.2}$$

$$\vec{B}(\vec{x}, t) = \vec{\nabla} \times \vec{A}(\vec{x}, t). \tag{8.3}$$

The field operators are Heisenberg picture operators satisfying the wave equation

$$\left(\frac{1}{c^2}\frac{\partial^2}{\partial t^2} - \Delta\right)\vec{A}(\vec{x}, t) = 0. \tag{8.4}$$

For the following discussion it will be helpful to decompose the radiation field into a complete set of mode functions $\vec{U}_\lambda(\vec{x})$ labelled by some index λ. They satisfy the Helmholtz equation

$$\left(\Delta + \frac{\omega_\lambda^2}{c^2}\right)\vec{U}_\lambda(\vec{x}) = 0 \tag{8.5}$$

with suitable boundary conditions, and are normalized such that

$$\int d^3x \, \vec{U}_\lambda^*(\vec{x}) \cdot \vec{U}_{\lambda'}(\vec{x}) = \delta_{\lambda\lambda'}. \tag{8.6}$$

The quantity ω_λ denotes the frequency of the mode λ. We further impose the condition that the modes are transverse,

$$\vec{\nabla} \cdot \vec{U}_\lambda(\vec{x}) = 0, \tag{8.7}$$

such that the completeness relation for the modes takes the form

$$\sum_\lambda U_\lambda^i(\vec{x})U_\lambda^{j*}(\vec{x}') = \delta_{\mathrm{T}}^{ij}(\vec{x}, \vec{x}') \equiv \left(\delta^{ij} - \frac{\partial^i\partial^j}{\Delta}\right)\delta(\vec{x} - \vec{x}'). \tag{8.8}$$

Here, $\delta_{\mathrm{T}}^{ij}(\vec{x}, \vec{x}')$ is the transverse δ-function which, applied to any vector field, projects onto the transverse component of the field.

---

[14]In this chapter we write explicitly all physical constants such as Planck's constant $\hbar$, the speed of light c, electron charge e and mass m. Further, we use Gaussian units such that the fine structure constant is given by $\alpha = e^2/\hbar c \approx 1/137$.

On using the mode functions we can represent the vector potential as,

$$\vec{A}(\vec{x}, t) = \sum_{\lambda} \sqrt{\frac{2\pi \hbar c^2}{\omega_\lambda}} \left(\vec{U}_\lambda(\vec{x}) e^{-i\omega_\lambda t} b_\lambda + \vec{U}_\lambda^*(\vec{x}) e^{i\omega_\lambda t} b_\lambda^\dagger \right), \qquad (8.9)$$

while the electric field operator takes the form

$$\vec{E}(\vec{x}, t) = i \sum_{\lambda} \sqrt{2\pi \hbar \omega_\lambda} \left(\vec{U}_\lambda(\vec{x}) e^{-i\omega_\lambda t} b_\lambda - \vec{U}_\lambda^*(\vec{x}) e^{i\omega_\lambda t} b_\lambda^\dagger \right). \qquad (8.10)$$

The operators $b_\lambda^\dagger$ and b_λ are the creation and annihilation operators for the photons of mode λ, satisfying bosonic commutation relations,

$$\left[b_\lambda, b_{\lambda'}^\dagger \right] = \delta_{\lambda\lambda'}, \qquad (8.11)$$

$$[b_\lambda, b_{\lambda'}] = \left[b_\lambda^\dagger, b_{\lambda'}^\dagger \right] = 0. \qquad (8.12)$$

In particular, we denote by

$$|\lambda\rangle = b_\lambda^\dagger |0\rangle \qquad (8.13)$$

the one-photon states, where $|0\rangle$ is the vacuum state of the field. Exploiting these commutation relations one obtains the equal-time commutation relations for the field operators, e.g.

$$\left[E^i(\vec{x}, t), A^j(\vec{x}', t) \right] = 4\pi i\hbar c\, \delta_{\mathrm{T}}^{ij}(\vec{x}, \vec{x}'). \qquad (8.14)$$

The projection onto the transverse component embodied in the transverse δ-function on the right-hand side of this equation ensures that the commutation relations are compatible with Gauss's law $\vec{\nabla} \cdot \vec{E} = 0$ for the free field. Finally, in the radiation gauge the free Hamiltonian of the transverse degrees of freedom of the field can be written as

$$H_B = \frac{1}{8\pi} \int d^3x \left[\vec{E}^2 + \vec{B}^2 \right] = \frac{1}{8\pi} \int d^3x \left[\vec{E}^2 + \vec{A}(-\Delta)\vec{A} \right]$$

$$= \sum_{\lambda} \hbar \omega_\lambda \left(b_\lambda^\dagger b_\lambda + \frac{1}{2} \right). \qquad (8.15)$$

The degrees of freedom of the electromagnetic radiation field will now be coupled to the electronic degrees of freedom of a bound quantum system, an atom, for example, whose self-Hamiltonian is denoted by H_S. The Hamiltonian H_I describing the interaction between the electrons of the system and the radiation field can be written as

$$H_I(t) = - \int d^3x \left\{ \frac{e}{c} \vec{j}(\vec{x}, t) \cdot \vec{A}(\vec{x}, t) + \frac{e}{mc} \vec{s}(\vec{x}, t) \cdot \vec{B}(\vec{x}, t) \right\}. \qquad (8.16)$$

We treat the matter degrees of freedom in the non-relativistic approximation and neglect the diamagnetic term which is proportional to the square of the vector potential. The quantity $\vec{j}(\vec{x}, t)$ is the paramagnetic electron current density,

$$\vec{j}(\vec{x}, t) = \sum_\alpha \frac{1}{2m} \left\{ \vec{p}^{(\alpha)}(t) \delta\left(\vec{x} - \vec{x}^{(\alpha)}(t)\right) + \delta\left(\vec{x} - \vec{x}^{(\alpha)}(t)\right) \vec{p}^{(\alpha)}(t) \right\}, \qquad (8.17)$$

whereas $\vec{s}(\vec{x}, t)$ is the electron spin density,

$$\vec{s}(\vec{x}, t) = \sum_\alpha \delta\left(\vec{x} - \vec{x}^{(\alpha)}(t)\right) \frac{\hbar}{2} \vec{\sigma}^{(\alpha)}(t). \qquad (8.18)$$

Later on we shall also use the expressions for the electron density

$$\varrho(\vec{x}, t) = \sum_\alpha \delta\left(\vec{x} - \vec{x}^{(\alpha)}(t)\right), \qquad (8.19)$$

and for the orbital angular momentum density

$$\begin{aligned}
\vec{l}(\vec{x}, t) &= m\vec{x} \times \vec{j}(\vec{x}, t) \\
&= \sum_\alpha \frac{1}{2} \left\{ \vec{L}^{(\alpha)}(t) \delta\left(\vec{x} - \vec{x}^{(\alpha)}(t)\right) + \delta\left(\vec{x} - \vec{x}^{(\alpha)}(t)\right) \vec{L}^{(\alpha)}(t) \right\}. \quad (8.20)
\end{aligned}$$

In these formulae the index α runs over the electrons of the system, and $\vec{x}^{(\alpha)}(t)$, $\vec{p}^{(\alpha)}(t)$, $\vec{L}^{(\alpha)}(t)$, and $\hbar\vec{\sigma}^{(\alpha)}(t)/2$ denote, respectively, position, linear momentum, orbital angular momentum, and spin operators of the particles in the interaction picture. Equation (8.16) thus describes the interaction of the electromagnetic field with the paramagnetic electron current density j and with the density of the magnetic moment associated with the electron spin density. Note that the interaction Hamiltonian $H_I(t)$ as given in (8.16) involves the interaction picture operators for the field and for the matter degrees of freedom, such that $H_I(t)$ is the interaction picture Hamiltonian.

8.1.2 Determination of the QED operation

Let us now derive the general expression for the operation pertaining to a continuous monitoring of the electromagnetic field variables. To this end, we first take an ideal case and suppose that we can perform a complete, orthogonal measurement of the one-photon states $|\lambda\rangle$. This means that we assume some measurement device which is able to detect the quanta radiated by the source together with a measurement of a complete set of quantum numbers λ of the photon.

Following our discussion in Sections 6.2 and 6.3 we consider again the field vacuum $|0\rangle$ acting as the probe state. In accordance with our previous notation we denote by τ the considered time interval of the interaction between object and probe system, that is, between source and radiation field. The arbitrarily chosen initial time is denoted by t_0, such that $t = t_0 + \tau$ is the final time at which the measurement on the radiation field is performed. Also, the initial density matrix of the source at time t_0 is written as $\tilde{\sigma}$.

Employing second-order perturbation theory we then find with the help of expression (8.16) for the interaction Hamiltonian that the corresponding operation is given by

$$\Phi_\lambda^{(1)}(\tilde{\sigma}) = \Omega_\lambda \tilde{\sigma} \Omega_\lambda^\dagger, \tag{8.21}$$

where the operator Ω_λ takes the form (cf. eqn (6.86))

$$\Omega_\lambda = \frac{-i}{\hbar} \int_{t_0}^{t} dt' \langle \lambda | H_I(t') | 0 \rangle = i \sqrt{\frac{2\pi c^2}{\hbar \omega_\lambda}} \int_{t_0}^{t} dt' e^{i\omega_\lambda t'}$$

$$\times \int d^3x \left\{ \frac{e}{c} \vec{j}(\vec{x}, t') \cdot \vec{U}_\lambda^*(\vec{x}) + \frac{e}{mc} \vec{s}(\vec{x}, t') \cdot \left[\vec{\nabla} \times \vec{U}_\lambda^*(\vec{x}) \right] \right\}. \tag{8.22}$$

The operation for the event that no photon is detected over the time interval τ can be written as

$$\Phi^{(0)}(\tilde{\sigma}) = \Omega_0 \tilde{\sigma} \Omega_0^\dagger, \tag{8.23}$$

where

$$\Omega_0 = I - \frac{1}{\hbar^2} \int_{t_0}^{t} dt' \int_{t_0}^{t'} dt'' \langle 0 | H_I(t') H_I(t'') | 0 \rangle. \tag{8.24}$$

Inserting a complete set of states, which may be restricted to the one-photon sector here, we get the equivalent expression

$$\Omega_0 = I - \frac{1}{\hbar^2} \int_{t_0}^{t} dt' \int_{t_0}^{t'} dt'' \sum_\lambda \langle \lambda | H_I(t') | 0 \rangle^\dagger \langle \lambda | H_I(t'') | 0 \rangle. \tag{8.25}$$

These relations will be used below to determine the operation for the detection scheme.

In order to deal with the time integration involved in the operations (8.22) and (8.25) we shall decompose the various densities into eigenoperators of the Hamiltonian H_S of the source. These eigenoperators are defined through (compare eqn (3.120))

$$\vec{j}(\vec{x}, \omega) = \sum_{E' - E = \hbar\omega} \Pi(E) \vec{j}(\vec{x}, 0) \Pi(E'), \tag{8.26}$$

with analogous relations for the particle density ϱ, the spin density $\vec{s}$ and the orbital momentum density $\vec{l}$. The operators $\vec{j}(\vec{x},0)$, $\varrho(\vec{x},0)$, $\vec{s}(\vec{x},0)$ and $\vec{l}(\vec{x},0)$ denote the corresponding Schrödinger picture operators taken at a fixed time $t = 0$. Furthermore, $\Pi(E)$ denotes the projector onto the eigenspace of H_S

corresponding to the energy eigenvalue E and the sums are extended over those energies E and E' which belong to a fixed energy difference of $\hbar\omega$. We assume that H_S describes a bound system such that the interaction picture densities are given sums taken over a discrete set of system frequencies ω,

$$\vec{j}(\vec{x},t) = \sum_{\omega} e^{-i\omega t}\vec{j}(\vec{x},\omega) + \text{h.c.}, \qquad (8.27)$$

$$\varrho(\vec{x},t) = \sum_{\omega} e^{-i\omega t}\varrho(\vec{x},\omega) + \text{h.c.}, \qquad (8.28)$$

$$\vec{s}(\vec{x},t) = \sum_{\omega} e^{-i\omega t}\vec{s}(\vec{x},\omega) + \text{h.c.}, \qquad (8.29)$$

$$\vec{l}(\vec{x},t) = \sum_{\omega} e^{-i\omega t}\vec{l}(\vec{x},\omega) + \text{h.c.} \qquad (8.30)$$

On using this decompositions into eigenoperators we can write the operator Ω_λ which determines the operation for the event that a photon with quantum numbers λ is detected as follows,

$$\Omega_\lambda = \sum_{\omega} \sqrt{\frac{2\pi c^2}{\hbar\omega_\lambda}} \exp[i(\omega_\lambda - \omega)t_0]\frac{\exp[i(\omega_\lambda - \omega)\tau] - 1}{\omega_\lambda - \omega}$$

$$\times \int d^3x \left\{ \frac{e}{c}\vec{j}(\vec{x},\omega) \cdot \vec{U}_\lambda^*(\vec{x}) + \frac{e}{mc}\vec{s}(\vec{x},\omega) \cdot \left[\vec{\nabla} \times \vec{U}_\lambda^*(\vec{x})\right] \right\}, \qquad (8.31)$$

Since the mode frequencies ω_λ are positive the sum over the system frequencies ω may be restricted to run over the positive frequencies only, which will always be assumed in the following.

We see from eqn (8.31) that the operation defined by Ω_λ leads, in general, to complicated short-time behaviour of the object system. In the following we assume that the discrete system frequencies ω are well separated with a minimal distance which is large compared to $\Delta \sim 1/\tau$, the energy uncertainty associated with the interaction time τ. Physically, this means that over interaction times of order τ we can distinguish the frequencies radiated by the system and identify uniquely the corresponding transition of the object system. In other words, we are considering a spectral detection of the object system with a finite resolution which is small compared to the widths of the transitions but large in comparison to the distance between their frequencies.

Under this assumption, we see that for a given frequency ω_λ of the field modes there is exactly one term which dominates the sum over ω in eqn (8.31). This is the term which fulfils the condition

$$\omega_\lambda \in I_\omega \equiv [\omega - \Delta, \omega + \Delta]. \qquad (8.32)$$

Thus, by keeping only the secular term in the operation we obtain for the QED operation (omitting irrelevant phase factors),

$$\Omega_\lambda = \sqrt{\frac{2\pi c^2}{\hbar \omega_\lambda}} \frac{\exp[i(\omega_\lambda - \omega)\tau] - 1}{\omega_\lambda - \omega}$$

$$\times \int d^3x \left\{ \frac{e}{c} \vec{j}(\vec{x}, \omega) \cdot \vec{U}_\lambda^*(\vec{x}) + \frac{e}{mc} \vec{s}(\vec{x}, \omega) \cdot \left[\vec{\nabla} \times \vec{U}_\lambda^*(\vec{x}) \right] \right\}, \quad (8.33)$$

where the system frequency ω is determined by condition (8.32).

8.1.3 Stochastic dynamics of multipole radiation

As an example let us discuss the case of the spectral and angular momentum measurement of the emitted photons. The complete set of quantum numbers λ measured by the device consists of the frequency of the photons, the square of the total angular momentum, the component of the angular momentum in a fixed direction $\vec{e}_z$, and the parity quantum number. Hence, the index λ is given by the set of quantum numbers

$$\lambda = (k, J, M, \pi), \quad (8.34)$$

where $J = 1, 2, 3, \ldots$, $M = -J, -J+1, \ldots, J-1, J$, and $\pi = 0, 1$. The one-photon states $|\lambda\rangle$ thus describe free photons with definite frequency $\omega_\lambda \equiv \omega_k = ck$. They are eigenstates of the square of the total angular momentum of the field, of the z-component of the total angular momentum, and of the parity operator with respective eigenvalues $\hbar^2 J(J+1)$, $\hbar M$, and $(-1)^{J-\pi}$. Our aim is to derive a stochastic differential equation for the density matrix $\sigma(t)$ of the source corresponding to this measurement scheme.

As is well known the mode functions $\vec{U}_\lambda(\vec{x})$ represent electric multipole fields for $\pi = 1$ and magnetic multipole fields for $\pi = 0$ (Akhiezer and Berestetskii, 1965). Coupling the orbital angular momentum and the spin of the photons, one can express the mode functions in terms of superpositions of products of vector spherical harmonics and Bessel functions, corresponding to their angular and radial part, respectively. They form a complete set and can be chosen to satisfy the normalization condition

$$\int d^3x \, \vec{U}_\lambda^*(\vec{x}) \cdot \vec{U}_{\lambda'}(\vec{x}) = \delta(k - k')\delta_{JJ'}\delta_{MM'}\delta_{\pi\pi'}. \quad (8.35)$$

We do not need here the explicit expressions for the mode functions; they may be found, for example, in Shore (1990). In the range of optical frequencies we may use the approximation $kd \ll 1$, that is we assume that the wavelength is large compared to the linear extension d of the source. Consequently, only the behaviour of the mode functions in the vicinity of the origin is relevant. Substituting the explicit expressions into (8.33) one obtains for the operation corresponding to the detection of a photon with quantum numbers $\lambda = (k, J, M, \pi)$,

$$\Omega_{kJM\pi} = \sqrt{\frac{\omega_k}{\pi\hbar}} \frac{k^J}{(2J-1)!!} \sqrt{\frac{J+1}{J(2J+1)}} \frac{\exp[i(\omega_k - \omega)\tau] - 1}{\omega_k - \omega} Q_{JM\pi}(\omega). \quad (8.36)$$

Here we have introduced the frequency components of the electric 2^J-pole moments,

$$Q_{JM,\pi=1}(\omega) = \sqrt{\frac{4\pi}{2J+1}} \int d^3x \, e\varrho(\vec{x},\omega) r^J Y_{JM}^*(\vec{x}), \tag{8.37}$$

and of the magnetic 2^J-pole moments,

$$Q_{JM,\pi=0}(\omega) = \sqrt{\frac{4\pi}{2J+1}} \int d^3x \frac{e}{mc} \left[\frac{1}{J+1} \vec{l}(\vec{x},\omega) + \vec{s}(\vec{x},\omega) \right] \cdot \vec{\nabla} \left[r^J Y_{JM}^*(\vec{x}) \right]. \tag{8.38}$$

In these formulae the $Y_{JM}(\vec{x})$ denote the ordinary spherical harmonics, $r \equiv |\vec{x}|$, and $n!! = 1 \cdot 3 \cdot 5 \cdots n$ for n odd.

It follows from eqn (8.36) that the spectral angular momentum measurement of the radiated photons leads to a PDP with the quantum jumps

$$\tilde{\sigma} \longrightarrow \sigma = \frac{Q_{JM\pi}(\omega)\tilde{\sigma}Q_{JM\pi}^\dagger(\omega)}{\mathrm{tr}_S \left\{ Q_{JM\pi}^\dagger(\omega)Q_{JM\pi}(\omega)\tilde{\sigma} \right\}}. \tag{8.39}$$

For each particular set of quantum numbers (ω, J, M, π) we thus have a certain jump which is given by the application of the corresponding transition frequency component $Q_{JM\pi}(\omega)$ of the 2^J-pole moment.

The rate corresponding to the jump (8.39) is given in the Markov and rotating wave approximation by the expression

$$\frac{1}{\tau} \int_0^\infty dk \, \mathrm{tr}_S \left\{ \Omega_{kJM\pi}^\dagger \Omega_{kJM\pi} \tilde{\sigma} \right\} \approx \gamma_J(\omega) \mathrm{tr}_S \left\{ Q_{JM\pi}^\dagger(\omega)Q_{JM\pi}(\omega)\tilde{\sigma} \right\}, \tag{8.40}$$

where

$$\gamma_J(\omega) \equiv \frac{2(J+1)}{J(2J+1)[(2J-1)!!]^2} \frac{k^{2J+1}}{\hbar}. \tag{8.41}$$

These are the well-known expressions for the transition rates of electric and magnetic multipole radiation.

Finally, we need to derive the expressions for the event of no photodetection. Invoking again the Markov and the rotating wave approximation, eqn (8.25) leads to

$$\Omega_0 \approx I - \tau \frac{1}{2} \sum_{\omega JM\pi} \gamma_J(\omega)Q_{JM\pi}^\dagger(\omega)Q_{JM\pi}(\omega), \tag{8.42}$$

where we have neglected the Lamb shift which yields a contribution to the Hamiltonian H_S of the source. Therefore, we get the following expression for the operation of the zero photon detection event,

$$\frac{\Omega_0\tilde{\sigma}\Omega_0^\dagger}{\mathrm{tr}_S \left\{ \Omega_0^\dagger\Omega_0\tilde{\sigma} \right\}} \approx I - \frac{i}{\hbar}\mathcal{G}(\tilde{\sigma})d\tau, \tag{8.43}$$

where the super-operator

$$
\mathcal{G}(\sigma) = -\frac{i\hbar}{2} \sum_{\omega J M \pi} \gamma_J(\omega) \left(Q^\dagger_{JM\pi}(\omega) Q_{JM\pi}(\omega)\sigma + \sigma Q^\dagger_{JM\pi}(\omega) Q_{JM\pi}(\omega) \right)
$$
$$
+ i\hbar \sum_{\omega J M \pi} \gamma_J(\omega) \mathrm{tr}_S \left\{ Q^\dagger_{JM\pi}(\omega) Q_{JM\pi}(\omega)\sigma \right\} \sigma \tag{8.44}
$$

represents the generator for the deterministic parts of the PDP.

Summarizing these results we have found that the spectral and angular momentum monitoring of the source results in a PDP which obeys the following stochastic differential equation for the density matrix $\sigma(t)$ in the interaction picture,

$$
d\sigma(t) = -\frac{i}{\hbar}\mathcal{G}(\sigma(t))dt \tag{8.45}
$$
$$
+ \sum_{\omega J M \pi} \left[\frac{Q_{JM\pi}(\omega)\sigma(t)Q^\dagger_{JM\pi}(\omega)}{\mathrm{tr}_S \left\{ Q^\dagger_{JM\pi}(\omega) Q_{JM\pi}(\omega)\sigma(t) \right\}} - \sigma(t) \right] dN_{\omega JM\pi}(t).
$$

The complete jump statistics of the PDP is embodied in the Poisson increments which satisfy

$$
dN_{\omega JM\pi}(t)\, dN_{\omega' J' M' \pi'}(t) = \delta_{\omega\omega'}\delta_{JJ'}\delta_{MM'}\delta_{\pi\pi'} dN_{\omega JM\pi}(t), \tag{8.46}
$$
$$
\mathrm{E}\left[dN_{\omega JM\pi}(t)\right] = \gamma_J(\omega)\mathrm{tr}_S \left\{ Q^\dagger_{JM\pi}(\omega) Q_{JM\pi}(\omega)\sigma(t) \right\} dt. \tag{8.47}
$$

It is immediately clear that eqn (8.45) leads to a stochastic process for pure states. As already discussed in Chapter 6, this is directly connected to the fact that we have performed a measurement of a complete system of observables of the photon, and because the back-action induced by these measurements depends on the photon frequency ω_k only through the frequency ω of the transition of the source.

8.1.4 Representation of incomplete measurements

Let us consider the specific case of electric dipole radiation ($J = 1$, $\pi = 1$). The jump operators may then be written as

$$
Q_{1M,\pi=1}(\omega) \equiv \vec{e}^*_M \cdot \vec{A}(\omega), \quad M = 0, \pm 1, \tag{8.48}
$$

where

$$
\vec{e}_0 = \vec{e}_z, \quad \vec{e}_{\pm 1} = \mp \frac{1}{\sqrt{2}} [\vec{e}_x \pm i\vec{e}_y], \tag{8.49}
$$

and

$$
\vec{A}(\omega) = \int d^3x\, e\varrho(\vec{x}, \omega)\vec{x} \tag{8.50}
$$

is an eigenoperator of the total dipole moment

$$\vec{D}(t) = \int d^3x \, e\varrho(\vec{x}, t)\vec{x} = \sum_\omega e^{-i\omega t} \vec{A}(\omega) + \text{h.c.} \qquad (8.51)$$

corresponding to the transition frequency ω.

A complete measurement for electric dipole radiation thus encompasses the detection of the transition frequency ω and of the angular momentum component $\hbar M$ of the photon along the fixed direction $\vec{e}_z$. The stochastic evolution equation (8.45) then directly yields the following stochastic equation for pure states,

$$d\psi(t) = -\frac{i}{\hbar}G(\psi(t))dt + \sum_\omega \sum_{M=0,\pm} \left[\frac{\vec{e}_M^* \cdot \vec{A}(\omega)\psi(t)}{||\vec{e}_M^* \cdot \vec{A}(\omega)\psi(t)||} - \psi(t) \right] dN_{\omega,M}(t), \quad (8.52)$$

where the generator of the deterministic pieces of the PDP is

$$G(\psi) = -\frac{i\hbar}{2} \sum_\omega \gamma(\omega) \left(\vec{A}^\dagger(\omega) \cdot \vec{A}(\omega) - \langle \vec{A}^\dagger(\omega) \cdot \vec{A}(\omega) \rangle_\psi \right) \psi$$

$$\equiv \hat{H}\psi + \frac{i\hbar}{2} \sum_\omega \gamma(\omega)\langle \vec{A}^\dagger(\omega) \cdot \vec{A}(\omega) \rangle_\psi \, \psi. \qquad (8.53)$$

The expectation values of the Poisson increments $dN_{\omega,M}(t)$ are given by

$$\text{E}\left[dN_{\omega,M}(t)\right] = \gamma(\omega)||\vec{e}_M^* \cdot \vec{A}(\omega)\psi(t)||^2 dt, \qquad (8.54)$$

where

$$\gamma(\omega) = \frac{4}{3}\frac{\omega^3}{\hbar c^3}, \qquad (8.55)$$

which gives the well-known transition rate for electric dipole radiation.

We now suppose that the measurement of the photon spin component $\hbar M$ is carried out on the non-selective level, that is the information on the photon angular momentum along $\vec{e}_z$ is thrown away. The stochastic process then loses the property of transforming pure states into pure states under the time evolution. The dynamics must therefore be described in terms of a stochastic evolution equation for the density matrix $\sigma(t)$ which takes the form

$$d\sigma(t) = -\frac{i}{\hbar}\mathcal{G}(\sigma(t))dt$$

$$+ \sum_\omega \left[\frac{\vec{A}(\omega)\sigma(t)\vec{A}^\dagger(\omega)}{\text{tr}_S\left\{ \vec{A}^\dagger(\omega) \cdot \vec{A}(\omega)\sigma(t) \right\}} - \sigma(t) \right] dN_\omega(t). \qquad (8.56)$$

As can be seen the operation pertaining to the detection of a photon with frequency ω now involves an incoherent sum over the angular momentum components,

$$\vec{A}(\omega)\sigma\vec{A}^\dagger(\omega) = \sum_{M=0,\pm} (\vec{e}_M^* \cdot \vec{A}(\omega))\sigma(\vec{e}_M^* \cdot \vec{A}(\omega))^\dagger. \qquad (8.57)$$

This expresses the fact that no information on the angular momentum of the photons is obtained during the monitoring of the field. As a result the statistics of the Poisson increments depends only on the mean transition rate obtained by averaging over the angular momentum components,

$$\mathrm{E}\left[dN_\omega(t)\right] = \gamma(\omega)\mathrm{tr}_S\left\{\vec{A}^\dagger(\omega)\cdot\vec{A}(\omega)\sigma(t)\right\}dt. \tag{8.58}$$

One can easily imagine a situation in which even the information on the frequencies of the photons is thrown away. In this case we obviously have only one type of quantum jump. These jumps are counted by a single Poisson increment $dN(t)$ which satisfies

$$\mathrm{E}\left[dN(t)\right] = \sum_\omega \gamma(\omega)\mathrm{tr}_S\left\{\vec{A}^\dagger(\omega)\cdot\vec{A}(\omega)\sigma(t)\right\}dt. \tag{8.59}$$

The stochastic differential equation for the resulting PDP thus reads

$$d\sigma(t) = -\frac{i}{\hbar}\mathcal{G}(\sigma(t))dt$$
$$+ \left[\frac{\sum_\omega \gamma(\omega)\vec{A}(\omega)\sigma(t)\vec{A}^\dagger(\omega)}{\mathrm{tr}_S\left\{\sum_\omega \gamma(\omega)\vec{A}^\dagger(\omega)\cdot\vec{A}(\omega)\sigma(t)\right\}} - \sigma(t)\right] dN(t). \tag{8.60}$$

In summary, we observe that by successively reducing the number of observables measured selectively in the detection scheme one obtains different stochastic evolution equations, involving different jump super-operators which contain an increasing number of summands. This results from the smaller amount of information which is extracted from the system: The coarse graining of the detection scheme leads to a back-action on the object system which describes a stronger increase of entropy.

On the other hand, as can be seen from the equations derived above, the generator $\mathcal{G}$ for the deterministic pieces of the PDPs is always the same. This is due to the fact that the deterministic part of the process corresponds to the event of zero photon detection and is, thus, not influenced by the specific form of the detection scheme. This can also be deduced directly from eqn (8.25) which shows that the operation for the zero detection event depends only on the one-photon subspace, but not on the specific basis of one-photon states chosen. The same situation already occurred in the case of homodyne photodetection (see Section 6.4).

8.2   Dark state resonances

In order to illustrate the foregoing discussion we investigate in this section a transition between two angular momentum manifolds (Mølmer, Castin and Dalibard, 1993). This example leads to the emergence of a certain trapping state of the process, known as a dark state, which will play an important rôle in the following section.

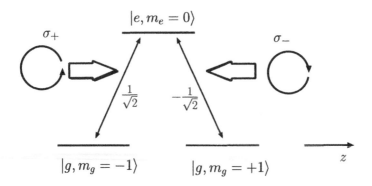

FIG. 8.1. Level scheme leading to a dark state resonance by a quantum interference effect. The figure shows the coupling between two $J_e = J_g = 1$
angular momentum manifolds through two counter-propagating laser beams
with different helicities $\sigma_\pm$.

8.2.1   Waiting time distribution and trapping state

We consider a two-level atom with transition frequency ω_0. Both levels are threefold degenerate forming manifolds with total angular momentum $J_e = 1$ (excited
level) and $J_g = 1$ (ground level). The level scheme is depicted in Fig. 8.1. We
introduce the energy eigenstates $|g, m_g\rangle$ and $|e, m_e\rangle$ which are simultaneously
eigenstates of the z-component of the atomic angular momentum operator with
eigenvalues $\hbar m_g$ and $\hbar m_e$, respectively. The frequency component $\vec{A}(\omega_0) \equiv \vec{A}$ of
the atomic dipole operator $\vec{D}$ can then be written as

$$\vec{A} = \sum_{m_g, m_e = 0, \pm 1} |g, m_g\rangle \langle g, m_g | \vec{D} | e, m_e \rangle \langle e, m_e|. \qquad (8.61)$$

Since $\vec{D}$ is a vector operator, $\vec{A}$ can easily be expressed in terms of the reduced
matrix element of the dipole operator, which will be denoted by d, and the
Clebsch–Gordan coefficients $\langle 1 m_e 1 M | 1 m_g \rangle$ which describe the coupling of the
atomic angular momentum and the photon angular momentum with components
$M = 0, \pm \hbar$.

In addition to the coupling to the vacuum of the radiation field the atom
is subjected to a resonant laser field with frequency $\omega_L = \omega_0$, which is linearly
polarized in the y-direction. One observes that the transition $m_e = 0 \leftrightarrow m_g = 0$ is
forbidden since the corresponding Clebsch–Gordan coefficient $\langle 1010 | 10 \rangle$ vanishes.
If we take some initial state in the manifold spanned by the states $|g, m_g = \pm 1\rangle$
we then find that the dynamics is confined to the subspace spanned by the basis
states

$$|e, m_e = 0\rangle, \quad |g, m_g = +1\rangle, \quad |g, m_g = -1\rangle. \qquad (8.62)$$

The representation for the frequency component of the dipole operator in this
subspace is found to be

$$A_x = \frac{d}{2} \left(|g, +1\rangle\langle e, 0| + |g, -1\rangle\langle e, 0| \right), \tag{8.63}$$

$$A_y = -\frac{id}{2} \left(|g, +1\rangle\langle e, 0| - |g, -1\rangle\langle e, 0| \right), \tag{8.64}$$

$$A_z = 0. \tag{8.65}$$

On using the resonance condition the Hamilton operator describing the linearly polarized laser field can be written in the interaction picture as

$$H_L = -E_0 \left(A_y + A_y^\dagger \right) = -\frac{i\hbar\Omega}{2} \begin{pmatrix} 0 & +1 & -1 \\ -1 & 0 & 0 \\ +1 & 0 & 0 \end{pmatrix}, \tag{8.66}$$

where $\Omega = dE_0/\hbar$ denotes the Rabi frequency and E_0 the laser field amplitude. The second equation shows the matrix representation in the basis (8.62). If we note further that

$$\vec{A}^\dagger \cdot \vec{A} = A_x^\dagger A_x + A_y^\dagger A_y = |d|^2 |e, 0\rangle\langle e, 0|, \tag{8.67}$$

we can write the non-Hermitian Hamiltonian $\hat{H}$ in eqn (8.53)

$$\hat{H} = H_L - \frac{i\hbar\gamma}{2} |e, 0\rangle\langle e, 0| = -\frac{i\hbar}{2} \begin{pmatrix} \gamma & \Omega & -\Omega \\ -\Omega & 0 & 0 \\ \Omega & 0 & 0 \end{pmatrix}, \tag{8.68}$$

where the atom–laser interaction has been added to $\hat{H}$. Note that we have absorbed the square of the reduced matrix element into γ which thus has the dimension of inverse time.

On using these results it is easy to write down analytical expressions for the non-Hermitian evolution and the cumulative distribution $F[\psi, t]$ for the random waiting time t for the PDP defined by the general equation (8.52). Since after any jump the state always ends up in the manifold spanned by the states $|g, \pm 1\rangle$, it suffices to determine these quantities for the states lying in this subspace, that is, for the states of the form

$$\varphi = \varphi_+ |g, +1\rangle + \varphi_- |g, -1\rangle \equiv \begin{pmatrix} 0 \\ \varphi_+ \\ \varphi_- \end{pmatrix}. \tag{8.69}$$

On diagonalizing $\hat{H}$ one easily finds that for these states

$$\exp\left(-\frac{i}{\hbar} \hat{H} t \right) \varphi = \frac{1}{2} \begin{pmatrix} 2 \left(\varphi_+ - \varphi_- \right) g(t) \\ \varphi_+ \left(1 - h(t) \right) + \varphi_- \left(1 + h(t) \right) \\ \varphi_+ \left(1 + h(t) \right) + \varphi_- \left(1 - h(t) \right) \end{pmatrix}, \tag{8.70}$$

whereas the waiting time distribution is given by

$$F[\varphi, t] = 1 - ||\exp(-i\hat{H}t/\hbar)\varphi||^2 \tag{8.71}$$

$$= 1 - \frac{1}{2}\left(1 + h^2(t) + 2g^2(t)\right) - \Re\left(\varphi_+^* \varphi_-\right)\left(1 - h^2(t) - 2g^2(t)\right).$$

Here we have introduced the abbreviations

$$g(t) = -\frac{\Omega}{\mu}e^{-\gamma t/4}\sin\left(\frac{\mu t}{2}\right),$$

$$h(t) = -e^{-\gamma t/4}\left[\cos\left(\frac{\mu t}{2}\right) + \frac{\gamma}{2\mu}\sin\left(\frac{\mu t}{2}\right)\right],$$

with

$$\mu \equiv \sqrt{2\Omega^2 - \frac{\gamma^2}{4}}. \tag{8.72}$$

In the following this quantity is assumed to be real and positive.

The important property of this example is that the non-Hermitian Hamiltonian (8.68) has a zero mode which is given by the state

$$\psi_{nc} = \frac{1}{\sqrt{2}}\left(|g, +1\rangle + |g, -1\rangle\right) = \frac{1}{\sqrt{2}}\begin{pmatrix} 0 \\ 1 \\ 1 \end{pmatrix}, \tag{8.73}$$

that is, we have

$$\hat{H}\psi_{nc} = 0. \tag{8.74}$$

It follows that $F[\psi_{nc}, t] \equiv 0$. The state ψ_{nc} is therefore a stable state of the process for any measurement scheme, which means that, once this state is reached, no further jumps take place and the state vector becomes independent of time. Moreover, since the other eigenvalues of $\hat{H}$ have negative imaginary parts, the state ψ_{nc} is attractive under deterministic evolution and may therefore be called a trapping state of the process.

From the physical point of view the emergence of the trapping state results from an interference effect as is easily seen by looking at the form of the non-Hermitian Hamiltonian. The linearly polarized laser field may be represented by the superposition of two counter-propagating fields in the z-direction (see Fig. 8.1). One component is right-circular polarized and one is left-circular polarized. The right-circular component induces the transitions $|g, -1\rangle \rightarrow |e, 0\rangle$, whereas the left-circular component yields the transitions $|g, +1\rangle \rightarrow |e, 0\rangle$. As can be see from $\hat{H}$ the amplitudes for these transitions interfere destructively and cancel exactly for the linear combination of the ground state manifold given the trapping state (8.73). Therefore, ψ_{nc} does not couple to the external laser field, thus representing a *dark state*.

8.2.2 *Measurement schemes and stochastic evolution*

Following Mølmer, Castin and Dalibard (1993) let us discuss two different measurement schemes, namely the measurement of the photon angular momentum

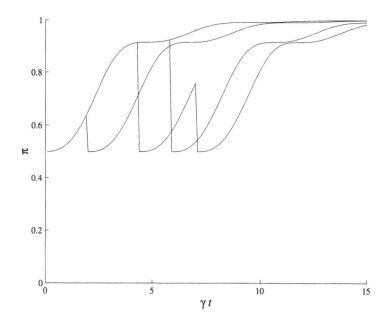

FIG. 8.2. Four realizations of the PDP for a two-level system driven by a resonant laser field polarized in the y-direction. For each realization $\psi(t)$ the
figure shows the population $\pi(t) = |\langle\psi_{nc}|\psi(t)\rangle|^2$ of the trapping state ψ_{nc}.
The stochastic process corresponds to the measurement of the photon angular
momentum along the z-axis. The initial state is $\psi(0) = |g, -1\rangle$ and $\Omega = \gamma$.

along the z-axis, and along the y-axis. For the measurement along the z-axis one
finds from eqns (8.63)–(8.65) two non-vanishing jump operators

$$\vec{e}_\pm^* \cdot \vec{A} = \mp \frac{d}{\sqrt{2}} |g, \mp 1\rangle\langle e, 0|, \tag{8.75}$$

where we have taken the polarization vectors $\vec{e}_0 = \vec{e}_z$ and $\vec{e}_\pm = \mp(\vec{e}_x \pm i\vec{e}_y)/\sqrt{2}$.
The jump operator $\vec{e}_\pm^* \cdot \vec{A}$ implies that the photon carries away the angular
momentum $\pm\hbar$ along the z-axis, such that the atomic transition involves a change
of the z-component of its total angular momentum given by $\mp\hbar$. We observe that
the trapping into the state ψ_{nc} results essentially from the non-unitary evolution
generated by $\hat{H}$. This is illustrated in Fig. 8.2 where we depict the occupation

$$\pi(t) \equiv |\langle\psi_{nc}|\psi(t)\rangle|^2 \tag{8.76}$$

of the trapping state for four realizations of the process corresponding to this
measurement scheme. As can be seen, the quantum jumps put the state vec-
tor into either $|g, +1\rangle$ or $|g, -1\rangle$, for which $\pi(t) = \frac{1}{2}$. One of the realizations
shown in the figure evolves continuously into the dark state ψ_{nc}. For the initial

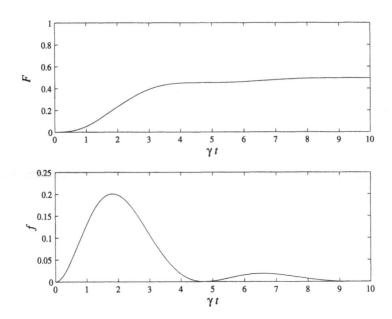

FIG. 8.3. The waiting time distribution function $F[|g,-1\rangle,t]$ and the corresponding density $f(t) = dF[|g,-1\rangle,t]/dt$ for the two-level system with a dark state and $\Omega = \gamma$.

state $\psi(0) = |g,-1\rangle$ chosen in the figure such a realization occurs with a finite probability which is given by the corresponding defect

$$\lim_{t\to\infty} F[|g,-1\rangle,t] = \frac{1}{2} \qquad (8.77)$$

of the waiting time distribution. Figure 8.3 shows a plot of the cumulative distribution function $F[|g,-1\rangle,t]$ and its density.

The situation is markedly different if we monitor the photon angular momentum along the y-axis. For this case we take the polarization vectors $\vec{e}_0 = \vec{e}_y$ and $\vec{e}_\pm = \mp(\vec{e}_z \pm i\vec{e}_x)/\sqrt{2}$ to obtain the jump operators

$$\vec{e}_0^* \cdot \vec{A} = -\frac{id}{\sqrt{2}}|\psi_c\rangle\langle e,0|, \qquad (8.78)$$

$$\vec{e}_\pm^* \cdot \vec{A} = +\frac{id}{2}|\psi_{nc}\rangle\langle e,0|, \qquad (8.79)$$

where we have introduced the state

$$\psi_c = \frac{1}{\sqrt{2}}\left(|g,+1\rangle - |g,-1\rangle\right), \qquad (8.80)$$

which is perpendicular to the trapping state. The jump operator $\vec{e}_0^* \cdot \vec{A}$ represents the back-action in the case that the photon angular momentum component along

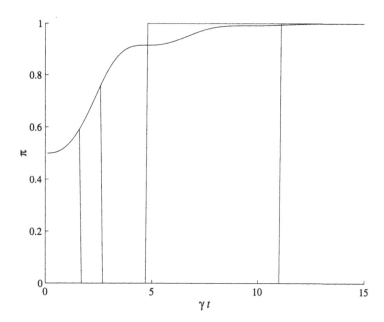

FIG. 8.4. The same as Fig. 8.2, but now the photon angular momentum is measured along the y-direction. The quantum jumps put the state vector either directly into the trapping state, in which case $\pi = 1$, or into the state perpendicular to it, such that $\pi = 0$.

the y-axis turns out to be zero. It puts the atom into the state ψ_c. By contrast, the expression for $\vec{e}_\pm^* \cdot \vec{A}$ shows that both measurement results $\pm\hbar$ for the angular momentum component put the atom into the trapping state ψ_{nc}. Thus, the trapping state may be reached directly through a quantum jump which is illustrated in Fig. 8.4.

   Both these measurement schemes lead, of course, to the same density matrix equation and cannot be distinguished if one looks only at the corresponding ensemble of type $\mathcal{E}_\rho$. As discussed extensively in Chapter 5, the two measurement schemes can, in fact, be distinguished if the measurement is performed on the selective level, that is, if we investigate the corresponding ensembles of type $\mathcal{E}_P$. To demonstrate this we plot in Figs. 8.5 and 8.6 the variances introduced in eqns (5.12), (5.13) and (5.14) pertaining to the observable

$$R = |\psi_{nc}\rangle\langle\psi_{nc}|, \tag{8.81}$$

which is just the projection onto the trapping state. The population $\pi(t)$ defined in eqn (8.76) is a random variable, which may be written as

$$\pi(t) = \langle\psi(t)|R|\psi(t)\rangle. \tag{8.82}$$

Thus we have

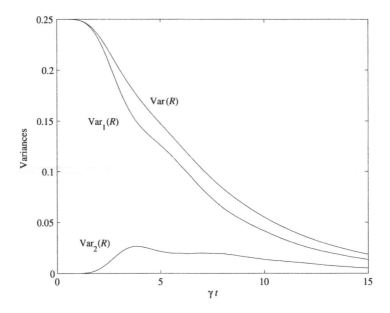

FIG. 8.5. The different variances of the observable $R = |\psi_{nc}\rangle\langle\psi_{nc}|$ as a function of time. The measurement scheme records the photon angular momentum along the z-axis. $\mathrm{Var}(R)$ has been determined from the solution of the density matrix equation, whereas $\mathrm{Var}_1(R)$ and $\mathrm{Var}_2(R)$ have been estimated from a sample of 10^4 realizations.

$$\mathrm{Var}_1(R) = \mathrm{E}\left[\pi(t)\right] - \mathrm{E}\left[\pi^2(t)\right], \tag{8.83}$$

$$\mathrm{Var}_2(R) = \mathrm{E}\left[\pi^2(t)\right] - \left(\mathrm{E}\left[\pi(t)\right]\right)^2, \tag{8.84}$$

$$\mathrm{Var}(R) = \mathrm{E}\left[\pi(t)\right] - \left(\mathrm{E}\left[\pi(t)\right]\right)^2. \tag{8.85}$$

Obviously, the sum of $\mathrm{Var}_1(R)$ and $\mathrm{Var}_2(R)$ is equal to $\mathrm{Var}(R)$ and is independent of the measurement scheme. The figures clearly exhibit that after some time $\mathrm{Var}_1(R)$ becomes considerably smaller for the measurements along the y-axis than for the measurement along the z-axis. This is easily understood in the context of quantum measurement theory. As we have seen above the measurements along the y-axis put the state vector into the trapping state ψ_{nc} or, else, into the orthogonal state ψ_c. This measurement scheme therefore corresponds to a measurement of R, that is, the scheme tends to generate a basis in which R is diagonal. After all realizations have performed their first jump the measurement scheme has produced an ensemble $\mathcal{E}_P$ consisting of eigenstates of R such that $\mathrm{Var}_1(R) = 0$. On the other hand, the measurements of the angular momentum along the z-axis project the state vector onto the states $|g, \pm 1\rangle$ for which $\mathrm{Var}_1(R) = \frac{1}{4}$. In this case it therefore takes much longer for $\mathrm{Var}_1(R)$ to decrease to zero.

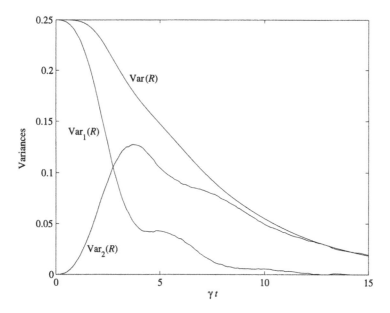

FIG. 8.6. The same as Fig. 8.5, but with the measurement scheme recording the photon angular momentum along the y-axis.

The variance $\mathrm{Var}_2(R)$ is a measure of the dispersion of $\pi(t)$ over the ensemble $\mathcal{E}_P$ which, for the simulation of the figures, was taken to consist of 10^4 realizations. Since $\mathrm{Var}_1(R)$ is larger for the measurement along the z-axis, $\mathrm{Var}_2(R)$ is smaller for this measurement scheme. The conclusion is that, in order to minimize the statistical fluctuations of $\pi(t) = \langle \psi(t)|R|\psi(t)\rangle$, one should try to obtain as less as possible information on the quantum observable R. If one is only interested in a numerically efficient simulation method, then one should use the measurement along the z-axis to get better statistics.

8.3   Laser cooling and Lévy processes

The dynamics involved in the cooling of the translational motion of atoms by means of appropriately designed laser fields provides important and interesting examples for stochastic processes in Hilbert space. In fact, this is one of the first examples for which the stochastic wave function dynamics has been used with great success (Cohen-Tannoudji, Bardou and Aspect, 1992; Castin and Mølmer, 1995). It turns out that the stochastic method enables one to deal with a series of problems in the deep quantum regime (Cohen-Tannoudji, 1992a) where an enormous variability of relevant time scales is involved, and where a treatment on the basis of the optical Bloch equations is difficult.

Let us consider for example the so-called Doppler cooling which employs the Doppler shift of the laser frequency as seen from the rest frame of the moving atoms. The basic physical principle underlying this cooling mechanism (Cohen-Tannoudji, 1992b) is the following. We take two laser beams counter-propagating along the z-axis. The laser frequency ω_L of both beams is tuned below the rest frame transition frequency ω of a simple two-level atom in the laser fields, such that the detuning $\delta \equiv \omega_L - \omega$ becomes negative. If the atom is moving into the positive z-direction, for example, it gets closer to resonance by the blue shift of the frequency of the laser field propagating in the negative z-direction. The atom is thus excited by the resonant absorption of photons from that beam and, by conservation of linear momentum, experiences a radiation pressure in the direction of the laser beam. The subsequent spontaneous emission of fluorescence radiation, however, is essentially isotropic in space which leads effectively to a dissipation of energy and a net friction force exerted on the atom. This friction force has the same direction as the laser beams. Thus, the two counter-propagating beams yield a damping of the z-component $p = p_z$ of the atomic momentum.

The dissipation of the atomic momentum is accompanied by fluctuating forces which are due to the random momentum kicks of size $\hbar k = \hbar\omega/c$ caused by the spontaneously emitted photons. It thus appears that the variance Δp^2 of the atomic momentum cannot be smaller than a fundamental limit which is given in terms of the recoil energy

$$E_R \equiv \frac{\hbar^2 k^2}{2M} = \frac{1}{2}k_B T_R \tag{8.86}$$

by the relation

$$\Delta p^2 > 2M E_R, \tag{8.87}$$

where M denotes the atomic mass. The recoil energy may be associated with a certain temperature, known as the recoil temperature

$$T_R = \frac{\hbar^2 k^2}{M k_B}. \tag{8.88}$$

As an example, for the transition $2^3 S_1 \leftrightarrow 2^3 P_1$ of helium one finds that $T_R \approx 4\mu K$.

This single photon recoil limit provided by the recoil temperature T_R seems to be a fundamental limit for any cooling mechanism which involves spontaneous photon emissions. In order to overcome this limit one has to suppress in some way spontaneous emission processes. As we have seen in the preceding section this is indeed possible if the atomic wave function is driven into a dark state which does not couple to the laser fields as a result of a quantum interference effect. This is precisely the sub-recoil cooling scheme which will be discussed in this section (Cohen-Tannoudji, 1992a, 1992b; Bardou et al., 1994). By using variants of this scheme or by invoking carefully designed pulses it has been demonstrated in a series of beautiful experiments that cooling temperatures in the nanokelvin range can be achieved (Kasevich and Chu, 1992; Lawall et al., 1994; Reichel et al., 1995; Lawall et al., 1995).

As will be discussed below the sub-recoil cooling mechanism leads to several interesting properties of the statistics of quantum jumps. The destructive interference of the transition amplitudes for the atom–laser interaction gives rise to long-tail waiting time distributions which behave asymptotically like certain stable Lévy distributions. In addition to important practical applications, the sub-recoil cooling scheme thus provides a nice example of the emergence of Lévy statistics as a result of quantum interference phenomena (Bardou, Bouchaud, Aspect and Cohen-Tannoudji, 2001).

8.3.1 *Dynamics of the atomic wave function*

Up to now we have completely neglected translational degrees of freedom. This means that the centre of mass $\vec{R}$ of the atom has been fixed at the origin of the coordinate system where, making use of the dipole approximation, the electric field is evaluated.

To derive the stochastic dynamics of the state vector for the cooling scheme we must take into account the motion of the centre of mass of the atoms (Cohen-Tannoudji, Dupont-Roc and Grynberg, 1998). The translational degrees of freedom will be treated fully quantum mechanically, that is, the conjugated variables $\vec{R}$ and atomic momentum $\vec{P}$ are regarded as operators satisfying canonical commutation relations,

$$[P_i, Q_j] = -i\hbar\delta_{ij}. \tag{8.89}$$

The Hamiltonian H_S describing the free atom now consists of two parts, namely the kinetic energy of the centre-of-mass motion and the energy of the internal degrees of freedom. For simplicity we consider a two-level atom and denote by Π_e and Π_g the projections onto the manifolds of excited and ground level, respectively. Hence we have

$$H_S = \frac{\vec{P}^2}{2M} + \hbar\omega\Pi_e, \tag{8.90}$$

where ω is the atomic transition frequency. The ground state energy has been taken to be zero.

The atom is subjected to two interactions: First, the interaction with a coherent laser field which will be described by a classical c-number field $\vec{E}_L(\vec{x}, t)$ and, second, the interaction with the quantized radiation field whose electric Schrödinger picture field operator will be denoted by $\vec{E}(\vec{x})$. Invoking the dipole approximation we may thus write the atom–laser interaction as

$$H_L(t) = -\vec{D} \cdot \vec{E}_L(\vec{R}, t), \tag{8.91}$$

whereas the interaction of the atom with the radiation field takes the form

$$H_I = -\vec{D} \cdot \vec{E}(\vec{R}). \tag{8.92}$$

The quantity $\vec{D}$ is the atomic dipole operator. Note that both electric fields are evaluated at the centre of mass of the atom. The total Hamiltonian for the atom

interacting with the coherent laser and the quantized radiation fields therefore
reads

$$H(t) = H_S + H_L(t) + H_I. \tag{8.93}$$

To be definite we consider in the following the cooling scheme sketched at
the beginning of this section, namely, the electric field $\vec{E}_L(\vec{R}, t)$ describes two
counter-propagating laser beams along the z-direction (see Fig. 8.1). The beam
propagating into the positive z-direction is right-circular polarized, the beam
propagating into the negative z-direction is left-circular polarized. Introducing
the corresponding polarization vectors,

$$\vec{e}_{\pm} = \mp \frac{1}{\sqrt{2}} \left(\vec{e}_x \pm i\vec{e}_y \right), \tag{8.94}$$

we can write

$$\vec{E}_L(\vec{R}, t) = \frac{E_0}{2} \left(\vec{e}_+ \exp\{+ik_L z\} + \vec{e}_- \exp\{-ik_L z\} \right) e^{-i\omega_L t} + \text{c.c.,} \tag{8.95}$$

where E_0 is the field amplitude. As in Section 8.2 we consider the case of a
$J_e = 1 \leftrightarrow J_g = 1$ transition. The lowering part $\vec{A}$ of the atomic dipole operator
is thus given by eqns (8.63)–(8.65). On using these expressions we find for the
atom–laser interaction in the rotating wave approximation

$$H_L(t) = \frac{\hbar\Omega}{2\sqrt{2}} \left(|e,0\rangle\langle g,+1| e^{-i(\omega_L t + k_L z)} - |e,0\rangle\langle g,-1| e^{-i(\omega_L t - k_L z)} \right) + \text{h.c.,} \tag{8.96}$$

where $\Omega = E_0 d/\hbar$ is the Rabi frequency.

Since the centre-of-mass coordinate $\vec{R}$ is canonically conjugated to the atomic
momentum $\vec{P}$ the exponential $\exp(-i\vec{k} \cdot \vec{R})$ acts as a shift in momentum space,
that is

$$\exp(-i\vec{k} \cdot \vec{R}) = \int d^3p \, |\vec{p}\rangle\langle\vec{p} + \hbar\vec{k}|, \tag{8.97}$$

where the states $|\vec{p}\rangle$ are momentum eigenstates of the centre-of-mass motion
which are normalized as

$$\langle\vec{p}|\vec{p}'\rangle = \delta(\vec{p} - \vec{p}'). \tag{8.98}$$

Hence we find

$$H_L(t) = \frac{\hbar\Omega}{2\sqrt{2}} \int d^3p \, |e,0,\vec{p}\rangle \left(\langle g,+1,\vec{p}+\hbar\vec{k}_L| - \langle g,-1,\vec{p}-\hbar\vec{k}_L| \right) e^{-i\omega_L t}$$
$$+\text{h.c.,} \tag{8.99}$$

where $\vec{k}_L \equiv k_L \vec{e}_z$ and we have introduced the product states

$$|g,\pm 1,\vec{p}\rangle \equiv |g,\pm 1\rangle \otimes |\vec{p}\rangle, \quad |e,0,\vec{p}\rangle \equiv |e,0\rangle \otimes |\vec{p}\rangle. \tag{8.100}$$

These are simultaneous eigenstates of the kinetic energy of the centre-of-mass
motion and of the Hamiltonian for the internal degrees of freedom.

Equation (8.99) has an obvious physical interpretation: The atom–laser interaction induces transitions from the ground to the excited state manifold. By the conservation of angular momentum, in the transition $|g, -1\rangle \rightarrow |e, 0\rangle$ the atom absorbs a right-circular polarized photon from the beam propagating in the positive z-direction. The conservation of linear momentum thus implies that the atomic momentum changes by an amount $+\hbar \vec{k}_L$. Likewise, in the transition $|g, +1\rangle \rightarrow |e, 0\rangle$ a left-circular polarized photon is absorbed from the beam propagating in the negative z-direction such that the atomic momentum changes by $-\hbar \vec{k}_L$.

The form (8.99) for the laser interaction suggests introducing, in analogy to the procedure of the preceding section (cf. eqns (8.73) and (8.80)), the states

$$|\psi_{nc}(\vec{p})\rangle = \frac{1}{\sqrt{2}} \left(|g, +1, \vec{p} + \hbar \vec{k}_L\rangle + |g, -1, \vec{p} - \hbar \vec{k}_L\rangle \right), \tag{8.101}$$

$$|\psi_c(\vec{p})\rangle = \frac{1}{\sqrt{2}} \left(|g, +1, \vec{p} + \hbar \vec{k}_L\rangle - |g, -1, \vec{p} - \hbar \vec{k}_L\rangle \right), \tag{8.102}$$

which satisfy the normalization conditions

$$\langle \psi_j(\vec{p}) | \psi_{j'}(\vec{p}') \rangle = \delta_{jj'} \delta(\vec{p} - \vec{p}'), \tag{8.103}$$

where j and j' stand for 'nc' or 'c'. In terms of these states we can write the atom–laser interaction Hamiltonian as

$$H_L(t) = \frac{\hbar \Omega}{2} \int d^3 p \, |e, 0, \vec{p}\rangle \langle \psi_c(\vec{p})| e^{-i\omega_L t} + \text{h.c.} \tag{8.104}$$

This form shows directly that the laser field couples only to the state $|\psi_c(\vec{p})\rangle$, whereas $|\psi_{nc}(\vec{p})\rangle$ is a zero mode of this interaction,

$$H_L(t) |\psi_c(\vec{p})\rangle = \frac{\hbar \Omega}{2} |e, 0, \vec{p}\rangle e^{-i\omega_L t}, \tag{8.105}$$

$$H_L(t) |\psi_{nc}(\vec{p})\rangle = 0. \tag{8.106}$$

Let us now turn to the interaction of the atom with the quantized radiation field. Using plane wave field modes normalized in a box of volume V with periodic boundary conditions,

$$U_\lambda(\vec{x}) = \frac{1}{\sqrt{V}} \vec{e}_M(\vec{k}) e^{i\vec{k} \cdot \vec{x}}, \tag{8.107}$$

we may write the electric field operator evaluated at the position $\vec{R}$ of the centre of mass as

$$\vec{E}(\vec{R}) = i \sum_{\vec{k}, M = \pm 1} \sqrt{\frac{2\pi \hbar \omega_k}{V}} \left(\vec{e}_M(\vec{k}) \exp \left[-i(\omega_k t - \vec{k} \cdot \vec{R}) \right] b_M(\vec{k}) \right.$$
$$\left. - \vec{e}_M^*(\vec{k}) \exp \left[+i(\omega_k t - \vec{k} \cdot \vec{R}) \right] b_M^\dagger(\vec{k}) \right). \tag{8.108}$$

Here $\hbar \vec{k}$ is the photon momentum, $\omega_k = ck$ the frequency, and the vectors $\vec{e}_M(\vec{k})$ denote the corresponding polarization vectors. We take here $M = \pm 1$, where

$M = +1$ corresponds to a right-circular, and $M = -1$ to a left-circular polarized photon (helicity $+1$ and -1, respectively).

The operators $b_M(\vec{k})$ and $b_M^\dagger(\vec{k})$ denote the annihilation and creation operators for photons with wavenumber $\vec{k}$ and polarization M. The operation describing the measurement of such a photon over the time interval $\tau = t - t_0$ is found to be

$$\Omega_{\vec{k},M} = -\frac{i}{\hbar} \int\limits_{t_0}^{t} dt' \langle \vec{k}, M | H_I(t') | 0 \rangle$$

$$= -i\sqrt{\frac{2\pi\omega_k}{\hbar V}} e^{i(\omega_k - \omega)t_0} \frac{e^{i(\omega_k - \omega)\tau} - 1}{\omega_k - \omega} \vec{e}_M^*(\vec{k}) \cdot \vec{A} e^{-i\vec{k}\cdot\vec{R}}. \quad (8.109)$$

Thus we deduce that the back-action for the photon measurement is given by (omitting an irrelevant phase factor)

$$\frac{\Omega_{\vec{k},M}\psi}{||\Omega_{\vec{k},M}\psi||} = \frac{\vec{e}_M^*(\vec{k}) \cdot \vec{A} e^{-i\vec{k}\cdot\vec{R}}\psi}{||\vec{e}_M^*(\vec{k}) \cdot \vec{A} e^{-i\vec{k}\cdot\vec{R}}\psi||}. \quad (8.110)$$

We see that the operation $\Omega_{\vec{k},M}$ acting on the atomic wave function ψ involves the operator $\exp(-i\vec{k}\cdot\vec{R})$ which induces a shift of the atomic momentum by $-\hbar\vec{k}$ (see eqn (8.97)). This shift describes the recoil of the atomic momentum caused by the emission of a photon with momentum $+\hbar\vec{k}$.

Thus we observe that the measurement of the linear momentum of the photon amounts to an indirect measurement of the recoil momentum of the centre of mass of the atom. This is also seen directly from the explicit form for the back-action which is obtained by making use of eqns (8.97) and (8.63). We find

$$\vec{e}_M^*(\vec{k}) \cdot \vec{A} e^{-i\vec{k}\cdot\vec{R}} = \frac{-Md}{2\sqrt{2}} \int d^3p \left(e^{-i\varphi}(\cos\theta - M)|g, +1, \vec{p} - \hbar\vec{k}\rangle\langle e, 0, \vec{p}| \right.$$

$$\left. + e^{+i\varphi}(\cos\theta + M)|g, -1, \vec{p} - \hbar\vec{k}\rangle\langle e, 0, \vec{p}| \right). \quad (8.111)$$

This equation shows that the excited state decays spontaneously into a linear superposition of the ground states of the atom whereby the atomic momentum changes by $-\hbar\vec{k}$. The amplitudes of the ground states depend on the direction of the photon momentum $\hbar\vec{k}$ through the polar coordinates (θ, φ) of $\vec{k}$.

With the help of eqn (8.111) we get for the rate $d\Gamma$ of the emission of a photon into the solid angle $d\Omega(\vec{k})$ with polarization M

$$d\Gamma(\vec{k}, M) = \frac{1}{\tau} \sum_{\vec{k}' \in d\Omega(\vec{k})} ||\Omega_{\vec{k}',M}\psi||^2 \quad (8.112)$$

$$= \frac{3}{32\pi}\gamma(1 + \cos^2\theta)\langle\psi|\Pi_e|\psi\rangle d\Omega(\vec{k}), \quad (8.113)$$

where $\gamma = 4\omega^3|d|^2/3\hbar c^3$. Equation (8.113) shows that the rate of spontaneous radiation is independent of M and of the azimuth φ. It only depends on the z-component

$$u \equiv \hbar k_z = \hbar k \cos\theta \tag{8.114}$$

of the photon momentum. According to eqn (8.113) the random variable u follows the distribution

$$q(u) = \frac{3}{8}\frac{1}{\hbar k}\left[1 + \left(\frac{u}{\hbar k}\right)^2\right] \tag{8.115}$$

which is normalized as

$$\int_{-\hbar k}^{+\hbar k} du\, q(u) = 1. \tag{8.116}$$

The above results yield the full three-dimensional structure of the process. An effective one-dimensional description is obtained if we average over the un-observed x and y-components of the atomic momentum and over the azimuth φ and the polarization M. This means that we now describe the process on the non-selective level with regard to these degrees of freedom. The jump operators describing the back-action on the atom then depend only on the z-component u of the photon momentum and project either on $|g, +1\rangle$ or on $|g, -1\rangle$, that is we have the following two types of jump operator

$$J_+(u) = \frac{1}{\sqrt{2}}\int dp|g, +1, p-u\rangle\langle e, 0, p|, \tag{8.117}$$

$$J_-(u) = \frac{1}{\sqrt{2}}\int dp|g, -1, p-u\rangle\langle e, 0, p|, \tag{8.118}$$

where $p \equiv p_z$ denotes the z-component of the atomic momentum.

Collecting our results we obtain the following stochastic differential equation for the PDP describing the dynamics of the internal degree of freedom and of the z-component of the translational motion of the atom,

$$d\psi(t) = -\frac{i}{\hbar}G(\psi)dt + \int_{-\hbar k}^{+\hbar k} du \sum_{s=\pm}\left[\frac{J_s(u)\psi}{||J_s(u)\psi||} - \psi\right]dN_{u,s}(t), \tag{8.119}$$

where the Poisson increments satisfy

$$E[dN_{u,s}(t)] = \gamma||J_s(u)\psi(t)||^2q(u)dt, \tag{8.120}$$

$$dN_{u,s}(t)dN_{u',s'}(t) = \delta_{ss'}\delta(u-u')dN_{u,s}(t). \tag{8.121}$$

One should note that $dN_{u,s}$ is a field of Poisson increments (see Section 1.5.4) indexed by a continuous variable u for the jump sizes. The generator $G(\psi)$ of the deterministic motion is given by

$$G(\psi) = \hat{H}\psi + \frac{i\hbar\gamma}{2}\langle\psi|\Pi_e|\psi\rangle\psi, \tag{8.122}$$

and the non-Hermitian Hamiltonian takes the form

$$\hat{H} = \frac{P^2}{2M} + \hbar\left(-\delta - \frac{i\gamma}{2}\right)\Pi_e + H_L. \tag{8.123}$$

The quantity $\delta = \omega_L - \omega$ is the detuning between atomic transition frequency ω and laser frequency ω_L, which will be taken to be zero in the following, i.e. $\delta = 0$. Here we have also performed a canonical transformation with the unitary operator $\exp\left(-i\omega_L\Pi_e t\right)$ in order to remove the explicit time dependence of the Hamiltonian, such that the atom–laser interaction reads

$$H_L = \frac{\hbar\Omega}{2}\int dp|e, 0, p\rangle\langle\psi_c(p)| + h.c. \tag{8.124}$$

These equations describe the PDP for the cooling mechanism.

8.3.2 Coherent population trapping

To understand the essential features of the stochastic dynamics of the cooling mechanism we introduce the family of the manifolds

$$\begin{aligned}\mathcal{M}(p) &= \text{span}\left\{|e, 0, p\rangle, \; |g, +1, p + \hbar k_L\rangle, \; |g, -1, p - \hbar k_L\rangle\right\}\\ &= \text{span}\left\{|e, 0, p\rangle, \; |\psi_{nc}(p)\rangle, \; |\psi_c(p)\rangle\right\},\end{aligned} \tag{8.125}$$

which are labelled by the momentum p. We denote by $\Pi(p)$ the orthogonal projection onto the manifold $\mathcal{M}(p)$. It is then easily verified that this projection commutes with the Hamiltonian $\hat{H}$, that is,

$$\left[\hat{H}, \Pi(p)\right] = 0. \tag{8.126}$$

This fact is connected to the conservation of the total linear momentum of the atom–laser system and implies that the manifolds $\mathcal{M}(p)$ are invariant under the deterministic evolution of the process, which means that, once the state vector has reached $\mathcal{M}(p)$ through a jump, it remains in $\mathcal{M}(p)$ until the next jump occurs. If we rewrite the jump operators as

$$J_\pm(u) = \frac{1}{2}\int dp\left(|\psi_{nc}(p - u \mp \hbar k_L)\rangle \pm |\psi_c(p - u \mp \hbar k_L)\rangle\right)\langle e, 0, p| \tag{8.127}$$

we see that a jump with $J_\pm(u)$ leads to a change of the manifold given by

$$\mathcal{M}(p) \longrightarrow \mathcal{M}(p - u \mp \hbar k_L). \tag{8.128}$$

Remember that u is a random number in the interval $[-\hbar k, +\hbar k]$, which is distributed according to the density $q(u)$ given in eqn (8.115). Physically, u describes the random recoil due to spontaneous emissions.

Let us take the states $|e, 0, p\rangle$, $|\psi_{nc}(p)\rangle$, $|\psi_c(p)\rangle$ as basis states of a certain manifold $\mathcal{M}(p)$. The non-Hermitian Hamiltonian $\hat{H}(p)$ which generates the deterministic evolution periods in $\mathcal{M}(p)$ can then be written as

$$\hat{H}(p) = \left(\frac{p^2}{2M} + E_R \right) I + \hat{V}(p), \tag{8.129}$$

where I denotes the 3×3 unit matrix, E_R is the recoil energy (8.86) and we have introduced the non-Hermitian 3×3 matrix

$$\hat{V}(p) = \begin{pmatrix} -E_R - i\hbar\gamma/2 & 0 & \hbar\Omega/2 \\ 0 & 0 & \hbar kp/M \\ \hbar\Omega/2 & \hbar kp/M & 0 \end{pmatrix}. \tag{8.130}$$

Note that the off-diagonal elements describe the various couplings, namely the atom–laser interaction

$$\langle e, 0, p' | \hat{H} | \psi_c(p) \rangle = \langle e, 0, p' | H_L | \psi_c(p) \rangle = \frac{\hbar\Omega}{2} \delta(p - p'), \tag{8.131}$$

and the coupling between the ground state levels,

$$\langle \psi_c(p') | \hat{H} | \psi_{nc}(p) \rangle = \langle \psi_c(p') | P^2/2M | \psi_{nc}(p) \rangle = \frac{\hbar kp}{M} \delta(p - p'), \tag{8.132}$$

leading to the Doppler energy $\hbar kp/M$. It is this last term which yields a coupling between the internal degree of freedom and the translational motion, called the motional coupling.

For simplicity we set $\hbar = 1$ in the following. As a specific example we shall investigate below the laser cooling on the transition $2^3 S_1 \leftrightarrow 2^3 P_1$ of helium for which we have $2\pi/k = 1.083$ μm, $\gamma/2\pi = 1.6$ MHz, and $E_R/\hbar\gamma = 0.027$. The Rabi frequency is chosen to be $\Omega = 0.3\gamma$. This was the first system for which the cooling mechanism described below has been demonstrated experimentally. All numerical values and simulations given below refer to this example.

The key point of the sub-recoil cooling process is the spectrum of the complex eigenvalues of the non-Hermitian Hamiltonian $\hat{V}(p)$ and the resulting structure of the waiting time distribution

$$F[\psi, \tau] = 1 - \| \exp(-i\hat{V}(p)\tau)\psi \|^2. \tag{8.133}$$

If $p = 0$ we immediately read off from eqn (8.130) that $|\psi_{nc}(0)\rangle$ is a zero mode of $\hat{V}(0)$. This is precisely the dark state which was already encountered in Section 8.2. Due to the atom–laser interaction and the motional coupling all three levels develop a finite width for non-vanishing momentum p. These widths are defined as twice the negative imaginary parts of the complex eigenvalues $\varepsilon_e(p)$, $\varepsilon_{nc}(p)$, and $\varepsilon_c(p)$ of $\hat{V}(p)$,

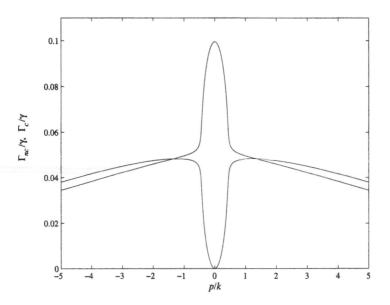

FIG. 8.7. Plot of the widths $\Gamma_{nc}(p)$ and $\Gamma_c(p)$ of the ground state manifolds as a function of the atomic momentum p. The parameters chosen correspond to the transition $2^3S_1 \leftrightarrow 2^3P_1$ of helium.

$$\Gamma_e(p) = -2\Im\,\varepsilon_e(p), \tag{8.134}$$

$$\Gamma_{nc}(p) = -2\Im\,\varepsilon_{nc}(p), \tag{8.135}$$

$$\Gamma_c(p) = -2\Im\,\varepsilon_c(p). \tag{8.136}$$

The corresponding exact eigenvectors of the non-Hermitian Hamiltonian $\hat{V}(p)$ will be denoted by $|\Phi_e(p)\rangle$, $|\Phi_{nc}(p)\rangle$, and $|\Phi_c(p)\rangle$. Here we have labelled the eigenstates such that their continuously connected states at $p = 0$ have the largest weight on the states $|\psi_e(0)\rangle$, $|\psi_{nc}(0)\rangle$, and $|\psi_c(0)\rangle$, respectively. In particular, $\Gamma_{nc}(p) \to 0$ and $|\Phi_{nc}(p)\rangle \to |\psi_{nc}(0)\rangle$ for $p \to 0$. In fact, it is easily found with the help of the characteristic eigenvalue equation that $\Gamma_{nc}(p)$ is given for small p by the expression

$$\Gamma_{nc}(p) = \gamma \left(\frac{4E_R}{\Omega} \right)^2 \left(\frac{p}{k} \right)^2 + \mathcal{O}(p^4). \tag{8.137}$$

The width of the state $|\Phi_{nc}(p)\rangle$ which has a large contribution from the state $|\psi_{nc}(p)\rangle$ thus vanishes as p^2. This is illustrated in Fig. 8.7 which shows a plot of $\Gamma_{nc}(p)$ and $\Gamma_c(p)$ as a function of the momentum p. The third width $\Gamma_e(p)$ may be found from the condition that the sum of the widths is identically equal to γ. We observe a dip in the curve for $\Gamma_{nc}(p)$ and a peak of the width $\Gamma_c(p)$ around $p = 0$. For large values of the momentum both widths are of the same order and decrease due to the Doppler shift.

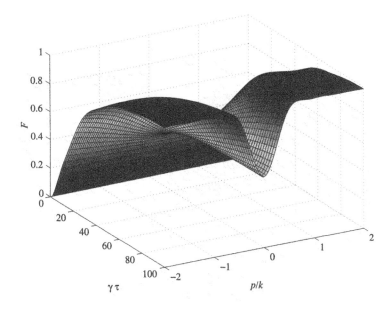

FIG. 8.8. The waiting time distribution $F[|g, +1, p + k\rangle, \tau]$ as a function of the waiting time τ and of the momentum p. As a result of the quantum interference in the laser excitation, with probability $\frac{1}{2}$ an atom which is initially in the state $|g, +1, k\rangle$ never leaves the manifold $\mathcal{M}(0)$.

The plot of the decay rates shows that around zero momentum $\Gamma_{nc}(p)$ and $\Gamma_c(p)$ differ by orders of magnitude. This fact has a drastic influence on the properties of the waiting time distribution (8.133). As we have seen the atomic state vector enters some manifold $\mathcal{M}(p)$ (eqn (8.125)) through a jump. If p is exactly equal to zero, then with a probability of $1/2$ the atom remains trapped forever in the manifold $\mathcal{M}(p = 0)$. This is due to the fact that in both states $|g, \pm 1, \pm k\rangle$ the amplitude for being in the non-coupling state $|\psi_{nc}(p = 0)\rangle$, which is an exact zero mode of $\hat{V}(p = 0)$, is equal to $1/\sqrt{2}$. We illustrate this in Figs. 8.8 and 8.9 which show the waiting time distribution (8.133) for the initial states $|g, +1, p + k\rangle$ and $|\psi_{nc}(p)\rangle$ as a function of the waiting time τ and of the momentum p.

For small but non-zero momentum p the atomic state vector leaves the manifold $\mathcal{M}(p)$ with certainty. However, the waiting time τ for the next jump can be extremely long: Both states $|g, \pm 1, p \pm k\rangle$ have a large contribution (of approximately 50%) from the state $|\Phi_{nc}(p)\rangle$ which is an exact eigenstate of $\hat{V}(p)$ with an extremely small width $\Gamma_{nc}(p)$. What happens is that during the deterministic evolution period following the jump into $\mathcal{M}(p)$ the atomic state vector is rapidly driven into the state $|\Phi_{nc}(p)\rangle$ where it remains trapped for a long time. The approach to the state $|\Phi_{nc}(p)\rangle$ is very fast since the widths $\Gamma_c(p)$ and $\Gamma_e(p)$ are by orders of magnitude larger than the width $\Gamma_{nc}(p)$.

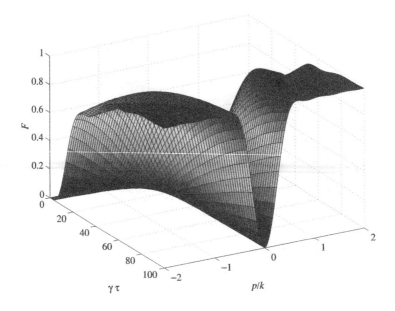

FIG. 8.9. The waiting time distribution $F[|\psi_{nc}(p)\rangle, \tau]$ for the initial state $|\psi_{nc}(p)\rangle$. Due to the quantum interference in the laser excitation, the probability of a jump out of the state $|\psi_{nc}(p)\rangle$ is strongly suppressed for small momenta p.

The above picture is confirmed by numerical simulations of the process defined by the stochastic differential equation (8.119). We show in Fig. 8.10 a single realization of the PDP describing the cooling process. The realization involves a total number of 10^4 quantum jumps. The picture shows a remarkable fact which is typical for the sample paths of the process, namely that the total evolution time is dominated by the waiting time τ for a single quantum jump of about $\gamma\tau = 5 \cdot 10^7$. Note that the momentum is constant between the jumps. As will be demonstrated in the next subsection the statistics of the waiting times is governed by a long-range distribution similar to a stable Lévy distribution.

It is also obvious from the figure that the long waiting times are interrupted by a series of very small ones. This is illustrated in Figs. 8.11 and 8.12 where we have depicted enlarged sections of Fig. 8.10. It is this proliferation of vastly different time scales which makes the dynamics so difficult for a treatment based on the optical Bloch equations. In the stochastic simulations, however, the treatment of the long-range waiting time distribution does not cause any problems: On using the inversion method, for example, even the longest waiting time period can be calculated in a single step. It is thus not necessary to adapt the numerical time step to the smallest time scale of the problem; the stochastic simulation algorithm chooses an appropriate time step automatically.

Since $|\Phi_{nc}(p)\rangle$ is very close to $|\psi_{nc}(p)\rangle$ for small p we can thus say that the

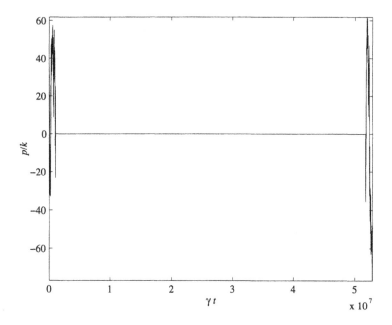

FIG. 8.10. Stochastic simulation of sub-recoil laser cooling. The plot shows the momentum $p(t)$ for a single realization of the stochastic process defined by the differential equation (8.119) involving 10^4 quantum jumps.

atoms that fall into a small region around $p = 0$ can be trapped coherently with a high probability for a long time and that an extremely large fraction of those atoms is with certainty in the state $|\psi_{nc}(p)\rangle$. In order to have really a cooling process we need, however, a further mechanism which brings at least a part of the atoms into the vicinity of $p = 0$. Such a mechanism is provided by the momentum diffusion due to the random kicks caused by the spontaneous emission processes. Thus we see that the cooling of the atomic momenta results from the interplay of two basic processes, namely from the diffusion of momentum, which is due to spontaneous emissions, and from the coherent population trapping, which stems from a quantum interference effect. One may interpret this by saying that the quantum interference effect gives rise to a diffusion in momentum space involving a diffusion coefficient which strongly depends on the momentum. In contrast to the usual scheme of Doppler cooling the basic physical mechanism is not an effective friction force, but rather a velocity selective coherent population trapping.

8.3.3 Waiting times and momentum distributions

To get further insights into the statistics of the random waiting times we fix a region $|p| < p_t$ of momenta smaller than some trapping momentum p_t. It is assumed here that p_t is chosen so small that the escape rate $\Gamma_{nc}(p)$ from the

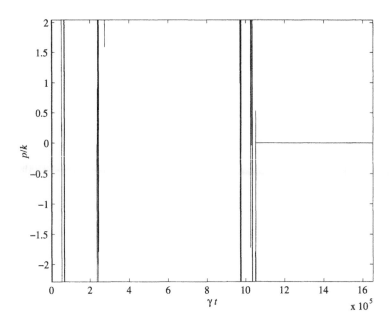

FIG. 8.11. Enlarged section of Fig. 8.10 showing a succession of a large number of jumps with small waiting times. These jumps finally drive the atom into the vicinity of zero momentum where it remains trapped for a long time.

states $|\Phi_{nc}(p)\rangle$ behaves to a good approximation as p^2 for $|p| < p_t$ (see eqn (8.137)), and that $p_t \ll k$. Suppose that an atom lands in this region through a quantum jump starting outside the trapping region with momentum $|p| > p_t$. Since the size of the jumps is of the order k it may be assumed further that the atoms land somewhere in the trapping region with a uniform momentum distribution.

As discussed in the preceding subsection, with a high probability the atomic state after the jump is driven nearly immediately into the state $\Phi_{nc}(p)$, from which it may escape with a rate $\Gamma_{nc}(p)$. Under the condition that a specific momentum p is given the density of the waiting time distribution for the next quantum jump is approximately

$$f(p,\tau) = \Gamma_{nc}(p) \exp\left(-\Gamma_{nc}(p)\tau\right). \qquad (8.138)$$

If we combine this with the above assumption of a uniform momentum distribution we get the following expression for the unconditioned probability density of the waiting times,

$$s(\tau) = \frac{1}{2p_t} \int_{-p_t}^{+p_t} f(p,\tau)dp = \frac{1}{2\Gamma_{nc}(p_t)^{1/2}\tau^{3/2}} \int_{0}^{\Gamma_{nc}(p_t)\tau} du\sqrt{u}e^{-u}. \qquad (8.139)$$

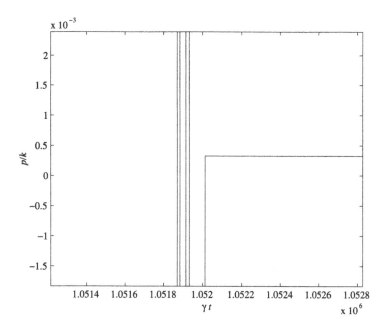

FIG. 8.12. Enlarged section of Fig. 8.11 showing that the atom lands in the trapping region with a very small momentum of less than $0.5 \cdot 10^{-3}k$.

For $\Gamma_{nc}(p_t)\tau \gg 1$ the integral converges to $\Gamma(3/2) = \sqrt{\pi}/2$ which leads to the asymptotic expression

$$s(\tau) \approx \frac{\sqrt{\pi}}{4} \frac{1}{\tau_t} \left(\frac{\tau_t}{\tau}\right)^{3/2}, \qquad (8.140)$$

where we have defined the characteristic trapping time τ_t associated with p_t,

$$\tau_t = [\Gamma_{nc}(p_t)]^{-1}. \qquad (8.141)$$

By virtue of eqn (8.137) this yields

$$\gamma\tau_t \approx \left(\frac{\Omega}{4E_R}\right)^2 \left(\frac{k}{p_t}\right)^2. \qquad (8.142)$$

The distribution (8.140) for the waiting time shows a slowly decaying power law behaviour $s(\tau) \sim \tau^{-3/2}$, which is characteristic of a stable Lévy distribution with scaling exponent $\alpha = \frac{1}{2}$. By an appropriate choice for the scaling parameter we find that the Lévy distribution

$$s_L(\tau) = \frac{\sqrt{\pi}}{4} \frac{1}{\tau_t} \left(\frac{\tau_t}{\tau}\right)^{3/2} \exp\left(-\frac{\pi^2}{16} \frac{\tau_t}{\tau}\right) \qquad (8.143)$$

has the same asymptotic behaviour as the waiting time distribution $s(\tau)$. The distribution (8.143) has already been encountered in eqn (1.264) as an example

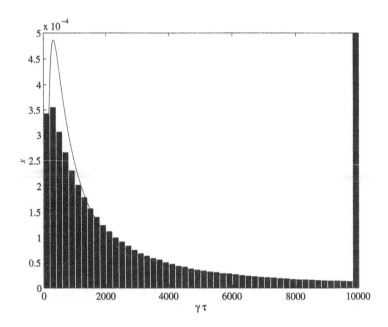

FIG. 8.13. Distribution $s(\tau)$ of the waiting times τ for initial states with momentum inside the trapping region $|p| < p_t$. The figure shows a histogram of a simulation with 10^6 realizations with $p_t = 0.1k$ corresponding to a trapping time of about $\gamma\tau_t = 771$. For comparison the figure also shows the stable Lévy distribution with scale parameter $\alpha = \frac{1}{2}$ (solid line). The last bin of the histogram contains all waiting times larger than those shown.

of a stable Lévy distribution whose explicit analytical expression is known. We have thus found that the quantum interference effect which leads to a vanishing width of the non-coupling state $|\psi_{nc}(p=0)\rangle$ gives rise to a long-range waiting time distribution (Bardou *et al.*, 1994).

The power law decay of the waiting time distribution is characteristic of the asymptotic behaviour of a random variable which follows asymptotically a stable Lévy distribution. Figure 8.13 shows the result of a numerical simulation of the waiting time distribution and compares it with the distribution (8.143). One clearly observes the slow decay of the distribution. In the simulations we have used $p_t = 0.1k$ corresponding to a trapping time of about $\gamma\tau_t = 771$ by virtue of eqn (8.143). As expected the simulation data are in good agreement with the distribution (8.143) for times much larger than this trapping time τ_t.

As explained in Section 1.6.3 the distribution (8.143) is infinitely divisible and stable with a scaling exponent of $\alpha = \frac{1}{2}$. It follows that the random variable τ following $s_L(\tau)$ has the following property. We take N independent copies $\tau_1, \tau_2, \ldots, \tau_N$ of τ and define the scaled sum

$$\bar{\tau} \equiv \frac{1}{N^{1/\alpha}} \sum_{i=1}^{N} \tau_i = \frac{1}{N^2} \sum_{i=1}^{N} \tau_i. \tag{8.144}$$

The new random variable $\bar{\tau}$ is then again distributed according to the Lévy distribution (8.143). Note that the scaling factor is $1/N^2$ and *not* $1/N$ as it would be in those cases where the central limit theorem could be applied. The quantity $\sum_{i=1}^{N} \tau_i/N$ scales as $N\bar{\tau}$ which shows that the ordinary mean value of a large number N of copies of τ is with a high probability larger than any given term of the sum. This demonstrates the drastic departure of the waiting time statistics from the usual behaviour described by the central limit theorem which is not applicable here since the moments of τ diverge.

The above scaling relations are exactly valid only for the Lévy distribution $s_L(\tau)$ and not for the true waiting time distribution $s(\tau)$. However, we can invoke the stability property of the Lévy distribution to derive universal scaling relations. To this end, we consider an infinite sequence τ_i, $i = 1, 2, 3, \ldots$, of copies of a random variable following the distribution $s(\tau)$ and take a fixed N. Equation (8.144) can then be viewed as defining a transformation

$$\mathcal{F}_N : \{\tau_i\} \mapsto \{\bar{\tau}_j\} \tag{8.145}$$

to a new sequence $\bar{\tau}_j$, $j = 1, 2, 3, \ldots$, of random numbers given explicitly by

$$\bar{\tau}_j = \frac{1}{N^{1/\alpha}} \sum_{k=1+N(j-1)}^{Nj} \tau_k. \tag{8.146}$$

The map $\mathcal{F}_N$ can be regarded as a renormalization transformation: The sequence τ_i is grouped into sets of N elements each, the elements of each set are summed and the sum is scaled by a factor of $N^{-1/\alpha}$. This yields a new, *coarse-grained* sequence of random variables $\bar{\tau}_j$.

It can be shown that under successive application of the renormalization transformation $\mathcal{F}_N$ the distributions of the coarse-grained sequences converge to a corresponding α-stable Lévy distribution, which is (8.143) in the present case. The renormalization transformation thus drives the waiting time distribution $s(\tau)$ into the $\frac{1}{2}$-stable Lévy distribution (8.143).

The mathematical reason for this behaviour is that the attractive fixed points of the renormalization transformation $\mathcal{F}_N$, considered as a map in the space of probability distributions, are given by the α-stable distributions. In this context the central limit theorem can be understood as expressing the existence of a fixed point characterized by $\alpha = 2$. The waiting time distribution $s(\tau)$ discovered above belongs to the basin of attraction of the fixed point characterized by $\alpha = 1/2$. This is the origin of a universal behaviour in the statistics of the waiting times. The renormalization transformation is illustrated in Fig. 8.14 for the case $N = 5$.

Let us finally look at the resulting distributions of the atomic momenta (Aspect *et al.*, 1989). Such a distribution may be defined by

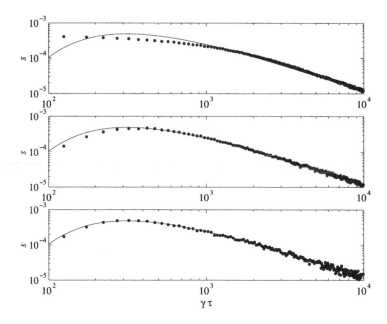

FIG. 8.14. Illustration of the renormalization transformation (8.145) of the random waiting time in laser cooling for $N = 5$. The figure shows double logarithmic plots for a simulation with a total number of 10^6 realization with $p_t = 0.1k$ (symbols) and the corresponding stable Lévy distribution (solid lines). Top: original data, middle: first application of $\mathcal{F}_5$, bottom: second application of $\mathcal{F}_5$.

$$w(p) = \langle p|\mathrm{tr}_{\mathrm{int}}(\rho)|p\rangle, \qquad (8.147)$$

where $\rho = \mathrm{E}[|\psi\rangle\langle\psi|]$ is the atomic density matrix and $\mathrm{tr}_{\mathrm{int}}$ denotes the trace over the internal degrees of freedom of the atom. This normalized density describes the distribution of p as it will be found if a momentum measurement is performed on a sample of atoms. Since in the trapping region an appreciable amount of the total number of atoms will be in states $|\psi_{nc}(p')\rangle$ with p' small and since $\langle p|\psi_{nc}(p')\rangle$ consists of two δ-peaks at $p = p' \pm k$ we expect that the distribution $w(p)$ contains two peaks at positions $p = \pm k$ over a broad background.

For a rough estimate of the width δp of these peaks we fix some interaction time t and ask for the range $|p| < \delta p$ of momenta such that the atoms remain trapped with appreciable probability during time t. Clearly, this range of momenta is found from the relation $\Gamma_{nc}(\delta p)t \sim 1$, which yields on using (8.137)

$$\frac{\delta p}{k} \sim \frac{\Omega}{4E_R}\frac{1}{\sqrt{\gamma t}}. \qquad (8.148)$$

This estimate shows that δp scales as $\Omega/\sqrt{t}$. It implies that there is, at least in principle, no lower limit for the width of the momentum distribution; for

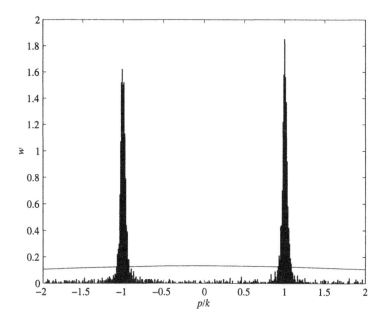

FIG. 8.15. The distribution $w(p)$ of the atomic momentum at time $\gamma t = 10^4$ obtained from a simulation with 10^4 realizations. The distribution shows two sharp peaks at $p = \pm k$ with a width δp which is much smaller than the recoil momentum k. The initial momenta have been drawn from a Gaussian distribution with a standard deviation of $3k$ (solid line).

sufficiently large interaction times the width δp becomes much smaller than the recoil momentum k. Consequently, the associated cooling temperature T_c defined by

$$\frac{1}{2}k_{\mathrm{B}}T_c = \frac{\delta p^2}{2M} \tag{8.149}$$

is much smaller than the recoil temperature T_R. This is illustrated in Fig. 8.15 which shows the distribution $w(p)$ of the atomic momenta at time $\gamma t = 10^4$ obtained from a stochastic simulation with 10^4 realizations.

Another momentum distribution may be defined as

$$v(p) = \langle \psi_{nc}(p) | \rho | \psi_{nc}(p) \rangle, \tag{8.150}$$

which yields the fraction of atoms that will be found on measurement in the state $|\psi_{nc}(p)\rangle$. Here we expect a single peak at $p = 0$ which is clearly seen in Fig. 8.16.

Concluding, it must be emphasized that it makes no sense to define the cooling temperature in terms of the mean kinetic energy of the atoms. The reason for choosing definition (8.149) for T_c is that the trapped atoms are with a high probability in the states $|\psi_{nc}(p)\rangle$ which represent *coherent* superpositions of

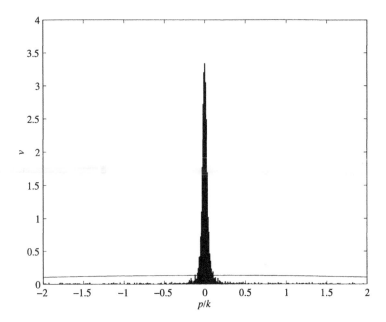

FIG. 8.16. The distribution $v(p)$ (eqn 8.150) at time $\gamma t = 10^4$ obtained from a simulation with 10^4 realizations. The distribution shows a single peak at $p = 0$ with a width much smaller than the recoil momentum k. The initial momenta have been drawn from a Gaussian distribution with a standard deviation of $3k$ (solid line).

momentum eigenstates. The definition of T_c thus takes into account the quantum correlations between the internal and the translational degrees of freedom of the atoms.

8.4 Strong field interaction and the Floquet picture

The situation encountered in the preceding sections was that of a bound quantum system which is coupled to an environment consisting of a continuum of electromagnetic field modes, and to a coherent external driving field. The procedure to treat this situation was simply to add the Hamiltonian describing the external driving to the coherent part of the equation of motion. However, following the derivation of the quantum optical master equation, one observes that this procedure is only justified if the external driving field represents a small perturbation and may be treated on an equal footing with the coupling between the reduced system and its environment. Adding the Hamiltonian of the external field to the coherent part of the dynamics without also changing the dissipative part of the dynamics, amounts to making the rotating wave approximation for the interaction between system and driving field. It is obvious that these conditions are violated if the external field is strong.

An appropriate strategy to deal with strong periodic driving fields is to treat the interaction between the open system and external driving field exactly by employing the Floquet basis of the open system, rather than the stationary eigenstates of the unperturbed system Hamiltonian. This idea leads to a Markovian quantum master equation for the reduced density matrix which is in Lindblad form when written in the Floquet basis. Such an approach has been applied to the description of Rydberg atoms in strong microwave fields under the influence of external noise (Blümel et al., 1991), and to a parametrically driven harmonic oscillator (Kohler, Dittrich and Hänggi, 1997).

In the present section we shall connect the idea of treating dissipative quantum systems in strong driving fields in the Floquet picture with the stochastic representation of the continuous monitoring of the environment. We derive the PDP describing the spectral detection in the presence of strong external driving. As will be seen the jumps of the PDP occur between manifolds spanned by Floquet states of the system, reflecting the back-action resulting from the emitted radiation. The observed frequencies of the radiation spectrum turn out to be the differences between Floquet eigenvalues. We will also discuss a simple example which serves to demonstrate the connection of the Floquet theory to the dressed atom picture of resonance flourescence.

8.4.1 *Floquet theory*

The free evolution of the reduced quantum system is generated by some time-independent Hamiltonian H_S^0. The coupling to the driving field is represented by a Hamiltonian $H_L(t)$ which is assumed to be periodic in time, that is

$$H_L(t + T_L) = H_L(t), \tag{8.151}$$

where $T_L \equiv 2\pi/\omega_L$ and ω_L denotes the driving frequency. Thus, the Hamiltonian that describes the reduced system without its interaction with the environment is given by the T_L-periodic operator

$$H_S(t) = H_S^0 + H_L(t). \tag{8.152}$$

The strategy is now to treat exactly that part of the dynamics which is described by the system Hamiltonian $H_S(t)$. By this procedure one avoids the rotating wave approximation for the interaction between system and external driving, as well as any perturbative treatment for this part of the dynamics. To this end, we invoke the Floquet theory of quantum mechanical systems with a time-periodic Hamiltonian (Shirley, 1965; Zeldovich, 1967). The Schrödinger equation

$$\frac{d}{dt}\psi(t) = -\frac{i}{\hbar}H_S(t)\psi(t) \tag{8.153}$$

is a differential equation with time-periodic coefficients. According to Floquet's theorem there exists a complete set of solutions $\psi_r(t)$ of eqn (8.153) which are labelled by an index r and which can be written in the form

$$\psi_r(t) = u_r(t) \exp\left\{-i\varepsilon_r t/\hbar\right\}. \qquad (8.154)$$

The form of these solutions is quite similar to that of the stationary states of a system with a time-independent Hamiltonian. The quantities ε_r appearing in the phase factors $\exp(-i\varepsilon_r t/\hbar)$ are time independent and are called quasi-energies or Floquet indices. The state vectors $u_r(t)$ depend periodically on time,

$$u_r(t + T_L) = u_r(t), \qquad (8.155)$$

and are referred to as Floquet states. The completeness of the Floquet solutions is expressed by the relation

$$\sum_r |u_r(t)\rangle\langle u_r(t)| = I, \qquad (8.156)$$

which shows that, for each fixed time t, the $u_r(t)$ form a complete set of basis states. Moreover these states can be chosen to be orthogonal,

$$\langle u_r(t)|u_{r'}(t)\rangle = \delta_{rr'}. \qquad (8.157)$$

Thus, the Floquet states form a time-dependent basis in the Hilbert space of the system under consideration. It follows that any solution $\psi(t)$ of the time-dependent Schrödinger equation (8.153) can be decomposed as

$$\psi(t) = \sum_r a_r u_r(t) \exp\left\{-i\varepsilon_r t/\hbar\right\}. \qquad (8.158)$$

The important point to note is that according to Floquet's theorem the amplitudes a_r are *time independent*.

Inserting eqn (8.154) into the time-dependent Schrödinger equation we observe that the Floquet wave functions $u_r(t)$ may be obtained from the eigenvalue equation

$$\mathbf{H}u_r(t) \equiv \left\{H_S(t) - i\hbar\partial_t\right\} u_r(t) = \varepsilon_r u_r(t) \qquad (8.159)$$

by imposing periodic boundary conditions of the form (8.155). The operator $\mathbf{H}$ in eqn (8.159) is the Floquet operator which has to be considered as an operator on the extended Hilbert space

$$\mathcal{H}_F = \mathcal{H}_S \otimes \mathcal{H}_{T_L}, \qquad (8.160)$$

where $\mathcal{H}_S$ is the Hilbert space of the open system, and $\mathcal{H}_{T_L}$ denotes the space of square-integrable T_L-periodic functions. The scalar product in the extended Hilbert space is given by

$$\langle\langle u|v\rangle\rangle \equiv \int_0^{T_L} \frac{dt}{T_L} \langle u(t)|v(t)\rangle, \qquad (8.161)$$

such that the expression $-i\hbar\partial_t$ in the eigenvalue eqn (8.159) becomes a Hermitian operator on $\mathcal{H}_F$.

It is important to remark the following fact. Given some solution $u_r(t)$ of the Floquet eigenvalue equation (8.159) with quasi-energy ε_r, then for any integer n also

$$u_{r,n}(t) \equiv u_r(t) \exp\{i\omega_L nt\} \qquad (8.162)$$

is a solution with quasi-energy

$$\varepsilon_{r,n} \equiv \varepsilon_r + \hbar\omega_L n. \qquad (8.163)$$

On the other hand, for all integers n the states (8.162) lead to one and the same solution of the time-dependent Schrödinger equation since

$$u_{r,n}(t) \exp\{-i\varepsilon_{r,n}t/\hbar\} = u_r(t) \exp\{-i\varepsilon_r t/\hbar\} . \qquad (8.164)$$

The class of states (8.162) belonging to a fixed r and to different integers n are therefore physically equivalent.

The time-evolution operator $U_S(t, t')$ corresponding to the Schrödinger equation (8.153) obeys

$$\frac{\partial}{\partial t}U_S(t,t') = -\frac{i}{\hbar}H_S(t)U_S(t,t'), \quad U_S(t',t') = I. \qquad (8.165)$$

By virtue of the Floquet representation (8.158) it takes the form

$$U_S(t,t') = \sum_r |u_r(t)\rangle\langle u_r(t')| \exp\left\{-i\varepsilon_r(t-t')/\hbar\right\} , \qquad (8.166)$$

as is easily checked with the help of eqn (8.159) and the completeness relation (8.156).

8.4.2 Stochastic dynamics in the Floquet picture

In order to derive a stochastic process for the dynamics of the reduced system coupled to the continuum of radiation modes we consider the dipole operator

$$\vec{D}(t) = U_S^\dagger(t,0)\vec{D}U_S(t,0). \qquad (8.167)$$

This is the exact Heisenberg picture dipole operator for the system plus external driving field. We are seeking a decomposition of $\vec{D}(t)$ into frequency components $\vec{A}(\omega)$ such that we can write

$$\vec{D}(t) = \sum_\omega \exp\{-i\omega t\}\vec{A}(\omega) + \text{h.c.}, \qquad (8.168)$$

where the sum is extended over a certain set of positive frequencies ω. Employing the Floquet decomposition (8.166) of the time evolution operator $U_S(t,0)$ one finds

$$\vec{D}(t) = \sum_{r,r'} |u_r(0)\rangle\langle u_r(t)|\vec{D}|u_{r'}(t)\rangle\langle u_{r'}(0)| \exp\{-i(\varepsilon_{r'} - \varepsilon_r)t/\hbar\}. \qquad (8.169)$$

Here, the dipole matrix element depends periodically on time and we may use the Fourier decomposition to get

$$\langle u_r(t)|\vec{D}|u_{r'}(t)\rangle = \sum_n \langle\langle u_{r,n}|\vec{D}|u_{r'}\rangle\rangle \exp\{i\omega_L nt\}, \qquad (8.170)$$

where we have used the states (8.162) and the scalar product (8.161) in the extended Hilbert space. Thus, we can write the dipole operator as

$$\vec{D}(t) = \sum_{r,r',n} |u_r(0)\rangle\langle\langle u_{r,n}|\vec{D}|u_{r'}\rangle\rangle\langle u_{r'}(0)| \exp\{-i(\varepsilon_{r'} - \varepsilon_r - \hbar\omega_L n)t/\hbar\}, \qquad (8.171)$$

which finally leads to

$$\vec{A}(\omega) = \sum_{r,r',n} |u_r(0)\rangle\langle\langle u_{r,n}|\vec{D}|u_{r'}\rangle\rangle\langle u_{r'}(0)|. \qquad (8.172)$$

The sum in this expression runs over all sets (r, r', n) of quantum numbers that satisfy the condition

$$\varepsilon_{r'} - \varepsilon_r - \hbar\omega_L n = \hbar\omega > 0. \qquad (8.173)$$

For a given frequency ω there are, in general, several sets (r, r', n) that fulfil this constraint. The frequency component $\vec{A}(\omega)$ of the dipole operator is different from zero if at least one of the corresponding matrix elements is different from zero, that is, if

$$\langle\langle u_{r,n}|\vec{D}|u_{r'}\rangle\rangle \equiv \int_0^{T_L} \frac{dt}{T_L} e^{-i\omega_L nt} \langle u_r(t)|\vec{D}|u_{r'}(t)\rangle \neq 0. \qquad (8.174)$$

This means that the n-th Fourier component of the oscillating matrix element of the dipole operator in the Floquet basis must be non-zero. Let us denote by $\Omega_{\rm rad}$ the set of positive frequencies ω given by eqn (8.173) and the additional selection rule (8.174). The set $\Omega_{\rm rad}$ determines the positions of the peaks in the radiation spectrum emitted by the system.

At this stage we can now proceed in a similar manner as in Section 8.1, whereby the frequency decomposition (8.51) of the interaction picture dipole operator is to be replaced by the decomposition (8.168) of the Heisenberg picture dipole operator. In order to be able to apply the rotating wave approximation with respect to the system–environment coupling the condition

$$|\omega - \omega'| \gg \gamma(\omega)|\langle\langle u_{r,n}|\vec{D}|u_{r'}\rangle\rangle|^2 \qquad (8.175)$$

must be satisfied for the different frequencies of the radiation spectrum. The result is then that the spectral detection leads to a PDP for the state vector of the source which is given by (Breuer and Petruccione, 1997; Breuer, Huber and Petruccione, 2000)

$$d\psi(t) = -\frac{i}{\hbar}G(\psi(t))dt + \sum_{\omega}\sum_{M=0,\pm}\left[\frac{\vec{e}_M^* \cdot \vec{A}(\omega)\psi(t)}{||\vec{e}_M^* \cdot \vec{A}(\omega)\psi(t)||} - \psi(t)\right]dN_{\omega,M}(t).$$

(8.176)

The expressions for the generator $G(\psi)$ and for the expectation values of the Poisson increments $dN_{\omega,M}(t)$ are formally identical to the ones derived in Section 8.1 (see eqns (8.53) and (8.54)). However, the new feature is that the jump operators $\vec{A}(\omega)$ are defined in terms of the Floquet representation, involving the exact time evolution due to the periodic driving force. According to eqn (8.172) the $\vec{A}(\omega)$ are lowering operators pertaining to the positive transition frequencies $\omega \in \Omega_{\text{rad}}$ of the source and describe quantum jumps between manifolds spanned by Floquet states.

Summarizing, we find that the positions of the peaks of the fluorescence spectrum are determined by the differences of the quasi-energies modulo integer multiples of $\hbar\omega_L$. A given frequency ω appears in the radiation spectrum if the matrix elements $\langle\langle u_{r,n}|\vec{D}|u_{r'}\rangle\rangle$ do not vanish, since otherwise the corresponding jump operator would be zero, according to the selection rule (8.174). Besides the usual selection rules for these matrix elements caused by symmetry, those selection rules that arise from the Fourier content of the Floquet states are important. For example, the condition for the generation of high harmonic radiation (L'Huillier, Schafer and Kulander, 1991), i.e. large n in eqn (8.173), is that high Fourier modes of the Floquet states are considerably excited. This condition is fulfilled in particular in the vicinity of near degeneracies or avoided crossings in the quasi-energy spectrum plotted as a function of the driving field amplitude (Breuer, Dietz and Holthaus, 1988).

The equation of motion for the reduced density matrix of the source is easily obtained by determining the equation of motion for the covariance of the PDP derived above. Transforming back to the Schrödinger picture we get the master equation

$$\frac{d}{dt}\rho_S(t) = -\frac{i}{\hbar}[H_S(t), \rho_S(t)] + \mathcal{D}_t(\rho_S(t)),$$

(8.177)

where

$$\mathcal{D}_t(\rho_S) \equiv \sum_{\omega}\gamma(\omega)\left(\vec{A}(\omega,t)\rho_S\vec{A}^\dagger(\omega,t)\right.$$

(8.178)

$$\left. -\frac{1}{2}\vec{A}^\dagger(\omega,t)\cdot\vec{A}(\omega,t)\rho_S - \frac{1}{2}\rho_S\vec{A}^\dagger(\omega,t)\cdot\vec{A}(\omega,t)\right)$$

is the dissipator involving the time-dependent operators

$$\vec{A}(\omega,t) = \sum_{r,r',n}|u_{r,n}(t)\rangle\langle\langle u_{r,n}|\vec{D}|u_{r'}\rangle\rangle\langle u_{r'}(t)|.$$

(8.179)

Again, the sum runs over those sets (r, r', n) that correspond to the given frequency ω of the radiation spectrum. For each fixed t the dissipator $\mathcal{D}_t$ is in Lindblad form. However, by contrast to the case of the quantum optical master equation, it depends explicitly on time. The physical reason for this fact is easily understood. The external driving leads to a strong distortion of the dipole moment. Since the system couples to the environment via its dipole moment, the driving field also strongly influences the dissipation mechanism.

8.4.3 *Spectral detection and the dressed atom*

Let us consider a two-level atom which is strongly driven by a resonant driving field. Assuming the rotating wave approximation to be valid, we may write the system Hamiltonian as

$$H_S(t) = \omega_0 \sigma_+ \sigma_- + \frac{\Omega}{2} \left(\sigma_+ e^{-i\omega_L t} + \sigma_- e^{i\omega_L t} \right), \tag{8.180}$$

where ω_0 denotes the level spacing, Ω is the Rabi frequency and we have set $\hbar = 1$. At resonance, that is for $\omega_L = \omega_0$, a basis of Floquet states is given by

$$u_\pm(t) = \frac{1}{\sqrt{2}} \begin{pmatrix} \pm e^{-i\omega_L t} \\ 1 \end{pmatrix}, \tag{8.181}$$

with corresponding quasi-energies

$$\varepsilon_\pm = \pm \frac{1}{2}\Omega. \tag{8.182}$$

Note that the Floquet states have been chosen such that their corresponding quasi-energy difference is equal to the Rabi frequency Ω. The Schrödinger picture dipole operator is given in the present notation by

$$D = \sigma_+ + \sigma_-. \tag{8.183}$$

Thus, in the weak driving case we have just one jump operator which is given by the lowering operator σ_-. The resulting PDP defined by means of this jump operator is precisely the process that has been used to simulate the quantum optical Bloch equation in Section 6.3.

To determine the jump operators in the strong driving case we employ the Floquet representation (8.166) of the time-evolution operator to obtain the Heisenberg picture dipole operator,

$$D(t) = U_S^\dagger(t,0) D U_S(t,0) = |u_+(0)\rangle\langle u_+(t)|D|u_+(t)\rangle\langle u_+(0)| \tag{8.184}$$
$$+ |u_-(0)\rangle\langle u_-(t)|D|u_-(t)\rangle\langle u_-(0)|$$
$$+ |u_+(0)\rangle\langle u_+(t)|D|u_-(t)\rangle\langle u_-(0)|e^{i(\varepsilon_+ - \varepsilon_-)t}$$
$$+ |u_-(0)\rangle\langle u_-(t)|D|u_+(t)\rangle\langle u_+(0)|e^{i(\varepsilon_- - \varepsilon_+)t}.$$

Since

$$\langle u_\pm(t)|D|u_\pm(t)\rangle = \pm\frac{1}{2}(e^{i\omega_L t} + e^{-i\omega_L t}), \qquad (8.185)$$

$$\langle u_+(t)|D|u_-(t)\rangle = \frac{1}{2}(e^{i\omega_L t} - e^{-i\omega_L t}), \qquad (8.186)$$

we get

$$D(t) = \frac{1}{2}(e^{i\omega_L t} + e^{-i\omega_L t})\left(|u_+(0)\rangle\langle u_+(0)| - |u_-(0)\rangle\langle u_-(0)|\right)$$

$$+\frac{1}{2}(e^{i(\omega_L+\Omega)t} - e^{-i(\omega_L-\Omega)t})|u_+(0)\rangle\langle u_-(0)|$$

$$-\frac{1}{2}(e^{i(\omega_L-\Omega)t} - e^{-i(\omega_L+\Omega)t})|u_-(0)\rangle\langle u_+(0)|. \qquad (8.187)$$

From this relation we immediately infer that we have three jump operators belonging to the positive frequencies ω_L and $\omega_L \pm \Omega$, namely

$$A(\omega_L) = \frac{1}{2}\left(|u_+(0)\rangle\langle u_+(0)| - |u_-(0)\rangle\langle u_-(0)|\right), \qquad (8.188)$$

$$A(\omega_L + \Omega) = \frac{1}{2}|u_-(0)\rangle\langle u_+(0)|, \qquad (8.189)$$

$$A(\omega_L - \Omega) = -\frac{1}{2}|u_+(0)\rangle\langle u_-(0)|. \qquad (8.190)$$

Thus, instead of one jump operator σ_- we get three jump operators in the strong driving case. The condition for strong driving is provided by (8.175) which leads here to the condition $\Omega \gg \gamma$.

To write the corresponding PDP defined by eqn (8.176) more explicitly we decompose the state vector in the Floquet basis as follows,

$$\psi(t) = \phi_+(t)|u_+(0)\rangle + \phi_-(t)|u_-(0)\rangle, \qquad (8.191)$$

and represent it through a two-component vector,

$$\psi(t) \longleftrightarrow \phi(t) \equiv \begin{pmatrix} \phi_+(t) \\ \phi_-(t) \end{pmatrix}. \qquad (8.192)$$

The jump operators given in eqn (8.188) are represented by Pauli matrices,

$$A(\omega_L) \longleftrightarrow \frac{1}{2}\sigma_3, \quad A(\omega_L \pm \Omega) \longleftrightarrow \pm\frac{1}{2}\sigma_\mp. \qquad (8.193)$$

Equation (8.176) therefore leads to the stochastic differential equation

$$d\phi(t) = \left[\frac{\sigma_3\phi(t)}{||\sigma_3\phi(t)||} - \phi(t)\right]dN_3(t) + \sum_{q=\pm}\left[\frac{\sigma_q\phi(t)}{||\sigma_q\phi(t)||} - \phi(t)\right]dN_q(t). \qquad (8.194)$$

We observe that the generator of the deterministic evolution periods of the PDP vanishes, such that the process becomes a pure jump process. The statistics of these jumps is determined by the Poisson increments which satisfy,

$$\mathrm{E}\,[dN_3(t)] = \frac{\gamma}{4}dt, \quad \mathrm{E}\,[dN_\pm(t)] = \frac{\gamma}{4}|\phi_\mp(t)|^2 dt. \tag{8.195}$$

It follows from the relative weight of the jump rates that, provided the system is known to jump at some time t, the jumps with $\sigma_\pm$ occur with the probabilities $\frac{1}{2}|\phi_\mp(t)|^2$, while the jump with σ_3 takes place with probability $\frac{1}{2}$.

The physical meaning of the PDP can be elucidated by drawing on the correspondence between the Floquet picture and a full quantum treatment of the driving mode (Shirley, 1965). This correspondence leads directly to an interpretation of the process in terms of the dressed atom picture (Cohen-Tannoudji and Reynaud, 1977, 1979). In accordance with the notation used in eqn (8.162) we define

$$u_{\pm,n}(t) = u_\pm(t)e^{in\omega_L t}. \tag{8.196}$$

These Floquet states have quasi-energies

$$\varepsilon_{\pm,n} = \pm\frac{1}{2}\Omega + n\omega_L. \tag{8.197}$$

For strong driving fields the Floquet states correspond to the dressed states of the atom, that is to the stationary eigenstates of the combined system consisting of atom and quantized driving mode. For large photon numbers N the eigenenergies $E_{\pm,N}$ of the dressed states $|\pm,N\rangle$ of the atom are approximately given by

$$E_{\pm,N} \approx \pm\frac{1}{2}\Omega + N\omega_L. \tag{8.198}$$

Comparing this expression with the quasi-energies (8.197) of the Floquet states (8.196) we see that we have the following correspondence

$$u_{\pm,n}(t) \longleftrightarrow |\pm, N+n\rangle \tag{8.199}$$

between the Floquet states and the dressed states. On using this correspondence we find that the transition with jump operator $A(\omega_L + \Omega)$ corresponds to the transition

$$|+, N+1\rangle \longrightarrow |-, N\rangle, \tag{8.200}$$

whereas the jump with the operator $A(\omega_L - \Omega)$ corresponds to the transition

$$|-, N+1\rangle \longrightarrow |+, N\rangle \tag{8.201}$$

between the dressed atomic states. Likewise, the transition induced by the jump operator $A(\omega_L)$ can be interpreted as the transition

$$\alpha|+, N+1\rangle + \beta|-, N+1\rangle \longrightarrow \alpha|+, N\rangle - \beta|-, N\rangle. \tag{8.202}$$

Thus we conclude that the Floquet representation yields a simple jump process describing quantum jumps between dressed atomic states. It is also obvious that the three types of quantum jump correspond to the three peaks of the strong-driving Mollow spectrum (3.306): We have a central peak at frequency ω_L and two sideband peaks at frequencies $\omega_L \pm \Omega$. Since in the stationary case both Floquet states are found with probability $\frac{1}{2}$, it follows from (8.195) that the integrated intensities of the two sideband peaks coincide and are equal to $\frac{1}{2}$ times the integrated intensity of the central peak.

References

Akhiezer, A. I. and Berestetskii, V. B. (1965). *Quantum Electrodynamics*. Interscience Publishers, John Wiley, New York.

Aspect, A., Arimondo, E., Kaiser, R., Vansteenkiste, N. and Cohen-Tannoudji, C. (1989). Laser cooling below the one-photon recoil energy by velocity-selective coherent population trapping: Theoretical analysis. *J. Opt. Soc. Am.*, **B6**, 2112–2124.

Bardou, F., Bouchaud, J.-P., Emile, O., Aspect, A. and Cohen-Tannoudji, C. (1994). Subrecoil laser cooling and Lévy flights. *Phys. Rev. Lett.*, **72**, 203–206.

Bardou, F., Bouchaud, J.-P., Aspect, A. and Cohen-Tannoudji, C. (2001). *Lévy Statistics and Laser Cooling*. Cambridge University Press, Cambridge.

Bjorken, J. D. and Drell, S. D. (1965). *Relativistic Quantum Fields*. McGraw-Hill, New York.

Blümel, R., Buchleitner, A., Graham, R., Sirko, L., Smilansky, U. and Walther, H. (1991). Dynamical localization in the microwave interaction of Rydberg atoms: The influence of noise. *Phys. Rev.*, **A44**, 4521–4540.

Breuer, H. P., Dietz, K. and Holthaus, M. (1988). Strong laser fields interacting with matter I. *Z. Phys.*, **D10**, 13–26.

Breuer, H. P., Huber, W. and Petruccione, F. (2000). Quasi-stationary distributions of dissipative nonlinear quantum oscillators in strong periodic driving fields. *Phys. Rev.*, **E61**, 4883–4889.

Breuer, H. P. and Petruccione, F. (1997). Dissipative quantum systems in strong laser fields: Stochastic wave-function method and Floquet theory. *Phys. Rev.*, **A55**, 3101–3116.

Castin, Y. and Mølmer, K. (1995). Monte Carlo wave-function analysis of 3d optical molasses. *Phys. Rev. Lett.*, **74**, 3772–3775.

Cohen-Tannoudji, C., Bardou, F. and Aspect, A. (1992). Review on fundamental processes in laser cooling. In *Proceedings of Laser Spectroscopy X* (eds. Ducloy, M., Giacobino, E. and Camy, G.), pp. 3–14, World Scientific, Singapore.

Cohen-Tannoudji, C. and Reynaud, S. (1977). Dressed-atom description of resonance fluorescence and absorption spectra of a multi-level atom in an intense laser beam. *J. Phys.*, **B10**, 345–363.

Cohen-Tannoudji, C. and Reynaud, S. (1979). Atoms in strong light-fields: Photon antibunching in single atom fluorescence. *Phil. Trans. R. Soc. Lond.*, **A293**, 223–237.

Cohen-Tannoudji, C. (1992a). Laser cooling and trapping of neutral atoms: Theory. *Phys. Rep.*, **219**, 153–164.

Cohen-Tannoudji, C. (1992b). *Atomic Motion in Laser Light*, Les Houches Session LIII on *Fundamental Systems in Quantum Optics*, pp. 1–164. North-Holland, Elsevier, Amsterdam.

Cohen-Tannoudji, C., Dupont-Roc, J. and Grynberg, G. (1998). *Atom–Photon Interactions*. John Wiley, New York.

Kasevich, M. and Chu, S. (1992). Laser cooling below a photon recoil with three-level atoms. *Phys. Rev. Lett.*, **69**, 1741–1744.

Kohler, S., Dittrich, T. and Hänggi, P. (1997). Floquet–Markovian description of the parametrically driven, dissipative harmonic oscillator. *Phys. Rev.*, **E55**, 300–313.

Lawall, J., Bardou, F., Saubamea, B., Shimizu, K., Leduc, M., Aspect, A. and Cohen-Tannoudji, C. (1994). Two-dimensional subrecoil laser cooling. *Phys. Rev. Lett.*, **73**, 1915–1918.

Lawall, J., Kulin, S., Saubamea, B., Bigelow, N., Leduc, M. and Cohen-Tannoudji, C. (1995). Three-dimensional laser cooling of helium beyond the single-photon recoil limit. *Phys. Rev. Lett.*, **75**, 4194–4197.

L'Huillier, A., Schafer, K. J. and Kulander, K. C. (1991). Theoretical aspects of intense field harmonic generation. *J. Phys.*, **B24**, 3315–3341.

Mølmer, K., Castin, Y. and Dalibard, J. (1993). Monte Carlo wave-function method in quantum optics. *J. Opt. Soc. Am.*, **B10**, 524–538.

Reichel, J., Bardou, F., Ben Dahan, M., Peik, E., Rand, S., Salomon, C. and Cohen-Tannoudji, C. (1995). Raman cooling of cesium below 3 nK: New approach inspired by Lévy flight statistics. *Phys. Rev. Lett.*, **75**, 4575–4578.

Shirley, J. H. (1965). Solution of the Schrödinger equation with a Hamiltonian periodic in time. *Phys. Rev.*, **138B**, 979–987.

Shore, B. W. (1990). *The Theory of Coherent Atomic Excitation*. Volume 2. *Multilevel Atoms and Incoherence*. John Wiley, New York.

Zeldovich, Ya. B. (1967). The quasienergy of a quantum-mechanical system subjected to a periodic action. *Sov. Phys. JETP*, **24**, 1006–1008.

Part IV

Non-Markovian quantum processes

9

PROJECTION OPERATOR TECHNIQUES

As discussed in Chapter 3 the laws describing the dynamics of open quantum systems can be derived from the unitary dynamics of the total system. In general, the reduction of the degrees of freedom in the effective description of the open system results in non-Markovian behaviour. It is the aim of this part to introduce the reader to some powerful techniques which allow a systematic description of the non-Markovian features of the dynamics of open systems.

A general framework to derive exact equations of motion for an open system is provided by projection operator techniques. These techniques were introduced by Nakajima (1958) and Zwanzig (1960) and independently by the Brussels school (Prigogine, 1962). They are widely used in non-equilibrium statistical mechanics (Haake, 1973; Balescu, 1975; Grabert, 1982; Kubo, Toda and Hashitsume, 1985).

The basic idea underlying the application of projection operator techniques to open quantum systems is to regard the operation of tracing over the environment as a formal projection $\rho \mapsto \mathcal{P}\rho$ in the state space of the total system. The superoperator $\mathcal{P}$ has the property of a projection operator, that is $\mathcal{P}^2 = \mathcal{P}$, and the density matrix $\mathcal{P}\rho$ is said to be the *relevant* part of the density ρ of the total system. Correspondingly, one defines a projection $\rho \mapsto \mathcal{Q}\rho$ onto the *irrelevant* part $\mathcal{Q}\rho$, where $\mathcal{P} + \mathcal{Q}$ is equal to the identity map. The aim is then to derive a closed equation of motion for the relevant part $\mathcal{P}\rho$.

We are going to discuss in this chapter two variants of projection operator techniques, the Nakajima–Zwanzig and the time-convolutionless technique. Both methods lead to an exact equation of motion for the relevant part $\mathcal{P}\rho$. In the case of the Nakajima–Zwanzig method this is an integro-differential equation involving a retarded time integration over the history of the reduced system, while the time-convolutionless equation of motion provides a first-order differential equation which is local in time.

The time-convolutionless projection operator technique leads to a time-local expansion of the equation of motion with respect to the strength of the system–environment coupling. It thus supports an investigation of non-Markovian effects beyond the Born approximation. To each order in the coupling the equation of motion involves a time-dependent but local generator. The rules for the perturbation expansion of the convolutionless generator will be developed. We are mainly concerned in this chapter with the derivation of the most important general results; specific physical applications will be studied in the next chapter.

9.1   The Nakajima–Zwanzig projection operator technique

We consider the general physical situation of an open system S coupled to an environment B. The dynamics of the density matrix $\rho(t)$ of the combined system is specified by some microscopic Hamiltonian of the form

$$H = H_0 + \alpha H_I, \tag{9.1}$$

where H_0 generates the uncoupled time evolution of the system and environment, H_I describes their interaction, and α denotes a dimensionless expansion parameter. When working in the interaction representation, the equation of motion for the density matrix reads

$$\frac{\partial}{\partial t}\rho(t) = -i\alpha[H_I(t), \rho(t)] \equiv \alpha\mathcal{L}(t)\rho(t), \tag{9.2}$$

where we have set $\hbar = 1$ and the interaction picture representation of the interaction Hamiltonian is defined by

$$H_I(t) = \exp(iH_0 t)H_I \exp(-iH_0 t). \tag{9.3}$$

The Liouville super-operator is denoted by $\mathcal{L}(t)$.

9.1.1   Projection operators

In order to derive an exact equation of motion for the reduced density matrix ρ_S of the open system it is convenient to define a super-operator $\mathcal{P}$ according to

$$\rho \mapsto \mathcal{P}\rho = \text{tr}_B\{\rho\} \otimes \rho_B \equiv \rho_S \otimes \rho_B, \tag{9.4}$$

where ρ_B is some fixed state of the environment. This super-operator projects on the relevant part of the density matrix ρ in the sense that $\mathcal{P}\rho$ gives the complete information required to reconstruct the reduced density matrix ρ_S of the open system. Accordingly, a complementary super-operator $\mathcal{Q}$,

$$\mathcal{Q}\rho = \rho - \mathcal{P}\rho, \tag{9.5}$$

may be introduced, which projects on the irrelevant part of the density matrix. The super-operators $\mathcal{P}$ and $\mathcal{Q}$ are maps in the state space of the combined system, that is in the space of density matrices of the total Hilbert space $\mathcal{H} = \mathcal{H}_S \otimes \mathcal{H}_B$. They have the obvious properties

$$\mathcal{P} + \mathcal{Q} = I, \tag{9.6}$$
$$\mathcal{P}^2 = \mathcal{P}, \tag{9.7}$$
$$\mathcal{Q}^2 = \mathcal{Q}, \tag{9.8}$$
$$\mathcal{P}\mathcal{Q} = \mathcal{Q}\mathcal{P} = 0, \tag{9.9}$$

which can be easily checked using the definitions (9.4) and (9.5) and assuming ρ_B to be normalized, $\text{tr}_B\rho_B = 1$.

The density matrix ρ_B used in definition (9.4) is an operator in $\mathcal{H}_B$. It may represent a quite arbitrary, but known environmental state, called the reference state. The choice of ρ_B strongly depends on the specific application one has in mind. In the following we shall suppose this state to be time independent. Typically, it is taken to be the stationary Gibbs state of the environment. In many cases it may also be assumed that the odd moments of the interaction Hamiltonian with respect to the reference state vanish

$$\text{tr}_B \left\{ H_I(t_1)H_I(t_2)\ldots H_I(t_{2n+1})\rho_B \right\} = 0, \tag{9.10}$$

which leads to the relation

$$\mathcal{P}\mathcal{L}(t_1)\mathcal{L}(t_2)\ldots\mathcal{L}(t_{2n+1})\mathcal{P} = 0 \tag{9.11}$$

for $n = 0, 1, 2, \ldots$. This technical assumption is not required for the derivation of the equation of motion. It will however be used later on in order to simplify the expressions of the perturbation expansion. It is important to remark that we do not demand any particular form for the initial conditions at this point. In particular we do not assume factorizing initial conditions.

9.1.2 The Nakajima–Zwanzig equation

Our aim is now to derive a closed equation for the relevant part $\mathcal{P}\rho(t)$, i.e. for the density matrix $\rho_S(t) = \text{tr}_B \rho(t)$ of the open system. By applying the projection operators $\mathcal{P}$ and $\mathcal{Q}$ to the Liouville–von Neumann equation (9.2) and by invoking the time independence of the reference state the following set of coupled differential equations for the relevant and the irrelevant part of the density matrix is obtained,

$$\frac{\partial}{\partial t}\mathcal{P}\rho(t) = \mathcal{P}\frac{\partial}{\partial t}\rho(t) = \alpha\mathcal{P}\mathcal{L}(t)\rho(t), \tag{9.12}$$

$$\frac{\partial}{\partial t}\mathcal{Q}\rho(t) = \mathcal{Q}\frac{\partial}{\partial t}\rho(t) = \alpha\mathcal{Q}\mathcal{L}(t)\rho(t). \tag{9.13}$$

On inserting the identity $I = \mathcal{P} + \mathcal{Q}$ between the Liouville operator and the density matrix ρ this may also be written as

$$\frac{\partial}{\partial t}\mathcal{P}\rho(t) = \alpha\mathcal{P}\mathcal{L}(t)\mathcal{P}\rho(t) + \alpha\mathcal{P}\mathcal{L}(t)\mathcal{Q}\rho(t), \tag{9.14}$$

$$\frac{\partial}{\partial t}\mathcal{Q}\rho(t) = \alpha\mathcal{Q}\mathcal{L}(t)\mathcal{P}\rho(t) + \alpha\mathcal{Q}\mathcal{L}(t)\mathcal{Q}\rho(t). \tag{9.15}$$

To get a closed equation for the relevant part of the density matrix we solve eqn (9.15) and insert the solution into eqn (9.14). The formal solution of eqn (9.15) corresponding to a given $\rho(t_0)$ at some initial time t_0 may be expressed as

$$\mathcal{Q}\rho(t) = \mathcal{G}(t,t_0)\mathcal{Q}\rho(t_0) + \alpha\int_{t_0}^{t} ds\, \mathcal{G}(t,s)\mathcal{Q}\mathcal{L}(s)\mathcal{P}\rho(s), \tag{9.16}$$

where we have introduced the propagator

$$\mathcal{G}(t, s) \equiv \mathrm{T}_{\leftarrow} \exp\left[\alpha \int_{s}^{t} ds' \mathcal{QL}(s')\right]. \tag{9.17}$$

As usual, the operator $\mathrm{T}_{\leftarrow}$ describes the chronological time ordering: It orders any product of super-operators such that the time arguments increase from right to left. The propagator $\mathcal{G}(t, s)$ thus satisfies the differential equation

$$\frac{\partial}{\partial t}\mathcal{G}(t, s) = \alpha \mathcal{QL}(t)\mathcal{G}(t, s) \tag{9.18}$$

with the initial condition

$$\mathcal{G}(s, s) = I. \tag{9.19}$$

Inserting the expression (9.16) for the irrelevant part of the density matrix into the equation of motion (9.14) for the relevant part we obtain the desired exact equation for the time evolution of the relevant part of the density matrix,

$$\frac{\partial}{\partial t}\mathcal{P}\rho(t) = \alpha \mathcal{PL}(t)\mathcal{G}(t, t_0)\mathcal{Q}\rho(t_0) + \alpha \mathcal{PL}(t)\mathcal{P}\rho(t)$$

$$+ \alpha^2 \int_{t_0}^{t} ds \mathcal{PL}(t)\mathcal{G}(t, s)\mathcal{QL}(s)\mathcal{P}\rho(s). \tag{9.20}$$

This equation is known as the Nakajima–Zwanzig equation. It is an exact equation for the relevant degrees of freedom of the reduced system. The right-hand side involves an inhomogeneous term $\mathcal{PL}(t)\mathcal{G}(t, t_0)\mathcal{Q}\rho(t_0)$ depending on the initial condition at time t_0, and an integral over the past history of the system in the time interval $[t_0, t]$. It thus describes completely non-Markovian memory effects of the reduced dynamics. If condition (9.11) is satisfied for $n = 0$, the second term in the Nakajima–Zwanzig equation vanishes and we may cast it into the compact form

$$\frac{\partial}{\partial t}\mathcal{P}\rho(t) = \int_{t_0}^{t} ds\, \mathcal{K}(t, s)\mathcal{P}\rho(s) + \alpha \mathcal{PL}(t)\mathcal{G}(t, t_0)\mathcal{Q}\rho(t_0). \tag{9.21}$$

The convolution or memory kernel

$$\mathcal{K}(t, s) = \alpha^2 \mathcal{PL}(t)\mathcal{G}(t, s)\mathcal{QL}(s)\mathcal{P} \tag{9.22}$$

represents a super-operator in the relevant subspace.

The integro-differential equation (9.21) is exact and holds for all initial conditions and for almost arbitrary systems and interactions. Unfortunately, the Nakajima–Zwanzig equation is usually as difficult to solve as the Liouville equation describing the dynamics of the total system. This means that perturbation expansions are needed in order to discuss the relevant dynamics in a way accessible to analytical or numerical computations. Obviously, the equation may be expanded in the coupling constant α, i.e. in powers of the interaction Hamiltonian H_I. Alternatively, it may be expanded around t in powers of the memory time, i.e. in the width of the kernel $\mathcal{K}(t,s)$, where, of course, for $\mathcal{K}(t,s) \approx \delta(t-s)$ in the absence of memory effects we obtain the Markovian description. Sometimes it might also be convenient to perform the perturbation expansion for the Laplace transform of $\rho_S(t)$ in the Schrödinger picture.

For a factorizing initial condition $\rho(t_0) = \rho_S(t_0) \otimes \rho_B$ we have $\mathcal{P}\rho(t_0) = \rho(t_0)$ and, therefore, $\mathcal{Q}\rho(t_0) = 0$. Hence the inhomogeneous term of the Nakajima–Zwanzig equation (9.21) vanishes and the exact equation for the relevant part of the density matrix reduces to

$$\frac{\partial}{\partial t}\mathcal{P}\rho(t) = \int_{t_0}^{t} ds \mathcal{K}(t,s)\mathcal{P}\rho(s). \tag{9.23}$$

To second order in the coupling strength α we obtain

$$\mathcal{K}(t,s) = \alpha^2 \mathcal{P}\mathcal{L}(t)\mathcal{Q}\mathcal{L}(s)\mathcal{P} + \mathcal{O}(\alpha^3), \tag{9.24}$$

which leads to an equation of motion of second order for $\mathcal{P}\rho(t)$

$$\frac{\partial}{\partial t}\mathcal{P}\rho(t) = \alpha^2 \int_{t_0}^{t} ds \mathcal{P}\mathcal{L}(t)\mathcal{L}(s)\mathcal{P}\rho(s), \tag{9.25}$$

where we again made use of $\mathcal{P}\mathcal{L}(t)\mathcal{P} = 0$. If we now introduce the explicit expressions for the projection operator $\mathcal{P}$ and for the generator $\mathcal{L}(t)$ we get the Born approximation of the master equation

$$\frac{\partial}{\partial t}\rho_S(t) = -\alpha^2 \int_{t_0}^{t} ds \, \mathrm{tr}_B[H_I(t), [H_I(s), \rho_S(s) \otimes \rho_B]], \tag{9.26}$$

which we already met in eqn (3.116).

This approach to the non-Markovian dynamics of open quantum systems has some practical disadvantages. The perturbative approximation of the memory kernel simplifies the derivation of the equations of motion, but unfortunately not their structure. The approximate equation of motion is again an integro-differential equation, whose numerical solution may be quite involved.

9.2   The time-convolutionless projection operator method

In practice the time convolution in the memory kernel of the Nakajima–Zwanzig equation is difficult to treat. In this section we show how to remove the time convolution in the master equation. This is achieved through a method which is known as the time-convolutionless projection operator technique. This technique has been developed by Shibata et al. (Shibata, Takahashi and Hashitsume, 1977; Chaturvedi and Shibata, 1979; Shibata and Arimitsu, 1980) and we are going to apply it here to the microscopic theory of an open quantum system which is coupled to an environment. The method yields a systematic expansion of the dynamics of the system of interest in terms of the coupling strength. In particular, we will develop expressions for the quantum master equation up to fourth order in the coupling for factorizing and for non-factorizing initial conditions.

9.2.1   The time-local master equation

The idea of the time-convolutionless projection operator technique is to eliminate the dependence of the future time evolution on the history of the system from the Nakajima–Zwanzig master equation and thus to derive an exact master equation for the open system which is local in time. In order to achieve this objective we proceed in the following way: The density matrix $\rho(s)$ on the right-hand side of eqn (9.16) is replaced by the expression

$$\rho(s) = G(t, s)(\mathcal{P} + \mathcal{Q})\rho(t), \tag{9.27}$$

where $G(t, s)$ is the backward propagator of the composite system, i.e. the inverse of the unitary time evolution of the total system. Formally, we may write

$$G(t, s) = \mathrm{T}_{\rightarrow} \exp\left[-\alpha \int_s^t ds' \, \mathcal{L}(s')\right], \tag{9.28}$$

where $\mathrm{T}_{\rightarrow}$ indicates the antichronological time-ordering.

With the help of the relation (9.27) the equation (9.16) for the irrelevant part of the density matrix may now be written as

$$\mathcal{Q}\rho(t) = \mathcal{G}(t, t_0)\mathcal{Q}\rho(t_0) + \alpha \int_{t_0}^t ds \, \mathcal{G}(t, s)\mathcal{Q}\mathcal{L}(s)\mathcal{P}G(t, s)(\mathcal{P} + \mathcal{Q})\rho(t). \tag{9.29}$$

Introducing the super-operator

$$\Sigma(t) = \alpha \int_{t_0}^t ds \, \mathcal{G}(t, s)\mathcal{Q}\mathcal{L}(s)\mathcal{P}G(t, s), \tag{9.30}$$

we can express the irrelevant part of the density matrix through

$$[1 - \Sigma(t)] \, \mathcal{Q}\rho(t) = \mathcal{G}(t, t_0)\mathcal{Q}\rho(t_0) + \Sigma(t)\mathcal{P}\rho(t). \tag{9.31}$$

Note that the super-operator $\Sigma(t)$ contains both propagators $\mathcal{G}$ and G, so that it does not specify a well-defined chronological order. $\Sigma(t)$ has the obvious properties $\Sigma(t_0) = 0$ and $\Sigma(t)|_{\alpha=0} = 0$. Hence, $1 - \Sigma(t)$ may be inverted for not too large couplings and in any case for small $t - t_0$. Thus, we get

$$\mathcal{Q}\rho(t) = [1 - \Sigma(t)]^{-1} \Sigma(t)\mathcal{P}\rho(t) + [1 - \Sigma(t)]^{-1} \mathcal{G}(t, t_0)\mathcal{Q}\rho(t_0). \tag{9.32}$$

This equation states that the irrelevant part $\mathcal{Q}\rho(t)$ of the density matrix can in principle be determined from the knowledge of the relevant part $\mathcal{P}\rho(t)$ at time t and from the initial condition $\mathcal{Q}\rho(t_0)$. The dependence on the history of the relevant part which occurs in the Nakajima–Zwanzig equation has thus been removed by the introduction of the exact backward propagator $G(t, s)$. It must be noted, however, that for strong couplings and/or large time intervals $t - t_0$ it may happen that eqn (9.31) cannot be solved uniquely for $\mathcal{Q}\rho(t)$ such that the inverse of $1 - \Sigma(t)$ does not exist. We are going to exemplify this situation in Section 10.1.2.

To complete the derivation of the time-convolutionless master equation, we insert eqn (9.32) into the equation of motion for the relevant part (9.14) and obtain the following exact time-convolutionless (TCL) form of the master equation,

$$\frac{\partial}{\partial t}\mathcal{P}\rho(t) = \mathcal{K}(t)\mathcal{P}\rho(t) + \mathcal{I}(t)\mathcal{Q}\rho(t_0), \tag{9.33}$$

with the time-local generator, called the TCL generator,

$$\mathcal{K}(t) = \alpha\mathcal{P}\mathcal{L}(t)[1 - \Sigma(t)]^{-1}\mathcal{P}, \tag{9.34}$$

and the inhomogeneity

$$\mathcal{I}(t) = \alpha\mathcal{P}\mathcal{L}(t)[1 - \Sigma(t)]^{-1}\mathcal{G}(t, t_0)\mathcal{Q}. \tag{9.35}$$

The equation of motion (9.33) is exact and local in time. Although the super-operators $\mathcal{K}(t)$ and $\mathcal{I}(t)$ are, in general, extremely complicated objects, eqn (9.33) can be used as a starting point of a systematic approximation method by expanding $\mathcal{K}(t)$ and $\mathcal{I}(t)$ in powers of the coupling strength α. This will be shown in the following subsections.

9.2.2   Perturbation expansion of the TCL generator

Of course, the super-operator $\mathcal{K}(t)$ only exists when it is possible to invert the operator $[1 - \Sigma(t)]$. Let us assume then that $\Sigma(t)$ may be expanded into a geometric series

$$[1 - \Sigma(t)]^{-1} = \sum_{n=0}^{\infty} [\Sigma(t)]^n. \tag{9.36}$$

On substituting this into the expression (9.34) one gets

$$\mathcal{K}(t) = \alpha\sum_{n=0}^{\infty}\mathcal{P}\mathcal{L}(t)[\Sigma(t)]^n\mathcal{P} = \sum_{n=1}^{\infty}\alpha^n\mathcal{K}_n(t). \tag{9.37}$$

To determine the contribution $\mathcal{K}_n(t)$ of n-th order in α to the TCL generator $\mathcal{K}(t)$ we also expand $\Sigma(t)$ in powers of α,

$$\Sigma(t) = \sum_{n=1}^{\infty} \alpha^n \Sigma_n(t), \tag{9.38}$$

insert this into eqn (9.37), and sort equal powers of α. For example, to fourth order in α this gives:

$$\mathcal{K}_1(t) = \mathcal{P}\mathcal{L}(t)\mathcal{P}, \tag{9.39}$$

$$\mathcal{K}_2(t) = \mathcal{P}\mathcal{L}(t)\Sigma_1(t)\mathcal{P}, \tag{9.40}$$

$$\mathcal{K}_3(t) = \mathcal{P}\mathcal{L}(t)\left\{[\Sigma_1(t)]^2 + \Sigma_2(t)\right\}\mathcal{P}, \tag{9.41}$$

$$\mathcal{K}_4(t) = \mathcal{P}\mathcal{L}(t)\left\{[\Sigma_1(t)]^3 + \Sigma_1(t)\Sigma_2(t) + \Sigma_2(t)\Sigma_1(t) + \Sigma_3(t)\right\}\mathcal{P}. \tag{9.42}$$

Finally, the contributions $\Sigma_n(t)$ are found with the help of eqns (9.30) and (9.38) by expanding also the propagators $\mathcal{G}(t,s)$ and $G(t,s)$ defined in eqns (9.17) and (9.28) in powers of α.

Let us determine more explicitly the first four terms of the expansion. To simplify the expressions we use condition (9.11) and take $t_0 = 0$. Equation (9.39) immediately gives

$$\mathcal{K}_1(t) = \mathcal{P}\mathcal{L}(t)\mathcal{P} = 0. \tag{9.43}$$

The first-order term $\Sigma_1(t)$ is given by

$$\Sigma_1(t) = \int_0^t dt_1 \mathcal{Q}\mathcal{L}(t_1)\mathcal{P}, \tag{9.44}$$

which yields

$$\mathcal{K}_2(t) = \int_0^t dt_1 \mathcal{P}\mathcal{L}(t)\mathcal{L}(t_1)\mathcal{P}. \tag{9.45}$$

The second-order term $\Sigma_2(t)$ is found to be

$$\Sigma_2(t) = \int_0^t dt_1 \int_0^{t_1} dt_2 \left[\mathcal{Q}\mathcal{L}(t_1)\mathcal{Q}\mathcal{L}(t_2)\mathcal{P} - \mathcal{Q}\mathcal{L}(t_2)\mathcal{P}\mathcal{L}(t_1)\right]. \tag{9.46}$$

Since $\mathcal{P}\mathcal{Q} = 0$ we conclude from eqn (9.44) that $[\Sigma_1(t)]^2 = 0$ and, therefore,

$$\mathcal{K}_3(t) = \mathcal{P}\mathcal{L}(t)\Sigma_2(t)\mathcal{P} = \int_0^t dt_1 \int_0^{t_1} dt_2 \mathcal{P}\mathcal{L}(t)\mathcal{L}(t_1)\mathcal{L}(t_2)\mathcal{P} = 0, \tag{9.47}$$

where we made use of condition (9.11) for $n = 0$ and $n = 1$. To find $\mathcal{K}_4(t)$ we first note that $[\Sigma_1(t)]^3 = \Sigma_1(t)\Sigma_2(t) = 0$ because of $\mathcal{P}\mathcal{Q} = 0$. Thus we have from eqn (9.42)

$$\mathcal{K}_4(t) = \mathcal{P}\mathcal{L}(t)\left[\Sigma_2(t)\Sigma_1(t) + \Sigma_3(t)\right]\mathcal{P}. \tag{9.48}$$

Invoking eqns (9.44) and (9.46) the first term is found to be

$$\mathcal{PL}(t)\Sigma_2(t)\Sigma_1(t)\mathcal{P} = -\int_0^t dt_1 \int_0^{t_1} dt_2 \int_0^{t_2} dt_3 \tag{9.49}$$

$$\times \mathcal{PL}(t) \left[\mathcal{L}(t_2)\mathcal{PL}(t_1)\mathcal{L}(t_3)\mathcal{P} + \mathcal{L}(t_3)\mathcal{PL}(t_2)\mathcal{L}(t_1)\mathcal{P} + \mathcal{L}(t_3)\mathcal{PL}(t_1)\mathcal{L}(t_2)\mathcal{P} \right].$$

Note that to get this expression the triple time integral has been brought into time-ordered form, $t \geq t_1 \geq t_2 \geq t_3 \geq 0$. Similarly, one finds

$$\mathcal{PL}(t)\Sigma_3(t)\mathcal{P} \tag{9.50}$$

$$= \int_0^t dt_1 \int_0^{t_1} dt_2 \int_0^{t_2} dt_3 \mathcal{PL}(t) \left[\mathcal{L}(t_1)\mathcal{QL}(t_2)\mathcal{L}(t_3)\mathcal{P} + \mathcal{L}(t_3)\mathcal{PL}(t_2)\mathcal{L}(t_1)\mathcal{P} \right].$$

Summarizing, the fourth-order contribution to the TCL generator takes the form,

$$\mathcal{K}_4(t) = \int_0^t dt_1 \int_0^{t_1} dt_2 \int_0^{t_2} dt_3 \Big(\mathcal{PL}(t)\mathcal{L}(t_1)\mathcal{L}(t_2)\mathcal{L}(t_3)\mathcal{P} - \mathcal{PL}(t)\mathcal{L}(t_1)\mathcal{PL}(t_2)\mathcal{L}(t_3)\mathcal{P}$$

$$- \mathcal{PL}(t)\mathcal{L}(t_2)\mathcal{PL}(t_1)\mathcal{L}(t_3)\mathcal{P} - \mathcal{PL}(t)\mathcal{L}(t_3)\mathcal{PL}(t_1)\mathcal{L}(t_2)\mathcal{P} \Big). \tag{9.51}$$

The second-order generator $\mathcal{K}_2(t)$ of the TCL master equation leads to the following equation for the reduced density matrix $\rho_S(t)$,

$$\frac{\partial}{\partial t}\rho_S(t) = -\alpha^2 \int_0^t ds \, \mathrm{tr}_B[H_I(t), [H_I(s), \rho_S(t) \otimes \rho_B]], \tag{9.52}$$

which should be contrasted to the corresponding second-order approximation (9.26) of the Nakajima–Zwanzig equation: Both equations are of second order and it is therefore to be expected that they approximate the exact dynamics with the same accuracy. This point will be illustrated in Section 10.1.2 with the help of a simple example. In practice, the TCL form is to be preferred because it involves a time-local generator instead of a convolution kernel.

The expressions (9.45) and (9.51) can be made more explicit if one writes the interaction Hamiltonian as a sum of products of Hermitian operators F_k and Q_k which act in the system's and the environment's Hilbert space, respectively,

$$H_I = \sum_k F_k \otimes Q_k. \tag{9.53}$$

Such a decomposition of the interaction into Hermitian operators is always possible. It is usually identical to the physically given form of the system–environment interaction.

Let us assume that the reference state ρ_B of the environment is a Gaussian state. This implies that all moments of H_I with respect to ρ_B can be expressed in terms of moments of second order. Thus, we define the correlation functions

$$\nu_{ij}(t_1, t_2) = \Re \, \text{tr}_B \left\{ Q_i(t_1) Q_j(t_2) \rho_B \right\}, \qquad (9.54)$$

$$\eta_{ij}(t_1, t_2) = \Im \, \text{tr}_B \left\{ Q_i(t_1) Q_j(t_2) \rho_B \right\}, \qquad (9.55)$$

where $\Re$ and $\Im$ denote the real and imaginary part, respectively. It will be convenient to introduce the following shorthand notation,

$$\hat{0} = F_{i_0}(t), \quad \hat{1} = F_{i_1}(t_1), \quad \hat{2} = F_{i_2}(t_2), \ldots, \qquad (9.56)$$

and

$$\nu_{01} = \nu_{i_0 i_1}(t, t_1), \quad \nu_{12} = \nu_{i_1 i_2}(t_1, t_2), \ldots, \qquad (9.57)$$

$$\eta_{01} = \eta_{i_0 i_1}(t, t_1), \quad \eta_{12} = \eta_{i_1 i_2}(t_1, t_2), \ldots. \qquad (9.58)$$

With these definitions any moment of H_I may be expressed in terms of the system operators F_i and of the correlation functions ν_{ij} and η_{ij} in a compact way. For example, we have for the second-order moments,

$$\text{tr}_B \left\{ H_I(t) H_I(t_1) \rho_S \otimes \rho_B \right\} = \sum_{i_0, i_1} (\nu_{01} + i\eta_{01}) \, \hat{0}\hat{1}\rho_S, \qquad (9.59)$$

and, by virtue of the Gaussian property, for the fourth-order moments,

$$\text{tr}_B \left\{ H_I(t) H_I(t_1) H_I(t_2) H_I(t_3) \rho_S \otimes \rho_B \right\}$$
$$= \sum_{i_0, i_2, i_3, i_3} \left[(\nu_{01} + i\eta_{01})(\nu_{23} + i\eta_{23}) + (\nu_{02} + i\eta_{02})(\nu_{13} + i\eta_{13}) \right.$$
$$\left. + (\nu_{03} + i\eta_{03})(\nu_{12} + i\eta_{12}) \right] \hat{0}\hat{1}\hat{2}\hat{3}\rho_S. \qquad (9.60)$$

Invoking the above shorthand notation we can write the second-order contribution (9.45) to the TCL generator as follows,

$$\mathcal{K}_2(t)\rho_S \otimes \rho_B = -\sum_{i_0, i_1} \int_0^t dt_1 \left(\nu_{01} \left[\hat{0}, [\hat{1}, \rho_S] \right] + i\eta_{01} \left[\hat{0}, \{\hat{1}, \rho_S\} \right] \right) \otimes \rho_B, \qquad (9.61)$$

while the fourth-order contribution (9.51) becomes

$$\mathcal{K}_4(t)\rho_S \otimes \rho_B = \sum_{i_0,i_1,i_2,i_3} \int_0^t dt_1 \int_0^{t_1} dt_2 \int_0^{t_2} dt_3$$

$$\times \Big(\nu_{02}\nu_{13} \left[\hat{0}, \left[[\hat{1},\hat{2}], [\hat{3},\rho_S]\right]\right] + i\nu_{02}\eta_{13} \left[\hat{0}, \left[[\hat{1},\hat{2}], \{\hat{3},\rho_S\}\right]\right]$$

$$+ i\eta_{02}\nu_{13} \left[\hat{0}, \left\{[\hat{1},\hat{2}], [\hat{3},\rho_S]\right\}\right] - \eta_{02}\eta_{13} \left[\hat{0}, \left\{[\hat{1},\hat{2}], \{\hat{3},\rho_S\}\right\}\right]$$

$$+ \nu_{03}\nu_{12} \left[\hat{0}, \left[[\hat{1},\hat{3}], [\hat{2},\rho_S]\right]\right] + i\nu_{03}\eta_{12} \left[\hat{0}, \left[[\hat{1},\hat{3}], \{\hat{2},\rho_S\}\right]\right]$$

$$+ i\eta_{03}\nu_{12} \left[\hat{0}, \left\{[\hat{1},\hat{3}], [\hat{2},\rho_S]\right\}\right] - \eta_{03}\eta_{12} \left[\hat{0}, \left\{[\hat{1},\hat{3}], \{\hat{2},\rho_S\}\right\}\right]$$

$$+ (\nu_{03}\nu_{12} - \eta_{03}\eta_{12}) \left[\hat{0}, \left[\hat{1}, [[\hat{2},\hat{3}], \rho_S]\right]\right]$$

$$+ i(\nu_{03}\eta_{12} + \eta_{03}\nu_{12}) \left[\hat{0}, \left[\hat{1}, \{[\hat{2},\hat{3}], \rho_S\}\right]\right] \Big) \otimes \rho_B. \tag{9.62}$$

These formulae will be employed in Chapter 10 to derive the TCL generators of quantum master equations for specific microscopic models.

9.2.3   The cumulant expansion

A general formula for the n-th-order contribution $\mathcal{K}_n(t)$ to the generator of the TCL master equation can be derived by employing a technique which was originally designed by van Kampen (1974a, 1974b) for the perturbation expansion of stochastic differential equations. Let us restrict to the homogeneous case $\mathcal{P}\rho(0) = \rho(0)$ and introduce the notation

$$\langle \mathcal{X} \rangle \equiv \mathcal{P}\mathcal{X}\mathcal{P} \tag{9.63}$$

for any super-operator $\mathcal{X}$. The formal solution of the Liouville–von Neumann equation (9.2) then leads to the representation

$$\mathcal{P}\rho(t) = \left\langle \mathrm{T}_{\leftarrow} \exp\left[\alpha \int_0^t ds \mathcal{L}(s)\right] \right\rangle \mathcal{P}\rho(0) \tag{9.64}$$

for the relevant part of the density matrix. Of course, this is just a shorthand notation for

$$\mathcal{P}\rho(t) = \left[1 + \alpha \int_0^t dt_1 \langle \mathcal{L}(t_1) \rangle + \alpha^2 \int_0^t dt_1 \int_0^{t_1} dt_2 \langle \mathcal{L}(t_1)\mathcal{L}(t_2) \rangle \right.$$

$$\left. + \alpha^3 \int_0^t dt_1 \int_0^{t_1} dt_2 \int_0^{t_2} dt_3 \langle \mathcal{L}(t_1)\mathcal{L}(t_2)\mathcal{L}(t_3) \rangle + \cdots \right] \mathcal{P}\rho(0). \tag{9.65}$$

Differentiating this equation with respect to time we get

$$\frac{\partial}{\partial t}\mathcal{P}\rho(t) = \left[\alpha\langle\mathcal{L}(t)\rangle + \alpha^2 \int\limits_0^t dt_1 \langle\mathcal{L}(t)\mathcal{L}(t_1)\rangle\right.$$

$$\left. + \alpha^3 \int\limits_0^t dt_1 \int\limits_0^{t_1} dt_2 \langle\mathcal{L}(t)\mathcal{L}(t_1)\mathcal{L}(t_2)\rangle + \cdots \right]\mathcal{P}\rho(0). \qquad (9.66)$$

The trick is now to invert the expansion on the right-hand side of eqn (9.65) to express $\mathcal{P}\rho(0)$ in terms of $\mathcal{P}\rho(t)$ and to insert the result into eqn (9.66). As was shown by van Kampen this procedure may be carried out in a systematic fashion to yield an expansion for the equation of motion in powers of α. Comparing this expansion with the TCL master equation one is led to the following result for the n-th-order contribution of the generator $\mathcal{K}(t)$,

$$\mathcal{K}_n(t) = \int\limits_0^t dt_1 \int\limits_0^{t_1} dt_2 \ldots \int\limits_0^{t_{n-2}} dt_{n-1} \langle\mathcal{L}(t)\mathcal{L}(t_1)\mathcal{L}(t_2)\ldots\mathcal{L}(t_{n-1})\rangle_{\mathrm{oc}}, \qquad (9.67)$$

where the quantities

$$\langle\mathcal{L}(t)\mathcal{L}(t_1)\mathcal{L}(t_2)\ldots\mathcal{L}(t_{n-1})\rangle_{\mathrm{oc}} \qquad (9.68)$$

$$\equiv \sum(-1)^q\mathcal{P}\mathcal{L}(t)\ldots\mathcal{L}(t_i)\mathcal{P}\mathcal{L}(t_j)\ldots\mathcal{L}(t_k)\mathcal{P}\mathcal{L}(t_l)\ldots\mathcal{L}(t_m)\mathcal{P}\ldots\mathcal{P}$$

are called *ordered cumulants*. They are defined by the following rules.

First, one writes down a string of the form $\mathcal{P}\mathcal{L}\ldots\mathcal{L}\mathcal{P}$ with n factors of $\mathcal{L}$ in between two $\mathcal{P}$s. Next one inserts an arbitrary number q of factors $\mathcal{P}$ between the $\mathcal{L}$s such that at least one $\mathcal{L}$ stands between two successive $\mathcal{P}$ factors. The resulting expression is multiplied by a factor $(-1)^q$ and all $\mathcal{L}$s are furnished with a time argument: The first one is always $\mathcal{L}(t)$. The remaining $\mathcal{L}$s carry any permutation of the time arguments $t_1, t_2, \ldots, t_{n-1}$ with the only restriction that the time arguments in between two successive $\mathcal{P}$s must be ordered chronologically. In eqn (9.68) we thus have $t \geq \ldots \geq t_i$, $t_j \geq \ldots \geq t_k$, $t_l \geq \ldots \geq t_m$, etc. Finally, the ordered cumulant is obtained by a summation over all possible insertions of $\mathcal{P}$ factors and over all allowed distributions of the time arguments.

For commuting $\mathcal{L}$s the ordered cumulants reduce to the ordinary cumulants. The reader may easily check that under the condition (9.11) all odd contributions $\mathcal{K}_{2n+1}(t)$ vanish and that these rules immediately yield the expressions (9.45) and (9.51) for the second and the fourth-order contribution of the TCL generator.

9.2.4 *Perturbation expansion of the inhomogeneity*

As in the Nakajima–Zwanzig equation the inhomogeneity $\mathcal{I}(t)\mathcal{Q}\rho(0)$ in the time-convolutionless quantum master equation (9.33) depends on the density matrix $\rho(0)$ at the initial time $t_0 = 0$. For factorizing initial conditions $\mathcal{Q}\rho(0)$ vanishes and the resulting exact equation of motion is homogeneous. In this section we discuss the effect of non-factorizing initial conditions of the form

$$\rho(0) = \sum_k O_k \rho_{\text{eq}} P_k \equiv \mathcal{A}\rho_{\text{eq}}, \tag{9.69}$$

where the operators O_k and P_k act in the Hilbert space $\mathcal{H}_S$ of the open system and ρ_{eq} denotes the equilibrium density matrix of the combined system. This type of initial condition arises for example in the determination of equilibrium correlation functions of system observables. It can also be used to describe the preparation of the system through a quantum measurement since the super-operator $\mathcal{A}$ defined in eqn (9.69) takes the form of a quantum operation if one chooses $P_k = O_k^\dagger$ (see Section 2.4.2).

The perturbation expansion of the inhomogeneity $\mathcal{I}(t)\mathcal{Q}\rho(0)$ may be performed in two steps: First, one expands the super-operator $\mathcal{I}(t)$ given in eqn (9.35) in powers of the coupling strength α, and, second, one determines $\mathcal{Q}\rho(0)$. The expansion of $\mathcal{I}(t)$,

$$\mathcal{I}(t) = \sum_{n=1}^\infty \alpha^n \mathcal{I}_n(t), \tag{9.70}$$

is very similar to the expansion of the generator $\mathcal{K}(t)$ and takes the form (Chang and Skinner, 1993)

$$\mathcal{I}_1(t) = \mathcal{P}\mathcal{L}(t)\mathcal{Q}, \tag{9.71}$$

$$\mathcal{I}_2(t) = \int_0^t dt_1 \mathcal{P}\mathcal{L}(t)\mathcal{L}(t_1)\mathcal{Q}, \tag{9.72}$$

$$\mathcal{I}_3(t) = \int_0^t dt_1 \int_0^{t_1} dt_2 \left[\mathcal{P}\mathcal{L}(t)\mathcal{L}(t_1)\mathcal{Q}\mathcal{L}(t_2)\mathcal{Q} - \mathcal{P}\mathcal{L}(t)\mathcal{L}(t_2)\mathcal{P}\mathcal{L}(t_1)\mathcal{Q}\right], \tag{9.73}$$

where we again use condition (9.11).

The next step consists in the determination of $\mathcal{Q}\rho(0) = \mathcal{Q}\mathcal{A}\rho_{\text{eq}}$. Since the super-operators $\mathcal{A}$ and $\mathcal{Q}$ commute, this amounts to the computation of $\mathcal{Q}\rho_{\text{eq}}$ which can be achieved by the following method (Breuer, Kappler and Petruccione, 2001). We define for any fixed $\tau > 0$ the density matrix $\rho_\tau(t)$ to be the density of a system which has been prepared in such a way that the state $\rho_\tau(-\tau)$ at time $t = -\tau$ is some factorizing state. Thus, we have $\mathcal{P}\rho_\tau(-\tau) = \rho_\tau(-\tau)$ and $\mathcal{Q}\rho_\tau(-\tau) = 0$. Assuming the total system to be ergodic, we get

$$\rho_{\text{eq}} = \lim_{\tau \to \infty} \rho_\tau(0). \tag{9.74}$$

This enables one to express $\mathcal{Q}\rho_{\text{eq}}$ as

$$\mathcal{Q}\rho_{\text{eq}} = \lim_{\tau \to \infty} \mathcal{Q}\rho_\tau(0). \tag{9.75}$$

On the other hand, eqn (9.32) yields a relation between the relevant part $\mathcal{P}\rho_\tau(0)$ and the irrelevant part $\mathcal{Q}\rho_\tau(0)$. Using this relation for $t_0 = -\tau$ we obtain

$$\mathcal{Q}\rho_\tau(0) = [1 - \Sigma_\tau(0)]^{-1}\Sigma_\tau(0)\mathcal{P}\rho_\tau(0), \tag{9.76}$$

where

$$\Sigma_\tau(0) = \alpha \int_{-\tau}^{0} ds\, \mathcal{G}(0,s)\mathcal{Q}\mathcal{L}(s)\mathcal{P}G(0,s). \tag{9.77}$$

Combining eqns (9.75) and (9.76), we get

$$\mathcal{Q}\rho_{\text{eq}} = \mathcal{R}\,\mathcal{P}\rho_{\text{eq}}, \tag{9.78}$$

where the operator $\mathcal{R}$ is defined as

$$\mathcal{R} = \lim_{\tau\to\infty} [1 - \Sigma_\tau(0)]^{-1}\Sigma_\tau(0) = \lim_{\tau\to\infty}\sum_{n=1}^{\infty}(\Sigma_\tau(0))^n. \tag{9.79}$$

The relevant part $\mathcal{P}\rho_{\text{eq}}$ of the equilibrium density matrix and the irrelevant part $\mathcal{Q}\rho_{\text{eq}}$ are therefore related by the exact equation (9.78).

Also the operator $\mathcal{R}$ may be expanded in powers of α, namely

$$\mathcal{R} = \sum_{n=1}^{\infty} \alpha^n \mathcal{R}_n. \tag{9.80}$$

Again, this expansion is accomplished by expanding $\Sigma_\tau(0)$ in powers of α, which yields

$$\mathcal{R}_1 = \int_{-\infty}^{0} dt_1 \mathcal{L}(t_1)\mathcal{P}, \tag{9.81}$$

$$\mathcal{R}_2 = \int_{-\infty}^{0} dt_1 \int_{-\infty}^{t_1} dt_2 \mathcal{Q}\mathcal{L}(t_1)\mathcal{L}(t_2)\mathcal{P}, \tag{9.82}$$

$$\mathcal{R}_3 = \int_{-\infty}^{0} dt_1 \int_{-\infty}^{t_1} dt_2 \int_{-\infty}^{t_2} dt_3 \big[\mathcal{L}(t_1)\mathcal{Q}\mathcal{L}(t_2)\mathcal{L}(t_3) \tag{9.83}$$

$$-\mathcal{L}(t_2)\mathcal{P}\mathcal{L}(t_1)\mathcal{L}(t_3) - \mathcal{L}(t_3)\mathcal{P}\mathcal{L}(t_1)\mathcal{L}(t_2)\big]\mathcal{P}.$$

The last step which completes the expansion of the inhomogeneity is the combination of the expansions for the operators $\mathcal{I}(t)$ and $\mathcal{R}$ which finally leads to

$$\mathcal{I}(t)\mathcal{Q}\rho(0) = \mathcal{I}(t)\mathcal{A}\mathcal{R}\mathcal{P}\rho_{\text{eq}} \equiv \mathcal{J}(t)\mathcal{P}\rho_{\text{eq}}. \tag{9.84}$$

Explicitly, the second and the fourth-order contributions to the super-operator $\mathcal{J}(t)$ are given by

$$\mathcal{J}_2(t) = \mathcal{I}_1(t)\mathcal{AR}_1 = \int\limits_{-\infty}^{0} dt_1 \mathcal{PL}(t)\mathcal{AL}(t_1)\mathcal{P}, \qquad (9.85)$$

and by

$$\mathcal{J}_4(t) = \mathcal{I}_1(t)\mathcal{AR}_3 + \mathcal{I}_2(t)\mathcal{AR}_2 + \mathcal{I}_3(t)\mathcal{AR}_1, \qquad (9.86)$$

with

$$\mathcal{I}_1(t)\mathcal{AR}_3 = \int\limits_{-\infty}^{0} dt_1 \int\limits_{-\infty}^{t_1} dt_2 \int\limits_{-\infty}^{t_2} dt_3 \Big[\mathcal{PL}(t)\mathcal{AL}(t_1)\mathcal{QL}(t_2)\mathcal{L}(t_3)\mathcal{P}$$

$$- \mathcal{PL}(t)\mathcal{AL}(t_2)\mathcal{PL}(t_1)\mathcal{L}(t_3)\mathcal{P} - \mathcal{PL}(t)\mathcal{AL}(t_3)\mathcal{PL}(t_1)\mathcal{L}(t_2)\mathcal{P} \Big],$$

$$\mathcal{I}_2(t)\mathcal{AR}_2 = \int\limits_{0}^{t} dt_1 \int\limits_{-\infty}^{0} dt_2 \int\limits_{-\infty}^{t_2} dt_3 \mathcal{PL}(t)\mathcal{L}(t_1)\mathcal{AQL}(t_2)\mathcal{L}(t_3)\mathcal{P},$$

$$\mathcal{I}_3(t)\mathcal{AR}_1 = \int\limits_{0}^{t} dt_1 \int\limits_{0}^{t_1} dt_2 \int\limits_{-\infty}^{0} dt_3 \Big[\mathcal{PL}(t)\mathcal{L}(t_1)\mathcal{QL}(t_2)\mathcal{AL}(t_3)\mathcal{P}$$

$$- \mathcal{PL}(t)\mathcal{L}(t_2)\mathcal{PL}(t_1)\mathcal{AL}(t_3)\mathcal{P} \Big]. \qquad (9.87)$$

Using the form (9.53) for the interaction Hamiltonian and employing the short-hand notation introduced in Section 9.2.2 we obtain for the second-order contribution,

$$\mathcal{J}_2(t)\rho_S \otimes \rho_B = -\sum_{i_0,i_1} \int\limits_{-\infty}^{0} dt_1 \left(\nu_{01} \left[\hat{0}, A\left[\hat{1}, \rho_S \right] \right] + i\eta_{01} \left[\hat{0}, A\{ \hat{1}, \rho_S \} \right] \right) \otimes \rho_B,$$

$$(9.88)$$

which is similar to the expression for $\mathcal{K}_2(t)$ (compare eqn (9.61)). Section 10.2.4 gives an example of the application of the above method to the calculation of equilibrium correlation functions.

9.2.5 *Error analysis*

The perturbation expansion of the generator $\mathcal{K}(t)$ and of the inhomogeneity $\mathcal{J}(t)$ to various orders in the coupling strength α can be used to obtain a computable estimation for the error introduced by a certain approximation. To this end, we consider the relative error

$$e_{\mathrm{r}}^{(2n)} = \frac{\left\| \mathcal{K}(t) - \sum\limits_{k=1}^{n} \alpha^{2k}\mathcal{K}_{2k}(t) \right\|}{\|\mathcal{K}(t)\|} \qquad (9.89)$$

of the approximation of order $2n$, where $\|\mathcal{K}\|$ denotes any appropriate norm for the super-operator $\mathcal{K}$. To leading order in α this can be written as

$$e_{\mathrm{r}}^{(2n)} = \alpha^{2n} \frac{\|\mathcal{K}_{2n+2}(t)\|}{\|\mathcal{K}_2(t)\|}. \tag{9.90}$$

Thus, the error of the approximation to second order can be related to the fourth-order approximation by

$$e_{\mathrm{r}}^{(2)} \sim \alpha^2 \frac{\|\mathcal{K}_4(t)\|}{\|\mathcal{K}_2(t)\|}. \tag{9.91}$$

If we assume that the coefficients $\mathcal{K}_2$ and $\mathcal{K}_4$ have the same order of magnitude, then this relation can be used to estimate the formal expansion parameter α by means of the known relative error, i.e.

$$\alpha^2 \sim e_{\mathrm{r}}^{(2)}. \tag{9.92}$$

Employing the same argument again, we find for the error of the fourth-order approximation

$$e_{\mathrm{r}}^{(4)} \sim \alpha^4 \frac{\|\mathcal{K}_6(t)\|}{\|\mathcal{K}_2(t)\|}. \tag{9.93}$$

However, if the approximation to sixth order is not known, we cannot compute $e_{\mathrm{r}}^{(4)}$ directly. To construct a computable error estimate we use the assumption that the orders of magnitude of $\mathcal{K}_2$, $\mathcal{K}_4$ and $\mathcal{K}_6$ are the same, and we obtain

$$e_{\mathrm{r}}^{(4)} \sim \alpha^4 \approx \left(e_{\mathrm{r}}^{(2)}\right)^2. \tag{9.94}$$

Thus, the relative error of the approximation to fourth order can be estimated by computing the square of the relative error of the approximation to second order. This procedure yields for a certain order of approximation a crude error estimate which is easy to compute since it only relies on the actual approximation and not on the evaluation of higher-order terms. A number of examples will be studied in the next chapter.

The same arguments can of course also be applied to the matrix elements of the super-operator $\mathcal{K}(t)$ and of the inhomogeneity $\mathcal{J}(t)$, yielding computable error estimates for these quantities as well.

9.3 Stochastic unravelling in the doubled Hilbert space

The most general master equation for the reduced density matrix $\rho_S(t)$ which results from the time-convolutionless projection operator technique takes the following form in the homogeneous case,

$$\frac{\partial}{\partial t}\rho_S(t) = A(t)\rho_S(t) + \rho_S(t)B^\dagger(t) + \sum_i C_i(t)\rho_S(t)D_i^\dagger(t), \tag{9.95}$$

with some time-dependent linear operators $A(t)$, $B(t)$, $C_i(t)$ and $D_i(t)$. This equation is linear in $\rho_S(t)$ and local in time, but it need not be in Lindblad form.

Nevertheless, it can be represented by a certain stochastic process for the wave functions of the reduced system (Breuer, Kappler and Petruccione, 1999).

In order to construct an unravelling of eqn (9.95) we follow the strategy which was already applied to the calculation of multi-time correlation functions in Section 6.1.4. Namely, we describe the state of the open system by a pair of stochastic wave functions

$$\theta(t) = \begin{pmatrix} \phi(t) \\ \psi(t) \end{pmatrix},$$ (9.96)

such that $\theta(t)$ becomes a stochastic process in the doubled Hilbert space $\tilde{\mathcal{H}} = \mathcal{H}_S \oplus \mathcal{H}_S$. Denoting the corresponding probability density functional by $\tilde{P}[\theta, t]$, we can define the reduced density matrix as

$$\rho_S(t) = \int D\theta D\theta^* \, \tilde{P}[\theta, t] \, |\phi\rangle\langle\psi|.$$ (9.97)

Consider now the following stochastic differential equation for the process $\theta(t)$ in the doubled Hilbert space,

$$d\theta(t) = -iG(\theta, t)dt + \sum_i \left(\frac{\|\theta(t)\|}{\|J_i(t)\theta(t)\|} J_i(t)\theta(t) - \theta(t) \right) dN_i(t),$$ (9.98)

where the Poisson increments $dN_i(t)$ satisfy

$$E[dN_i(t)] = \frac{\|J_i(t)\theta(t)\|^2}{\|\theta(t)\|^2} dt,$$ (9.99)

$$dN_i(t)dN_j(t) = \delta_{ij}dN_i(t),$$ (9.100)

and the non-linear operator $G(\theta, t)$ is defined as

$$G(\theta, t) = i \left(F(t) + \frac{1}{2} \sum_i \frac{\|J_i(t)\theta(t)\|^2}{\|\theta(t)\|^2} \right) \theta(t),$$ (9.101)

with the time-dependent operators

$$F(t) = \begin{pmatrix} A(t) & 0 \\ 0 & B(t) \end{pmatrix}, \quad J_i(t) = \begin{pmatrix} C_i(t) & 0 \\ 0 & D_i(t) \end{pmatrix}.$$ (9.102)

Equation (9.98) describes a PDP whose deterministic pieces are solutions of the differential equation

$$i\frac{\partial}{\partial t}\theta(t) = G(\theta, t),$$ (9.103)

and whose jumps take the form

$$\theta(t) \longrightarrow \frac{\|\theta(t)\|}{\|J_i\theta(t)\|} J_i(t)\theta(t) = \frac{\|\theta(t)\|}{\|J_i\theta(t)\|} \begin{pmatrix} C_i(t)\phi(t) \\ D_i(t)\psi(t) \end{pmatrix}.$$ (9.104)

Employing the rules of the calculus for PDPs one easily demonstrates that the covariance matrix

$$\widetilde{\rho}(t) = \int D\theta D\theta^* \widetilde{P}[\theta, t] \, |\theta\rangle\langle\theta| \tag{9.105}$$

of the process $\theta(t)$ satisfies the equation of motion

$$\frac{\partial}{\partial t}\widetilde{\rho}(t) = F(t)\widetilde{\rho}(t) + \widetilde{\rho}(t)F^\dagger(t) + \sum_i J_i(t)\widetilde{\rho}(t)J_i^\dagger(t). \tag{9.106}$$

Note that, in general, this equation is also *not* in Lindblad form. The important point is, however, that the component (9.97) of $\widetilde{\rho}(t)$ obeys eqn (9.95), as is easily verified using the block structure of the operators $F(t)$ and $J_i(t)$. This provides the sought unravelling of eqn (9.95) in terms of a process in the doubled Hilbert space. A specific example will be studied in Section 10.1.3.

References

Balescu, R. (1975). *Equilibrium and Nonequilibrium Statistical Mechanics*. John Wiley, New York.

Breuer, H. P., Kappler, B. and Petruccione, F. (1999). Stochastic wave-function method for non-Markovian quantum master equations. *Phys. Rev.*, **A59**, 1633–1643.

Breuer, H. P., Kappler, B. and Petruccione, F. (2001). The time-convolutionless projection operator technique in the quantum theory of dissipation and decoherence. *Ann. Phys. (N.Y.)*, **291**, 36–70.

Chang, T.-M. and Skinner, J. L. (1993). Non-Markovian population and phase relaxation and absorption lineshape for a two-level system strongly coupled to a harmonic quantum bath. *Physica*, **A193**, 483–539.

Chaturvedi, S. and Shibata, F. (1979). Time-convolutionless projection operator formalism for elimination of fast variables. Application to Brownian motion. *Z. Phys.*, **B35**, 297–308.

Grabert, H. (1982). *Projection Operator Techniques in Nonequilibrium Statistical Mechanics*, Volume 95 of *Springer Tracts in Modern Physics*. Springer-Verlag, Berlin.

Haake, F. (1973). *Statistical treatment of open systems by generalized master equations*. Springer Tracts in Modern Physics, **66**, 98–168.

Kubo, R., Toda, M. and Hashitsume, N. (1985). *Statistical Physics II. Nonequilibrium Statistical Mechanics*. Springer-Verlag, Berlin.

Nakajima, S. (1958). On quantum theory of transport phenomena. *Progr. Theor. Phys.*, **20**, 948–959.

Prigogine, I. (1962). *Non-Equilibrium Statistical Mechanics*. Interscience Publishers, New York.

Shibata, F. and Arimitsu, T. (1980). Expansion formulas in nonequilibrium statistical mechanics. *J. Phys. Soc. Jap.*, **49**, 891–897.

Shibata, F. and Takahashi, Y. and Hashitsume, N. (1977). A generalized stochastic Liouville equation. Non-Markovian versus memoryless master equations. *J. Stat. Phys.*, **17**, 171–187.

van Kampen, N. G. (1974*a*). A cumulant expansion for stochastic linear differential equations. I. *Physica*, **74**, 215–238.

van Kampen, N. G. (1974*b*). A cumulant expansion for stochastic linear differential equations. II. *Physica*, **74**, 239–247.

Zwanzig, R. (1960). Ensemble method in the theory of irreversibility. *J. Chem. Phys.*, **33**, 1338–1341.

10

NON-MARKOVIAN DYNAMICS IN PHYSICAL SYSTEMS

In this chapter we apply the time-convolutionless (TCL) projection operator technique developed in the preceding chapter to some typical examples. The first example is that of the spontaneous decay of a two-level system into a reservoir with arbitrary spectral density. For this simple problem the TCL expansion can be carried out to all orders in the system–reservoir coupling. The model therefore serves to examine the performance of the TCL method.

As specific examples we treat the damped Jaynes–Cummings model and the spontaneous decay into a photonic band gap. The damped Jaynes–Cummings model on resonance is also used to exemplify the breakdown of the TCL expansion in the strong coupling regime. In the off-resonant case the model leads to a TCL generator which is not in Lindblad form and involves a time-dependent decay rate that can take on negative values. The corresponding unravelling through a stochastic process in the doubled Hilbert space designed in Section 9.3 will be illustrated by means of numerical simulations.

The last two sections are devoted to quantized Brownian motion, where non-Markovian effects play an important rôle for certain parameters. The first example is the damped harmonic oscillator, the second one is the spin-boson model.

The purpose of this chapter is to illustrate that the TCL projection operator technique has several advantages in practical applications. Although the technique is clearly perturbative, it does not rely on an expansion of a given physical quantity, but rather on an expansion of the equation of motion for the reduced density matrix which may then be solved either analytically or numerically. From this solution the desired physical quantities are determined. In many cases the range of validity of this procedure is much larger than that achieved by a direct perturbation expansion of the desired quantity to the same order.

The TCL technique is applicable to a large class of physical systems: It does not rely on a specific form of the interaction, of the initial state of the combined total system, or of the spectral density of the environment, nor does it employ specific symmetry or scaling properties. Moreover, the perturbation expansion of the TCL generator provides a systematic way to go beyond the Markovian approximation. For many systems that arise in the theory of open quantum systems the analysis of the Markovian approximation is relatively easy to perform and often gives a first qualitative picture of the long-time dynamics. The expansion technique then offers the possibility of studying non-Markovian phenomena in a systematic way, or to judge whether or not a non-Markovian treatment is necessary.

10.1   Spontaneous decay of a two-level system

The first example of the application of the TCL projection operator technique is an exactly solvable model of a two-level system which decays spontaneously into the field vacuum. This system has been studied already in Section 3.4.2 employing the Born–Markov approximation which led to an equation of motion for the reduced density matrix in Lindblad form. Here we treat the model in the rotating wave approximation to all orders in the coupling between system and reservoir. The exact solutions for various forms of the spectral density will be compared with the results of the time-convolutionless master equation derived in Section 9.2.1 and with the corresponding perturbation expansion of the TCL generator $\mathcal{K}(t)$ developed in Sections 9.2.2 and 9.2.3.

The Hamiltonian of the total system is given by

$$H = H_S + H_B + H_I = H_0 + H_I, \tag{10.1}$$

where

$$H_0 = \omega_0 \sigma_+ \sigma_- + \sum_k \omega_k b_k^\dagger b_k, \tag{10.2}$$

$$H_I = \sigma_+ \otimes B + \sigma_- \otimes B^\dagger \text{ with } B = \sum_k g_k b_k. \tag{10.3}$$

The transition frequency of the two-level system is denoted by ω_0, and $\sigma_\pm$ are the raising and lowering operators. The index k labels the different field modes of the reservoir with frequencies ω_k, creation and annihilation operators $b_k^\dagger$, b_k and coupling constants g_k.

10.1.1   *Exact master equation and TCL generator*

Let us introduce the states (Garraway, 1997)

$$\psi_0 = |0\rangle_S \otimes |0\rangle_B, \tag{10.4}$$

$$\psi_1 = |1\rangle_S \otimes |0\rangle_B, \tag{10.5}$$

$$\psi_k = |0\rangle_S \otimes |k\rangle_B, \tag{10.6}$$

where $|0\rangle_S = \sigma_-|1\rangle_S$ and $|1\rangle_S = \sigma_+|0\rangle_S$ indicate the ground and excited state of the system, respectively, the state $|0\rangle_B$ denotes the vacuum state of the reservoir, and $|k\rangle_B = b_k^\dagger|0\rangle_B$ denotes the state with one photon in mode k. In the interaction picture the state $\phi(t)$ of the total system obeys the Schrödinger equation

$$\frac{d}{dt}\phi(t) = -iH_I(t)\phi(t), \tag{10.7}$$

where

$$H_I(t) = \sigma_+(t)B(t) + \sigma_-(t)B^\dagger(t) \tag{10.8}$$

is the interaction picture Hamiltonian with

$$\sigma_\pm(t) = \sigma_\pm \exp(\pm i\omega_0 t), \tag{10.9}$$

and

$$B(t) = \sum_k g_k b_k \exp(-i\omega_k t). \tag{10.10}$$

It is easy to check that the total Hamiltonian (10.1) commutes with the 'particle number' operator

$$N = \sigma_+\sigma_- + \sum_k b_k^\dagger b_k, \tag{10.11}$$

i.e. we have

$$[H, N] = 0. \tag{10.12}$$

Thus, N is a conserved quantity. It follows that any initial state of the form

$$\phi(0) = c_0\psi_0 + c_1(0)\psi_1 + \sum_k c_k(0)\psi_k \tag{10.13}$$

evolves after time t into the state

$$\phi(t) = c_0\psi_0 + c_1(t)\psi_1 + \sum_k c_k(t)\psi_k. \tag{10.14}$$

The amplitude c_0 is constant since $H_I(t)\psi_0 = 0$, while the amplitudes $c_1(t)$ and $c_k(t)$ are time dependent. The time development of these amplitudes is governed by a system of differential equations which is easily derived from the Schrödinger equation (10.7),

$$\dot{c}_1(t) = -i\sum_k g_k \exp[i(\omega_0 - \omega_k)t]c_k(t), \tag{10.15}$$

$$\dot{c}_k(t) = -ig_k^* \exp[-i(\omega_0 - \omega_k)t]c_1(t). \tag{10.16}$$

Assuming that $c_k(0) = 0$, i.e. there are no photons in the initial state, we solve the second equation and insert the solution into the first to get a closed equation for $c_1(t)$,

$$\dot{c}_1(t) = -\int_0^t dt_1 f(t - t_1)c_1(t_1). \tag{10.17}$$

This procedure was already used in the classic paper by Weisskopf and Wigner (1930) on the determination of the natural line width. The kernel $f(t - t_1)$ is given by the correlation function

$$f(t - t_1) = \mathrm{tr}_B\{B(t)B^\dagger(t_1)\rho_B\} \exp[i\omega_0(t - t_1)], \tag{10.18}$$

where $\rho_B = (|0\rangle\langle 0|)_B$ is the vacuum state of the reservoir. As usual, this kernel may be expressed in terms of the spectral density $J(\omega)$ of the reservoir as follows,

$$f(t - t_1) = \int d\omega J(\omega) \exp[i(\omega_0 - \omega)(t - t_1)]. \tag{10.19}$$

The exact equation (10.17) can now be solved by means of a Laplace transformation. In subsequent sections we will present solutions for different forms of the spectral density.

With the help of the probability amplitudes c_0, $c_1(t)$ and $c_k(t)$ we can now express the reduced density matrix $\rho_S(t)$ of the two-state system as

$$\rho_S(t) = \mathrm{tr}_B\{|\phi(t)\rangle\langle\phi(t)|\} = \begin{pmatrix} |c_1(t)|^2 & c_0^* c_1(t) \\ c_0 c_1^*(t) & 1 - |c_1(t)|^2 \end{pmatrix}. \tag{10.20}$$

Differentiating this expression with respect to time and recalling that $\dot{c}_0 = 0$ we get

$$\frac{d}{dt}\rho_S(t) = \begin{pmatrix} \frac{d}{dt}|c_1(t)|^2 & c_0^* \dot{c}_1(t) \\ c_0 \dot{c}_1^*(t) & -\frac{d}{dt}|c_1(t)|^2 \end{pmatrix}. \tag{10.21}$$

Introducing further the quantities

$$S(t) = -2\Im\left\{\frac{\dot{c}_1(t)}{c_1(t)}\right\}, \tag{10.22}$$

$$\gamma(t) = -2\Re\left\{\frac{\dot{c}_1(t)}{c_1(t)}\right\}, \tag{10.23}$$

we can rewrite eqn (10.21) as

$$\frac{d}{dt}\rho_S(t) = -\frac{i}{2}S(t)[\sigma_+\sigma_-, \rho_S(t)] \tag{10.24}$$

$$+\gamma(t)\left\{\sigma_-\rho_S(t)\sigma_+ - \frac{1}{2}\sigma_+\sigma_-\rho_S(t) - \frac{1}{2}\rho_S(t)\sigma_+\sigma_-\right\}.$$

This is an exact master equation for the reduced system dynamics. Obviously, $S(t)$ plays the rôle of a time-dependent Lamb shift and $\gamma(t)$ that of a time-dependent decay rate.

The remarkable point of the master equation (10.24) is that it is already of the form of a time-convolutionless master equation,

$$\frac{d}{dt}\rho_S(t) = \mathcal{K}_S(t)\rho_S(t), \tag{10.25}$$

where the generator $\mathcal{K}_S(t)$ is connected to the TCL generator $\mathcal{K}(t)$ by means of the relation

$$\mathcal{K}_S(t)\rho_S(t) = \mathrm{tr}_B\{\mathcal{K}(t)\rho_S(t) \otimes \rho_B\}. \tag{10.26}$$

We observe that the structure of $\mathcal{K}_S(t)$ is similar to that of a Lindblad generator. However, due to the time dependence of the coefficients $S(t)$ and $\gamma(t)$ eqn (10.24) does not provide, in general, a quantum dynamical semigroup. We will also see below that the time-dependent rate $\gamma(t)$ may even become negative which violates the complete positivity of the generator.

In Sections 9.2.2 and 9.2.3 a perturbation expansion for the TCL generator in powers of the coupling strength α was developed. Let us demonstrate that this expansion produces in all orders a master equation of the form (10.24) and that it generates a perturbation expansion of the coefficients $\gamma(t)$ and $S(t)$ in the master equation which takes the form

$$\gamma(t) = \sum_{n=1}^{\infty} \alpha^{2n} \gamma_{2n}(t), \tag{10.27}$$

$$S(t) = \sum_{n=1}^{\infty} \alpha^{2n} S_{2n}(t). \tag{10.28}$$

To get this expansion we introduce an expansion parameter α by replacing the correlation function $f(t)$ by $\alpha^2 f(t)$, since the spectral density is to be considered as a quantity of second order in the coupling.

We first note that $\sigma_+ = (|1\rangle\langle 0|)_S$ is an eigenoperator of the generator $\mathcal{K}_S(t)$, i.e. we have

$$\mathcal{K}_S(t)\sigma_+ = -\frac{1}{2}[\gamma(t) + iS(t)]\sigma_+. \tag{10.29}$$

On the other hand, with the help of (10.26) the general formula (9.68) for the expansion of the TCL generator into ordered cumulants leads to

$$\mathcal{K}_S(t)\sigma_+ = \sum_{n=1}^{\infty} \alpha^{2n} \int_0^t dt_1 \int_0^{t_1} dt_2 \cdots \int_0^{t_{2n-2}} dt_{2n-1} \tag{10.30}$$
$$\times \mathrm{tr}_B \left\{ \langle \mathcal{L}(t)\mathcal{L}(t_1) \cdots \mathcal{L}(t_{2n-1}) \rangle_{\mathrm{oc}} \sigma_+ \otimes \rho_B \right\}.$$

Note that only even orders of α appear in the expansion because relation (9.11) is satisfied here. Employing $\mathcal{L}(t)\rho = -i[H_I(t), \rho]$ one easily verifies that $\sigma_+ \otimes \rho_B$ is an eigenoperator of the super-operator $\mathcal{L}(t)\mathcal{L}(t_1)$ corresponding to the eigenvalue $-f(t - t_1)$, i.e.

$$\mathcal{L}(t)\mathcal{L}(t_1)\sigma_+ \otimes \rho_B = -f(t - t_1)\sigma_+ \otimes \rho_B. \tag{10.31}$$

Since further $\mathcal{P}\sigma_+ \otimes \rho_B = \sigma_+ \otimes \rho_B$ the expansion (10.30) can be restated as follows,

$$\mathcal{K}_S(t)\sigma_+ = \sum_{n=1}^{\infty} \alpha^{2n} \int_0^t dt_1 \int_0^{t_1} dt_2 \cdots \int_0^{t_{2n-2}} dt_{2n-1} \tag{10.32}$$
$$\times (-1)^n \langle f(t - t_1)f(t_2 - t_3) \cdots f(t_{2n-2} - t_{2n-1}) \rangle_{\mathrm{oc}} \sigma_+.$$

Comparing the last expression with eqn (10.29) we immediately get the TCL expansion for the coefficients of the exact quantum master equation

$$\gamma_{2n}(t) + iS_{2n}(t) = \int\limits_0^t dt_1 \int\limits_0^{t_1} dt_2 \ldots \int\limits_0^{t_{2n-2}} dt_{2n-1} \tag{10.33}$$

$$\times 2(-1)^{n+1} \langle f(t-t_1)f(t_2-t_3)\ldots f(t_{2n-2}-t_{2n-1})\rangle_{\mathrm{oc}}.$$

The rules for the construction of the quantity $\langle f(t-t_1)\ldots f(t_{2n-2}-t_{2n-1})\rangle_{\mathrm{oc}}$ can be stated as follows: Write down the corresponding expression for the ordered cumulant of order $2n$ according to the rules given below eqn (9.68). Omit all factors of $\mathcal{P}$ and replace the pairs $\mathcal{L}(t_i)\mathcal{L}(t_j)$ of successive $\mathcal{L}$-factors by the correlation function $f(t_i - t_j)$.

Following these rules eqn (9.45) immediately yields the second-order contribution for the coefficients of the master equation,

$$\gamma_2(t) + iS_2(t) = 2\int\limits_0^t dt_1 f(t-t_1), \tag{10.34}$$

while (9.51) leads to the fourth-order contribution,

$$\gamma_4(t) + iS_4(t) = 2\int\limits_0^t dt_1 \int\limits_0^{t_1} dt_2 \int\limits_0^{t_2} dt_3 [f(t-t_2)f(t_1-t_3) + f(t-t_3)f(t_1-t_2)]. \tag{10.35}$$

It is convenient to define the real functions $\Phi(t)$ and $\Psi(t)$ through

$$2f(t) = \Phi(t) + i\Psi(t). \tag{10.36}$$

In terms of these functions the second-order contributions to the decay rate $\gamma(t)$ and to the energy shift $S(t)$ determining the Born approximation read

$$\gamma_2(t) = \int\limits_0^t dt_1 \Phi(t-t_1), \tag{10.37}$$

$$S_2(t) = \int\limits_0^t dt_1 \Psi(t-t_1), \tag{10.38}$$

and hence the Markovian decay rate γ_{M} and the Markovian Lamb shift S_{M} are found by extending the integration to infinity,

$$\gamma_{\mathrm{M}} = \int\limits_0^\infty ds\, \Phi(s), \tag{10.39}$$

$$S_{\mathrm{M}} = \int\limits_0^\infty ds\, \Psi(s). \tag{10.40}$$

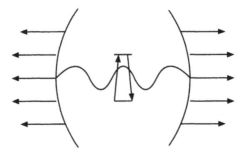

FIG. 10.1. Schematic representation of the damped Jaynes–Cummings model.

The fourth-order TCL contributions to the decay rate and to the energy shift may be expressed as

$$\gamma_4(t) = \frac{1}{2} \int\limits_0^t dt_1 \int\limits_0^{t_1} dt_2 \int\limits_0^{t_2} dt_3 \Big[\Phi(t - t_2)\Phi(t_1 - t_3) + \Phi(t - t_3)\Phi(t_1 - t_2)$$

$$-\Psi(t - t_2)\Psi(t_1 - t_3) - \Psi(t - t_3)\Psi(t_1 - t_2) \Big], \qquad (10.41)$$

and

$$S_4(t) = \frac{1}{2} \int\limits_0^t dt_1 \int\limits_0^{t_1} dt_2 \int\limits_0^{t_2} dt_3 \Big[\Psi(t - t_2)\Phi(t_1 - t_3) + \Phi(t - t_2)\Psi(t_1 - t_3)$$

$$+\Psi(t - t_3)\Phi(t_1 - t_2) + \Phi(t - t_3)\Psi(t_1 - t_2) \Big]. \qquad (10.42)$$

Thus we see that the form of the exact master equation (10.24) is preserved in all orders of the coupling and that the TCL expansion amounts to an expansion of the coefficients $S(t)$ and $\gamma(t)$. In the following subsection we study the non-Markovian behaviour of some specific models involving different spectral densities (Breuer, Kappler and Petruccione, 1999).

10.1.2   Jaynes–Cummings model on resonance

The damped Jaynes–Cummings model describes the coupling of a two-level atom to a single cavity mode which in turn is coupled to a reservoir consisting of harmonic oscillators in the vacuum state. A schematic representation of the Jaynes–Cummings model is depicted in Fig. 10.1. If we restrict ourselves to the case of a single excitation in the atom–cavity system, the cavity mode can be eliminated in favour of an effective spectral density of the form

$$J(\omega) = \frac{1}{2\pi} \frac{\gamma_0 \lambda^2}{(\omega_0 - \omega)^2 + \lambda^2}, \qquad (10.43)$$

where ω_0 is the transition frequency of the two-level system. The parameter λ defines the spectral width of the coupling, which is connected to the reservoir correlation time τ_B by the relation $\tau_B = \lambda^{-1}$ and the time scale τ_R on which the state of the system changes is given by $\tau_R = \gamma_0^{-1}$.

In order to compute the exact probability amplitude $c_1(t)$ we evaluate the reservoir-correlation function $f(t)$ using the spectral density (10.43),

$$f(t) = \frac{1}{2}\gamma_0\lambda\exp(-\lambda|t|). \tag{10.44}$$

For this $f(t)$ the differential equation (10.17) for the probability amplitude $c_1(t)$ can easily be solved to give the exact solution

$$c_1(t) = c_1(0)e^{-\lambda t/2}\left[\cosh\left(\frac{dt}{2}\right) + \frac{\lambda}{d}\sinh\left(\frac{dt}{2}\right)\right], \tag{10.45}$$

where $d = \sqrt{\lambda^2 - 2\gamma_0\lambda}$. This yields the time-dependent population of the excited state

$$\rho_{11}(t) = \rho_{11}(0)e^{-\lambda t}\left[\cosh\left(\frac{dt}{2}\right) + \frac{\lambda}{d}\sinh\left(\frac{dt}{2}\right)\right]^2. \tag{10.46}$$

Using eqns (10.22) and (10.23) we therefore obtain a vanishing Lamb shift, $S(t) \equiv 0$, and the time-dependent decay rate

$$\gamma(t) = \frac{2\gamma_0\lambda\sinh(dt/2)}{d\cosh(dt/2) + \lambda\sinh(dt/2)}. \tag{10.47}$$

In Fig. 10.2(a) we show this time-dependent decay rate $\gamma(t)$ ('exact') together with the Markovian decay rate $\gamma_M = \gamma_0$ ('Markov') for $\tau_R = 5\tau_B$. For short times, i.e. for times of the order of τ_B, the exact decay rate grows linearly with t, which leads to the correct quantum mechanical short-time behaviour of the transition probability. In the long-time limit the decay rate saturates at a value larger than the Markovian decay rate, which represents corrections to the golden rule. The population of the excited state is depicted in Fig. 10.2(b): For short times, the exact population decreases quadratically and is larger than the Markovian population, which is simply given by $\rho_{11}(0)\exp(-\gamma_0 t)$, whereas in the long-time limit the exact population is slightly less than the Markovian population.

Next, we determine the solution of the generalized quantum master equation (9.26) in the Born approximation. To this end, we insert the spectral density of the coupling strength (10.43) into eqn (10.36) to obtain $\Psi(t) \equiv 0$ and

$$\Phi(t) = \gamma_0\lambda\exp(-\lambda t). \tag{10.48}$$

This leads to the following form of the generalized master equation,

$$\frac{d}{dt}\rho_S(t) = \gamma_0\lambda\int_0^t ds\, e^{-\lambda(t-s)}\left[\sigma_-\rho_S(s)\sigma_+ - \frac{1}{2}\sigma_+\sigma_-\rho_S(s) - \frac{1}{2}\rho_S(s)\sigma_+\sigma_-\right]. \tag{10.49}$$

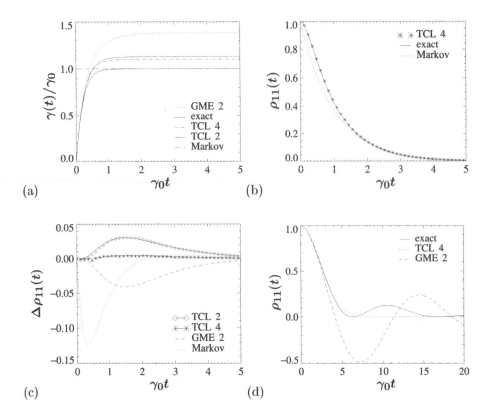

FIG. 10.2. The damped Jaynes–Cummings model on resonance. Exact solution (exact), time-convolutionless master equation to second (TCL 2) and fourth order (TCL 4), generalized master equation to second order (GME 2), and the Markovian quantum master equation (Markov): (a) Decay rate of the excited state population, (b) the population of the excited state, and (c) deviation of the approximate solutions from the exact result, for $\tau_R = 5\tau_B$ (moderate coupling). The symbols show the results of a stochastic simulation of the time-convolutionless quantum master equation using 10^5 realizations. (d) Population of the excited state for $\tau_R = 0.2\tau_B$ (strong coupling).

We differentiate with respect to t and obtain

$$\ddot{\rho}_S(t) = -\lambda\dot{\rho}_S(t) + \gamma_0\lambda\left[\sigma_-\rho_S(t)\sigma_+ - \frac{1}{2}\sigma_+\sigma_-\rho_S(t) - \frac{1}{2}\rho_S(t)\sigma_+\sigma_-\right]. \quad (10.50)$$

Due to the exponential memory kernel, this equation of motion is an ordinary differential equation which is local in time, and contains only $\rho_S(t)$, $\dot{\rho}_S(t)$ and $\ddot{\rho}_S(t)$. The solution leads to the following time evolution of the population of the upper level,

$$\tilde{\rho}_{11}(t) = \tilde{\rho}_{11}(0)e^{-\lambda t/2}\left[\cosh\left(\frac{d't}{2}\right) + \frac{\lambda}{d'}\sinh\left(\frac{d't}{2}\right)\right], \tag{10.51}$$

where $d' = \sqrt{\lambda^2 - 4\gamma_0\lambda}$. From this expression, we can determine the time-dependent decay rate

$$\tilde{\gamma}(t) = -\frac{\dot{\tilde{\rho}}_{11}(t)}{\tilde{\rho}_{11}(t)} = \frac{2\gamma_0\lambda\sinh(d't/2)}{d'\cosh(d't/2) + \lambda\sinh(d't/2)}, \tag{10.52}$$

the structure of which is similar to the exact decay rate (10.47). Note, however, the difference between the parameters d and d' which can also be seen in Fig 10.2(a) where we have also plotted the decay rate $\tilde{\gamma}(t)$ ('GME 2'): For short times, the decay rate $\tilde{\gamma}(t)$ is in good agreement with $\gamma(t)$, but in the long-time limit, $\tilde{\gamma}(t)$ is too large.

Finally, the time-convolutionless decay rate can be determined from eqn (10.41). The second and fourth-order contributions are given by

$$\gamma_2(t) = \gamma_0\left(1 - e^{-\lambda t}\right), \tag{10.53}$$

and

$$\gamma_4(t) = \frac{\gamma_0^2}{\lambda}\left[\sinh(\lambda t) - \lambda t\right]e^{-\lambda t}. \tag{10.54}$$

This corresponds to a Taylor expansion of the exact decay rate $\gamma(t)$ in powers of the expansion parameter $\alpha^2 = \gamma_0/\lambda = \tau_B/\tau_R$. Thus we see that the TCL expansion provides an expansion in the ratio of the reservoir correlation time τ_B to the system's relaxation time τ_R. Figure 10.2(a) clearly shows that the second-order contribution $\gamma_2(t)$ to the TCL expansion as well as the contribution up to fourth order $\gamma^{(4)}(t) = \gamma_2(t) + \gamma_4(t)$ approximate the exact decay rate very well for short times, and that the fourth-order contribution $\gamma^{(4)}$ is also a good approximation in the long-time limit.

The time evolution of the population of the excited state can be obtained by integrating the rate $\gamma^{(4)}(t)$ with respect to t. This yields

$$\rho_{11}^{(4)}(t) = \rho_{11}(0)\exp\left(-\int_0^t ds\,\gamma^{(4)}(s)\right). \tag{10.55}$$

In order to compare the quality of the different approximation schemes, we show the difference between the approximated populations and the exact population in Fig. 10.2(c). In addition to the analytical solutions of the generalized master equation and the time-convolutionless master equation, we have also performed a stochastic simulation of the time-convolutionless quantum master equations with 10^5 realizations. Since the time-dependent rates $\gamma_2(t)$ and $\gamma^{(4)}(t)$ are positive for all t, the corresponding generators are in Lindblad form, and we can use the stochastic unravelling developed in Section 6.1.1. Figure 10.2(c) shows that the stochastic simulation is in very good agreement with the corresponding analytical solutions. Moreover, we see that the difference between the solution of the time-convolutionless master equation to fourth order and the exact master equation is small (see also Fig. 10.2(b)), whereas the errors of the generalized master equation and of the time-convolutionless master equation to second order are larger and of the same order of magnitude. In fact, the Markov approximation even leads to a slight improvement of the accuracy, compared to the Born approximation, which is surprising if we consider the heuristic derivation of the quantum master equation.

The approximation schemes used here are perturbative and hence rely on the assumption that the coupling is not too strong. But what happens if the system approaches the strong coupling regime? We will investigate this question by means of the damped Jaynes–Cummings model on resonance, where the explicit expressions of the quantities of interest are known.

First, let us take a look at the exact expression for the population of the excited state (10.46): In the strong coupling regime, that is for $\gamma_0 > \lambda/2$ or $\tau_R < 2\tau_B$, the parameter d is purely imaginary. Defining $\hat{d} = -id$ we can write the exact population as

$$\rho_{11}(t) = \rho_{11}(0)e^{-\lambda t}\left[\cos\left(\frac{\hat{d}t}{2}\right) + \frac{\lambda}{\hat{d}}\sin\left(\frac{\hat{d}t}{2}\right)\right]^2, \qquad (10.56)$$

which is an oscillating function that has discrete zeros at

$$t = \frac{2}{\hat{d}}\left(n\pi - \arctan\frac{\hat{d}}{\lambda}\right), \qquad (10.57)$$

where $n = 1, 2, \ldots$. Hence, the rate $\gamma(t)$ diverges at these points (see eqn (10.23)). Obviously, $\gamma(t)$ can only be an analytical function for $t \in [0, t_0[$, where t_0 is the smallest positive zero of $\rho_{11}(t)$.

On the other hand, as we have just seen, the time-convolutionless quantum master equation corresponds basically to a Taylor expansion of $\gamma(t)$ in powers of γ_0, and the radius of convergence of this series is given by the region of analyticity of $\gamma(t)$. For $\gamma_0 < \lambda/2$, this is the whole positive real axis, but for $\gamma_0 > \lambda/2$ the perturbation expansion only converges for $t < t_0$. This behaviour can be clearly seen in Fig. 10.2(d), where we depict $\rho_{11}(t)$ and $\rho_{11}^{(4)}(t)$ for $\tau_R = \tau_B/5$, i.e. for strong coupling: The perturbation expansion converges to $\rho_{11}(t)$ for $t \lesssim t_0 \approx 6.3/\gamma_0$, but fails to converge for $t > t_0$.

The solution of the generalized master equation to second order shows a quite distinct behaviour, but also fails in the strong-coupling regime: For $\gamma_0 > \lambda/4$ the population $\rho_{11}(t)$ starts to oscillate and even takes negative values, which is unphysical (see Fig. 10.2(d)).

The 'failure' of the time-convolutionless master equation at $t = t_0$ can also be understood from a more intuitive point of view. The time-convolutionless equation of motion states that the evolution of the reduced density matrix only

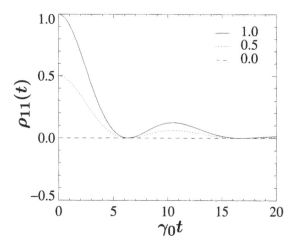

FIG. 10.3. The damped Jaynes–Cummings model on resonance. Exact population for the three different initial conditions $\rho_{11}(0) = 1.0$, 0.5, 0.0 in the
strong-coupling regime ($\tau_R = 0.2\tau_B$).

depends on the actual value of $\rho_S(t)$ and on the TCL generator. However, at
$t = t_0$ the time evolution also depends on the initial value of the density matrix.
This fact is illustrated in Fig. 10.3, where we have plotted the population $\rho_{11}(t)$
for three different initial conditions, namely $\rho_{11}(0) = 1.0$, 0.5, 0.0. At $t = t_0$,
the corresponding density matrices coincide, regardless of the initial condition.
However, the future time evolution for $t > t_0$ is different for these trajectories.
It is therefore intuitively clear that a time-convolutionless form of the equation
of motion which is local in time ceases to exist for $t > t_0$. The formal reason for
this fact is that at $t = t_0$ the operator $1 - \Sigma(t)$ (see Section 9.2.1) is not invertible
and hence the generator $\mathcal{K}(t)$ does not exist at this point.

10.1.3 Jaynes–Cummings model with detuning

In this section we treat the damped Jaynes–Cummings model with detuning,
i.e. the same setup as in the preceding example but the centre frequency of the
cavity is detuned by an amount Δ against the atomic transition frequency ω_0.
In this case the spectral density takes the form

$$J(\omega) = \frac{1}{2\pi} \frac{\gamma_0 \lambda^2}{(\omega_0 - \Delta - \omega)^2 + \lambda^2}, \tag{10.58}$$

and thus the functions $\Phi(t)$ and $\Psi(t)$ are given by

$$\Phi(t) = \gamma_0 \lambda e^{-\lambda t} \cos(\Delta t), \tag{10.59}$$

$$\Psi(t) = \gamma_0 \lambda e^{-\lambda t} \sin(\Delta t). \tag{10.60}$$

With these functions, the time-dependent Lamb shift $S^{(4)}(t) = S_2(t) + S_4(t)$ and
decay rate $\gamma^{(4)}(t) = \gamma_2(t) + \gamma_4(t)$ to fourth order in the coupling can be calculated

using eqns (10.42) and (10.41). The integrals can be evaluated exactly and lead to the expressions

$$S^{(4)}(t) = \frac{\gamma_0 \lambda \Delta}{\lambda^2 + \Delta^2} \left[1 - e^{-\lambda t} \left(\cos(\Delta t) + \tfrac{\lambda}{\Delta} \sin(\Delta t) \right) \right] \tag{10.61}$$
$$- \frac{\gamma_0^2 \lambda^2 \Delta^3 e^{-\lambda t}}{2(\lambda^2 + \Delta^2)^3} \left\{ \left[1 - 3 \left(\tfrac{\lambda}{\Delta} \right)^2 \right] \left(e^{\lambda t} - e^{-\lambda t} \cos(2\Delta t) \right) \right.$$
$$- 2 \left[1 - \left(\tfrac{\lambda}{\Delta} \right)^4 \right] \Delta t \sin(\Delta t) + 4 \left[1 + \left(\tfrac{\lambda}{\Delta} \right)^2 \right] \lambda t \cos(\Delta t)$$
$$\left. - \tfrac{\lambda}{\Delta} \left[3 - \left(\tfrac{\lambda}{\Delta} \right)^2 \right] e^{-\lambda t} \sin(2\Delta t) \right\}$$

and

$$\gamma^{(4)}(t) = \frac{\gamma_0 \lambda^2}{\lambda^2 + \Delta^2} \left[1 - e^{-\lambda t} \left(\cos(\Delta t) - \tfrac{\Delta}{\lambda} \sin(\Delta t) \right) \right] \tag{10.62}$$
$$+ \frac{\gamma_0^2 \lambda^5 e^{-\lambda t}}{2(\lambda^2 + \Delta^2)^3} \left\{ \left[1 - 3 \left(\tfrac{\Delta}{\lambda} \right)^2 \right] \left(e^{\lambda t} - e^{-\lambda t} \cos(2\Delta t) \right) \right.$$
$$- 2 \left[1 - \left(\tfrac{\Delta}{\lambda} \right)^4 \right] \lambda t \cos(\Delta t) + 4 \left[1 + \left(\tfrac{\Delta}{\lambda} \right)^2 \right] \Delta t \sin(\Delta t)$$
$$\left. + \tfrac{\Delta}{\lambda} \left[3 - \left(\tfrac{\Delta}{\lambda} \right)^2 \right] e^{-\lambda t} \sin(2\Delta t) \right\}.$$

In Fig. 10.4(a) we have depicted $\gamma^{(4)}(t)$ together with the exact decay rate, which can be calculated by solving the differential equation for the probability amplitude $c_1(t)$ for $\Delta = 8\lambda$ and $\lambda = 0.3\gamma_0$. Note that the spontaneous decay rate is severely suppressed compared to the spontaneous decay on resonance. This can also be seen by computing the Markovian decay rate γ_M which is given by

$$\gamma_M = \frac{\gamma_0 \lambda^2}{\lambda^2 + \Delta^2} \approx 0.015 \gamma_0. \tag{10.63}$$

However, this strong suppression is most effective in the long-time limit. For short times, $\gamma(t)$ oscillates with a large amplitude and can even take negative values, which leads to an increase of the population. This is due to photons which have been emitted by the atom and are reabsorbed at a later time. Hence, the exact quantum master equation as well as its time-convolutionless approximation are not in Lindblad form, but conserve the positivity of the reduced density matrix. This is of course not a contradiction to the Lindblad theorem, since a basic assumption of this theorem is that the reduced system dynamics constitutes a dynamical semigroup. This assumption is obviously violated here.

Although the time-convolutionless master equation is not in Lindblad form and involves negative transition rates, it can be represented through an appropriate stochastic process as was shown in Section 9.3. The dynamics of the stochastic wave function $\theta(t)$, which is an element of the doubled Hilbert space

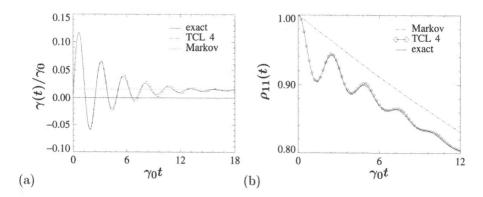

FIG. 10.4. The damped Jaynes–Cummings model with detuning. Exact solution (exact), time-convolutionless master equation to fourth order (TCL 4),
and the Markovian quantum master equation (Markovian): (a) Decay rate
of the excited state population, and (b) the population of the excited state,
including a stochastic simulation of the time-convolutionless quantum master
equation with 10^5 realizations (symbols), for $\lambda = 0.3\gamma_0$ and $\Delta = 8\lambda$.

$\tilde{\mathcal{H}} = \mathcal{H}_S \oplus \mathcal{H}_S$, is governed by the stochastic differential equation (9.98), where
the operator $F(t)$ is given by

$$F(t) = -\frac{1}{2}\gamma^{(4)}(t)\begin{pmatrix} \sigma_+\sigma_- & 0 \\ 0 & \sigma_+\sigma_- \end{pmatrix}, \tag{10.64}$$

and we have a single jump operator

$$J(t) = \begin{pmatrix} \gamma^{(4)}(t)\sigma_- & 0 \\ 0 & \sigma_- \end{pmatrix}. \tag{10.65}$$

The jumps induce instantaneous transitions of the form

$$\theta(t) \longrightarrow \frac{\|\theta(t)\|}{\|J\theta(t)\|}J\theta(t) \sim \begin{pmatrix} \gamma^{(4)}(t)|0\rangle_S \\ |0\rangle_S \end{pmatrix}. \tag{10.66}$$

If the rate $\gamma^{(4)}(t)$ is positive these transitions lead to a positive contribution to
the ground state population $\rho_{00}(t)$, while a negative rate results in a decrease of
$\rho_{00}(t)$.

In Fig. 10.4(b), we show the results of a stochastic simulation for 10^5 realizations, together with the analytical solution of the time-convolutionless quantum
master equation and the exact solution. Obviously, the agreement of all three
curves is good and the stochastic simulation algorithm works excellently even
for negative decay rates. In addition, we also show the solution of the Markovian
quantum master equation which clearly underestimates the decay for short times
and does not show oscillations.

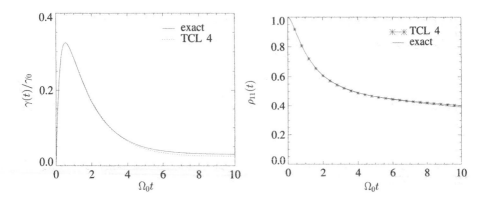

FIG. 10.5. Spontaneous decay in a photonic band gap: Exact solution (exact), and time-convolutionless master equation to fourth order (TCL 4). Left: Decay rate of the excited state population. Right: The population of the excited state, including a stochastic simulation of the time-convolutionless quantum master equation with 10^5 realizations (symbols), for $W_1 = 1.1$, $W_2 = 0.1$, $\Gamma_1/\Omega_0 = 10$, and $\Gamma_2/\Omega_0 = 1$.

10.1.4 Spontaneous decay into a photonic band gap

As our final example, we treat a simple model for the spontaneous decay of a two-level system in a photonic band gap which was introduced by Garraway (1997). To this end, we consider a spectral density of the form

$$J(\omega) = \frac{\Omega_0^2}{2\pi} \left(\frac{W_1 \Gamma_1}{(\omega - \omega_0)^2 + (\Gamma_1/2)^2} - \frac{W_2 \Gamma_2}{(\omega - \omega_0)^2 + (\Gamma_2/2)^2} \right), \qquad (10.67)$$

where Ω_0^2 describes the overall coupling strength, Γ_1 the bandwidth of the 'flat' background continuum, Γ_2 the width of the gap, and W_1 and W_2 the relative strength of the background and the gap, respectively. In Fig. 10.5 we show the excited state's decay rate $\gamma(t)$ for the parameters $\Gamma_1/\Omega_0 = 10$, $\Gamma_2/\Omega_0 = 1$, $W_1 = 1.1$, and $W_2 = 0.1$. For short times, $\gamma(t)$ increases linearly on the time scale Γ_1^{-1} and then attains a maximum value. This phase of the dynamics stems from transitions into the 'flat' background continuum. For longer times, i.e. $t \gg \Gamma_2^{-1}$, the system resolves the structure of the spectral density and transitions are strongly suppressed due to the presence of the gap. The decay rate therefore becomes smaller and smaller until it reaches its final value. Thus, the population of the excited state decreases rapidly for times of order Γ_2^{-1}, and slowly in the long-time limit (see Fig. 10.5).

The time-dependent Lamb shift $S^{(4)}(t)$ and the decay rate $\gamma^{(4)}(t)$ of the time-convolutionless quantum master equation to fourth order can be computed by inserting the spectral density of the coupling strength $J(\omega)$ into eqn (10.36). This yields $\Psi(t) \equiv 0$ and

$$\Phi(t) = 2\Omega_0^2 \left(W_1 e^{-\Gamma_1 t/2} - W_2 e^{-\Gamma_2 t/2} \right), \tag{10.68}$$

which can be inserted into eqns (10.42) and (10.41). Since $\Psi(t) \equiv 0$ the Lamb shift $S^{(4)}(t)$ vanishes; the time-dependent decay rate $\gamma^{(4)}(t)$ can be computed explicitly, and is in good agreement with the exact decay rate for our choice of parameters (see Fig. 10.5).

10.2 The damped harmonic oscillator

The damped harmonic oscillator has already been studied at various places in Chapter 3, where we discussed the master equation (3.307) in the quantum optical limit and the high-temperature Brownian motion master equation (3.410). It provides one of the few open systems which can be solved analytically – for example by solving the exact Heisenberg equations of motion (Section 3.6.3) or by means of the Feynman–Vernon path integral technique (Section 3.6.4). Therefore, the damped harmonic oscillator is particularly suited to investigate the parameter regime where the perturbation expansion of the time-convolutionless master equation yields reliable results beyond the Born–Markov limit. We will show that the perturbative treatment is in good agreement with the exact solution for sufficiently high temperatures at any coupling strength and for low temperatures at weak or moderate couplings.

10.2.1 *The model and frequency renormalization*

We consider the Caldeira–Leggett model for a harmonic oscillator, linearly coupled to a bath of harmonic oscillators (see Chapter 3.6). The Hamiltonian of the composite system thus takes the form

$$H = H_S + H_B + H_I, \tag{10.69}$$

where

$$H_S = \frac{1}{2}P^2 + \frac{1}{2}\omega_b^2 X^2, \tag{10.70}$$

$$H_B = \sum_n \left(\frac{1}{2m_n}p_n^2 + \frac{1}{2}m_n\omega_n^2 x_n^2 \right), \tag{10.71}$$

$$H_I = -X \sum_n \kappa_n x_n = -X \otimes B. \tag{10.72}$$

The Hamiltonians H_S and H_B generate the free time evolution of the system and the bath, respectively, and H_I denotes their interaction. Note that we have chosen units such that $\hbar = 1$ and that we have rescaled position and momentum variables according to $x \to X/\sqrt{m}$ and $p \to P\sqrt{m}$, so that the mass of the oscillator is normalized to one in the subsequent expressions.

The frequency ω_b which enters the definition of the system Hamiltonian (10.70) is not the physically observable frequency of the oscillator but the bare frequency, since the coupling to the environment induces a frequency shift which depends on the cutoff frequency Ω (see Section 3.6.2.1). The bare frequency is connected to the renormalized frequency ω_0 by the relation

$$\omega_b^2 = \omega_0^2 + \omega_c^2, \qquad (10.73)$$

with

$$\omega_c^2 = \sum_n \frac{\kappa_n^2}{m_n \omega_n^2} \simeq 2 \int_0^\infty d\omega \frac{J(\omega)}{\omega} = 2\gamma\Omega, \qquad (10.74)$$

where $J(\omega)$ denotes the Ohmic spectral density with a Lorentz–Drude cutoff function (see eqn (3.392)).

The frequency ω_c^2 depends on the coupling strength and is of second order α^2 in the coupling. Hence, the bare frequency ω_b is a function containing terms of order α^2. On the other hand, the generator $\mathcal{K}(t) = \mathcal{K}(t; \omega_b)$ and the inhomogeneity $\mathcal{J}(t) = \mathcal{J}(t; \omega_b)$ (see eqn (9.84)) depend, among other parameters, on the bare frequency ω_b. In order to obtain a consistent expansion in terms of the coupling strength, we thus have to take into account explicitly the potential renormalization. To second order this yields

$$\mathcal{K}_2(t) = \mathcal{K}_2(t; \omega_0) \quad \text{and} \quad \mathcal{J}_2(t) = \mathcal{J}_2(t; \omega_0), \qquad (10.75)$$

while the fourth-order contributions read

$$\mathcal{K}_4(t) = \mathcal{K}_4(t; \omega_0) + \frac{\omega_c^2}{2\omega_0} \frac{\partial \mathcal{K}_2(t; \omega_0)}{\partial \omega_0}, \qquad (10.76)$$

$$\mathcal{J}_4(t) = \mathcal{J}_4(t; \omega_0) + \frac{\omega_c^2}{2\omega_0} \frac{\partial \mathcal{J}_2(t; \omega_0)}{\partial \omega_0}, \qquad (10.77)$$

where the expressions $\mathcal{K}_n(t)$ and $\mathcal{J}_n(t)$ denote the n-th order contributions to the generator and the inhomogeneity as derived in Sections 9.2.2 and 9.2.4.

Thus, for the computation of the second-order contribution we simply use the renormalized frequency ω_0 instead of the bare frequency ω_b. To fourth order, we have to consider two terms: The first term is again obtained by replacing the bare frequency with the renormalized frequency in the usual expressions for $\mathcal{K}_4$ and $\mathcal{J}_4$. The second term is given by differentiating $\mathcal{K}_2$ and $\mathcal{J}_2$ with respect to the renormalized frequency.

10.2.2   *Factorizing initial conditions*

In this section we investigate the relaxation of an initially factorizing state into the equilibrium state. To this end, we determine the generator $\mathcal{K}(t)$ to fourth order using eqns (9.61) and (9.62). After some algebra we obtain in the Schrödinger representation the following quantum master equation with time-dependent coefficients:

$$\frac{d}{dt}\rho_S(t) = -i\left[H_S, \rho_S(t)\right] - \frac{i}{2}\Delta(t)\left[X^2, \rho_S(t)\right]$$
$$-i\lambda(t)[X, \{P, \rho_S(t)\}] - D_{PP}(t)\left[X, [X, \rho_S(t)]\right]$$
$$+2D_{PX}(t)\left[X, [P, \rho_S(t)]\right]. \tag{10.78}$$

This equation has precisely the same structure as the exact quantum master equation (Haake and Reibold, 1985; Hu, Paz and Zhang, 1992; Karrlein and Grabert, 1997) and it can be shown that it is not in Lindblad form (see, e.g. Săndulescu and Scutaru, 1987).

The coefficient $\Delta(t)$ leads to a time-dependent energy shift, $\lambda(t)$ is the classical damping term and $D_{PP}(t)$ and $D_{PX}(t)$ are diffusion terms. The physically observable frequency of the oscillator is given by

$$\omega_p^2(t) = \omega_b^2 + \Delta(t) = \omega_0^2 + \omega_c^2 + \Delta(t). \tag{10.79}$$

As we will see below, in the long-time limit ω_c^2 and $\Delta(t)$ tend to compensate each other, such that the observable frequency $\omega_p(t)$ is close to the renormalized frequency ω_0.

To second order in the coupling strength, the coefficients entering the master equation read

$$\Delta^{(2)}(t) = -\int_0^t ds\, D(s)\cos(\omega_0 s), \tag{10.80}$$

$$\lambda^{(2)}(t) = \frac{1}{2\omega_0}\int_0^t ds\, D(s)\sin(\omega_0 s), \tag{10.81}$$

$$D_{PX}^{(2)}(t) = \frac{1}{4\omega_0}\int_0^t ds\, D_1(s)\sin(\omega_0 s), \tag{10.82}$$

$$D_{PP}^{(2)}(t) = \frac{1}{2}\int_0^t ds\, D_1(s)\cos(\omega_0 s), \tag{10.83}$$

which is in agreement with the results obtained by Hu, Paz and Zhang. We recall that the dissipation and the noise kernel $D(s)$ and $D_1(s)$ were introduced in eqns (3.393) and (3.394), respectively. The Markovian limit is obtained in the limit $t \to \infty$, yielding for Ohmic dissipation

$$\Delta^M = -\frac{2\gamma\Omega^3}{\Omega^2 + \omega_0^2}, \tag{10.84}$$

$$\lambda^M = \frac{\gamma\Omega^2}{\Omega^2 + \omega_0^2}, \tag{10.85}$$

$$D_{PX}^{\mathrm{M}} = \frac{\gamma k_{\mathrm{B}} T \Omega^2}{\Omega^2 + \omega_0^2} \sum_{n=-\infty}^{\infty} \frac{\omega_0^2 - \Omega|\nu_n|}{(\nu_n^2 + \omega_0^2)(|\nu_n| + \Omega)}, \tag{10.86}$$

$$D_{PP}^{\mathrm{M}} = \frac{\gamma \omega_0 \Omega^2}{\Omega^2 + \omega_0^2} \coth\left(\frac{\omega_0}{2k_{\mathrm{B}} T}\right). \tag{10.87}$$

In the high-temperature limit $k_{\mathrm{B}} T \gg \Omega \gg \omega_0$ this yields the coefficients of the well-known Caldeira–Leggett master equation (3.410), namely

$$\Delta^{\mathrm{CL}} = -2\gamma\Omega, \tag{10.88}$$

$$\lambda^{\mathrm{CL}} = \gamma, \tag{10.89}$$

$$D_{PX}^{\mathrm{CL}} = \frac{\gamma k_{\mathrm{B}} T}{\Omega}, \tag{10.90}$$

$$D_{PP}^{\mathrm{CL}} = 2\gamma k_{\mathrm{B}} T. \tag{10.91}$$

Note that the coefficient D_{PX}^{CL} is small in comparison to the other coefficients and may thus be set equal to zero. The explicit expressions for the coefficients to fourth order can be found in (Breuer, Kappler and Petruccione, 2001). The energy shift $\Delta(t)$ and the damping coefficient $\lambda(t)$ only depend on the dissipation kernel $D(t)$. Hence, these quantities are independent of the temperature of the reservoir. In Fig. 10.6 we show the time dependence of the physically observable frequency $\omega_p(t)$ and of the coefficient $\lambda(t)$ for moderate coupling ($\gamma = \omega_0$, $\Omega = 20\omega_0$). Both functions decay on a time scale which is of the order of magnitude of the inverse cutoff frequency Ω^{-1} and approach a constant value for long times. In the limit $\Omega \to \infty$ the fourth-order contribution becomes negligible and the constants are given by the Markovian limits: $\omega_p \to \omega_0$ and $\lambda(t) \to \lambda^{\mathrm{CL}}$.

The time dependence of the temperature-dependent diffusion coefficients $D_{PP}(t)$ and $D_{PX}(t)$ to fourth order is depicted in Fig. 10.7 for three different temperature regimes. For high temperatures, i.e. $k_{\mathrm{B}} T \gg \Omega$, the diffusion coefficients vary on a time scale which is of the order of the inverse cutoff frequency (top). For intermediate temperatures, i.e. $\Omega \gg k_{\mathrm{B}} T \gg \omega_0$, the coefficients vary on a time scale of the order of the thermal correlation time $1/k_{\mathrm{B}} T$ (middle), and for low temperatures the time dependence of the diffusion coefficients is governed by the inverse system frequency ω_0^{-1} (bottom). From this behaviour, we expect the perturbation approximation to be in good agreement with the exact solution for high temperatures if $\gamma \ll \Omega$, for intermediate temperatures in the regime $\gamma \ll k_{\mathrm{B}} T$ and for low temperatures in the regime $\gamma \ll \omega_0$. This point will be discussed in more detail in Section 10.2.3, where we compute the stationary density matrix of the reduced system.

Let us now consider the relative error of the coefficients of the quantum master equation, which we define for $\lambda(t)$ as

$$e_{\mathrm{r}}^{(2n)} = \lim_{t \to \infty} \left| \frac{\lambda^{(2n)}(t) - \lambda(t)}{\lambda(t)} \right|, \tag{10.92}$$

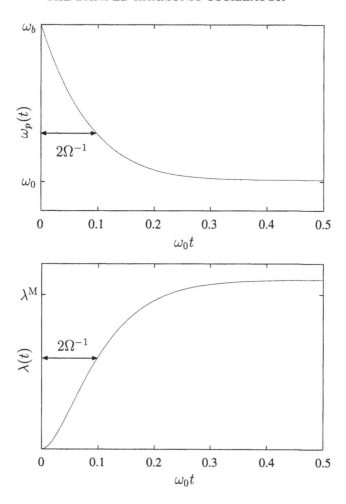

FIG. 10.6. Time dependence of the physically observable frequency $\omega_p(t)$ and of the damping coefficient $\lambda(t)$ of the harmonic oscillator to fourth order in the coupling strength. The parameters are: $\gamma = \omega_0$, $\Omega = 20\,\omega_0$.

and in a similar way for the other coefficients. This type of error was discussed in Section 9.2.5. By using the expansion of the coefficients to fourth order we obtain the following estimates for these errors:

$$\hat{e}_r^{(2)} = \lim_{t \to \infty} \left| \frac{\lambda^{(4)}(t) - \lambda^{(2)}(t)}{\lambda^{(2)}(t)} \right|, \qquad \hat{e}_r^{(4)} = \left(\hat{e}_r^{(2)} \right)^2, \tag{10.93}$$

and similar expressions can be found for the other coefficients. These estimates are compared with the actual errors in Fig. 10.8. The figure shows that $\hat{e}_r^{(2)}$ provides a very good estimate of the actual error $e_r^{(2)}$ of the perturbation expansion. This is not surprising since this estimate is based on the fourth-order

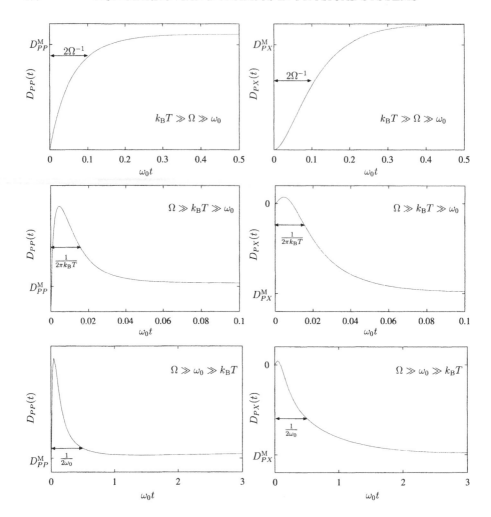

FIG. 10.7. Time dependence of the diffusion terms $D_{PP}(t)$ and $D_{PX}(t)$ to fourth order in the coupling strength. We distinguish three parameter regimes. High temperatures, intermediate coupling (top): $\gamma = \omega_0$, $\Omega = 20\,\omega_0$, $k_B T = 1000\,\omega_0$. Intermediate temperatures, intermediate coupling (middle): $\gamma = \omega_0$, $\Omega = 200\,\omega_0$, $k_B T = 10\,\omega_0$. Low temperatures, weak coupling (bottom): $\gamma = 0.1\,\omega_0$, $\Omega = 20\,\omega_0$, $k_B T = 0.01\,\omega_0$.

contribution. On the other hand, the estimate $\hat{e}_r^{(4)}$ does not involve higher-order terms but only relies on the second and fourth-order contribution. Nevertheless it provides a good means for getting the order of magnitude of the actual error which is enough for most applications. Thus, except for the second order, the error of the perturbation expansion can be reliably estimated without computing higher-order terms of the expansion.

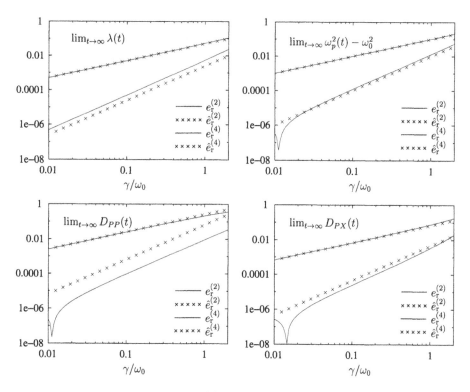

FIG. 10.8. Exact relative errors $e_r^{(2)}$ and $e_r^{(4)}$ and their estimates $\hat{e}_r^{(2)}$ and $\hat{e}_r^{(4)}$ to second and fourth order for the coefficients $\lambda(t)$, $\omega_p(t)^2 - \omega_0^2$, $D_{PP}(t)$ and $D_{PX}(t)$.

10.2.3 The stationary state

In this section we investigate the stationary state of the time-convolutionless master equation in order to find the parameter regime in which the perturbation expansion yields reliable numerical results. Since the stationary state of the harmonic oscillator is Gaussian, it is completely determined by the first and second moments of the observables X and P. The dynamics of these moments is governed by the adjoint master equation

$$
\begin{aligned}
\frac{d}{dt}\langle \mathcal{O}\rangle_t &= i\,\langle [H_S, \mathcal{O}]\rangle_t + \frac{i}{2}\Delta(t)\,\langle [X^2, \mathcal{O}]\rangle_t \\
&\quad + i\lambda(t)\,\langle \{P, [X, \mathcal{O}]\}\rangle_t - D_{PP}(t)\,\langle [X, [X, \mathcal{O}]]\rangle_t \\
&\quad + 2D_{PX}(t)\,\langle [X, [P, \mathcal{O}]]\rangle_t\,,
\end{aligned}
\tag{10.94}
$$

where $\mathcal{O}$ is some arbitrary system operator and $\langle \mathcal{O}\rangle_t$ denotes the time-dependent expectation value $\mathrm{tr}_S\{\mathcal{O}\rho_S(t)\}$ of the observable $\mathcal{O}$ with respect to the reduced density matrix $\rho_S(t)$.

For the observables X and P eqn (10.94) yields the classical equations of motion of a damped harmonic oscillator, namely

$$\frac{d}{dt}\langle X\rangle_t = \langle P\rangle_t, \tag{10.95}$$

$$\frac{d}{dt}\langle P\rangle_t = -\omega_p^2(t)\langle X\rangle_t - 2\lambda(t)\langle P\rangle_t, \tag{10.96}$$

and for the second moments we obtain

$$\frac{d}{dt}\langle X^2\rangle_t = \langle\{X,P\}\rangle_t, \tag{10.97}$$

$$\frac{d}{dt}\langle P^2\rangle_t = -\omega_p^2(t)\langle\{X,P\}\rangle_t - 4\lambda(t)\langle P^2\rangle + 2D_{PP}(t), \tag{10.98}$$

$$\frac{d}{dt}\langle\{X,P\}\rangle_t = -2\omega_p^2(t)\langle X^2\rangle_t + 2\langle P^2\rangle_t,$$
$$-2\lambda(t)\langle\{X,P\}\rangle_t + 4D_{PX}(t). \tag{10.99}$$

The stationary solution of these equations of motion reads (Săndulescu and Scutaru, 1987)

$$\langle X\rangle = \langle P\rangle = \langle\{X,P\}\rangle = 0, \tag{10.100}$$

$$\langle X^2\rangle = \lim_{t\to\infty}\frac{1}{2\omega_p^2(t)}\left[\frac{D_{PP}(t)}{\lambda(t)} + 4D_{PX}(t)\right], \tag{10.101}$$

$$\langle P^2\rangle = \lim_{t\to\infty}\frac{D_{PP}(t)}{2\lambda(t)}. \tag{10.102}$$

In the weak coupling limit, i.e. in the limit $\gamma \to 0$, the stationary state can be computed by inserting the Markovian coefficients defined in eqns (10.84)–(10.87), which lead to the expressions

$$\langle X^2\rangle = \frac{1}{2\omega_0}\coth\left(\frac{\omega_0}{2k_BT}\right), \tag{10.103}$$

$$\langle P^2\rangle = \frac{\omega_0}{2}\coth\left(\frac{\omega_0}{2k_BT}\right). \tag{10.104}$$

The variances ΔX^2 and ΔP^2 are (up to a scaling factor ω_0) identical and thus, the state of the harmonic oscillator is not squeezed. In the high-temperature regime the variances are proportional to the temperature, i.e. the oscillator behaves like a classical oscillator. On the other hand, in the low-temperature regime the variances are bound by the Heisenberg uncertainty relation and the oscillator behaves quantum mechanically.

For $\gamma > 0$ we can approximate the stationary solution by using the expressions for the coefficients of the master equation in fourth order. This yields corrections to the second moments $\langle X^2\rangle$ and $\langle P^2\rangle$ which are linear in γ. In Fig. 10.9 we compare the approximated with the exact variances for $\gamma = 0.25\,\omega_0$ and $\gamma = 0.5\,\omega_0$ in

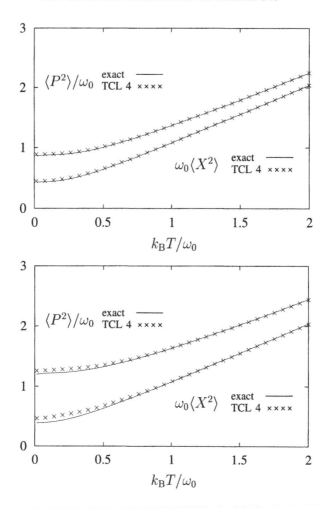

FIG. 10.9. Low-temperature behaviour of the stationary variances $\langle X^2 \rangle$ and $\langle P^2 \rangle$ for $\gamma = 0.25 \, \omega_0$ (top) and for $\gamma = 0.5 \, \omega_0$ (bottom). 'TCL 4' denotes the solution of the quantum master equation to fourth order and 'exact' the solution given in (Grabert, Schramm and Ingold, 1988). The cutoff is chosen to be $\Omega = 20 \, \omega_0$.

the low-temperature regime. Both solutions show qualitatively the same behaviour: The variance in the position decreases, while the variance in the momentum increases for increasing temperatures – the state of the oscillator is squeezed in this parameter regime. Figure 10.9 also indicates that the approximation of the master equation to fourth order yields reliable numerical results for $\gamma \lesssim \omega_0/2$.

A more detailed quantitative analysis of the error of the approximation is depicted in Fig. 10.10. For a fixed value of the cutoff frequency Ω this figure shows the regions in the parameter space where the relative error in the variance of the

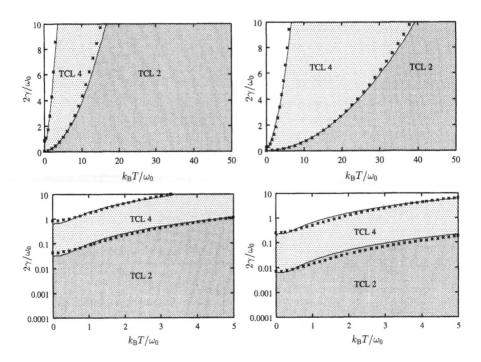

FIG. 10.10. The regions in parameter space, where the relative error of the stationary value of $\langle P^2 \rangle$ is less than 5% (left side) and less than 1% (right side). TCL 2 (TCL 4) denotes the solution of the time-convolutionless master equation to second (fourth) order. The crosses mark the boundaries of the regions where eqn (10.105) holds. The bottom shows an enlarged section of the low-temperature and weak-coupling regime. The cutoff is chosen to be $\Omega = 20\,\omega_0$.

momentum is less than 5% and 1%, respectively. Obviously, the perturbation expansion is in good agreement with the exact results for $\gamma \lesssim k_B T / 2$ in the high-temperature regime, and for $\gamma \lesssim \omega_0 / 2$ in the limit $T \to 0$. Numerical investigations show that for $\gamma \lesssim \Omega / 2$ this behaviour may be summarized by the condition

$$\frac{2\omega_0 \gamma}{\omega_0^2 + (k_B T)^2} \leq c_n(\Omega, e_r) \tag{10.105}$$

where the constant c_n depends on the cutoff frequency Ω, the desired accuracy e_r, and the order n of the approximation, but is independent of the temperature. For $\Omega = 20\,\omega_0$ and $e_r = 5\%$ we have, for example, $c_2 = 0.042$ and $c_4 = 0.84$, while for $e_r = 1\%$ one gets $c_2 = 0.0068$ and $c_4 = 0.25$. In Fig. 10.10 we have also marked the boundary of the parameter regime for which eqn (10.105) holds, which is in very good agreement with that region where the relative error is, in fact, less than the prescribed value of e_r.

The fact that the perturbation expansion also yields numerically reliable results in the low-temperature regime seems to contradict the fact that the thermal correlation time $\tau_B \sim 1/k_B T$ becomes infinite in the limit $T \to 0$. On the other hand, as discussed in Section 10.2.2, the time dependence of the temperature-dependent diffusion coefficients is dominated by the time scale ω_0^{-1}. Thus, the memory time in the low-temperature regime is the inverse system frequency and not the thermal correlation time. In fact, as we saw in Section 3.4 a perturbative treatment is extremely accurate in the quantum optical regime even in the limit $T \to 0$. This explains why the Markovian master equation is particularly useful in that regime and why it provides an approximation which is uniform in T and holds even in the vacuum optical case $(T = 0)$.

10.2.4 Non-factorizing initial conditions

In this section we want to determine the equilibrium position autocorrelation function $\langle X(t)X \rangle$ which is defined by

$$\langle X(t)X \rangle \equiv \mathrm{tr}\left\{ e^{iHt} X e^{-iHt} X \rho_{\mathrm{eq}} \right\}, \tag{10.106}$$

where the trace is taken over the system and reservoir. This definition can be transformed into the interaction picture,

$$\langle X(t)X \rangle = \mathrm{tr}\left\{ X(t)V(t) \right\}, \tag{10.107}$$

where $X(t) = \exp(iH_S t)X \exp(-iH_S t)$ is the interaction picture position operator. The operator $V(t)$ introduced above is a solution of the equation of motion

$$\frac{d}{dt}V(t) = -i[H_I(t), V(t)] \equiv \mathcal{L}(t)V(t), \tag{10.108}$$

corresponding to the initial condition

$$V(0) = X\rho_{\mathrm{eq}} \equiv \mathcal{A}\rho_{\mathrm{eq}}. \tag{10.109}$$

In general, this initial value neither provides a true density matrix nor does it represent a factorizing state. However, it does belong to the class of initial conditions (9.69) discussed in Section 9.2.4.

Since the operator X only acts in the Hilbert space of the open system it commutes with the projection operator $\mathcal{P}$ and we can write the position autocorrelation function as follows,

$$\langle X(t)X \rangle = \mathrm{tr}\left\{ X(t)\mathcal{P}V(t) \right\}. \tag{10.110}$$

The equation of motion for the correlation function is now obtained by differentiating both sides of this equation with respect to t and making use of eqns (9.33) and (9.84),

$$\frac{d}{dt}\langle X(t)X \rangle = i\,\mathrm{tr}\left\{ [H_S, X(t)]\mathcal{P}V(t) \right\} \tag{10.111}$$

$$+ \mathrm{tr}\left\{ X(t)\mathcal{K}(t)\mathcal{P}V(t) \right\} + \mathrm{tr}\left\{ X(t)\mathcal{J}(t)\mathcal{P}\rho_{\mathrm{eq}} \right\}.$$

In a similar way we can derive an equation of motion for the momentum–position correlation function, namely

$$\frac{d}{dt}\langle P(t)X \rangle = i\,\mathrm{tr}\left\{[H_S, P(t)]\mathcal{P}V(t)\right\} \tag{10.112}$$
$$+\mathrm{tr}\left\{P(t)\mathcal{K}(t)\mathcal{P}V(t)\right\} + \mathrm{tr}\left\{P(t)\mathcal{J}(t)\mathcal{P}\rho_{\mathrm{eq}}\right\}.$$

Inserting the expression for $\mathcal{K}(t)$ for the harmonic oscillator derived in Section 10.2.2 we obtain a closed set of differential equations for $\langle X(t)X \rangle$ and $\langle P(t)X \rangle$:

$$\frac{d}{dt}\langle X(t)X \rangle = \langle P(t)X \rangle + I_{XX}(t), \tag{10.113}$$

$$\frac{d}{dt}\langle P(t)X \rangle = -\omega_p^2(t)\langle X(t)X \rangle - 2\lambda(t)\langle P(t)X \rangle + I_{PX}(t), \tag{10.114}$$

where the inhomogeneities are defined as

$$I_{XX}(t) = \mathrm{tr}\left\{X(t)\mathcal{J}(t)\mathcal{P}\rho_{\mathrm{eq}}\right\}, \tag{10.115}$$
$$I_{PX}(t) = \mathrm{tr}\left\{P(t)\mathcal{J}(t)\mathcal{P}\rho_{\mathrm{eq}}\right\}. \tag{10.116}$$

The equation of motion for the correlation function $\langle P(t)X \rangle$ is, apart from the inhomogeneity, identical to the equation of motion for the observables X and P (see eqn (10.96)). This is the basic assertion of the quantum regression theorem. Thus, in the weak coupling limit, where the effect of the inhomogeneity can be neglected, the quantum regression theorem holds to the same level of accuracy, as the master equation. For stronger couplings the equation of motion has to be supplemented by the inhomogeneous term. It is important to note that this result is not in contradiction to the reported failure of the quantum regression theorem (Grabert, 1982; Talkner, 1995; Ford and O'Connell, 1996), since the description of the reduced dynamics by means of a convolutionless equation of motion does *not* constitute a quantum Markov process possessing the semigroup property.

By using the definition of the super-operator $\mathcal{I}(t)$ (see eqn (9.35)) it can be shown that the inhomogeneous term $I_{XX}(t)$ vanishes to all orders in the coupling strength. This can also be seen directly from the exact Heisenberg equation of motion $\dot{X}_H(t) = P_H(t)$ which leads to $I_{XX}(t) \equiv 0$ in eqn (10.113). The term $I_{PX}(t)$ is evaluated by making use of the perturbation expansion of the super-operator $\mathcal{J}(t)$ given in Section 9.2.4. To second order in the coupling strength we find

$$I_{PX}^{(2)} = \langle X^2 \rangle \int_{-\infty}^{0} ds\, D(t-s)\cos(\omega_0 s) - \frac{1}{2\omega_0}\int_{-\infty}^{0} ds\, D_1(t-s)\sin(\omega_0 s). \tag{10.117}$$

The fourth-order contribution $I_{PX}^{(4)}$ is given by Breuer, Kappler and Petruccione (2001).

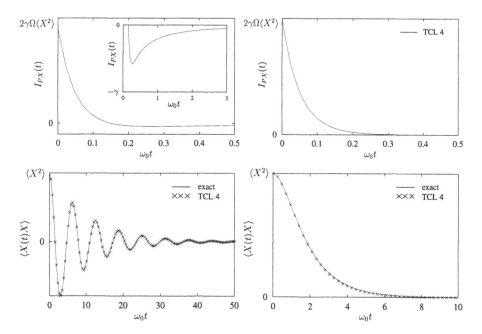

FIG. 10.11. Time dependence of the inhomogeneous term $I_{PX}(t)$ (top) and the real part of the position autocorrelation function $\langle X(t)X \rangle$ (bottom) for low temperatures and weak couplings, i.e. $k_B T = 0.01\,\omega_0$ and $\gamma = 0.1\,\omega_0$ (left) and for high temperatures and intermediate couplings, i.e. $k_B T = 1000\,\omega_0$ and $\gamma = \omega_0$ (right). The cutoff is chosen to be $\Omega = 20\,\omega_0$. The inset shows the long-time behaviour of $I_{PX}(t)$ for low temperatures after the initial jolt has decayed (note the different units).

The time dependence of $I_{PX}(t)$ is depicted in Fig. 10.11. In the low-temperature regime, there are two different contributions to $I_{PX}(t)$: For small times, i.e. for times of the order of the inverse cutoff frequency, $I_{PX}(t)$ takes large positive values. For intermediate times, i.e. for times of the order of the inverse system frequency, $I_{PX}(t)$ is negative. In the limit $t \to \infty$ the inhomogeneous term $I_{PX}(t)$ vanishes and the master equation becomes homogeneous. In contrast to this behaviour, $I_{PX}(t)$ simply decays exponentially on a time scale Ω^{-1} in the high-temperature regime.

This behaviour has important consequences for the time evolution of the position autocorrelation function. To see this, we consider the equation of motion for $\langle P(t)X \rangle$, eqn (10.114). The first term, $-\omega_p^2(t)\langle X(t)X \rangle$, gives rise to a large negative contribution in the short-time behaviour due to the initial jolts in the physically observable frequency $\omega_p(t)$ (see Fig. 10.6), which would lead to a fast decay of $\langle P(t)X \rangle$ on a time scale Ω^{-1}. However, this contribution is compensated by the inhomogeneous term $I_{PX}(t)$ which takes large positive values for all temperatures. Thus the somewhat artificial initial jolts, which are induced by

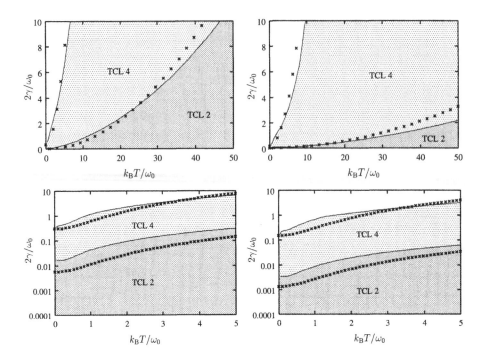

FIG. 10.12. The regions in parameter space where the relative error of the position autocorrelation function $\langle X(t)X \rangle$ is less than 5% (left side) and less than 1% (right side). TCL 2 (TCL 4) denotes the solution of the time-convolutionless master equation to second (fourth) order. The crosses mark the boundaries of the regions where eqn (10.105) holds. The cutoff is chosen to be $\Omega = 200\,\omega_0$.

the short-time behaviour of the generator $\mathcal{K}(t)$, are compensated by the inhomogeneity $\mathcal{J}(t)$.

Figure 10.11 also shows the real part of the autocorrelation function $\langle X(t)X \rangle$ in the underdamped and in the overdamped regime. The lines indicate the exact solution as given by Grabert, Schramm and Ingold (1988), and the symbols denote the solution of the perturbation expansion of the equations of motion (10.114). The agreement of both expressions is very good for the considered parameters. A systematic investigation of the range of validity of the perturbation expansion to second and fourth order is depicted in Fig. 10.12. This figures show the regions in parameter space where the relative error

$$e_r = \frac{\max_{t>0}\left\{|C_{\mathrm{ex}}(t) - C_n(t)|\right\}}{\langle X^2 \rangle} \tag{10.118}$$

for a certain order of approximation n is less than 5% and 1%, respectively. The function $C_{\mathrm{ex}}(t)$ denotes the exact autocorrelation function and $C_n(t)$ its

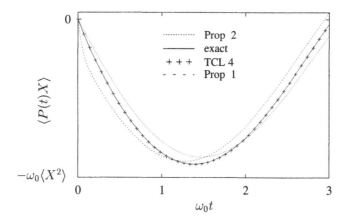

FIG. 10.13. Approximation of the correlation function $\langle P(t)X \rangle$ determined by neglecting the inhomogeneity (see text for details). The parameters read: $k_B T/\omega_0 = 0.01$, $\gamma/\omega_0 = 0.1$, and $\Omega = 20\,\omega_0$.

nth order approximation. In this figure we have also included the boundaries of the regions where eqn (10.105) holds for $c_2 = 0.0055$, $c_4 = 0.3$ ($e_r = 5\%$) and $c_2 = 0.0013$, $c_4 = 0.15$ ($e_r = 1\%$). As in the case of the relative error of the stationary variance $\langle P^2 \rangle$, this gives a good estimate of the error of the perturbation expansion. Note, however, that the quality of this estimate also depends on the ratio of the coupling strength γ to the cutoff frequency Ω, since the approximation of the physically observable frequency $\omega_p(t)$ and of the classical damping coefficient $\lambda(t)$ depend on this ratio.

10.2.5 *Disregarding the inhomogeneity*

The explicit computation of the inhomogeneity $\mathcal{J}(t)$ for non-factorizing initial conditions can become rather tedious since it consists of more terms than the generator $\mathcal{K}(t)$. Thus, there naturally arises the question if this effort is really necessary or if it is possible to disregard the inhomogeneity. In this section we will briefly discuss two proposals of this kind which yield the correct results in the weak-coupling limit and show their failure for certain parameters.

The first possibility we want to investigate is to simply ignore the correlations between the system and the environment and to assume a factorizing initial state. Thus, the equation of motion for the correlation function $\langle P(t)X \rangle$ would take the form

$$\frac{d}{dt}\langle P(t)X \rangle = -\omega_p^2(t)\langle X(t)X \rangle - 2\lambda(t)\langle P(t)X \rangle. \tag{10.119}$$

However, as we discussed in Section 10.2.4 the short-time dynamics of the physically observable frequency ω_p is dominated by a large initial jolt which leads to an enhanced decay of the autocorrelation function. This behaviour is depicted in Fig. 10.13 ('Prop 1').

The second proposal we want to discuss avoids this difficulty. In this approach we also ignore the initial correlations, but in contrast to the first approach the time evolution of the master equation is generated by the stationary value of $\mathcal{K}(t)$. Thus the equation of motion for $\langle P(t)X \rangle$ would take the form

$$\frac{d}{dt}\langle P(t)X \rangle = -\tilde{\omega}_p^2 \langle X(t)X \rangle - 2\tilde{\lambda}\langle P(t)X \rangle, \qquad (10.120)$$

where

$$\tilde{\omega}_p = \lim_{t\to\infty} \omega_p(t), \quad \tilde{\lambda} = \lim_{t\to\infty} \lambda(t). \qquad (10.121)$$

As demonstrated in Fig. 10.13 (see 'Prop 2') this approach also leads to significant deviations from a calculation which takes into account the inhomogeneity. These deviations are due to the negative contribution to $I_{PX}(t)$ (see the inset of Fig. 10.11).

Thus, it is in general inevitable to explicitly include the inhomogeneity in the case of non-factorizing initial states. An attempt to approximate the initial state by a factorizing state can lead to additional errors, which are of the order of the corrections one makes by expanding the generator to higher orders.

10.3 The spin-boson system

The investigation of the spin-boson system, i.e. a two-level system interacting with a bath of harmonic oscillators, is of particular interest in the theory of open quantum systems. First of all many physically interesting systems can be approximated by a two-level system in the low-temperature limit by neglecting higher excitations. This point has been discussed in detail by Leggett *et al.* (1987) in the context of the problem of tunnelling between two potential wells. Another reason for the wide interest in the spin-boson system is that it provides an excellent model for the investigation of certain approximation methods.

10.3.1 *Microscopic model*

The spin-boson system describes a two-level system which is linearly coupled to a bath of harmonic oscillators. The essential difference to the model treated in Section 10.1 is that the system–reservoir coupling is not treated within the rotating wave approximation. The Hamiltonian of the composite system thus takes the form

$$H = H_S + H_B + H_I, \qquad (10.122)$$

$$H_S = \frac{1}{2}\omega_0 \sigma_z, \qquad (10.123)$$

$$H_B = \sum_n \left(\frac{1}{2m_n}p_n^2 + \frac{1}{2}m_n\omega_n^2 x_n^2 \right), \qquad (10.124)$$

$$H_I = -\frac{1}{2}\sigma_x \otimes B, \quad B = \sum_n \kappa_n x_n, \qquad (10.125)$$

where H_S and H_B generate the free time evolution of the system and the bath, respectively, and H_I denotes their interaction. As in the case of the harmonic oscillator (see Section 10.2) the reduced system dynamics is characterized by the dissipation kernel $D(t) = i[B(t), B]$ and by the noise kernel $D_1(t) = \langle\{B(t), B\}\rangle$ (see eqns (3.385) and (3.386), respectively). The spectral density $J(\omega)$ of the model is again taken to be of the Lorentz–Drude form,

$$J(\omega) = \frac{\gamma}{\pi}\frac{\omega}{\omega_0}\frac{\Omega^2}{\Omega^2 + \omega^2}. \tag{10.126}$$

With this definition of the damping constant γ the dimension of $D(t)$ and $D_1(t)$ is equal to the square of an energy as it should be for an interaction Hamiltonian of the form (10.125).

10.3.2 Relaxation of an initially factorizing state

In this section we compute the perturbation expansion of the generator of the quantum master equation for the spin-boson system and determine its stationary solution. The quantum master equation for the reduced density matrix is most conveniently written in terms of the Bloch vector $\langle\vec{\sigma}(t)\rangle$ (see eqn (3.223)) as follows,

$$\frac{d}{dt}\langle\vec{\sigma}(t)\rangle = A(t)\langle\vec{\sigma}(t)\rangle + 2\vec{b}(t), \tag{10.127}$$

where $A(t)$ is a time-dependent 3×3 matrix and the inhomogeneity $\vec{b}(t)$ is a 3-vector. These quantities can be computed by using the expressions for the generator $\mathcal{K}(t)$ given in Section 9.2.2. In the Schrödinger picture the perturbation expansion of $A(t)$ and $\vec{b}(t)$ takes the form

$$A(t) = \begin{pmatrix} 0 & -\omega_0 & 0 \\ \omega_0 + a_{yx}(t) & a_{yy}(t) & 0 \\ 0 & 0 & a_{zz}(t) \end{pmatrix}, \quad \vec{b}(t) = \begin{pmatrix} 0 \\ 0 \\ b_z(t) \end{pmatrix}. \tag{10.128}$$

To second order in the coupling strength we obtain

$$a_{yx}^{(2)}(t) = \frac{1}{2}\int_0^t ds\, D_1(s)\sin(\omega_0 s), \tag{10.129}$$

$$a_{yy}^{(2)}(t) = a_{zz}^{(2)}(t) = -\frac{1}{2}\int_0^t ds\, D_1(s)\cos(\omega_0 s), \tag{10.130}$$

$$b_z^{(2)}(t) = -\frac{1}{4}\int_0^t ds\, D(s)\sin(\omega_0 s), \tag{10.131}$$

which is similar to the coefficients of the quantum master equation of the harmonic oscillator (see Section 10.2.2). Note however that the term $\Delta(t)$ which

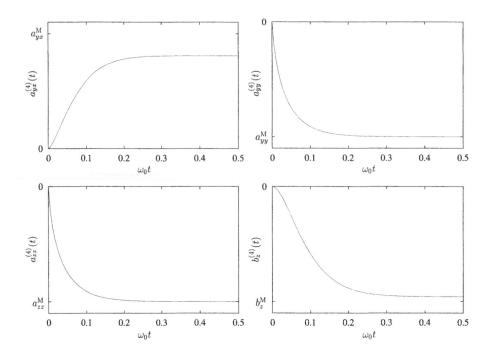

FIG. 10.14. Time dependence of the dephasing coefficients $a_{yx}(t)$ and $a_{yy}(t)$ and of the relaxation coefficients $a_{zz}(t)$ and $b_z(t)$ to fourth order in the coupling strength in the intermediate-coupling regime, i.e. $\gamma = 0.3\,\omega_0$, $\Omega = 20\,\omega_0$, $k_B T = 10\,\omega_0$.

leads to a large cutoff dependent frequency shift is not present in this model. Hence, there is no need for an explicit frequency renormalization procedure in this case. The fourth-order contributions can be found in (Breuer, Kappler and Petruccione, 2001).

The coefficients $a_{yx}(t)$ and $a_{yy}(t)$ are responsible for the dephasing of the two-level system and the coefficients $a_{zz}(t)$ and $b_z(t)$ describe the relaxation to the equilibrium state. The time dependence of these coefficients to fourth order in the coupling is depicted in Fig. 10.14 for an intermediate-coupling strength. Note that the stationary values of $a_{yx}(t)$ and $b_z(t)$ show a significant deviation from the Markovian limit whereas the approximation of $a_{yy}(t)$ and $a_{zz}(t)$ to second order is quite accurate.

In the Markovian regime and in the high-temperature limit $k_B T \gg \Omega \gg \omega_0$ the matrix A and the inhomogeneity $\vec{b}$ can be approximated by

$$A^M = \begin{pmatrix} 0 & -\omega_0 & 0 \\ \omega_0 + \frac{\gamma k_B T}{\Omega} & -\frac{\gamma k_B T}{\omega_0} & 0 \\ 0 & 0 & -\frac{\gamma k_B T}{\omega_0} \end{pmatrix}, \quad \vec{b}^M = \begin{pmatrix} 0 \\ 0 \\ -\frac{\gamma}{4} \end{pmatrix}. \qquad (10.132)$$

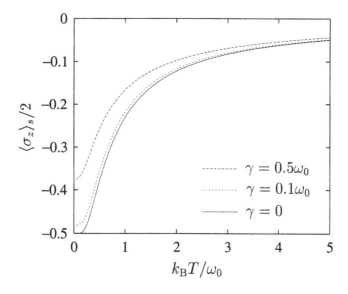

FIG. 10.15. Temperature dependence of the stationary value of $\langle\sigma_z\rangle_s$ for different coupling strengths. The population of the excited state is $\rho_{11} = (1 + \langle\sigma_z\rangle_s)/2$.

It is important to note that in this limit the relaxation rate as well as the dephasing rate are proportional to the temperature, in contrast to the harmonic oscillator where the temperature of the environment only has an effect on the diffusive behaviour of the system but not on dissipation.

The stationary state of the two-level system is readily obtained by making use of eqns (10.127) and (10.128),

$$\langle\sigma_x\rangle_s = \langle\sigma_y\rangle_s = 0, \quad \langle\sigma_z\rangle_s = -\lim_{t\to\infty}\frac{2b_z(t)}{a_{zz}(t)}. \tag{10.133}$$

The temperature dependence of $\langle\sigma_z\rangle_s$ is depicted in Fig. 10.15. For $\gamma = 0$ we obtain the thermodynamic limit in which the reduced density matrix is determined by the Boltzmann distribution. For $\gamma > 0$ we find deviations from the Boltzmann distribution. In particular the population of the excited state does not vanish in the limit $T \to 0$ in contrast to the predictions of a calculation to second order. These results are qualitatively similar to the results obtained by De Raedt and De Raedt (1984).

The order of magnitude of the relative error of the stationary state can be determined by estimating the relative error of the coefficients a_{zz} and b_z using the procedure outlined in Section 9.2.5. Thus the relative error to second and fourth order can be estimated by

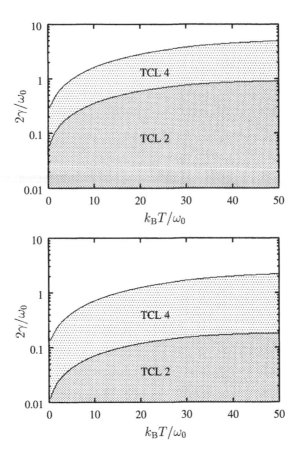

FIG. 10.16. The regions in parameter space where the estimated relative errors $\hat{e}_r^{(2)}$ and $\hat{e}_r^{(4)}$ of the stationary value $\langle \sigma_z \rangle_s$ are less than 5% (top) and less than 1% (bottom). The cutoff is chosen to be $\Omega = 20\,\omega_0$.

$$\hat{e}_r^{(2)} \approx \lim_{t \to \infty} \left(\left| \frac{a_{zz}^{(4)}(t) - a_{zz}^{(2)}(t)}{a_{zz}^{(2)}(t)} \right| + \left| \frac{b_z^{(4)}(t) - b_z^{(2)}(t)}{b_z^{(2)}(t)} \right| \right) \qquad (10.134)$$

$$\hat{e}_r^{(4)} \approx \lim_{t \to \infty} \left(\left| \frac{a_{zz}^{(4)}(t) - a_{zz}^{(2)}(t)}{a_{zz}^{(2)}(t)} \right|^2 + \left| \frac{b_z^{(4)}(t) - b_z^{(2)}(t)}{b_z^{(2)}(t)} \right|^2 \right). \qquad (10.135)$$

As in the case of the harmonic oscillator, we expect that the estimate for the relative error to second order provides a quantitatively correct result, whereas the error estimate to fourth order only yields the correct order of magnitude.

In Fig. 10.16 we show the regions in parameter space where the error estimates lower than 5% and 1%, respectively. For low and intermediate temperatures the behaviour of the error estimate is essentially the same as for the harmonic oscillator. However, in the high-temperature limit $k_BT \gg \Omega$ the maximum coupling strength is proportional to $1/T$. This is due to the fact that the relaxation coefficient $a_{zz}(t)$ and the dephasing coefficient $a_{yy}(t)$ are proportional to the temperature in this limit. Thus, when we consider the high-temperature limit we have to fix the product γT in order to obtain meaningful results (Chang and Skinner, 1993). This behaviour is very different from that of the harmonic oscillator, where the temperature has an influence only on diffusion and not on damping.

10.3.3 Equilibrium correlation functions

In this section we compute the equilibrium correlation function $\langle \sigma_x(t)\sigma_x \rangle$. To this end, we consider the equations of motion for the correlation functions $\langle \sigma_x(t)\sigma_x \rangle$ and $\langle \sigma_y(t)\sigma_x \rangle$, which can be derived from the adjoint master equation. Following the strategy described in Section 10.2.4 we obtain

$$\frac{d}{dt}\langle \sigma_x(t)\sigma_x \rangle = -\omega_0 \langle \sigma_y(t)\sigma_x \rangle + I_{XX}(t), \qquad (10.136)$$

$$\frac{d}{dt}\langle \sigma_y(t)\sigma_x \rangle = [\omega_0 + a_{yx}(t)]\langle \sigma_x(t)\sigma_x \rangle + a_{yy}(t)\langle \sigma_y(t)\sigma_x \rangle + I_{YX}(t), \quad (10.137)$$

where the inhomogeneities are defined as

$$I_{XX}(t) = \text{tr}\left\{ \sigma_x(t)\mathcal{J}(t)\mathcal{P}\rho_{eq} \right\}, \qquad (10.138)$$
$$I_{YX}(t) = \text{tr}\left\{ \sigma_y(t)\mathcal{J}(t)\mathcal{P}\rho_{eq} \right\}. \qquad (10.139)$$

Using the appropriate expansion for $\mathcal{J}(t)\mathcal{P}\rho_{eq}$ (see Section 9.2.4) we obtain $I_{XX}(t) \equiv 0$ to all orders in the coupling strength. This is, of course, an immediate consequence of the exact Heisenberg equation of motion $\dot{\sigma}_x = -\omega_0\sigma_y$. The second-order contribution to the inhomogeneity $I_{YX}(t)$ is found to be

$$I_{YX}^{(2)} = \frac{i}{2}\langle \sigma_z \rangle_s \int\limits_{-\infty}^{0} ds\, D_1(t-s)\cos(\omega_0 s) - \frac{i}{2}\int\limits_{-\infty}^{0} ds\, D(t-s)\sin(\omega_0 s). \quad (10.140)$$

The fourth-order contribution to I_{YX} may be found in (Breuer, Kappler and Petruccione, 2001).

Figure 10.17 shows the time dependence of the correlation function $\langle \sigma_x(t)\sigma_x \rangle$ in the low and high-temperature regimes. The Markovian approximation yields quantitatively good results for the real part of the correlation functions, whereas the fourth-order contribution introduces significant corrections in the imaginary part. Note that in the low-temperature regime the parameters are chosen in such a way that the system is underdamped, whereas in the high-temperature regime we have chosen a coupling strength for which the system is overdamped. In the latter case, the Markovian approximation reveals that the dynamics of the

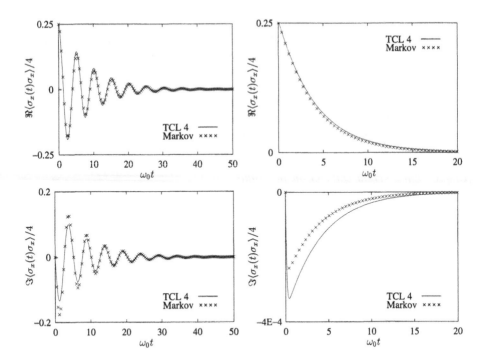

FIG. 10.17. Real and imaginary parts of the correlation function $\langle \sigma_x(t)\sigma_x \rangle$ in the low-temperature regime, i.e. $k_B T/\omega_0 = 0.1$, $\gamma = 0.5\,\omega_0$ (left) and in the high-temperature regime, i.e. $k_B T/\omega_0 = 50$, $\gamma = 0.2\,\omega_0$ (right). The cutoff is $\Omega = 20\,\omega_0$.

correlation function is bi-exponential where the fast and slow decay rates can be approximated by

$$r_{\text{fast}} = \frac{\gamma k_B T}{\omega_0}, \quad r_{\text{slow}} = \frac{\omega_0^3}{\gamma k_B T}. \tag{10.141}$$

10.3.4   Transition from coherent to incoherent motion

Especially in the context of macroscopic quantum coherence the cross-over from coherent to incoherent motion is very important: quantum coherence can only be observed for a coupling strength which is well below the critical damping γ_c where this cross-over takes place.

As pointed out by Egger, Grabert and Weiss (1997) the critical damping strength γ_c depends on the criterion used to define coherent motion. Since we are only interested in the order of magnitude of the critical damping strength we take a simple criterion which is independent of the initial preparation, i.e. which is valid for the expectation value $\langle \sigma_x(t) \rangle$ as well as for the correlation function $\langle \sigma_x(t)\sigma_x(0) \rangle$. This criterion is based on the eigenvalues of the matrix

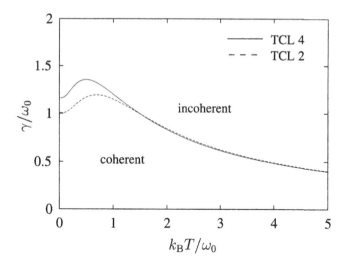

FIG. 10.18. The cross-over from coherent to incoherent motion for the perturbative approximation to second and fourth order. The cutoff is chosen to be
$\Omega = 20\,\omega_0$.

$$\tilde{A} = \lim_{t \to \infty} \begin{pmatrix} 0 & -\omega_0 \\ \omega_0 + a_{yx}(t) & a_{yy}(t) \end{pmatrix} \tag{10.142}$$

which determines the long-time dynamics of $\langle \sigma_x(t) \rangle$ and $\langle \sigma_x(t)\sigma_x(0) \rangle$ (see eqns
(10.127) and (10.136)). If both eigenvalues are real, then the quantities under
consideration decay exponentially in the long-time limit whereas they oscillate
if both eigenvalues are complex. Thus we define the critical damping γ_c through
the condition

$$\lim_{t \to \infty} \left[a_{yy}^2(t) - 4\omega_0(\omega_0 + a_{yx}(t)) \right] = 0. \tag{10.143}$$

Figure 10.18 shows the critical damping strength for the approximations to second and fourth order as a function of temperature. In the limit $T \to 0$ the
estimated relative error of $a_{yx}^{(4)}(t)$ is less than 1% whereas the estimated error of
$a_{yy}^{(4)}(t)$ is 18%. The relative error of γ_c is expected to be of the same order of
magnitude.

References

Breuer, H. P., Kappler, B. and Petruccione, F. (1999). Stochastic wave-function
method for non-Markovian quantum master equations. *Phys. Rev.*, **A59**, 1633–
1643.

Breuer, H. P., Kappler, B. and Petruccione, F. (2001). The time-convolutionless
projection operator technique in the quantum theory of dissipation and decoherence. *Ann. Phys. (N.Y.)*, **291**, 36–70.

Chang, T.-M. and Skinner, J. L. (1993). Non-Markovian population and phase relaxation and absorption lineshape for a two-level system strongly coupled to a harmonic quantum bath. *Physica A*, **193**, 483–539.

De Raedt, B. and De Raedt, H. (1984). Thermodynamics of a two-level system coupled to bosons. *Phys. Rev.*, **B29**, 5325–5336.

Egger, R., Grabert, H. and Weiss, U. (1997). Crossover from coherent to incoherent dynamics in damped quantum systems. *Phys. Rev.*, **E55**, R3809–R3812.

Ford, G. W. and O'Connell, R. F. (1996). There is no quantum regression theorem. *Phys. Rev. Lett.*, **77**, 798–801.

Garraway, B. M. (1997). Nonperturbative decay of an atomic system in a cavity. *Phys. Rev.*, **A55**, 2290–2303.

Grabert, H. (1982). Nonlinear relaxation and fluctuations of damped quantum systems. *Z. Phys.*, **B49**, 161–172.

Grabert, H., Schramm, P. and Ingold, G.-L. (1988). Quantum Brownian motion: The functional integral approach. *Phys. Rep.*, **168**, 115–207.

Haake, F. and Reibold, R. (1985). Strong damping and low-temperature anomalies for the harmonic oscillator. *Phys. Rev.*, **A32**, 2462–2475.

Hu, B. L., Paz, J. P. and Zhang, Y. (1992). Quantum Brownian motion in a general environment: Exact master equation with nonlocal dissipation and colored noise. *Phys. Rev.*, **D45**, 2843–2861.

Karrlein, R. and Grabert, H. (1997). Exact time evolution and master equations for the damped harmonic oscillator. *Phys. Rev.*, **E55**, 153–164.

Leggett, A. J., Chakravarty, S., Dorsey, A. T., Fisher, M. P. A., Garg, A. and Zwerger, W. (1987). Dynamics of the dissipative two-state system. *Rev. Mod. Phys.*, **59**, 1–85.

Săndulescu, A. and Scutaru, H. (1987). Open quantum systems and the damping of collective modes in deep ineleastic collisions. *Ann. Phys. (N.Y.)*, **173**, 277–317.

Talkner, P. (1995). The failure of the quantum regression hypothesis. *Ann. Phys. (N.Y.)*, **167**, 390–436.

Weisskopf, V. and Wigner, E. (1930). Berechnung der natürlichen Linienbreite auf Grund der Diracschen Lichttheorie. *Z. Phys.*, **63**, 54–73.

Part V

Relativistic quantum processes

11

MEASUREMENTS IN RELATIVISTIC QUANTUM MECHANICS

In the relativistic domain the quantum theory of measurement exhibits several new and interesting features which are not encountered in the non-relativistic theory (Landau and Peierls, 1931; Bohr and Rosenfeld, 1933; Bloch, 1967). The present chapter is devoted to a systematic study of measurement and state reduction in relativistic quantum mechanics. An elementary introduction to the problem may be found in (Breuer and Petruccione, 1999).

The basic idea underlying the presentation of this chapter is to consider the state of a quantum system as a functional on the set of spacelike hypersurfaces in Minkowski space. This idea has been developed in the early days of relativistic quantum field theory by Dirac, Schwinger and Tomonaga with the aim to give a manifest covariant formulation for the time-evolution equation of the state vector, known as the Schwinger–Tomonaga equation. The same concept turns out to be very useful in a systematic treatment of measurement and state reduction in relativistic quantum mechanics, as has been pointed out in a series of papers by Aharonov and Albert (1980, 1981, 1984a, 1984b). In fact, a covariant state reduction postulate is obtained if one regards the state reduction as taking place along those spacelike hypersurfaces which cross the classical chance event provided by the readout of a local measurement. As is demonstrated in the present chapter, this concept leads to a consistent formulation for the state evolution of a quantum system which is conditioned on a single or on multiple local measurements arbitrarily distributed in space-time, and for the corresponding joint probability distributions containing arbitrary local and non-local quantum correlations. We also develop in this context the relativistic formulation of continuous measurements in terms of a covariant PDP for the state vector.

An important point which must be taken into account in the relativistic quantum theory of measurement is the possibility of non-local measurements. It has been shown by Aharonov, Albert and Vaidman (1986) that it is possible to carry out the measurement of certain non-local quantities without measuring local properties of the system. Such measurements are performed as indirect measurements invoking entangled quantum probes. We shall discuss here in detail the measurement of non-local observables and the verification of non-local, entangled quantum states.

As will be seen the possibility of the measurement of non-local quantities is strongly restricted by the causality principle. These restrictions can be formulated with the help of a theorem due to Popescu and Vaidman (1994) which states, essentially, that non-local measurements must necessarily erase local information in order to be compatible with the causality principle. The present chapter contains a proof of this theorem and a detailed discussion of its physical implications. In particular, one can deduce that only a certain class of non-local operators allows a quantum non-demolition measurement which is in agreement with the von Neumann–Lüders postulate.

The quantum theory of measurement thus leads to important consequences for the notions of observables and states if it is combined with the requirements of special relativity. Several examples for the application of the general theory will be presented. Furthermore, the preparation of states, the notion of exchange measurements and the instantaneous transfer of a coherent quantum state, known as quantum teleportation, will be discussed.

11.1   The Schwinger–Tomonaga equation

In the following we write $x^\mu = (x^0, \vec{x})$ for the space-time coordinates of a point x in Minkowski space $\mathbb{R}^4$, where μ runs from 0 to 3. The first component $x^0 = t$ is the time coordinate and $\vec{x} = (x^1, x^2, x^3)$ denotes the space coordinates. We choose units such that $\hbar = c = 1$, where c is the speed of light. The Lorentz invariant inner product of two 4-vectors x and y is defined by

$$xy = x^\mu y_\mu = g_{\mu\nu} x^\mu y^\nu = x^0 y^0 - \vec{x} \cdot \vec{y}, \tag{11.1}$$

where $g_{\mu\nu}$ denotes the metric tensor, and repeated indices are summed.

In the interaction picture the time evolution of the state vector $|\Psi(t)\rangle$ is governed by the Schrödinger equation

$$\frac{\partial}{\partial t} |\Psi(t)\rangle = -iH_I(t)|\Psi(t)\rangle, \tag{11.2}$$

where

$$H_I(t) = \int d^3x \, \mathcal{H}(t, \vec{x}) \tag{11.3}$$

is the interaction picture Hamiltonian and $\mathcal{H}(x) = \mathcal{H}(t, \vec{x})$ denotes the Hamiltonian density of the theory. For simplicity we investigate here theories without derivative couplings and the Hamiltonian density is assumed to transform as a scalar under Lorentz transformations.

11.1.1   *States as functionals of spacelike hypersurfaces*

Given a fixed coordinate system, $|\Psi(t)\rangle$ characterizes the state of a quantum mechanical system at a fixed time $x^0 = t$ and thus allows the evaluation of the expectation values for all observables which are localized on the three-dimensional hypersurface given by $x^0 = t = $ constant. The relativistically invariant generalization of this concept is that of a state vector which is associated with a general, three-dimensional spacelike hypersurface σ. Such a hypersurface is defined to be a three-dimensional manifold in Minkowski space which extends to infinity in all spacelike directions and which has at each point $x \in \sigma$ a unit, timelike normal vector $n^\mu(x)$ satisfying

$$n_\mu(x)n^\mu(x) = 1, \quad n^0(x) \geq 1. \tag{11.4}$$

The state vector then becomes a functional

$$|\Psi\rangle = |\Psi(\sigma)\rangle \tag{11.5}$$

on the space of such hypersurfaces. The same holds for the density matrix of the system which again yields a functional

$$\rho = \rho(\sigma) \tag{11.6}$$

on the set of spacelike hypersurfaces.

The corresponding generalization of eqn (11.2) is the Schwinger–Tomonaga equation for the state vector (Tomonaga, 1946, 1947; Schwinger, 1948; Schweber, 1961),

$$\frac{\delta |\Psi(\sigma)\rangle}{\delta \sigma(x)} = -i\mathcal{H}(x)|\Psi(\sigma)\rangle, \tag{11.7}$$

or, else, for the density matrix,

$$\frac{\delta \rho(\sigma)}{\delta \sigma(x)} = -i\left[\mathcal{H}(x), \rho(\sigma)\right]. \tag{11.8}$$

The Schwinger–Tomonaga equation is a functional differential equation. It may be considered as a differential equation in a continuous family of time variables. Each point $x \in \sigma$ may be represented in a special coordinate system as $x^\mu = (x^0(\vec{x}), \vec{x})$, such that each space point $\vec{x}$ has its own time variable $x^0 = x^0(\vec{x})$. Equations (11.7) and (11.8) involve a variation $\delta\sigma$ of the hypersurface σ in which these time variables are varied independently, with the restriction that $\sigma + \delta\sigma$ is again a spacelike hypersurface.

Formally, the functional derivative $\delta/\delta\sigma(x)$ in eqns (11.7) and (11.8) is defined as follows (see Fig. 11.1). Consider some point $x \in \sigma$ and an infinitesimal variation $\sigma \to \sigma + \delta\sigma$ of the hypersurface around x. The volume of the four-dimensional space-time region enclosed by σ and $\sigma + \delta\sigma$ will be denoted by $\Omega(x)$. Then we define for any functional $F(\sigma)$,

$$\frac{\delta F(\sigma)}{\delta \sigma(x)} \equiv \lim_{\Omega(x) \to 0} \frac{F(\sigma + \delta\sigma) - F(\sigma)}{\Omega(x)}. \tag{11.9}$$

To give an explicit example for this kind of derivative we consider some vector field $\Gamma_\mu(x)$ and define the functional

$$F(\sigma) \equiv \int_\sigma d\sigma(x)n^\mu(x)\Gamma_\mu(x). \tag{11.10}$$

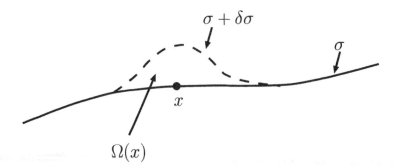

FIG. 11.1. Illustration of definition (11.9) of the functional derivative $\delta/\delta\sigma(x)$. The figure shows some small variation $\delta\sigma$ of the spacelike hypersurface σ around the point x such that σ and $\sigma + \delta\sigma$ enclose a four-dimensional space–time volume $\Omega(x)$.

Here, the integration is performed over the hypersurface σ with the help of the Lorentz-invariant surface element $d\sigma(x)$ on σ which is induced by the Lorentz metric. Explicitly one has

$$d\sigma(x) = \frac{d^3x}{n^0(x)}, \qquad (11.11)$$

where $d^3x = dx^1 dx^2 dx^3$ and $n^0(x)$ is the time component of the unit normal vector $n^\mu(x)$ at the point x on σ. Thus, $F(\sigma)$ is just the flow of the vector field $\Gamma_\mu(x)$ through the hypersurface σ. Using definition (11.9) we find the functional derivative,

$$\frac{\delta F(\sigma)}{\delta\sigma(x)} = \lim_{\Omega(x)\to 0} \frac{1}{\Omega(x)} \left[\int_{\sigma+\delta\sigma} d\sigma(x')n^\mu(x')\Gamma_\mu(x') - \int_\sigma d\sigma(x')n^\mu(x')\Gamma_\mu(x') \right]$$

$$= \lim_{\Omega(x)\to 0} \frac{1}{\Omega(x)} \int_{\Omega(x)} d^4x' \partial^\mu \Gamma_\mu(x')$$

$$= \partial^\mu\Gamma_\mu(x), \qquad (11.12)$$

where we have used Gauss's theorem in the second step. The four-dimensional space-time integral is extended over the region enclosed by the hypersurfaces σ and $\sigma + \delta\sigma$. This obvious result states that the change in the flow through σ which results from an infinitesimal variation of σ around x is given by the divergence of the vector field at x. For a vector field satisfying $\partial^\mu\Gamma_\mu(x) = 0$ this means that the corresponding flow $F(\sigma)$ is independent of σ.

The Schwinger–Tomonaga equation (11.7) or (11.8) is obviously covariant since both the Hamiltonian density $\mathcal{H}(x)$ and the functional derivative $\delta/\delta\sigma(x)$

FIG. 11.2. Sketch of the second-order variation involved in the integrability condition (11.13). Lying on the same hypersurface σ, the points x and y are separated by a spacelike distance. For the Schwinger–Tomonaga equation to be integrable the order of the variations around x and y must be immaterial which is ensured by microcausality.

transform as scalars under Lorentz transformations. Being an equation in a continuous family of variables, the Schwinger–Tomonaga equation must fulfil a certain integrability condition in order for a solution to exist (Schweber, 1961). In direct analogy to the case of a partial differential equation in a finite number of dimensions, a necessary integrability condition for the Schwinger–Tomonaga equation takes the form

$$\frac{\delta^2 \rho(\sigma)}{\delta\sigma(x)\delta\sigma(y)} - \frac{\delta^2 \rho(\sigma)}{\delta\sigma(y)\delta\sigma(x)} = 0, \tag{11.13}$$

where the points x and y are located on the same hypersurface σ and are thus separated by a spacelike interval (see Fig. 11.2). The integrability condition is a direct consequence of the requirement of microcausality for the Hamiltonian density, which states that $\mathcal{H}(x)$ and $\mathcal{H}(y)$ commute for spacelike separations (Weinberg, 1995),

$$[\mathcal{H}(x), \mathcal{H}(y)] = 0, \quad \text{for } (x - y)^2 < 0. \tag{11.14}$$

Namely, using the Jacobi identity we immediately get from the microcausality condition

$$\frac{\delta^2 \rho(\sigma)}{\delta\sigma(x)\delta\sigma(y)} - \frac{\delta^2 \rho(\sigma)}{\delta\sigma(y)\delta\sigma(x)} = [[\mathcal{H}(x), \mathcal{H}(y)], \rho(\sigma)] = 0. \tag{11.15}$$

The integrability condition ensures that the Schwinger–Tomonaga equation has a unique solution $\rho(\sigma)$ once an appropriate initial density matrix $\rho(\sigma_0)$ has been specified on an initial hypersurface σ_0. Formally, this solution can be written as

$$\rho(\sigma) = U(\sigma, \sigma_0)\rho(\sigma_0)U^\dagger(\sigma, \sigma_0), \tag{11.16}$$

where we have introduced the unitary evolution operator

$$U(\sigma, \sigma_0) = \mathrm{T}_\leftarrow \exp\left[-i \int_{\sigma_0}^{\sigma} d^4x\, \mathcal{H}(x)\right]. \tag{11.17}$$

As usual, $\mathrm{T}_\leftarrow$ denotes the chronological time-ordering operator.

11.1.2   Foliations of space-time

A foliation of Minkowski space is defined to be a smooth one-parameter family

$$\mathcal{F} = \{\sigma(\tau)\} \tag{11.18}$$

of spacelike hypersurfaces $\sigma(\tau)$ with the property that each space-time point x is located on precisely one hypersurface of the family. If we denote the latter by σ_x, this means that we have $\sigma_x = \sigma(\tau)$ for exactly one parameter value τ.

A given foliation $\sigma(\tau)$ gives rise to a corresponding family of state vectors

$$|\Psi(\tau)\rangle = |\Psi(\sigma(\tau))\rangle. \tag{11.19}$$

The Schwinger–Tomonaga equation can then be formulated as an integral equation,

$$|\Psi(\tau)\rangle = |\Psi(0)\rangle - i \int_{\sigma_0}^{\sigma(\tau)} d^4x\, \mathcal{H}(x)|\Psi(\sigma_x)\rangle. \tag{11.20}$$

The four-dimensional integration is extended over the region enclosed by an initial hypersurface $\sigma_0 = \sigma(\tau = 0)$ and the hypersurface $\sigma(\tau)$ of the family which lies entirely in the future of σ_0.

The hypersurfaces $\sigma(\tau)$ of a foliation can be conveniently defined with the help of an implicit equation of the form

$$f(x, \tau) = 0, \tag{11.21}$$

where $f(x, \tau)$ is a smooth scalar function. With an appropriate normalization of f we may assume that the unit normal vector $n_\mu(x)$ at the point $x \in \sigma(\tau)$ is given by

$$n_\mu(x) = \frac{\partial f(x, \tau)}{\partial x^\mu}. \tag{11.22}$$

It follows from eqns (11.20), (11.21) and (11.22) that $|\Psi(\tau)\rangle$ obeys the equation of motion

$$\frac{d}{d\tau}|\Psi(\tau)\rangle = -i \int_{\sigma(\tau)} d\sigma(x) \left|\frac{\partial f}{\partial \tau}\right| \mathcal{H}(x)|\Psi(\tau)\rangle \equiv -iH(\tau)|\Psi(\tau)\rangle, \tag{11.23}$$

where the integration is performed over the hypersurface $\sigma(\tau)$ of the foliation. This is a manifest covariant form of the Schrödinger equation (11.2). To prove it we first note that for two hypersurfaces of the foliation corresponding to two infinitesimally separated parameter values τ and $\tau + d\tau$ eqn (11.20) yields

$$d|\Psi(\tau)\rangle = |\Psi(\tau + d\tau)\rangle - |\Psi(\tau)\rangle = -i \int_{\sigma(\tau)}^{\sigma(\tau+d\tau)} d^4x\, \mathcal{H}(x)|\Psi(\tau)\rangle. \tag{11.24}$$

On using $d^4x = d\sigma(x)|n_0 \partial x_0/\partial \tau|d\tau = d\sigma(x)|\partial f/\partial \tau|d\tau$ one is immediately led to eqn (11.23).

As an example, consider a foliation by means of a family of parallel hyper-surfaces $\sigma(\tau)$ given through the equation

$$f(x, \tau) \equiv nx - \tau = 0, \tag{11.25}$$

with a constant unit normal vector n^μ. Such a foliation can be associated with an observer O moving along the straight world line $y(\tau) = n\tau$ with constant velocity $\vec{v}$ such that

$$n = \frac{dy}{d\tau} = (\gamma, \gamma\vec{v}) \tag{11.26}$$

is the 4-velocity of O with

$$\gamma = \frac{1}{\sqrt{1 - |\vec{v}|^2}}. \tag{11.27}$$

The parameter τ denotes the proper time of the observer, that is, the time of a clock attached to O. At each fixed τ the time axis in an observer's rest frame is given by the unit vector n, whereas instantaneous 3-space at that time is given by the flat, spacelike hypersurface $\sigma(\tau)$ which is orthogonal to n and contains the point $y(\tau)$, i.e. which is defined by the equation

$$n(x - y(\tau)) \equiv nx - \tau = 0. \tag{11.28}$$

The hypersurface $\sigma(\tau)$ is therefore the set of those space-time points x to which observer O assigns one and the same time coordinate τ. We have $|\partial f / \partial \tau| = 1$ and, hence, eqn (11.23) takes the form (Jauch and Rohrlich, 1980)

$$\frac{d}{d\tau} |\Psi(\tau)\rangle = -i \int_{\sigma(\tau)} d\sigma(x) \, \mathcal{H}(x) |\Psi(\tau)\rangle \equiv -iH(\tau) |\Psi(\tau)\rangle. \tag{11.29}$$

In particular, in the special coordinate system in which the unit normal vector n coincides with the time axis, $n^\mu = (1, 0, 0, 0)$, this equation becomes identical to the Schrödinger equation (11.2). Analogous equations hold, of course, for the density matrix $\rho(\sigma)$.

11.2   The measurement of local observables

This section deals with the measurement of local quantities. We begin our discussion by constructing the operation for an indirect measurement at a single point (or at a single localized space-time region) and formulate an appropriate relativistic state reduction postulate. The requirements of relativistic covariance lead to the prescription that the state reduction must be performed on all spacelike hypersurfaces which pass the classical event given by the measurement outcome.

This prescription for the state vector reduction yields the result that the probability amplitude becomes a multivalued function on the space-time continuum. Thus, in the relativistic domain the state vector (or the density matrix) must be regarded, in general, as a functional on the set of spacelike hypersurfaces. This concept has already been used in the previous section for the formulation of the Schwinger–Tomonaga equation.

Having discussed single measurements, we turn to the description of multiple measurements carried out at a set of points which may be arbitrarily distributed in space and time. The evolution of the state vector conditioned on the readouts of the various measurements will be formulated in terms of a relativistically covariant stochastic process. It is shown further that the measurement outcomes can be described by a consistent family of Lorentz-invariant joint probability distributions. The latter are shown to contain all local and non-local quantum correlations. As an example we briefly discuss EPR-type correlations which are embodied in non-local, entangled quantum states. We further discuss here the formulation of relativistic PDPs describing continuous measurements.

11.2.1   The operation for a local measurement

We consider some observable $A(\sigma_m)$ which is associated with a spacelike hypersurface σ_m and which is given in terms of an integral over some local, Hermitian field $\varphi(x)$,

$$A(\sigma_m) = \int_{\sigma_m} d\sigma(x) G(x) \varphi(x). \qquad (11.30)$$

Here, $G(x)$ denotes some smooth function with compact support around some point x_m located on σ_m. In the following we shall describe the local measurement of $A(\sigma_m)$ as an indirect measurement. More specifically, we assume in analogy to the procedure in Section 2.4.6 that the quantum field $\varphi(x)$ is coupled linearly to the generalized momentum P of some probe system. The interaction between the field and the quantum probe is further assumed to be localized in some small space-time region containing the support of $G(x)$ (see Fig. 11.3). After the interaction the generalized coordinate Q, canonically conjugated to P, is measured on the probe system. The corresponding readout q of the Q measurement and the initial probe state $|\phi\rangle$ then lead to an, in general approximate, measurement of $A(\sigma_m)$.

We consider two hypersurfaces σ and σ' which are separated from the interaction region as indicated in Fig. 11.3. Our basic assumption will be that locally any observable can be measured with the help of such an indirect measurement scheme and that, at least in principle, the interaction time can be made arbitrarily small such that the free evolution of both the field and the quantum probe can be neglected over this time. We then get the following expression for the unitary evolution operator which takes the state of the total system (object plus quantum probe) from the hypersurface σ' to the hypersurface σ,

$$V(\sigma, \sigma') = \exp\left[-iA(\sigma_m)P\right]. \qquad (11.31)$$

As indicated in Fig. 11.3 the hypersurface σ' is taken to intersect the backward light cone $V_-(x_i)$ based at x_i which is defined by

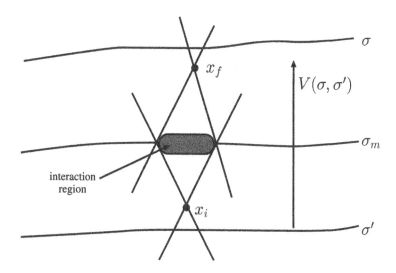

FIG. 11.3. Schematic picture for the measurement of an observable $A(\sigma_m)$ with
compact support on the hypersurface σ_m. The forward light cone of x_i and
the backward light cone of x_f are tangential to the interaction region as
indicated. The unitary operator $V(\sigma, \sigma')$ maps the state vector on σ' to the
state vector on σ.

$$V_-(x_i) = \left\{ x \in \mathbb{R}^4 \mid (x - x_i)^2 > 0, \ x^0 < x_i^0 \right\}, \tag{11.32}$$

while σ intersects the forward light cone $V_+(x_f)$ based at x_f,

$$V_+(x_f) = \left\{ x \in \mathbb{R}^4 \mid (x - x_f)^2 > 0, \ x^0 > x_f^0 \right\}. \tag{11.33}$$

The points x_i and x_f are chosen such that these light cones are tangential to
the interaction region. It is clear that σ' and σ do not intersect the interaction
region. The Q measurement is carried out on a surface such as σ which crosses
the forward light cone based at x_f.

Likewise, the initial state of the field and probe must be given on a hypersurface such as σ' which crosses the backward light cone of x_i. This state is given
by a product state $|\Psi(\sigma')\rangle \otimes |\phi\rangle$, where $|\Psi(\sigma')\rangle$ denotes the state of the quantum
field. The final state after the interaction between quantum field and probe and
the subsequent ideal Q measurement with the result $Q = q$ then takes the form

$$|q\rangle\langle q|V(\sigma, \sigma')\left(|\Psi(\sigma')\rangle \otimes |\phi\rangle\right) \equiv |q\rangle \Omega(q)|\Psi(\sigma')\rangle. \tag{11.34}$$

Here we have applied the state projection postulate to the ideal Q measurement
which projects the state vector onto the eigenstate $|q\rangle$ of Q corresponding to the
eigenvalue q. The above expression implies that the indirect measurement device
is described by the operation

$$\Omega(q) = \langle q| \exp\left[-iA(\sigma_m)P\right]|\phi\rangle = \phi\left(q - A(\sigma_m)\right), \tag{11.35}$$

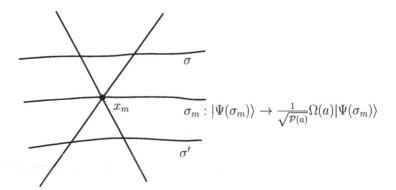

FIG. 11.4. Idealized picture for the measurement at a single space-time point x_m. The hypersurface σ_m is chosen to cross x_m, whereas σ' crosses the backward light cone of x_m and σ its forward light cone. The operation $\Omega(a)$ describes the reduction of the state vector conditioned on the outcome a of the indirect measurement.

which acts on the states of the quantum field. Here we have introduced the wave function $\phi(q) = \langle q|\phi \rangle$ of the initial probe state in the Q representation.

As an example we take the initial wave function of the probe in the Q representation to be a Gaussian function (Diòsi, 1991) with variance η^2,

$$\phi(q) = \left(2\pi\eta^2\right)^{-1/4} \exp\left[-\frac{q^2}{4\eta^2}\right]. \tag{11.36}$$

The operation is then found to be

$$\Omega(q) = \left(2\pi\eta^2\right)^{-1/4} \exp\left[-\frac{(q - A(\sigma_m))^2}{4\eta^2}\right]. \tag{11.37}$$

This demonstrates that a probe which has been prepared initially in a Gaussian state with variance η^2 enables the approximate measurement of the observable $A(\sigma_m)$ to an error of the order η. We also observe that the readout q leads to an inferred value for the observable $A(\sigma_m)$ which is given by $q = a$. This is due to the fact that the above Gaussian has zero mean value and, therefore, $q = a$ is a bias-free estimate for the observable. In the following we shall always assume that the mean value of the probe states is zero (which can always be achieved by subtraction of the bias, of course) and identify the readout q with the inferred value a.

11.2.2   Relativistic state reduction

The whole situation developed so far can now be idealized by shrinking the total interaction region to a single point x_m as illustrated in Fig. 11.4. The device then acts as an indirect measurement of the local observable $A(x_m)$.

Since all hypersurfaces crossing the forward (backward) light cone based at x_m are unitarily equivalent we are thus led to the following state reduction postulate for local measurements. Consider some foliation $\mathcal{F} = \{\sigma(\tau)\}$ of space-time. Then the state reduction

$$|\Psi(\sigma_m)\rangle \longrightarrow \frac{1}{\sqrt{\mathcal{P}(a)}} \Omega(a)|\Psi(\sigma_m)\rangle \qquad (11.38)$$

occurs along that hypersurface σ_m of the foliation which crosses the point x_m. The operation for the local measurement at x_m now takes the form

$$\Omega(a) = \phi\left(a - A(x_m)\right). \qquad (11.39)$$

The state reduction (11.38) is conditioned on the readout $q = a$ which is distributed according to the probability density

$$\mathcal{P}(a) = ||\Omega(a)|\Psi(\sigma_m)\rangle||^2 = \langle\Psi(\sigma_m)|\Omega^\dagger(a)\Omega(a)|\Psi(\sigma_m)\rangle. \qquad (11.40)$$

We recall that $\Omega^\dagger(a)\Omega(a)$ is a Hermitian and positive operator,

$$\Omega^\dagger(a)\Omega(a) = |\phi(a - A(x_m))|^2 \geq 0, \qquad (11.41)$$

which satisfies

$$\int da\,\Omega^\dagger(a)\Omega(a) = \int dq\,|\phi(a - A(x_m))|^2 = I, \qquad (11.42)$$

such that the density $\mathcal{P}(a)$ is normalized,

$$\int da\,\mathcal{P}(a) = \int da\,\langle\Psi(\sigma_m)|\Omega^\dagger(a)\Omega(a)|\Psi(\sigma_m)\rangle = \langle\Psi(\sigma_m)|\Psi(\sigma_m)\rangle = 1. \quad (11.43)$$

Analogous relations are of course valid for mixed states $\rho(\sigma_m)$.

For different foliations, the corresponding surfaces that intersect the point x_m can, of course, be different. The above state reduction postulate thus amounts to the prescription that the state vector reduction occurs along all spacelike hypersurfaces which cross the point x_m in which the local measurement is being performed. This is the state reduction postulate first formulated by Aharonov and Albert (1984b). A dynamical model for this relativistic reduction postulate has been developed by Breuer and Petruccione (1998). For an extended region given by the support of $G(x)$ we can say that the state reduction takes place along all spacelike hypersurfaces which coincide on the support of $G(x)$ (see Figs. 11.5 and 11.6).

The above state reduction postulate is obviously covariant since it is formulated without reference to any specific coordinate system. Additionally, the probability distribution $\mathcal{P}(a)$ for the readout a as well as for the operation $\Omega(a)$ applied to the state vector does not depend on the specific surface σ as long

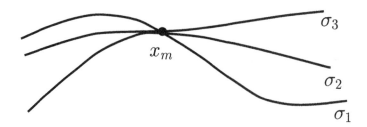

FIG. 11.5. Illustration of the state vector reduction postulate. Different foliations lead to different hypersurfaces crossing the point x_m. The state reduction thus occurs along *all* spacelike hypersurfaces passing x_m.

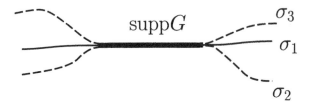

FIG. 11.6. For the measurement of a local observable with compact support on
some hypersurface the state reduction takes place along all spacelike hypersurfaces which coincide on the support of the measured quantity.

as it crosses the point x_m. Being a local observable, $A(x_m)$ commutes with the
Hamiltonian density $\mathcal{H}(x)$ for spacelike separations, and, hence, $\mathcal{H}(x)$ commutes
with the operation:

$$[\mathcal{H}(x), \Omega(a)] = 0, \quad \text{for } (x - x_m)^2 < 0. \qquad (11.44)$$

This is the condition of microcausality for the operation. Considering then some
hypersurface which crosses x_m and varying this surface around any point x on
it, keeping x_m fixed, we get with the help of eqns (11.40) and (11.44),

$$\frac{\delta}{\delta\sigma(x)}\mathcal{P}(a) = i\langle\Psi(\sigma)|\left[\mathcal{H}(x), \Omega^\dagger(a)\Omega(a)\right]|\Psi(\sigma)\rangle = 0. \qquad (11.45)$$

This proves that $\mathcal{P}(a)$ is, in fact, independent of σ as long as σ crosses x_m. The
same conclusion holds for an extended region given by the support of $G(x)$, with
the restriction, of course, that the variations leave invariant the common support
of $G(x)$.

In view of the above state vector reduction postulate, a given foliation $\mathcal{F} = \{\sigma(\tau)\}$ now gives rise to a corresponding stochastic process $|\Psi(\sigma(\tau))\rangle$ for the state
vector. Starting from a state $|\Psi(\sigma_0)\rangle$ on an initial hypersurface $\sigma_0 = \sigma(\tau = 0)$ the
state vector evolves continuously according to the Schwinger–Tomonaga equation

until the (uniquely determined) hypersurface σ_m which crosses x_m is reached. Then, conditioned on the readout a, the state reduction (11.38) occurs along σ_m and the state vector evolves continuously again. If $\sigma(\tau)$ crosses the future light cone of x_m the state vector on $\sigma(\tau)$ resulting from the readout a can be written as

$$|\Psi(\sigma(\tau))\rangle = \frac{1}{\sqrt{\mathcal{P}(a)}} \mathrm{T}_{\leftarrow} \left(\Omega(a) \exp\left[-i \int\limits_{\sigma_0}^{\sigma(\tau)} d^4x\, \mathcal{H}(x) \right] \right) |\Psi(\sigma_0)\rangle. \qquad (11.46)$$

Here, the chronological time ordering operator $\mathrm{T}_{\leftarrow}$ acts on the exponential involving the local Hamiltonian density $\mathcal{H}(x)$ as well as on the operation $\Omega(a)$ which contains the local observable $A(x_m)$.

The stochastic state vector evolution formulated above is obviously Markovian. Note also that different foliations lead to different state vector evolutions. However, for identical readouts the state vector (11.46) depends only on the initial surface σ_0 and on the final surface $\sigma(\tau)$. The process is therefore integrable in the same sense as the purely unitary evolution according to the Schwinger–Tomonaga equation: We take two different foliations $\mathcal{F}_1 = \{\sigma_1(\tau_1)\}$ and $\mathcal{F}_2 = \{\sigma_2(\tau_2)\}$ with common initial and final hypersurfaces σ_0 and σ, respectively. These foliations yield corresponding state vectors $|\Psi_1(\sigma_1(\tau_1))\rangle$ and $|\Psi_2(\sigma_2(\tau_2))\rangle$. Integrability then means that both foliations will lead to one and the same state vector on the common final surface σ, namely we have

$$|\Psi_1(\sigma)\rangle = |\Psi_2(\sigma)\rangle, \qquad (11.47)$$

provided that we start from the same initial state vector on σ_0 and that we have identical readouts a. The integrability of the process is an immediate consequence of the property of microcausality. It implies that the state vector is, in fact, a functional on the set of spacelike hypersurfaces.

It is important to realize that the state vector history as given in eqn (11.46) is conditioned on the outcome a of the measurement at x_m. It thus depends on the classical event that the observable Q takes on the value a. Let us suppose that the measurement of Q is performed immediately after the interaction between the field and the quantum probe and that the result is communicated via a classical light signal. The readout a is then available everywhere in the forward light cone based at x_m. The latter is defined as the set of points x satisfying $(x - x_m)^2 > 0$ and $x^0 > x_m^0$ and will be denoted by $V_+(x_m)$ (see eqn (11.33)). Consider an observer O moving along a world line $y = y(\tau)$ where τ denotes the proper time. If that world line intrudes into $V_+(x_m)$ at the proper time τ_1, say, the observer knows the readout a of the measurement on the probe and can thus set up the state vector history (11.46) depending on a. It is clear that the state reduction (11.38) takes place in the observer's past, that is for $\tau_m < \tau_1$, which means that it occurs (possibly a long time) before the observer gets the information on the readout.

11.2.3 *Multivalued space-time amplitudes*

As we have seen the state vector reduction expressed by eqn (11.38) yields a covariant stochastic state vector history (11.46) associated with each foliation of space-time. This fact has been demonstrated to be a simple consequence of the causality principle. However, it must be emphasized that the state reduction postulate leads to the important conclusion that the probability amplitudes which are determined by the functional $|\Psi(\sigma)\rangle$ need not represent single-valued functions on Minkowski space (Aharonov and Albert, 1984b).

We illustrate this point with the help of an example. To this end, we first construct a simple device for the effective position measurement carried out on a single-electron state. Consider the observable

$$A = \int d^3x\, G(\vec{x})\psi^\dagger(x)\psi(x), \quad x = (t_m, \vec{x}), \tag{11.48}$$

where $\psi_\alpha^\dagger(x)$ and $\psi_\alpha(x)$ denote the field operators of the electron field, α being a spinor index. They create and annihilate, respectively, an electron at x and satisfy Fermionic anticommutation relations,

$$\{\psi_\alpha(t,\vec{x}), \psi_\beta(t,\vec{x}\,')\} = \{\psi_\alpha^\dagger(t,\vec{x}), \psi_\beta^\dagger(t,\vec{x}\,')\} = 0, \tag{11.49}$$

$$\{\psi_\alpha(t,\vec{x}), \psi_\beta^\dagger(t,\vec{x}\,')\} = \delta(\vec{x} - \vec{x}\,')\delta_{\alpha\beta}. \tag{11.50}$$

Furthermore, $G(\vec{x})$ is a smooth function with compact support. We suppose that $G(\vec{x})$ is equal to 1 in a small region $\mathcal{G}$ of space around x_m and falls rapidly to zero outside $\mathcal{G}$.

Taking $|\Psi(\sigma_m)\rangle$ to be a one-electron state, we consider the amplitude

$$\chi(x) = \langle 0|\psi(x)|\Psi(\sigma_m)\rangle \tag{11.51}$$

for the electron to be at the space-time point x, where $|0\rangle$ denotes the ground state vacuum of the electron field. Our aim is to determine how the operation $\Omega(q)$ pertaining to the measurement of A acts on this amplitude $\chi(x)$. We find

$$\langle 0|\psi(x)\Omega(q)|\Psi(\sigma_m)\rangle = \langle q|\langle 0|\psi(x)e^{-iAP}|\Psi(\sigma_m)\rangle|\phi\rangle$$
$$= \langle q|\langle 0|e^{iAP}\psi(x)e^{-iAP}|\Psi(\sigma_m)\rangle|\phi\rangle, \tag{11.52}$$

where we have used $A|0\rangle = 0$ in the second step. The anticommutation relations yield

$$[A, \psi(x)] = -G(\vec{x})\psi(x), \tag{11.53}$$

from which it follows that

$$e^{iAP}\psi(x)e^{-iAP} = e^{-iG(\vec{x})P}\psi(x). \tag{11.54}$$

Therefore we get

$$\langle 0|\psi(x)\Omega(q)|\Psi(\sigma_m)\rangle = \langle q|\langle 0|e^{-iG(\vec{x})P}\psi(x)|\Psi(\sigma_m)\rangle|\phi\rangle$$
$$= \phi(q - G(\vec{x}))\chi(x). \tag{11.55}$$

This shows that the operation acts as follows on the one-electron amplitude,

$$\chi(x) \longrightarrow \phi(q - G(\vec{x}))\chi(x), \tag{11.56}$$

and that the probability density for the readout q becomes

$$\mathcal{P}(q) = \int d^3x |\phi(q - G(\vec{x}))|^2 |\chi(x)|^2. \tag{11.57}$$

In order to enable a sufficiently accurate measurement the probe wave function $\phi(q)$ must be a function which is sharply peaked around $q = 0$, with a width much smaller than 1. Equation (11.57) clearly shows that our device allows the approximate measurement of the random variable $q = G(\vec{x})$.

Suppose now that the amplitude $\chi(x)$ consists of two localized wave packets $\chi^{(1)}(x)$ and $\chi^{(2)}(x)$, such that the support of $\chi^{(1)}(x)$ is contained in the region $\mathcal{G}$, where $G(\vec{x})$ is equal to 1. The support of $\chi^{(2)}(x)$ and that of $G(\vec{x})$ are assumed to be disjoint. We then have

$$\mathcal{P}(q) = |\phi(q - 1)|^2 \int d^3x |\chi^{(1)}(x)|^2 + |\phi(q)|^2 \int d^3x |\chi^{(2)}(x)|^2. \tag{11.58}$$

This means that we get the readout $q \approx 1$ with probability $\int d^3x |\chi^{(1)}(x)|^2$, showing that the particle is in $\mathcal{G}$, and the readout $q \approx 0$ with probability $\int d^3x |\chi^{(2)}(x)|^2$, showing that the particle is not in $\mathcal{G}$. As can be inferred from eqn (11.56), in the first case the operation projects the amplitude $\chi(x)$ onto $\chi^{(1)}(x)$, in the second case onto $\chi^{(2)}(x)$,

$$\chi(x) \longrightarrow \chi^{(1)}(x)/\|\chi^{(1)}\|, \quad \text{for } q \approx 1, \tag{11.59}$$
$$\chi(x) \longrightarrow \chi^{(2)}(x)/\|\chi^{(2)}\|, \quad \text{for } q \approx 0. \tag{11.60}$$

Let us now consider the following situation (Aharonov and Albert, 1984b) which will lead to the conclusion that the amplitude $\chi(x)$ is not a single-valued function on the space-time continuum (see Fig. 11.7). Suppose that we have prepared on an initial hypersurface σ_0 some one-electron state $|\Psi(\sigma_0)\rangle$ such that the amplitude $\langle 0|\psi(x)|\Psi(\sigma_0)\rangle$ represents a superposition of two wave packets $\chi^{(1)}(x)$ and $\chi^{(2)}(x)$,

$$\langle 0|\psi(x)|\Psi(\sigma_0)\rangle = \chi^{(1)}(x) + \chi^{(2)}(x), \quad x \in \sigma_0. \tag{11.61}$$

The wave packets are supposed to be localized in small regions of space around $\vec{x}^{(1)}$ and $\vec{x}^{(2)}$, respectively, and follow their world tubes with zero mean velocity. For simplicity we may neglect the extension as well as the spreading of the wave packets. These effects can easily be taken into account, but do not change the argument.

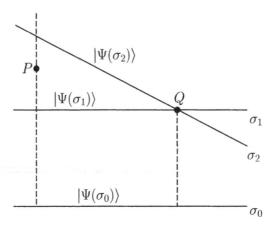

FIG. 11.7. The figure illustrates that space-time amplitudes are, in general, multivalued functions: The state vectors associated with two different hypersurfaces σ_1 and σ_2 may yield different amplitudes at a common point Q if at the point P a position measurement is carried out.

At some space-time point P a position measurement is performed with the help of a device of the type considered above. We assume that the measurement leads to the result that the electron is at P (the case $q = 1$ above). Given such a situation we may consider two flat spacelike hypersurfaces σ_1 and σ_2 which intersect at the space-time point Q. Both hypersurfaces appear as equal-time hypersurfaces in appropriately chosen coordinate frames K_1 and K_2, that is there are observers O_1 and O_2 at rest in K_1 and K_2, respectively, such that σ_1 is an equal-time hypersurface for O_1, and σ_2 is an equal-time hypersurface for O_2. The important difference between both observers is that for O_2 the measurement has already taken place, whereas for O_1 it has not. Consequently, both observers assign different amplitudes to one and the same objective space-time point Q. Namely, the state reduction following the measurement of the electron at P yields

$$\langle 0|\psi(Q)|\Psi(\sigma_2)\rangle = 0 \tag{11.62}$$

on σ_2. On the other hand, on σ_1 the state reduction has not yet occurred so that we have

$$\langle 0|\psi(Q)|\Psi(\sigma_1)\rangle = \chi^{(2)}(Q) \neq 0. \tag{11.63}$$

Thus we find that the amplitudes differ at the point Q where both hypersurfaces intersect,

$$\langle 0|\psi(Q)|\Psi(\sigma_1)\rangle \neq \langle 0|\psi(Q)|\Psi(\sigma_2)\rangle. \tag{11.64}$$

This relation clearly demonstrates our claim, namely that the one-particle amplitude $\langle 0|\psi(x)|\Psi(\sigma)\rangle$ is a multivalued function on space-time: The value of this amplitude at x depends, in general, on the hypersurface σ which crosses x. In other words, the amplitude depends on the foliation and, thus, on the complete state vector history.

Although there is obviously a Lorentz transformation which maps the plane σ_1 to σ_2, the corresponding one-particle amplitudes are clearly not related by a unitary transformation. This is due to the fact that the measurement on the system makes it an open system such that in situations like the one considered above there is no unitary representation of the Lorentz group for the reduced system. The same conclusion holds also if we consider the corresponding non-selective measurement: In that case we have a mixed state on σ_2, whereas we have a pure state on σ_1.

11.2.4   The consistent hierarchy of joint probabilities

Generalizing the foregoing analysis we now consider a sequence of local measurements performed at the points

$$x^{(1)}, x^{(2)}, \ldots, x^{(K)}, \tag{11.65}$$

which may be arbitrarily distributed in space and time. At each $x^{(k)}$, where $k = 1, 2, \ldots, K$, a local observable $A^{(k)} = A(x^{(k)})$ is measured indirectly through the coupling to a quantum probe in the initial state $|\phi^{(k)}\rangle$. The coupling between the field and the k-th quantum probe involves the generalized momentum $P^{(k)}$ and a direct measurement of the canonically conjugated observable $Q^{(k)}$ is performed after the interaction. We assume here that the K quantum probes act completely independently on the field, that is, we may write

$$|\phi\rangle = |\phi^{(1)}\rangle \otimes |\phi^{(2)}\rangle \otimes \ldots \otimes |\phi^{(K)}\rangle \tag{11.66}$$

for the total state of the K probe particles. We also assume as before that the mean values $\langle \phi^{(k)} | Q^{(k)} | \phi^{(k)} \rangle$ vanish, such that the inferred values $a^{(k)}$ for the observables $A^{(k)}$ are simply given by $a^{(k)} = q^{(k)}$. Writing $\phi^{(k)}(q^{(k)})$ for the $Q^{(k)}$ representation of the initial state of the k-th probe particle we thus obtain the operations

$$\Omega^{(k)}(a^{(k)}) = \phi^{(k)} \left(a^{(k)} - A^{(k)} \right), \tag{11.67}$$

which describe the change of the system state conditioned on the outcome $a^{(k)}$ at $x^{(k)}$.

As mentioned before, the points $x^{(k)}$ may be distributed arbitrarily in space and time. Thus, we may have spacelike as well as timelike separations between them. Since the $A^{(k)}$ are assumed to be local observables, the causality principle ensures that for spacelike separations the operations commute with themselves and with the Hamiltonian density, that is we have

$$\left[\Omega^{(k)}(a^{(k)}), \Omega^{(l)}(a^{(l)}) \right] = 0, \quad \text{for } (x^{(k)} - x^{(l)})^2 < 0, \tag{11.68}$$

$$\left[\mathcal{H}(x), \Omega^{(k)}(a^{(k)}) \right] = 0, \quad \text{for } (x - x^{(k)})^2 < 0. \tag{11.69}$$

The results of the previous subsection are now readily generalized to the above collection of local measurements (Breuer and Petruccione, 1999). Taking some foliation with initial surface σ_0 we find for the state vector on $\sigma(\tau)$ (compare with eqn (11.46)),

$$|\Psi(\sigma(\tau))\rangle = \mathcal{N} \cdot \mathrm{T}_{\leftarrow} \left(\prod_{k=1}^{K} \Omega^{(k)}(a^{(k)}) \exp \left[-i \int_{\sigma_0}^{\sigma(\tau)} d^4x\, \mathcal{H}(x) \right] \right) |\Psi(\sigma_0)\rangle. \quad (11.70)$$

Again, the time-ordering operator acts on the exponential as well as on the operations $\Omega^{(k)}$. The normalization factor $\mathcal{N}$ is given by

$$\mathcal{N} = \left[\mathcal{P}_K(a^{(1)}, \ldots, a^{(K)}) \right]^{-1/2}, \quad (11.71)$$

where $\mathcal{P}_K(a^{(1)}, \ldots, a^{(K)})$ denotes the joint probability of the readouts,

$$\mathcal{P}_K(a^{(1)}, \ldots, a^{(K)}) = \left\| \mathrm{T}_{\leftarrow} \left(\prod_{k=1}^{K} \Omega^{(k)}(a^{(k)}) \exp \left[-i \int_{\sigma_0}^{\sigma(\tau)} d^4x\, \mathcal{H}(x) \right] \right) |\Psi(\sigma_0)\rangle \right\|^2.$$

$$(11.72)$$

The surface $\sigma(\tau)$ may be chosen arbitrarily with the only restriction that all the $x^{(k)}$ must be located in the past of it, which means that $\sigma(\tau)$ must cross the forward light cones of all the $x^{(k)}$.

Equation (11.70) associates with each foliation a unique stochastic process $|\Psi(\sigma(\tau))\rangle$. Each realization of the process represents a state vector history which is conditioned on the readouts that follow the joint probability (11.72). It should be clear that this process is Markovian and integrable. In particular, the joint probability $\mathcal{P}_K(a^{(1)}, \ldots, a^{(K)})$ represents a Lorentz-invariant expression and does not depend on the foliation. The reason is that the time-ordering operator $\mathrm{T}_{\leftarrow}$ is defined in an invariant fashion by virtue of the causality conditions (11.68) and (11.69). We also note that due to the completeness relations

$$\int da^{(k)} \Omega^{(k)\dagger}(a^{(k)}) \Omega^{(k)}(a^{(k)}) = I \quad (11.73)$$

the joint probability is normalized,

$$\int da^{(1)} \ldots \int da^{(K)} \mathcal{P}_K(a^{(1)}, \ldots, a^{(K)}) = 1. \quad (11.74)$$

Let us illustrate in Fig. 11.8 how eqn (11.70) works for the specific case of three measurement points (a similar situation has been considered by Bloch (1967)). The distances between $x^{(1)}$ and $x^{(2)}$ and between $x^{(1)}$ and $x^{(3)}$ are spacelike, whereas $x^{(2)}$ and $x^{(3)}$ are separated by a timelike interval. The figure

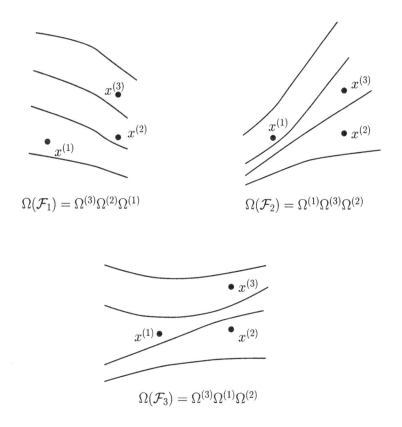

FIG. 11.8. An example for eqn (11.70) involving three measurement points with two spacelike and one timelike distance. The figure shows three different foliations which all lead to one and the same total operation.

shows three different foliations $\mathcal{F}_1$, $\mathcal{F}_2$, and $\mathcal{F}_3$. According to eqn (11.70) these foliations lead to the three total operations

$$\Omega(\mathcal{F}_1) = \Omega^{(3)}\Omega^{(2)}\Omega^{(1)}, \qquad (11.75)$$
$$\Omega(\mathcal{F}_2) = \Omega^{(1)}\Omega^{(3)}\Omega^{(2)}, \qquad (11.76)$$
$$\Omega(\mathcal{F}_3) = \Omega^{(3)}\Omega^{(1)}\Omega^{(2)}, \qquad (11.77)$$

which describe the state vector on the final hypersurface of the corresponding foliation (for simplicity, the Hamiltonian density may be set equal to zero). $\Omega^{(1)}$ commutes with $\Omega^{(2)}$ and with $\Omega^{(3)}$, but $\Omega^{(2)}$ does not commute with $\Omega^{(3)}$, of course. As is easily seen all three total operations are identical. This shows the integrability of the process as well as the relativistic invariance of the joint probability distribution.

It must be emphasized that the state vector history (11.70) is conditioned on the total readout $(a^{(1)}, \ldots, a^{(K)})$. We consider again an observer O moving

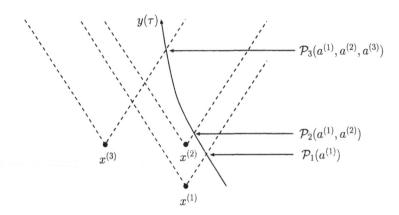

Fig. 11.9. The hierarchy of joint probabilities (11.78) seen by an observer moving along some world line $y(\tau)$: Each time the observer intrudes into a new light cone of the measurement points $x^{(k)}$, a new member of the family is generated.

along some world line $y = y(\tau)$. As before we suppose that the results of the measurements at the points $x^{(k)}$ are communicated via light signals such that the outcomes of the single measurements are available in the respective future light cones $V_+(x^{(k)})$. Moving along $y(\tau)$ the observer O successively intrudes into these light cones. Suppose that at the proper time τ observer O has intruded into the light cones of the points $x^{(k_l)}$, where $l = 1, 2, \ldots, L$, and $L \leq K$. Observer O then defines for any foliation the state vector history which is given by eqn (11.70) with the only modification that the product $\prod_k$ now extends only over the points $x^{(k_l)}$ in the forward light cones of which O is located, that is only over those measurements for which O knows the outcome.

It should be clear that in order to set up the state vector history observer O need not know the outcomes of the measurements in advance, nor does O have to know in advance which observable is measured at $x^{(k)}$ or if any measurement at this point has been performed at all. The information on what is measured and where the measurements took place can be communicated, of course, together with the corresponding readouts.

Let us, for ease of notation, label the measurement points in such a way that observer O intrudes first into the light cone of $x^{(1)}$, then the light cone of $x^{(2)}$, ..., and finally the light cone of $x^{(K)}$. Moving along, observer O thus generates a family of joint probabilities (see Fig. 11.9),

$$\mathcal{P}_1(a^{(1)}), \quad \mathcal{P}_2(a^{(1)}, a^{(2)}), \quad \ldots \quad \mathcal{P}_K(a^{(1)}, \ldots, a^{(K)}), \tag{11.78}$$

which are defined by

$$\mathcal{P}_k(a^{(1)},\ldots,a^{(k)}) = \left\| \mathrm{T}_\leftarrow \left(\prod_{l=1}^{k} \Omega^{(l)}(a^{(l)}) \exp\left[-i \int_{\sigma_0}^{\sigma(\tau)} d^4x\, \mathcal{H}(x) \right] \right) |\Psi(\sigma_0)\rangle \right\|^2.$$

(11.79)

The points $x^{(1)},\ldots,x^{(k)}$ are located in the space-time region enclosed by the hypersurfaces σ_0 and $\sigma(\tau)$. As we have already noted all joint probabilities (11.78) are normalized and do not depend on the specific chosen foliation. The latter property implies that any two observers with the same information on the readouts agree completely on the corresponding joint probabilities. The following consistency condition can also be verified easily,

$$\mathcal{P}_k(a^{(1)},\ldots,a^{(k)}) = \int da^{(k+1)} \mathcal{P}_{k+1}(a^{(1)},\ldots,a^{(k)},a^{(k+1)}).$$

(11.80)

This shows that each new readout $a^{(k+1)}$ which is communicated to observer O when she or he intrudes into the light cone $V_+(x^{(k+1)})$ is compatible with the joint probabilities of lower order in the hierarchy. Thus we conclude that (11.78) forms a consistent hierarchy of joint probability distributions.

11.2.5 EPR correlations

It is important to note that the above formulae have been derived without restrictions on the initial state vector $|\Psi(\sigma_0)\rangle$. They are thus capable of describing arbitrary quantum correlations between timelike and spacelike separated points. Quantum correlations at spacelike separated points arise if $|\Psi(\sigma_0)\rangle$ represents a non-local, entangled state. As an example we consider Bohm's formulation (Bohm, 1951) of the famous EPR *gedanken* experiment developed by Einstein, Podolsky and Rosen (1935): A two-particle system in a state of total spin 0,

$$|\Psi(\sigma_0)\rangle = \frac{1}{\sqrt{2}} \left(|+\rangle^{(1)}|-\rangle^{(2)} - |-\rangle^{(1)}|+\rangle^{(2)} \right)$$

(11.81)

disintegrates into two particles with spin $\frac{1}{2}$ flying with opposite velocities along the x-direction. In the initial state (11.81) the states $|\pm\rangle^{(k)}$ are eigenstates of the z-components $\sigma_z^{(k)}$ of the spin operator at two points $x^{(k)}$, $k=1,2$, with eigenvalues ±1. At the points $x^{(k)}$, being spacelike separated, the spin projections $\vec{\alpha}\cdot\vec{\sigma}^{(1)}$ and $\vec{\beta}\cdot\vec{\sigma}^{(2)}$ are measured. Here $\vec{\alpha}$ and $\vec{\beta}$ denote unit vectors lying in the (y,z)-plane and forming angles α and β with the z-axis. Neglecting any particle interaction we now have for the joint probability of the measurements

$$\mathcal{P}_2(a^{(1)},a^{(2)}) = \|\Omega^{(1)}(a^{(1)})\Omega^{(2)}(a^{(2)})|\Psi(\sigma_0)\rangle\|^2.$$

(11.82)

Assuming that the measurements represent ideal measurements of the spin projections we have two possible readouts $a^{(k)} = \pm1$, whereas the operations are given by the projections on the corresponding eigenvectors. We then find for the single measurements at $x^{(1)}$ and $x^{(2)}$ the mean values

$$\langle \Psi(\sigma_0) | \vec{\alpha} \cdot \vec{\sigma}^{(1)} | \Psi(\sigma_0) \rangle = \langle \Psi(\sigma_0) | \vec{\beta} \cdot \vec{\sigma}^{(2)} | \Psi(\sigma_0) \rangle = 0, \qquad (11.83)$$

and the variances

$$\mathrm{Var}\left(\vec{\alpha} \cdot \vec{\sigma}^{(1)} \right) = \langle \Psi(\sigma_0) | \left(\vec{\alpha} \cdot \vec{\sigma}^{(1)} \right)^2 | \Psi(\sigma_0) \rangle = 1, \qquad (11.84)$$

$$\mathrm{Var}\left(\vec{\beta} \cdot \vec{\sigma}^{(2)} \right) = \langle \Psi(\sigma_0) | \left(\vec{\beta} \cdot \vec{\sigma}^{(2)} \right)^2 | \Psi(\sigma_0) \rangle = 1. \qquad (11.85)$$

The correlation coefficient (see eqn (1.40)) is easily found to be

$$\begin{aligned}
\mathrm{Cor}(\alpha, \beta) &= \sum_{a^{(1)}, a^{(2)}} a^{(1)} a^{(2)} \mathcal{P}_2(a^{(1)}, a^{(2)}) \\
&= \langle \Psi(\sigma_0) | \left(\vec{\alpha} \cdot \vec{\sigma}^{(1)} \right) \left(\vec{\beta} \cdot \vec{\sigma}^{(2)} \right) | \Psi(\sigma_0) \rangle \\
&= -\cos(\alpha - \beta).
\end{aligned} \qquad (11.86)$$

The correlation coefficient describes the quantum correlations embodied in the joint probability distribution $\mathcal{P}_2(a^{(1)}, a^{(2)})$. For $\alpha = \beta$ we have $\mathrm{Cor}(\alpha, \beta) = -1$ from which it follows that the readouts are perfectly anticorrelated, namely that $a^{(1)} = -a^{(2)}$.

For general measurements on the particles we have the relations

$$\mathcal{P}_1(a^{(1)}) = || \Omega^{(1)}(a^{(1)}) | \Psi(\sigma_0) \rangle ||^2, \qquad (11.87)$$

$$\mathcal{P}_1(a^{(1)}) = \int da^{(2)} \mathcal{P}_2(a^{(1)}, a^{(2)}). \qquad (11.88)$$

Equation (11.87) reveals that the probabilities for the single, local measurement at $x^{(1)}$ are completely independent of the measurement at $x^{(2)}$: They do not depend on the readout $a^{(2)}$, nor do they depend on the observable being measured there, that is on the angle β, for example. The probabilities for the single measurement at $x^{(1)}$ do not even depend on where a possible second measurement is performed, as long as the latter is spacelike separated from $x^{(1)}$, of course. As pointed out already, eqn (11.88) means that the unconditioned probabilities for the single measurement at $x^{(1)}$ are not changed when the local observer gets any information on the measurement at a second point $x^{(2)}$, which clearly reveals the consistency of the joint probabilities for the present example.

11.2.6 *Continuous measurements*

With the help of the idea of the state vector as a function on the set of spacelike hypersurfaces, the stochastic wave function representation of continuous, selective measurements can be given a relativistically covariant form. As an example we construct here a piecewise deterministic process for the state vector of the source of an optical cavity which is covariant under Lorentz transformations and which describes the stochastic dynamics induced by a continuous monitoring of the radiated photons through a moving detector. The generalization to the general case of a quantum dynamical semigroup generated by an arbitrary number of Lindblad operators is straightforward; further details of the theory may be found in (Breuer and Petruccione, 2001).

We consider the output signal of a two-level atom in an optical cavity which is detected by a photocounter. The photodetector moves with velocity $\vec{v} = v\vec{e}$ relative to the cavity in the direction $\vec{e}$ of the output signal, thereby following a world line $y(\tau)$ in Minkowski space. Again, the parameter τ denotes the proper time of the detector, that is, the time of a clock attached to the detector. In the following we allow for an accelerated motion for which the 4-velocity

$$n(\tau) \equiv \frac{dy}{d\tau} = (\gamma, \gamma\vec{v}) \qquad (11.89)$$

is not a constant, where $n^\mu(\tau)n_\mu(\tau) = 1$, and $\gamma \equiv (1 - v^2)^{-1/2}$. For each τ the equation

$$n(\tau)(x - y(\tau)) = 0 \qquad (11.90)$$

defines a flat, spacelike hypersurface $\sigma(\tau)$. This hypersurface is the set of those space-time points x to which an observer O moving with the detector assigns one and the same time coordinate τ. The family of hypersurfaces $\mathcal{F} = \{\sigma(\tau)\}$ determined by the path $y(\tau)$ represents a foliation of a certain space-time region. More precisely, this is the space-time region in which different hypersurfaces of the family do not intersect. The corresponding condition is $l(x) < g^{-1}$, where $l(x)$ is the distance of a point x from the world line $y(\tau)$ and g is the acceleration of the detector measured in its own rest frame (Misner, Thorne and Wheeler, 1973). The family $\{\sigma(\tau)\}$ then provides a foliation of the set of space-time points x which fulfil this condition. As we shall see, beyond a distance g^{-1} from the detector's world line a complete, continuous measurement and, therefore, a stochastic representation of the source dynamics ceases to exist, in general.

Our aim is now to construct a relativistically covariant, stochastic representation of the source dynamics in terms of a state vector $|\Psi(\tau)\rangle = |\Psi(\sigma(\tau))\rangle$ which results from the continuous monitoring of the radiated photons through the moving photodetector. Each time a photon is detected the wave function of the source undergoes an instantaneous change. We assume, as in the non-relativistic theory (see Section 6.3), that this change is obtained through the application of some Lindblad operator L to the state vector. Due to the lack of an absolute time the term *instantaneous change of the wave function* is not a relativistically covariant one. However, by the very principles of quantum mechanics, source, radiation field, and detector have to be regarded as a whole. The important conclusion is that a Lorentz transformation affects the quantum object and the probe as well as the detector and the hypersurfaces $\sigma(\tau)$ of the foliation of observer O. Thus we do get a covariant prescription for the state vector reduction if we postulate that the state vector reduction occurs instantaneously in the detector's rest frame, that is, along a certain spacelike hypersurface $\sigma(\tau)$ of the foliation of the observer O associated to it.

As a consequence the jump operator now becomes a function $L(\tau) = L(\sigma(\tau))$ of the hypersurfaces of the foliation. In direct analogy to the non-relativistic formulation we obtain on the basis of our state reduction postulate the following Markovian stochastic state vector equation describing a piecewise deterministic process:

$$
\begin{aligned}
d|\Psi(\tau)\rangle = {} & -iH(\tau)|\Psi(\tau)\rangle d\tau \\
& -\frac{1}{2}\left(L^\dagger(\tau)L(\tau) - \langle L^\dagger(\tau)L(\tau)\rangle\right)|\Psi(\tau)\rangle da \\
& + \left(\frac{L(\tau)|\Psi(\tau)\rangle}{\sqrt{\langle L^\dagger(\tau)L(\tau)\rangle}} - |\Psi(\tau)\rangle\right) dN(a),
\end{aligned}
\tag{11.91}
$$

where we have introduced the abbreviation

$$
\langle L^\dagger(\tau)L(\tau)\rangle \equiv \langle \Psi(\tau)|L^\dagger(\tau)L(\tau)|\Psi(\tau)\rangle.
\tag{11.92}
$$

The first term on the right-hand side of eqn (11.91) represents the unitary dynamics (see eqn (11.23)), while the second and the third term yield the irreversible part of the evolution induced by the continuous monitoring of the quantum object. The structure of these terms is similar to the one encountered in Section 6.1.1. The differential da plays the rôle of an invariant time increment which will be determined below. The quantity $dN(a)$ is the increment of a Poisson process which obeys

$$
\mathrm{E}\left[dN(a)\right] = \langle L^\dagger(\tau)L(\tau)\rangle da, \quad \left[dN(a)\right]^2 = dN(a),
\tag{11.93}
$$

where E denotes the expectation value of the process. The second relation in (11.93) tells us that $dN(a)$ takes on the values 0 or 1. As long as no photon is detected we have $dN(a) = 0$ and, thus, the second term of eqn (11.91) gives the evolution of the state vector conditioned on the outcome that no photon is detected. If a photon is detected we have $dN(a) = 1$, such that the third term of eqn (11.91) yields the corresponding jump of the state vector given by

$$
|\Psi(\sigma)\rangle \longrightarrow \frac{L(\sigma)|\Psi(\sigma)\rangle}{\sqrt{\langle L^\dagger(\sigma)L(\sigma)\rangle}}.
\tag{11.94}
$$

Thus, the Poisson process $N(a)$ simply counts the number of photon detection events.

According to eqn (11.94) the stochastic jumps of the state vector occur along the hypersurfaces $\sigma = \sigma(\tau)$ of the foliation given by the detector path. If a photon has been detected at a certain proper time τ the state vector reduction has to be performed at the corresponding retarded proper time

$$
\tau_{\mathrm{ret}} = \tau - \frac{R}{c},
\tag{11.95}
$$

taking into account the time required for the light signal to propagate from the source to the detector, where R denotes the instantaneous distance from the source to the detector. This follows directly from the fact that the detected signal yields information on the state of the source at the retarded time. Thus, the precise prescription for the state vector reduction takes the following form: The reduction of the state vector occurs along the spacelike hypersurface $\sigma(\tau_{\mathrm{ret}})$ at the retarded proper time τ_{ret} which corresponds to the proper time τ of the actual measuring event. Thus, eqn (11.91) gives rise to a stochastic equation of motion for the source wave function $|\Psi(\tau_{\mathrm{ret}})\rangle = |\Psi(\sigma(\tau_{\mathrm{ret}}))\rangle$.

In order to determine the invariant time parameter a used in eqn (11.91) we first observe that, according to eqn (11.93), the photocurrent as measured in the rest frame of the detector is given by

$$J = \langle L^{\dagger}(\tau)L(\tau)\rangle \frac{da}{d\tau}. \qquad (11.96)$$

J is the average number of photon counts per unit of the proper time interval $d\tau$. Due to the Lorentz invariant nature of the scalar product we may simply set $\langle L^{\dagger}(\tau)L(\tau)\rangle = \gamma_0$, where γ_0 is an invariant emission rate characteristic of the source. Thus we have

$$J = \gamma_0 \frac{da}{d\tau}. \qquad (11.97)$$

If the detector is at rest with respect to the source ($v = 0$), the detected photocurrent must be $J_0 = \gamma_0$. Thus, a must be equal to the proper time of the source, i.e. a is the time of a clock fixed at a position in the vicinity of the source. To see that this conclusion is correct we consider the case of a moving detector ($v \neq 0$). It is easy to show with the help of the transformation laws for the electromagnetic field strength tensor that the photocurrrent J as measured in the rest frame of the detector is given by

$$J = \gamma_0 \sqrt{\frac{1 - v}{1 + v}}, \qquad (11.98)$$

where we do not assume that v is constant. Now, with the above choice for the quantity a we find

$$\frac{da}{d\tau} = \sqrt{\frac{1 - v}{1 + v}}, \qquad (11.99)$$

which proves our claim.

As mentioned earlier the source must lie within a distance $l < g^{-1}$ from the world line of the detector. This condition implies that $da/d\tau_{\mathrm{ret}} > 0$ and that, therefore, eqn (11.91) represents a sensible stochastic state vector equation with a positive increment da. If this condition is violated, obviously no complete continuous monitoring by the detector is possible. As an example one might think of a detector in hyperbolic motion, in which case the observer can outrun the photons radiated by the source.

The covariance of our stochastic state vector equation (11.91) under Lorentz transformations is obvious. Since da is an invariant, the quantity $dN(a)$ is an

invariant stochastic process. Furthermore, since the quantum expectation value
$\langle L^{\dagger}(\sigma)L(\sigma) \rangle$ transforms as a Lorentz scalar both the dissipative and the stochastic term of (11.91) transform covariantly. It is important to emphasize that the transformation laws also involve a transformation of the jump operator, namely

$$L'(\sigma') = UL(\sigma)U^{\dagger}, \tag{11.100}$$

where U denotes the unitary representation of the Lorentz transformation. Physically, this means that the quantum object as well as the environment and the measuring apparatus have to be Lorentz transformed in order to obtain covariance of the stochastic process. In view of the physical meaning of the process as a continuous measurement, this is a plausible prescription which is in full agreement with both quantum mechanics and special relativity.

11.3   Non-local measurements and causality

In the analysis of the preceding section we have assumed that the total initial state of the quantum probe is given by a simple product state of the form (11.66). A more general class of measurements can be constructed with the help of quantum probes which are in entangled initial states. We are going to demonstrate in the present section that entangled probe states enable one to perform the measurement of non-local observables and states as first shown by Aharonov and Albert (1980, 1981, 1984a, 1984b).

The possibility of non-local measurements has several important consequences for the notions of states and observables in relativistic quantum theory. As before we will assume that only local interactions between object and probe system are involved. The dynamics of the total system will therefore be compatible with the causality principle, of course. However, if causality is combined with certain conditions on the properties of the measuring device and on the behaviour of the state vector of the object system alone, one is led to important restrictions on the measurability of observable and states (Aharonov, Albert and Vaidman, 1986; Sorkin, 1993).

The conditions imposed by the causality principle will be derived in this section. For example, we shall demonstrate below that the projections on non-local, entangled states are not measurable in the conventional sense of quantum mechanics. It will also be shown that only specific classes of entangled, non-local states and only certain types of non-local observables with entangled eigenstates allow a quantum non-demolition (QND) measurement.

Although the possibility of QND measurements of non-local observables and states is strongly restricted by causality, the *preparation* of non-local states is not. Indeed, as will be shown any non-local entangled state could in principle be prepared without conflict with causality. There is also a quite different type of measurement, known as an exchange measurement, which never leads to any contradiction to causality and which will also be discussed. Finally, as an interesting application we investigate the transmission of an unknown quantum state with the help of a classical communication channel and a quantum channel provided by an EPR entangled quantum state. This so-called quantum teleportation nicely illustrates some of the features of relativistic quantum measurements.

11.3.1 *Entangled quantum probes*

We consider the case that the probe system has been prepared initially in some entangled state $|\phi\rangle$. In the $Q^{(k)}$ representation we write for the wave function of the total probe system

$$\phi(q^{(1)}, \ldots, q^{(K)}) = \langle q^{(1)}, \ldots, q^{(K)}|\phi\rangle. \tag{11.101}$$

Our previous result (11.72) on the joint probability distribution can then be immediately generalized to yield

$$\mathcal{P}_K(a^{(1)}, \ldots, a^{(K)}) \tag{11.102}$$

$$= \left\| T_\leftarrow \left(\Omega(a^{(1)}, \ldots, a^{(K)}) \exp\left[-i \int\limits_{\sigma_0}^{\sigma(\tau)} d^4x\, \mathcal{H}(x) \right] \right) |\Psi(\sigma_0)\rangle \right\|^2,$$

where, again, the $x^{(1)}, \ldots, x^{(K)}$ are located in the space-time region enclosed by the hypersurfaces σ_0 and $\sigma(\tau)$. The total operation may be written

$$\Omega(a^{(1)}, \ldots, a^{(K)}) = \phi(a^{(1)} - A^{(1)}, \ldots, a^{(K)} - A^{(K)}). \tag{11.103}$$

It should be noted that, in general, this is only a formal expression since any two operators $A^{(k)}$ and $A^{(l)}$ need not commute if the corresponding points $x^{(k)}$ and $x^{(l)}$ are separated by a timelike distance. However, under the time-ordering operator $T_\leftarrow$ in eqn (11.102) the operation is unambiguously defined.

The usage of an entangled quantum probe leads to decisive consequences for the conditioned evolution of the state of the object system. The most important feature is that for entangled probe states the evolution does not, in general, transform pure states into pure states and that the process becomes non-Markovian. This can be seen with the help of the simplest case, namely that of a measurement at two spacelike separated points $x^{(1)}$ and $x^{(2)}$. Since the corresponding operators $A^{(1)}$ and $A^{(2)}$ commute, the operation

$$\Omega(a^{(1)}, a^{(2)}) = \phi(a^{(1)} - A^{(1)}, a^{(2)} - A^{(2)}) \tag{11.104}$$

is unambiguously defined without the time-ordering operator, of course.

We suppose that the probe state $|\phi\rangle$ is an entangled state. To be specific we may regard $|\phi\rangle$ as the state of a two-particle system which is prepared in an entangled state in some localized region. These probe particles are then brought to the space-time points $x^{(1)}$ and $x^{(2)}$, respectively, where they interact with the object system. Afterwards, the observables $Q^{(1)}$ and $Q^{(2)}$ are measured on the probe particles to yield the readouts $a^{(1)}$ and $a^{(2)}$, as indicated in Fig. 11.10.

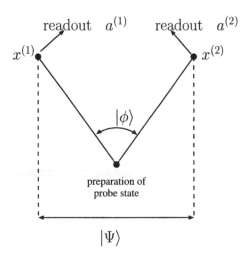

FIG. 11.10. General scheme for an indirect measurement device which uses an entangled probe state $|\phi\rangle$.

Consider now two different foliations $\mathcal{F}$ and $\mathcal{F}'$ by parallel surfaces as indicated in Fig. 11.11. As before we may regard the hypersurfaces belonging to these foliations as equal-time hypersurfaces of two observers O and O' moving with different 4-velocities n and n'. Clearly, n is the unit normal vector of $\mathcal{F}$ and n' is that of $\mathcal{F}'$.

In the case that the probe represents an entangled state one finds that O and O' associate completely different state histories to their foliations. Let us first look at the situation from the viewpoint of O. Observer O starts from some initial state $|\Psi(\sigma_0)\rangle$ on σ_0. As indicated in the figure, there is some surface σ_1 of the foliation $\mathcal{F}$ which crosses both points $x^{(1)}$ and $x^{(2)}$, which means that both measurements are simultaneous for observer O. This implies that the state reduction in $\mathcal{F}$ occurs along σ_1 and that the readout $a^{(1)}$, $a^{(2)}$ gives the final state

$$|\Psi(\sigma_2)\rangle = \mathcal{N} \, \Omega(a^{(1)}, a^{(2)})|\Psi(\sigma_0)\rangle \qquad (11.105)$$

on the surface σ_2 of the foliation of O, $\mathcal{N}$ being some normalization factor. Thus, observer O describes the state history as an evolution of pure states, the reduction taking place at a single instant of time in the rest frame of O.

The state history associated with the foliation $\mathcal{F}'$ is completely different. In this foliation there is no hypersurface which crosses both points $x^{(1)}$ and $x^{(2)}$, that is for observer O' the two measurements are not simultaneous: There exists an intermediate region of time in which the first measurement at $x^{(2)}$ and the corresponding state reduction already took place whereas the one at $x^{(1)}$ has not. For an entangled quantum probe this leads to the result that in this intermediate region object and probe are, in general, in an entangled total state which is given by

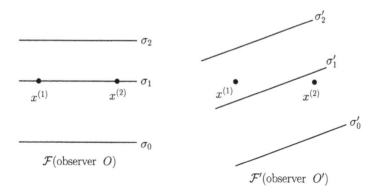

FIG. 11.11. Two different foliations $\mathcal{F}$ and $\mathcal{F}'$ corresponding to two observers O and O' who describe the non-local measurement depicted in Fig. 11.10 in their respective coordinate frames.

$$|\Phi(\sigma_1')\rangle = \mathcal{N} \int da^{(1)} |a^{(1)}\rangle \otimes |a^{(2)}\rangle \otimes \langle a^{(1)}, a^{(2)}| e^{-iA^{(2)} P^{(2)}} |\phi\rangle |\Psi(\sigma_0')\rangle. \quad (11.106)$$

By contrast, if the probe is in a direct product state $|\phi\rangle = |\phi^{(1)}\rangle \otimes |\phi^{(2)}\rangle$, the intermediate state also becomes a direct product

$$|\Phi(\sigma_1')\rangle = \mathcal{N}|\phi^{(1)}\rangle \otimes |a^{(2)}\rangle \otimes \langle a^{(2)}| e^{-iA^{(2)} P^{(2)}} |\phi^{(2)}\rangle |\Psi(\sigma_0')\rangle, \quad (11.107)$$

as it was in those cases considered in the preceding section. It is also clear that after the second measurement at $x^{(1)}$ the total object-plus-probe system is always in a product state again which takes the form

$$|\Phi(\sigma_2')\rangle = \mathcal{N}|a^{(1)}\rangle \otimes |a^{(2)}\rangle \otimes \Omega(a^{(1)}, a^{(2)}) |\Psi(\sigma_0')\rangle. \quad (11.108)$$

The entanglement of the object–probe system shows that the state referring to the observables of the object system represents, in general, a mixture in the intermediate region of the foliation $\mathcal{F}'$,

$$\rho(\sigma_1') = \frac{1}{\mathcal{P}_1(a^{(2)})} \int da^{(1)} \phi(a^{(1)}, a^{(2)} - A^{(2)}) \rho(\sigma_0') \phi^\dagger(a^{(1)}, a^{(2)} - A^{(2)}), \quad (11.109)$$

where

$$\mathcal{P}_1(a^{(2)}) = \int da^{(1)} \langle \Psi(\sigma_0')| \phi^\dagger \phi(a^{(1)}, a^{(2)} - A^{(2)}) |\Psi(\sigma_0')\rangle \quad (11.110)$$

is the unconditioned probability density for the readout $a^{(2)}$ of the measurement at $x^{(2)}$. Thus we see that observer O' associates the following history for the object state with her/his foliation $\mathcal{F}'$,

$$\rho(\sigma_0') \text{ (pure state)} \longrightarrow \rho(\sigma_1') \text{ (mixture)} \longrightarrow \rho(\sigma_2') \text{ (pure state)}. \quad (11.111)$$

In the first step the pure state $\rho(\sigma_0')$ is transformed into a mixture $\rho(\sigma_1')$ by the first measurement at $x^{(2)}$. In the second step this mixture then transforms into a pure state $\rho(\sigma_2')$ again: The destruction of the pure state by the first measurement is completely undone by the second measurement.

The above description clearly reveals the non-Markovian character of the state history. The final state $\rho(\sigma_2')$ of the object system after the second measurement is obtained by applying the operation $\Omega(a^{(1)}, a^{(2)})$ to the initial state $\rho(\sigma_0')$, that is by applying it to an object state which was given at a finite time interval $\Delta\tau$ prior to the second measurement. We remark that this time interval corresponding to the intermediate region of time between the two measurements can be made arbitrarily large. For example, if $x^{(1)} = (0, \vec{x}^{(1)})$ and $x^{(2)} = (0, \vec{x}^{(2)})$ are the coordinates of the measurement events in the rest frame of O and if O' moves in the direction of $\vec{x}^{(1)} - \vec{x}^{(2)}$ the time interval between the first and the second measurement is found to be

$$\Delta\tau = \gamma v \Delta x, \tag{11.112}$$

where $\gamma = (1 - v^2)^{-1/2}$, v is the speed of O' relative to O, and $\Delta x = |\vec{x}^{(1)} - \vec{x}^{(2)}|$.

The non-Markovian feature of the process is due to the fact that the information which is required to restore the pure object state in the second step is contained in the probe state during the time interval $\Delta\tau$. This information is lost by tracing over the variables of the quantum probe. As we have seen, the process describing the behaviour of the total object-plus-probe system is, of course, Markovian.

11.3.2   Non-local measurement by EPR probes

In this subsection we shall demonstrate that the use of entangled probe states allows the quantum non-demolition measurement of certain non-local observables. As in the preceding subsection our quantum probe constitutes a two-particle system with canonical coordinates $Q^{(1)}$ and $Q^{(2)}$ and corresponding conjugated momenta $P^{(1)}$ and $P^{(2)}$. The initial probe state is taken to be an EPR-type entangled state which may be defined with the help of the relations

$$\left(P^{(1)} + P^{(2)}\right)|\phi\rangle = 0, \tag{11.113}$$

$$\left(Q^{(1)} - Q^{(2)}\right)|\phi\rangle = 0. \tag{11.114}$$

Note that the total canonical momentum

$$P = P^{(1)} + P^{(2)} \tag{11.115}$$

commutes with the relative coordinate

$$\tilde{Q} = Q^{(1)} - Q^{(2)}, \tag{11.116}$$

such that P and $\tilde{Q}$ can take simultaneously sharp values. Introducing also the average of the coordinates

$$Q = \frac{1}{2}\left(Q^{(1)} + Q^{(2)}\right), \tag{11.117}$$

as well as the relative momentum

$$\tilde{P} = \frac{1}{2}\left(P^{(1)} - P^{(2)}\right) \tag{11.118}$$

we have two new pairs of canonically conjugated coordinates and momenta,

$$[P, Q] = [\tilde{P}, \tilde{Q}] = -i, \tag{11.119}$$
$$[P, \tilde{Q}] = [\tilde{P}, Q] = 0. \tag{11.120}$$

Using the mixed $(P, \tilde{Q})$ representation we can thus define the probe state by means of

$$|\phi\rangle = |p = 0, \tilde{q} = 0\rangle. \tag{11.121}$$

Let us determine the operation describing the readouts $a^{(1)} = q^{(1)}$ and $a^{(2)} = q^{(2)}$ for the local measurement of $Q^{(1)}$ and $Q^{(2)}$ after the object–probe interaction. Employing the above relations we find

$$\begin{aligned}
\Omega(a^{(1)}, a^{(2)}) &= \langle a^{(1)}, a^{(2)}| \exp[-iA^{(1)}P^{(1)} - iA^{(2)}P^{(2)}]|\phi\rangle \\
&= \langle a^{(1)}, a^{(2)}| \exp[-i(A^{(1)} - A^{(2)})\tilde{P} - \frac{i}{2}(A^{(1)} + A^{(2)})P]|\phi\rangle \\
&= \langle a^{(1)}, a^{(2)}| \exp[-i(A^{(1)} - A^{(2)})\tilde{P}]|\phi\rangle,
\end{aligned} \tag{11.122}$$

where we have used the fact that P commutes with $\tilde{P}$ and that $P|\phi\rangle = 0$. Consider now the non-local operator

$$A = A^{(1)} - A^{(2)}, \tag{11.123}$$

and its spectral decomposition

$$A = \sum_a a \, \Pi_A(a), \tag{11.124}$$

which is assumed to be discrete for simplicity. Here and in the following we write $\Pi_A(a)$ for the projection onto the eigenspace of the operator A belonging to the eigenvalue a.

If we introduce the spectral decomposition of A into the expression (11.122) for the operation we get

$$\begin{aligned}
\Omega(a^{(1)}, a^{(2)}) &= \sum_a \langle a^{(1)}, a^{(2)}|e^{-ia\tilde{P}}|p = 0, \tilde{q} = 0\rangle \, \Pi_A(a) \\
&= \sum_a \langle a^{(1)}, a^{(2)}|p = 0, \tilde{q} = a\rangle \, \Pi_A(a).
\end{aligned} \tag{11.125}$$

In the second step we have used the fact that $\exp(-ia\tilde{P})$ is a translation operator which shifts $\tilde{q}$ by the amount a, since $\tilde{P}$ is canonically conjugated to $\tilde{Q}$. Now, the remaining matrix element vanishes unless $a^{(1)} - a^{(2)} = a$,

$$\langle a^{(1)}, a^{(2)} | p = 0, \tilde{q} = a \rangle = \delta(a^{(1)} - a^{(2)} - a), \tag{11.126}$$

which gives

$$\Omega(a^{(1)}, a^{(2)}) = \sum_a \delta(a^{(1)} - a^{(2)} - a) \, \Pi_A(a). \tag{11.127}$$

This equation clearly shows that the operation depends only on the difference of the readouts and that the possible values for this difference coincide with the eigenvalues of A. The correctly normalized operation pertaining to the outcome $a = a^{(1)} - a^{(2)}$ can therefore be written as follows,

$$\Omega(a) = \Pi_A(a), \quad a \in \mathrm{spec}(A), \tag{11.128}$$

showing that the operation is just equal to the corresponding projection of the spectral family of the operator A.

Thus, we see that the EPR entangled probe state allows an ideal quantum non-demolition measurement of the non-local observable A. The operation describes the back action on the quantum object which is in full agreement with the von Neumann–Lüders projection postulate: The readouts yield the eigenvalues of A and if the initial state of the object system was is an eigenstate of A then this state is not changed by the measurement.

To appreciate what has been achieved it is important to realize the following properties of the above measurement scheme. First, it must be emphasized that the measurement device does not yield any information on the local observables $A^{(1)}$ and $A^{(2)}$. This is connected to the fact that neither $Q^{(1)}$ nor $Q^{(2)}$ are sharply defined in the initial probe state, since these observables do not commute with P. Only the relative coordinate $\tilde{Q}$ which does commute with P is sharply defined. The entangled probe state $|\phi\rangle$ is thus not appropriate for a measurement of the local observables $A^{(1)}$ and $A^{(2)}$. Consider, for example, the local measurement at $x^{(2)}$ and the foliation $\mathcal{F}'$ of the previous subsection (Fig. 11.11). If we introduce the spectral decomposition of the local observable $A^{(2)}$,

$$A^{(2)} = \sum_\nu a_\nu^{(2)} \Pi_\nu^{(2)} \tag{11.129}$$

we find for the density matrix describing the quantum object after the first measurement at $x^{(2)}$,

$$\rho(\sigma_1') = \sum_\nu \Pi_\nu^{(2)} \rho(\sigma_0') \Pi_\nu^{(2)}. \tag{11.130}$$

This is precisely the density matrix as it would be obtained in a *non-selective* measurement of the local observable $A^{(2)}$. It follows that the measurement of $Q^{(2)}$ does not give any information on $A^{(2)}$: The quantum object behaves as if we measure $A^{(2)}$ and immediately erase that information. In particular, the possible outcomes $a^{(2)}$ are not given by the eigenvalues of $A^{(2)}$, but are uniformly distributed and independent of the initial object state.

Thus we see that the device allows the measurement of $A = A^{(1)} - A^{(2)}$ without measuring the local quantities $A^{(1)}$, $A^{(2)}$. Of course one can also measure any linear combination of the form

$$A = \alpha A^{(1)} + \beta A^{(2)}. \tag{11.131}$$

This is achieved by an appropriate replacement of the local observable, $A^{(1)} \to \alpha A^{(1)}$, $A^{(2)} \to -\beta A^{(2)}$. Recall that our basic assumption is that, at least in principle, we can measure any local observable. Similarly, one can also measure any product

$$A = A^{(1)} \cdot A^{(2)} \tag{11.132}$$

provided both local observables have a definite sign, for example $A^{(1)} > 0$, $A^{(2)} > 0$. Namely, in that case we may replace $A^{(1)} \to \ln A^{(1)}$, $A^{(2)} \to \ln A^{(2)}$. However, as we shall demonstrate later, it is not possible to measure all observables belonging to the algebra of operators of the object system.

The second important point to be noted is that the device allows the measurement of non-local observables which cannot be measured locally. Consider the observable $A = A^{(1)} + A^{(2)}$. Suppose first that A is non-degenerate. In that case we can measure A also locally by simply measuring $A^{(1)}$ and $A^{(2)}$ separately (with the help of a probe which is in a direct product state). Such a measurement will then also be a QND measurement of A: The corresponding readouts $a^{(1)}$, $a^{(2)}$ are the eigenvalues of $A^{(1)}$ and $A^{(2)}$, respectively, and $a = a^{(1)} + a^{(2)}$ is an eigenvalue of A. Moreover, all eigenstates of A are also eigenstates of $A^{(1)}$ and of $A^{(2)}$, which follows from the fact that the local observables commute and that A is non-degenerate. Therefore, the separate measurement of $A^{(1)}$ and $A^{(2)}$ leaves unchanged all eigenstates of A.

The situation changes completely if A is degenerate. The separate measurement will then, in general, not be a QND measurement of A. To see this we consider some degenerate eigenvalue a of A and two corresponding orthogonal eigenstates $|\chi_1\rangle$ and $|\chi_2\rangle$. The non-local measurement procedure constructed above clearly has the property that it leaves invariant the total subspace spanned by $|\chi_1\rangle$ and $|\chi_2\rangle$. However, we can always choose these states such that they are simultaneous eigenstates of $A^{(1)}$ and $A^{(2)}$. We suppose that the corresponding eigenvalues are different and consider the initial state

$$|\chi\rangle = \frac{1}{\sqrt{2}} \left(|\chi_1\rangle + |\chi_2\rangle \right). \tag{11.133}$$

What happens then during separate local measurements of $A^{(1)}$ and $A^{(2)}$ is that this initial state goes over with probability $\frac{1}{2}$ to either $|\chi_1\rangle$ or $|\chi_2\rangle$. The separate measurement of the local quantities is therefore not a QND measurement of their sum.

As an example, which will be used several times later on, we consider a system of two particles with spin $\frac{1}{2}$. One particle interacts locally with the device at $x^{(1)}$, the other at $x^{(2)}$. The total Hilbert space is, of course, given by the tensor product

$\mathcal{H} = \frac{1}{2} \otimes \frac{1}{2}$. As our local observables we take the spin components of the particles along the z-direction,

$$A^{(1)} = \frac{1}{2}\sigma_z^{(1)}, \quad A^{(2)} = \frac{1}{2}\sigma_z^{(2)}, \tag{11.134}$$

and consider the measurement of the sum of the spins along that direction

$$A = \frac{1}{2}\left(\sigma_z^{(1)} + \sigma_z^{(2)}\right) \equiv J_z. \tag{11.135}$$

A basis of eigenstates of J_z may be written in an obvious notation as

$$|j = 0, m = 0\rangle = \frac{1}{\sqrt{2}}\left(|+\rangle^{(1)}|-\rangle^{(2)} - |-\rangle^{(1)}|+\rangle^{(2)}\right), \tag{11.136}$$

$$|j = 1, m = 0\rangle = \frac{1}{\sqrt{2}}\left(|+\rangle^{(1)}|-\rangle^{(2)} + |-\rangle^{(1)}|+\rangle^{(2)}\right), \tag{11.137}$$

$$|j = 1, m = +1\rangle = |+\rangle^{(1)}|+\rangle^{(2)}, \tag{11.138}$$

$$|j = 1, m = -1\rangle = |-\rangle^{(1)}|-\rangle^{(2)}, \tag{11.139}$$

where m denotes the eigenvalue of J_z. These eigenstates have been chosen to be simultaneous eigenstates of the square of the total spin

$$\vec{J} = \frac{1}{2}\left(\vec{\sigma}^{(1)} + \vec{\sigma}^{(2)}\right), \tag{11.140}$$

that is $\vec{J}^2|j, m\rangle = j(j+1)|j, m\rangle$. The states $|j = 0, m = 0\rangle$ and $|j = 1, m = 0\rangle$ span the two-fold degenerate eigenspace belonging to the eigenvalue 0 of J_z. The non-local measurement using an EPR-type probe state allows the QND measurement of J_z. In a separate measurement of the local spins of the particles, however, the singlet state $|j = 0, m = 0\rangle$, for example, will be transformed into either $|+\rangle^{(1)}|-\rangle^{(2)}$ or $|-\rangle^{(1)}|+\rangle^{(2)}$.

Let us now demonstrate that besides linear combinations and certain products we can also measure arbitrary modular sums of local observables. The fact that this type of non-local observable is measurable at least in principle, will play an important rôle in the further development of the theory.

Obviously, it suffices to construct an explicit measuring device for the quantity

$$B = A \bmod z \equiv \left(A^{(1)} - A^{(2)}\right) \bmod z, \tag{11.141}$$

where $z > 0$ is some real number. To measure B we employ the following initial probe state,

$$|\phi\rangle = \mathcal{N} \sum_n |p = 0, \tilde{q} = nz\rangle, \tag{11.142}$$

where the sum runs over an appropriate subset of the set of integers and $\mathcal{N}$ is some irrelevant normalization factor. The probe state is a superposition of

EPR entangled states. In contrast to the probe used before, not only are the local coordinates $Q^{(1)}$ and $Q^{(2)}$ undetermined in this state, but also the relative coordinate $\tilde{Q}$ is only determined up to a multiple of z, that is $(\tilde{Q} \bmod z)|\phi\rangle = 0$.

Proceeding in precisely the same manner as above we now get the following expression for the operation,

$$\Omega(a^{(1)}, a^{(2)}) = \sum_n \sum_a \langle a^{(1)}, a^{(2)} | p = 0, \tilde{q} = a + nz \rangle \Pi_A(a). \qquad (11.143)$$

Correctly normalized, the operation takes the form

$$\Omega(a^{(1)}, a^{(2)}) = \sum_{a+nz=a^{(1)}-a^{(2)}} \Pi_A(a). \qquad (11.144)$$

Here the sum extends over all a and n with the constraint of a fixed value for the quantity $a + nz = a^{(1)} - a^{(2)}$. This shows that the operation depends, in fact, only on the quantity $a \bmod z \equiv b$ and we may write the operation as

$$\Omega(b) = \sum_{a \bmod z = b} \Pi_A(a) = \Pi_B(b). \qquad (11.145)$$

Hence, the operation $\Omega(b)$ is equal to the projection onto the eigenspace belonging to the eigenvalue b of the modular sum B. The use of the probe state (11.142) therefore allows the measurement of the modular sum of local observables. We remark that it is not necessary that the sum over n in the initial probe state extends over all integers. In the following we shall apply the measurement of modular sums only to bounded operators. In that case it suffices if the sum runs over a finite number of integers, namely those which project the eigenvalues of A into the fundamental interval $[0, z)$.

11.3.3 Quantum state verification

The discussion of the preceding subsection raises the question of whether we can measure all non-local observables by an appropriate design of quantum probes and by employing various local interactions between probe and object system. Surprisingly, the answer to this question is that we cannot measure all non-local observables, that is we cannot design an ideal, quantum non-demolition measurement for each observable belonging to the algebra of operators of the object system. In fact, as we shall see the measurement of most non-local observables is incompatible with the causality principle. In the following sections we shall formulate the restrictions on the measurability of observables which are imposed by causality and investigate some classes of non-local operators and states which are measurable.

For a clear and systematic treatment of the subject the concept of a so-called state verification measurement turns out to be extremely useful. This concept is more general than the usual notion for the measurement of states which is used in quantum mechanics. In the present subsection we define state verification measurements and describe them in terms of operations and effects within the framework of generalized measurement theory developed in Section 2.4.2.

A state verification measurement of a given state $|\Psi_0\rangle$ is defined to be a measuring device which performs a Yes/No-decision experiment with the following property. Let $|\Psi\rangle$ be any state of the underlying Hilbert space and

$$|\Psi\rangle = \alpha|\Psi_0\rangle + \beta|\Psi_\perp\rangle \qquad (11.146)$$

its decomposition into the component parallel to $|\Psi_0\rangle$ and the component perpendicular to it, such that $\langle\Psi_0|\Psi_\perp\rangle = 0$ and $|\alpha|^2 + |\beta|^2 = 1$. If prior to the measurement the system is in the state $|\Psi\rangle$ then the measuring device responds with the result Yes with probability $|\alpha|^2$, and it responds with the result No with probability $|\beta|^2$.

Analogously we define a QND state verification as a state verification measurement with the following additional property. If the result is Yes then the system is in the state $|\Psi_0\rangle$ after the measurement. This implies that the initial state $|\Psi_0\rangle$ is left unchanged by the measurement since the result Yes is then obtained with certainty. In the case that the measurement yields No the system is in a state $|\Psi'_\perp\rangle$ which is orthogonal to $|\Psi_0\rangle$, but which is not necessarily equal to the orthogonal component $|\Psi_\perp\rangle$ of the initial state. The latter property implies that on repeating the experiment we again find No with certainty.

It should be clear that quantum state verification measurements do not, in general, represent measurements of the projection

$$\Pi_0 = |\Psi_0\rangle\langle\Psi_0| \qquad (11.147)$$

onto the initial state in the conventional sense of quantum mechanics. In the case of a QND state verification the orthogonal component is allowed to change during the measurement, and for a general state verification measurement no assumption at all is made regarding the behaviour of the state vector of the system.

We now describe the form of a general state verification using the language of operations and effects. According to its definition the device has a set $\{a\}$ of Yes-readouts and a set $\{b\}$ of No-readouts. Correspondingly we have a collection of Yes-operations $\Omega_{\mathrm{Yes}}(a)$ and of No-operations $\Omega_{\mathrm{No}}(b)$. The behaviour of the density matrix of the object system as a result of the non-selective state verification can then be described by

$$\rho \longrightarrow \rho' = \sum_a \Omega_{\mathrm{Yes}}(a)\rho\Omega^\dagger_{\mathrm{Yes}}(a) + \sum_b \Omega_{\mathrm{No}}(b)\rho\Omega^\dagger_{\mathrm{No}}(b). \qquad (11.148)$$

As we saw in Section 2.4.2 this is, apart from the assumption of discreteness of the sets of the readouts, the most general setting. In particular it includes the possibility of quantum probes in mixed states and those of incomplete, approximate, or non-selective measurement of the final probe states.

On introducing the Yes- and the No-effect,

$$F_{\text{Yes}} = \sum_a \Omega_{\text{Yes}}^\dagger(a)\Omega_{\text{Yes}}(a), \qquad (11.149)$$

$$F_{\text{No}} = \sum_b \Omega_{\text{No}}^\dagger(b)\Omega_{\text{No}}(b), \qquad (11.150)$$

we get from the conservation of probability

$$F_{\text{Yes}} + F_{\text{No}} = I. \qquad (11.151)$$

Let us consider the initial state $|\Psi\rangle$ and its decomposition (11.146). We can write the probability for the outcome *Yes* of the state verification of $|\Psi_0\rangle$ as follows

$$\begin{aligned}
\mathcal{P}_{\text{Yes}}(\Psi) &= \langle\Psi|F_{\text{Yes}}|\Psi\rangle \\
&= |\alpha|^2\langle\Psi_0|F_{\text{Yes}}|\Psi_0\rangle + |\beta|^2\langle\Psi_\perp|F_{\text{Yes}}|\Psi_\perp\rangle + (\alpha\beta^*\langle\Psi_\perp|F_{\text{Yes}}|\Psi_0\rangle + \text{c.c}) \\
&= |\alpha|^2.
\end{aligned} \qquad (11.152)$$

The last equality expresses the condition for a state verification measurement and must be true for all α, β, and for all $|\Psi_\perp\rangle$. Setting first $\beta = 0$ and then $\alpha = 0$ one easily deduces the relations

$$\langle\Psi_0|F_{\text{Yes}}|\Psi_0\rangle = 1, \quad \langle\Psi_\perp|F_{\text{Yes}}|\Psi_\perp\rangle = 0. \qquad (11.153)$$

If these relations are inserted into eqn (11.152) one also finds that

$$\langle\Psi_\perp|F_{\text{Yes}}|\Psi_0\rangle = 0. \qquad (11.154)$$

It follows that for any state verification measurement the Yes-effect is equal to the projection onto the verified state,

$$F_{\text{Yes}} = \Pi_0 \equiv |\Psi_0\rangle\langle\Psi_0|. \qquad (11.155)$$

With the help of eqns (11.153) one is led to the following equations,

$$\Omega_{\text{Yes}}(a)|\Psi_\perp\rangle = 0, \qquad (11.156)$$

and

$$\Omega_{\text{No}}(b)|\Psi_0\rangle = 0. \qquad (11.157)$$

Equation (11.156) holds for all a and for all states perpendicular to the verified state, while eqn (11.157) is valid for all b.

Summarizing we see that for any state verification measurement all Yes-operations annihilate the No-states, all No-operations annihilate the Yes-state, and the Yes-effect is equal to the projection onto the Yes-state. With the help of the conservation of probability it follows also that the No-effect is equal to the projection onto the orthogonal complement,

$$F_{\text{No}} = I - \Pi_0 \equiv \Pi_\perp. \tag{11.158}$$

In order for the state verification to be a QND measurement two further conditions must be satisfied. First, when applied to the state $|\Psi_0\rangle$ all Yes-operations must yield a state which is proportional to $|\Psi_0\rangle$. Hence, only a single Yes-operation is required which must be equal to the projection onto the verified state,

$$\Omega_{\text{Yes}} = \Pi_0. \tag{11.159}$$

The second additional condition for a QND state verification is that all No-operations must map the subspace orthogonal to $|\Psi_0\rangle$ into itself.

These results will be applied in the next subsection to the state verification of non-local states.

11.3.4   Non-local operations and the causality principle

In the present and the following subsection we study the measurement of some quantum system which consists of two localized parts (1) and (2) described by the Hilbert spaces $\mathcal{H}^{(1)}$ and $\mathcal{H}^{(2)}$, respectively. The total Hilbert space of the quantum object is the tensor product $\mathcal{H} = \mathcal{H}^{(1)} \otimes \mathcal{H}^{(2)}$. We investigate the state verification of an entangled state $|\Psi_0\rangle$ in this space. Introducing appropriate local bases $|\chi_i^{(1)}\rangle$ and $|\chi_i^{(2)}\rangle$ in both parts of the system a given state $|\Psi_0\rangle$ can be represented in terms of its Schmidt decomposition as follows (see Section 2.2.2),

$$|\Psi_0\rangle = \sum_{i=1}^{D} \alpha_i |\chi_i^{(1)}\rangle \otimes |\chi_i^{(2)}\rangle. \tag{11.160}$$

We assume here and in the following that $|\Psi_0\rangle$ is normalized and that its Schmidt number, denoted by D, is finite. Thus, the Schmidt decomposition consists of a finite number of terms and we assume that the complex numbers α_i, $i = 1, \ldots, D$, are different from zero,

$$\sum_{i=1}^{D} |\alpha_i|^2 = 1, \quad \alpha_i \neq 0. \tag{11.161}$$

For $D \geq 2$ the state $|\Psi_0\rangle$ is an entangled state. We recall that $|\Psi_0\rangle$ is called maximally entangled if the absolute values of all non-vanishing coefficients in the Schmidt decomposition are equal to each other, that is if $|\alpha_i| = 1/\sqrt{D}$ for all $i = 1, \ldots, D$.

We further denote by $\tilde{\mathcal{H}}^{(1)}$ and $\tilde{\mathcal{H}}^{(2)}$ the local subspaces spanned by those basis vectors that occur in the Schmidt decomposition of $|\Psi_0\rangle$ with a non-zero coefficient,

$$\tilde{\mathcal{H}}^{(1)} = \text{span}\left\{|\chi_i^{(1)}\rangle\right\}_{i=1,\ldots,D}, \tag{11.162}$$

$$\tilde{\mathcal{H}}^{(2)} = \text{span}\left\{|\chi_i^{(2)}\rangle\right\}_{i=1,\ldots,D}. \tag{11.163}$$

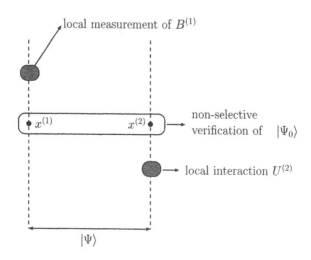

FIG. 11.12. Illustration of the causality principle expressed by eqn (11.164). After a verification measurement of $|\Psi_0\rangle$ a local observable $B^{(1)}$ is measured on part (1) of the system. Prior to the verification measurement a local interaction described by the unitary operator $U^{(2)}$ may be applied to part (2) of the system. The measurement of $B^{(1)}$ and the interaction $U^{(2)}$ are separated by a spacelike distance such that the expectation value of $B^{(1)}$ must be independent of whether the interaction $U^{(2)}$ is or is not applied to the system. The state verification itself is described on the non-selective level.

The tensor product $\tilde{\mathcal{H}}^{(1)} \otimes \tilde{\mathcal{H}}^{(2)}$ is thus a D^2-dimensional space.

11.3.4.1 *Formulation of the causality principle* We consider the setup sketched in Fig. 11.12. The initial state of the system may be any state $|\Psi\rangle$. At two spacelike separated points $x^{(1)}$ and $x^{(2)}$ a state verification measurement of the state $|\Psi_0\rangle$ is performed. In the localized region belonging to part (1) of the system we perform a measurement of some observable $B^{(1)}$ referring to the local variables in that part. In the following we denote by $\mathrm{E}\left[B^{(1)}|\Psi\right]$ the expectation value of $B^{(1)}$ for this local measurement performed after the state verification, under the condition that the initial state is $|\Psi\rangle$.

Suppose now that some local interaction may be applied on part (2) of the system prior to the state verification measurement. This can be described by some unitary operator $U^{(2)}$ acting on the Hilbert space $\mathcal{H}^{(2)}$. For example, if the quantum object constitutes a spin system one may regard $U^{(2)}$ as describing the interaction of the system with a magnetic field by which an experimenter can flip the spin of the particle in part (2). Since we assume that anything can be measured locally we may also assume that any unitary transformation can be realized by an appropriately chosen local interactions in each local part of the system.

We assume that the measurement of $B^{(1)}$ is carried out in a space-time region which is separated from the interaction region of $U^{(2)}$ by a spacelike interval. The causality principle then implies that the result of the local measurement carried out in part (1) *after* the state verification, does not depend in any way on the local interaction which is applied to the system in part (2) *before* the state verification. This means that the expectation value of any observable $B^{(1)}$ determined after the state verification must be the same regardless of whether $U^{(2)}$ is applied to the system or not, that is we must have

$$E[B^{(1)}|U^{(2)}\Psi] = E[B^{(1)}|\Psi].$$ (11.164)

This equation expresses the requirement of the causality principle on the state verification measurement.

It should be clear that the state verification must be described here as a non-selective measurement. The reason is that any local observer in part (1) does not know the result of the verification of the non-local state $|\Psi_0\rangle$ at the time when $B^{(1)}$ is measured. However, if eqn (11.164) were not always true a local observer in part (2) could transfer information from part (2) to part (1) with a superluminal speed. Thus, the expectation value is given by the expression,

$$
\begin{aligned}
E[B^{(1)}|\Psi] = \quad & \sum_a \langle\Psi|\Omega_{\mathrm{Yes}}^{\dagger}(a)B^{(1)}\Omega_{\mathrm{Yes}}(a)|\Psi\rangle \\
& + \sum_b \langle\Psi|\Omega_{\mathrm{No}}^{\dagger}(b)B^{(1)}\Omega_{\mathrm{No}}(b)|\Psi\rangle.
\end{aligned}
$$ (11.165)

The causality principle formulated by eqns (11.164) and (11.165) leads to important consequences for the measurability of non-local quantities. It must be emphasized that eqn (11.164) is trivially satisfied if we assume, as we of course do, that only local object–probe interactions are involved in the measurements. In fact, as is easily seen eqn (11.164) becomes trivial if we write it in terms of the state vector (or density matrix) for the total object–probe system. However, important conclusions can be drawn from the above formulation of the causality principle if it is combined with the conditions for a non-local state verification measurement and with certain properties of the underlying Hilbert space of the combined object system.

11.3.4.2 *Theorem on the erasing of local information*   In the following we prove a theorem by Popescu and Vaidman (1994) which can be derived from the causality principle (11.164) and which leads to several interesting conclusions on the possibilities of non-local measurements. This theorem states that

$$E[B^{(1)}|\Psi] = E[B^{(1)}|\Psi_0] \quad \text{for all} \quad |\Psi\rangle \in \tilde{\mathcal{H}}^{(1)} \otimes \mathcal{H}^{(2)}.$$ (11.166)

This means that the expectation values for all local observables in part (1) are independent of the initial state $|\Psi\rangle$ prior to the state verification measurement. In other words, after the state verification a local observer in part (1) cannot find out by the measurement of local observables $B^{(1)}$ the initial state of the system prior to the state verification. After the state verification there is thus no trace of the initial state in the local density matrix referring to the variables of part (1) of the system. This property is called the erasing of local information by the state verification measurement.

It must be stressed that the erasing of local information refers to initial states $|\Psi\rangle$ which belong to the subspace $\tilde{\mathcal{H}}^{(1)} \otimes \mathcal{H}^{(2)}$. For an arbitrary $|\Psi\rangle \in \mathcal{H}^{(1)} \otimes \mathcal{H}^{(2)}$ eqn (11.166) is, in general, wrong. As an example we take $|\Psi_0\rangle = |\chi^{(1)}\rangle \otimes |\chi^{(2)}\rangle$. This product state can obviously be verified by performing separate local measurements of the projections onto the states $|\chi^{(1)}\rangle$ and $|\chi^{(2)}\rangle$ in part (1) and part (2) of the system. Consider further the state $|\Psi\rangle = |\psi^{(1)}\rangle \otimes |\chi^{(2)}\rangle$, where $|\psi^{(1)}\rangle$ is orthogonal to $|\chi^{(1)}\rangle$. Since the chosen state verification is an ideal quantum measurement we have the conditional expectations $E[B^{(1)}|\Psi_0] = \langle\chi^{(1)}|B^{(1)}|\chi^{(1)}\rangle$ and $E[B^{(1)}|\Psi] = \langle\psi^{(1)}|B^{(1)}|\psi^{(1)}\rangle$ which are, in general, not equal to each other.

To prove (11.166) we first evaluate the expression for the conditional expectation value (11.165). Using the decomposition (11.146) of $|\Psi\rangle$ and taking into account the properties (11.156) and (11.157) of the state verification measurement we obtain,

$$E[B^{(1)}|\Psi]$$
$$= |\alpha|^2 \sum_a \langle\Psi_0|\Omega_{\text{Yes}}^{\dagger}(a)B^{(1)}\Omega_{\text{Yes}}(a)|\Psi_0\rangle + |\beta|^2 \sum_b \langle\Psi_{\perp}|\Omega_{\text{No}}^{\dagger}(b)B^{(1)}\Omega_{\text{No}}(b)|\Psi_{\perp}\rangle$$
$$= |\alpha|^2 E[B^{(1)}|\Psi_0] + |\beta|^2 E[B^{(1)}|\Psi_{\perp}]. \tag{11.167}$$

Thus we find

$$E[B^{(1)}|\Psi] = E[B^{(1)}|\Psi_0] + |\beta|^2 \left(E[B^{(1)}|\Psi_{\perp}] - E[B^{(1)}|\Psi_0] \right). \tag{11.168}$$

If $|\Psi\rangle$ has a non-vanishing orthogonal component, that is if $\beta \neq 0$ in the decomposition (11.146), the last equation tells us that $E[B^{(1)}|\Psi]$ is equal to $E[B^{(1)}|\Psi_0]$ if and only if $E[B^{(1)}|\Psi_{\perp}]$ is equal to $E[B^{(1)}|\Psi_0]$,

$$E[B^{(1)}|\Psi] = E[B^{(1)}|\Psi_0] \Longleftrightarrow E[B^{(1)}|\Psi_{\perp}] = E[B^{(1)}|\Psi_0]. \tag{11.169}$$

For the proof of eqn (11.166) the following strategy is used. In a first step one demonstrates that any state $|\Psi\rangle \in \tilde{\mathcal{H}}^{(1)} \otimes \mathcal{H}^{(2)}$ can be represented in the form,

$$|\Psi\rangle = \sum_{r=1}^{N} c_r U_r^{(2)} |\Psi_0\rangle, \tag{11.170}$$

where the c_r are c-numbers and the $U_r^{(2)}$ are unitary operators which act on the local space $\mathcal{H}^{(2)}$ of part (2) of the system. This means that any such state vector can be represented as a superposition of a finite number of states which can be generated from $|\Psi_0\rangle$ by the application of unitary transformations which act only on part (2) of the system. In the second step of the proof one demonstrates that any state of the form (11.170) obeys eqn (11.166). In this second step one uses the causality principle (11.164) as well as the property (11.169).

The existence of the representation (11.170) will be demonstrated by an explicit construction of the local, unitary operators $U_r^{(2)}$. The space $\tilde{\mathcal{H}}^{(1)} \otimes \mathcal{H}^{(2)}$ is spanned by the basis vectors

$$|\chi_k^{(1)}\rangle \otimes |\chi_l^{(2)}\rangle, \quad k = 1, \ldots, D, \quad l = 1, 2, \ldots. \tag{11.171}$$

Note that the restriction $k \leq D$ stems from our requirement that $|\Psi\rangle$ belongs to the space $\tilde{\mathcal{H}}^{(1)} \otimes \mathcal{H}^{(2)}$. As we have remarked already the theorem is not true for all state vectors belonging to $\mathcal{H}^{(1)} \otimes \mathcal{H}^{(2)}$. Obviously, it suffices to demonstrate that any basis vector (11.171) can be represented in the form (11.170). This is achieved with the help of two unitary transformations $U_1^{(2)}$ and $U_2^{(2)}$ which act in the space $\mathcal{H}^{(2)}$ and which are defined by

$$U_1^{(2)}|\chi_k^{(2)}\rangle = |\chi_l^{(2)}\rangle, \tag{11.172}$$

$$U_1^{(2)}|\chi_l^{(2)}\rangle = |\chi_k^{(2)}\rangle, \tag{11.173}$$

$$U_1^{(2)}|\chi_i^{(2)}\rangle = |\chi_i^{(2)}\rangle, \quad \text{for } i \neq k, l, \tag{11.174}$$

and by

$$U_2^{(2)}|\chi_k^{(2)}\rangle = -|\chi_l^{(2)}\rangle, \tag{11.175}$$

$$U_2^{(2)}|\chi_l^{(2)}\rangle = |\chi_k^{(2)}\rangle, \quad \text{for } k \neq l, \tag{11.176}$$

$$U_2^{(2)}|\chi_i^{(2)}\rangle = |\chi_i^{(2)}\rangle, \quad \text{for } i \neq k, l. \tag{11.177}$$

Thus, $U_1^{(2)}$ simply exchanges the states $|\chi_k^{(2)}\rangle$ and $|\chi_l^{(2)}\rangle$, and $U_2^{(2)}$ exchanges these states and introduces an additional minus sign. For $k = l$ the operator $U_1^{(2)}$ is equal to the identity, whereas $U_2^{(2)}$ multiplies the k-th basis vector by -1. Applying these operators to $|\Psi_0\rangle$ we get for $k \neq l$

$$U_1^{(2)}|\Psi_0\rangle = +\alpha_k|\chi_k^{(1)}\rangle|\chi_l^{(2)}\rangle + \alpha_l|\chi_l^{(1)}\rangle|\chi_k^{(2)}\rangle + \sum_{i \neq k,l} \alpha_i|\chi_i^{(1)}\rangle|\chi_i^{(2)}\rangle, \tag{11.178}$$

$$U_2^{(2)}|\Psi_0\rangle = -\alpha_k|\chi_k^{(1)}\rangle|\chi_l^{(2)}\rangle + \alpha_l|\chi_l^{(1)}\rangle|\chi_k^{(2)}\rangle + \sum_{i \neq k,l} \alpha_i|\chi_i^{(1)}\rangle|\chi_i^{(2)}\rangle. \tag{11.179}$$

It follows that
$$U_1^{(2)}|\Psi_0\rangle - U_2^{(2)}|\Psi_0\rangle = 2\alpha_k|\chi_k^{(1)}\rangle \otimes |\chi_l^{(2)}\rangle, \tag{11.180}$$

which gives a representation of the required form, namely

$$|\chi_k^{(1)}\rangle \otimes |\chi_l^{(2)}\rangle = \left(\frac{1}{2\alpha_k}U_1^{(2)} - \frac{1}{2\alpha_k}U_2^{(2)}\right)|\Psi_0\rangle. \tag{11.181}$$

As is easily seen, the case $k = l$ leads to the same equation. Note that $\alpha_k \neq 0$ for $k = 1, \ldots, D$ due to our assumptions on $|\Psi\rangle$ and $|\Psi_0\rangle$. This concludes the proof of eqn (11.170).

We now have to show that any $|\Psi\rangle$ of the form (11.170) satisfies eqn (11.166). This is done by induction over N, the number of terms involved in the representation. For $N = 1$ we may set $c_1 = 1$ since $|\Psi\rangle$ is normalized. Equation (11.166) then follows immediately from the causality principle (11.164).

Assume that (11.166) holds for all states with a decomposition of the form (11.170) involving N terms. Consider some normalized state whose representation of this form contains $N + 1$ terms,

$$|\Psi\rangle = \sum_{r=1}^{N+1} c_r U_r^{(2)} |\Psi_0\rangle. \tag{11.182}$$

For brevity we write $\mathrm{E}[\Psi]$ instead of $\mathrm{E}[B^{(1)}|\Psi]$ in what follows. According to the causality principle (11.164) we have

$$\mathrm{E}[\Psi] = \mathrm{E}\left[\left(U_{N+1}^{(2)}\right)^{-1} \Psi\right], \tag{11.183}$$

where

$$\left(U_{N+1}^{(2)}\right)^{-1} |\Psi\rangle = \sum_{r=1}^{N} c_r \left(U_{N+1}^{(2)}\right)^{-1} U_r^{(2)} |\Psi_0\rangle + c_{N+1} |\Psi_0\rangle. \tag{11.184}$$

If the sum over r in (11.184) is zero we have already proved the theorem. Let us therefore assume that it is non-zero and introduce the normalized state

$$|\tilde{\Psi}\rangle = \mathcal{N} \sum_{r=1}^{N} c_r \left(U_{N+1}^{(2)}\right)^{-1} U_r^{(2)} |\Psi_0\rangle, \tag{11.185}$$

with an appropriate normalization factor $\mathcal{N}$. The induction assumption then gives

$$\mathrm{E}[\tilde{\Psi}] = \mathrm{E}[\Psi_0]. \tag{11.186}$$

Next we decompose the state $|\tilde{\Psi}\rangle$,

$$|\tilde{\Psi}\rangle = \alpha |\Psi_0\rangle + \beta |\Psi_\perp\rangle, \tag{11.187}$$

where we may assume that $\beta \neq 0$ since otherwise the theorem follows immediately. Equation (11.186) together with the property (11.169) yields

$$\mathrm{E}[\Psi_\perp] = \mathrm{E}[\Psi_0]. \tag{11.188}$$

In view of eqns (11.184), (11.185), and (11.187) we also have the decomposition,

$$\left(U_{N+1}^{(2)}\right)^{-1}|\Psi\rangle = \left(\frac{\alpha}{\mathcal{N}} + c_{N+1}\right)|\Psi_0\rangle + \frac{\beta}{\mathcal{N}}|\Psi_\perp\rangle \equiv \alpha'|\Psi_0\rangle + \beta'|\Psi_\perp\rangle. \quad (11.189)$$

Since $\beta' \neq 0$ (see above) we may now use eqn (11.188) to conclude with the help of (11.169) that

$$\mathrm{E}\left[\left(U_{N+1}^{(2)}\right)^{-1}\Psi\right] = \mathrm{E}[\Psi_0], \quad (11.190)$$

from which the theorem follows by making use of eqn (11.183).

It should be noted that the theorem has been proven under quite general circumstances. As physical assumptions we have merely used the existence of the operations and effects describing a state verification measurement as well as the causality principle.

11.3.5 *Restrictions on the measurability of operators*

Let us draw some important conclusions from the theorem on the erasing of local information with regard to the measurability of non-local operators in the combined Hilbert space of the system.

The first conclusion is that the projections $\Pi_0 = |\Psi_0\rangle\langle\Psi_0|$ onto entangled states $|\Psi_0\rangle$ are not measurable. To prove this statement we derive a contradiction to the causality principle. Let

$$|\Psi_0\rangle = \sum_{i=1}^{D} \alpha_i |\chi_i^{(1)}\rangle \otimes |\chi_i^{(2)}\rangle \quad (11.191)$$

be the Schmidt decomposition of $|\Psi_0\rangle$. Since we assume that this state is entangled, at least two of the coefficients α_i are different from zero, that is we have $\alpha_k, \alpha_l \neq 0$, for some pair of indices $k \neq l$. A QND measurement of the projection Π_0 means that one has just two operations, namely $\Omega_{\mathrm{Yes}} = \Pi_0$ and $\Omega_{\mathrm{No}} = \Pi_\perp = I - \Pi_0$. It follows that not only $|\Psi_0\rangle$ but also any state $|\Psi_\perp\rangle$ orthogonal to it must be left unchanged by the measurement, that is

$$\mathrm{E}[B^{(1)}|\Psi_0] = \langle\Psi_0|B^{(1)}|\Psi_0\rangle, \quad (11.192)$$
$$\mathrm{E}[B^{(1)}|\Psi_\perp] = \langle\Psi_\perp|B^{(1)}|\Psi_\perp\rangle. \quad (11.193)$$

The erasing of local information expressed by eqn (11.166) then implies that

$$\langle\Psi_\perp|B^{(1)}|\Psi_\perp\rangle = \langle\Psi_0|B^{(1)}|\Psi_0\rangle \quad (11.194)$$

must hold for all states which are orthogonal to $|\Psi_0\rangle$ and which belong to the space $\tilde{\mathcal{H}}^{(1)} \otimes \mathcal{H}^{(2)}$. Now we take $|\Psi_\perp\rangle = |\chi_k^{(1)}\rangle \otimes |\chi_l^{(2)}\rangle$ and choose $B^{(1)} = |\chi_l^{(1)}\rangle\langle\chi_l^{(1)}|$. This leads to

$$\langle\Psi_\perp|B^{(1)}|\Psi_\perp\rangle = \langle\chi_k^{(1)}|B^{(1)}|\chi_k^{(1)}\rangle = 0, \quad (11.195)$$
$$\langle\Psi_0|B^{(1)}|\Psi_0\rangle = \sum_i |\alpha_i|^2 \langle\chi_i^{(1)}|B^{(1)}|\chi_i^{(1)}\rangle = |\alpha_l|^2 \neq 0, \quad (11.196)$$

which is incompatible with eqn (11.194). This shows that a measurement of the projections onto an entangled state would contradict the causality principle.

We note that a contradiction to causality only emerges if at least two of the Schmidt coefficients α_i are non-vanishing, that is only if $|\Psi_0\rangle$ is an entangled state. Of course, all projections onto product states $|\chi_k^{(1)}\rangle \otimes |\chi_l^{(2)}\rangle$ are measurable.

As an example one may think of the square of the total spin operator $\vec{J}$ (see eqn (11.140)) for a non-local system of two spin-$\frac{1}{2}$ particles. The operator $\vec{J}^2$ has a three-fold degenerate eigenspace belonging to the eigenvalue 2, and a non-degenerate eigenstate corresponding to the eigenvalue 0. The non-degenerate eigenstate, namely the singlet state $|j = 0, m = 0\rangle$, is an entangled state. A measurement of $\vec{J}^2$ is thus equivalent to a measurement of the projection onto an entangled state. Thus, the operator $\vec{J}^2$ is not measurable.

These considerations can be generalized as follows. Let A be some measurable observable in the space $\mathcal{H} = \mathcal{H}^{(1)} \otimes \mathcal{H}^{(2)}$ with at least one non-degenerate eigenvalue a and a corresponding entangled eigenstate $|\Psi_0\rangle$ of the form (11.191) with Schmidt number $D \geq 2$. As before we define $\tilde{\mathcal{H}}^{(1)}$ and $\tilde{\mathcal{H}}^{(2)}$ to be the subspaces spanned by the respective basis vectors $|\chi_i^{(1)}\rangle$ and $|\chi_i^{(2)}\rangle$, where $i = 1, \ldots, D$. If we further suppose that A leaves invariant the subspace $\tilde{\mathcal{H}}^{(1)} \otimes \tilde{\mathcal{H}}^{(2)}$ the causality principle leads to the following conclusion: The non-degenerate eigenstate $|\Psi_0\rangle$ as well as all other eigenstates of A in the space $\tilde{\mathcal{H}}^{(1)} \otimes \tilde{\mathcal{H}}^{(2)}$ must be maximally entangled.

To prove this statement we introduce a basis $|\Psi_\nu\rangle$, $\nu = 0, \ldots, D^2 - 1$, of eigenvectors of A belonging to the space $\tilde{\mathcal{H}}^{(1)} \otimes \tilde{\mathcal{H}}^{(2)}$. The state given by $\nu = 0$ is just the eigenstate $|\Psi_0\rangle$ corresponding to the non-degenerate eigenvalue a.

Now, any measurement of A is necessarily a state verification of $|\Psi_0\rangle$. To cast the causality condition into an appropriate form we introduce the density $\rho^{(1)}(\Psi)$ which is defined to be the local density matrix describing the variables of part (1) after the state verification, under the condition that the initial state was $|\Psi\rangle$. This density is obtained by taking the partial trace over part (2) of the density matrix of the total system after the state verification,

$$\rho^{(1)}(\Psi) = \mathrm{tr}^{(2)} \left\{ \sum_a \Omega_{\mathrm{Yes}}(a)|\Psi\rangle\langle\Psi|\Omega_{\mathrm{Yes}}^\dagger(a) + \sum_b \Omega_{\mathrm{No}}(b)|\Psi\rangle\langle\Psi|\Omega_{\mathrm{No}}^\dagger(b) \right\}.$$
(11.197)

The erasing of local information expressed by eqn (11.166) can then be re-formulated as

$$\rho^{(1)}(\Psi) = \rho^{(1)}(\Psi_0), \quad \text{for all} \quad |\Psi\rangle \in \tilde{\mathcal{H}}^{(1)} \otimes \mathcal{H}^{(2)}.$$
(11.198)

Since all $|\Psi_\nu\rangle$ belong to the space $\tilde{\mathcal{H}}^{(1)} \otimes \mathcal{H}^{(2)}$ we have for all ν

$$\rho^{(1)}(\Psi_0) = \rho^{(1)}(\Psi_\nu).$$
(11.199)

The A measurement leaves unchanged all eigenstates. Thus we conclude

$$\mathrm{tr}^{(2)} \left\{ |\Psi_0\rangle\langle\Psi_0| \right\} = \mathrm{tr}^{(2)} \left\{ |\Psi_\nu\rangle\langle\Psi_\nu| \right\}.$$
(11.200)

Summing this equation over ν and taking into account that the $|\Psi_\nu\rangle$ form a basis in the space $\tilde{\mathcal{H}}^{(1)} \otimes \tilde{\mathcal{H}}^{(2)}$ we immediately get,

$$D^2 \cdot \text{tr}^{(2)} \left\{ |\Psi_0\rangle\langle\Psi_0| \right\} = \text{tr}^{(2)} \left\{ \sum_{\nu=0}^{D^2-1} |\Psi_\nu\rangle\langle\Psi_\nu| \right\}$$
$$= \text{tr}^{(2)} \left\{ I_{\tilde{\mathcal{H}}^{(1)} \otimes \tilde{\mathcal{H}}^{(2)}} \right\}$$
$$= D \cdot I_{\tilde{\mathcal{H}}^{(1)}}, \tag{11.201}$$

from which it follows by using the Schmidt decomposition of $|\Psi_0\rangle$,

$$\text{tr}^{(2)} \left\{ |\Psi_0\rangle\langle\Psi_0| \right\} = \sum_{i=1}^{D} |\alpha_i|^2 |\chi_i^{(1)}\rangle\langle\chi_i^{(1)}| = \frac{1}{D} I_{\tilde{\mathcal{H}}^{(1)}}. \tag{11.202}$$

This proves that $|\alpha_i| = 1/\sqrt{D}$ for all $i = 1, \ldots, D$, that is, that the non-degenerate eigenstate $|\Psi_0\rangle$ is maximally entangled. Equation (11.200) now tells us that this must be true also for the other eigenstates $|\Psi_\nu\rangle$. Thus we see that, in fact, all eigenstates of A in the subspace $\tilde{\mathcal{H}}^{(1)} \otimes \tilde{\mathcal{H}}^{(2)}$ have a Schmidt decomposition of the form

$$|\Psi_\nu\rangle = \sum_{i=1}^{D} \beta_{\nu,i} |\chi_{\nu,i}^{(1)}\rangle \otimes |\chi_{\nu,i}^{(2)}\rangle, \quad \text{where} \quad |\beta_{\nu,i}| = \frac{1}{\sqrt{D}}, \tag{11.203}$$

with appropriate local basis vectors $|\chi_{\nu,i}^{(1)}\rangle$ and $|\chi_{\nu,i}^{(2)}\rangle$.

As our final application of the causality principle we shall give a complete characterization of all non-degenerate, measurable observables for our previous example, namely for a non-local system of two spin-$\frac{1}{2}$ particles (Popescu and Vaidman, 1994). Thus we let A be a non-degenerate observable in the composite Hilbert space of these particles. Obviously, there are two possible cases: Either all eigenstates are direct products, or else there is at least one entangled eigenstate.

If all eigenstates are direct products one can easily verify that it is possible to choose appropriate local bases such that the eigenstates of A take on the following form,

$$|\Psi_1\rangle = |+z\rangle|+z'\rangle, \tag{11.204}$$
$$|\Psi_2\rangle = |-z\rangle|+z'\rangle, \tag{11.205}$$
$$|\Psi_3\rangle = |+z''\rangle|-z'\rangle, \tag{11.206}$$
$$|\Psi_4\rangle = |-z''\rangle|-z'\rangle, \tag{11.207}$$

where we denote by $|\pm n\rangle$ the eigenstate of the spin component along the direction n, with corresponding eigenvalue $\pm\frac{1}{2}$. The causality principle then leads to the conclusion that the direction z'' must be parallel or antiparallel to z, which means that the eigenstates of A can always be cast into the form,

$$|\Psi_1\rangle = |+z\rangle|+z'\rangle, \tag{11.208}$$
$$|\Psi_2\rangle = |-z\rangle|+z'\rangle, \tag{11.209}$$

$$|\Psi_3\rangle = |+z\rangle| - z'\rangle, \tag{11.210}$$
$$|\Psi_4\rangle = |-z\rangle| - z'\rangle. \tag{11.211}$$

For the proof of this statement we take the local observable $B^{(1)} = |-z\rangle\langle -z|$. Since $|\Psi_1\rangle$ is an eigenstate of A,

$$E[B^{(1)}|\Psi_1] = \langle \Psi_1|B^{(1)}|\Psi_1\rangle = 0. \tag{11.212}$$

For the state $|\Phi\rangle = |+z\rangle| - z'\rangle$ the causality principle yields

$$E[B^{(1)}|\Phi] = E[B^{(1)}|\Psi_1] = 0. \tag{11.213}$$

On the other hand, since A is assumed to be a measurable operator we must have

$$
\begin{aligned}
E[B^{(1)}|\Phi] &= \sum_{i=1}^{4} |\langle\Phi|\Psi_i\rangle|^2 \langle\Psi_i|B^{(1)}|\Psi_i\rangle \\
&= |\langle +z| + z''\rangle|^2 \cdot |\langle -z| + z''\rangle|^2 + |\langle +z| - z''\rangle|^2 \cdot |\langle -z| - z''\rangle|^2 \\
&= 0,
\end{aligned} \tag{11.214}
$$

showing that indeed $z'' = \pm z$, as claimed.

Let us now consider the case that at least one eigenstate is an entangled state. Our general considerations then reveal that all four eigenstates of A must be maximally entangled. By a suitable choice of local basis vectors one can therefore always cast these eigenstates into the following form,

$$|\Psi_1\rangle = \frac{1}{\sqrt{2}} \left(|+z\rangle| - z'\rangle - |-z\rangle| + z'\rangle \right), \tag{11.215}$$

$$|\Psi_2\rangle = \frac{1}{\sqrt{2}} \left(|+z\rangle| - z'\rangle + |-z\rangle| + z'\rangle \right), \tag{11.216}$$

$$|\Psi_3\rangle = \frac{1}{\sqrt{2}} \left(|+z\rangle| + z'\rangle - |-z\rangle| - z'\rangle \right), \tag{11.217}$$

$$|\Psi_4\rangle = \frac{1}{\sqrt{2}} \left(|+z\rangle| + z'\rangle + |-z\rangle| - z'\rangle \right). \tag{11.218}$$

A non-degenerate operator with eigenstates of this form is called a Bell-state operator. Thus we have shown that the set of all measurable, non-degenerate operators decomposes into two classes: Either all eigenstates are direct products of the form (11.208) or else all eigenstates are maximally entangled states of the form (11.215). What we have demonstrated is that a measurement of any non-degenerate operator which does not belong to one of these two classes would contradict causality. It remains to be shown, however, how those operators whose eigenstates *are* of the above form *can* be measured.

It is obvious that any non-degenerate operator with eigenstates of the form (11.208) can be measured. In fact, this is achieved by two separate measurements carried out locally in both parts of the system. We now demonstrate that

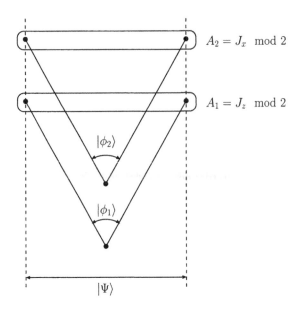

$A_2 = J_x \mod 2$

$A_1 = J_z \mod 2$

$|\phi_2\rangle$

$|\phi_1\rangle$

$|\Psi\rangle$

FIG. 11.13. A measurement device which enables the measurement of the Bell-s-tate operator with the non-degenerate eigenstates given by eqn (11.215). One uses two quantum probes by which the non-local quantities $A_1 = J_z \mod 2$ and $A_1 = J_x \mod 2$ are measured.

also any Bell-state operator, that is any non-degenerate operator A with eigenstates of the form (11.215) can be measured. This will be done with the help of the measurement of two non-local observables A_1 and A_2 (see Fig. 11.13). We construct these observables and verify that they represent operators which we already know to be measurable.

The A measurement must be a QND state verification of all eigenstates $|\Psi_i\rangle$. We denote by Π_0 the subspace spanned by $|\Psi_1\rangle$ and $|\Psi_2\rangle$, and by Π_1 the subspace spanned by $|\Psi_3\rangle$ and $|\Psi_4\rangle$. With the help of the first measurement of A_1 we want to find out whether the initial state belongs to Π_0 or to Π_1. To this end, consider the operator

$$J_z = \frac{1}{2}\left(\sigma_z^{(1)} + \sigma_{z'}^{(2)}\right). \tag{11.219}$$

In the subspace Π_0 it has the eigenvalue 0, whereas Π_1 is spanned by the eigenstates corresponding to the eigenvalues ± 1. Thus we have $J_z^2 = 0$ in Π_0, and $J_z^2 = 1$ in Π_1. The relation $J_z^2 = J_z \mod 2$ tells us that J_z^2 belongs to the class of measurable operators. Thus, our first measurement is the measurement of the quantity

$$A_1 = J_z \mod 2. \tag{11.220}$$

Through the second measurement we must be able to distinguish the states $|\Psi_{1,2}\rangle$ in Π_0 as well as the states $|\Psi_{3,4}\rangle$ in Π_1. To find an appropriate observable to achieve this we first note that

$$\sigma_x^{(1)}|\pm z\rangle = |\mp z\rangle, \quad \sigma_{x'}^{(2)}|\pm z'\rangle = |\mp z'\rangle, \tag{11.221}$$

which shows that the operator $\sigma_x^{(1)}\sigma_{x'}^{(2)}$ flips the spin of both particles. Now, $|\Psi_1\rangle$ and $|\Psi_3\rangle$ are odd, whereas $|\Psi_2\rangle$ and $|\Psi_4\rangle$ are even under a spin flip of both particles. Thus, a measurement of $\sigma_x^{(1)}\sigma_{x'}^{(2)}$ together with the previous measurement of A_1 enables us to verify all four eigenstates, which constitutes a QND measurement of the Bell-state operator A. To show that $\sigma_x^{(1)}\sigma_{x'}^{(2)}$ is indeed measurable we write it as follows,

$$\sigma_x^{(1)}\sigma_{x'}^{(2)} = 2J_x^2 - 1, \tag{11.222}$$

where

$$J_x = \frac{1}{2}\left(\sigma_x^{(1)} + \sigma_{x'}^{(2)}\right). \tag{11.223}$$

Thus, the measurement of $\sigma_x^{(1)}\sigma_{x'}^{(2)}$ is equivalent to the measurement of J_x^2. Again, we have $J_x^2 = J_x \bmod 2$ and, hence, the second measurement to be performed is that of the observable

$$A_2 = J_x \bmod 2. \tag{11.224}$$

Thus we have constructed explicitly a scheme which allows the measurement of the Bell-state operator A. It is clear that the device constitutes a QND measurement of A: Take any eigenspace of A_1 and any eigenspace of A_2. Then both eigenspaces are two-dimensional, orthogonal to each other, and have a common one-dimensional subspace which represents an eigenstate of A. It follows that all eigenstates of A are left unchanged by the measurement. This can also be seen by noting that the non-locally measured quantities commute, $[A_1, A_2] = 0$, as is easily verified.

11.3.6   *QND verification of non-local states*

In the previous section we have discussed the measurability of operators in the tensor product space $\mathcal{H}^{(1)} \otimes \mathcal{H}^{(2)}$. Let us turn to the QND verification of entangled states in this space, that is we now consider the verification of single state vectors, rather than the measurement of operators. We would like to answer the question of which type of entangled states allows a QND verification and how such measurements could be carried out. Again we invoke the causality principle which enables one to exclude the measurability of a large class of entangled states.

We consider the following general measurement scheme (see Fig. 11.14). The measurement consists of a sequence of N measurements of the (in general non-local) observables A_1, A_2, ..., A_N. Each A_n is measured with the help of an appropriate (possibly entangled) quantum probe as described in Section 11.3.1. We suppose that all A_n are of the form of those operators which were shown to

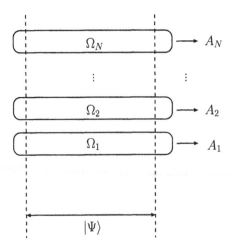

FIG. 11.14. General measurement device for a QND state verification measurement. The scheme involves the measurement of non-local quantities $A_1, A_2, \ldots, A_N$ described by the operations $\Omega_1(a_1), \Omega_2(a_2), \ldots, \Omega_N(a_N)$.

be measurable. Denoting the operation corresponding to the nth readout a_n by $\Omega_n(a_n)$ the total operation for the device takes the form

$$\Omega(a_1, \ldots, a_N) = \Omega_N(a_N)\Omega_{N-1}(a_{N-1}) \ldots \Omega_1(a_1), \qquad (11.225)$$

where the time-ordering is understood. As for any measurement we have

$$\sum_{a_1, \ldots, a_N} \Omega^\dagger(a_1, \ldots, a_N)\Omega(a_1, \ldots, a_N) = I, \qquad (11.226)$$

which follows immediately from the time-ordering and with the help of the property,

$$\sum_{a_n} \Omega_n^\dagger(a_n)\Omega_n(a_n) = I. \qquad (11.227)$$

For all measuring procedures considered here we require that the operation is a function of the measured quantity. This implies that all operations commute with their adjoint,

$$[\Omega_n(a_n), \Omega_n^\dagger(a_n)] = 0. \qquad (11.228)$$

This equation, in turn, yields that in addition to (11.226) we also have

$$\sum_{a_1, \ldots, a_N} \Omega(a_1, \ldots, a_N)\Omega^\dagger(a_1, \ldots, a_N) = I, \qquad (11.229)$$

as is easily verified. Physically, this means that an initial density matrix which is proportional to the identity is not changed by the measurement.

We now regard the measurement device as a state verification of some entangled state $|\Psi_0\rangle$. Introducing the corresponding Yes/No-operations we thus have, in addition to eqn (11.151),

$$\sum_a \Omega_{\text{Yes}}(a)\Omega_{\text{Yes}}^{\dagger}(a) + \sum_b \Omega_{\text{No}}(b)\Omega_{\text{No}}^{\dagger}(b) = I. \qquad (11.230)$$

We need one further condition. We will assume here that the operations pertaining to the measurement leave invariant the space $\tilde{\mathcal{H}}^{(1)} \otimes \tilde{\mathcal{H}}^{(2)}$, that is the space which is spanned by the basis vectors with a non-vanishing coefficient in the Schmidt decomposition. In physical terms this means that the quantities being measured do not excite new basis vectors, beside those already present in the Schmidt decomposition. In view of this condition we may restrict the whole discussion in the following to the subspace $\tilde{\mathcal{H}}^{(1)} \otimes \tilde{\mathcal{H}}^{(2)}$.

Our goal is to prove the following statement. For any entangled state $|\Psi_0\rangle$ which allows a QND state verification of the type considered above the causality principle requires that all coefficients α_i of its Schmidt decomposition (11.160) must necessarily be equal to each other, that is $|\alpha_i| = 1/\sqrt{D}$. This means that only maximally entangled states can be verified by such QND measurements without contradiction to causality. In addition we are going to demonstrate that and how these maximally entangled can be measured.

For the proof of the above statement we use the causality condition in the form (11.198). Explicitly, that condition can be written with the help of the Yes/No-operations as

$$\text{tr}^{(2)}\{|\Psi_0\rangle\langle\Psi_0|\} \qquad (11.231)$$

$$= \text{tr}^{(2)}\left\{\sum_a \Omega_{\text{Yes}}(a)|\Psi\rangle\langle\Psi|\Omega_{\text{Yes}}^{\dagger}(a) + \sum_b \Omega_{\text{No}}(b)|\Psi\rangle\langle\Psi|\Omega_{\text{No}}^{\dagger}(b)\right\}.$$

Our strategy is similar to the one used in the previous Section 11.3.5: We choose a set of orthonormal basis vectors $|\Psi_\nu\rangle$, $\nu = 0, 1, \ldots, D^2-1$, which span the space $\tilde{\mathcal{H}}^{(1)} \otimes \tilde{\mathcal{H}}^{(2)}$, insert these basis vectors into eqn (11.231), and sum over ν. The essential difference to the case of a QND operator measurement is that we cannot assume here that all basis vectors $|\Psi_\nu\rangle$ are unchanged through the measurement: In general, only $|\Psi_0\rangle$ does not change in the QND state verification. Thus, instead of eqn (11.201) we now have

$$D^2 \cdot \text{tr}^{(2)}\{|\Psi_0\rangle\langle\Psi_0|\} \qquad (11.232)$$

$$= \text{tr}^{(2)}\left\{\sum_a \Omega_{\text{Yes}}(a)I_{\tilde{\mathcal{H}}^{(1)}\otimes\tilde{\mathcal{H}}^{(2)}}\Omega_{\text{Yes}}^{\dagger}(a) + \sum_b \Omega_{\text{No}}(b)I_{\tilde{\mathcal{H}}^{(1)}\otimes\tilde{\mathcal{H}}^{(2)}}\Omega_{\text{No}}^{\dagger}(b)\right\}.$$

However, in view of eqn (11.230) and of our requirement that the operations leave invariant the space $\tilde{\mathcal{H}}^{(1)} \otimes \tilde{\mathcal{H}}^{(2)}$ we can conclude from the last equation,

$$D^2 \cdot \text{tr}^{(2)}\{|\Psi_0\rangle\langle\Psi_0|\} = \text{tr}^{(2)}\left\{I_{\tilde{\mathcal{H}}^{(1)}\otimes\tilde{\mathcal{H}}^{(2)}}\right\} = D \cdot I_{\tilde{\mathcal{H}}^{(1)}}, \qquad (11.233)$$

from which we obtain

$$\text{tr}^{(2)}\left\{|\Psi_0\rangle\langle\Psi_0|\right\} = \frac{1}{D} I_{\tilde{\mathcal{H}}^{(1)}}. \tag{11.234}$$

This proves our statement, namely that in order to be measurable without contradicting the causality condition, the state $|\Psi_0\rangle$ must be maximally entangled. Stated differently, after the QND state verification the local density matrices referring to the local variables in part (1) of the system must necessarily be proportional to the identity in the space $\tilde{\mathcal{H}}^{(1)}$. This clearly expresses once again the erasing of local information: After the state verification there is no trace of the initial state in the local mixtures, and the latter describe states of maximal entropy.

What has to be demonstrated finally is that the maximally entangled states can, at least in principle, be verified by some QND measurement. This will be done by an explicit construction of a measuring device (Aharonov, Albert and Vaidman, 1986). To this end, we first note that by an appropriate choice of the phases of the local basis vectors one can always put a maximally entangled state into the form

$$|\Psi_0\rangle = \frac{1}{\sqrt{D}} \sum_{i=1}^{D} |\chi_i^{(1)}\rangle \otimes |\chi_i^{(2)}\rangle. \tag{11.235}$$

A verification of this state can be carried out with the help of two successive, non-local measurements, similar to the device constructed to measure the Bell-state operator.

The first measurement is that of the observable

$$A = A^{(1)} + A^{(2)}, \tag{11.236}$$

where the local observables $A^{(1)}$ and $A^{(2)}$ are defined through

$$A^{(1)}|\chi_i^{(1)}\rangle = -i|\chi_i^{(1)}\rangle, \tag{11.237}$$

$$A^{(2)}|\chi_i^{(2)}\rangle = +i|\chi_i^{(2)}\rangle, \tag{11.238}$$

with $i = 1, 2, \ldots, D$. Obviously, $|\Psi_0\rangle$ is an eigenstate of A with eigenvalue 0. Moreover, the vectors $|\Psi\rangle$ belonging to the D-dimensional eigenspace $A = 0$ have the general form

$$|\Psi\rangle = \sum_{i=1}^{D} \beta_i |\chi_i^{(1)}\rangle \otimes |\chi_i^{(2)}\rangle. \tag{11.239}$$

This shows that by the outcome $A = 0$ we verify that the state has a Schmidt decomposition in the given local basis vectors $|\chi_i^{(1)}\rangle$, $|\chi_i^{(2)}\rangle$.

By the second measurement we want to verify that the coefficients β_i in (11.239) are all equal to each other. To find an appropriate observable we consider the local unitary operators $U^{(1)}$, $U^{(2)}$ defined through

$$U^{(1)}|\chi_i^{(1)}\rangle = |\chi_{i+1}^{(1)}\rangle, \quad i = 1, \ldots, D-1, \tag{11.240}$$

$$U^{(1)}|\chi_D^{(1)}\rangle = |\chi_1^{(1)}\rangle, \tag{11.241}$$

$$U^{(2)}|\chi_i^{(2)}\rangle = |\chi_{i+1}^{(2)}\rangle, \quad i = 1, \ldots, D-1, \tag{11.242}$$

$$U^{(2)}|\chi_D^{(2)}\rangle = |\chi_1^{(2)}\rangle. \tag{11.243}$$

These operators induce a cyclic shift of the index i. Introducing the unitary operator $U = U^{(1)}U^{(2)}$ we find that for any state (11.239) in the space $A = 0$,

$$U|\Psi\rangle = \sum_{i=1}^{D} \beta_{i-1}|\chi_i^{(1)}\rangle \otimes |\chi_i^{(2)}\rangle, \tag{11.244}$$

where $\beta_0 \equiv \beta_D$. Thus we see that $U|\Psi\rangle = |\Psi\rangle$ if and only if all β_i are equal to each other. It follows that $|\Psi_0\rangle$ is the only eigenstate of U corresponding to the eigenvalue 1 in the space $A = 0$.

Since $U^{(1)}$, $U^{(2)}$ are local unitary operators we can introduce local observables $B^{(1)}$, $B^{(2)}$ by means of

$$U^{(1)} = \exp\left\{iB^{(1)}\right\}, \quad U^{(2)} = \exp\left\{iB^{(2)}\right\}, \tag{11.245}$$

such that, since $B^{(1)}$ and $B^{(2)}$ commute,

$$U = \exp\left\{i\left(B^{(1)} + B^{(2)}\right)\right\}. \tag{11.246}$$

From the above property of U we deduce that $|\Psi_0\rangle$ is the only eigenstate of the operator

$$B = \left(B^{(1)} + B^{(2)}\right) \bmod 2\pi \tag{11.247}$$

belonging to the eigenvalue 0 in the space $A = 0$. As was demonstrated in Section 11.3.2 such modular sums are measurable.

This completes the construction of the measuring device: The state $|\Psi_0\rangle$ is verified by the outcomes $A = 0$ and $B = 0$ of two non-local measurements. The initial state $|\Psi_0\rangle$ is obviously left unchanged by the measurement. Moreover, since A and B commute it is easy to show that the device constitutes a QND state verification measurement.

11.3.7 *Preparation of non-local states*

We have seen above that causality imposes strong restrictions on the measurability of non-local observables and states. In particular, we have found that a normalized, entangled and non-local state with a Schmidt decomposition of the form

$$|\Psi\rangle = \sum_{i=1}^{D} \alpha_i |\chi_i^{(1)}\rangle \otimes |\chi_i^{(2)}\rangle \tag{11.248}$$

is not measurable by our QND state verification device, unless all $|\alpha_i|$ are equal to each other, that is unless the state is maximally entangled. However, it is important to observe that all states of the form (11.248) can be *prepared* by an appropriate, non-local measuring device without contradiction to causality.

A device for the preparation of the states (11.248) can be designed as follows (Aharonov, Albert and Vaidman, 1986). First we prepare locally the (normalized) states

$$|\psi^{(1)}\rangle = \sum_{i=1}^{D} \alpha_i |\chi_i^{(1)}\rangle, \tag{11.249}$$

$$|\psi^{(2)}\rangle = \frac{1}{\sqrt{D}} \sum_{i=1}^{D} |\chi_i^{(2)}\rangle, \tag{11.250}$$

which are then combined to form the initial state

$$|\Phi\rangle = |\psi^{(1)}\rangle \otimes |\psi^{(2)}\rangle = \frac{1}{\sqrt{D}} \sum_{i,j=1}^{D} \alpha_i |\chi_i^{(1)}\rangle \otimes |\chi_j^{(2)}\rangle \tag{11.251}$$

in the tensor product space $\mathcal{H}^{(1)} \otimes \mathcal{H}^{(2)}$. Both parts (1) and (2) of the combined system may be separated by a spacelike distance.

In the second step we now measure the non-local observable $A = A^{(1)} + A^{(2)}$, where the local quantities $A^{(1)}$ and $A^{(2)}$ are given through eqn (11.237). The outcome $A = 0$ of that measurement projects the initial state $|\Phi\rangle$ onto the state $|\Psi\rangle$ whose Schmidt decomposition is of the desired form in the given local bases $|\chi_i^{(1)}\rangle$ and $|\chi_i^{(2)}\rangle$, namely we have

$$|\Phi\rangle \longrightarrow |\Psi\rangle = \sum_{i=1}^{D} \alpha_i |\chi_i^{(1)}\rangle \otimes |\chi_i^{(2)}\rangle. \tag{11.252}$$

Of course, there is no conflict with the causality principle since the measurement of A involves only local interactions. However, the preparation of $|\Psi\rangle$ is successful only with a certain probability which is given by

$$\mathcal{P}(A = 0) = |\langle \Psi | \Phi \rangle|^2 = \frac{1}{D} \sum_{i=1}^{D} |\alpha_i|^2 = \frac{1}{D}. \tag{11.253}$$

The important conclusion to be drawn from the above considerations is that for any measurement device one has to distinguish carefully between the states that are *measurable* and those states that can be *prepared* by the device. It is this possibility of preparation which justifies regarding the vectors (11.248) in the tensor product space really as *states* of the combined system.

11.3.8 *Exchange measurements*

At several places we have derived contradictions to the causality principle by means of the assumption that a QND state verification measurement leaves invariant the state $|\Psi_0\rangle$ which is to be verified, as well as its orthogonal complement. It may be seen from the formulation of the causality condition expressed by eqn (11.166) that no contradiction will arise if one performs a demolition measurement such that in all cases the object system will end up in one and the same state. Denoting the latter by $|\Psi_f\rangle$ and assuming for simplicity that we have only a single Yes-operation we have the following Yes/No-operations for the device,

$$\Omega_{\text{Yes}} = |\Psi_f\rangle\langle\Psi_0|, \tag{11.254}$$

$$\Omega_{\text{No}}(b) = |\Psi_f\rangle\langle\Psi_\perp(b)|. \tag{11.255}$$

The states $|\Psi_\perp(b)\rangle$ are orthogonal to $|\Psi_0\rangle$ and may, of course, depend on b, the only necessary restriction being the normalization condition

$$F_{\text{Yes}} + F_{\text{No}} = \Omega^\dagger_{\text{Yes}}\Omega_{\text{Yes}} + \sum_b \Omega^\dagger_{\text{No}}(b)\Omega_{\text{No}}(b)$$

$$= |\Psi_0\rangle\langle\Psi_0| + \sum_b |\Psi_\perp(b)\rangle\langle\Psi_\perp(b)| = I. \tag{11.256}$$

The above Yes/No-operations satisfy all conditions of a state verification measurement. The final state $|\Psi_f\rangle$ may or may not be equal to $|\Psi_0\rangle$. For example, it may be given by the vacuum state $|0\rangle$ of the electromagnetic field. This case occurs, e.g. if the device measures the quanta of the field which are annihilated after registration.

An important example for such demolition measurements are so-called exchange measurements. We define an exchange measurement as a device involving local interactions with each part of the system. It is assumed that the Hilbert space $\mathcal{H}^{(k)}$ of the local part (k) of the object system and the Hilbert space $\mathcal{H}_P^{(k)}$ of the corresponding part of the probe are isomorphic. Introducing local basis vectors $|\chi_i^{(k)}\rangle$ and $|\phi_i^{(k)}\rangle$ in $\mathcal{H}^{(k)}$ and $\mathcal{H}_P^{(k)}$, respectively, the local object–probe interaction in part (k) is supposed to take the form

$$|\chi_i^{(k)}\rangle \otimes |\phi_j^{(k)}\rangle \longrightarrow |\chi_j^{(k)}\rangle \otimes |\phi_i^{(k)}\rangle. \tag{11.257}$$

The interaction thus acts simply by exchanging the labels of the basis vectors of object and probe.

It is clear that such an interaction if applied to any non-local state of the quantum object leads to a final object state which is isomorphic to the initial probe state and vice versa. For example, taking the initial object state

$$|\Psi_0\rangle = \sum_i \alpha_i |\chi_i^{(1)}\rangle \otimes |\chi_i^{(2)}\rangle \tag{11.258}$$

and the initial probe state

$$|\Phi_0\rangle = \sum_j \beta_j |\phi_j^{(1)}\rangle \otimes |\phi_j^{(2)}\rangle \tag{11.259}$$

we find as a result of the interactions in both parts (1) and (2) of the total system

$$|\Psi_0\rangle \otimes |\Phi_0\rangle \longrightarrow \left(\sum_j \beta_j |\chi_j^{(1)}\rangle \otimes |\chi_j^{(2)}\rangle \right) \otimes \left(\sum_i \alpha_i |\phi_i^{(1)}\rangle \otimes |\phi_i^{(2)}\rangle \right). \quad (11.260)$$

This shows that the final state $|\Psi_f\rangle$ of the object system is isomorphic to the initial probe state $|\Phi_0\rangle$, namely

$$|\Psi_f\rangle = \sum_j \beta_j |\chi_j^{(1)}\rangle \otimes |\chi_j^{(2)}\rangle, \quad (11.261)$$

whereas the final probe state is isomorphic to the initial object state. Thus the rôles of object and probe state have been exchanged as a result of the local interactions in both parts of the system.

It must be noted, however, that with the local exchange interactions the measurement is not yet complete: After the local interactions the various parts of the probe state must be brought together to one place, at which a local measurement on the probe state can be carried out. Any non-local measurement on the probe state is, of course, subjected to the same causality restrictions as discussed in the previous subsections.

11.4   Quantum teleportation

Non-local entangled states describe quantum correlations which are expressed through the joint probability for measurements performed on separated parts of a combined quantum system. An interesting application is to employ the properties of entangled states for the coherent transfer of an unknown quantum state from one part of a system to another, spacelike separated part. We shall study here this transfer of a quantum state, known as quantum teleportation (Bennett *et al.*, 1993; Vaidman, 1994).

11.4.1   *Coherent transfer of quantum states*

Let us first define precisely what is meant by the teleportation of a quantum state (see Fig. 11.15). We consider a system which is composed of three Hilbert spaces $\mathcal{H}^{(1)}$, $\mathcal{H}^{(2)}$ and $\mathcal{H}^{(3)}$. In the following these spaces are supposed to be isomorphic, that is to have the same dimension D. The total space of the composite system is the threefold tensor product

$$\mathcal{H} = \mathcal{H}^{(1)} \otimes \mathcal{H}^{(2)} \otimes \mathcal{H}^{(3)}. \quad (11.262)$$

The spaces $\mathcal{H}^{(1)}$ and $\mathcal{H}^{(2)}$ describe the degrees of freedom of some localized part of the total system which will be referred to as part (A). The third space $\mathcal{H}^{(3)}$ represents a second localized part of the system, denoted by part (B), which is separated from part (A) by a spacelike distance. We may associate a local observer, called Alice, with part (A), and a local observer, called Bob, with

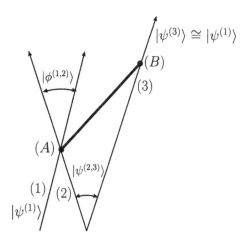

FIG. 11.15. Schematic space-time picture of quantum teleportation. The incoming state $|\psi^{(1)}\rangle$ in the Hilbert space of Alice (A) is transferred to an isomorphic copy $|\psi^{(3)}\rangle$ of the Hilbert space of Bob (B). The quantum channel is provided by a maximally entangled state $|\psi^{(2,3)}\rangle$. The bold line represents a classical communication channel through which Alice communicates her measurement result to Bob. If Alice carries out a QND measurement she is left with an entangled state $|\phi^{(1,2)}\rangle$ which is isomorphic to $|\psi^{(2,3)}\rangle$, provided her measurement has been successful.

part (B). Thus, Alice can measure all local observables pertaining to the space $\mathcal{H}^{(1)} \otimes \mathcal{H}^{(2)}$, whereas the local observables in the space $\mathcal{H}^{(3)}$ are available to Bob.

Consider now some state

$$|\psi^{(1)}\rangle = \sum_{i=0}^{D-1} \alpha_i |\chi_i^{(1)}\rangle \qquad (11.263)$$

belonging to the first Hilbert space $\mathcal{H}^{(1)}$.[15] Alice aims to transfer this state instantaneously and coherently to Bob, without measuring or destroying it: The absolute values as well as the phase relations of the amplitudes α_i are left unchanged after the transfer process.

Such a transfer process can in fact be achieved with the help of an entangled state $|\psi^{(2,3)}\rangle$ in the space $\mathcal{H}^{(2)} \otimes \mathcal{H}^{(3)}$. To be definite we take this state to be a certain maximally entangled state,

$$|\psi^{(2,3)}\rangle = \frac{1}{\sqrt{D}} \sum_{j=0}^{D-1} |\chi_j^{(2)}\rangle \otimes |\chi_j^{(3)}\rangle. \qquad (11.264)$$

---

[15]Note that in contrast to our previous convention the index i which labels the basis vectors now runs from 0 to $D-1$.

The state $|\psi^{(2,3)}\rangle$ can be prepared in some localized space-time region. It is then separated into two parts, described by the Hilbert spaces $\mathcal{H}^{(2)}$ and $\mathcal{H}^{(3)}$, which are brought to Alice and Bob, respectively. Thus, the quantum correlations embodied in $|\psi^{(2,3)}\rangle$ may be detected through joint measurements carried out by Alice and Bob. In the following the state $|\psi^{(2,3)}\rangle$ will be referred to as a quantum channel.

The initial state of the combined system is now given by

$$|\Psi_0\rangle = |\psi^{(1)}\rangle \otimes |\psi^{(2,3)}\rangle = \frac{1}{\sqrt{D}} \sum_{i,j=0}^{D-1} \alpha_i |\chi_i^{(1)}\rangle \otimes |\chi_j^{(2)}\rangle \otimes |\chi_j^{(3)}\rangle. \qquad (11.265)$$

Suppose then that Alice performs a certain verification measurement of the state

$$|\phi^{(1,2)}\rangle = \frac{1}{\sqrt{D}} \sum_{k=0}^{D-1} |\chi_k^{(1)}\rangle \otimes |\chi_k^{(2)}\rangle, \qquad (11.266)$$

which belongs to the Hilbert space available to Alice. If the state verification yields a positive result the initial state $|\Psi_0\rangle$ of the total system is projected onto the state $|\phi^{(1,2)}\rangle$. As is easily deduced from the above equations we have

$$\langle \phi^{(1,2)} | \Psi_0 \rangle = \frac{1}{D} \sum_{i=0}^{D-1} \alpha_i |\chi_i^{(3)}\rangle \equiv \frac{1}{D} |\psi^{(3)}\rangle. \qquad (11.267)$$

This shows that Alice finds a positive result with a probability which is given by

$$\mathcal{P} = ||\langle \phi^{(1,2)} | \Psi_0 \rangle||^2 = \frac{1}{D^2}. \qquad (11.268)$$

Conditioned on this result the state prior to the measurement is transformed into the final state after the measurement as follows,

$$|\Psi_0\rangle = |\psi^{(1)}\rangle \otimes |\psi^{(2,3)}\rangle \longrightarrow |\phi^{(1,2)}\rangle \otimes |\psi^{(3)}\rangle. \qquad (11.269)$$

Thus we see that the state reduction conditioned on a positive outcome for the measurement of Alice leads to the desired teleportation process: The final state $|\psi^{(3)}\rangle$ in Bob's Hilbert space $\mathcal{H}^{(3)}$ given by eqn (11.267) represents a copy of Alice's initial state $|\psi^{(1)}\rangle$ (eqn (11.263)) in $\mathcal{H}^{(1)}$.

The following facts should be noted. Being induced by a state reduction following the measurement in one part of the system, the teleportation process (11.269) can be viewed as occurring instantaneously in all Lorentz frames. Namely, as we have already discussed it can be regarded as taking place along all spacelike hypersurfaces which cross the classical event produced by Alice's measurement.

Since the teleportation of a state involves only local interactions in one part of the system there is, of course, no conflict with causality and no transfer of information with a speed greater than the speed of light. Indeed, the reduced density matrix which refers to Bob's Hilbert space $\mathcal{H}^{(3)}$ is proportional to the unit matrix in this space, which is due to the fact that the quantum channel (11.264) represents a maximally entangled state. Since the operations describing Alice's measurement are local and act only in $\mathcal{H}^{(1)} \otimes \mathcal{H}^{(2)}$, Bob's reduced density matrix is not affected by Alice's measurement, if the latter is described on the non-selective level. Bob's density matrix which describes a state of maximal ignorance is thus unchanged during Alice's measurement.

However, if the result of Alice's measurement is communicated via a classical channel to Bob, he can use that information to conclude that his state $|\psi^{(3)}\rangle$ is isomorphic to $|\psi^{(1)}\rangle$, provided the result of Alice's measurement was positive. Note that in this case neither Alice nor Bob have measured the transferred state. After her measurement Alice's state $|\phi^{(1,2)}\rangle$ is an entangled state which is isomorphic to the quantum channel $|\psi^{(2,3)}\rangle$. The teleportation of the given state works although it remains completely unknown to both Alice and Bob.

11.4.2 Teleportation and Bell-state measurement

If the teleportation process is carried out on an ensemble of initial states in the manner described above it is, in general, successful only for a fraction $1/D^2$ of the ensemble. The reason for this fact is that a negative outcome of Alice's state verification destroys, as a rule, the initial state $|\psi^{(1)}\rangle$. However, Alice can, at least in principle, always design a device which yields a teleportation rate of 100%. For this purpose Alice performs a measurement which projects the initial state onto an orthogonal basis which consists of maximally entangled states, that is Alice performs a measurement of a Bell-state operator in $\mathcal{H}^{(1)} \otimes \mathcal{H}^{(2)}$.

To explain this point in more detail we first note that an orthogonal basis of maximally entangled states in $\mathcal{H}^{(1)} \otimes \mathcal{H}^{(2)}$ is provided by the following expression,

$$|\phi_\nu^{(1,2)}\rangle = \sum_{k=0}^{D-1} \beta_{r,k} |\chi_{(k+n) \bmod D}^{(1)}\rangle \otimes |\chi_k^{(2)}\rangle. \tag{11.270}$$

The index $\nu = 0, 1, \ldots, D^2 - 1$ labels the basis vectors $|\phi_\nu^{(1,2)}\rangle$. Each ν has a unique representation of the form

$$\nu = r + nD, \tag{11.271}$$

where

$$r = \nu \bmod D = 0, 1, \ldots, D - 1, \tag{11.272}$$

and $n = 0, 1, \ldots, D - 1$. The amplitudes $\beta_{r,k}$ in eqn (11.270) are defined to be

$$\beta_{r,k} = \frac{1}{\sqrt{D}} \exp\left(-2\pi i \frac{kr}{D}\right). \tag{11.273}$$

Let us demonstrate that eqn (11.270) yields, in fact, a basis of the desired form. First, it is immediately clear that the $|\phi_\nu^{(1,2)}\rangle$ are maximally entangled.

Namely, by an appropriate relabelling of the basis vectors in $\mathcal{H}^{(1)}$, corresponding to the first factor in the tensor product, we can always achieve the condition that (11.270) takes on the standard form of the Schmidt decomposition. Since all $\beta_{r,k}$ are equal in absolute value,

$$|\beta_{r,k}| = \frac{1}{\sqrt{D}}, \tag{11.274}$$

we find that all $|\phi_\nu^{(1,2)}\rangle$ are maximally entangled.

It therefore remains to be shown that the $|\phi_\nu^{(1,2)}\rangle$ are orthogonal and normalized. To this end, one first verifies

$$\sum_{k=0}^{D-1} \beta_{r',k}^* \beta_{r,k} = \delta_{r'r}, \tag{11.275}$$

which is easily done with the help of the summation formula for the geometric series. Setting $\nu = r + nD$, and $\nu' = r' + n'D$ we find for the scalar product,

$$\langle \phi_{\nu'}^{(1,2)} | \phi_\nu^{(1,2)} \rangle = \sum_{k,l=0}^{D-1} \beta_{r',k}^* \beta_{r,l} \langle \chi_{(k+n') \bmod D}^{(1)} | \chi_{(l+n) \bmod D}^{(1)} \rangle \langle \chi_k^{(2)} | \chi_l^{(2)} \rangle$$

$$= \sum_{k=0}^{D-1} \beta_{r',k}^* \beta_{r,k} \langle \chi_{(k+n') \bmod D}^{(1)} | \chi_{(k+n) \bmod D}^{(1)} \rangle$$

$$= \sum_{k=0}^{D-1} \beta_{r',k}^* \beta_{r,k} \delta_{n'n} = \delta_{r'r} \delta_{n'n}. \tag{11.276}$$

In the second and the third step we have used the orthogonality of the basis vectors $|\chi_k^{(1)}\rangle$ and $|\chi_k^{(2)}\rangle$, whereas in the last step eqn (11.275) has been used. Thus we finally obtain the desired orthogonality of the basis vectors

$$\langle \phi_{\nu'}^{(1,2)} | \phi_\nu^{(1,2)} \rangle = \delta_{\nu'\nu}, \tag{11.277}$$

which concludes the proof. Applying eqn (11.270) to the special case $D = 2$ (spin-$\frac{1}{2}$ particles) one finds the Bell states given by eqn (11.215).

Returning to quantum teleportation we now decompose the total initial state (11.265) of the device as follows,

$$|\Psi_0\rangle = \sum_{\nu=0}^{D^2-1} |\phi_\nu^{(1,2)}\rangle \langle \phi_\nu^{(1,2)} | \Psi_0 \rangle. \tag{11.278}$$

From eqns (11.265) and (11.270) we get

$$\langle \phi_\nu^{(1,2)} | \Psi_0 \rangle = \frac{1}{D} \sum_{i=0}^{D-1} \alpha_i \sqrt{D} \beta_{r,(i-n) \bmod D}^* |\chi_{(i-n) \bmod D}^{(3)}\rangle. \tag{11.279}$$

For each ν we introduce a corresponding unitary operator $U_\nu^{(3)}$ which acts in Bob's Hilbert space $\mathcal{H}^{(3)}$. We define these operators by their action on Bob's local basis,

$$U_\nu^{(3)}|\chi_i^{(3)}\rangle = \sqrt{D}\beta_{r,(i-n)\ \mathrm{mod}\ D}^*|\chi_{(i-n)\ \mathrm{mod}\ D}^{(3)}\rangle, \quad i = 0, 1, \ldots, D-1. \quad (11.280)$$

By virtue of eqn (11.274) these operators are indeed unitary. In addition to the multiplication by phase factors, they simply induce a cyclic permutation of the basis states $|\chi_i^{(3)}\rangle$.

This enables us to rewrite eqn (11.278) as follows,

$$|\Psi_0\rangle = |\psi^{(1)}\rangle \otimes |\psi^{(2,3)}\rangle = \frac{1}{D}\sum_{\nu=0}^{D^2-1} |\phi_\nu^{(1,2)}\rangle \otimes U_\nu^{(3)}|\psi^{(3)}\rangle, \quad (11.281)$$

where $|\psi^{(3)}\rangle$ is given by eqn (11.267). From this equation we read off that a teleportation rate of 100% is indeed possible if Alice carries out a measurement of the complete basis $|\phi_\nu^{(1,2)}\rangle$. All D^2 possible outcomes of that measurement occur with probability $1/D^2$. Receiving the outcome ν of Alice's measurement via the classical communication channel, Bob can then apply the inverse of the corresponding unitary operator $U_\nu^{(3)}$ to his state which yields in all cases an isomorphic copy $|\psi^{(3)}\rangle$ of Alice's initial state $|\psi^{(1)}\rangle$.

As an example, let us apply the above construction to the case $D = 2$, that is, to the spin-$\frac{1}{2}$ example studied previously. We then find that Bob has to apply one of the following unitary operators

$$U_0^{(3)} = I, \quad U_1^{(3)} = \sigma_z, \quad U_2^{(3)} = \sigma_x, \quad U_3^{(3)} = i\sigma_y, \quad (11.282)$$

depending on the outcome of Alice's measurement.

11.4.3 Experimental realization

We conclude our discussion with the discussion of an experimental realization of quantum teleportation performed by Bouwmeester et al. (1997). The experimental setup is sketched in Fig. 11.16. It works with three electromagnetic field modes b_1, b_2 and b_3. We write $b_{i\lambda}^\dagger$ and $b_{i\lambda}$ for the corresponding creation and annihilation operators, where $i = 1, 2, 3$ and $\lambda = \leftrightarrow, \updownarrow$ denotes the photon polarization. The quantum channel is provided by an EPR source which generates the entangled two-photon state

$$|\psi^{(2,3)}\rangle = \frac{1}{\sqrt{2}}\left(b_{2\leftrightarrow}^\dagger b_{3\updownarrow}^\dagger - b_{2\updownarrow}^\dagger b_{3\leftrightarrow}^\dagger\right)|0\rangle \quad (11.283)$$

by parametric down-conversion (Michler et al., 1996). A careful design of the device guarantees that all three modes contain a single photon. The total Hilbert space is then isomorphic to the tensor product of the Hilbert spaces corresponding to the three modes.

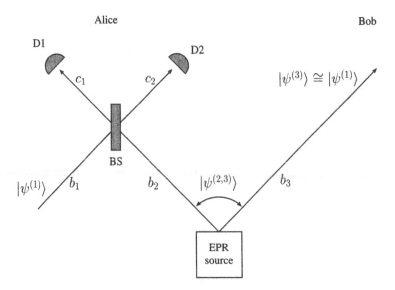

FIG. 11.16. Sketch of the setup used for an experimental realization of quantum teleportation. Alice's state analysis is realized by the superposition of the two field modes b_1 and b_2 at a beam splitter BS and the subsequent measurement of the outgoing photons in modes c_1, c_2 through detectors D1 and D2. The successful teleportation to the field mode b_3 has been demonstrated by a state analysis of that mode (not shown).

The state to be transferred is given by

$$|\psi^{(1)}\rangle = \left(\alpha b_{1\leftrightarrow}^\dagger + \beta b_{1\updownarrow}^\dagger\right)|0\rangle, \quad |\alpha|^2 + |\beta|^2 = 1, \qquad (11.284)$$

such that we have the following three-photon initial state,

$$|\Psi_0\rangle = \frac{1}{\sqrt{2}}\left(\alpha b_{1\leftrightarrow}^\dagger + \beta b_{1\updownarrow}^\dagger\right)\left(b_{2\leftrightarrow}^\dagger b_{3\updownarrow}^\dagger - b_{2\updownarrow}^\dagger b_{3\leftrightarrow}^\dagger\right)|0\rangle. \qquad (11.285)$$

Alice performs a state verification measurement on her photons in modes b_1 and b_2 using a beam splitter of transmittivity $\frac{1}{2}$ (see the discussion on homodyne photodetection in Section 6.4). She measures with the help of her detectors D1 and D2 the quanta annihilated by the operators $c_{1\lambda}$ and $c_{2\lambda}$. With a suitable choice of the phases for these operators we have

$$c_{1\lambda} = \frac{1}{\sqrt{2}}\left(b_{1\lambda} + b_{2\lambda}\right), \qquad (11.286)$$

$$c_{2\lambda} = \frac{1}{\sqrt{2}}\left(b_{1\lambda} - b_{2\lambda}\right). \qquad (11.287)$$

A crucial point of the experiment is to ensure that the photons in modes b_1 and b_2 arrive simultaneously at the beam splitter, since otherwise they could be distinguished through their arrival times. A more refined theoretical analysis (Braunstein and Mann, 1995) is also required which takes into account the space-time structure of the states in the various modes. We neglect such considerations since they do not modify the principal statements.

In order to find the possible coincidence events measured through detectors D1 and D2 we solve relations (11.286) and (11.287) for $b_{1\lambda}$ and $b_{2\lambda}$ and substitute into eqn (11.285), which leads to

$$
\begin{aligned}
|\Psi_0\rangle = & + \frac{1}{\sqrt{8}} \left(c_{1\leftrightarrow}^\dagger c_{2\uparrow}^\dagger - c_{1\uparrow}^\dagger c_{2\leftrightarrow}^\dagger \right) \left(\alpha b_{3\leftrightarrow}^\dagger + \beta b_{3\uparrow}^\dagger \right) |0\rangle \\
& - \frac{1}{\sqrt{8}} \left(c_{1\uparrow}^\dagger c_{1\leftrightarrow}^\dagger - c_{2\uparrow}^\dagger c_{2\leftrightarrow}^\dagger \right) \left(\alpha b_{3\leftrightarrow}^\dagger - \beta b_{3\uparrow}^\dagger \right) |0\rangle \\
& - \frac{1}{\sqrt{8}} \left(\left[c_{1\uparrow}^\dagger \right]^2 - \left[c_{2\uparrow}^\dagger \right]^2 \right) \beta b_{3\leftrightarrow}^\dagger |0\rangle \\
& + \frac{1}{\sqrt{8}} \left(\left[c_{1\leftrightarrow}^\dagger \right]^2 - \left[c_{2\leftrightarrow}^\dagger \right]^2 \right) \alpha b_{3\uparrow}^\dagger |0\rangle .
\end{aligned}
\tag{11.288}
$$

From this equation we infer that we may distinguish four different cases of coincidence events. In case 1, which is described by the first term on the right-hand side of eqn (11.288), both detectors register a single photon. Note that in this case the polarizations of the photons are necessarily different from each other, since terms proportional to $c_{1\leftrightarrow}^\dagger c_{2\leftrightarrow}^\dagger$ or $c_{1\uparrow}^\dagger c_{2\uparrow}^\dagger$ do not appear in the above decomposition. This is an interference effect: Due to the beam splitter there are two different paths which contribute to the event of registering a single photon in both detectors. The amplitudes for these paths interfere destructively such that only that component of the initial two-photon state contributes which contains different polarizations in the modes b_1, b_2. It follows from the first line that case 1 occurs with a probability of $\frac{1}{4}$ and leads to the following state of mode b_3,

$$
|\psi^{(3)}\rangle = \left(\alpha b_{3\leftrightarrow}^\dagger + \beta b_{3\uparrow}^\dagger \right) |0\rangle ,
\tag{11.289}
$$

which represents a copy of the state $|\psi^{(1)}\rangle$ given by eqn (11.284). This case is investigated in the Zeilinger experiment (Bouwmeester *et al.*, 1997) where the teleportation process is demonstrated with the help of an additional state analysis carried out on mode b_3.

Case 2, which also occurs with probability $\frac{1}{4}$ is described by the second line of eqn (11.288). It consists of those events where two photons are detected by either D1 or by D2 with different polarizations. Case 3 (third line) and case 4 (fourth line) represent those events where either D1 or D2 registers two photons with both polarizations equal to $\uparrow$ or both equal to $\leftrightarrow$, respectively. The total probability for case 3 is equal to $|\beta|^2/2$ and equal to $|\alpha|^2/2$ for case 4.

In case 2 Bob gets the final state $(\alpha b_{3\leftrightarrow}^\dagger - \beta b_{3\uparrow}^\dagger)|0\rangle$ in mode b_3. By a simple application of σ_z he again obtains a copy of $|\psi^{(1)}\rangle$. Thus, if Alice could distinguish faithfully the cases 1 and 2 from the cases 3 and 4, a teleportation rate of 50% could be achieved in principle with the above scheme. This is the upper limit since in cases 3 and 4 the (normalized) final state of mode b_3 is either $b_{3\leftrightarrow}^\dagger|0\rangle$ or $b_{3\updownarrow}^\dagger|0\rangle$ such that the phase relations of Alice's initial state $|\psi^{(1)}\rangle$ are irreversibly lost.

Quantum teleportation offers completely new perspectives for experiments on the foundations of quantum mechanics. It could be used, for example, to transmit quantum states over large distances without destruction by decoherence effects. Note also that the teleported state itself could be part of an entangled state in some larger Hilbert space. It is thus possible to transfer the entanglement between particles and to carry out tests of the Bell inequalities (Bell, 1964; Clauser and Shimony, 1978) on particles which did not directly interact in the past.

In addition to new experiments on the foundations of quantum mechanics several further applications of teleportation devices in quantum information processing and communication have been suggested (Bouwmeester *et al.*, 1997). For example, they could be used to store the information carried by photons on trapped ions, or to teleport a quantum state over large distances even if the available quantum channels are of very poor quality, employing entanglement purification. Moreover, quantum teleportation schemes could also serve as links between quantum computers or to protect states in quantum computers from their environment.

We remark finally that the teleportation scheme studied above provides a nice example of the concept of states as functionals on the set of spacelike hypersurfaces, which was introduced in Section 11.1.1. Let us denote by $x^{(1)}$ and $x^{(2)}$ the space-time points corresponding to the photodetection events at detectors D1 and D2, respectively. Above we have tacitly assumed that these events occur simultaneously in Alice's coordinate frame. However, in principle D1 and D2 could be arbitrarily far apart and we may consider some spacelike hypersurface σ which crosses the forward light cone based at $x^{(1)}$ and the backward light cone based at $x^{(2)}$. If σ is flat, it may therefore be viewed as an equal-time hypersurface in the coordinate frame of a moving observer O for whom the detection at D1 takes place earlier than that at D2. According to our discussion in Sections 11.2.1 and 11.2.3 we find that observer O associates a state to the hypersurface σ which is *not* related by any unitary representation of the Lorentz group to any state in the foliation of Alice. In fact, according to eqn (11.288) we have on σ a state vector which is conditioned on the outcome of the first measurement at D1 and which is given by one of six possible states, namely two two-photon states,

$$c_{2\updownarrow}^\dagger \left(\alpha b_{3\leftrightarrow}^\dagger + \beta b_{3\updownarrow}^\dagger \right)|0\rangle, \quad c_{2\leftrightarrow}^\dagger \left(\alpha b_{3\leftrightarrow}^\dagger + \beta b_{3\updownarrow}^\dagger \right)|0\rangle, \tag{11.290}$$

three one-photon states,

$$b_{3\updownarrow}^\dagger|0\rangle, \quad b_{3\leftrightarrow}^\dagger|0\rangle, \quad \left(\alpha b_{3\leftrightarrow}^\dagger - \beta b_{3\updownarrow}^\dagger \right)|0\rangle, \tag{11.291}$$

and one three-photon state,

$$\frac{1}{\sqrt{3}} \left(\beta \left[c_{2\updownarrow}^{\dagger} \right]^2 b_{3\leftrightarrow}^{\dagger} - \alpha \left[c_{2\leftrightarrow}^{\dagger} \right]^2 b_{3\updownarrow}^{\dagger} + c_{2\leftrightarrow}^{\dagger} c_{2\updownarrow}^{\dagger} \left[\alpha b_{3\leftrightarrow}^{\dagger} - \beta b_{3\updownarrow}^{\dagger} \right] \right) |0\rangle. \qquad (11.292)$$

Note also that the mixture which describes the measurement on the non-selective level has no definite photon number. This illustrates once again that a state vector (or a density matrix) must be represented as a functional on the set of spacelike hypersurfaces, which means that it must be linked to the various foliations of space-time.

References

Aharonov, Y. and Albert, D. Z. (1980). States and observables in relativistic quantum field theories. *Phys. Rev.*, **D21**, 3316–3324.

Aharonov, Y. and Albert, D. Z. (1981). Can we make sense out of the measurement process in relativistic quantum mechanics? *Phys. Rev.*, **D24**, 359–370.

Aharonov, Y. and Albert, D. Z. (1984a). Is the usual notion of time evolution adequate for quantum-mechanical system? I. *Phys. Rev.*, **D29**, 223–227.

Aharonov, Y. and Albert, D. Z. (1984b). Is the usual notion of time evolution adequate for quantum mechanical system? II. Relativistic considerations. *Phys. Rev.*, **D29**, 228–234.

Aharonov, Y., Albert, D. Z. and Vaidman, L. (1986). Measurement process in relativistic quantum theory. *Phys. Rev.*, **D34**, 1805–1813.

Bell, J. S. (1964). On the Einstein–Podolsky–Rosen paradox. *Physics*, **1**, 195–200.

Bennett, C. H., Brassard, G., Crépeau, C., Jozsa, R., Peres, A. and Wootters, W. K. (1993). Teleporting an unknown quantum state via dual classical and Einstein–Podolsky–Rosen channels. *Phys. Rev. Lett.*, **70**, 1895–1899.

Bloch, I. (1967). Some relativistic oddities in the quantum theory of observation. *Phys. Rev.*, **156**, 1377–1384.

Bohm, D. (1951). *Quantum Theory*. Prentice-Hall, New York.

Bohr, N. and Rosenfeld, L. (1933). *Zur Frage der Meßbarkeit der elektromagnetischen Feldgrößen*. Det Kgl. Danske Videnskabernes Selskab., Mathematisk-fysiske Meddelelser **XII**, 8, Kopenhagen.

Bouwmeester, D., Pan, J.-W., Mattle, K., Eibl, M., Weinfurter, H. and Zeilinger, A. (1997). Experimental quantum teleportation. *Nature*, **390**, 575–579.

Braunstein, S. L. and Mann, A. (1995). Measurement of the Bell operator and quantum teleportation. *Phys. Rev.*, **A51**, R1727–1730.

Breuer, H. P. and Petruccione, F. (1998). Relativistic formulation of quantum-state diffusion. *J. Phys. A: Math. Gen.*, **31**, 33–52.

Breuer, H. P. and Petruccione, F. (1999). State vector reduction in relativistic quantum mechanics: An introducion. In *Open Systems and Measurement in Relativistic Quantum Theory* (eds. Breuer, H. P. and Petruccione, F.), Volume 526 of *Lecture Notes in Physics*, pp. 1–30. Springer-Verlag, Berlin.

Breuer, H. P. and Petruccione, F. (2001). Relativistic theory of continuous measurements. In *Chance in Physics. Foundations and Perspectives* (eds. Bricmont, J., Dürr, D., Galavotti, M. C., Ghirardi, G., Petruccione, F. and Zanghi, N.), Volume 574 of *Lecture Notes in Physics*, pp. 195–212, Springer-Verlag, Berlin.

Clauser, J. F. and Shimony, A. (1978). Bell's theorem: Experimental tests and implications. *Rep. Prog. Phys.*, **41**, 1881–1927.

Diòsi, L. (1991). Covariant formulation of multiple localized quantum measurements. *Phys. Rev.*, **A43**, 17–21.

Einstein, A., Podolsky, B. and Rosen, N. (1935). Can quantum-mechanical description of physical reality be considered complete? *Phys. Rev.*, **47**, 777–780.

Jauch, J. M. and Rohrlich, F (1980). *The Theory of Photons and Electrons.* Springer-Verlag, New York.

Landau, L. and Peierls, R. (1931). Erweiterung des Unbestimmtheitsprinzips für die relativistische Quantentheorie. *Z. Phys.*, **69**, 56–69.

Michler, M., Mattle, K., Weinfurter, H. and Zeilinger, A. (1996). Interferometric Bell-state analysis. *Phys. Rev.*, **A53**, R1209–R1212.

Misner, C. W., Thorne, K. S. and Wheeler, J. A. (1973). *Gravitation.* Freeman, San Francisco.

Popescu, S. and Vaidman, L. (1994). Causality constraints on nonlocal quantum measurements. *Phys. Rev.*, **A49**, 4331–4338.

Schweber, S. S. (1961). *An Introduction to Relativistic Quantum Field Theory.* Row, Peterson and Company, Evanston, Illinois.

Schwinger, J. (1948). Quantum electrodynamics. I. A covariant formulation. *Phys. Rev.*, **74**, 1439–1461.

Sorkin, R. D. (1993). *Impossible measurements on quantum fields.* In *Directions in General Relativity* (eds. Hu, B. L. and Jacobson, T. A.), Volume 2. Cambridge University Press, Cambridge.

Tomonaga, S. (1946). On a relativistically invariant formulation of the quantum theory of wave fields. I. *Prog. Theor. Phys.*, **1**, 27–42.

Tomonaga, S. (1947). On a relativistically invariant formulation of the quantum theory of wave fields. II. *Prog. Theor. Phys.*, **2**, 101–116.

Vaidman, L. (1994). Teleportation of quantum states. *Phys. Rev.*, **A49**, 1473–1476.

Weinberg, S. (1995). *The Quantum Theory of Fields*, Volume I. Cambridge University Press, Cambridge.

12

OPEN QUANTUM ELECTRODYNAMICS

In quantum electrodynamics the matter degrees of freedom are coupled to the radiation field $A_\mu(x)$ through a local, gauge-invariant interaction density of the form $j^\mu(x)A_\mu(x)$, where $j^\mu(x)$ is the conserved current density of the charged matter. Due to the linear structure of this coupling, the problem of constructing a complete formal representation of the reduced matter dynamics can be solved exactly if the electromagnetic radiation field is initially in a state describable by a Gaussian characteristic functional. An appropriate strategy to achieve this goal is the use of functional methods from field theory. In this chapter we combine such methods with the super-operator technique to derive an exact, relativistic representation for the reduced density matrix of the matter degrees of freedom.

The reduced dynamics involves a certain time-ordered functional of current density super-operators which completely describes the influence of the electromagnetic radiation field on the matter dynamics. This functional has already been employed in our study of quantum Brownian motion in Section 3.6.4 for the derivation of the Feynman–Vernon influence functional and of the path integral representation for the propagator function of the Caldeira–Leggett model.

The functional super-operator technique developed in this chapter can be used as a starting point for the derivation of the master equations encountered in previous chapters. Here, we illustrate this technique by means of a specific application. Namely, we investigate the suppression of the quantum coherence of charged particles through the emission of radiation in typical interference devices. To this end, the degree of decoherence will be characterized through a relativistic, gauge-invariant decoherence functional. An appropriate technique allowing the explicit determination of the decoherence functional for simple interference devices will be developed. It turns out that the relative motion of two interfering wave packets leads to a loss of coherence which is caused by the emission of soft, low-frequency bremsstrahlung. This decoherence mechanism provides a quite fundamental process in quantum electrodynamics since it dominates for short times and because it is at work even in the electromagnetic field vacuum at zero temperature.

Finally, we investigate the possibility of the destruction of coherence of the superposition of many-particle states. It will be argued that, while the decoherence effect is small for single electrons at non-relativistic speed, it is strongly amplified for certain superpositions of many-particle states.

12.1   Density matrix theory for QED

In contrast to the convention in previous chapters we shall use in this chapter rationalized Heaviside–Lorentz units such that $\hbar = c = 1$ and the fine structure constant is given by

$$\alpha = \frac{e^2}{4\pi} \approx \frac{1}{137}, \tag{12.1}$$

where e is the electron charge and c the speed of light. As in Chapter 11 space-time coordinates of 4-vectors in Minkowski space are denoted by $x^\mu = (x^0, \vec{x})$ and we use a metric tensor with signature $(+1, -1, -1, -1)$. Accordingly, the Lorentz scalar product of two 4-vectors $x^\mu = (x^0, \vec{x})$ and $y^\mu = (y^0, \vec{y})$ is written as

$$xy = x^\mu y_\mu = g_{\mu\nu} x^\mu y^\nu = x^0 y^0 - \vec{x} \cdot \vec{y}, \tag{12.2}$$

where we use the summation convention. Occasionally, we will reintroduce factors of c and $\hbar$.

12.1.1   *Field equations and correlation functions*

To be specific we perform our study in the Coulomb or radiation gauge. In this gauge the elimination of the radiation degrees of freedoms is easily carried out. Although we lose manifest Lorentz covariance by taking a fixed gauge, the Lorentz covariance and the gauge invariance of the final results are easily established.

12.1.1.1   *Maxwell equations and commutation relations*   In the Coulomb gauge the free radiation field is described by the vector potential $\vec{A}$ satisfying the Coulomb gauge condition

$$\vec{\nabla} \cdot \vec{A} = 0. \tag{12.3}$$

The electric and magnetic fields are given in terms of $\vec{A}$ by[16]

$$\vec{E}_{\mathrm{T}} = -\frac{\partial \vec{A}}{\partial t}, \tag{12.4}$$

$$\vec{B} = \vec{\nabla} \times \vec{A}. \tag{12.5}$$

The free Maxwell equations,

$$\vec{\nabla} \cdot \vec{E}_{\mathrm{T}} = 0, \tag{12.6}$$

$$\vec{\nabla} \cdot \vec{B} = 0, \tag{12.7}$$

$$\vec{\nabla} \times \vec{E}_{\mathrm{T}} = -\frac{\partial \vec{B}}{\partial t}, \tag{12.8}$$

$$\vec{\nabla} \times \vec{B} = \frac{\partial \vec{E}_{\mathrm{T}}}{\partial t}, \tag{12.9}$$

---

[16]We put an index T on the electric radiation field $\vec{E}_{\mathrm{T}}$ to emphasize its transverse character.

then lead to the wave equation for the transverse vector potential,

$$\Box \vec{A} \equiv \left(\frac{\partial^2}{\partial t^2} - \Delta \right) \vec{A} = 0. \tag{12.10}$$

The wave equation (12.10) may be solved with the help of the plane wave decomposition of the vector potential,

$$\vec{A}(x) = \int \frac{d^3 k}{\sqrt{2(2\pi)^3 \omega}} \sum_{\lambda=1,2} \left[\vec{e}_\lambda(\vec{k}) b_\lambda(\vec{k}) e^{-ikx} + \vec{e}_\lambda(\vec{k}) b_\lambda^\dagger(\vec{k}) e^{+ikx} \right]. \tag{12.11}$$

As usual, the plane wave modes are characterized by a wave vector $\vec{k}$ and two unit polarization vectors $\vec{e}_\lambda(\vec{k})$, such that

$$\vec{k} \cdot \vec{e}_\lambda(\vec{k}) = 0, \tag{12.12}$$

$$\vec{e}_\lambda(\vec{k}) \cdot \vec{e}_{\lambda'}(\vec{k}) = \delta_{\lambda\lambda'}, \tag{12.13}$$

$$\sum_{\lambda=1,2} e_\lambda^i(\vec{k}) e_\lambda^j(\vec{k}) = \delta_{ij} - \frac{k_i k_j}{|\vec{k}|^2} \equiv P_{ij}(\vec{k}), \quad i, j = 1, 2, 3. \tag{12.14}$$

The tensor $P_{ij}(\vec{k})$ is the projector onto the transverse component in k-space. In position space it takes the form

$$P_{ij} = \delta_{ij} - \frac{\partial_i \partial_j}{\Delta}. \tag{12.15}$$

The field operators $b_\lambda(\vec{k})$ and $b_\lambda^\dagger(\vec{k})$ represent the destruction and creation of a photon with wave vector $\vec{k}$ and polarization $\vec{e}_\lambda(\vec{k})$, satisfying bosonic commutation relations

$$\left[b_\lambda(\vec{k}), b_{\lambda'}(\vec{k}') \right] = \left[b_\lambda^\dagger(\vec{k}), b_{\lambda'}^\dagger(\vec{k}') \right] = 0, \tag{12.16}$$

$$\left[b_\lambda(\vec{k}), b_{\lambda'}^\dagger(\vec{k}') \right] = \delta_{\lambda\lambda'} \delta(\vec{k} - \vec{k}'). \tag{12.17}$$

The momentum 4-vector of the photon is written as

$$k^\mu = (k^0, \vec{k}) = (\omega, \vec{k}), \quad \omega = |\vec{k}|, \tag{12.18}$$

and $kx \equiv k^\mu x_\mu = \omega x^0 - \vec{k} \cdot \vec{x}$.

With the help of the above relations it is easy to determine the commutation relation for the free vector potential at arbitrary times,

$$D^{\mathrm{T}}(x - x')_{ij} \equiv i \left[A_i(x), A_j(x') \right] = -P_{ij} D(x - x'). \tag{12.19}$$

The function $D(x - x')$ is the Pauli–Jordan commutator function of the electromagnetic field. Its Fourier representation takes the form

$$D(x - x') = -i \int \frac{d^3 k}{2(2\pi)^3 \omega} \left(e^{-ik(x-x')} - e^{ik(x-x')} \right). \qquad (12.20)$$

Explicitly, one finds

$$D(x - x') = -\frac{1}{4\pi r} \left[\delta(r - t) - \delta(r + t) \right] \qquad (12.21)$$

$$= -\frac{1}{2\pi} \text{sign}(x_0 - x_0') \delta[(x - x')^2], \qquad (12.22)$$

where

$$r \equiv |\vec{x} - \vec{x}'|, \quad t \equiv x_0 - x_0'. \qquad (12.23)$$

We observe that $D(x - x')$ is an antisymmetric function and satisfies

$$D(x - x')|_{x_0 = x_0'} = 0, \qquad (12.24)$$

$$\partial_0 D(x - x')|_{x_0 = x_0'} = -\delta(\vec{x} - \vec{x}'). \qquad (12.25)$$

Moreover, $D(x - x')$ solves the homogeneous wave equation

$$\Box D(x - x') = 0, \qquad (12.26)$$

and is obviously invariant under Lorentz transformations[17]

$$x^\mu \longrightarrow \Lambda^\mu{}_\nu x^\nu, \qquad (12.27)$$

that is, we have

$$D(\Lambda[x - x']) = D(x - x'). \qquad (12.28)$$

The commutator function $D(x - x')$ is a singular function on Minkowski space which vanishes outside the surface of the light cone defined by $(x - x')^2 = 0$ and has a δ-type singularity on it. It is connected to the retarded Green function $D_{\text{ret}}(x - x')$ and to the advanced Green function $D_{\text{adv}}(x - x')$ of the wave equation through the relations

$$D(x - x') = -\left(D_{\text{ret}}(x - x') - D_{\text{adv}}(x - x') \right), \qquad (12.29)$$

$$\text{sign}(x_0 - x_0') D(x - x') = -\frac{1}{2\pi} \delta[(x - x')^2] \qquad (12.30)$$

$$= -\left(D_{\text{ret}}(x - x') + D_{\text{adv}}(x - x') \right).$$

Finally, we remark that the commutation relation (12.19) immediately yields by virtue of eqn (12.25) the equal-time commutation relation between the vector potential and the transverse electric field operator,

$$\left[\vec{E}_i^{\text{T}}(t, \vec{x}), A_j(t, \vec{x}') \right] = +i P_{ij} \delta(\vec{x} - \vec{x}') \equiv i \delta_{ij}^{\text{T}}(\vec{x} - \vec{x}'). \qquad (12.31)$$

---

[17]More precisely, it is invariant under orthochronous Lorentz transformations which satisfy $\Lambda^0{}_0 \geq 1$.

12.1.1.2 *The anticommutator function*   As we have already seen in our study of quantum Brownian motion, in addition to the commutator function a further correlation function plays an important rôle in the description of the reduced system dynamics. This is the anticommutator correlation function which is defined by

$$D_1^{\mathrm{T}}(x-x')_{ij} \equiv \langle\{A_i(x),A_j(x')\}\rangle_f\,. \tag{12.32}$$

The angular brackets denote the average with respect to the radiation field in a thermal equilibrium state at a certain temperature T,

$$\langle\mathcal{O}\rangle_f \equiv \mathrm{tr}_f\left\{\mathcal{O}\frac{1}{Z_f}\exp\left[-H_f/k_{\mathrm{B}}T\right]\right\}, \tag{12.33}$$

where

$$H_f = \int d^3k \sum_{\lambda=1,2} \omega b_\lambda^\dagger(\vec{k})b_\lambda(\vec{k}) \tag{12.34}$$

represents the Hamiltonian of the free radiation field, tr_f denotes the trace over the radiation degrees of freedom, and Z_f is a normalization factor, the partition function of the field.

With the help of

$$\left\langle\left\{b_\lambda(\vec{k}),b_{\lambda'}(\vec{k}')\right\}\right\rangle_f = \left\langle\left\{b_\lambda^\dagger(\vec{k}),b_{\lambda'}^\dagger(\vec{k}')\right\}\right\rangle_f = 0, \tag{12.35}$$

$$\left\langle\left\{b_\lambda(\vec{k}),b_{\lambda'}^\dagger(\vec{k}')\right\}\right\rangle_f = \delta_{\lambda\lambda'}\delta(\vec{k}-\vec{k}')\coth\left(\omega/2k_{\mathrm{B}}T\right), \tag{12.36}$$

one finds

$$D_1^{\mathrm{T}}(x-x')_{ij} = P_{ij}D_1(x-x'), \tag{12.37}$$

with the anticommutator function

$$\begin{aligned}
D_1(x-x') &= \int \frac{d^3k}{2(2\pi)^3\omega}\left(e^{-ik(x-x')}+e^{ik(x-x')}\right)\coth\left(\omega/2k_{\mathrm{B}}T\right)\\
&\equiv D_1^{\mathrm{vac}}(x-x') + D_1^{\mathrm{th}}(x-x').
\end{aligned} \tag{12.38}$$

In the last line we have decomposed D_1 into a vacuum and a thermal part. The vacuum part is easily found to be

$$\begin{aligned}
D_1^{\mathrm{vac}}(x-x') &= \int \frac{d^3k}{2(2\pi)^3\omega}\left(e^{-ik(x-x')}+e^{ik(x-x')}\right)\\
&= \frac{1}{4\pi^2 r}\left[\mathrm{P}\frac{1}{r-t}+\mathrm{P}\frac{1}{r+t}\right]\\
&= -\frac{1}{2\pi^2}\mathrm{P}\frac{1}{(x-x')^2}.
\end{aligned} \tag{12.39}$$

P denotes the Cauchy principal value. Like the commutator function $D(x-x')$ the vacuum contribution $D_1^{\mathrm{vac}}(x-x')$ of the anticommutator function is Lorentz invariant and satisfies the free wave equation. While $D(x - x')$ is antisymmetric, $D_1^{\text{vac}}(x - x')$ is obviously symmetric. We note that $D_1^{\text{vac}}(x - x')$ does not vanish outside the light cone and exhibits a principal value singularity on the light cone.

To determine the thermal part of the anticommutator function one first carries out the angular integration of the k-integral in (12.38) to obtain

$$D_1^{\text{th}}(x - x') = \frac{1}{4\pi^2 r} \int\limits_0^\infty d\omega \left[\coth\left(\omega/2k_B T\right) - 1\right] \left[\sin\omega(r - t) + \sin\omega(r + t)\right].$$

(12.40)

With the help of the formula (4.57) this yields

$$D_1^{\text{th}}(x - x') = \frac{1}{4\pi^2 r} \left[\frac{1}{\tau_B}\coth\left(\frac{r - t}{\tau_B}\right) - \frac{1}{r - t} + \frac{1}{\tau_B}\coth\left(\frac{r + t}{\tau_B}\right) - \frac{1}{r + t}\right].$$

(12.41)

This expression involves the thermal correlation time

$$\tau_B \equiv \frac{1}{\pi k_B T}$$

(12.42)

which was already introduced in eqn (4.58).

One observes that $D_1^{\text{th}}(x - x')$ is a regular function on Minkowski space. If we combine the above expression for $D_1^{\text{th}}(x - x')$ with expression (12.39) for the vacuum contribution we see that the contributions involving $1/(r \pm t)$ cancel each other. With the convention that the singularity on the light cone $(x - x')^2 = 0$ is to be treated as a principal value singularity we can write the total anticommutator function as follows,

$$D_1(x - x') = \frac{1}{4\pi^2 r \tau_B} \left[\coth\left(\frac{r - t}{\tau_B}\right) + \coth\left(\frac{r + t}{\tau_B}\right)\right]$$

$$= \frac{1}{4\pi^2 r \tau_B} \frac{\sinh(2r/\tau_B)}{\sinh[(r - t)/\tau_B]\sinh[(r + t)/\tau_B]}.$$

(12.43)

It must be noted that the anticommutator function is not Lorentz invariant, which is connected to the fact that the thermal radiation field singles out a certain frame. Of course, the full anticommutator function is symmetric and satisfies the free wave equation. According to eqn (12.43), $D_1(x - x')$ changes sign if one crosses the surface of the light cone, that is $D_1(x - x') > 0$ for space-like distances $((x - x')^2 < 0)$ and $D_1(x - x') < 0$ for time-like distances $((x - x')^2 > 0)$.

The dipole approximation of the anticommutator function is obtained by taking the limit $r \longrightarrow 0$ in eqn (12.43) which yields

$$D_1(t, 0) = -\frac{1}{2\pi^2 \tau_B^2 \sinh^2(t/\tau_B)}.$$

(12.44)

In the limit $|t| \ll \tau_B$ this function diverges as t^{-2}, while it decays exponentially as $\exp(-2|t|/\tau_B)$ for times $|t| \gg \tau_B$.

On the other hand, by taking the limit $t \longrightarrow 0$ of eqn (12.43) we find

$$D_1(0,r) = \frac{1}{2\pi^2 r \tau_B} \coth(r/\tau_B). \qquad (12.45)$$

This shows that $D_1(0,r)$ diverges as r^{-2} for small distances ($r \ll \tau_B$) but decays only algebraically as r^{-1} for large distances ($r \gg \tau_B$).

12.1.1.3 *The correlation function of the electric field* A simple but interesting application is the determination of the correlation function of the transverse electric field $\vec{E}_{\mathrm{T}}(x)$. Invoking homogeneity in space and time, one can see that it suffices to calculate the quantity

$$\left\langle \left\{ E_i^{\mathrm{T}}(x), E_j^{\mathrm{T}}(0) \right\} \right\rangle_f = \left\langle \left\{ \dot{A}_i(x), \dot{A}_j(0) \right\} \right\rangle_f$$
$$= -P_{ij} \partial_0^2 D_1(t,r)$$
$$= -\left(\delta_{ij} \Delta - \partial_i \partial_j \right) D_1(t,r), \qquad (12.46)$$

where $t = x^0$ and $r = |\vec{x}|$. In the last step we have used the fact that D_1 satisfies the homogeneous wave equation. By use of the explicit expression (12.43) this gives

$$\left\langle \left\{ E_i^{\mathrm{T}}(x), E_j^{\mathrm{T}}(0) \right\} \right\rangle_f = (\delta_{ij} - \hat{x}_i \hat{x}_j) \qquad (12.47)$$
$$\times \frac{-1}{2\pi^2 \tau_B^3 r} \left[\frac{\cosh[(r-t)/\tau_B]}{\sinh^3[(r-t)/\tau_B]} + \frac{\cosh[(r+t)/\tau_B]}{\sinh^3[(r+t)/\tau_B]} \right],$$

where $\hat{x}_i = x_i/r$. Summing over $i = j$ we are led to the expression

$$\left\langle \left\{ \vec{E}_{\mathrm{T}}(x), \vec{E}_{\mathrm{T}}(0) \right\} \right\rangle_f = \frac{-1}{\pi^2 \tau_B^3 r} \left[\frac{\cosh[(r-t)/\tau_B]}{\sinh^3[(r-t)/\tau_B]} + \frac{\cosh[(r+t)/\tau_B]}{\sinh^3[(r+t)/\tau_B]} \right]. \qquad (12.48)$$

Let us investigate first the limit $r \longrightarrow 0$ of eqn (12.48) which is given by

$$\left\langle \left\{ \vec{E}_{\mathrm{T}}(t,0), \vec{E}_{\mathrm{T}}(0) \right\} \right\rangle_f = \frac{1}{\pi^2 \tau_B^4} \frac{6 + 4\sinh^2(t/\tau_B)}{\sinh^4(t/\tau_B)}. \qquad (12.49)$$

For times which are short compared to the thermal correlation time this yields

$$\left\langle \left\{ \vec{E}_{\mathrm{T}}(t,0), \vec{E}_{\mathrm{T}}(0) \right\} \right\rangle_f \approx \frac{6}{\pi^2 t^4}, \quad |t| \ll \tau_B, \qquad (12.50)$$

which corresponds to the vacuum contribution of the correlation function. Thus, the vacuum part dominates for short times and diverges as t^{-4}.

The thermal contribution to the correlation function of the electric field is given by

$$\left\langle\left\{\vec{E}_{\mathrm{T}}(t,0),\vec{E}_{\mathrm{T}}(0)\right\}\right\rangle_{\mathrm{th}} = \frac{1}{\pi^2\tau_B^4}\left[\frac{6+4\sinh^2(t/\tau_B)}{\sinh^4(t/\tau_B)} - \frac{6}{(t/\tau_B)^4}\right]. \tag{12.51}$$

For $|t|/\tau_B \longrightarrow 0$ the term within the square brackets approaches the value $2/15$, such that we may write

$$\left\langle\left\{\vec{E}_{\mathrm{T}}(t,0),\vec{E}_{\mathrm{T}}(0)\right\}\right\rangle_{\mathrm{th}} = \frac{2}{15\pi^2\tau_B^4}g\left(\frac{t}{\tau_B}\right), \tag{12.52}$$

where the function

$$g(\tau) \equiv \frac{15}{2}\left[\frac{6+4\sinh^2\tau}{\sinh^4\tau} - \frac{6}{\tau^4}\right] \tag{12.53}$$

satisfies $g(0) = 1$. The energy density of the field in thermal equilibrium is thus found to be

$$u_{\mathrm{th}} = \frac{1}{2}\left\langle\left\{\vec{E}_{\mathrm{T}}(0),\vec{E}_{\mathrm{T}}(0)\right\}\right\rangle_{\mathrm{th}} = \frac{1}{15\pi^2\tau_B^4} = \frac{\pi^2(k_{\mathrm{B}}T)^4}{15},$$

which is the Stefan–Boltzmann law of black-body radiation.

On the other hand, in the case $|t| \gg \tau_B$ eqn (12.49) yields

$$\left\langle\left\{\vec{E}_{\mathrm{T}}(t,0),\vec{E}_{\mathrm{T}}(0)\right\}\right\rangle_f \approx \frac{16}{\pi^2\tau_B^4}e^{-2|t|/\tau_B}. \tag{12.54}$$

The correlation function thus decays exponentially for times large compared to the correlation time τ_B.

Finally, let us investigate the limit $|t| \longrightarrow 0$ in eqn (12.48). This limit gives

$$\left\langle\left\{\vec{E}_{\mathrm{T}}(0,r),\vec{E}_{\mathrm{T}}(0)\right\}\right\rangle_f = \frac{-2}{\pi^2\tau_B^3 r}\frac{\cosh(r/\tau_B)}{\sinh^3(r/\tau_B)}, \tag{12.55}$$

so that we have

$$\left\langle\left\{\vec{E}_{\mathrm{T}}(0,r),\vec{E}_{\mathrm{T}}(0)\right\}\right\rangle_f \approx -\frac{2}{\pi^2 r^4}, \quad \text{for} \quad r \ll \tau_B, \tag{12.56}$$

$$\left\langle\left\{\vec{E}_{\mathrm{T}}(0,r),\vec{E}_{\mathrm{T}}(0)\right\}\right\rangle_f \approx -\frac{16}{\pi^2\tau_B^3 r}e^{-2r/\tau_B}, \quad \text{for} \quad r \gg \tau_B. \tag{12.57}$$

The correlation function of the electric field thus diverges as r^{-4} for short distances, $r \ll \tau_B$. By contrast to the behaviour of the anticommutator function D_1 in this limit, it decays exponentially for distances large compared to τ_B.

12.1.2   The reduced density matrix

We now turn to the reduced dynamics of the matter degrees of freedom which are coupled to the radiation field. In the Coulomb gauge the Hamiltonian density in the interaction picture takes the form (Weinberg, 1995; Jauch and Rohrlich, 1980)

$$\mathcal{H}(x) = \mathcal{H}_{\mathrm{C}}(x) + \mathcal{H}_{\mathrm{T}}(x). \tag{12.58}$$

Here,

$$\mathcal{H}_{\mathrm{T}}(x) = -\vec{j}(x) \cdot \vec{A}(x) \equiv j^{\mu}(x) A_{\mu}(x) \tag{12.59}$$

represents the density of the interaction of the matter current density $j^{\mu}(x)$ with the transverse radiation field $A^{\mu} = (0, \vec{A})$. The matter current density is assumed to satisfy the continuity equation

$$\partial_{\mu} j^{\mu}(x) = 0, \tag{12.60}$$

expressing the local conservation of charge. The Coulomb energy density is given by

$$\mathcal{H}_{\mathrm{C}}(x^0, \vec{x}) = \frac{1}{2} \int d^3y \, \frac{j^0(x^0, \vec{x}) j^0(x^0, \vec{y})}{4\pi |\vec{x} - \vec{y}|}, \tag{12.61}$$

such that

$$H_{\mathrm{C}}(x^0) = \frac{1}{2} \int d^3x \int d^3y \frac{j^0(x^0, \vec{x}) j^0(x^0, \vec{y})}{4\pi |\vec{x} - \vec{y}|} \tag{12.62}$$

is the instantaneous Coulomb energy of the charge distribution given by $j^0(x)$.

For notational convenience, we have set in eqn (12.59) the scalar potential of the radiation field equal to zero, $A^0 = 0$. This allows us to write the correlation functions of the field as

$$D^{\mathrm{T}}(x - x')_{\mu\nu} = i\,[A_{\mu}(x), A_{\nu}(x')] = -P_{\mu\nu} D(x - x'), \tag{12.63}$$
$$D_1^{\mathrm{T}}(x - x')_{\mu\nu} = \langle\{A_{\mu}(x), A_{\nu}(x')\}\rangle_f = +P_{\mu\nu} D_1(x - x'), \tag{12.64}$$

where $P_{\mu\nu}$ is defined by eqn (12.15), and by $P_{00} = P_{0i} = P_{i0} = 0$ for $i = 1, 2, 3$. It must be remembered, however, that these and all other correlation functions to be introduced later involve the transverse projection $P_{\mu\nu}$.

As in the preceding section all fields are taken to be in the interaction picture. The interaction picture density matrix $\rho(t)$ for the coupled system then satisfies the Liouville–von Neumann equation which we write as

$$\frac{\partial}{\partial t}\rho(t) = \int d^3x \, \mathcal{L}(t, \vec{x})\rho(t). \tag{12.65}$$

The Liouville super-operator $\mathcal{L}(x)$ is given by $\mathcal{L}(x) = \mathcal{L}_{\mathrm{C}}(x) + \mathcal{L}_{\mathrm{T}}(x)$, where we introduce the Liouville super-operators pertaining to the densities of the Coulomb field and of the transverse field,

$$\mathcal{L}_{\mathrm{C}}(x)\rho \equiv -i[\mathcal{H}_{\mathrm{C}}(x), \rho], \qquad (12.66)$$

$$\mathcal{L}_{\mathrm{T}}(x)\rho \equiv -i[\mathcal{H}_{\mathrm{T}}(x), \rho]. \qquad (12.67)$$

Integrating the Liouville–von Neumann equation formally with the help of the chronological time-ordering operator and taking the trace over the radiation field, we obtain the following equation,

$$\rho_m(t_f) = \mathrm{tr}_f \left\{ \mathrm{T}_{\leftarrow} \exp\left[\int_{t_i}^{t_f} d^4x\, \mathcal{L}(x) \right] \rho(t_i) \right\}. \qquad (12.68)$$

This equation relates the density matrix $\rho_m(t_f)$ describing the matter degrees of freedom at some final time t_f to the density matrix $\rho(t_i)$ of the combined matter–field system at some initial time t_i. It will be the starting point for the derivation of the influence functional in the next section.

12.2   The influence functional of QED

Employing functional techniques of field theory, we derive in this section a super-operator representation for the influence functional of QED. The influence functional will be given in the form of a functional of super-operators of the matter current density, involving certain Green functions of the radiation field. The resulting expression will be discussed physically. In particular, it will be related to well-known expressions for the vacuum-to-vacuum amplitude of the electromagnetic radiation field in the presence of a classical current density, and to the classical formulation of a system of charged particles in terms of a non-local action principle (Breuer and Petruccione, 2001).

12.2.1   *Elimination of the radiation degrees of freedom*

Our aim is to eliminate the variables of the electromagnetic radiation field to obtain an exact representation for the reduced density matrix ρ_m of the matter degrees of freedom. To this end, we invoke the formal representation (12.68). The derivation is performed in two steps. First, we eliminate the time-ordering of the electromagnetic variables and, second, we determine the remaining trace over the field variables employing the Gaussian property of the field state.

12.2.1.1   *Eliminating the time ordering of the field variables*   Our first step is a decomposition of the chronological time-ordering operator $\mathrm{T}_{\leftarrow}$ into a time-ordering operator $\mathrm{T}^j_{\leftarrow}$ for the matter current and a time-ordering operator $\mathrm{T}^A_{\leftarrow}$ for the electromagnetic field variables as $\mathrm{T}_{\leftarrow} = \mathrm{T}^j_{\leftarrow}\mathrm{T}^A_{\leftarrow}$. This enables one to write eqn (12.68) as

$$\rho_m(t_f) = \mathrm{T}^j_{\leftarrow}\left(\mathrm{tr}_f \left\{ \mathrm{T}^A_{\leftarrow} \exp\left[\int_{t_i}^{t_f} d^4x\, (\mathcal{L}_{\mathrm{C}}(x) + \mathcal{L}_{\mathrm{T}}(x)) \right] \rho(t_i) \right\} \right). \qquad (12.69)$$

The currents $j^\mu(x)$ commute under the time-ordering $T_{\leftarrow}^j$. We may therefore treat them formally as commuting c-number fields under the time-ordering symbol. Since the super-operator $\mathcal{L}_C(x)$ contains only matter variables, the corresponding contribution can be pulled out of the trace. Hence, we have

$$\rho_m(t_f) = T_{\leftarrow}^j \left(\exp\left[\int_{t_i}^{t_f} d^4x\, \mathcal{L}_C(x) \right] \mathrm{tr}_f \left\{ T_{\leftarrow}^A \exp\left[\int_{t_i}^{t_f} d^4x\, \mathcal{L}_T(x) \right] \rho(t_i) \right\} \right).$$
(12.70)

We now proceed by eliminating the time-ordering of the A fields. With the help of the Wick theorem (Itzykson and Zuber, 1980) we get

$$T_{\leftarrow}^A \exp\left[\int_{t_i}^{t_f} d^4x\, \mathcal{L}_T(x) \right]$$
(12.71)

$$= \exp\left[\frac{1}{2} \int_{t_i}^{t_f} d^4x \int_{t_i}^{t_f} d^4x' [\mathcal{L}_T(x), \mathcal{L}_T(x')]\theta(t-t') \right] \exp\left[\int_{t_i}^{t_f} d^4x\, \mathcal{L}_T(x) \right].$$

In order to determine the commutator of the Liouville super-operators we invoke the Jacobi identity which yields for an arbitrary test density ρ,

$$\begin{aligned}
[\mathcal{L}_T(x), \mathcal{L}_T(x')]\rho &= \mathcal{L}_T(x)\mathcal{L}_T(x')\rho - \mathcal{L}_T(x')\mathcal{L}_T(x)\rho \\
&= -[\mathcal{H}_T(x), [\mathcal{H}_T(x'), \rho]] + [\mathcal{H}_T(x'), [\mathcal{H}_T(x), \rho]] \\
&= -[[\mathcal{H}_T(x), \mathcal{H}_T(x')], \rho].
\end{aligned}$$
(12.72)

The commutator of the transverse energy densities may be simplified to read

$$[\mathcal{H}_T(x), \mathcal{H}_T(x')] = j^\mu(x)j^\nu(x')[A_\mu(x), A_\nu(x')],$$
(12.73)

since the contribution involving the commutator of the currents vanishes by virtue of the time-ordering operator $T_{\leftarrow}^j$. Thus, it follows from eqns (12.72) and (12.73) that the commutator of the Liouville super-operators may be written as

$$[\mathcal{L}_T(x), \mathcal{L}_T(x')]\rho = -[A_\mu(x), A_\nu(x')][j^\mu(x)j^\nu(x'), \rho].$$
(12.74)

It is useful to introduce current super-operators $J_+(x)$ and $J_-(x)$ by means of

$$J_+^\mu(x)\rho \equiv j^\mu(x)\rho, \qquad J_-^\mu(x)\rho \equiv \rho j^\mu(x).$$
(12.75)

Thus, $J_+(x)$ is defined to be the current density acting from the left, while $J_-(x)$ acts from the right on an arbitrary density. With the help of these definitions we bring the expression for the commutator of the Liouville super-operators into the form

$$[\mathcal{L}_T(x), \mathcal{L}_T(x')] = -[A_\mu(x), A_\nu(x')]\left(J_+^\mu(x)J_+^\nu(x') - J_-^\mu(x)J_-^\nu(x')\right). \quad (12.76)$$

Inserting this result into eqn (12.71), we can write eqn (12.70) as

$$\rho_m(t_f) = T_\leftarrow^j\left(\exp\left[\int_{t_i}^{t_f} d^4x\,\mathcal{L}_C(x)\right.\right.$$
$$-\frac{1}{2}\int_{t_i}^{t_f} d^4x \int_{t_i}^{t_f} d^4x'\theta(t-t')[A_\mu(x), A_\nu(x')]J_+^\mu(x)J_+^\nu(x')$$
$$\left.+\frac{1}{2}\int_{t_i}^{t_f} d^4x \int_{t_i}^{t_f} d^4x'\theta(t-t')[A_\mu(x), A_\nu(x')]J_-^\mu(x)J_-^\nu(x')\right]$$
$$\left.\times \mathrm{tr}_f\left\{\exp\left[\int_{t_i}^{t_f} d^4x\,\mathcal{L}_T(x)\right]\rho(t_i)\right\}\right). \quad (12.77)$$

This is an exact formal representation for the reduced density matrix of the matter variables. The time-ordering of the radiation degrees of freedom has been removed and the latter enter eqn (12.77) only through the functional

$$W[J_+, J_-] \equiv \mathrm{tr}_f\left\{\exp\left[\int_{t_i}^{t_f} d^4x\,\mathcal{L}_T(x)\right]\rho(t_i)\right\}, \quad (12.78)$$

since the commutator of the A fields is a c-number function.

12.2.1.2 *The influence super-operator* The functional (12.78) involves an average over the field variables with respect to the initial state $\rho(t_i)$ of the combined matter–field system. It therefore contains all correlations in the initial state of the total system. In the following our central goal is to investigate how correlations are built up through the interaction between matter and radiation field. We therefore consider an initial state of low entropy which is given by a product state of the form

$$\rho(t_i) = \rho_m(t_i) \otimes \rho_f, \quad (12.79)$$

where $\rho_m(t_i)$ is the density matrix of the matter at the initial time and the density matrix ρ_f of the radiation field describes the thermal equilibrium state at temperature T. The influence of the special choice (12.79) for the initial condition may be eliminated by pushing t_i to $-\infty$ and by switching on the interaction adiabatically. This is the usual procedure used in quantum field theory in order to define asymptotic states and the S-matrix. The matter and the field variables are then described as *in*-fields, obeying free field equations with renormalized mass.

For an arbitrary initial condition $\rho(t_i)$ the functional $W[J_+, J_-]$ can be determined by means of a cumulant expansion (see Section 9.2.3). Since the initial state (12.79) is Gaussian with regard to the field variables and since the Liouville super-operator $\mathcal{L}_T(x)$ is linear in the radiation field, the cumulant expansion terminates after the second-order term. In addition, a linear term does not appear in the expansion because of $\langle A_\mu(x)\rangle_f = 0$. Thus we immediately obtain

$$W[J_+, J_-] = \exp\left[\frac{1}{2}\int_{t_i}^{t_f} d^4x \int_{t_i}^{t_f} d^4x' \langle \mathcal{L}_T(x)\mathcal{L}_T(x')\rangle_f\right] \rho_m(t_i). \qquad (12.80)$$

Inserting the definition of the Liouville super-operator $\mathcal{L}_T(x)$ into the exponent of this expression one finds after some algebra,

$$\frac{1}{2}\int_{t_i}^{t_f} d^4x \int_{t_i}^{t_f} d^4x' \langle \mathcal{L}_T(x)\mathcal{L}_T(x')\rangle_f \rho_m(t_i) \qquad (12.81)$$

$$\equiv -\frac{1}{2}\int_{t_i}^{t_f} d^4x \int_{t_i}^{t_f} d^4x' \operatorname{tr}_f[\mathcal{H}_T(x),[\mathcal{H}_T(x'),\rho(t_i)]]$$

$$= -\frac{1}{2}\int_{t_i}^{t_f} d^4x \int_{t_i}^{t_f} d^4x' \qquad (12.82)$$

$$[\langle A_\nu(x')A_\mu(x)\rangle_f J_+^\mu(x)J_+^\nu(x') + \langle A_\mu(x)A_\nu(x')\rangle_f J_-^\mu(x)J_-^\nu(x')$$
$$-\langle A_\nu(x')A_\mu(x)\rangle_f J_+^\mu(x)J_-^\nu(x') - \langle A_\mu(x)A_\nu(x')\rangle_f J_-^\mu(x)J_+^\nu(x')] \rho_m(t_i).$$

On using this result together with eqn (12.80), we can cast eqn (12.77) into the following form,

$$\rho_m(t_f) = \mathrm{T}_\leftarrow^j \exp\left[\int_{t_i}^{t_f} d^4x \mathcal{L}_C(x)\right. \qquad (12.83)$$

$$+\frac{1}{2}\int_{t_i}^{t_f} d^4x \int_{t_i}^{t_f} d^4x'\{ -(\theta(t-t')[A_\mu(x), A_\nu(x')] + \langle A_\nu(x')A_\mu(x)\rangle_f)J_+^\mu(x)J_+^\nu(x')$$

$$+(\theta(t-t')[A_\mu(x), A_\nu(x')] - \langle A_\mu(x)A_\nu(x')\rangle_f)J_-^\mu(x)J_-^\nu(x')$$

$$\left. +\langle A_\nu(x')A_\mu(x)\rangle_f J_+^\mu(x)J_-^\nu(x') + \langle A_\mu(x)A_\nu(x')\rangle_f J_-^\mu(x)J_+^\nu(x')\}\right]\rho_m(t_i).$$

At this point it is useful to introduce some new notation for the correlation functions of the electromagnetic field, namely the Feynman propagator and its complex conjugate ($T_\rightarrow$ denotes the antichronological time-ordering),

$$iD_F(x - x')_{\mu\nu} \equiv \langle T_\leftarrow[A_\mu(x)A_\nu(x')]\rangle_f \tag{12.84}$$
$$= \theta(t - t')[A_\mu(x), A_\nu(x')] + \langle A_\nu(x')A_\mu(x)\rangle_f,$$
$$iD_F^*(x - x')_{\mu\nu} \equiv -\langle T_\rightarrow[A_\mu(x)A_\nu(x')]\rangle_f \tag{12.85}$$
$$= \theta(t - t')[A_\mu(x), A_\nu(x')] - \langle A_\mu(x)A_\nu(x')\rangle_f,$$

as well as the two-point correlation functions

$$D_+(x - x')_{\mu\nu} \equiv \langle A_\mu(x)A_\nu(x')\rangle_f, \tag{12.86}$$
$$D_-(x - x')_{\mu\nu} \equiv \langle A_\nu(x')A_\mu(x)\rangle_f. \tag{12.87}$$

As is easily verified these functions are related through the identity

$$-iD_F(x - x')_{\mu\nu} + iD_F^*(x - x')_{\mu\nu} + D_+(x - x')_{\mu\nu} + D_-(x - x')_{\mu\nu} = 0. \tag{12.88}$$

With the help of this notation the density matrix of the matter can now be represented in the compact form

$$\rho_m(t_f) = T_\leftarrow^j \exp(i\Phi[J_+, J_-])\, \rho_m(t_i), \tag{12.89}$$

where we have introduced the *influence phase functional*

$$i\Phi[J_+, J_-] = \int_{t_i}^{t_f} d^4x\, \mathcal{L}_C(x) + \frac{1}{2}\int_{t_i}^{t_f} d^4x \int_{t_i}^{t_f} d^4x' \tag{12.90}$$
$$\times \left\{-iD_F(x - x')_{\mu\nu}J_+^\mu(x)J_+^\nu(x') + iD_F^*(x - x')_{\mu\nu}J_-^\mu(x)J_-^\nu(x')\right.$$
$$\left. + D_-(x - x')_{\mu\nu}J_+^\mu(x)J_-^\nu(x') + D_+(x - x')_{\mu\nu}J_-^\mu(x)J_+^\nu(x')\right\}.$$

Equation (12.89) provides an exact representation for the matter density matrix which takes on the desired form: It involves the electromagnetic field variables only through the various two-point correlation functions introduced above. One observes that the dynamics of the matter variables is given by a time-ordered *influence super-operator*, that is a time-ordered exponential function whose exponent is a bilinear functional of the current super-operators $J_\pm(x)$.

It should be remarked that the influence phase $\Phi[J_+, J_-]$ is both a functional of the quantities $J_\pm(x)$ and a super-operator which acts in the space of density matrices of the matter degrees of freedom. There are several alternative methods which may be used to arrive at an expression of the form (12.90) as, for example, path integral techniques (Feynman and Vernon, 1963) or Schwinger's closed time-path method (Chou, Su, Hao and Yu, 1985; Diòsi, 1990).

For the study of decoherence phenomena another equivalent formula for the influence phase functional will be useful. To obtain this formula we use the commutator and the anticommutator functions which have been introduced and determined explicitly in Section 12.1. These functions are related to the above correlation functions through

$$D_+(x - x')_{\mu\nu} = \frac{1}{2}D_1^{\mathrm{T}}(x - x')_{\mu\nu} - \frac{i}{2}D^{\mathrm{T}}(x - x')_{\mu\nu}, \tag{12.91}$$

$$D_-(x - x')_{\mu\nu} = \frac{1}{2}D_1^{\mathrm{T}}(x - x')_{\mu\nu} + \frac{i}{2}D^{\mathrm{T}}(x - x')_{\mu\nu}, \tag{12.92}$$

$$iD_F(x - x')_{\mu\nu} = \frac{1}{2}D_1^{\mathrm{T}}(x - x')_{\mu\nu} - \frac{i}{2}\mathrm{sign}(t - t')D^{\mathrm{T}}(x - x')_{\mu\nu}, \tag{12.93}$$

$$-iD_F^*(x - x')_{\mu\nu} = \frac{1}{2}D_1^{\mathrm{T}}(x - x')_{\mu\nu} + \frac{i}{2}\mathrm{sign}(t - t')D^{\mathrm{T}}(x - x')_{\mu\nu}. \tag{12.94}$$

Correspondingly, we define a commutator super-operator $J_c(x)$ and an anticommutator super-operator $J_a(x)$ by means of

$$J_c^\mu(x)\rho \equiv [j^\mu(x), \rho], \tag{12.95}$$

$$J_a^\mu(x)\rho \equiv \{j^\mu(x), \rho\}, \tag{12.96}$$

which are related to the previously introduced super-operators $J_\pm^\mu(x)$ by

$$J_c^\mu(x) = J_+^\mu(x) - J_-^\mu(x), \tag{12.97}$$

$$J_a^\mu(x) = J_+^\mu(x) + J_-^\mu(x). \tag{12.98}$$

In terms of these quantities the influence phase functional becomes

$$i\Phi[J_c, J_a] = \int_{t_i}^{t_f} d^4x\, \mathcal{L}_{\mathrm{C}}(x) \tag{12.99}$$

$$+ \int_{t_i}^{t_f} d^4x \int_{t_i}^{t} d^4x' \left\{ \frac{i}{2}D^{\mathrm{T}}(x - x')_{\mu\nu} J_c^\mu(x) J_a^\nu(x') - \frac{1}{2}D_1^{\mathrm{T}}(x - x')_{\mu\nu} J_c^\mu(x) J_c^\nu(x') \right\}.$$

This form of the influence phase functional will be particularly useful later on. It represents the influence of the radiation field on the matter dynamics in terms of the two fundamental two-point correlation functions $D(x - x')$ and $D_1(x - x')$. Note that the double space-time integral in eqn (12.99) is already a time-ordered integral since the integration over x_0' extends over the time interval from t_i to $t = x_0$.

The result expressed by eqns (12.89) and (12.99) has been used in Section 3.6.4 for the determination of the influence functional of the Caldeira–Leggett model (see eqns (3.507) and (3.508)). In fact, for the Caldeira–Leggett model we have to use a coupling of the form $H_I = -xB$ (see eqn (3.377)). Going through the above derivation we see that in this case the influence functional (12.99) takes on precisely the form given in (3.508): The commutator and anticommutator super-operators J_c^μ, J_a^μ must be replaced by the super-operators x_c and

x_a defined in eqn (3.510), while the counter-term (3.509) plays a similar rôle to the Coulomb term $\mathcal{L}_C(x)$. Furthermore, the correlation functions $D^T(x - x')_{\mu\nu}$ and $D_1^T(x - x')_{\mu\nu}$ must be replaced by the correlation functions of the Caldeira–Leggett model defined in eqns (3.385) and (3.386).

12.2.2 Vacuum-to-vacuum amplitude

It is instructive to compare eqns (12.89) and (12.90) with the structure of the Markovian quantum master equation in Lindblad form containing a set of Lindblad operators A_i (see Section 3.2.2). One observes that the terms of the influence phase functional involving the current super-operators in the combinations J_+J_- and J_-J_+ correspond to the gain terms in the Lindblad equation having the form $A_i\rho_m A_i^\dagger$. These terms may be interpreted as describing the back-action on the reduced system of the matter degrees of freedom induced by 'real' processes in which photons are absorbed or emitted. The presence of these terms leads to a transformation of pure states into statistical mixtures. Namely, if we disregard the terms containing the combinations J_+J_- and J_-J_+ the right-hand side of (12.89) can be written as $U(t_f, t_i)\rho_m(t_i)U^\dagger(t_f, t_i)$, where

$$U(t_f, t_i) = \mathrm{T}_\leftarrow^j \exp\left[-i\int_{t_i}^{t_f} d^4x\,\mathcal{H}_C(x)\right.$$
$$\left. -\frac{i}{2}\int_{t_i}^{t_f} d^4x\int_{t_i}^{t_f} d^4x'\,D_F(x - x')_{\mu\nu}j^\mu(x)j^\nu(x')\right]. \tag{12.100}$$

This shows that the contributions involving the Feynman propagators and the combinations J_+J_+ and J_-J_- of super-operators preserve the purity of states.

Taking the limits $t_i \longrightarrow -\infty$ and $t_f \longrightarrow +\infty$ the operator $U(t_f, t_i)$ becomes the time-ordered product of the functional

$$A[j] = \exp\left[-i\int d^4x\,\mathcal{H}_C(x) - \frac{i}{2}\int d^4x\int d^4x'\,D_F(x - x')_{\mu\nu}j^\mu(x)j^\nu(x')\right]$$
$$\equiv \exp\left[i\left(S^{(1)} + iS^{(2)}\right)\right]. \tag{12.101}$$

At zero temperature this expression is the vacuum-to-vacuum amplitude in the presence of a classical current density $j^\mu(x)$ (Feynman and Hibbs, 1965). It yields the amplitude for the field to start in the vacuum at $t_i = -\infty$ and to end up in the vacuum at $t_f = +\infty$. In the second equation of (12.101) we have decomposed the exponent into real and imaginary parts, using the corresponding decomposition of the Feynman propagator as given in eqn (12.93).

The exponent of $A[j]$ provides a complex action functional $S = S^{(1)} + iS^{(2)}$. Its imaginary part is found to be

$$S^{(2)} = \frac{1}{4} \int d^4x \int d^4x' D_1(x - x') \vec{j}_T(x) \cdot \vec{j}_T(x')$$

$$= -\frac{1}{4} \int d^4x \int d^4x' D_1(x - x') j_\mu(x) j^\mu(x'). \qquad (12.102)$$

In the first line $\vec{j}_T$ denotes the transverse component of the current density, satisfying $\vec{\nabla} \cdot \vec{j}_T = 0$. The coupling to the transverse component is due to the fact that $D_1^T(x - x')_{\mu\nu}$ contains the transverse projection $P_{\mu\nu}$. In the second line of (12.102) we have used the current conservation (12.60) to cast the expression into covariant form. To see how this form emerges we transform into Fourier space, using (12.38) and the Fourier transform of the current density,

$$j^\mu(k) = \int d^4x e^{-ikx} j^\mu(x). \qquad (12.103)$$

This leads to

$$S^{(2)} = \frac{1}{2} \int \frac{d^3k}{2(2\pi)^3\omega} \coth(\omega/2k_BT) \left(\vec{j}(k) \cdot \vec{j}^*(k) - \frac{(\vec{k} \cdot \vec{j}(k))(\vec{k} \cdot \vec{j}(k))^*}{|\vec{k}|^2} \right)$$

$$= \frac{1}{2} \int \frac{d^3k}{2(2\pi)^3\omega} \coth(\omega/2k_BT) \left(-j^\mu(k) j_\mu^*(k) \right), \qquad (12.104)$$

where we have used

$$\vec{j}(k) \cdot \vec{j}^*(k) - \frac{(\vec{k} \cdot \vec{j}(k))(\vec{k} \cdot \vec{j}(k))^*}{|\vec{k}|^2} = \vec{j}(k) \cdot \vec{j}^*(k) - j_0(k) j_0^*(k)$$

$$= -j^\mu(k) j_\mu^*(k), \qquad (12.105)$$

which follows from current conservation, $k_\mu j^\mu(k) = 0$. Thus we have

$$|A[j]|^2 = \exp\left[-2S^{(2)} \right] \qquad (12.106)$$

$$= \exp\left[-\int \frac{d^3k}{2(2\pi)^3\omega} \coth(\omega/2k_BT) \left(-j^\mu(k) j_\mu^*(k) \right) \right].$$

This is the no-photon emission probability, that is the probability that the current density does not emit any photons (Itzykson and Zuber, 1980). We see that for finite temperatures, $T > 0$, the no-photon emission probability is reduced in comparison to the vacuum case. This is just the effect of induced emission and absorption processes.

By use of the relation (12.30) and of current conservation the real part of the action functional if found to be

$$S^{(1)} = -\frac{1}{8\pi} \int d^4x \int d^4x' \left\{ \frac{\delta(x_0 - x_0')}{|\vec{x} - \vec{x}'|} j_0(x) j_0(x') - \delta\left[(x - x')^2 \right] \vec{j}_T(x) \cdot \vec{j}_T(x') \right\}$$

$$= -\frac{1}{8\pi} \int d^4x \int d^4x' \delta\left[(x - x')^2 \right] j_\mu(x) j^\mu(x'). \qquad (12.107)$$

This is the classical Wheeler–Feynman action (Wheeler and Feynman, 1945, 1949; Rohrlich, 1965). It allows the description of the interaction of a system of

charged classical particles by means of a non-local action principle. The degrees
of freedom of the electromagnetic field have been completely eliminated under
the boundary condition of no net emission of radiation (which is known as the
complete absorber theory). As a result one is left with a non-local action functional which contains both the retarded and the advanced Green function (see
eqn (12.30)).

12.2.3 *Second-order equation of motion*

The representation (12.89) immediately yields the following second-order equation of motion for the density matrix of the matter degrees of freedom,

$$\frac{d}{dt}\rho_m(t) = -i\left[H_C(t), \rho_m(t)\right] \tag{12.108}$$

$$-\frac{i}{2}\int d^3x \int d^3x' \int_{t_i}^{t} dx_0' D(x - x')\left[\vec{j}_{\mathrm{T}}(x), \left\{\vec{j}_{\mathrm{T}}(x'), \rho_m(t)\right\}\right]$$

$$-\frac{1}{2}\int d^3x \int d^3x' \int_{t_i}^{t} dx_0' D_1(x - x')\left[\vec{j}_{\mathrm{T}}(x), \left[\vec{j}_{\mathrm{T}}(x'), \rho_m(t)\right]\right],$$

where $x = (t, \vec{x})$ and $x' = (x_0', \vec{x}')$. This equation is just the time-convolutionless
master equation to second order in the coupling. As we know it provides a non-
Markovian master equation since it involves the non-local dissipation and noise
kernels $D(x - x')$ and $D_1(x - x')$, respectively. The various master equations
encountered in the previous chapters can be derived from this equation. For
example, the quantum optical master equation is obtained from it by performing
the Markovian and the rotating wave approximation.

12.2.3.1 *The quantum optical limit* It might be instructive to sketch how the
quantum optical master equation can be derived from eqn (12.108). To this
end, we insert the Fourier representations of the Green functions $D(x - y)$ and
$D_1(x-y)$ (see eqns (12.20) and (12.38)) into eqn (12.108). Using then the Fourier
components of the transverse current density which are defined by

$$\vec{j}_{\mathrm{T}}(x_0, \vec{k}) = \int d^3x\, \vec{j}_{\mathrm{T}}(x_0, \vec{x}) e^{-i\vec{k}\cdot\vec{x}}, \tag{12.109}$$

the space integrals over $\vec{x}$ and $\vec{x}'$ can easily be evaluated. Next, one performs
the Markov approximation which consists in the replacement

$$\int_{t_i}^{t} dx_0' \longrightarrow \int_{-\infty}^{t} dx_0' = \int_{0}^{\infty} d\tau, \tag{12.110}$$

where we have substituted $\tau \equiv t - x_0'$. Finally, we introduce the decomposition
of the current into eigenoperators of the system Hamiltonian, such that

$$\vec{j}_T(x_0, \vec{k}) = \sum_{\omega_n} e^{-i\omega_n x_0} \vec{j}_T(\omega_n, \vec{k}).$$ (12.111)

The sum runs over the frequencies ω_n of the system defined by the energy differences of the unperturbed system. This representation enables us to employ the rotating wave approximation in the quantum optical limit. The latter consists in keeping only the secular terms in double sums over the system frequencies ω and is equivalent to an averaging procedure over the rapidly oscillating terms. Putting it all together we obtain the following master equation,

$$\frac{d}{dt}\rho_m(t) = -i\left[H_C(t), \rho_m(t)\right] - i\left[H_{LS} + H_{SS}, \rho_m(t)\right]$$ (12.112)

$$+ \sum_{\omega_n>0} \gamma_-(\omega_n) \int d\Omega(\vec{k}) \left(\vec{j}_T(\omega_n, \vec{k})\rho_m \vec{j}_T^\dagger(\omega_n, \vec{k}) - \frac{1}{2}\left\{\vec{j}_T^\dagger(\omega_n, \vec{k})\vec{j}_T(\omega_n, \vec{k}), \rho_m\right\}\right)$$

$$+ \sum_{\omega_n>0} \gamma_+(\omega_n) \int d\Omega(\vec{k}) \left(\vec{j}_T^\dagger(\omega_n, \vec{k})\rho_m \vec{j}_T(\omega_n, \vec{k}) - \frac{1}{2}\left\{\vec{j}_T(\omega_n, \vec{k})\vec{j}_T^\dagger(\omega_n, \vec{k}), \rho_m\right\}\right),$$

where $d\Omega(\vec{k})$ denotes the element of the solid angles in the direction of the wavevector $\vec{k}$.

The last two terms on the right-hand side in (12.112) represent the dissipator of the master equation describing incoherent processes, namely induced absorption processes, the rates of which are proportional to $\gamma_+(\omega_n) = \omega_n N(\omega_n)/8\pi^2$, as well as induced emission and spontaneous processes taking place with a rate which is proportional to $\gamma_-(\omega_n) = \omega_n[N(\omega_n) + 1]/8\pi^2$.

The coherent Hamiltonian part of the dynamics contains the Coulomb interaction H_C. Additionally, it is modified through the presence of the radiation field which leads to contributions from the Lamb shift Hamiltonian

$$H_{LS} = \sum_{\omega_n} P \int \frac{d^3k}{2(2\pi)^3\omega} \frac{1}{\omega_n - \omega} \vec{j}_T^\dagger(\omega_n, \vec{k})\vec{j}_T(\omega_n, \vec{k})$$ (12.113)

and from the Stark shift Hamiltonian

$$H_{SS} = \sum_{\omega_n} P \int \frac{d^3k}{2(2\pi)^3\omega} \frac{N(\omega)}{\omega_n - \omega} \left(\vec{j}_T^\dagger(\omega_n, \vec{k})\vec{j}_T(\omega_n, \vec{k}) - \vec{j}_T(\omega_n, \vec{k})\vec{j}_T^\dagger(\omega_n, \vec{k})\right).$$

(12.114)

It is easy to check that the density matrix equation (12.112) leads to the various forms for the quantum optical master equation used in previous chapters (note the change of units here).

12.2.3.2 Mass renormalization and Lamb shift

Let us have a closer look at the Lamb shift Hamiltonian (12.113). The contribution from H_{LS} is formally infinite and must be renormalized according to the renormalization procedure of QED (see, e.g. Weinberg, 1995). For our purposes it suffices to make a few remarks on the non-relativistic treatment of this term.

Invoking the dipole approximation and introducing the spectral decomposition of the system Hamiltonian through $H_S = \sum_n E_n |n\rangle\langle n|$, we find for the shift ΔE_n of the n-th level induced by H_{LS}:

$$\Delta E_n = -\frac{e^2}{6\pi^2 m^2} \sum_{m \neq n} \mathrm{P} \int_0^{\Omega_{\max}} d\omega \frac{\omega}{\omega - \omega_{mn}} |\langle n|\vec{p}|m\rangle|^2, \qquad (12.115)$$

where $\omega_{mn} = E_m - E_n$ and $\Omega_{\max}$ is an ultraviolet frequency cutoff. To leading order, the frequency integral diverges linearly with the cutoff. To split off the linearly divergent part we write the integrand as

$$\frac{\omega}{\omega - \omega_{mn}} = 1 + \frac{\omega_{mn}}{\omega - \omega_{mn}}. \qquad (12.116)$$

Correspondingly, the energy shift consists of two parts,

$$\Delta E_n = \Delta E_n' + \Delta E_n''. \qquad (12.117)$$

The first part is given by

$$\Delta E_n' = -\frac{e^2}{6\pi^2 m^2} \sum_{m \neq n} \int_0^{\Omega_{\max}} d\omega |\langle n|\vec{p}|m\rangle|^2 = -\frac{e^2 \Omega_{\max}}{6\pi^2 m^2} \langle n|\vec{p}^2|n\rangle. \qquad (12.118)$$

This term may be interpreted as the matrix element which stems from a renormalization of the mass in the kinetic energy of the system Hamiltonian. In fact, if we write the physical mass $m = m_0 + \delta m$ as a sum of a bare mass m_0 and an electromagnetic mass contribution δm, the total kinetic energy reads

$$\frac{\vec{p}^2}{2m} = \frac{\vec{p}^2}{2(m_0 + \delta m)} \approx \frac{\vec{p}^2}{2m_0} - \frac{\delta m}{2m_0^2}\vec{p}^2. \qquad (12.119)$$

Comparing this with (12.118) we see that the mass correction is given by

$$\delta m = \frac{e^2 \Omega_{\max}}{3\pi^2}. \qquad (12.120)$$

This is the same mass renormalization as found from the classical Abraham–Lorentz equation for the electron (see Section 12.3.4). If we take the cutoff $\Omega_{\max}$ to be on the order of the electron mass, we find that the mass correction is small, $\delta m/m = e^2/3\pi^2 \approx 0.0031$. We note that in a full relativistic treatment δm only depends logarithmically on the cutoff, that is we have (Bjorken and Drell, 1964)

$$\delta m = \frac{3e^2 m}{8\pi^2} \ln\left(\frac{\Omega_{\max}}{m}\right). \qquad (12.121)$$

The second part of the energy shift in (12.117) is given by

$$\Delta E_n'' = -\frac{e^2}{6\pi^2 m^2} \sum_{m\neq n} \mathrm{P} \int\limits_0^{\Omega_{\max}} d\omega \frac{\omega_{mn}}{\omega - \omega_{mn}} |\langle n|\vec{p}|m\rangle|^2. \tag{12.122}$$

In our present non-relativistic calculation this expression gives the main contribution to the observed atomic level shift caused by the vacuum fluctuations of the electromagnetic field. The sum over intermediate states can be worked out and yields the famous Bethe formula for the Lamb shift in hydrogen (Bjorken and Drell, 1964).

12.3 Decoherence by emission of bremsstrahlung

As a further application of expression (12.99) for the influence phase functional we investigate in this section the destruction of quantum coherence through the emission of bremsstrahlung. To this end, we investigate a simple, prototypical interference device for a charged particle and ask for the loss of coherence which is induced by the interaction with the radiation field. The interference device involves two possible paths of the charged particle: The initial wave packet is split into two components which first move apart and which are then recombined to measure locally their capability to interfere. This type of interference experiment thus involves a relative motion of two spatially separated components of the wave function. It turns out that the possibility of the emission of radiation leads to a (partial) destruction of quantum coherence which results in a reduction of the interference contrast.

An appropriate measure for the decoherence in the interference experiment is a certain relativistically covariant and gauge-invariant functional, the *decoherence functional*, which will be derived and determined explicitly in the present section. It will also be demonstrated that the obtained decoherence functional involves the typical features of the emission of bremsstrahlung. Explicit analytical expressions for the vacuum and the thermal contribution to the decoherence functional and for the corresponding coherence lengths are determined. These expressions reveal that bremsstrahlung leads to a fundamental decoherence mechanism which dominates for short times and which is present even in the electromagnetic field vacuum at zero temperature. The influence of bremsstrahlung on the centre-of-mass coordinate of a systems of many identical charged particles is also studied and shown to lead to a strong suppression of quantum coherence.

12.3.1 *Introducing the decoherence functional*

Let us consider an interference device of the type sketched in Fig. 12.1. A charged particle, say an electron, is emitted by the source Q and can move along two different world lines y_1 and y_2 to reach a screen at S, where an interference pattern is observed. These paths represent two quantum alternatives whose probability amplitudes may be described by two wave packets $|\Psi_1(t_i)\rangle$ and $|\Psi_2(t_i)\rangle$. With the help of the superposition principle we find that the wave function

$$|\Psi(t_i)\rangle = |\Psi_1(t_i)\rangle + |\Psi_2(t_i)\rangle \tag{12.123}$$

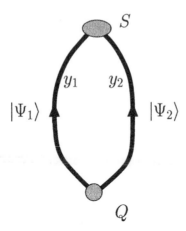

FIG. 12.1. Sketch of a prototypical interference device which is employed to introduce the decoherence functional. An electron emerges from the source Q and can follow two possible paths y_1 and y_2, which leads, in general, to an interference pattern observed on a screen at S. The two quantum alternatives may be described through the wave packets $|\Psi_1\rangle$ and $|\Psi_2\rangle$.

describes the physical situation depicted in the figure. Alternatively, the electron state can be represented in terms of the density matrix $\rho_m(t_i) = |\Psi(t_i)\rangle\langle\Psi(t_i)|$ which may be written as

$$\rho_m(t_i) = \rho_{11}(t_i) + \rho_{22}(t_i) + \rho_{12}(t_i) + \rho_{21}(t_i), \qquad (12.124)$$

where $\rho_{\alpha\beta}(t_i) = |\Psi_\alpha(t_i)\rangle\langle\Psi_\beta(t_i)|$ with $\alpha, \beta = 1, 2$. One observes the emergence of the interference term $\rho_{12}(t_i) + \rho_{21}(t_i)$. Remember that we are working in the interaction picture and that we therefore have $\rho_m(t) = \rho_m(t_i)$ for all times in the case of a vanishing coupling between matter and electromagnetic field.

Our aim is to determine with the help of the influence super-operator the structure of the electron density matrix ρ_m in the presence of the electromagnetic radiation field. An essential simplification is achieved if the matter current density can be treated as a classical current. This approximation can be justified under the following conditions. First, we assume that the wavelength $\bar{\lambda} = c/\omega$ of the photons emitted by the currents is large compared to the Compton wavelength $\bar{\lambda}_{\rm C}$ of the electron,

$$\bar{\lambda} \gg \bar{\lambda}_{\rm C} = \frac{\hbar}{mc}, \qquad (12.125)$$

and, thus, also large in comparison to the classical electron radius $r_e = \alpha\bar{\lambda}_{\rm C} \approx 2.8 \times 10^{-15}$m. This requirement is equivalent to $\hbar\omega \ll mc^2$. In this low-energy regime one may neglect pair creation and annihilation amplitudes and treat the matter current density as a given classical field (Jauch and Rohrlich, 1980; Cohen-Tannoudji, Dupont-Roc and Grynberg, 1998). The same procedure is used, for example, in the non-perturbative analysis of radiative corrections in the low-frequency limit (see below). In an experiment of the type sketched in Fig. 12.1 the paths involve an acceleration of the electron through a certain field of force. This force gives rise to a certain characteristic acceleration time τ_p. We define τ_p as the inverse of the highest frequency in the power spectrum of the force acting on the electron. In the following we call τ_p the preparation time since it can be interpreted as the time required to set into motion the interfering wave packets. As a consequence of the existence of such a characteristic time we have a natural upper cutoff $\Omega_{\max}$ for the frequency spectrum of the emitted radiation which is of the order

$$\Omega_{\max} \sim \frac{1}{\tau_p} = \frac{c}{\sigma_0}, \tag{12.126}$$

where the length scale σ_0 represents the order of the minimal wavelength of the radiation. Our above requirement thus takes the form

$$\sigma_0 \gg \bar{\lambda}_C. \tag{12.127}$$

This also implies that the characteristic acceleration time τ_p is large compared to r_e/c. It is known from classical electrodynamics that this condition ensures that the energy radiated is small compared to the kinetic energy of the particle and that therefore radiative damping effects are small (Jackson, 1999) (see also Section 12.3.4).

The second condition is that the motion of the current can be reasonably described within a semiclassical approximation. This leads to the requirement $\Delta v/v \ll 1$, where v is a typical velocity and Δv the velocity uncertainty. Assuming that the wave packets represent states of minimal uncertainty with spatial width Δx one is led to the condition

$$\frac{\Delta v}{v} \sim \frac{\hbar}{mv\Delta x} \ll 1, \tag{12.128}$$

or, equivalently,

$$\frac{\bar{\lambda}_{dB}}{\Delta x} \ll 1, \tag{12.129}$$

where $\bar{\lambda}_{dB} = \hbar/mv$ is the de Broglie wavelength. This is the typical condition for a semiclassical treatment.

In view of these conditions we now assume that $\rho_m(t_i)$ represents a state which is an approximate eigenstate of the current density. Thus, if $\rho_m(t_i)$ is a pure state,

$$\rho_m(t_i) = |\Psi(t_i)\rangle\langle\Psi(t_i)|, \tag{12.130}$$

we suppose that

$$j^\mu(x)|\Psi(t_i)\rangle \approx s^\mu(x)|\Psi(t_i)\rangle, \tag{12.131}$$

where $s^\mu(x)$ is a classical current density. Hence, we also have to the same degree of accuracy,

$$J_c^\mu(x)\rho_m(t_i) = [j^\mu(x), \rho_m(t_i)] \approx 0. \tag{12.132}$$

The initial state $\rho_m(t_i)$ does not necessarily have to be a pure state. It suffices to require (12.132), where

$$\langle j^\mu(x) \rangle = \mathrm{tr}_m \{ j^\mu(x)\rho_m(t_i) \} = s^\mu(x) \tag{12.133}$$

is the expectation value of the current density. In any case we immediately obtain with the help of (12.132) and expression (12.99) for the influence phase

$$\rho_m(t_f) \approx \rho_m(t_i). \tag{12.134}$$

This equation states that the system is essentially unaffected by the radiation field, i.e. by virtue of our assumption that the initial state is an approximate current eigenstate, the dynamics of the density matrix is nearly the same as that of the free system. The same conclusion can be drawn from an investigation of the dynamics of a Gaussian wave packet under the influence of the radiation field by use of the exact analytical expression for the propagator function in the dipole approximation (see Section 12.3.4).

Let us now return to the interference device and assume that the superposition (12.123) consists of two approximate current eigenstates,

$$j^\mu(x)|\Psi_1(t_i)\rangle \approx s_1^\mu(x)|\Psi_1(t_i)\rangle, \tag{12.135}$$
$$j^\mu(x)|\Psi_2(t_i)\rangle \approx s_2^\mu(x)|\Psi_2(t_i)\rangle, \tag{12.136}$$

where $s_1(x)$ and $s_2(x)$ are classical current densities. These currents are assumed to be concentrated within two world tubes around the paths y_1 and y_2 of the interference device, respectively (see Fig. 12.2). By virtue of eqn (12.135) we have

$$J_c^\mu(x)\rho_{11}(t_i) \approx J_c^\mu(x)\rho_{22}(t_i) \approx 0, \tag{12.137}$$
$$J_c^\mu(x)\rho_{12}(t_i) \approx (s_1^\mu(x) - s_2^\mu(x))\,\rho_{12}(t_i), \tag{12.138}$$
$$J_a^\mu(x)\rho_{12}(t_i) \approx (s_1^\mu(x) + s_2^\mu(x))\,\rho_{12}(t_i), \tag{12.139}$$

and

$$\mathcal{L}_C(x)\rho_{11}(t_i) \approx \mathcal{L}_C(x)\rho_{22}(t_i) \approx 0, \tag{12.140}$$
$$\mathcal{L}_C(x)\rho_{12}(t_i) \approx -i\,(\mathcal{H}_{C1}(x) - \mathcal{H}_{C2}(x))\,\rho_{12}(t_i), \tag{12.141}$$

where

$$\mathcal{H}_{C1,2}(x) = \frac{1}{2}\int d^3y\, \frac{s_{1,2}^0(x^0, \vec{x})s_{1,2}^0(x^0, \vec{y})}{4\pi|\vec{x} - \vec{y}|} \tag{12.142}$$

are the Coulomb energy densities associated with the current densities $s_1^\mu(x)$ and $s_2^\mu(x)$, respectively. We may suppose that the corresponding Coulomb energies

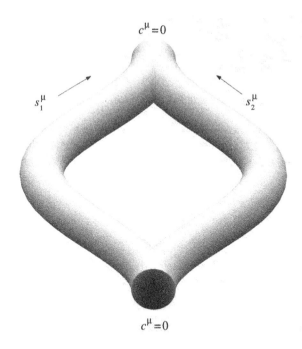

$$c^\mu = 0$$

$$s_1^\mu \qquad s_2^\mu$$

$$c^\mu = 0$$

FIG. 12.2. A closed current world tube indicating the support of the current difference density $c^\mu = (s_1^\mu - s_2^\mu)/\sqrt{2}$. The world tube is located around the closed path $C = y_1 - y_2$ formed by following y_1 in the positive and y_2 in the negative direction (see Fig. 12.1).

for both possible paths are equal to each other. The expression (12.99) for the influence phase functional now immediately leads to

$$\rho_m(t_f) \approx \rho_{11}(t_i) + \rho_{22}(t_i) + \exp(i\Phi)\rho_{12}(t_i) + \exp(-i\Phi^*)\rho_{21}(t_i), \qquad (12.143)$$

where

$$i\Phi \equiv i\Phi[s_1, s_2] \qquad\qquad\qquad\qquad (12.144)$$

$$= \int_{t_i}^{t_f} d^4x \int_{t_i}^{t} d^4x' \frac{i}{2} D^{\mathrm{T}}(x - x')_{\mu\nu} \left(s_1^\mu(x) - s_2^\mu(x)\right)\left(s_1^\nu(x') + s_2^\nu(x')\right)$$

$$- \frac{1}{4}\int_{t_i}^{t_f} d^4x \int_{t_i}^{t_f} d^4x' D_1^{\mathrm{T}}(x - x')_{\mu\nu}\left(s_1^\mu(x) - s_2^\mu(x)\right)\left(s_1^\nu(x') - s_2^\nu(x')\right).$$

We observe that the electromagnetic field affects the interference terms through a complex phase $\Phi = \Phi[s_1, s_2]$ which is a functional of the two possible classical paths y_1 and y_2, or, more precisely, of the associated current densities $s_1^\mu(x)$ and

$s_2^\mu(x)$. The real part of Φ leads to a distortion of the interference pattern. The imaginary part of Φ, on the other hand, yields a suppression of the interference terms in (12.143) given by the factor

$$|\exp(i\Phi)| = \exp\Gamma. \qquad (12.145)$$

The exponent Γ will be referred to as the decoherence functional (compare the discussion in Section 4.1). From the second term in eqn (12.144) we infer that Γ represents a bilinear functional of the current difference

$$c^\mu(x) = \frac{1}{\sqrt{2}}\left(s_1^\mu(x) - s_2^\mu(x)\right) \qquad (12.146)$$

and may be written as follows,

$$\Gamma[c] = -\frac{1}{2}\int\limits_{t_i}^{t_f} d^4x \int\limits_{t_i}^{t_f} d^4x'\, D_1^{\mathrm{T}}(x-x')_{\mu\nu} c^\mu(x) c^\nu(x'). \qquad (12.147)$$

12.3.2 *Physical interpretation*

To bring $\Gamma[c]$ into a more convenient form let us define t_0 to be the time corresponding to Q (see Fig. 12.1), that is the time at which the wave packet is separated into two components, while t_f denotes the final time when both packets are recombined at S. It is obvious that the current difference $c^\mu(x)$ vanishes for times prior to t_0 and for times later than the final time t_f. In fact, the support of the current difference $c^\mu(x)$ lies in the interior of a closed world tube around the loop

$$C = y_1 - y_2 \qquad (12.148)$$

which is formed by following y_1 in the positive and y_2 in the negative direction, as illustrated in Fig. 12.2. Current conservation, $\partial_\mu c^\mu = 0$, therefore enables us to write the decoherence functional as (compare the discussion following eqn (12.102))

$$\Gamma[c] = -\frac{1}{2}\int d^4x \int d^4x'\, D_1(x-x')\left[-c^\mu(x) c_\mu(x')\right]. \qquad (12.149)$$

We observe that $\Gamma[c]$ represents a relativistically covariant and gauge invariant functional. Moreover, $\Gamma[c]$ is Lorentz invariant in the vacuum case as may be seen explicitly from the fact that $D_1^{\mathrm{vac}}(x-x')$ is given by the invariant function (12.39).

Equation (12.149) suggests several interesting physical interpretations. With the help of the Feynman propagator (12.84) and the two-point correlation functions (12.86) and (12.87) we can express the complex phase factor $\exp(i\Phi)$ in eqn (12.143) as follows,

$$\exp(i\Phi) = A[s_1]A[s_2]^* \tag{12.150}$$

$$\times \exp\left[\frac{1}{2}\int d^4x \int d^4x' D_-(x-x')_{\mu\nu}s_1^\mu(x)s_2^\nu(x')\right.$$

$$\left. +\frac{1}{2}\int d^4x \int d^4x' D_+(x-x')_{\mu\nu}s_2^\mu(x)s_1^\nu(x')\right],$$

where $A[j]$ is defined in eqn (12.101). This form for the phase factor was first derived by Ford (1993) for the case of zero temperature with the aim of studying the influence of conducting boundaries on electron coherence. As we have seen, at zero temperature $A[j]$ is the vacuum-to-vacuum amplitude in the presence of a classical current density $j^\mu(x)$. The first term on the right-hand side in (12.150) is thus the product of the vacuum-to-vacuum amplitudes in the presence of the current densities s_1^μ and s_2^μ. This contribution describes virtual processes in which photons are emitted and reabsorbed by either the currents s_1^μ and s_2^μ. Correspondingly, the exponential on the right-hand side of eqn (12.150) is the contribution of the emission of real photons. These processes also contribute to the decoherence functional since a photon can be emitted by both currents and carries away information on the path taken by the electron. Moreover, at finite temperatures thermally induced emission and absorption processes occur.

Expression (12.150) also answers the question as to whether virtual vacuum processes or real photon emissions are responsible for the decoherence effect expressed by the functional $\Gamma[c]$. We see that the decoherence factor (12.145) depends on virtual processes through the Feynman propagator $D_F(x-x')_{\mu\nu}$ (which is contained in the vacuum-to-vacuum amplitudes) as well as on real processes described by the Green's functions $D_\pm(x-x')_{\mu\nu}$. Thus, it is the combined effect of virtual and real processes which leads to decoherence. The physical reason is that the mere possibility of real photon emissions also reduces the vacuum-to-vacuum amplitudes which can already lead to a reduction of the interference contrast. For example, one can think of an interference device in which the path y_1 describes a uniform motion of the electron, whereas the other path y_2 is strongly bent, involving a large acceleration of the electron. Suppose we observe the photons emitted in the experiment. If we know that no photon has been emitted it is then very likely that the electron has taken the path y_1, which results in a suppression of the interference pattern formed by the electrons of the corresponding sub-ensemble. This situation is similar to that of a two-level atom which is initially in a superposition of the excited state and the ground state. If we find that the atom did not emit a photon during a large time interval (large compared to the inverse of the emission rate) we have effectively measured the state of the atom to be the ground state. Thus, the off-diagonal terms of the atomic density matrix approximately vanish without the emission of real photons.

It is also interesting to express the decoherence factor directly in terms of the amplitude $A[c]$. Invoking (12.91) we find

$$\exp\left(\Gamma[c]\right) = |A[c]|^2 .$$ (12.151)

Obviously, we have $\Gamma[c] \leq 0$, and $\Gamma[c] = 0$ for $s_1^\mu = s_2^\mu$, that is for a vanishing current difference, $c^\mu = 0$. Equation (12.151) gives rise to another interesting interpretation: The decoherence factor which multiplies the interference term is given by the no-photon emission probability in the presence of the current density c^μ. This current is the same as the current which would be created by two particles with opposite charges $\pm e/\sqrt{2}$, one moving along y_1 and the other along y_2. The smaller the vacuum-to-vacuum amplitude for this current density the larger the reduction of the interference contrast. This must have been expected since it is the difference between the currents s_1 and s_2 which determines the extent to which the two possible paths can be distinguished, and, thus, the degree of the loss of coherence.

These interpretations in terms of the emitted photons must be taken, however, with some care. The reason is that we consider here processes on a finite time scale and not transitions between asymptotic states. It is well known that certain matter currents emit an infinite number of long-wavelength (soft) photons whose frequencies approach zero, while their total energy adds up to a finite value. This is the so-called infrared catastrophe (Weinberg, 1995; Jauch and Rohrlich, 1980) which arises in the perturbative calculation of radiative corrections to any process involving charged matter. The complete removal of infrared divergences requires a non-perturbative treatment in which the amplitudes for the emission of real and virtual soft photons are summed to all orders, such that the processes involving real and virtual photons become indistinguishable in the low-frequency limit. Infrared divergences can be shown to cancel provided a finite resolution $\Omega_{\min}$ for the photodetection is introduced: Insisting on the perturbative picture, one could say that there is always an infinite number of quanta, namely those whose frequency is lower than $\Omega_{\min}$, which escapes undetected and cannot be observed in principle.

Our analysis treats the matter current classically but it is non-perturbative (Jauch and Rohrlich, 1980). In view of the above considerations it is obvious that the decoherence functional $\Gamma[c]$ does not lead to infrared divergences since it describes a process taking place in the finite time interval between the splitting of the wave packet at t_0 and the recombination at t_f. This gives rise to a natural frequency resolution of the order

$$\Omega_{\min} \sim \frac{1}{t_f - t_0} .$$ (12.152)

The emergence of this effective infrared cutoff will be seen explicitly in the calculations of the next subsection, where it will be demonstrated that the arising integrals over the photon frequencies converge at the lower limit $\omega \longrightarrow 0$. In addition we also have an ultraviolet cutoff $\Omega_{\max}$ which has already been introduced in eqn (12.126). This cutoff can be accounted for by the introduction of a finite width σ_0 characterizing the current world tube, as will be seen in the next subsection.

12.3.3 Evaluation of the decoherence functional

We wish to evaluate here the decoherence functional for some specific situations. On using eqn (12.149) and the Fourier representation (12.38) for the anticommutator function $D_1(x - x')$ we find

$$\Gamma[c] = - \int \frac{d^3k}{2(2\pi)^3\omega} \coth(\omega/2k_\mathrm{B}T) \left[-c^\mu(k)c_\mu^*(k) \right], \qquad (12.153)$$

where we have introduced the Fourier transform

$$c^\mu(k) \equiv \int d^4x \exp(-ikx)c^\mu(x) \qquad (12.154)$$

of the current difference $c^\mu(x)$.

Let us first show explicitly how a finite width of the current world tubes gives rise to an ultraviolet cutoff scale. To this end, the currents $s_1^\mu(x)$ and $s_2^\mu(x)$ are taken to be concentrated within world tubes of spatial extent σ_0 around the world lines $y_1 = y_1(\tau)$ and $y_2 = y_2(\tau)$. The world lines are parametrized by their proper time τ, such that

$$u_{1,2}^\mu(\tau) = \frac{dy_{1,2}^\mu(\tau)}{d\tau} \qquad (12.155)$$

are the corresponding 4-velocities. To be specific we write

$$s_{1,2}^\mu(x) = e \int d\tau\, u_{1,2}^\mu(\tau)\delta_{\sigma_0}(x - y_{1,2}(\tau)), \qquad (12.156)$$

where

$$\delta_{\sigma_0}(x - x') = \delta(x_0 - x_0') \frac{1}{(2\pi\sigma_0^2)^{3/2}} \exp\left[-\frac{(\vec{x} - \vec{x}')^2}{2\sigma_0^2} \right] \qquad (12.157)$$

is a smeared δ-function described by a Gaussian with width σ_0. We remark that we neglect the spin contribution to the current density. This is justified as long as the length scales involved in the problem under consideration are large compared to the Compton wavelength (Ford, 1993).

On using eqn (12.156) in eqn (12.154) the Fourier transform of the current difference is found to be

$$c^\mu(k) = \frac{e}{\sqrt{2}} \left[\int d\tau\, u_1^\mu(\tau) \exp(-iky_1(\tau)) - \int d\tau\, u_2^\mu(\tau) \exp(-iky_2(\tau)) \right]$$
$$\times \exp\left[-\frac{1}{2}\sigma_0^2\omega^2 \right]. \qquad (12.158)$$

We see that the finite width σ_0 of the current world tubes yields an effective ultraviolet cutoff $\Omega_{\max} \sim \sigma_0^{-1}$ as given in eqn (12.126). Our main interest is an estimation of the decoherence functional for some specific situations. We

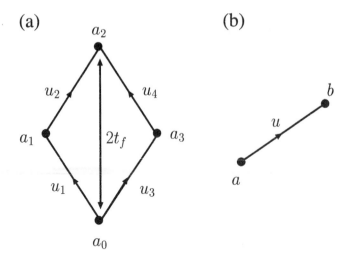

FIG. 12.3. (a) The closed loop $C = y_1 - y_2$ used for an explicit determination of the decoherence functional $\Gamma[c]$. The loop consists of four straight world line segments with 4-velocities u_1, u_2, u_3, u_4. The vertices are located at the space-time points a_0, a_1, a_2, a_3. As indicated, the loop corresponds to the total time $2t_f$. (b) A single line segment with initial point a, endpoint b and 4-velocity u.

therefore ignore in the following the specific form of the cutoff function in (12.158) and work with a sharp cutoff at the maximal frequency $\omega = \Omega_{\max}$. It will be seen below that the final expression for $\Gamma[c]$ depends on $\Omega_{\max}$ only through $\ln \Omega_{\max}$. This extremely weak logarithmic dependence shows that the precise value of σ_0 or of the preparation time τ_p is rather irrelevant. The important point to note here is that the emergence of an effective ultraviolet cutoff has a clear physical origin.

Thus, we now write eqn (12.153) as

$$\Gamma[c] = -\frac{1}{16\pi^3} \int\limits_0^{\Omega_{\max}} d\omega \, \omega \coth(\omega/2k_{\mathrm{B}}T) \int d\Omega(\hat{k}) \left[-c^\mu(k) c_\mu^*(k) \right], \qquad (12.159)$$

where $d\Omega(\hat{k})$ denotes the element of the solid angle in the direction of the unit vector $\hat{k} \equiv \vec{k}/|\vec{k}|$. The Fourier transform of the current difference can be expressed as a loop integral over the closed loop $C = y_1 - y_2$,

$$c^\mu(k) = \frac{e}{\sqrt{2}} \oint\limits_C dx^\mu \exp(-ikx). \qquad (12.160)$$

For simplicity let us consider the case that the loop C consists of four straight world line segments (see Fig. 12.3(a)). The vertices of the loop are denoted by a_0, a_1, a_2, a_3, whereas the corresponding 4-velocities are given by

$$u_n^\mu = \gamma_n(1, \vec{v}_n), \quad \gamma_n = (1 - |\vec{v}_n|^2)^{-1/2}. \tag{12.161}$$

We further assume that the arrangement is symmetric, that is we suppose that $u_1 = u_4$, $u_2 = u_3$, and that $a_1 - a_0 = a_2 - a_3$, $a_2 - a_1 = a_3 - a_0$. For a single line segment with initial point a, endpoint b and 4-velocity u (see Fig. 12.3(b)) we obtain

$$\int_a^b dx \exp(-ikx) = i\frac{u}{ku} \left[e^{-ikb} - e^{-ika}\right]. \tag{12.162}$$

With the help of this formula the Fourier transform (12.160) of the current difference is found to be,

$$c(k) = \frac{ie}{\sqrt{2}} \left\{ +\frac{u_1}{ku_1} \left[e^{-ika_1} - e^{-ika_0}\right] + \frac{u_2}{ku_2} \left[e^{-ika_2} - e^{-ika_1}\right] \right.$$
$$\left. -\frac{u_3}{ku_3} \left[e^{-ika_3} - e^{-ika_0}\right] - \frac{u_4}{ku_4} \left[e^{-ika_2} - e^{-ika_3}\right] \right\}. \tag{12.163}$$

It should be noted that $k_\mu c^\mu(k) = 0$ as required by current conservation. Note further that $c^\mu(k)$ shows the correct behaviour under Lorentz transformations. In particular, one finds that $c^\mu(k)$ transforms into $c^\mu(k) \exp(-ikb)$ under a space-time translations by the 4-vector b. If we now use the symmetry properties of the loop we arrive at

$$c(k) = \frac{ie}{\sqrt{2}} \left[\frac{u_2}{ku_2} - \frac{u_4}{ku_4}\right] \mathcal{G}(k), \tag{12.164}$$

where we have introduced

$$\mathcal{G}(k) = e^{-ika_2} \left(1 - e^{ik(a_2-a_3)}\right) \left(1 - e^{ik(a_2-a_1)}\right). \tag{12.165}$$

Using these results one is led to the following expression for the decoherence functional,

$$\Gamma[c] = \frac{\alpha}{8\pi^2} \int_0^{\Omega_{\text{max}}} \frac{d\omega}{\omega} \coth(\omega/2k_B T) \int d\Omega(\hat{k})\omega^2 \left[\frac{u_2}{ku_2} - \frac{u_4}{ku_4}\right]^2 |\mathcal{G}(k)|^2. \tag{12.166}$$

We denote the time interval associated with a single line segment of the loop by t_f, that is we set $t_0 = -t_f$ and $t_f - t_0 = 2t_f$ (see Fig. 12.3(a)). Then we have

$$k(a_2 - a_3) = \omega t_f(1 - \hat{k} \cdot \vec{v}_4), \tag{12.167}$$
$$k(a_2 - a_1) = \omega t_f(1 - \hat{k} \cdot \vec{v}_2). \tag{12.168}$$

In order to estimate further the expression (12.166) we approximate

$$k(a_2 - a_3) \approx k(a_2 - a_1) \approx \omega t_f, \tag{12.169}$$

which leads to

$$|\mathcal{G}(k)|^2 = 8\left[(1 - \cos\omega t_f) - \frac{1}{4}(1 - \cos 2\omega t_f)\right]. \qquad (12.170)$$

This allows us to estimate the decoherence functional (12.166) as follows,

$$\Gamma[c] \approx \frac{\alpha}{\pi^2} \int\limits_0^{\Omega_{\text{max}}} \frac{d\omega}{\omega} \coth(\omega/2k_BT) \left[(1 - \cos\omega t_f) - \frac{1}{4}(1 - \cos 2\omega t_f)\right]$$

$$\times \int d\Omega(\hat{k})\omega^2 \left[\frac{u_2}{ku_2} - \frac{u_4}{ku_4}\right]^2. \qquad (12.171)$$

Let us first concentrate on the integral over the photon frequencies ω, that is on the first integral on the right-hand side of (12.171). The integrand is proportional to ω^{-1}, which is a typical signature for the spectrum of bremsstrahlung (Jackson, 1999). In addition to vacuum bremsstrahlung there may be thermally induced emission and absorption processes (Itzykson and Zuber, 1980), which are embodied in the factor $\coth(\omega/2k_BT)$. At zero temperature (vacuum field) this factor may be replaced by 1.

In order to calculate the frequency integral it turns out to be useful to decompose it into a vacuum contribution and a thermal contribution which vanishes for $T = 0$. We therefore write

$$F \equiv \int\limits_0^{\Omega_{\text{max}}} \frac{d\omega}{\omega} \coth(\omega/2k_BT)\,(1 - \cos\omega t_f) \equiv F_{\text{vac}} + F_{\text{th}}, \qquad (12.172)$$

where

$$F_{\text{vac}} \equiv \int\limits_0^{\Omega_{\text{max}}} \frac{d\omega}{\omega}\,(1 - \cos\omega t_f) \qquad (12.173)$$

is the vacuum contribution, while

$$F_{\text{th}} \equiv \int\limits_0^{\Omega_{\text{max}}} \frac{d\omega}{\omega}\,[\coth(\omega/2k_BT) - 1]\,(1 - \cos\omega t_f) \qquad (12.174)$$

is the thermal contribution. The frequency integral F_{vac} can be evaluated as follows. Substituting $x = \omega t_f$ we get

$$F_{\text{vac}} = \int\limits_0^{\Omega_{\text{max}}t_f} \frac{dx}{x}\,(1 - \cos x) = \ln(g\Omega_{\text{max}}t_f) + O\left(\frac{1}{\Omega_{\text{max}}t_f}\right), \qquad (12.175)$$

where $\ln g \approx 0.577$ is Euler's constant (Gradshteyn and Ryzhik, 1980). For $\Omega_{\text{max}}t_f \gg 1$ we thus have asymptotically

$$F_{\text{vac}} \approx \ln\left(g\Omega_{\text{max}}t_f\right).$$ (12.176)

This relation demonstrates that the vacuum integral over the photon frequencies converges at the lower limit $\omega \longrightarrow 0$ and that it gives rise to an effective infrared cutoff of order $\Omega_{\text{min}} \sim 1/t_f$, as discussed in the previous subsection (see eqn (12.152)). We also observe that the vacuum frequency integral increases weakly with the logarithm of $\Omega_{\text{max}}t_f$. As indicated in eqn (12.176) we keep for simplicity in the following only the leading contribution in our expressions. It should be kept in mind, however, that one can include without difficulty the terms of higher order which vanish in the limit $\Omega_{\text{max}}t_f \longrightarrow \infty$. The integral (12.173) has already been evaluated in Section 4.2.2, where we have used, however, an exponential cutoff. A comparison with the formula (4.54) shows that a change of the cutoff shape introduced only a small change in the value of the integral.

The thermal contribution F_{th} has also been evaluated in Section 4.2.2 (see eqn (4.55)) under the condition $k_B T \ll \hbar\Omega_{\text{max}}$, which will also be assumed here. To give an example for this condition we take the ultraviolet cutoff $\Omega_{\text{max}} \sim 10^{19}$ s^{-1}, corresponding to a length scale of order $100\bar{\lambda}_C$. The above condition then means that $T \ll 10^8$ K. Thus we obtain with the help of (4.55)

$$F_{\text{th}} \approx \ln\left(\frac{\sinh\left(t_f/\tau_B\right)}{t_f/\tau_B}\right).$$ (12.177)

On using the results (12.176) and (12.177) we can now determine the frequency integral in eqn (12.171),

$$\int_0^{\Omega_{\text{max}}} \frac{d\omega}{\omega}\coth(\beta\omega/2)\left[(1-\cos\omega t_f) - \frac{1}{4}(1-\cos 2\omega t_f)\right]$$ (12.178)

$$\approx \frac{3}{4}\ln\left(g\Omega_{\text{max}}t_f\right) + \ln\left(\frac{\sinh(t_f/\tau_B)}{t_f/\tau_B}\right) - \frac{1}{4}\ln\left(\frac{\sinh(2t_f/\tau_B)}{2t_f/\tau_B}\right).$$

It remains to calculate the frequency-independent angular integral in eqn (12.171), that is we have to evaluate integrals of the form

$$I(u_n, u_m) \equiv \int d\Omega(\hat{k})\omega^2 \frac{u_n u_m}{(k u_n)(k u_m)}$$ (12.179)

$$= \int d\Omega(\hat{k})\frac{1 - \vec{v}_n \cdot \vec{v}_m}{(1 - \hat{k}\cdot\vec{v}_n)(1 - \hat{k}\cdot\vec{v}_m)},$$ (12.180)

where $n, m = 2, 4$. We note that the combination $d\Omega(\hat{k})\omega^2$ is an invariant quantity, such that $I(u_n, u_m)$ is a Lorentz-invariant integral. To determine this integral we may therefore transform to a coordinate system in which the second velocity is equal to zero, that is $\vec{v}_m = 0$. In this system the magnitude $v_n = |\vec{v}_n|$ of the first velocity is equal to the relative velocity

$$v_{nm} \equiv \sqrt{1 - \frac{1}{(u_n u_m)^2}}, \tag{12.181}$$

which is, by definition, a relativistic invariant quantity. Thus we arrive at

$$I(u_n, u_m) = \int d\Omega(\hat{k}) \frac{1}{1 - \hat{k} \cdot \vec{v}_n} = \frac{4\pi}{v_{nm}} \tanh^{-1} v_{nm}. \tag{12.182}$$

This formula is correct also for the case $u_n = u_m$, giving $I(u_n, u_n) = 4\pi$, as may be seen directly from the expansion of $\tanh^{-1}(x)$ for small arguments,

$$\tanh^{-1} x = x + \frac{1}{3}x^3 + \frac{1}{5}x^5 + \dots . \tag{12.183}$$

Thus we find,

$$\int d\Omega(\hat{k})\omega^2 \left[\frac{u_2}{ku_2} - \frac{u_4}{ku_4} \right]^2 = I(u_2, u_2) + I(u_4, u_4) - 2I(u_2, u_4)$$

$$= -8\pi \left(\frac{1}{v_{24}} \tanh^{-1} v_{24} - 1 \right). \tag{12.184}$$

Substituting (12.184) and (12.178) into eqn (12.171) we finally obtain

$$\Gamma[c] = \Gamma_{\text{vac}} + \Gamma_{\text{th}}, \tag{12.185}$$

where

$$\Gamma_{\text{vac}} \approx -\frac{6\alpha}{\pi} \ln\left(g\Omega_{\max} t_f\right) \left(\frac{1}{v_{24}} \tanh^{-1} v_{24} - 1 \right). \tag{12.186}$$

is the vacuum decoherence functional and

$$\Gamma_{\text{th}} \approx -\frac{8\alpha}{\pi} \left[\ln\left(\frac{\sinh(t_f/\tau_B)}{t_f/\tau_B} \right) - \frac{1}{4} \ln\left(\frac{\sinh(2t_f/\tau_B)}{2t_f/\tau_B} \right) \right]$$

$$\times \left(\frac{1}{v_{24}} \tanh^{-1} v_{24} - 1 \right) \tag{12.187}$$

is the thermal contribution to the decoherence functional. As expected, we see from these expressions that $\Gamma[c]$ strongly depends on the relative velocity v_{24} which is due to the fact that the decoherence is caused by the emission of bremsstrahlung. The larger v_{24} the larger is the involved acceleration of the charged particle which creates the radiation field.

An important result is that bremsstrahlung leads to a partial destruction of coherence even at zero temperature. The magnitude of the vacuum contribution

Γ_{vac} is seen to increase as the logarithm of the time t_f if the relative velocity is held fixed. This weak dependence is connected to the effective infrared resolution $\Omega_{\text{min}} \sim 1/t_f$ of the interference device: For increasing t_f photons of lower and lower frequencies could in principle be detected and, thus, more and more information is lost on tracing over the photon field. On the other hand, the term within the square brackets in (12.187) approaches $t_f/2\tau_B$ for $t_f \gg \tau_B$. Thus, keeping fixed the relative velocity the magnitude of the thermal contribution Γ_{th} increases linearly with t_f for times $t_f \gg \tau_B$. This describes the decohering influence of absorption and emission processes induced by the thermal field. It follows that for short times the vacuum contribution dominates, while decoherence is mainly due to thermally induced processes for large times. The time t_f^* corresponding to the cross-over between these two regimes is determined by the relation

$$\ln \left(g\Omega_{\text{max}} t_f^* \right) = \frac{2}{3} \frac{t_f^*}{\tau_B}. \tag{12.188}$$

Taking $\Omega_{\text{max}} \sim 10^{19}$ s^{-1} and $T = 1$ K we find from this condition that the cross-over time is of order

$$t_f^* \sim 30 \, \tau_B \sim 10^{-10} \text{ s}. \tag{12.189}$$

This means that for the given example the vacuum decoherence dominates for times small compared to 10^{-10} s.

To facilitate the further discussion, let us investigate the case of opposite velocities with equal magnitude, that is $\vec{v}_1 = \vec{v}_4 = -\vec{v}_2 = -\vec{v}_3$ (see Fig. 12.3(a)). The relative velocity is then found to be

$$v_{24} = \frac{2v}{1 + v^2}, \tag{12.190}$$

where

$$v = \frac{|\vec{a}_1 - \vec{a}_3|}{2t_f}. \tag{12.191}$$

This situation corresponds to the case of two wave packets in a superposition which first move apart with opposite mean velocities $\vec{v}_1$ and $\vec{v}_3 = -\vec{v}_1$, respectively, and, having reached their maximal distance $|\vec{a}_1 - \vec{a}_3|$, approach each other again with velocities $\vec{v}_2$ and $\vec{v}_4 = -\vec{v}_2$. For non-relativistic velocities we have $v_{24} \approx 2v$ and we may use the expansion (12.183) to obtain

$$\Gamma_{\text{th}} \approx -\frac{16\alpha}{3\pi} \frac{t_f}{\tau_B} v^2, \quad t_f \gg \tau_B. \tag{12.192}$$

One can then ask the following question: Given a fixed electron energy, that is a fixed velocity v, how far can we coherently separate the components of the electronic state without exceeding a given threshold $|\Gamma_0|$ for the decoherence? Provided the thermal contribution dominates, eqn (12.192) leads to the condition

$$\frac{16\alpha}{3\pi}\frac{t_f}{\tau_B}v^2 = |\Gamma_0|, \qquad (12.193)$$

from which we obtain the maximal possible separation

$$d_{\max} = 2vt_f = \frac{3\pi|\Gamma_0|}{8\alpha}\frac{c\tau_B}{v/c}. \qquad (12.194)$$

Choosing $|\Gamma_0| = 0.01$, which corresponds to a threshold of 1% decoherence, we find that the maximal distance at $T = 300$ K is given by

$$d_{\max} \approx \frac{4\mu\text{m}}{v/c}. \qquad (12.195)$$

This shows that one can achieve rather large coherent separations for non-relativistic electrons. For example, in the experiment performed by Hasselbach, Kiesel and Sonnentag (2000) an electronic beam was coherently separated by a lateral distance of about $d = 100\,\mu\text{m}$. To compare this experiment with our results we take an electron energy of $1\,\text{keV}$ and use a fixed value of $10\,\text{cm}$ for the distance corresponding to the time interval from t_0 to t_f. For $|\Gamma_0| = 0.01$ condition (12.193) then yields $d_{\max} \approx 4.5\,\text{cm}$ at $T = 1\,\text{K}$ and $d_{\max} \approx 0.26\,\text{cm}$ at $T = 300\,\text{K}$. Note that the quantity v in (12.193) represents the magnitude of the lateral component of the electron velocity in the experiment, which is due to the fact that it is the relative velocity that enters the formula for the decoherence functional. The values obtained for $d_{\max}$ are large compared with the lateral distance d of the experiment, demonstrating that our theory is in full agreement with experiment.

The result expressed through eqns (12.186) and (12.187) can also be discussed from another point of view. Namely, instead of keeping fixed the velocity v (eqn (12.191)), we consider a fixed maximal spatial distance $|\vec{a}_1 - \vec{a}_3|$ between the paths. Thus, for increasing t_f the velocity v becomes smaller and smaller and, consequently, the decoherence effect through bremsstrahlung becomes smaller and smaller. For large enough times v is non-relativistic such that the vacuum and the thermal contribution to the decoherence functional are given by

$$\Gamma_{\text{vac}} \approx -\frac{2\alpha}{\pi}\ln\left(g\Omega_{\max}t_f\right)\frac{|\vec{a}_1 - \vec{a}_3|^2}{t_f^2}, \qquad (12.196)$$

and

$$\Gamma_{\text{th}} \approx -\frac{8\alpha}{3\pi}\left[\ln\left(\frac{\sinh(t_f/\tau_B)}{t_f/\tau_B}\right) - \frac{1}{4}\ln\left(\frac{\sinh(2t_f/\tau_B)}{2t_f/\tau_B}\right)\right]\frac{|\vec{a}_1 - \vec{a}_3|^2}{t_f^2}. \qquad (12.197)$$

According to eqn (12.196) the magnitude of Γ_{vac} decreases essentially as t_f^{-2}, while eqn (12.197) shows that the magnitude of the thermal contribution Γ_{th} decreases as t_f^{-1} for $t_f \gg \tau_B$,

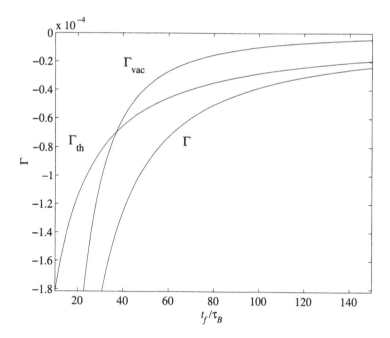

FIG. 12.4. The vacuum contribution Γ_{vac} and the thermal contribution Γ_{th} of the decoherence functional Γ according to eqns (12.186) and (12.187). For a fixed maximal distance $|\vec{a}_1 - \vec{a}_3| = c\tau_B$ between the paths the two contributions are plotted against the time t_f which is measured in units of the thermal correlation time τ_B. Parameters: $T = 1$ K and $\Omega_{\text{max}} = 10^{19}$ s^{-1}.

$$\Gamma_{\text{th}} \approx -\frac{4\alpha}{3\pi} \frac{|\vec{a}_1 - \vec{a}_3|^2}{t_f \tau_B}. \tag{12.198}$$

We again observe the cross-over between two regimes of times: For short times the vacuum decoherence dominates, whereas the thermally induced decoherence dominates for large times. This can be seen in Fig. 12.4 where we have plotted the expressions (12.186) and (12.187) as a function of t_f for a fixed value of $|\vec{a}_1 - \vec{a}_3|$.

The expressions (12.196) and (12.198) suggest we define a vacuum and a thermal coherence length by means of

$$\Gamma_{\text{vac}} \equiv -\frac{|\vec{a}_1 - \vec{a}_3|^2}{2L(t_f)_{\text{vac}}^2}, \tag{12.199}$$

$$\Gamma_{\text{th}} \equiv -\frac{|\vec{a}_1 - \vec{a}_3|^2}{2L(t_f)_{\text{th}}^2}. \tag{12.200}$$

This leads to (reintroducing factors of c)

$$L(t_f)_{\text{vac}} = \sqrt{\frac{\pi}{4\alpha \ln(g\Omega_{\text{max}}t_f)}} c \cdot t_f \approx \frac{10.4}{\sqrt{\ln(g\Omega_{\text{max}}t_f)}} c \cdot t_f, \tag{12.201}$$

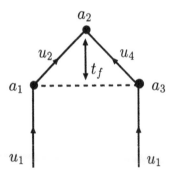

FIG. 12.5. A current loop involving two line segments which meet in the infinite past (compare with Fig. 12.3(a)). This situation corresponds to an interference device in which the relative velocity v_{13} between the interfering wave packets vanishes initially.

and

$$L(t_f)_{\text{th}} = \sqrt{\frac{3\pi}{8\alpha}}\sqrt{c^2\tau_B t_f} \approx 12.7\sqrt{c^2\tau_B t_f} \propto t_f^{1/2} \cdot T^{-1/2}. \qquad (12.202)$$

Equation (12.201) implies that the vacuum coherence length is roughly of order

$$L(t_f)_{\text{vac}} \sim c \cdot t_f. \qquad (12.203)$$

This simple result means that for a given time t_f the radiation field does not destroy quantum coherence on length scales which are small compared to the distance that light travels during this time. This also explains why the radiation field is quite ineffective in destroying quantum coherence of single, localized electrons.

We close this section by considering briefly another interference device which is depicted in Fig. 12.5: Here we suppose that the wave packets brought to interference are at rest initially. As indicated in the figure this case corresponds to a situation in which the line segments described by the 4-velocities u_1 and u_3 meet each other in the infinite past. Thus we set $u_1 = u_3$ in eqn (12.163) to obtain,

$$c(k) = \frac{ie}{\sqrt{2}}\left\{ \frac{u_1}{ku_1}\left[e^{-ika_1} - e^{-ika_3}\right]\right. \qquad (12.204)$$

$$\left. + \frac{u_2}{ku_2}\left[e^{-ika_2} - e^{-ika_1}\right] - \frac{u_4}{ku_4}\left[e^{-ika_2} - e^{-ika_3}\right]\right\}.$$

Performing here the approximation (12.169) we are again led to expression (12.166), where now

$$|\mathcal{G}(k)|^2 = 2(1 - \cos\omega t_f), \qquad (12.205)$$

which immediately yields the expression

$$\Gamma[c] = -\frac{2\alpha}{\pi} \left[\ln g\Omega_{\max} t_f + \ln \left(\frac{\sinh(t_f/\tau_B)}{t_f/\tau_B} \right) \right] \left(\frac{1}{v_{24}} \tanh^{-1} v_{24} - 1 \right). \quad (12.206)$$

In the non-relativistic limit we use the expansion (12.183) to obtain

$$\Gamma[c] \approx -\frac{8\alpha}{3\pi} \left[\ln (g\Omega_{\max} t_f) + \ln \left(\frac{\sinh(t_f/\tau_B)}{t_f/\tau_B} \right) \right] v^2, \quad (12.207)$$

where we have again assumed $\vec{v}_2 = -\vec{v}_4$.

Decoherence through bremsstrahlung exhibits a highly non-Markovian character since the decoherence functional depends on the whole paths of the interference device. This can be illustrated by a comparison of eqn (12.206) with eqns (12.186) and (12.187): After the time corresponding to the maximal distance between the wave packets we have in both cases two wave packets approaching each other with opposite velocities of the same magnitude v. The difference between the decoherence functionals obtained in the two cases shows that the suppression of quantum coherence through bremsstrahlung depends on the total history of the process and that the memory time is of the order of the total time t_f of the experiment.

12.3.4 Path integral approach

An exact analytical representation for the problem of a single electron interacting with the radiation field which takes into account the finite width as well as the spreading of the wave packets can be obtained in the (non-relativistic) dipole approximation (Barone and Caldeira, 1991; Dürr and Spohn, 2000). This simple case is of special interest for it allows an analytical treatment for several interesting cases which already exhibit the basic physical mechanism leading to the decoherence mechanism through bremsstrahlung (Breuer and Petruccione, 2000).

Invoking the dipole approximation we may replace the commutator function (eqns (12.19) and 12.20)) and the anticommutator function (eqns (12.37) and (12.38)) with their space-independent expressions as follows,

$$D^{\mathrm{T}}(x - x')_{ij} \rightarrow \delta_{ij} D(t - t') \equiv \delta_{ij} \int_0^\infty d\omega J(\omega) \sin \omega(t - t'), \quad (12.208)$$

$$D_1^{\mathrm{T}}(x - x')_{ij} \rightarrow \delta_{ij} D_1(t - t') \equiv \delta_{ij} \int_0^\infty d\omega J(\omega) \coth\left(\omega/2k_{\mathrm{B}}T\right) \cos \omega(t - t'),$$

$$(12.209)$$

where the spectral density is given by

$$J(\omega) = \frac{e^2}{3\pi^2} \omega \, \theta(\Omega_{\max} - \omega). \quad (12.210)$$

As before we have introduced here an ultraviolet cutoff Ω_{max}. One observes that the spectral density increases with the first power of the frequency ω. Had we used the dipole form $-e\vec{x} \cdot \vec{E}_T$ for the coupling of the electron coordinate $\vec{x}$ to the electric field (which is obtained through a canonical transformation), the corresponding spectral density would be proportional to the third power of the frequency. This means that the coupling to the radiation field in the dipole approximation may be described as a special case of the Caldeira–Leggett model discussed in Section 3.6. In the language of the theory of quantum Brownian motion the radiation field constitutes a super-Ohmic environment (Barone and Caldeira, 1991; Anglin, Paz and Zurek, 1997). Note also that we now include the factor e^2 into the definition of the correlation function.

Within the non-relativistic approximation we may replace the current density by

$$\vec{j}(t, \vec{x}) \longrightarrow \frac{\vec{p}(t)}{m_0}\delta(\vec{x} - \vec{x}(t)) + \delta(\vec{x} - \vec{x}(t))\frac{\vec{p}(t)}{m_0}, \qquad (12.211)$$

where $\vec{p}(t)$ and $\vec{x}(t)$ denote the momentum and position operator of the electron in the interaction picture with respect to the Hamiltonian

$$H_m = \frac{\vec{p}^2}{2m_0} + V(\vec{x}) \qquad (12.212)$$

for the electron, $V(\vec{x})$ being an external potential. The bare electron mass is denoted by m_0 (see below).

We are thus led to the following non-relativistic approximation (neglecting the spin degree of freedom) for the influence functional representation of the single electron density matrix (compare with eqn (12.99)),

$$\rho_m(t_f) = \mathrm{T}_\leftarrow \left(\exp\left[\int_{t_i}^{t_f} dt \int_{t_i}^{t} dt' \left\{ \frac{i}{2}D(t - t')\frac{\vec{p}_c(t)}{m_0}\frac{\vec{p}_a(t')}{m_0} \right. \right.\right. \qquad (12.213)$$
$$\left.\left.\left. -\frac{1}{2}D_1(t - t')\frac{\vec{p}_c(t)}{m_0}\frac{\vec{p}_c(t')}{m_0} \right\} \right] \right) \rho_m(t_i).$$

In accordance with the definitions (12.95) and (12.96) $\vec{p}_c$ is a commutator super-operator and $\vec{p}_a$ an anticommutator super-operator.

12.3.4.1 *Path integral representation and classical equations of motion* As we know from Section 3.6.4.2 the reduced density matrix given in eqn (12.213) admits an equivalent path integral representation. We introduce new coordinates through the relations $\vec{q} = \vec{x} - \vec{x}'$ and $\vec{r} = \frac{1}{2}(\vec{x} + \vec{x}')$, and set, for simplicity, the initial time equal to zero, $t_i = 0$. The propagator function J is then defined by

$$\rho_m(\vec{r}_f, \vec{q}_f, t_f) = \int d^3r_i \int d^3q_i\, J(\vec{r}_f, \vec{q}_f, t_f; \vec{r}_i, \vec{q}_i)\rho_m(\vec{r}_i, \vec{q}_i, 0), \qquad (12.214)$$

and admits a path integral representation of the form (see eqn (3.527))

$$J(\vec{r}_f, \vec{q}_f, t_f; \vec{r}_i, \vec{q}_i) = \int D\vec{r} \int D\vec{q} \, \exp\{i\mathcal{A}[\vec{r}, \vec{q}]\}. \qquad (12.215)$$

This is a double path integral taken over all paths $\vec{r}(t)$, $\vec{q}(t)$ which satisfy the boundary conditions,

$$\vec{r}(0) = \vec{r}_i, \qquad \vec{r}(t_f) = \vec{r}_f, \qquad \vec{q}(0) = \vec{q}_i, \qquad \vec{q}(t_f) = \vec{q}_f. \qquad (12.216)$$

The weight factor for the paths is defined in terms of an effective action functional $\mathcal{A}$ given by (see eqn (3.531))

$$\mathcal{A}[\vec{r}, \vec{q}] = \int_0^{t_f} dt \left(m_0 \ddot{\vec{r}}\dot{\vec{q}} - V(\vec{r} + \vec{q}/2) + V(\vec{r} - \vec{q}/2) \right)$$

$$+ \int_0^{t_f} dt \int_0^{t_f} dt' \theta(t - t') D(t - t') \dot{\vec{q}}(t) \dot{\vec{r}}(t')$$

$$+ \frac{i}{4} \int_0^{t_f} dt \int_0^{t_f} dt' D_1(t - t') \dot{\vec{q}}(t) \dot{\vec{q}}(t'). \qquad (12.217)$$

It is instructive to investigate first the classical equations of motion determined by the effective action functional. The first variation of $\mathcal{A}$ with respect to $\vec{r}(t)$ and $\vec{q}(t)$ leads to the classical equations of motion,

$$m_0 \ddot{\vec{r}}(t) + \frac{1}{2} \vec{\nabla}_{\vec{r}} \left(V(\vec{r} + \vec{q}/2) + V(\vec{r} - \vec{q}/2) \right) + \frac{d}{dt} \int_0^t dt' D(t - t') \dot{\vec{r}}(t')$$

$$= -\frac{i}{2} \frac{d}{dt} \int_0^{t_f} dt' D_1(t - t') \dot{\vec{q}}(t'), \qquad (12.218)$$

and

$$m_0 \ddot{\vec{q}}(t) + 2\vec{\nabla}_{\vec{q}} \left(V(\vec{r} + \vec{q}/2) + V(\vec{r} - \vec{q}/2) \right) + \frac{d}{dt} \int_t^{t_f} dt' D(t' - t) \dot{\vec{q}}(t') = 0. \qquad (12.219)$$

The real part of the equation of motion (12.218), which is obtained by setting the right-hand side equal to zero, yields the famous Abraham–Lorentz equation for the electron (Jackson, 1999). It describes the radiation damping through the damping kernel $D(t - t')$. To see this we write the real part of eqn (12.218) as

$$m_0 \ddot{\vec{r}}(t) + \frac{d}{dt} \int_0^t dt' D(t - t') \dot{\vec{r}}(t') = \vec{F}(t), \qquad (12.220)$$

where $\vec{F}(t)$ denotes the external force derived from the potential V. The damping kernel can be written as

$$D(t - t') = \int\limits_{0}^{\Omega_{\max}} d\omega \frac{e^2}{3\pi^2} \omega \sin \omega(t - t') = \frac{e^2}{3\pi^2} \frac{d}{dt'} f(t - t'), \qquad (12.221)$$

where we have introduced the function

$$f(t) \equiv \frac{\sin \Omega_{\max} t}{t}. \qquad (12.222)$$

To be specific the UV-cutoff $\Omega_{\max}$ is taken to be $\hbar \Omega_{\max} = mc^2$, which yields $\Omega_{\max} \approx 0.78 \times 10^{21} \, \mathrm{s}^{-1}$. The term of the equation of motion (12.220) involving the damping kernel can be written as follows,

$$\frac{d}{dt} \int\limits_{0}^{t} dt' D(t - t') \dot{\vec{r}}(t') = \frac{e^2}{3\pi^2} \frac{d}{dt} \int\limits_{0}^{t} dt' \left[\frac{d}{dt'} f(t - t') \right] \dot{\vec{r}}(t') \qquad (12.223)$$

$$= \frac{e^2}{3\pi^2} \frac{d}{dt} \left[-\int\limits_{0}^{t} dt' f(t - t') \ddot{\vec{r}}(t') + f(0) \dot{\vec{r}}(t) - f(t) \dot{\vec{r}}(0) \right].$$

For times t such that $\Omega_{\max} t \gg 1$, i.e. $t \gg 10^{-21}$ s, we may replace $f(t)$ by $\pi \delta(t)$, and approximate $f(t) \approx 0$, while $f(0) = \Omega_{\max}$. Thus we obtain,

$$\frac{d}{dt} \int\limits_{0}^{t} dt' D(t - t') \dot{\vec{r}}(t') = \frac{e^2}{3\pi^2} \frac{d}{dt} \left[-\frac{\pi}{2} \ddot{\vec{r}}(t) + \Omega_{\max} \dot{\vec{r}}(t) \right], \qquad (12.224)$$

which finally leads to the equation of motion,

$$\left(m_0 + \frac{e^2 \Omega_{\max}}{3\pi^2} \right) \dot{\vec{v}}(t) - \frac{e^2}{6\pi} \ddot{\vec{v}}(t) = \vec{F}(t), \qquad (12.225)$$

where $\vec{v} = \dot{\vec{r}}$ is the velocity. This equation is known as Abraham–Lorentz equation (Jackson, 1999). The above derivation, which is similar to the one given by Barone and Caldeira (1991), shows that the damping kernel leads to two contributions. The first one provides a dressing of the electron mass by an electromagnetic mass contribution $\delta m = e^2 \Omega_{\max}/3\pi^2$. This renormalization of the electron mass is the same as the one found in Section 12.2.3 (see eqn (12.120)). The second contribution introduced by the damping kernel is proportional to the third derivative of $\vec{r}(t)$ and describes the damping of the electron motion through the emitted radiation. This term is independent of the cutoff.

The equation of motion (12.225) can be obtained heuristically by means of the Larmor formula for the power radiated by an accelerated charge. More rigorously, it has been derived by Abraham and by Lorentz from the conservation law for the field momentum, assuming a spherically symmetric charge distribution and that the momentum is of purely electromagnetic origin (Jackson, 1999). The decomposition $m = m_0 + \delta m$ of the electron mass into a bare mass m_0 and an electromagnetic mass contribution δm is unphysical since the electron is never observed without its self-field and the associated field momentum. In other words, we identify m with the observed physical mass which enables us to write eqn (12.225) as

$$m\left(\dot{\vec{v}}(t) - \tau_0 \ddot{\vec{v}}(t)\right) = \vec{F}(t),\tag{12.226}$$

where we have defined the characteristic time constant

$$\tau_0 \equiv \frac{e^2}{6\pi m} = \frac{2}{3}r_e \approx 0.6 \times 10^{-23}\,\text{s}.\tag{12.227}$$

As is well known, eqn (12.226), being a classical equation of motion for the electron, leads to the problem of exponentially increasing runaway solutions and to an apparently acausal classical behaviour known as pre-acceleration (Jackson, 1999). The time τ_0 represents a characteristic radiation time scale of the electron motion in the following sense (see also the corresponding discussion in Section 12.3.1). Suppose the electron is at rest initially. If the external force $\vec{F}(t)$ acts upon the electron for a short period of time τ the condition $\tau \gg \tau_0$ implies that the kinetic energy of the electron is large in comparison to the radiated energy according to Larmor's formula. This implies that the influence of radiative damping may be neglected provided $\vec{F}(t)$ changes only slightly over times of the order of τ_0. As in Section 12.3.1 we shall use this condition in the following. It allows us to discard the term involving the damping kernel altogether in the classical equations of motion. In the language of Brownian motion we could say that the short-time behaviour is that of a strongly underdamped particle and that decoherence is entirely due to the noise kernel in the electron propagator function.

It might be useful to formulate this condition also for an electron which moves in a harmonic potential

$$V(\vec{x}) = \frac{1}{2}m\omega_0^2\vec{x}^2.\tag{12.228}$$

In this case the classical equation of motion reads

$$\frac{d^2}{dt^2}\vec{r} + \omega_0^2\vec{r} - \tau_0\frac{d^3}{dt^3}\vec{r} = 0.\tag{12.229}$$

With the help of the ansatz $\vec{r}(t) = \vec{r}_0 \exp(zt)$ we are led to a cubic equation for the characteristic frequencies,

$$z^2 + \omega_0^2 - \tau_0 z^3 = 0.\tag{12.230}$$

For vanishing coupling to the radiation field ($\tau_0 = 0$) the solutions are located at $z_\pm = \pm i\omega_0$, describing the free motion of a harmonic oscillator with frequency

ω_0. For $\tau_0 > 0$ eqn (12.230) has three roots, one is real and the other two are complex conjugated to each other. The real root corresponds to the runaway solution and must be discarded. Let us assume that the period of the oscillator is large compared to the radiation time,

$$\tau_0 \ll \frac{1}{\omega_0}. \tag{12.231}$$

Because of (12.227) this assumption is well satisfied even in the regime of optical frequencies. We may thus determine the complex roots to lowest order in $\omega_0\tau_0$,

$$z_\pm = \pm i\omega_0 - \frac{1}{2}\tau_0\omega_0^2. \tag{12.232}$$

The purely imaginary roots $\pm i\omega_0$ of the undisturbed harmonic oscillator are thus shifted into the negative half-plane under the influence of the radiation field. The negative real part describes the radiative damping. In fact, we see that $\vec{r}(t)$ decays as $\exp(-\gamma t/2)$, where

$$\gamma = \tau_0\omega_0^2 = \frac{2}{3}\alpha\frac{\hbar\omega_0^2}{mc^2} \tag{12.233}$$

is the damping constant for the radiative damping of the oscillator. If we consider time intervals τ of the order of magnitude of one oscillator period, $\omega_0\tau \sim 1$, we have $\gamma\tau = (\omega_0\tau_0)(\omega_0\tau) \sim \omega_0\tau_0 \ll 1$. Thus we see again that we may neglect effects of radiative damping provided the radiation time τ_0 is small compared to the typical time scale ω_0^{-1} of the undisturbed mechanical motion.

12.3.4.2 *Determination of the propagator function* We can now determine the propagator function for the electron density matrix explicitly for an arbitrary quadratic potential. The procedure is essentially the same as the one used in Section 3.6.4.2 for the Caldeira–Leggett model. We therefore sketch only briefly the result for a free electron moving in the radiation field. The details of the calculation may be found in (Breuer and Petruccione, 2000).

Under the above conditions the propagator function for the electron is found to be

$$J(\vec{r}_f, \vec{q}_f, t_f; \vec{r}_i, \vec{q}_i) = \left(\frac{m}{2\pi t_f}\right)^3 \exp\left\{\frac{im}{t_f}(\vec{r}_f - \vec{r}_i)(\vec{q}_f - \vec{q}_i) + \Gamma(\vec{q}_f, \vec{q}_i, t_f)\right\}. \tag{12.234}$$

As expected, depending only on the difference $\vec{r}_f - \vec{r}_i$, the propagator function is invariant under space translations. Furthermore, one easily recognizes that the contribution

$$G_0(\vec{r}_f - \vec{r}_i, \vec{q}_f - \vec{q}_i, t_f) \equiv \left(\frac{m}{2\pi t_f}\right)^3 \exp\left\{\frac{im}{t_f}(\vec{r}_f - \vec{r}_i)(\vec{q}_f - \vec{q}_i)\right\} \tag{12.235}$$

is simply the propagator function for the electron's density matrix in the case of a vanishing coupling to the radiation field.

The function $\Gamma(\vec{q}_f, \vec{q}_i, t_f)$ in eqn (12.234) describes the influence of the radiation field and may be written as

$$\Gamma(\vec{q}_f, \vec{q}_i, t_f) \approx -\frac{2\alpha}{3\pi} \left[\ln g\Omega_{\text{max}} t_f + \ln\left(\frac{\sinh(t_f/\tau_B)}{t_f/\tau_B} \right) \right] \frac{(\vec{q}_f - \vec{q}_i)^2}{(ct_f)^2}$$

$$\equiv -\frac{(\vec{q}_f - \vec{q}_i)^2}{2L(t_f)^2}. \qquad (12.236)$$

Here, we have introduced the quantity $L(t_f)$ which is defined by

$$L(t_f)^2 \equiv \frac{3\pi}{4\alpha} \left[\ln g\Omega_{\text{max}} t_f + \ln\left(\frac{\sinh(t_f/\tau_B)}{t_f/\tau_B} \right) \right]^{-1} \cdot (ct_f)^2, \qquad (12.237)$$

and which can be interpreted as a time-dependent coherence length.

12.3.4.3 Wave packet dynamics With the help of eqn (12.234) we can study the time-evolution of electronic wave packets in order to estimate the influence of the finite width and of the spreading on the decohrence mechanism. To this end, we investigate an initial state given by the superposition of two Gaussian wave packets separated by a distance $2a$. We assume that both packets have a width σ_0 and that they are centred initially at $\vec{x} = \pm\vec{a} = \pm(a, 0, 0)$. The packets are supposed to approach each other with the average speed $v = k_0/m > 0$. For simplicity the motion is assumed to occur along the x-axis. Thus we have the initial state

$$\psi_0(\vec{x}) = A_1 \left(\frac{1}{2\pi\sigma_0^2} \right)^{3/4} \exp\left[-\frac{(\vec{x} - \vec{a})^2}{4\sigma_0^2} - i\vec{k}_0(\vec{x} - \vec{a}) \right]$$

$$+ A_2 \left(\frac{1}{2\pi\sigma_0^2} \right)^{3/4} \exp\left[-\frac{(\vec{x} + \vec{a})^2}{4\sigma_0^2} + i\vec{k}_0(\vec{x} + \vec{a}) \right], \qquad (12.238)$$

where $\vec{k}_0 = (k_0, 0, 0)$ and A_1, A_2 are complex amplitudes. Our aim is to determine the interference pattern that arises in the moment $t_f = a/v$ of the collision of the centres of the packets at $\vec{x} = 0$. The position space density at the time t_f is found with the help of the formula

$$\rho_m(\vec{r}_f, t_f) \equiv \rho_m(\vec{r}_f, \vec{q}_f = 0, t_f) \qquad (12.239)$$

$$= \int d^3r_i \int d^3q_i \left(\frac{m}{2\pi t_f} \right)^3 \exp\left[-\frac{im}{t_f}(\vec{r}_f - \vec{r}_i)\vec{q}_i - \frac{\vec{q}_i^2}{2L(t_f)^2} \right]$$

$$\times \psi_0(\vec{r}_i + \frac{1}{2}\vec{q}_i)\psi_0^*(\vec{r}_i - \frac{1}{2}\vec{q}_i).$$

Performing the Gaussian integrations we get the result

$$\rho_m(\vec{r}_f, t_f) = \left(\frac{1}{2\pi\sigma(t_f)^2} \right)^{3/2} \exp\left[-\frac{\vec{r}_f^2}{2\sigma(t_f)^2} \right] \qquad (12.240)$$

$$\times \left\{ |A_1|^2 + |A_2|^2 + 2\Re\left(A_1 A_2^* \exp\left[i\varphi(\vec{r}_f) + \Gamma(t_f) \right] \right) \right\}.$$

We recognize a Gaussian envelope centred at $\vec{r}_f = 0$ with width $\sigma(t_f)$, an incoherent sum $|A_1|^2 + |A_2|^2$, and an interference term proportional to $A_1 A_2^*$. The interference term involves the phase

$$\varphi(\vec{r}_f) = -2\vec{k}_0\vec{r}_f(1 - \varepsilon) \qquad (12.241)$$

as well as the decoherence function

$$\Gamma(t_f) = -\frac{2a^2}{L(t_f)^2}(1 - \varepsilon). \qquad (12.242)$$

The quantity ε is given by

$$\varepsilon = \left(1 + \frac{L(t_f)^2}{4\sigma_0^2} + \frac{m^2\sigma_0^2 L(t_f)^2}{t_f^2}\right)^{-1} \qquad (12.243)$$

The term $-2\vec{k}_0\vec{r}_f$ of the phase $\varphi(\vec{r}_f)$ describes the interference pattern as it would be obtained for a free Schrödinger particle, while the term $2\vec{k}_0\vec{r}_f\varepsilon$ leads to a modification of the period of the pattern. Moreover, without the contribution proportional to ε the decoherence function (12.242) exactly coincides with the expression (12.207). Thus we see that the influence of the finite width and of the spreading of the wave packets brought to interference may indeed be neglected provided the condition $\varepsilon \ll 1$ holds. This condition is always satisfied in the present approximation. Since ε attains the maximum value

$$\varepsilon_{\max} = \frac{1}{1 + mL(t_f)^2/t_f} \qquad (12.244)$$

one is led to the requirement $L(t_f)^2 \gg \hbar t_f/m$, which is always fulfilled for times t_f and temperatures T satisfying $ct_f \gg \lambda_C$ and $k_B T \ll mc^2$.

12.4   Decoherence of many-particle states

In the previous section we have obtained several expressions for the decoherence functional which describes the loss of coherence in an interference device. We address here the question as to whether these results could explain the absence of the coherence of certain superpositions of states that can be considered as macroscopically distinct.

For single electrons the vacuum decoherence through bremsstrahlung turns out to be small at non-relativistic speeds. For example, taking $\Omega_{\max} \sim c/\lambda_C$ and t_f of the order of $1\,\mathrm{s}$, and using a velocity v which is already as large as $1/10$ of the speed of light, one finds that $|\Gamma_{\mathrm{vac}}| \sim 10^{-2}$, which corresponds to a 1% suppression of interference. By virtue of the weak logarithmic dependence on the cutoff scale this estimate is true also for other particles carrying an elementary charge.

Matters could however be different for many-particle states. To investigate this case it is important to specify clearly the structure of the superposition under consideration. We distinguish two extreme classes of many-particle states (Joos, 1996). If $|\varphi_1\rangle$ and $|\varphi_2\rangle$ are two states containing only a small number of particles, one may consider a superposition of the form

$$|\Psi\rangle = (\alpha|\varphi_1\rangle + \beta|\varphi_2\rangle))^N . \qquad (12.245)$$

As our notation indicates, the state $|\Psi\rangle$ is an N-fold tensor product of a superposition of the few-particle states $|\varphi_1\rangle$ and $|\varphi_2\rangle$. Therefore, the decoherence factor for $|\Psi\rangle$ is expected to be very small.

The situation changes, however, substantially if one considers another class of states, namely those of the form

$$|\Psi\rangle = \alpha|\varphi_1\rangle^N + \beta|\varphi_2\rangle^N. \qquad (12.246)$$

In contrast to (12.245) this is a superposition of two N-fold tensor products. It is this class of states that will be investigated in the following regarding their decoherence properties. Our aim is to derive the dependence of the decoherence functional on the particle number N.

Let us consider a system which is composed of N identical particles with mass m and charge e. Our aim is to construct an effective master equation for the density $\rho_{\text{cm}}(\vec{R}, \vec{R}')$ of the centre-of-mass coordinate

$$\vec{R} = \frac{1}{N} \sum_{i=1}^{N} \vec{x}_i \qquad (12.247)$$

for such a system. Here, the $\vec{x}_i$ are the particle coordinates and we suppress, for simplicity, the spin degree of freedom. We introduce the relative coordinates $\vec{r}_i$ through

$$\vec{x}_i = \vec{R} + \vec{r}_i(q). \qquad (12.248)$$

Since the $\vec{r}_i$ sum up to zero, they are functions of $3N - 3$ internal variables which we denote collective by q. Let us suppose that the state of the N-particle system is described in the position representation by a density matrix of the form

$$\rho_m = \rho_{\text{cm}}(\vec{R}, \vec{R}')\rho_{\text{int}}(q, q'), \qquad (12.249)$$

where ρ_{cm} and ρ_{int} are separately normalized to 1,

$$\int d^3R\, \rho_{\text{cm}}(\vec{R}, \vec{R}) = \int dq\, \rho_{\text{int}}(q, q) = 1. \qquad (12.250)$$

The density matrix ρ_{cm} describes the centre-of-mass coordinate, while ρ_{int} represents the state of the internal degrees of freedom. For example, one finds that the quantity

$$w(\vec{x} - \vec{R}) = \frac{1}{N} \int dq \sum_{i=1}^{N} \delta\left(\vec{x} - \vec{R} - \vec{r}_i(q)\right) \rho_{\text{int}}(q, q) \qquad (12.251)$$

is the density of finding a particle at $\vec{x}$ under the condition that the centre-of-mass coordinate is at $\vec{R}$. This function is obviously normalized as

$$\int d^3 x \, w(\vec{x}) = 1. \tag{12.252}$$

If the system described by the state (12.249) performs a translational motion it is reasonable to assume that its total current density can be approximated by an effective current density of the form

$$\vec{j}_{\text{cm}}(\vec{x}) = \frac{Ne}{2M} \left\{ \vec{P} w(\vec{x} - \vec{R}) + w(\vec{x} - \vec{R}) \vec{P} \right\}, \tag{12.253}$$

where $\vec{P} = -i\partial/\partial\vec{R}$ is the total momentum canonically conjugated to the centre-of-mass coordinate $\vec{R}$, Ne is the total charge and $M = Nm$ the total mass. The expression (12.253) implies that the current density of the internal degrees of freedom vanishes. In particular, it excludes the possibility that the whole systems is in a rotational state which would require us to introduce three further collective coordinates as, for example, the three Euler angles.

Equation (12.253) shows that the case of an N-particle system can be dealt with by using the replacements $e \longrightarrow Ne$ and $m \longrightarrow M = Nm$, and by interpreting the length scale σ_0, which appears in the UV cutoff scale $\Omega_{\text{max}} \sim 1/\sigma_0$, as the linear extension of the one-particle density $w(\vec{x})$. A representation for the density matrix $\rho_{\text{cm}}(\vec{R}, \vec{R}')$ is then obtained from (12.89) by substituting the effective current (12.253) into the functional (12.99). Invoking the non-relativistic (dipole) approximation one is led to the following representation for the centre-of-mass density,

$$\rho_{\text{cm}}(t_f) = \text{T}_{\leftarrow} \left(\exp \left[\int_{t_i}^{t_f} dt \int_{t_i}^{t_f} dt' \left\{ \frac{i}{2} D(t - t') \frac{\vec{P}_c(t)}{M} \frac{\vec{P}_a(t')}{M} \right. \right. \right. \tag{12.254}$$

$$\left. \left. \left. - \frac{1}{2} D_1(t - t') \frac{\vec{P}_c(t)}{M} \frac{\vec{P}_c(t')}{M} \right\} \right] \right) \rho_{\text{cm}}(t_i),$$

where $\vec{P}_c(t)$ and $\vec{P}_a(t)$ denote the interaction picture commutator and anticommutator super-operators for the total momentum. The dissipation and the noise kernel are given by the expressions (12.213), where the spectral density now takes the form

$$J(\omega) = \frac{N^2 e^2}{3\pi^2} \omega \, \theta(\Omega_{\text{max}} - \omega). \tag{12.255}$$

The results of Section 12.3.4 can now immediately be transferred to the present case with the help of the replacements given above. It follows that the vacuum decoherence functional for an N-particle state scales with the square N^2 of the particle number,

$$\Gamma_{\text{vac}} \sim -N^2 \frac{8\alpha}{3\pi} \ln\left(g\Omega_{\text{max}}t_f\right) v^2. \qquad (12.256)$$

This scaling with the particle number obviously leads to a strong amplification of the decoherence effect. To give a very extreme example we take $N = 10^{22}$ which corresponds to $\sigma_0 \sim 1$ cm for typical free electron densities in metals. Let us ask for the maximal speed v leading to 1% decoherence. With the help of (12.256) we find $v \sim 10^{-14}\,\text{m}\,\text{s}^{-1}$. For a distance of 1 m this implies, for example, that a successful interference experiment would take about three million years!

The scaling (12.256) of Γ with the square N^2 of the particle number can be traced back to two facts. First, the radiative back-action is proportional to the square of the total charge since the emitted radiation adds coherently in the limit of long wavelength. Second, the decoherence functional only depends on the logarithm of the cutoff Ω_{max}, which means that it depends only very weakly on the total mass or on the spatial extent of the N-particle state. In the cases discussed here one must expect, of course, a large radiative damping in addition to the decoherence effect. It is of great fundamental interest to investigate also the influence of these phenomena for composite neutral objects.

References

Anglin, J. R., Paz, J. P. and Zurek, W. H. (1997). Deconstructing decoherence. *Phys. Rev.*, **A55**, 4041–4053.

Barone, P. M. V. B. and Caldeira, A. O. (1991). Quantum mechanics of radiation damping. *Phys. Rev.*, **A43**, 57–63.

Bjorken, J. D. and Drell, S. D. (1964). *Relativistic Quantum Mechanics*. McGraw-Hill, New York.

Breuer, H. P. and Petruccione, F. (2000). Radiation damping and decoherence in quantum electrodynamics. In *Relativistic Quantum Measurement and Decoherence* (eds. Breuer, H. P. and Petruccione, F.), Volume 559 of *Lecture Notes in Physics*, pp. 31–65. Springer-Verlag, Berlin.

Breuer, H. P. and Petruccione, F. (2001). Destruction of quantum coherence through emission of bremsstrahlung. *Phys. Rev.*, **A63**, 032102–1(18).

Chou, K.-c., Su, Z.-b., Hao, B.-l. and Yu, L. (1985). Equilibrium and nonequilibrium formalisms made unified. *Phys. Rep.*, **118**, 1–131.

Cohen-Tannoudji, C., Dupont-Roc, J. and Grynberg, G. (1998). *Atom–Photon Interactions*. John Wiley, New York.

Diòsi, L. (1990). Landau's density matrix in quantum electrodynamics. *Found. Phys.*, **20**, 63–70.

Dürr, D. and Spohn, H. (2000). Decoherence through coupling to the radiation field. In *Decoherence: Theoretical, Experimental, and Conceptual Problems* (eds. Blanchard, P., Giulini, D., Joos, E., Kiefer, C. and Stamatescu, I.-O.), Volume 538 of *Lecture Notes in Physics*, pp. 77–86. Springer-Verlag, Berlin.

Feynman, R. P. and Vernon, F. L. (1963). The theory of a general quantum system interacting with a linear dissipative system. *Ann. Phys. (N.Y.)*, **24**, 118–173.

Feynman, R. P. and Hibbs, A. R. (1965). *Quantum Mechanics and Path Integrals*. McGraw-Hill, New York.

Ford, L. H. (1993). Electromagnetic vacuum fluctuations and electron coherence. *Phys. Rev.*, **D47**, 5571–5580.

Gradshteyn, I. S. and Ryzhik, I. M. (1980). *Table of Integrals, Series, and Products*. Academic Press, New York.

Hasselbach, F., Kiesel, H. and Sonnentag, P. (2000). Exploration of the fundamentals of quantum mechanics by charged particle interferometry. In *Decoherence: Theoretical, Experimental, and Conceptual Problems* (eds. Blanchard, P., Giulini, D., Joos, E., Kiefer, C. and Stamatescu, I.-O.), Volume 538 of *Lecture Notes in Physics*, pp. 201–212. Springer-Verlag, Berlin.

Itzykson, C. and Zuber, J.-B. (1980). *Quantum Field Theory*. McGraw-Hill, New York.

Jackson, J. D. (1999). *Classical Electrodynamics* (third edition). John Wiley, New York.

Jauch, J. M. and Rohrlich, F. (1980). *The Theory of Photons and Electrons*. Springer-Verlag, New York.

Joos, E. (1996). Decoherence through interaction with the environment. In *Decoherence and the Appearence of a Classical World in Quantum Theory* (eds. Giulini, D., Joos, E., Kiefer, C., Kupsch, J., Stamatescu, I.-O. and Zeh, H. D.), pp. 35–136. Springer-Verlag, Berlin.

Rohrlich, F. (1965). *Classical Charged Particles*. Addison-Wesley, Reading, Massachusetts.

Weinberg, S. (1995). *The Quantum Theory of Fields*, Volume I. Cambridge University Press, Cambridge.

Wheeler, J. A. and Feynman, R. P. (1945). Interaction with the absorber as the mechanism of radiation. *Rev. Mod. Phys.*, **17**, 157–181.

Wheeler, J. A. and Feynman, R. P. (1949). Classical electrodynamics in terms of direct interparticle action. *Rev. Mod. Phys.*, **21**, 425–433.

INDEX